Fundamentals of Sequential and Parallel Algorithms

Kenneth A. Berman and Jerome L. Paul

University of Cincinnati

PWS Publishing Company

I(T)P An International Thomson Publishing Company

Boston • Albany • Bonn • Cincinnati • Detroit • London • Madrid • Melbourne • Mexico City New York • Pacific Grove • Paris • San Francisco • Singapore • Tokyo • Toronto • Washington

PWS Publishing Company
20 Park Plaza, Boston, MA 02116-4324

To our wives, Peiyan and Ruta

I Ⓣ P
International Thomson Publishing
The trademark ITP is used under license

For more information, contact:

PWS Publishing Company
20 Park Plaza
Boston, MA 02116

International Thomson Publishing Europe
Berkshire House I68-I73
High Holborn
London WC1V 7AA
England

Thomas Nelson Australia
102 Dodds Street
South Melbourne, 3205
Victoria, Australia

Nelson Canada
1120 Birchmount Road
Scarsborough, Ontario
Canada M1K 5G4

International Thomson Editores
Campos Eliseos 385, Piso 7
Col. Polanco
11560 Mexico D.F., Mexico

International Thomson Publishing GmbH
Königswinterer Strasse 418
53227 Bonn, Germany

International Thomson Publishing Asia
221 Henderson Road
#05-10 Henderson Building
Singapore 0315

International Thomson Publishing Japan
Hirakawacho Kyowa Building, 31
2-2-1 Hirakawacho
Chiyoda-ku, Tokyo 102
Japan

Executive Editor: Michael J. Sugarman
Developmental Editor: Mary T. Stone
Editorial Assistant: Susan Garland
Production Editor: Pamela Rockwell
Interior Design: Abigail Heim
Cover Design: Julia Gecha

Cover Art: Ray Paul, "Totem," 1993
(acrylic, collage on canvas)
Marketing Manager: Nathan Wilbur
Cover Printer: John Pow Company
Text Printer and Binder: Quebecor/Martinsburg

Library of Congress Cataloging-in-Publication Data

Berman, Kenneth A.
Fundamentals of sequential and parallel algorithms / Kenneth A. Berman, Jerome L. Paul
p. cm.
Includes bibliographical references and index.
ISBN 0-534-94674-7 (alk. paper)
1. Computer algorithms. 2. Sequential processing (Computer science) 3. Parallel processing (Electronic computers) I. Paul, Jerome L. II. Title.
QA76.9.A43B47 1996
005.1--dc20 96-22995
CIP

Printed and bound in the United States of America. 96 97 98 99 00 — 10 9 8 7 6 5 4 3 2 1

Contents

PART THREE Major Design Strategies

Preface

The aim of this book is to provide comprehensive coverage of the traditional and current topics in sequential algorithms, while at the same time providing a solid foundation in the theory of parallel algorithms. The theory of parallel algorithms is becoming increasingly important, so that every computer science graduate should be exposed to the theory of parallel algorithms as part of his or her undergraduate training. The book covers parallel algorithms and architectures in sufficient detail to prepare for independent study or for taking a more advanced course devoted entirely to parallel algorithms.

Another goal of this book is to provide a large toolkit of known sequential and parallel algorithmic solutions to frequently encountered problems such as searching, sorting, matrix manipulations, scientific computations, Fourier analysis, and so forth. A third goal is to provide the reader with the mathematical tools necessary to analyze the correctness and performance of algorithms, and to be able to judge how close an algorithm is to being optimal for a given problem. These tools enable the student to choose the best algorithm to use in a given circumstance from among several algorithms that might be available for the problem. Perhaps the most important goal, however, is to enhance the reader's ability to create new algorithms or to modify existing algorithms in order to solve new problems.

Organization of the Book

Material on sequential and parallel algorithms is treated in a naturally integrated fashion. Several advantages result from integrating the sequential and parallel coverage. Often a good parallel algorithm results from looking for inherent parallelism that might be present in a sequential algorithm for a given problem. For example, algorithms based on the divide-and-conquer strategy usually have a naturally inherent parallelism. Another advantage of an integration of the sequential and parallel coverage is that any mathematical features of a given problem that are common to both a sequential and a parallel approach can be isolated and solved while the problem is still fresh in the student's mind.

An additional organizational feature of the book is its division into four parts: introductory material on sequential and parallel algorithms; development of more advanced tools; major design strategies; and special topics. Part One begins with a brief history of the subject of algorithms and a review of some elementary concepts from mathematics and data structures. The fundamental measures of the performance of sequential algorithms are established, along with the notion of asymptotic order. The discussion of sequential searching and sorting algorithms illustrate various general issues relating to the design and analysis of algorithms.

Parallel algorithms and architectures are then introduced, including a careful development of pseudocode for parallel algorithms. A number of constraints arise in the design of parallel algorithms that are not present in the design of sequential algorithms. These constraints are highlighted, together with a development of various performance measures for parallel algorithms. Searching and sorting are then revisited from the parallel point of view, which serves to illustrate these constraints and helps to delineate the differences between the design of a sequential algorithm and that of a parallel algorithm.

Most chapters in Parts Two, Three and Four contain a closing section where problems solved earlier in the chapter using sequential algorithms are revisited from the point of view of parallel algorithms. Since the treatments of sequential and parallel algorithms are kept in separate sections, an instructor is able to cover as much of the parallel material as desired.

Part Two expands the basic algorithmic toolkit, and shows how these tools can be applied. We gathered these expanded tools into Part Two so that in Part One we could quickly introduce the student to some of the main ideas of sequential and parallel algorithms. However, the organization allows an instructor to look ahead and cover some of the material in Part Two when discussing some of the examples in Part One.

In Part Three we give a complete overview of the standard major design strategies for algorithms, including the greedy method, divide-and-conquer, dynamic programming, backtracking, and branch-and-bound. In Part Four we discuss some special design strategies such as heuristic searching and probabilistic algorithms, plus special topics such as balanced search trees, matching and flow problems in graph theory, NP-complete problems for sequential algorithms, and NC and P-complete problems for parallel algorithms.

In lieu of a bibliography, we have included references and suggestions for further reading at the end of each chapter. For the most part, these references are to textbooks and to survey articles, thereby providing the reader with ample pointers to the literature.

Possible Course Outlines

The book is intended to serve as a text for a core-curriculum undergraduate course or first year graduate course in the design and analysis of algorithms. The book could also be used for independent study by readers wishing to learn more of the theory and practice of algorithms. We assume that most students have some familiarity with elementary data structures and basic concepts from discrete mathematics, although both of these topics are reviewed in enough detail to make the book relatively self-contained. The inclusion of an extended treatment of parallel algorithms, together with a number of special topics, provides enough material to serve as the basis for a two-semester course in algorithms. However, the organization of the book allows the instructor the flexibility to readily adapt the material to a one-semester course. For example, in a one-semester course the instructor might assign Chapters 1 and 2 as reading assignments, cover a sampling of the sorting algorithms in Chapter 4, and omit Section 6.5 on sorting networks. Some of the more specialized topics in Part Two could be omitted, such as interpolating asymptotic behavior and Sections 7.5–7.7. The instructor might then cover a representative sample of the algorithms illustrating the design strategies in Part Three, and pick and choose from among the topics in Part Four as time allows.

Material in Part Four may be integrated earlier in the text. For example, the material on balanced search trees in Chapter 17 could be covered immediately after binary search trees have been discussed. Algorithms for strongly connected components and biconnected components discussed in Chapter 19 could be covered immediately after covering depth-first search in Chapter 8. Some of the material on probabilistic algorithms in Chapter 18 such as randomizing deterministic algorithms could be discussed immediately after covering the material on probabilistic analysis of algorithms in Chapter 9. It would also be natural to cover some of the material in Chapter 20 on NP-completeness anytime after covering lower bound theory in Chapter 10.

The authors would appreciate receiving correspondence from readers pointing out errors in the text, or making comments or suggestions for improvement. Write to either author at:

Department of Electrical & Computer Engineering and Computer Science
University of Cincinnati
Cincinnati, OH 45221-0030

or e-mail: ken.berman@uc.edu. jerry.paul@uc.edu

Acknowledgments

This book is based on notes that have been developed since 1990 for a course in sequential and parallel algorithms at the University of Cincinnati. We wish to thank the many students whose suggestions and corrections over the years contributed greatly to the final product. We also thank all of our colleagues who have class tested these notes. Particular thanks go to Fred Annexstein, John Franco, Ken Meyer, John Schlipf, and Dieter Schmidt for their many helpful suggestions. Special thanks goes to Yizong Cheng who kindly provided the figures on the U.S. freeway system given in Chapter 16.

This material has also undergone a number of external reviews, which have been invaluable. The reviewers include:

Andrew Astromoff
San Francisco State University

Theodore Brown
Queens College-City University of New York

Donald Burlingame
Potsdam College

Adair Dingle
Lehigh University

Barry J. Donahue
Central Washington University

Rex A. Dwyer
North Carolina State University

Keith Harrow
Brooklyn College-City University of New York

T.C. Hu
University of California-San Diego

David J. John
Wake Forest University

Arkady Kanevsky
Texas A & M University

Sampath Kannan
University of Pennsylvania

Sheau-Doug Lang
University of Central Florida

Udi Manber
University of Arizona

Harold Martin
Northern Michigan University

Tim McGuire
Texas A & M University

Jerry Potter
Kent State University

Michael J. Quinn
University of New Hampshire

Robert S. Roos
Smith College

Richard Salter
Oberlin College

Cliff Shaffer
Virginia Polytechnic Institute and State University

Greg Starling
University of Arkansas

Robert E. Tarjan
AT&T Bell Laboratories

David Teague
Western Carolina University

Roger L. Wainright
University of Tulsa

Lynn R. Ziegler
St. John's University

We would also like to thank Frank Ruggirello who originally signed us up for this project, and Mike Sugarman, Mary Thomas-Stone, and Pam Rockwell at PWS for their patience, encouragement, and excellent technical advice.

We cannot overestimate our debt of gratitude to our wives, Peiyan and Ruta, for their constant encouragement, love, and support. Without their considerable sacrifices this book would not have been possible. We dedicate this book to them and to our children, Rachel and Sarah, and Ray and Renee.

K.A.B and *J.L.P*

Pseudocode Conventions for Sequential Algorithms

We include pseudocode for almost all of the algorithms that we discuss in this book. Our pseudocode resembles Pascal but is a hybrid of several high-level languages. To delineate blocks, we use matching keywords such as **if endif**, **for endfor**, **while endwhile**, **case endcase**, and so forth. Also, our priority rules for evaluating Boolean expressions differ from Pascal.

Our algorithms are always implemented as subprograms, which come in two flavors: functions and procedures. In the case of a function, we include a **return** statement for returning the value of the function. Information is passed to and from functions and procedures by argument-parameter correspondence, or by global variables. We always make explicit which parameters are input parameters, and which are output parameters. For procedures, a parameter can be an input parameter, an output parameter, or both. However, it is our assumption that all parameters for a function are strictly input parameters. *We assume that all input parameters (that are not also output parameters) are passed* **by value**, *so that the corresponding argument is not changed even if the input parameter is altered by the procedure or function, thereby avoiding unwanted side-effects.*

The opening syntax for a procedure has the following form, where information enclosed in pointed brackets <> is supplied by the designer (the pointed brackets themselves are not written in the pseudocode):

procedure <Name> (<List of parameters>)

Input: <parameter 1> (<description>)

.

.

.

<parameter *k*> (<description>)

Continued

Output: <parameter k+1> (<description>)

.
.
.

<parameter m> (<description>)

dcl: <Declaration of local variables and global variables>

The opening syntax for a function is similar, except that the **Output** statement then describes the value returned. The following example of opening syntax is from the procedure *InsertMaxHeap* given in Chapter 4:

procedure *InsertMaxHeap*(A[1:m],*HeapSize*,X)

Input: A[1:m] (an array of reals containing a heap in positions 1, . . . ,*HeapSize*)
HeapSize (the number of elements in the heap)
X (a value to be inserted into heap)

Output: A[1:m] (array altered by addition of X and heap maintained)

HeapSize (incremented by 1 from input value)

Note that we indicate arrays by using square brackets for index ranges. For brevity of notation, when we include an array A[1:m] in the parameter list of a function or procedure such as *InsertMaxHeap*, we are passing not only the array, but the variable m as well. Comments in the body of the pseudocode for an algorithm are enclosed in curly brackets {comments}. These comments are distinct from the descriptions of parameters in **Input** or **Output** statements that are enclosed in round brackets (descriptions). *To keep the pseudocode clean and easy to read, we avoid including extensive commentary in the pseudocode itself. Instead, we save such commentary for the text describing the algorithm.*

For brevity of code, the declare (**dcl**) *statement is usually omitted, since the role of local variables are usually clear from context.* Also, in most cases we are not explicit about the variable type (integer, real, character, etc.). Either the variable type is obvious from context, or the algorithm in question does not depend on the particular variable type being used (as in comparison-based sorting, searching, and so forth). We do, however, use a **dcl** statement when global variables are used, to distinguish between local and global variables.

What's in a name? That which we call a rose
By any other word would smell as sweet.

—*Romeo and Juliet, Act II, Scene II*
W. Shakespeare

A rose is a rose is a rose.

—Gertrude Stein

Echoing the quotation from Shakespeare, using the name *IMH* instead of *InsertMaxHeap* would certainly not affect the performance of the algorithm. However, we usually use names for algorithms (functions or procedures) that contain a reference to the problem being solved. The name of an algorithm might also contain a reference to the basic method used to solve the problem, like *InsertionSort* or *MergeSort*. While these naming conventions sometimes lead to rather long names, it makes remote references to the algorithm in the text much more descriptive.

In spite of Stein's truism, an *American Beauty* rose is not a *Flaming Crimson* rose is not a *Primrose Pink* rose. Similarly, a *PRAM* is not a 2-dimensional mesh is not a hypercube. Therefore, the names we give parallel algorithms will usually contain a reference to the particular parallel architecture for which the algorithm is designed. Again, this can lead to a lengthy name like *BitonicMergeSortPRAM*, but the price we pay for the length of the name is more than made up by descriptive nature of the name.

We now describe the form of our pseudocode constructs for branching and looping. We will not describe the action of these constructs here, since they are standard and found in any modern book on programming.

We use two branching constructs in our pseudocode, **if endif**, and **case endcase**. The **if endif** construct has the form:

if <Boolean expression> **then**
    <statement 1>
        .
        .
        .
    <statement k>
else
    <statement k + 1>
        .
        .
        .
    <statement m>
endif

The **else** clause is omitted when an alternative action is not to be executed if the boolean expression is false. When there is only one statement in the body of the **if** clause, and there is no **else** clause, then we often write the construct on one line as follows:

if <Boolean expression> **then** statement **endif**.

The **case endcase** construct with n branches has the form:

case
  :<Boolean expression 1>**:**
    <statement 1>
        .
        .
        .
    <statement k_1>

```
    :<Boolean expression 2>:
        <statement k_1+1>
            .
            .
            .
        <statement k_2>
            .
            .
            .
  :otherwise:
        <statement k_{n-1}+1>
            .
            .
            .
        <statement k_n>
  endcase
```

The **otherwise** clause is omitted if no action is required when all of the Boolean expressions are false. We assume that the Boolean expressions are evaluated in order, and that we branch out of the construct after executing the statements associated with the first Boolean expression having the value **.true.**, or after executing the code associated with the **otherwise** clause (if included) when all the Boolean expressions are false. Similar to our **if endif** construct, when the code associated with one of the **case** clauses consists of a single statement, then we write the statement on the same line as the clause itself.

We adopt the following priority rules when evaluating arithmetic or Boolean expressions. We list them from highest priority to lowest, with operations on the same level having equal priority. Priority between two operations at the same level proceeds left to right (we avoid using the exponentiation operator ∗∗).

1. Parentheses (evaluate parenthesized expressions from inside out, if nested)
2. Multiplication and Division ∗, /
3. Addition and Subtraction +, -
4. Relational operations <, >, ≤ , ≥ , =, ≠
5. Boolean connectives **.not.** **.and.** **.or.**

Thus, the Boolean expression:

$$(x + y)*z \le y - 1 \textbf{ .or. } z < 3 \textbf{ .and. .not. } x/z > s \textbf{ .or. } x = 5$$

is equivalent to the more fully parenthesized version:

$$(((x + y)*z \le y - 1) \textbf{ .or. } (z < 3 \textbf{ .and. } (\textbf{.not. } (x/z > s)))) \textbf{ .or. } (x = 5)$$

We assume the following *short-circuit* convention for the evaluation of a compound Boolean expression of the form E_1 **.and.** E_2, where E_1 and E_2 are Boolean expressions. If E_1 has the value **.false.**, then E_1 **.and.** E_2 automatically has the value **.false.**, and E_2 (which might not be defined) is not evaluated. For example, consider the compound Boolean expression:

$$i \geq 1 \textbf{ .and. } L[i] < Current,$$

where $L[1{:}n]$ is an array whose indices vary from 1 to n, inclusive. Suppose i has the value 0, so that the Boolean expression $i \geq 1$ is **.false.**. By our convention, this automatically implies that the compound Boolean expression is false, and the undefined $L[0]$ is thus not referenced. An analogous convention holds for compound expressions of the form E_1 **.or.** E_2.

We use three loop control constructions in our pseudocode:

```
for <variable> := <initial> to (downto) <final> do
    <statement 1>
        .
        .
        .
    <statement k>
endfor

while<Boolean expression> do
    <statement 1>
        .
        .
        .
    <statement k>
endwhile

repeat
    <statement 1>
        .
        .
        .
    <statement k>
until <Boolean expression>
```

We use **:=** for assignments versus = for logical comparison. We also use mathematical notation for the other logical comparison operators such as ≤, ≥, ≠, and so forth. For pointers, we use the notation $P\rightarrow$ to indicate the dynamic variable pointed to by the pointer P. For example, suppose the nodes of a linked list are stored in dynamic variables, which are records of the following type:

```
Node = record
           Info: Infotype
           Next: →Node
       end
```

If P is a pointer variable containing the address of *Node*, then $P \rightarrow Info$ denotes the contents of the information field of the node, and $P \rightarrow Next$ denotes the contents of the pointer field of the node.

The following function *Sum*, which computes the sum of n numbers stored in an array $X[1{:}n]$, illustrates some of our pseudocode conventions.

```
function Sum(X[1:n])

Input:  X[1:n] (array of n real numbers x_1, x_2, ..., x_n)
Output: x_1 + x_2 + ··· + x_n

   Summation := X[1]
   for i := 2 to n do
      Summation := Summation + X[i]
   endfor
   return(Summation)

end Sum
```

When writing pseudocode for sequential algorithms, we assume that we have a sequential (serial) computer having a single processor capable of executing all of the standard arithmetical operations, logical branching instructions, assignment statements, and so forth. We assume an internal random access memory (RAM), with each memory location being accessible with equal speed. Since our main concern is with high-level design strategies, most of our algorithms will assume that we have sufficient RAM to store all the variables used by the algorithm. Thus, we do not make much use of external memory devices accessed by **read** and **write** statements. In actual practice, however, it is sometimes necessary to design algorithms that input or output data from and to external devices such as disks and tapes, often in an "on line" fashion. An important example is provided by *B*-trees that contain keys from externally stored databases, where the complexity of the algorithms for searching, inserting, and deleting are dominated by the **read** and **write** operations required.

Pseudocode conventions for parallel algorithms involve additional complications due to the various models of parallel architectures that are currently in use. We postpone the discussion of pseudocode conventions for parallel algorithms until the necessary background is established.

Built-in Functions and Procedures

We use a number of built-in mathematical functions, such as the ceiling function $\lceil x \rceil$, the floor function $\lfloor x \rfloor$, **abs**(x), **odd**(n), **even**(n), and so forth, in our pseudocode for algorithms. A discussion of these built-in functions is included in Chapter 1. In the text portion, we usually use the notation $|x|$ to denote the absolute value **abs**(x) of a real number x. Given a finite set X, we will also use the notation $|X|$ to denote the cardinality (number of elements) of X (but this is not a built-in function). We also assume that there is a built-in procedure **interchange**(A,B), which interchanges the values of A and B.

In the statements of theorems, propositions, and exercises, we occasionally use the abbreviation "iff" for the phrase "if, and only if."

Part One

Introduction to Sequential and Parallel Algorithms

1

Introduction and Preliminaries

The use of algorithms did not begin with the introduction of computers. In fact, people have been using algorithms as long as they have been solving problems systematically. Informally, an algorithm can be described as a finite sequence of rules that solves a problem. While a completely rigorous definition of an algorithm uses the notion of a Turing Machine, the following definition will suffice for our purposes.

DEFINITION 1.1 An *algorithm* is a complete, step-by-step procedure for solving a specific problem. Each step must be unambiguously expressed in terms of a finite number of rules and guaranteed to terminate in a finite number of applications of these rules. A rule typically calls for the execution of one or more operations.

A *sequential* algorithm performs these operations one at a time in sequence, whereas a *parallel* algorithm performs many operations simultaneously.

In this text, the operations allowed in a sequential algorithm are restricted to instructions found in a typical high-level procedural computer language. These instructions include, for example, arithmetical operations, logical comparisons, and transfer of control. A parallel algorithm uses the same operations as a sequential algorithm, but a given operation can be performed on multiple data instances simultaneously. For instance, a parallel algorithm might add 1 to each element in an array of numbers in a single parallel step.

When designing parallel algorithms, it is important to consider the various models and architectures of parallel computers. For example, some parallel computers do not allow two different operations, such as a multiplication and an addition, to be performed at the same time (single-instruction, multiple-data), while others allow a mix of operations to be performed simultaneously (multiple-instruction, multiple-data).

In this chapter we provide some historical background for the study of sequential and parallel algorithms. Some elementary mathematical concepts that are needed for the analysis of algorithms are reviewed. The notion of recursion, a fundamental tool in the design and analysis of algorithms, is also reviewed in this chapter.

1.1 Algorithms from Ancient to Modern Times

The algorithms discussed in this section solve some classical problems in arithmetic. Some algorithms that are commonly executed on today's computers were originally developed more than three thousand years ago. The examples we present illustrate the fact that the most straightforward algorithm for solving a given problem often is not the most efficient.

1.1.1 Evaluating Powers

An ancient problem in arithmetic is the efficient evaluation of integer powers of a number x. The naive approach to evaluating x^n is to repeatedly multiply x by itself $n - 1$ times, yielding the following algorithm.

```
function NaivePowers(x,n)

Input:  x (a real number), n (a positive integer)
Output: x^n

  Product := x
  for i := 1 to n – 1 do
    Product := Product * x
  endfor
  return(Product)

end NaivePowers
```

Clearly, *NaivePowers* performs $n - 1$ multiplications. In the case that n is a power of two, say $n = 2^m$, an improved algorithm based on successive squaring was already in use as the basis for multiplication in ancient Egypt (c. 1800 B.C.). For example, to calculate x^8, first square x, then square the result x^2 yielding x^4, and finally square x^4. This method of successive squaring can be applied to calculate x^n when n is any power of 2: It uses $\log_2 n$ multiplications rather than the $n - 1$ used by *NaivePowers*.

```
function EgyptianPowers(x,n)

Input:  x (a real number), n (a power of two, n = 2^m)
Output: x^n

  j := 1
  Product := x
  while j < n do
    Product := Product * Product
    j := j + j
  endwhile
  return(Product)

end EgyptianPowers
```

The value of j is doubled during each pass of the **while** loop in *EgyptianPowers*, so that after the first k passes have been completed, j has the value 2^k. Therefore, after $k = m$ passes, j has the value $2^m = n$, and the **while** loop terminates. Since one multiplication is performed in each pass, *EgyptianPowers* performs a total of $m = \log_2 n$ multiplications. For large n, $\log_2 n$ is significantly smaller than $n - 1$ (see Figure 1.1), so that *EgyptianPowers* is a dramatic improvement in efficiency over *NaivePowers*.

EgyptianPowers is a special case of a squaring algorithm, called the *binary method*, that works for *any* integer n (not just when n is a power of 2). This binary method was known before the year 200 B.C. and was included in Pingala's classic Chandah-sutra.[1] The method is based on the following simple observation: If n is even, then x^n is the square of $x^n/2$, whereas if n is odd, then x^n is the product of x and the square of $x^{(n-1)/2}$. A recursive function *PowersRec* based on this observation is given in Section 1.3.2.

1.1.2 The Euclidean Algorithm

One of the oldest problems in number theory is to determine the *greatest common divisor* gcd(a,b) of two positive integers a and b. The greatest common divisor of a and b is the largest positive integer k that divides both a and b with no remainder. The problem of calculating gcd(a,b) was already known to the ancient Greek mathematicians. A naive algorithm computes the prime factorization of a and b and collects common prime powers whose product is then equal to gcd(a,b). However, for large a and b, computing the prime factorizations is very time consuming, even on today's fastest computers. A more efficient algorithm was published in *Euclid's Elements* (c. 300 B.C.). Euclid's algorithm is a refinement of an algorithm that was known 200 years earlier. This earlier algorithm is based on the observation that for $a \geq b$, an integer divides both a and b if, and only if, it divides $a - b$ and b. Thus,

$$\gcd(a,b) = \gcd(a-b,b), \quad a \geq b, \quad \gcd(a,a) = a. \qquad (1.1.1)$$

Figure 1.1 Table of values of $\log_2 n$ versus $n - 1$

$n = 2^m$	$m = \log_2 n$	$n - 1$
1	0	0
2	1	1
16	4	15
128	7	127
1,024	10	1,023
1,048,576	20	1,048,575
562,949,953,421,312	48	562,949,953,421,311

[1]See B. Datta and A. S. Singh, *History of Hindu Mathematics*, 1 (Bombay: 1935), 76.

Formula (1.1.1) yields the following algorithm for computing gcd(a,b).

```
function GCD(a,b)
Input:  a,b (two positive integers)
Output: gcd(a,b) (the greatest common divisor of a and b)
  while a ≠ b do
     if a > b then
        a := a – b
     else
        b := b – a
     endif
  endwhile
  return(a)
end GCD
```

After each iteration of the **while** loop in *GCD*, the larger of the previous values of a and b is replaced by a strictly smaller positive number. Hence, *GCD* eventually terminates, having calculated the gcd of the original a and b.

Euclid's gcd algorithm refines the algorithm *GCD* by utilizing the fact that if $a > b$ and $a - b$ is still greater than b, then $a - b$ in turn is replaced by $a - 2b$, and so forth. Hence, if a is not a multiple of b, then a is eventually replaced by $r = a - qb$, where r is the remainder when a is divided by b. Thus, all these successive subtractions can be replaced by the single invocation a **mod** b, where **mod** is the built-in function defined by

$$a \textbf{ mod } b = a - b\left\lfloor \frac{a}{b} \right\rfloor, \quad a \text{ and } b \text{ integers}, \quad b \neq 0,$$

where $\lfloor x \rfloor$ denotes the largest integer less than or equal to x.

For example, when calculating gcd(108,8), the thirteen subtractions 108 – 8, 100 – 8, 92 – 8, ... , 12 – 8 executed by the algorithm *GCD* can be replaced by the single calculation 108 **mod** 8 = 4.

The preceding discussion leads to an algorithm based on the following formula:

$$\gcd(a,b) = \gcd(b, a \textbf{ mod } b). \tag{1.1.2}$$

Note that when a is a multiple of b, gcd(a,b) = b, and a **mod** b = 0. So the usual convention gcd(b,0) = b shows that (1.1.2) remains valid when a is a multiple of b.

Euclid's description of the gcd algorithm based on (1.1.2) was complicated by the fact that the algebraic concept of zero was not yet formalized. The following is a modern version of Euclid's algorithm.

```
function EuclidGCD(a,b)

Input:   a,b (two nonnegative integers)
Output: gcd(a,b) (the greatest common divisor of a and b)

  while b ≠ 0 do
    Remainder := a mod b
    a := b
    b := Remainder
  endwhile
  return(a)

end EuclidGCD
```

We illustrate *EuclidGCD* for input $a = 10724$, $b = 864$. We have

$$\gcd(10724,864) = \gcd(864,356) = \gcd(356,152) = \gcd(152,52) = \gcd(52,48) = \gcd(48,4) = \gcd(4,0) = 4.$$

1.1.3 Babylonian Square Roots

Another mathematical problem gained special significance in the sixth century B.C. when the Pythagorean school of geometers made the startling discovery that the length of the hypotenuse of a right triangle with legs both equal to 1 cannot be expressed as the ratio of two integers. This conclusion is equivalent to saying that $\sqrt{2}$ is not a rational number and therefore its decimal expansion can never be completely calculated. Long before this discovery of irrational numbers people were interested in calculating the square root of a given positive number a to any desired degree of accuracy. A square root algorithm was already known to the Babylonians by 1500 B.C. and is perhaps the first nontrivial mathematical algorithm.

The Babylonian method for calculating $\sqrt{a}$ is based on averaging two points on either side of $\sqrt{a}$. The Babylonians may have discovered this algorithm by considering the problem of laying out a square plot of a given area. For example, consider an area of 5. As a first approximation, they may have considered a rectangular plot of dimensions 1 by 5. If they replaced one dimension by the average of the previous two dimensions, they would obtain a "more square" plot of dimension 3 by 5/3. If they next replaced one of the new dimensions by the average of 3 and 5/3, the dimensions of the plot would be 7/3 by 5/(7/3) (roughly 2.33 by 2.14). More repetitions of this technique leads to plots having sides that are better and better approximations to $\sqrt{5}$ (see Figure 1.2, page 8).

We can use the Babylonian square root algorithm to calculate the square root of any positive number a. Start with an initial guess $x = x_1$ for $\sqrt{a}$; any guess will do, but a good initial guess leads to more rapid convergence. We calculate successive approximations $x_2, x_3, \ldots$ to $\sqrt{a}$ using the formula

$$x_i = x_{i-1} + \frac{(a/x_{i-1})}{2}, \ i = 2,3, \ldots.$$

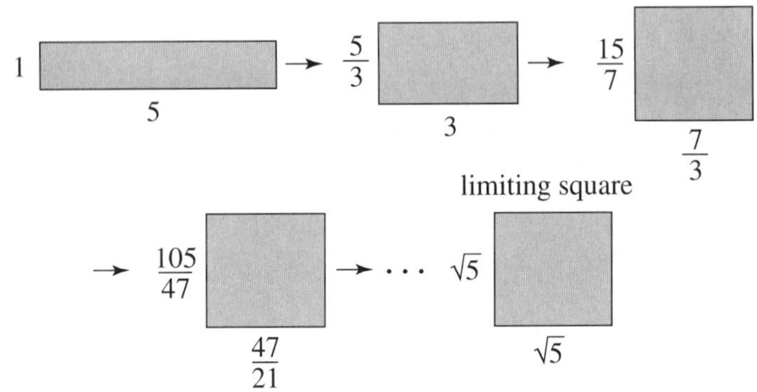

Figure 1.2 Increasing better approximations of $\sqrt{5}$

We write the Babylonian square root algorithm as a function whose input parameters are the number a and a positive real number *error* measuring the desired accuracy for computing $\sqrt{a}$. For simplicity, we use a as our initial approximation to $\sqrt{a}$.

```
function BabylonianSQRT(a,error)

Input:  a (a positive number), error (a positive real number)
Output: √a accurate to within error

   x := a
   while |x – a/x| > error do
      x := (x + a/x)/2
   endwhile
   return(x)

end BabylonianSQRT
```

Finding square roots is a special instance of the problem of determining the (approximate) roots of a polynomial (note that $\sqrt{a}$ is the positive root of the polynomial $x^2 - a$). More generally, suppose $f(x)$ is a real-valued function, and we wish to find a value of x (called a *zero of f*) such that $f(x) = 0$. If f is a continuous function, then the intermediate value theorem of calculus guarantees that a zero occurs in any closed interval $[a, b]$ where $f(a)$ and $f(b)$ have opposite signs. An algorithm (called the *bisection method*) for determining a zero of f proceeds by bisecting such an interval $[a, b]$ in half, then narrowing the search for a zero to one of the two subintervals where a sign change of f occurs. By repeating this process n times, an approximation to a zero is computed that is no more than $(b - a)/2^n$ from an actual zero. We leave the pseudocode for the bisection method to the exercises.

For the case when $f(x)$ is a differentiable function, a more efficient method for finding zeros of f was developed by Sir Isaac Newton in the seventeenth century. Newton's method is based on constructing the tangent line to the graph of f at an initial guess for a zero x_1. The point x_2 where this tangent line crosses the x axis is taken as the second approximation to a zero of f. This process is then repeated at x_2,

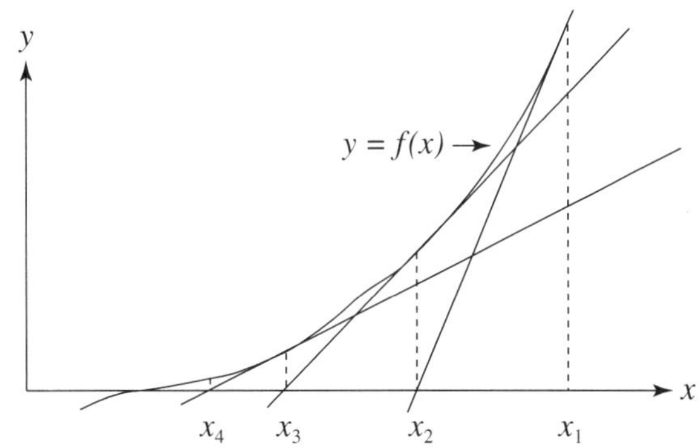

Figure 1.3 Newton's method for finding a zero of a differentiable function f

yielding a third approximation x_3 (see Figure 1.3). Successive iterations yield points $x_2, x_3, \dots$ given by the formula (see Exercise 1.10)

$$x_i = x_{i-1} - \frac{f(x_{i-1})}{f'(x_{i-1})}, \quad i = 2,3,\dots. \tag{1.1.3}$$

Curiously, when applied to the polynomial $x^2 - a$ Newton's method yields exactly the Babylonian square root algorithm. In general, certain conditions need to be imposed on the function f and starting point x_1 to guarantee that the points x_i given by (1.1.3) actually converge to a zero.

1.1.4 Evaluating Polynomials

A basic problem in mathematics is the problem of evaluating a polynomial $p(x) = a_n x^n + a_{n-1}x^{n-1} + \cdots + a_1 x + a_0$ at a particular value v of x. The most straightforward solution to the problem would be to compute each term $a_i x^i$ independently, $i = 1, \dots, n$, and sum the individual terms. However, when computing each power v^i, $i = 1, \dots, n$, it is more efficient to obtain v^i by multiplying the already calculated v^{i-1} by v. This simple observation leads to the following algorithm.

```
function PolyEval(a[0:n],v)

Input:  a[0:n] (an array of real numbers), v (a real number)
Output: the value of the polynomial a_n x^n + a_{n-1} x^{n-1} + ... + a_1 x + a_0   at x = v

  Sum := a[0]
  Product := 1
  for i := 1 to n do
     Product := Product * v
     Sum := Sum + a[i] * Product
  endfor
  return(Sum)

end PolyEval
```

PolyEval clearly does $2n$ multiplications and n additions, and this might seem the best that we can do. However, there is a simple algorithm for polynomial evaluation that cuts the number of multiplications in half. This algorithm goes under the name of Horner's Rule, since W. G. Horner popularized the method in 1819, but the algorithm was actually devised by Sir Isaac Newton in 1699. Horner's Rule for polynomial evaluation is based on a clever parenthesizing of the polynomial. For example, a fourth-degree polynomial $a_4x^4 + a_3x^3 + a_2x^2 + a_1x + a_0$ is rewritten as

$$(((a_4*x + a_3)*x + a_2)*x + a_1)*x + a_0.$$

This rewriting can be done for a polynomial of *any* degree n, yielding the following algorithm.

```
function HornerEval(a[0:n],v)

Input:  a[0:n] (an array of real numbers), v (a real number)
Output: the value of the polynomial a_n x^n + a_{n-1} x^{n-1} + ··· + a_1 x + a_0   at x = v

    Sum := a[n]
    for i := n – 1 downto 0 do
        Sum := Sum * v + a[i]
    endfor
    return(Sum)

end HornerEval
```

Clearly, *HornerEval* performs n multiplications and n additions when evaluating a polynomial of degree n. It can be proved that any algorithm for evaluating an nth-degree polynomial that only uses multiplications or divisions and additions or subtractions must perform at least n multiplications or divisions and at least n additions or subtractions. Hence, *HornerEval* is an optimal algorithm for evaluating polynomials.

Algorithms for various other mathematical problems have been developed over the years. These old algorithms were mostly concerned with algebraic rules for calculating numbers and solving arithmetic equations. Indeed, the word *algorithm* itself is taken from the name of the ninth-century Arabic mathematician and astronomer Al-Khowarizmi, who wrote a famous book (*Al Jabr*, c. 825 A.D.) on the manipulation of numbers and equations. One of Al-Khowarizmi's accomplishments involved converting Jewish-calendar dates to Islamic dates. In fact, many early algorithms involved calendars and fixing the days for feasts.

1.1.5 Toward a Modern Theory of Algorithms

Up until the twentieth century, the study of algorithms was largely done on an ad hoc basis, and no general theory of algorithms had emerged. However, the mechanical implementation of algorithms was already conceived by Charles Babbage as early as 1834, as evidenced by his work on the "Analytical Engine." This machine was to be a more general-purpose mechanical computational device than his earlier special-purpose "Difference Engine" for doing mathematical computations. Unlike the Difference Engine, the Analytical Engine was never built. However, it is clear from Babbage's writings that he recognized that a set of punched cards for controlling his engine would constitute a representation of an algorithm. The algorithmic nature of the Analytical Engine was further discussed in the writings of Ada Augusta Byron, Countess of Lovelace. Countess Lovelace actually wrote algorithms suitable for implementation on the Analytical Engine. She is considered to be the first computer programmer (the modern programming language ADA is named in her honor).

The formal mathematical concept of an algorithm was precisely formulated by mathematicians in the 1930s. Perhaps the most influential work in this area was done by the English mathematician Alan Turing, who described a simple but very general theoretical model (called a Turing Machine) of the mechanism by which computing machines can solve problems. Turing also described the stored-program paradigm that is the basis of modern general-purpose computers. This paradigm is usually attributed to John von Neumann, who formally demonstrated the practicality of encoding instructions to a "real" computer in the same (binary) language used for the data it processed, with both instructions and data being stored in the computer's memory. Of course, the advent of the electronic computer in the 1940s ushered in a veritable explosion of interest in the subject of algorithms.

Early computers were all single-processor machines and could only accept programs that implemented sequential algorithms. General algorithm design strategies for sequential algorithms then evolved, forming a theoretical framework for algorithmic problem solving. Effective computer implementation of algorithms led to the creation of the theory of data structures. The theory of structured programming set guidelines for translating algorithms into code having a high degree of modularity and easily understood logical flow. Modern programming languages were designed to support these data structures and structured programming concepts.

Using appropriate algorithms, computers can solve problems that were once inaccessible. Space travel, robotics, graphics, simulations of large and complex systems, and medical imaging including CAT scans have all been made possible by the development of efficient algorithms implemented on computers with large memories and extremely fast processors.

Since 1984, parallel computers containing tens, hundreds, or even thousands of processors working in parallel have been developed and marketed, creating a need for algorithms that exploit this parallelism. The subject of parallel algorithms has attracted an enormous amount of interest and research in recent years.

1.1.6 Serial and Parallel Computers

The sequential algorithms presented in this book are suitable for implementation on a serial computer having a single processor capable of executing all of the standard arithmetic operations, logical branching instructions, and assignment statements. We assume an internal random access memory (RAM), with each memory location being accessible with equal speed. Since our main concern is with high-level design strategies, most of our algorithms assume that we have sufficient RAM to store all the variables used by the algorithm. Thus, we only make occasional use of external memory devices accessed by **read** and **write** statements.

While a single-processor computer can only execute one instruction at a time, a parallel computer with p processors $P_1, \ldots, P_p$ can execute p instructions simultaneously, with each processor operating on (possibly) different data. Models of parallel computation that restrict the processors to performing the *same* instruction on multiple data are called SIMD (single-instruction, multiple-data) machines. Parallel computers that allow *different* instructions to be performed at the same time on multiple data are called MIMD (multiple-instruction, multiple-data) machines. The serial computer can be viewed in this terminology as a SISD (single-instruction, single-data) machine.

The notion of RAM memory for serial computers can be extended in two main and very different ways to parallel computers: *shared* memory versus *distributed* memory. The shared-memory model assumes that *all* processors have access to a single shared RAM memory. Parallel computers based on this model are referred to as *Parallel Random Access Machines*, or *PRAMs*. PRAMs fall into four main categories, depending on how conflicts are resolved when two processors attempt to read or write from the same memory location simultaneously. The distributed-memory model assumes that each processor has its own local memory, and information is passed between processors by sending messages along an interconnection network. Parallel algorithms must be designed according to the particular type of PRAM or interconnection network model used.

1.2 Mathematical Notation and Background

The design and analysis of algorithms requires familiarity with certain concepts from mathematics such as functions, sets, various summation formulas, and so forth. In this section we review some of these commonly used concepts and establish appropriate notation. Certain functions occur frequently in the design and analysis of algorithms and are usually implemented as built-in functions in most high-level languages. Two of the most commonly used functions are the ceiling $\lceil x \rceil$ and floor $\lfloor x \rfloor$, defined to be the smallest integer greater than or equal to x and the largest integer smaller than or equal to x, respectively. For example, $\lceil 1.23 \rceil = 2$, and $\lfloor 1.23 \rfloor = 1$. For a positive x, $\lfloor x \rfloor$ is obtained from x by truncating its decimal part. Note that $x - 1 < \lfloor x \rfloor \leq x$, and $x \leq \lceil x \rceil < x + 1$ (see Figure 1.4).

When $\lceil x \rceil$ and $\lfloor x \rfloor$ are invoked in our pseudocode as built-in functions, we retain the mathematical notation for such invocations rather than using names such as **ceiling**(x) and **floor**(x), respectively. On the other hand, for functions such as $\sqrt{x}$, we will use the name **sqrt**. The following is a list of names and definitions for some of the more commonly used built-in functions.

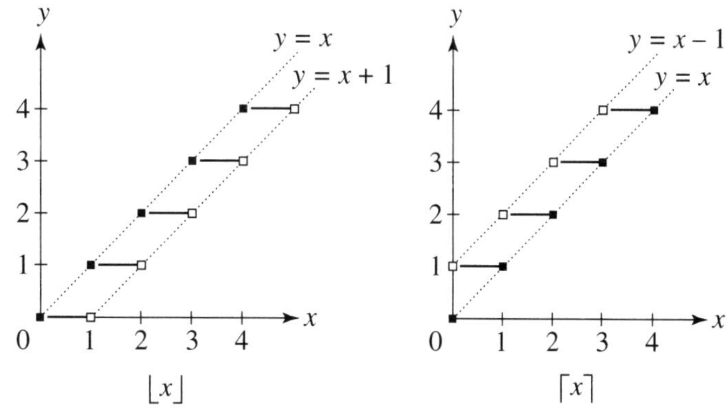

Figure 1.4 Graphs of the functions $y = \lfloor x \rfloor$, $y = x$, and $y = \lceil x \rceil$, shown for $x \geq 0$.

sqrt$(x) = \sqrt{x}$, x a nonnegative real number.

abs $(x) = |x| = \begin{cases} x & x \geq 0, \\ -x & \text{otherwise}, \end{cases}$ x a real number.

a **mod** $b = a - b\lfloor a/b \rfloor$, a and b integers, $b \neq 0$.

odd(n) = **.true.** if, and only if, n **mod** $2 = 1$, n an integer.

even(n) = **.true.** if, and only if, n **mod** $2 = 0$, n an integer.

Note that for positive integers a and b, a **mod** b is the remainder when a is divided by b.

Another important function in algorithms is the logarithm function. Given a base $b > 1$, the logarithm of x to the base b, denoted by $\log_b x$, is defined to be the functional inverse of the exponential function b^x. In other words, $\log_b x$, $x > 0$, is defined to be the power to which the base b must be raised to equal x, so that

$$b^{\log_b x} = x. \tag{1.2.1}$$

We illustrate the functional inverse relationship of 2^x and $\log_2 x$ in Figure 1.5 (page 14). Note that their graphs are the reflections of one another about the line $y = x$. Since 2^x grows very rapidly, it follows that $\log_2 x$ grows very slowly.

The differences between logarithmic growth ($y = \log_2 x$), linear growth ($y = x$), and exponential growth ($y = 2^x$) are dramatic (see Figure 1.6, page 14). The first few entries for x in Figure 1.6 were obtained by doubling the previous value. Note that doubling the input to the function $\log_2 x$ only increases its output by one. On the other hand, doubling the input to the function 2^x results in squaring its output. This is indeed a dramatic difference!

The exponential function b^x has the following fundamental properties:

$$b^{x+y} = b^x b^y, \tag{1.2.2}$$

$$(b^x)^y = (b^y)^x = b^{xy}. \tag{1.2.3}$$

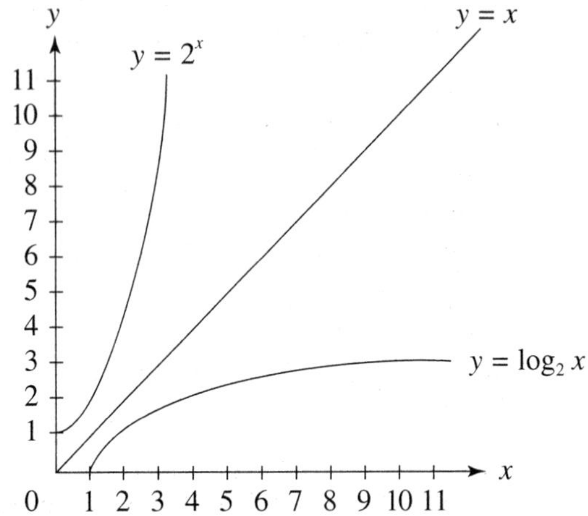

Figure 1.5 Graphs of $y = \log_2 x$, $y = x$, and $y = 2^x$

The following two properties of $\log_b x$ corresponding to (1.2.2) and (1.2.3) are immediately obtained by raising both sides to the power b and using (1.2.1).

$$\log_b xy = \log_b x + \log_b y, \tag{1.2.4}$$

$$\log_b x^y = y \log_b x. \tag{1.2.5}$$

The following useful formula allows us to change from one base to another (see Exercise 1.18):

$$\log_a x = \frac{\log_b x}{\log_b a}. \tag{1.2.6}$$

The most commonly used bases are 2, e, and 10. When the base is e, the logarithm is referred to as the *natural* logarithm and is denoted by $\ln x$.

Summations involving arithmetic and geometric progressions occur frequently in the analysis of algorithms. Given real numbers a, d, and a positive integer n, the sum of the first n terms of the arithmetic progression with leading term a and difference d is given by

Figure 1.6 Table of values of $y = x$, $y = \log_2 x$, and $y = 2^x$

x	$\log_2 x$	2^x
1	0	2
2	1	4
4	2	16
8	3	256
16	4	65,536
48	5.58496	562,949,953,421,312
562,949,953,421,312	48	eeeeeeeeenormous

$$a + (a + d) + (a + 2d) + \cdots + (a + (n-1)d) = \frac{n(2a + (n-1)d)}{2}. \qquad (1.2.7)$$

Formula (1.2.7) is easily proved by adding the progression to a copy of itself, but where we add the ith term in the original progression to the $(n - i + 1)$st term in the copy. Then each of the resulting n summands equals $(2a + (n - 1)d)$. In algorithm analysis, the formula (1.2.7) occurs most often with $a = d = 1$, in which case it reduces to

$$1 + 2 + \cdots + n = \frac{n(n+1)}{2}. \qquad (1.2.8)$$

There are several interesting geometrical-combinatorial interpretations of (1.2.8). One such interpretation and "proof without words" results from counting the points in a suitably drawn $n \times (n + 1)$ parallelogram of points, as illustrated by Figure 1.7 where $n = 8$.

Given a real number x and a positive integer n, the sum of the first n terms of the geometric progression $1, x, x^2, \ldots, x^p, \ldots$ is given by

$$1 + x + x^2 + \cdots + x^{n-1} = (x^n - 1)/(x - 1). \qquad (1.2.9)$$

Formula (1.2.9) follows easily by multiplying both sides by $x - 1$ and simplifying the left-hand side. An important special case in the analysis of algorithms occurs when $x = 2$. Then (1.2.9) reduces to

$$1 + 2 + 2^2 + \cdots + 2^{n-1} = 2^n - 1. \qquad (1.2.10)$$

Another way of interpreting formula (1.2.10) is to note that the base-two expansion of the number $2^n - 1$ consists of a sequence of n ones.

A formula that is useful when analyzing the average behavior of algorithms is obtained by replacing n by $n + 1$ in (1.2.9) and then differentiating both sides of the equation, yielding

$$1 + 2x + 3x^2 + \cdots + nx^{n-1} = \frac{nx^{n+1} - (n+1)x^n + 1}{(x-1)^2}. \qquad (1.2.11)$$

Figure 1.7 Counting the points in an $n \times (n + 1)$ parallelogram of points

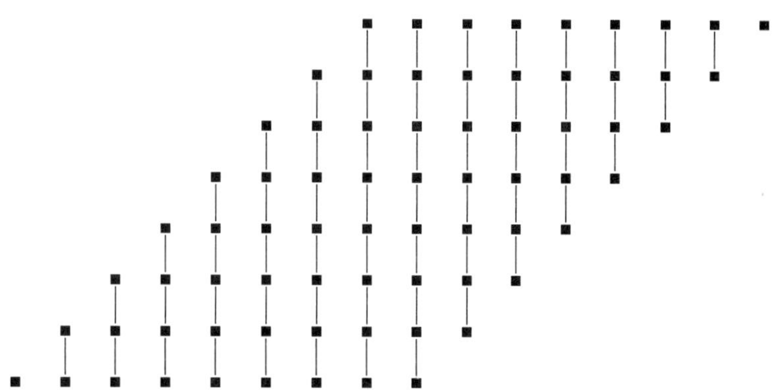

$1 + 2 + 3 + 4 + 5 + 6 + 7 + 8 + 8 + 7 + 6 + 5 + 4 + 3 + 2 + 1$
$= 2(1 + 2 + \cdots + 8) =$ number of points in parallelogram $= 8 \times 9$

For x a real number, $-1 < x < 1$, taking the limit of both sides of (1.2.9) and (1.2.11) as n approaches infinity yields the following two useful power series formulas.

$$\sum_{n=0}^{\infty} x^n = \frac{1}{1-x} \tag{1.2.12}$$

$$\sum_{n=0}^{\infty} (n+1)\,x^n = \frac{1}{(1-x)^2} \tag{1.2.13}$$

Of course, (1.2.13) also can be obtained from (1.2.12) by differentiating both sides of the equation.

We now consider two combinatorial quantities that arise often in counting arguments and in probability computations. Suppose we have a finite set S having n elements. For any k between 0 and n, the number of ways to make an *ordered* choice of k elements from S is given by the product

$$n^{(k)} = n(n-1)\ \dots\ (n-k+1). \tag{1.2.14}$$

In particular, note that $n! = n^{(n)}$. The following inequality will be useful (see Exercise 1.19):

$$\frac{n^{(k)}}{m^{(k)}} < \left(\frac{n}{m}\right)^k, \quad 0 < k \le n < m. \tag{1.2.15}$$

Let $C(n, k)$ denote the number of ways to make an *unordered* choice of k elements from S. It then follows that

$$C(n, k) = \frac{n^{(k)}}{k!}, \quad \text{for } k = 0, \dots, n, \tag{1.2.16}$$

since $n^{(k)}$ stood for the number of ordered choices of k elements, and there are $k!$ different orderings of these k elements. Hence, using (1.2.14) and (1.2.16), we have

$$C(n, k) = \frac{n(n-1)\dots(n-k+1)}{k!} = \frac{n!}{(n-k)!k!} \quad \text{for } k = 0, \dots, n. \tag{1.2.17}$$

We refer to $C(n, k)$ as "n choose k." We often use the alternate notation $\binom{n}{k}$ for $C(n, k)$. As an illustration, consider the set $S = \{a, b, c, d\}$. There are twelve ways to choose an ordered subset of two elements from S, namely,

ab, ba, ac, ca, ad, da, bc, cb, bd, db, cd, dc,

but only $12/2! = 6$ ways to choose an unordered two-element subset, namely,

$$\{a, b\}, \{a, c\}, \{a, d\}, \{b, c\}, \{b, d\}, \{c, d\}.$$

Thus, $4^{(2)} = 12$, and $C(4, 2) = \binom{4}{2} = 6$.

The number $C(n, k) = \binom{n}{k}$ is also called a *binomial coefficient* because of its appearance in the binomial expansion:

$$(a+x)^n = \sum_{i=0}^{n} \binom{n}{i} a^{n-i} x^i \tag{1.2.18}$$

$$= \binom{n}{0} a^n x^0 + \binom{n}{1} a^{n-1} x^1 + \binom{n}{2} a^{n-2} x^2 + \cdots + \binom{n}{i} a^{n-i} x^i + \cdots + \binom{n}{n} a^0 x^n.$$

Taking the partial derivative of both sides of (1.2.18) with respect to x and then multiplying by x yields the following useful identity.

$$nx(a+x)^{n-1} = \sum_{i=0}^{n} \binom{n}{i} a^{n-i} x^i i \tag{1.2.19}$$

The binomial coefficients give us yet another interpretation of the formula $1 + 2 + \cdots + n = n(n + 1)/2$. Combining (1.2.8) and (1.2.17) yields

$$1 + 2 + \cdots + n = \binom{n+1}{2}. \tag{1.2.20}$$

A direct verification of (1.2.20) can be given as follows. Let $\{1, 2, \ldots, n + 1\}$ be the $(n + 1)$-element set. Then there are n ways to choose a two-element subset containing 1, $n - 1$ ways to choose a two-element subset containing 2 but not 1, $n - 2$ ways to choose a two-element subset containing 3 but not 1 and 2, and so forth. Further properties of binomial coefficients are discussed in Chapter 7.

The binomial expansion (1.2.18) has an important generalization (due to Newton) to the case where $a = 1$ and $|x| < 1$ and n is an arbitrary real number c.

$$(1+x)^c = \sum_{i=0}^{\infty} \binom{c}{i} x^i, \tag{1.2.21}$$

where the generalized binomial coefficient is defined by

$$\binom{c}{k} = c(c-1)\ldots(c-k+1)/k!. \tag{1.2.22}$$

Various definitions and notation from the elementary theory of sets are needed in our discussion of asymptotic behavior given in Chapter 3. These notions from set theory are also required in our discussion of elementary probability theory given in Chapter 9. In addition to utilizing sets in the mathematical analysis of algorithms, there are many important algorithms that implement union and find operations for disjoint sets.

Whenever we talk about a set, we assume that the set is contained in a certain *universal set* U. The set U is always clear from context (all edges in a graph, all real-valued functions defined on the nonnegative integers, all outcomes of an experiment, etc.). The set of real numbers and the set of nonnegative integers occur often, and we denote them by $\mathbb{R}$ and $\mathbb{N}$, respectively.

Given a set A, we indicate that x is a member of (belongs to) A by $x \in A$. We write $x \notin A$ when x is not a member of (does not belong to) A. The set having no members, called the *empty set*, is denoted by $\varnothing$. Given two sets A and B, we say that A is a *subset* of B, denoted by $A \subseteq B$, if every member of A is also a member of B. We also say that A *is contained in* B. If $A \subseteq B$ and $A \neq B$, then we say that A is a *proper* subset of B, denoted by $A \subset B$. We also say that A is *strictly* contained in B. The *union* of A and B, denoted by $A \cup B$, is the set of all elements that are members of either A or B

(or both). The *intersection* of A and B, denoted by $A \cap B$, is the set of all elements that are members of both A and B. The sets A and B are called *disjoint*, if $A \cap B = \varnothing$. The *difference* between A and B, denoted by $A \setminus B$, is the set of all members of A that are not members of B. The *complement* of A, denoted by A^c, is the set of all members of the universal set U that are not members of A, that is $A^c = U \setminus A$. The *Cartesian product* $A \times B$ is the set of all ordered pairs $\{(a,b): a \in A, b \in B\}$.

Figure 1.8 gives *Venn diagrams* illustrating subset, union, intersection, difference, and complement.

1.3 Recursion

Recursion is a very powerful tool in the design and analysis of algorithms and is utilized throughout the text. In many situations, the solution to a given problem can be expressed naturally in terms of a solution (or solutions) to a smaller or simpler input (or set of inputs) to the same problem. This expression often takes the form of a *recurrence relation*, which includes an *initial condition* stating the known solution for a initial set of inputs. Exploiting such recurrence relations is the essence of *recursive* algorithmic strategy. Not only do recurrence relations yield elegant algorithms for solving many problems, recurrence relations also arise frequently in the analysis of the complexity of algorithms.

1.3.1 Recursive Functions and Procedures

A *recursive function* is simply a function whose evaluation at (noninitial) inputs involves invoking the function itself at another input (or set of inputs). A *recursive procedure* is defined similarly. For our first example, consider the function $Factorial(n) = n! = 1 \times 2 \times \cdots \times n$, where $0! = 1$ by convention. Clearly, $Factorial(n)$ satisfies the recurrence relation

$$Factorial(n) = n \times Factorial(n-1), \quad \textbf{init. cond. } Factorial(1) = 1. \qquad (1.3.1)$$

Figure 1.8 Venn diagrams for set operations

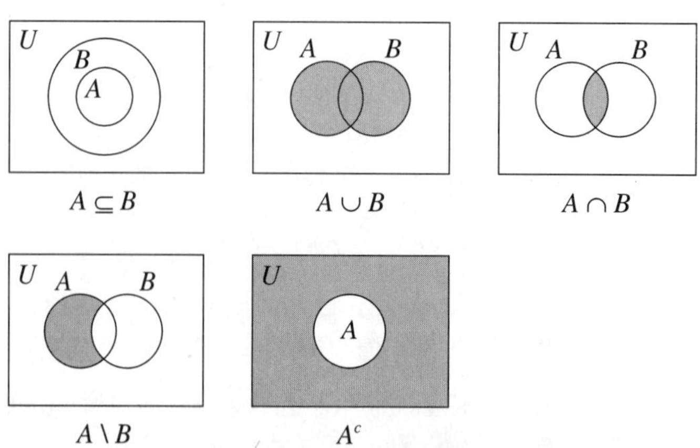

Employing recurrence relation (1.3.1), we obtain the following recursive function for calculating $n!$.

```
function Factorial(n) recursive
Input:  n (a nonnegative integer)
Output: n!
  if n ≤ 1 then
    return(1)
  else
    return (n*Factorial(n – 1))
  endif
end Factorial
```

What makes this recursive function work is that each invocation of *Factorial* brings us one step closer to the known solution for $n = 1$. Such progress is essential to any successful recursive function to ensure that the function resolves (terminates). In other words, the function must not invoke itself endlessly.

Consider again the problem of computing the greatest common divisor of two nonnegative integers a and b (gcd(a,b)). In Section 1.2 we described the nonrecursive algorithm *EuclidGCD* for solving this problem. A recursive version of this algorithm can be based on the following recurrence relation 1.3.2 (see also (1.1.2)).

$$\gcd(a,b) = \gcd(b, a \textbf{ mod } b), \quad \textbf{init. cond. } \gcd(a,0) = a \tag{1.3.2}$$

Since $b < a$ **mod** b, each application of (1.3.2) leads us to an input whose second argument is strictly smaller than the second argument of the previous input. Thus, the following recursive function *GCDRec* based on (1.3.2) resolves after finitely many invocations with a final value of 0 for the second input parameter.

```
function EuclidGCDRec(a,b) recursive
Input:  a,b (nonnegative integers)
Output: gcd(a,b)
  if b = 0 then
    return(a)
  else
    return(EuclidGCDRec(b,a mod b))
  endif
end EuclidGCDRec
```

Note that *Factorial* and *EuclidGCDRec* contain only one recursive invocation (reference) in the code and are therefore examples of *single-reference* recursion. A recursive function or procedure may contain several recursive reference statements in the code and still be single-reference recursion. For example, the following code for a function *f* containing the two recursive invocation statements (and no others)

```
if condition A then
   invoke f with given parameters
else
   invoke f with other parameters
endif
```

is still considered single-reference recursion, since only one of the two recursive references is executed by any given *current* invocation of the function *f*. The recursive function *PowersRec* given as our next example contains this type of construction. Recursive functions or procedures, which not only contain two or more recursive references in the code but actually execute at least two of these references in a given *current* invocation or call, are referred to as *multiple-reference* recursion. Recursive algorithms given later in this section for computing Fibonacci numbers and for solving the Towers of Hanoi problem are examples of multiple-reference recursive functions and procedures, respectively.

1.3.2 Exponentiation: A Glimpse of Cryptology

The algorithm *EgyptianPowers* given in Section 1.1 computed x^n when n was a power of 2 by repeated squaring. We now give pseudocode for *PowersRec*, which computes x^n for a general positive integer n. *PowersRec* is based on the following recurrence relation.

$$x^n = \begin{cases} (x*x)^{n/2} & \text{if } n \text{ is even,} \\ x*(x*x)^{(n-1)/2} & \text{otherwise,} \end{cases} \quad \textbf{init. cond. } x^1 = x \tag{1.3.3}$$

The pseudocode for *PowersRec* is based directly on (1.3.3).

```
function PowersRec(x,n) recursive
Input:  x (a real number), n (a positive integer)
Output: x^n
   if n = 1 then
      Pow = x                                  {initial condition}
   else
      if even(n) then                          {n even}
         Pow := PowersRec(x*x,n/2)
      else                                     {n odd}
         Pow := x * PowersRec(x*x,(n – 1)/2)
      endif
   endif
   return(Pow)
end PowersRec
```

For input (x,n), *PowersRec* performs between $\lfloor \log_2 n \rfloor$ and $2\lfloor \log_2 n \rfloor$ multiplications (see Exercise 1.24).

In the case of *Factorial* and *GCDRec*, nonrecursive versions are easily obtained directly from their respective recurrence relations. Coming up with the nonrecursive version *Powers* of *PowersRec* requires a little more effort.

```
function Powers(x,n)
Input:  x (a real number), n (a nonnegative integer)
Output: x^n (returns the value of x^n)
    if n = 0 then return(1) endif
    AccumPowers := 1
    Pow := x
    while n > 1 do
       if even(n) then
          n := n/2
       else
          AccumPowers := AccumPowers*Pow
          n := (n – 1)/2
       endif
       Pow := Pow*Pow
    endwhile
    return(Pow*AccumPowers)
end Powers
```

The ability to efficiently compute x^n has many applications. For example, various communication protocols in cryptology utilize efficient exponentiation. Consider the following problem: Alice and Bob wish to send confidential information to one another over some communication medium (e-mail, telephone conversation, fax, and so forth). However, they know that Eve Dropper has the ability to monitor all communication sent over this medium. Moreover, Alice and Bob have not shared in advance any secret information that could be used for encrypting information. We now describe a communication protocol based on exponentiation modulo p that can be used by Alice and Bob to share secrets.

Alice and Bob first send messages, monitored by Eve, in which they agree to use a conventional cryptographic system for communication. The cryptographic system, also assumed to be known by Eve, requires that they exchange a secret key value, which we assume is an arbitrary positive integer. While the key value is arbitrary, it must be known by both Alice and Bob. The following question arises: Can Alice and Bob publicly (that is, known to Eve) exchange information that allows them to efficiently determine a key, but at the same time makes it infeasible for Eve to discern the value of this key?

The answer lies in the notion of *one-way* functions. One-way functions are efficiently computable, but their inverses are infeasible to compute. In 1976, Diffie and Hellman first pointed out the utility of one-way functions in cryptology and identified exponentiation modulo p, where p is an integer with several hundred decimal digits, as an example. We now describe how efficient exponentiation yields an effective method of exchanging a secret key.

Alice and Bob send messages agreeing on the value of p, as well as on an integer b between 2 and $p - 1$. Of course, then Eve also knows the values of p and b. Next, Alice and Bob randomly choose integers m and n, respectively, between 1 and $p - 1$. Alice does not know n, and Bob does not know m (Eve knows neither n nor m). Using a modification *PowersRecModp* of *PowersRec* in which all calculations are carried out modulo p (see Exercise 1.27b), Alice then calculates and sends the integer b^m **mod** p to Bob. Similarly, Bob calculates and sends the integer b^n **mod** p to Alice. Here comes the secret: Using *PowersRecModp* again, Alice calculates (b^n **mod** $p)^m$ **mod** p, and Bob calculates (b^m **mod** $p)^n$ **mod** p. Lo and behold, they have just calculated the same number b^{mn} **mod** p, which is the secret key they now share.

Because Eve has monitored all exchanges between Alice and Bob, she now knows p, b, b^m **mod** p, and b^n **mod** p. Thus, theoretically Eve can determine the secret key b^{mn} **mod** p by calculating m from her knowledge of b^m **mod** p and then calculating (b^n **mod** $p)^m$ **mod** p. The problem of determining m from the known values of b, p, and b^m **mod** p is (a special case of) the *discrete logarithm* problem. The naive algorithm solving the discrete logarithm problem is to compute b^i **mod** p to b^m **mod** p as i takes on successive values 1, 2, 3, … , until we reach an i such that b^i **mod** $p = b^m$ **mod** p. Since this would take $p/2$ iterations on average, the naive algorithm is obviously infeasible. (Recall that p is an integer having more than a hundred digits!) While there are better algorithms for the discrete logarithm problem, unfortunately for Eve no efficient algorithm is known.

The algorithm *PowersRecModp* which Alice and Bob used to calculate powers of b requires special techniques for multiplying large integers. A classical divide-and-conquer method for multiplying large integers is discussed in Section 13.5 of Chapter 13. Since all calculations are reduced modulo p, *PowersRecModp* always multiplies or squares numbers less than p, so that each such operation takes time bounded above by a constant $C(p)$ depending only on p. Thus, if Alice and Bob choose p, m, and n having 200 decimal digits each, then they can exchange their secret key after performing 3000 multiplications of 200-digit numbers and 3000 reductions of a 400-digit number modulo p. All these computations can be done within a reasonable time.

1.3.3 Towers of Hanoi

For our first illustration of multiple-reference recursion, we consider the famous Towers of Hanoi puzzle. Suppose we have three pegs A, B, C, with n disks. The n disks are of different diameters and are initially placed on peg A so that a larger disk is always below a smaller disk (see Figure 1.9). The object is to move all n disks to peg C using peg B as auxiliary. A move consists in removing a single disk at the top of a peg and moving it to another peg. However, a given disk may never be placed on top of a smaller one. An elegant solution to the puzzle can be described easily using recursion.

Figure 1.9 The Towers of Hanoi puzzle

The problem is immediately solved for $n = 1$: Simply move the disk on peg A to peg C. Although it is not clear how an explicit solution to the puzzle might be obtained for general n, we can easily state a solution for the n disks in terms of a solution for the $n - 1$ disks:

1. Move the top $n - 1$ disks from A to B, using C as auxiliary.
2. Move the remaining disk from A to C.
3. Move the $n - 1$ disks from B to C, using A as auxiliary.

The following pseudocode for the recursive procedure *Towers* amounts to nothing more than a direct translation of these three steps. The only trick is to be careful to use the appropriate arguments in the two recursive calls. The original call to *Towers* is with three character arguments '*A*', '*B*', '*C*', and integer argument *n*.

```
procedure Towers(First,Auxiliary,Last,n) recursive

Input:  First,Auxiliary,Last (names of three pegs), n (a positive integer)
Output: a solution to the Towers of Hanoi puzzle

  if n = 1 then
     write ('Move disk',1,'from peg',First,'to peg',Last)
  else
     call Towers(First,Last,Auxiliary,n – 1)
     write ('Move disk',n,'from peg',First,'to peg',Last)
     call Towers(Auxiliary,First,Last,n – 1)
  endif

end Towers
```

Using mathematical induction (see Chapter 7), it is easily proved that *Towers* solves the Towers of Hanoi puzzle using the fewest possible moves. Let $t(n)$ denote the number of moves performed by *Towers* for n disks. Clearly, $t(n)$ satisfies the recurrence relation

$$t(n) = 2t(n - 1) + 1, \quad \textbf{init. cond. } t(1) = 1. \tag{1.3.4}$$

One way of determining a closed formula for $t(n)$ is to generate a few terms in the sequence $t(1)$, $t(2)$, … and conjecture a formula for $t(n)$ based on the observed pattern. To test the conjecture, one needs only to substitute the formula into (1.3.4) to check its validity. Recurrence relation (1.3.4) generates the sequence 1, 3, 7, 15, 31, 63, … . A good guess for $t(n)$ based on the beginning terms of this sequence is $t(n) = 2^n - 1$. It is easy to verify that $t(n)$ given by this formula satisfies (1.3.4).

It is often not so easy to guess the correct formula from the pattern exhibited by the first few terms in a sequence generated by a recurrence relation. In Chapter 7 we develop general methods for solving various types of recurrence relations. By way of illustration, we solve (1.3.4) along the lines of one of these general methods. Instead of working our way from the bottom up through the sequence, we can obtain our closed formula by unwinding the terms in a top-down fashion.

We first unwind the relation (1.3.4) one level by replacing the term $t(n - 1)$ with the term $2t(n - 2) + 1$, which results from replacing n by $n - 1$ in (1.3.4). We obtain

$$\begin{aligned} t(n) &= 2t(n - 1) + 1 \\ &= 2(2t(n - 2) + 1) + 1 = 2^2t(n - 2) + 2 + 1. \end{aligned}$$

Unwinding another level by substituting $2t(n - 3) + 1$ for $t(n - 2)$ yields

$$t(n) = 2^3t(n - 3) + 2^2 + 2 + 1.$$

Continuing this unwinding, we finally arrive at

$$\begin{aligned} t(n) &= 2^{n-1}t(1) + 2^{n-2} + 2^{n-3} + \cdots + 2 + 1 \\ &= 2^{n-1} + 2^{n-2} + 2^{n-3} + \cdots + 2 + 1 \text{ (since } t(1) = 1). \end{aligned}$$

The latter expression for $t(n)$, while still not in closed form, nevertheless is a geometric progression $1 + x + x^2 + \cdots + x^{n-1}$, with $x = 2$. Since $1 + x + x^2 + \cdots + x^{n-1} = (x^n - 1)/(x - 1)$ (see 1.2.10), it follows that $t(n) = 2^n - 1$.

1.3.4 Top-Down versus Bottom-Up Implementations of Recursion

To illustrate the difference between top-down and bottom-up transformations of a recursive solution into an algorithm, consider the problem of designing an algorithm to compute the nth Fibonacci number based on the recurrence relation

$$\text{fib}(n) = \text{fib}(n - 1) + \text{fib}(n - 2), \quad \textbf{init. cond. } \text{fib}(0) = 0, \text{fib}(1) = 1. \tag{1.3.5}$$

A top-down approach based directly on (1.3.5) yields the following recursive algorithm.

```
function Fib1(n) recursive

Input:  n (a nonnegative integer)
Output: the nth Fibonacci number

  if n ≤ 1 then
     return(n)
  else
     return(Fib1(n – 1) + Fib1(n – 2))
  endif

end Fib1
```

The top-down nature of *Fib*1 is evident when we trace the *tree of recursive calls* corresponding to the recursive invocations. Figure 1.10 shows how the tree of recursive calls is generated dynamically for $n = 4$. Each node in the tree represents an invocation of *Fib*1.

A left branch in the tree in Figure 1.10 (pages 26–27) corresponds to invoking *Fib*1 with argument $n - 1$, and a right branch corresponds to invoking *Fib*1 with argument $n - 2$. Downward arrows indicate invocations and upward arrows indicate resolutions.

In Figure 1.12a (page 28), we illustrate the complete tree of recursive invocations of *Fib*1 for $n = 5$. We have drawn the *recursive path around the tree* that traces the sequence of recursive invocations and resolutions. Starting at the top (root) and following this path generates a dynamic trace similar to that illustrated in Figure 1.10. *Fib*1 is extremely inefficient in that it performs many redundant recalculations (see Exercise 1.29).

Because it eliminates redundant recalculations, the function *Fib*2, based on the bottom-up approach, is much more efficient than *Fib*1. Working bottom-up, we first compute fib(2) by summing fib(1) and fib(0), then compute fib(3) by summing fib(2) and fib(1), and so forth (see Figure 1.12b, page 28). The following purely iterative pseudocode for *Fib*2 accomplishing these bottom-up computations utilizes the variables *Previous*, *Current*, and *Next*.

```
function Fib2(n)

Input:  n (a nonnegative integer)
Output: the nth Fibonacci number

  if n ≤ 1 then
     return(n)
  else
     Previous := 0
     Current := 1
     for i := 2 to n do
        Next := Previous + Current
        Previous := Current
        Current := Next
     endfor
  endif
  return(Current)

end Fib2
```

The three variables *Previous*, *Current*, and *Next* used in *Fib*2 mutually update each other and propagate from bottom to top, as shown in Figure 1.11 (page 27).

While both *Fib*1 and *Fib*2 utilize the recurrence relation (1.3.5), Figure 1.12 (page 28) shows how the bottom-up solution of this relation used by *Fib*2 is dramatically more efficient than the top-down solution used by *Fib*1. As we have seen, *Fib*2 avoids recalculations performed by *Fib*1. Another way to avoid recalculations, and at the same time retain a recursive algorithm that closely mimics the Fibonacci

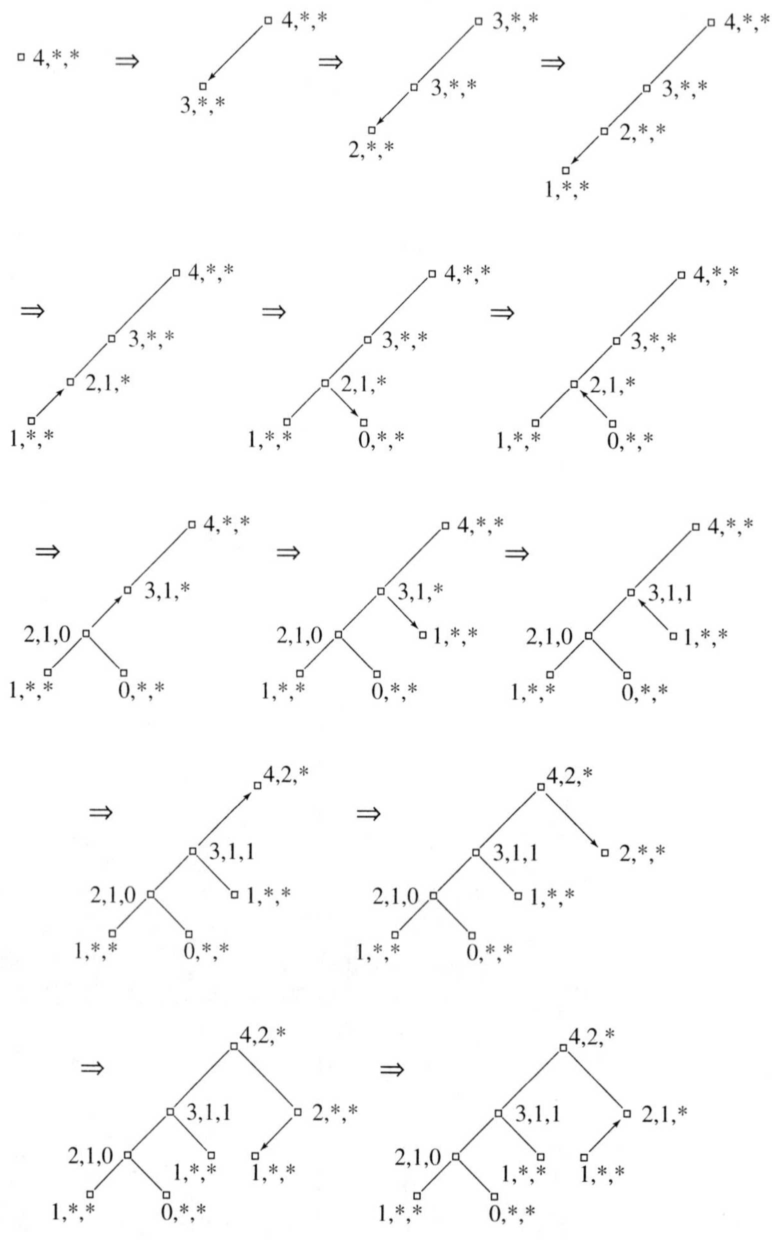

Figure 1.10 Recursive calls to *Fib*1 for $n = 4$. The values of n, *Fib*1$(n-1)$, *Fib*1$(n-2)$ are shown outside each node. (A * indicates that the corresponding functional value is yet to be resolved.)

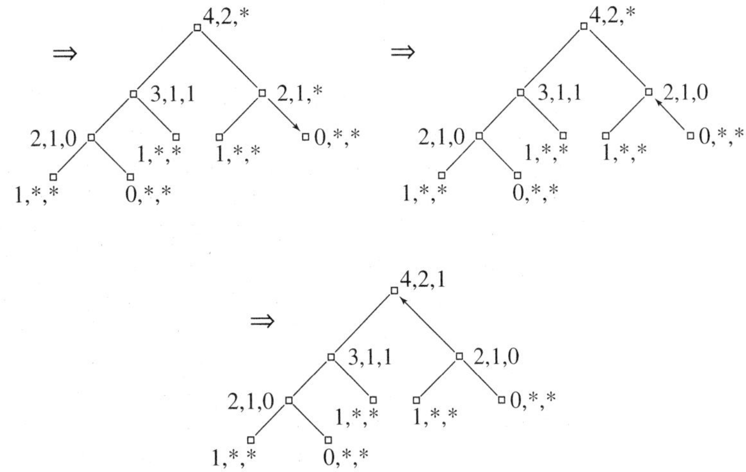

Figure 1.10 Continued

recurrence relation, is to use a global table to store previously calculated values. Recursive invocations are only generated for values not already stored in the table by the algorithm. Moreover, this canonical technique of using a table to store previously calculated values generalizes to any situation where recursion leads to recalculations. Thus, the top-down recursive nature of the algorithm is maintained, but a pinch of bottom-up strategy is mixed in. By saving the values that we calculate when we hit bottom in the tree of recursive invocations, we subsequently do not have to go beneath any node whose value has already been calculated.

We rewrite the pseudocode for *Fib*1 using this idea of saving values in a table *Table* [1:*n*]. We initialize all the entries of the array *Table* [1:*n*] to – 1 (a value different from any Fibonacci number). This initialization is global to the resulting recursive function, called *FibRec*3, which computes the *n*th Fibonacci number. In the following pseudocode, *Fib*3 is a dummy nonrecursive function, which initializes the table and then invokes *FibRec*3.

Figure 1.11 Bottom-up action of *Fib*2 for $n = 5$

						Next	8	*Current*	8
				Next	5	*Current*	5	*Previous*	5
		Next	3	*Current*	3	*Previous*	3		3
Next	2	*Current*	2	*Previous*	2		2		2
Current	1	*Previous*	1		1		1		1
Previous	0		0		0		0		0
$i = 2$		$i = 3$		$i = 4$		$i = 5$		**return**(*Current*)	

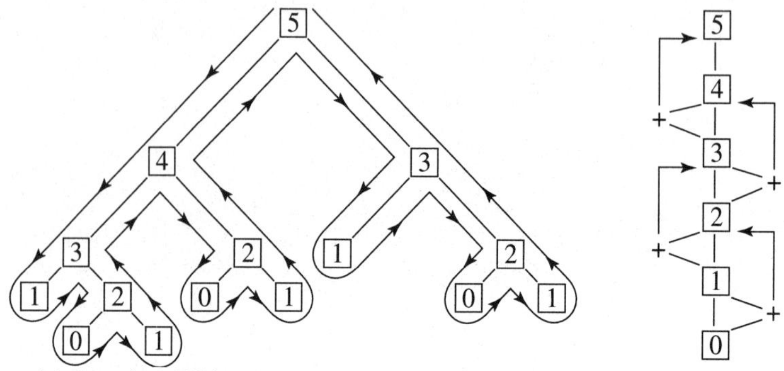

(a) function *Fib*1 (b) function *Fib*1

Figure 1.12 (a) Top-down (*Fib*1) versus (b) bottom-up (*Fib*2) solutions to the problem of computing the *n*th Fibonacci number for $n = 5$. Value of input *i* to *Fib*1 and *Fib*2 is shown in the nodes.

```
function Fib3(n)

Input:   n (a nonnegative integer)
Output: the nth Fibonacci number

  dcl Table[1:n] (array)

  for i := 1 to n do
     Table[i] := -1
  endfor
  return(Fib3Rec(n))

  function Fib3Rec(n) recursive

     if n ≤ 1 then
        return (n)
     else
        if Table[n] = – 1 then
           Table[n] := Fib3Rec(n – 1) + Fib3Rec(n – 2)
        endif
     endif
     return(Table[n])

  end Fib3Rec

end Fib3
```

In Figure 1.13, we illustrate the tree of recursive invocations of *Fib3Rec* for $n = 5$. A leaf node corresponding to returning a Fibonacci number already computed and stored in the table is labeled with a **T** and is referred to as a *table look-up*.

The recursive algorithms considered in this section each gave an elegant solution to the problem at hand. We will see many more examples of the power and elegance

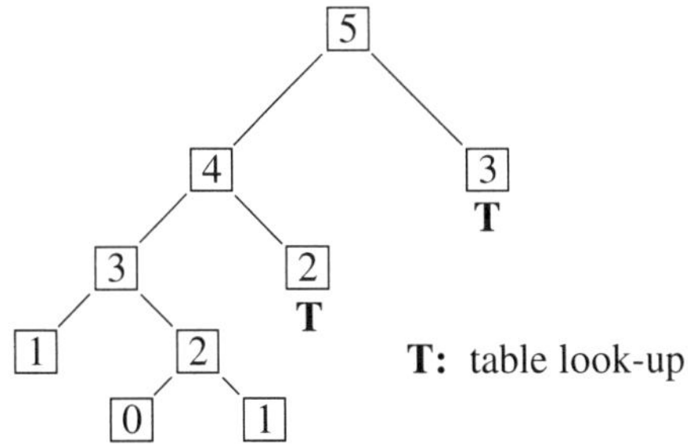

Figure 1.13 Tree of recursive invocations for *Fib3Rec* for $n = 5$

of recursion throughout the text. In fact, recursion is the basis of two important general design strategies, divide-and-conquer and dynamic programming (see Part Three).

1.4 Guidelines for Algorithm Design

Designing an algorithm for solving a complicated task may involve finding algorithms to solve many subtasks. The standard top-down approach to algorithm design involves outlining the major steps to a solution of the given problem and then performing a process of stepwise refinement into simpler and more manageable building blocks. The key to solving the original problem then boils down to finding algorithms for building-block problems that are conceptually self-contained such as evaluating polynomials, computing powers, searching, sorting, and so forth. In this text we are concerned with finding algorithms for solving such building-block problems.

The following may be regarded as a set of general guidelines for designing and analyzing an algorithm for solving a given problem.

Guidelines for Designing an Algorithm for Solving a Given Problem

1. Formulate the problem, taking into account the resources available such as the amount of memory and the number of processors.
2. Identify the appropriate abstract data types to model the problem.
3. Identify those parts of the problem that are already solved by (possibly modified) known algorithms (don't reinvent the wheel). Using an ad hoc method or one of the major design strategies, find algorithms for the remaining parts.
4. Once an algorithm for solving the given problem has been found, verify its correctness and analyze its performance.
5. Decide whether a search for a more efficient algorithm is worthwhile.

1.5 Closing Remarks

We have discussed several problems having a long history and presented some algorithms for solving these problems. These examples illustrated the fact that the most straightforward and simple algorithm for solving a given problem is often not the most efficient. Designing more efficient algorithms frequently requires utilizing more sophisticated data structures than were used in this chapter. The next chapter will give a review of some standard data structures, and other data structures will be introduced as needed throughout the text.

Measuring the performance of an algorithm was discussed informally in this chapter. In Chapter 3 we will formalize performance measures by introducing the notions of best-case, worst-case, and average complexities as well as the notion of asymptotic growth rates of functions.

Until relatively recently, algorithms were approached from the sequential point of view since only single-processor computers were available. Today there are computers that utilize hundreds and even thousands of processors. A major challenge facing computer scientists today is the design of algorithms and algorithm strategies that exploit this parallelism. In Chapters 5 and 6, we give the basic foundation of parallel algorithms and architectures and then integrate parallel algorithm discussion throughout the remainder of the text.

References and Suggestions for Further Reading

Knuth, D. E. *The Art of Computer Programming, Volume I: Fundamental Algorithms*, 2nd Ed. Reading, MA: Addison-Wesley, 1973. An excellent reference for further details concerning the early development of algorithms, this text also contains a detailed discussion of the binary method for evaluating powers.

Pollack, S. E. "The Development of Computer Science," in *Studies in Computer Science*, S. Pollack, ed. The Mathematical Association of America, 1982. This interesting article traces the evolution of computing machinery and the development of computer science.

Exercises

Section 1.1 Algorithms from Ancient to Modern Times

1.1 Trace the action of the algorithm *GCD* for the following input pairs.

a. (24,108)

b. (23,108)

c. (89,144)

d. (1953,1937)

1.2 Repeat Exercise 1.1 for the algorithm *EuclidGCD*.

1.3 Note that for the pair (89,144) given in Exercise 1.1 the algorithms *GCD* and *EuclidGCD* performed identical calculations. This phenomenon holds for infinitely many pairs of integers. One such collection can be obtained as successive pairs in the Fibonacci sequence fib(n) defined in Section 1.3.4 using recurrence relation (1.3.5).

a. Verify that (except for the last step) *GCD* and *EuclidGCD* perform identically on any input pair (fib (n – 1),fib(n)).

b. Obtain a formula for the number of steps required by *EuclidGCD* to compute gcd(fib(n – 1), fib(n)).

1.4 Show that the longest number of steps required by *EuclidGCD* for a pair of integers each having m digits or less is achieved by a suitable pair (fib (n – 1),fib(n)).

1.5 Trace the action of the algorithm *BabylonianSQRT* for the following input values of a and *error* = 0.001.

a. $a = 6$

b. $a = 23$

c. $a = 16$

1.6 A continuous function $f(x)$ such that $f(a)$ and $f(b)$ have opposite signs must have a zero (a point x such that $f(x) = 0$) in the interval (a, b). By checking whether $f(a)$ and $f((a + b)/2)$ have opposite signs, we can determine whether the zero occurs in the subinterval $[a, (a + b)/2]$ or the subinterval $[(a + b)/2, b]$. The *bisection method* for approximating a zero of $f(x)$ is based on repeating this process until a zero is obtained to within a predescribed error. Give pseudocode for the bisection method algorithm *Bisec(f(x),a,b,error)* for finding an approximation to a zero of a continuous function $f(x)$ in the interval $[a, b]$ accurate to within *error*.

1.7 **a.** Show how to use the bisection method of Exercise 1.6 to compute $\sqrt{c}$.

b. For $c = 6$ and *error* = 0.00001, compare the efficiency (number of iterations) of *BabylonianSQRT* for computing $\sqrt{c}$ with the bisection method algorithm on the interval [1, 6]. (You might want to write a program for this.)

1.8 For each of the following, do three iterations of the bisection method.

a. $f(x) = x^3 - 6$, initial interval [1, 3]

b. $f(x) = x^3 - 29$, initial interval [1, 3]

c. $f(x) = 2^x - 10$, initial interval [3, 5]

1.9 Show that Newton's method reduces to *BabylonianSQRT* in the special case when $f(x) = x^2 - a$.

1.10 Use calculus and the description given in Figure 1.3 to derive Newton's formula (1.1.3).

1.11 In practice, when writing a program requiring interactive input of numeric data, it is useful to check whether the user has entered any "bad characters" (that is, characters not corresponding to a digit between 0 and 9, inclusive). To prevent the program from crashing, we can input the data as a character string, test for bad characters, and convert it to an integer if none are found. Devise an algorithm based on Horner's Rule for converting a string of alphanumeric digits to its numeric value. Assume a function *ConvertDigit* exists for converting an alphanumeric digit to its integer equivalent (for example, *ConvertDigit* ('5') = 5). Also assume that the alphanumeric digits are input one at a time in an *online* fashion, so that the total number of digits is not known in advance.

1.12 To evaluate a polynomial at v and $-v$, one could simply call *HornerEval* twice, involving $2n$ multiplications and $2n$ additions. Describe a modification of *HornerEval* that solves this particular evaluation problem using only n multiplications and $n + 1$ additions. (A generalization of this process is the basis of the Fast Fourier Transform.)

Section 1.2 Mathematical Notation and Background

1.13 The function **round**(x) is the integer nearest to x, with ties going to the larger.

a. Express **round**(x) in terms of $\lfloor x \rfloor$.

b. Express $\lfloor x \rfloor$ in terms of **round**(x).

1.14 Suppose that the function **round**(x) is defined as in Exercise 1.13, except that ties go to the smaller.

a. Repeat Exercise 1.13 with $\lceil x \rceil$ and this altered definition of **round**(x).

b. What is the problem if (a) is attempted with the definition of **round**(x) given in Exercise 1.13?

1.15 Give a pseudocode for an algorithm *Digits* that uses the floor function, $\lfloor x \rfloor$, and the **mod** function to compute all the digits (base 10) of a given integer n. Store the digits in an array.

1.16 Give a closed formula for the sum $m + (m + 1) + \cdots + n, \quad 1 \le m \le n$.

1.17 **a.** Suppose A and B are two finite sets. Show that

$$|A \cup B| = |A| + |B| - |A \cap B|.$$

b. Generalize the formula in (a) to n sets.

1.18 Prove the following formula for changing from one base to another.

$$\log_a x = \frac{\log_b x}{\log_b a}$$

1.19 Show that

$$n^{(k)}/m^{(k)} < (n/m)^k \quad \text{for } 0 < k \le n < m.$$

1.20 Prove the following identity.

$$\sum_{k=0}^{n} \binom{n}{k} = 2^n$$

1.21 By taking an appropriate derivative in the binomial formula (1.2.18), prove the identity

$$n2^{n-1} = \sum_{k=0}^{n} k\binom{n}{k}.$$

1.22 Show that the number of nonempty subsets of even cardinality of a set n elements is exactly one less than the number of subsets of odd cardinality.

Section 1.3 Recursion

1.23 Solve the following recurrence relations.

a. $t(n) = 3t(n-1) + 5$, **init. cond.** $t(0) = -2$

b. $t(n) = t(n-1) + 3$, **init. cond.** $t(-2) = 0$

1.24 Show that for input (x,n), *PowersRec* performs between $\lfloor \log_2 n \rfloor$ and $2\lfloor \log_2 n \rfloor$ multiplications.

1.25 Verify the following classical recurrence relation.

$$C(n, k) = C(n-1, k-1) + C(n-1, k), \quad \textbf{init. cond. } C(n, 0) = C(n, n) = 1$$

1.26 Find a recurrence relation for $b_n = \binom{2n}{n}$.

1.27 **a.** Show that

$$(x + y) \textbf{ mod } p = ((x \textbf{ mod } p) + (y \textbf{ mod } p)) \textbf{ mod } p$$

$$(xy) \textbf{ mod } p = ((x \textbf{ mod } p)(y \textbf{ mod } p)) \textbf{ mod } p.$$

b. Give pseudocode for a modification *PowersRecModp* of *PowersRec* in which all calculations are carried out modulo p. The correctness of *PowersRecModp* follows from (a).

1.28 In Chapter 7 we derive an exact formula for fib(n). Argue that

$$(1.5)^{n-2} \le \text{fib}(n) \le 2^{n-1}, \quad n \ge 1.$$

1.29 We saw that *Fib*1 is extremely inefficient, since it performs many redundant recalculations. How many times is fib(k) computed by *Fib*1 when *Fib*1 is invoked with input n, $k = 1, 2, \ldots, n$?

2

Elementary Data Structures

In the algorithms discussed in Chapter 1, we assumed knowledge of various data items (integers, reals, Booleans, characters) and their organization into elementary data structures such as arrays. In this chapter, we will review other basic abstract data types (ADTs) such as linked lists, stacks, queues, and binary trees. We will discuss more advanced ADTs in the context of an applicable algorithm.

The performance of an algorithm often depends on the choice of data organization. An ADT is concerned with a theoretical description of the organization of the data and the operations to be performed on this data. The implementation of an ADT determines a *data structure*.

2.1 Linear ADTs

Many algorithms discussed in this text use a list as one of the primary abstract data types. Operations associated with a list include, for example, insertion, deletion, and accessing a particular element in the list. For convenience, when describing algorithms we often implement the list ADT as an array. However, in practice there are a number of situations where a linked list implementation would improve performance.

2.1.1 Implementing the List ADT: Arrays versus Linked Lists

The array implementation of a list has the important advantage of allowing quick and *direct* access to any element of the list. However, there are some disadvantages in using an array. For example, inserting an element at position i of a list stored in positions $1, \ldots, n$ of an array requires $n - i + 2$ array assignments. Deleting an element also requires many array assignments.

Another disadvantage relates to inefficient use of space. A particular list may occupy only a small portion of an array that has been declared to handle much larger lists. A linked list avoids these disadvantages at the expense of direct access.

A linked list is a set of nodes, where each node contains an information field and a pointer field. The information field contains data whose type depends on the application, and the pointer field contains a *pointer* that specifies the location of the next node in the linked list. In addition to the pointers contained in each node, there is a pointer to the beginning of the linked list (see Figure 2.1).

Most high-level languages support pointers and dynamic variables, so, except for additional space used by the pointers, we can implement lists without wasted storage. Nodes can be allocated and disposed of as needed. The nodes of a linked list can then be stored in dynamic variables, which are records of the following type.

ListNode = **record**
Info: *InfoType*
Next: → *ListNode*
end

InfoType represents a generic type, which could be an integer, a real, a string, or even a record containing many information fields.

Given a pointer to an appropriate position in the list, insertion and deletion of elements in a linked list is particularly efficient, since only two pointers need be altered. Suppose *Place* is a pointer to a node of the linked list and *NewNode* is a pointer to a node not in the linked list. The node pointed to by *NewNode* can be inserted into the linked list immediately after the node pointed to by *Place* and the linked list maintained as follows:

NewNode→*Next* := *Place*→*Next*
Place→*Next* := *NewNode*

If *Place* points to the first node of the linked list, then replace *Place* by *Header* in the above pseudocode. The action of this pseudocode is illustrated in Figure 2.2.

2.1.2 Stack ADT

A *stack* is a list in which all insertions and deletions are made at one end called the *top*. Thus, the last element inserted is the first element removed, so that a stack is also commonly referred to as a LIFO (last-in, first-out) list. The operation of inserting an element into the stack is called *push*, and the operation of removing an element is called *pop*. Stacks show up everywhere in the field of computer science. In addition to the use of stacks in algorithm design, computer systems use stacks to

Figure 2.1 A linked list

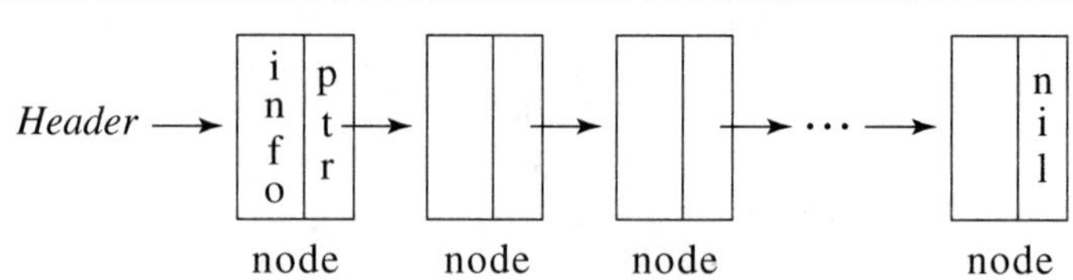

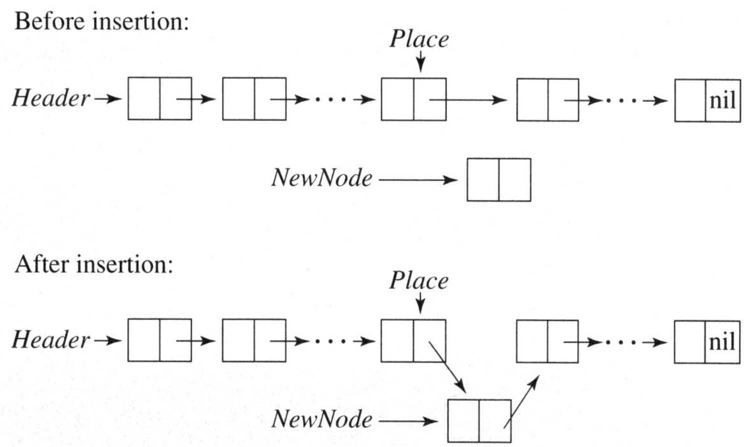

Figure 2.2 Inserting an element into a linked list using only two pointer assignments

maintain parameters, local variables, return addresses, and so forth, for function invocations and procedure calls.

Since a stack is a list, it can be implemented using an array or a linked list. In the array implementation, we use an array *Stack*[1:*MaxSize*] and the auxiliary variable *Top* to hold the index of the array element that is at the top of the stack. An empty stack corresponds to *Top* = 0, whereas a full stack corresponds to *Top* = *MaxSize*. Checking whether a stack is empty or full, respectively, are additional operations of the stack ADT.

Pushing (inserting) an element x on the stack can be performed using the following unrefined pseudocode.

```
if Top = MaxSize then {Stack is full}
   take appropriate action on stack full condition
else
   Top := Top + 1
   Stack[Top] := x
endif
```

Popping the stack and assigning the popped element to x is accomplished by the following unrefined pseudocode.

```
if Top = 0 then {Stack is empty}
   take appropriate action on stack empty condition
else
   x := Stack[Top]
   Top := Top – 1
endif
```

Pseudocode for pushing and popping a stack implemented as a linked list is left to the exercises.

2.1.3 Queue ADT

A *queue* is a list in which all insertions are made at one end of the list, called the *rear*, and all deletions are made at the other end, called the *front*. In a queue, the first element inserted is the first element removed, so that a queue is also referred to as a FIFO (first-in, first-out) list. Queues occur naturally in computer science. Examples include queues of tasks waiting for the line printer, access to disk storage, use of CPU, and so forth.

Since a queue is a list, it can be implemented by using an array or a linked list. Consider an array implementation *Queue*[1:*MaxSize*] that mimics, for example, a queue of people waiting in a checkout line. When an element is dequeued, all the remaining elements would be bumped up one position in the array, so that the element in the front of the queue will always have index one (*Queue*[1]). A variable *Rear* contains the index of the element at the rear of the queue and is simply incremented by 1 when enqueueing an element. However, dequeueing is very inefficient, since it involves *Rear* – 1 index updates (and a decrementing of *Rear* by one).

A more efficient (but still troublesome) array implementation would be to use *Front* and *Rear* as variables containing indices of the elements at the front and one past the rear of the queue, respectively. Both *Front* and *Rear* might be initialized to 1. We increment *Rear* by 1 for enqueueing and increment *Front* by 1 for dequeueing. The condition *Front* = *Rear* signals that the queue is empty, whereas *Rear* = *MaxSize* + 1 signals that the queue is full. As long as we never perform more than *MaxSize* enqueues, everything is fine. Otherwise, an obvious problem can arise: We might end up with the queue being signaled as full when it is practically empty. For example, starting from an initially empty queue (*Front* = *Rear* = 1), an alternating sequence of *MaxSize* enqueues and *MaxSize* – 1 dequeues results in *Front* = *MaxSize* and *Rear* = *MaxSize* + 1. An attempt now to enqueue an element would result in the queue being signaled as full even though the array contains a single element!

An array implementation of a queue that avoids the problems arising in the previous two implementations is the *circular array*, which is an array where the indices are viewed in a circular fashion. Another efficient way to implement a queue is to use a linked list. We leave the details of both the circular array and linked list implementations to the exercises.

2.1.4 Priority Queue ADT

A *priority queue* is an abstract data type similar to a queue, except that a real number value, called a *priority*, is associated with each element. The main difference between a queue and a priority queue occurs in the removal operation. Rather than dequeue the elements based on a FIFO rule, the elements are dequeued based on their priority values. The element dequeued is the one having the highest priority.

Numerous important applications of priority queues arise in practice. For example, jobs submitted to time-sharing computers are prioritized according to such things as time and space resource requests. A priority queue is implemented that maintains a dynamic schedule for execution of the jobs. When a particular job is completed or its execution is interrupted, execution of the job with the highest priority amongst the pending jobs is either begun or resumed. Completed jobs are removed from the priority queue, and interrupted jobs or new jobs are inserted into the priority queue. Another application of priority queues is time-dependent event simulation.

There are two obvious ways to implement a priority queue using the list ADT. In one implementation, we maintain the elements of the list in sorted order. Then the deletion of an element can be performed in constant time. However, inserting an element into a list of size n may require as many as n comparison steps. The second implementation of a priority queue does not maintain a sorted list. Instead, elements are simply inserted at one end of the list. However, to then delete an element the list must be scanned for the element of highest priority. Such a scan may require as many as n comparisons.

One way of obtaining a more efficient implementation of a priority queue is to use the heap ADT. Both the insertion and deletion of an element in a priority queue of n elements can be accomplished using at most $\log_2 n$ comparison steps when the priority queue is implemented as a heap. The heap ADT is discussed in detail in Chapter 4, where it is used in the design of the sorting algorithm *HeapSort*.

2.2 Tree ADTs

I think that I shall never see, a poem as lovely as a tree.

—Joyce Kilmer

Computer scientists, too, are enamored with trees. *Trees*, as defined in computer science and mathematics, sprout up everywhere in the subject of algorithms. Trees are often explicitly implemented in code for an algorithm or are implicitly present in the hierarchical logical organization of an algorithm. We begin our discussion of trees with the very important special type of tree known as a *binary tree*.

2.2.1 Binary Tree ADT

There are many equivalent definitions of a binary tree, but the following recursive definition is one of the most elegant.

DEFINITION 2.2.1 A *binary tree T* consists of a set of nodes, which is either empty or has the following properties:

1. One of the nodes, say R, is designated the root node.
2. The remaining nodes (if any) are partitioned into two disjoint subsets, called the *left subtree* and *right subtree*, respectively, each of which is a binary tree.

The roots of the left and right subtrees described in property (2) of the definition are called the *left child* and *right child* of R, respectively. Some examples of binary trees are given in Figure 2.3. A node V is called a *leaf* if it has no children. Just as in

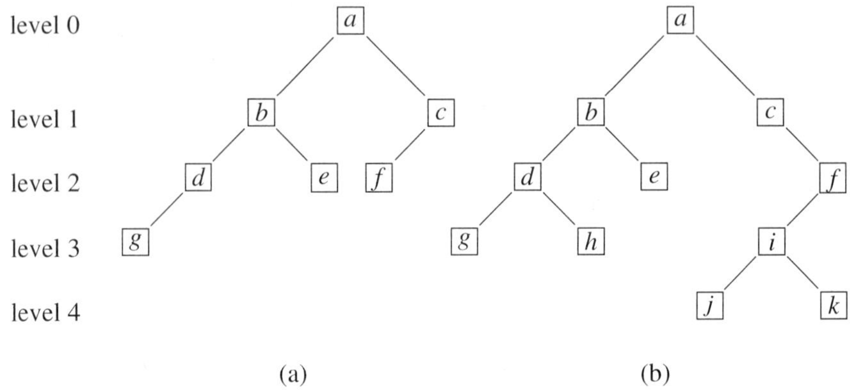

Figure 2.3 Sample binary trees of depths 3 and 4

family trees, we utilize genealogical notation such as *parent*, *grandchildren*, *great-grandchildren*, *sibling*, *descendant*, *ancestor*, and so forth.

The nodes of T can be partitioned into disjoint sets (*levels*) depending on their (genealogical) distance from the root R. In particular, the root R is at level 0. The *depth* (sometimes called *height*) of T is the maximum distance from R to a leaf. The trees shown in Figure 2.3a and 2.3b have depths 3 and 4, respectively.

For any nonnegative integer n, let b_n denote the number of different binary trees on n nodes. Then it turns out (see Section 7.3) that b_n is given by

$$b_n = \frac{1}{n+1}\binom{2n}{n}, \quad n \geq 1 \tag{2.2.1}$$

(also known as the nth *Catalan number*). Formula (2.2.1), together with the lower bound estimate

$$\binom{2n}{n} \geq \frac{4^n}{(2n+1)}, \quad n \geq 1$$

(see Exercise 7.31 of Chapter 7), shows that the number of binary trees on n nodes grows exponentially with n.

Implementation of Binary Trees Binary trees can be implemented using pointer and dynamic variables as follows. The nodes of the tree are represented by a dynamic variable having the following structure.

BinaryTreeNode = **record**
    Info: *InfoType*
    LeftChild: → *BinaryTreeNode*
    RightChild: → *BinaryTreeNode*
end *BinaryTreeNode*

In addition to the dynamic variables representing the nodes of the binary tree, we have a pointer variable *Root* pointing to the root node. The implementation of the binary tree from Figure 2.3b using the pointer and dynamic variables is given in Figure 2.4.

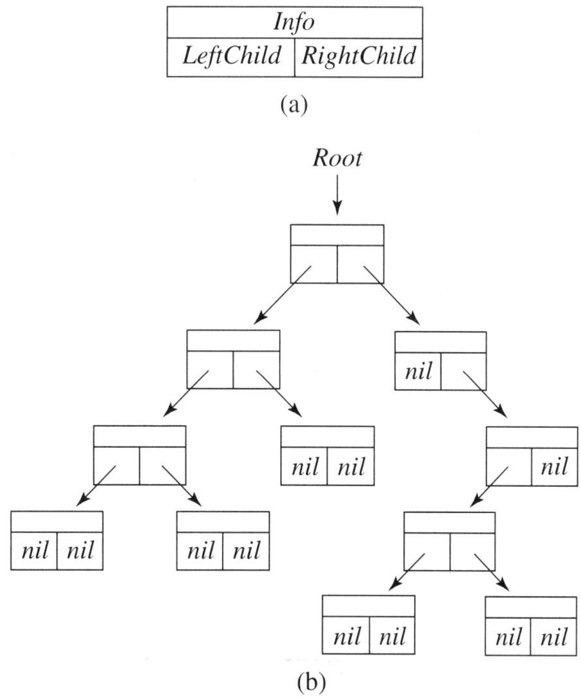

Figure 2.4 (a) Representation of a node binary tree; (b) implementation of a binary tree using pointers and dynamic variables

2.2.2 General Tree ADT and Forest ADT

There are a thousand hacking at the leaves of evil to one who is striking at the root.

—Henry Thoreau

The design of many algorithms requires more general trees than binary trees. General trees are often implicitly present in the logical organization of an algorithm. The following definition extends the concept of a (rooted) binary tree given by Definition 2.2.1 to a general (rooted) tree.

DEFINITION 2.2.2 A *rooted tree* T consists of a set of nodes (also called vertices), which is either empty or has the following properties:

1. One of the nodes, say R, is designated the *root* node.
2. The remaining nodes (if any) are partitioned into j disjoint subsets $T_1, T_2, \ldots, T_j$, each of which is a rooted tree (called subtrees).

For convenience, we use the term *tree* for the more precise term *rooted tree*. Given the root R, the roots of the subtrees described in (2) are called *children* of R. A node v is called a *leaf* if it has no children. Just as for binary trees, we utilize genea-

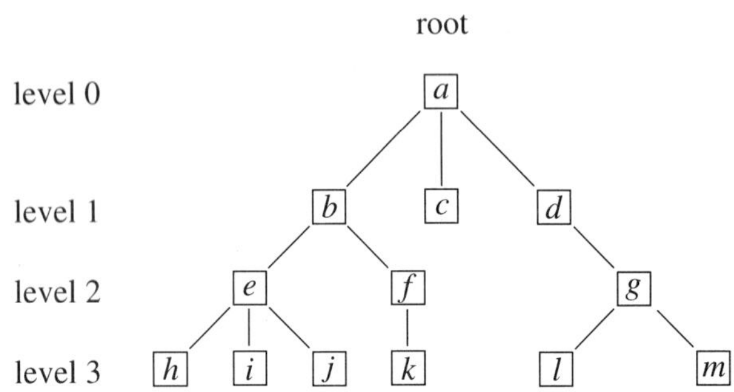

Figure 2.5 A sample tree of depth 3

logical terminology such as *parent*, *grandchildren*, *great-grandchildren*, *sibling*, *descendant*, *ancestor*, and so forth. We also have the notions of *levels* and *depth* of a tree. We illustrate a tree of depth 3 in Figure 2.5.

Child-Sibling Representation of a General Tree A child-sibling representation of a tree *T* can be implemented using pointer and dynamic variables as follows. The nodes of the tree are represented by the following structure.

TreeNode = **record**
  Info: *InfoType*
  LeftmostChild: →*TreeNode*
  NextSibling: →*TreeNode*
end *TreeNode*

As with binary trees, we keep a pointer variable *Root* that points to the root node of the tree. For example, the tree from Figure 2.5 is redrawn in Figure 2.6 using pointers and dynamic variables.

Remark This implementation of a general tree leads naturally to a transformation (known as the Knuth transformation) from a general tree to a binary tree. We simply think of the *LeftmostChild* as the left child of the node and the *NextSibling* as the right child. Indeed, Figure 2.6, showing the general tree implementation of the tree from Figure 2.5, is actually a binary tree.

Parent Representation of Trees and Forests Another representation of a general rooted tree, called the *parent representation*, keeps track of the parent of a node instead of its children. The parent representation has the advantage of efficiently generating the path in the tree from any given node of the tree to the root. However, the parent representation does not facilitate traversals of the tree.

The parent representation can actually be used to implement a slightly more general ADT known as a rooted forest. A *rooted forest* is a union *F* of (pairwise disjoint) rooted trees. A natural way to implement the parent representation of a (rooted) for-

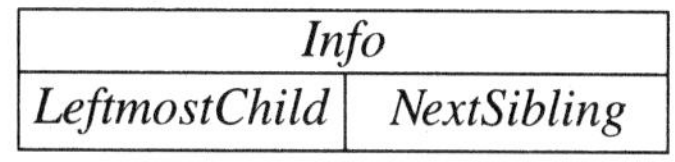

(a)

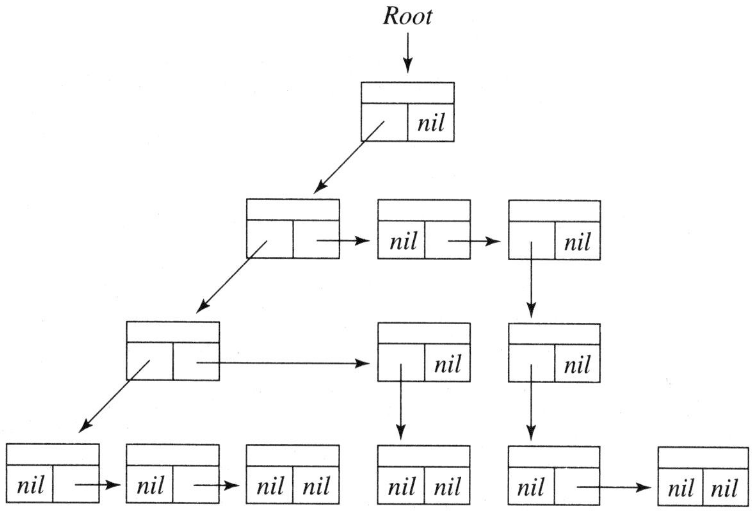

(b)

Figure 2.6 (a) Representation of a node of a general tree; (b) child-sibling implementation of a tree using pointers and dynamic variables

est F uses an array *Parent*[1:n]. For convenience, we assume that F has n nodes labeled 1,2, . . . , n. When referring to these nodes, we do not distinguish between a node and its label. For each $i \in \{1, 2, \ldots, n\}$, *Parent*[$i$] is zero, if i is a root of one of the trees in F, and *Parent*[i] is the parent of node i, otherwise. Information associated with the node i, $i = 1, \ldots, n$, can be stored in a second array *Info*[1:n]. In Figure 2.7*a* (page 44) the values of the two arrays *Parent* and *Info* are given for the tree from Figure 2.5. The number beside each node of the tree in the diagram is the label of that node, and the character inside each node is the information stored in that node. In Figure 2.7*b* the parent array representation of a sample forest is given.

The parent representation of a forest can also be implemented using pointers and dynamic variables. The nodes of the forest in this representation are given by the following structure.

ForestNode = **record**
    Info: *InfoType*
    Parent: →*ForestNode*
end *ForestNode*

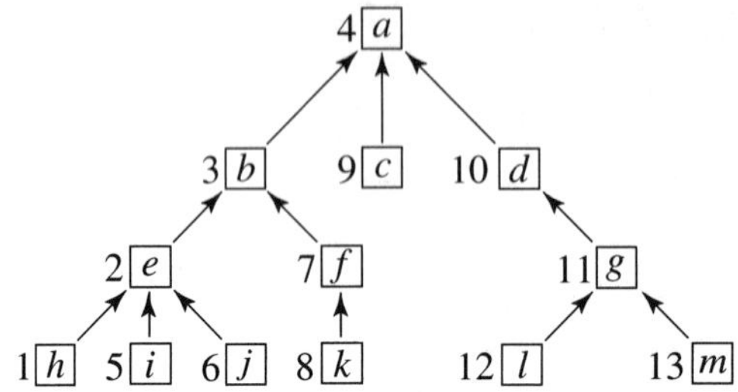

i	1	2	3	4	5	6	7	8	9	10	11	12	13
Info[i]	h	e	b	a	i	j	f	k	c	d	g	l	m
Parent[i]	2	3	4	0	2	2	3	7	4	4	10	11	11

(a)

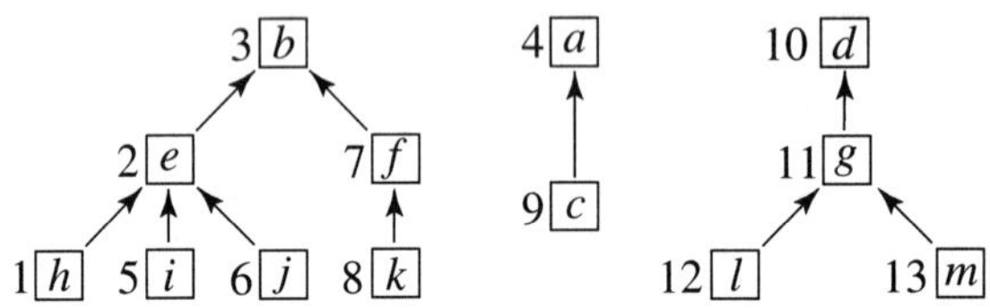

i	1	2	3	4	5	6	7	8	9	10	11	12	13
Info[i]	h	e	b	a	i	j	f	k	c	d	g	l	m
Parent[i]	2	3	0	0	2	2	3	7	4	0	10	11	11

(b)

Figure 2.7 (a) Parent array implementation of a tree; (b) parent array implementation of a forest

The parent field of a node contains a pointer to the parent of that node if it has a parent. If the node corresponds to a root of one of the trees in the forest, then it contains the value **nil**.

The parent implementation of trees and forests has many important applications. Many of these applications exploit the fact that the parent implementation of a tree T allows efficient generation of the *path* P in T from any given node u of T to the root r. (The path in T from u to r is the sequence of nodes $u_0, u_1, \ldots, u_p$, where $u_0 = u$ and $u_p = r$, such that u_i is the parent of u_{i-1}, $i = 1, 2, \ldots, p$.) For example, in Figure 2.7a we generate the path in the indicated tree T from vertex 8 to the root 4 as follows:

$$8, \mathit{Parent}[8] = 7, \mathit{Parent}[7] = 3, \mathit{Parent}[3] = 4$$

2.3 Tree Traversal

When a tree is implemented as a data structure in an algorithm, the nodes of the tree contain information (data) germane to the application in question. In many applications it is necessary to *access* the fields in every node of the tree. During the execution of an algorithm, the current value of a variable (pointer variable, index variable, and so forth) that refers to a node in the tree thereby makes the fields in a particular node accessible. A node is *visited* when the node is accessed and some operation is performed on the information field(s) contained in the node. For example, visiting a node might consist in simply printing an information field. A tree is *traversed* if all its nodes are visited. The number of node accesses performed by a tree-traversal algorithm is used to measure the efficiency of the algorithm. (Normally, a node is visited only once in a traversal, but may be accessed several times.)

2.3.1 Traversing Binary Trees

We discuss three standard traversals of a binary tree T, namely, the *preorder*, *inorder*, and *postorder* traversals of T. Each traversal can be defined recursively, as illustrated by the following definition of preorder traversal.

DEFINITION 2.3.1 The *preorder traversal* of T is defined recursively as follows:

If T is not empty, then

1. visit the root node R,
2. perform a preorder traversal of the left subtree LT of the root,
3. perform a preorder traversal of the right subtree RT of the root.

Note that no action is performed in the preorder traversal of an empty tree T. The inorder and postorder traversals of T differ from the preorder traversal only in the order in which the root is visited. In the inorder traversal the root is visited after the left subtree tree has been traversed. In the postorder traversal the root is visited last (see Figure 2.8).

In Figure 2.9, the information in each node of the binary tree is a single alphabetical character. Assuming that visiting a node consists of simply printing the informa-

Figure 2.8 Preorder, inorder, and postorder traversals of binary T

preorder:	visit R	traverse LT	traverse RT
inorder:	traverse LT	visit R	traverse RT
postorder:	traverse LT	traverse RT	visit R

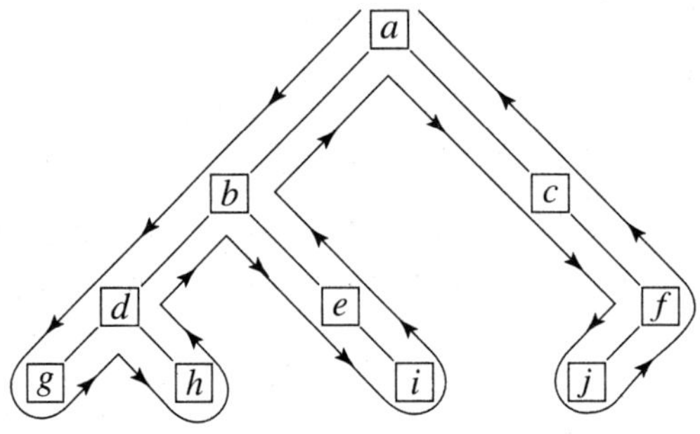

preorder: abdgheicfj inorder: gdhbeiacjf postorder: ghdiebjfca

Figure 2.9 Here visiting a node consists of printing the single character information field. All three traversals have the same *path around the tree,* as indicated.

tion stored in the node, then the resulting character strings printed out when the tree is traversed in preorder, inorder, and postorder are *abdghecfij*, *gdhbeacifj*, and *ghdebijfca*, respectively.

All three traversals have the same "path around the tree," as shown in Figure 2.9. The difference between them is the order in which visits occur. Figure 2.10 illustrates the sequence of accesses and visits for the tree given in Figure 2.9 for each of the three traversals.

Figure 2.10 Accessing versus visiting for the binary tree given in Figure 2.9 for preorder, inorder, and postorder traversals

preorder traversal		inorder traversal		postorder traversal	
access		access		access	
a	visit	*a*		*a*	
b	visit	*b*		*b*	
d	visit	*d*		*d*	
g	visit	*g*	visit	*g*	visit
d		*d*	visit	*d*	
h	visit	*h*	visit	*h*	visit
d		*d*		*d*	visit
b		*b*	visit	*b*	
e	visit	*e*	visit	*e*	
i	visit	*i*	visit	*i*	visit
e		*e*		*e*	visit
b		*b*		*b*	visit
a		*a*	visit	*a*	
c	visit	*c*	visit	*c*	
f	visit	*f*		*f*	
j	visit	*j*	visit	*j*	visit
f		*f*	visit	*f*	visit
c		*c*		*c*	visit
a		*a*		*a*	visit

The traversals preorder, inorder, and postorder have many applications. For example, preorder and postorder traversals of algebraic expression trees generate *prefix* and *postfix* expressions, respectively (see Exercise 2.7). Later in this section we discuss how a preorder traversal of an appropriate binary tree can be used to generate a binary coding for a given alphabet. Inorder traversals are particularly useful in connection with binary search trees. In particular, an inorder traversal of a binary search tree visits the nodes in sorted order.

We now give a recursive procedure for a preorder traversal of the binary tree T implemented using pointers and dynamic variables, where the pointer variable *Root* points to the root of T. The procedure *Visit* performs some action with the data in *Info*, such as printing it out.

```
procedure PreOrder(Root) recursive

Input:  Root (pointer to the root node of a binary tree T)
Output: the preorder traversal of T

   if Root ≠ nil then
      call Visit(Root→Info)
      call PreOrder(Root→LeftChild)
      call PreOrder(Root→RightChild)
   endif

end PreOrder
```

The procedures *InOrder* and *PostOrder* have pseudocode identical to *PreOrder* except for the order of the call to *Visit*.

Slightly different pseudocode for the tree traversals can be given when it is assumed that the tree T is nonempty. To illustrate, we give pseudocode for the altered version of *InOrder*. By assuming that *InOrder*2 is only invoked with nonempty trees, we avoid the need to check whether *Root* is **nil** at the beginning of the code. Before attempting to traverse a subtree, we add a check to see whether the subtree is nonempty. The depth of the recursion is reduced by one in the altered versions. However, it turns out that *InOrder* and *InOrder*2 perform the same number of comparisons of a pointer variable to **nil**.

```
procedure InOrder2(Root) recursive

Input:  Root (pointer to the root node of a tree T) {T is assumed nonempty}
Output: the inorder traversal of T

   if Root→LeftChild ≠ nil then
      call InOrder2(Root→LeftChild)
   endif
   call Visit(Root→Info)
```

Continued

```
    if Root→ RightChild ≠ nil then
        call InOrder2(Root→ RightChild)
    endif

end InOrder2
```

We now analyze the performance of *InOrder* (a similar analysis holds for *PreOrder* and *PostOrder*). The two main operations performed by *InOrder* are visiting a node (calls to *Visit*) and accessing a node. Since *InOrder* visits each node of T exactly once, the total number of node visits is n. During the performance of *InOrder*, each edge (line joining two nodes) in the tree T is traversed exactly twice, in opposite directions. For each edge traversal, the endpoint node (in the direction of the traversal) of the edge is accessed. It will be shown in Chapter 8 (Proposition 8.1.5) that the number of edges of a tree is one less than the number of nodes, so that T has $n - 1$ edges. It follows that *InOrder* performs $2(n - 1)$ node accesses, plus one additional initial access of the root. Therefore, the total number of node accesses performed by *InOrder* (or by *PreOrder* or *PostOrder*) is

$$1 + 2(n - 1) = 2n - 1.$$

2.3.2 Inorder Successors and Improving the Efficiency of Inorder Traversals

Clearly, any algorithm that performs a traversal of a tree having n nodes must access each node at least once. Hence, any traversal algorithm must perform at least n node accesses. Since *InOrder* performs $2n - 1$ accesses, when searching for a better algorithm we can only hope to narrow the gap between $2n - 1$ accesses and the lower bound of n accesses. We now show how to eliminate the reaccessing of nodes performed by *InOrder* when returning to the inorder successor of a node v having no right child. (The *inorder successor* of v is the node visited immediately after v in the inorder traversal.)

Consider the high-level description of an inorder traversal given in Figure 2.11.

Figure 2.11 High-level description of an inorder traversal of a binary tree T. Initially, the root is the current node.

1. Move left down the tree as far as possible, ending at a node now called the current node.
2. Visit the current node.
3. If the current node has no right child, then move back up the tree to the first unvisited node. If there is no such node, then terminate; otherwise this node becomes the current node and we repeat step 2.
4. Move to the right child of the current node, which now becomes the current node, and repeat step 1.

When translating Figure 2.11 into pseudocode, we maintain a stack of the vertices encountered as we move down the tree in step 1. If *v* is the current node having no right child, then the inorder successor of *v* is directly accessible since it is on the top of the stack.

```
procedure InOrder3(Root)

Input:  Root (pointer to the root node of a tree T) {T is assumed nonempty}
Output: the inorder traversal of T

  dcl Stack       {initialize empty stack}
  Terminate := .false.
  while .not. Terminate do {repeat steps 1–4 in Figure 2.11 until termination}

            {step 1}

    while Root→LeftChild ≠ nil do
      call Push(Stack,Root)
      Root := Root→LeftChild
    endwhile

            {step 2}

    call Visit(Root→Info)

            {step 3}

    while Root→RightChild = nil .and. .not. Terminate do
            {move back up tree to first unvisited node}
        if Empty(Stack) then          {if no such node exists, terminate}
          Terminate := .true.
        else
          call Pop(Stack,Root)
          call Visit(Root→Info)       {repeat step 2}
        endif
    endwhile

            {step 4}

    if .not. Terminate then
      Root := Root→RightChild
    endif
  endwhile

end InOrder3
```

During the execution of *InOrder*3, whenever *Stack* is popped the inorder successor of the node last visited is immediately accessed and visited. Going directly from *v* to its inorder successor for those *v* having no right child usually results in a considerable savings in the number of node accesses. For example, in Figure 2.12 we show

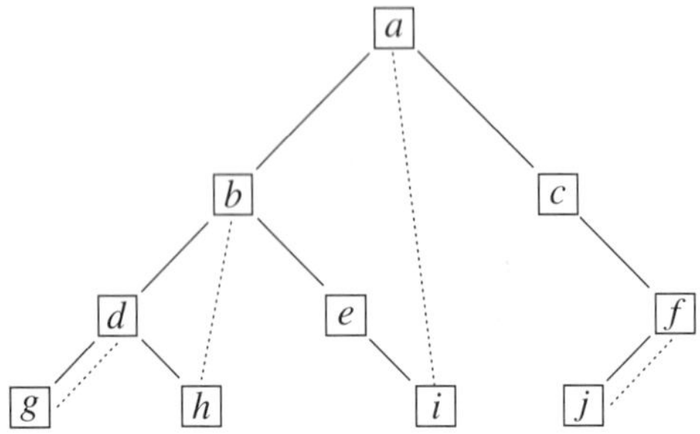

Root→*Info*			
a		*e*	visit
b		*i*	visit
d		*a*	visit
g	visit	*c*	visit
d	visit	*f*	
h	visit	*j*	visit
b	visit	*f*	visit

Figure 2.12 Sequence of accesses and visits performed by *InOrder*3 for the same binary tree given in Figure 2.9

the new sequence of node accesses that result from calling *InOrder*3 with the same tree used in Figure 2.9. The dotted arcs illustrate the immediate access to the relevant inorder successor that results from popping *Stack*. Note that *InOrder*3 performs 14 node accesses for the tree in Figure 2.12, whereas *InOrder* performs 19 node accesses for the same tree. The implicit threading saves three accesses (*d*,*e*,*b*), and two accesses (*c*,*a*) are saved by the termination condition.

2.3.3 Threaded Binary Trees

The dotted arcs shown in Figure 2.12 motivate the notion of a *threaded* binary tree. We concentrate here on the idea of an *inorder threading*, but the ideas are similar for *preorder* and *postorder threadings*.

In an inorder threading of a binary tree, we have a means of directly accessing the inorder successor (if any) of each node in the tree not having a right child. (The last node visited in an inorder traversal has the nil inorder successor.) This direct access to the inorder successor is the threading, as illustrated in Figure 2.12.

In a *symmetric* threading, we also have a means of directly accessing the inorder predecessor of nodes having no left child. However, we restrict attention here to inorder successors.

Assuming that a binary tree is implemented with dynamic variables and pointers, an inorder threading is obtained by altering the right child pointer in nodes v having no right child to point to the inorder successor. An additional single-bit Boolean field *Thread* is added to each node to distinguish between the case where *RightChild* points to a right child (*Thread* = 0) and where *RightChild* points to an inorder successor (*Thread* = 1). In Figure 2.13 we illustrate the pointer and dynamic variable implementation of the threaded tree in Figure 2.12. For convenience, we show the information field of the node pointed to by *LeftChild* and *RightChild*. Of course, in practice *LeftChild* and *RightChild* contain the addresses of the indicated nodes. In Figure 2.13, the *LeftChild* and *RightChild* fields are left blank when they contain the **nil** pointer.

We can create the original threading with a simple modification of the algorithm *InOrder*3. Given the threading, an algorithm for inorder traversal utilizing the

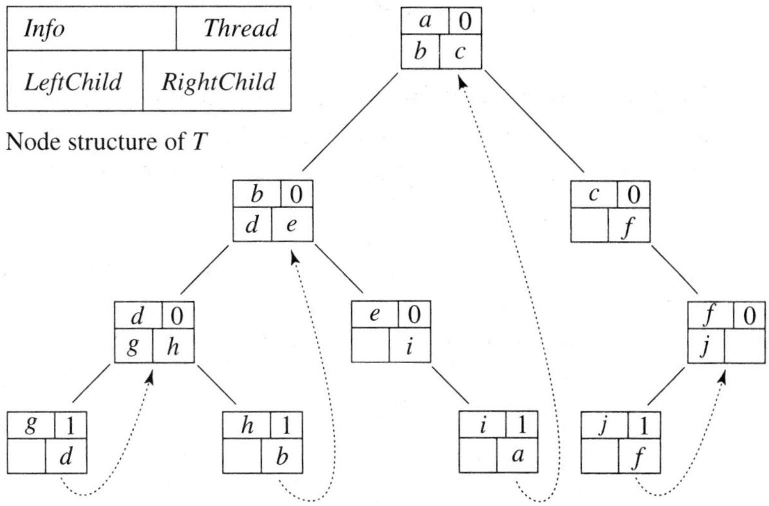

Figure 2.13 A binary tree T with an inorder-successor threading

threading is easily obtained. An inorder traversal based on the threading does not save any accesses over *InOrder*3, but it has the advantage that no stack is needed. Of course, inorder traversal algorithms based on threading have the additional overhead of checking whether *RightChild* is a pointer to a child or is a thread.

There are other ways to traverse binary trees without using (implicit or explicit) stacks. For example, parent fields can be introduced, or they can be simulated by dynamically reversing and resetting child pointers during the course of the traversal. We discuss these ideas further in the exercises.

2.3.4 Binary Prefix Codes

In each of the three standard traversal methods, we assume that a call to the procedure *Visit* with a given input node v involves only operations on the information fields of v itself. In practice, situations arise when visiting a node v might involve performing operations on the information fields of the children of v as well (called a *generalized visit*). A natural example of traversing a tree where we call a generalized visit procedure occurs in the context of binary coding and data compression.

For a given *alphabet* of symbols $A = \{a_1, a_2, \ldots, a_n\}$, one approach to the problem of data compression involves encoding each symbol in the alphabet with a binary string. We refer to the set of binary strings encoding the symbols in A as a *binary code* for A. Given a binary code for A, we obtain a binary encoding of any string consisting of symbols from A. We simply replace each symbol in the string with its binary encoding. For example, suppose $A = \{a, b, c\}$ and symbols a, b, c are encoded as 1, 00, 01, respectively. Then the binary encoding for the string *cabbc* is 011000001.

A commonly used binary code for storing alphabetical symbols in the computer is the ASCII code, which associates a 1-byte (8-bit) binary string with each symbol. By allowing variable length binary strings for binary codes, we can save space when storing text made up of alphabetical symbols. If we know the frequency of the occurrence of the various symbols, we can minimize the space required to store text by encoding the more frequently occurring symbols with shorter codes. We solve this optimization problem in Chapter 13, when we discuss Huffman codes. In this section we show how a generalized preorder traversal of binary trees can be used to generate unambiguous binary codes—that is, codes for which a well-defined decoding procedure exists.

To illustrate ambiguous codes, suppose a, b, c are encoded as 0, 1, 01, respectively. Then the binary string 011 can be decoded as either *abb* or *cb*. Note that the binary string 0 is a prefix for the binary string 01. It is easily verified that a binary code allows for unambiguous (unique) decoding if, and only if, no binary string in the code is a prefix of any other binary string in the code.

Binary codes having the property that no binary string in the code is a prefix of any other binary string in the code are called *prefix codes*. We can readily create a prefix code by utilizing a 2-tree T, which is a binary tree where every nonleaf node has exactly two children. The leaf nodes of T correspond to the n symbols in A. A high-level description of the process is as follows. Label each edge corresponding to a left child with a 0 and each edge corresponding to a right child with a 1. Assign to each symbol a_i the binary string obtained by reading the labels on the edges when following a path from the root to the leaf node corresponding to a_i. The binary strings generated by a sample tree for the alphabet $A = \{a, b, c, d, e, f\}$ are shown in Figure 2.14.

For each node N of the 2-tree T, let $B(N)$ denote the binary string corresponding to following a path from the root to N in the manner described earlier. Let L and R denote the left and right children of N, respectively. Note that $B(L)$ and $B(R)$ can

Figure 2.14 Encoding of alphabet $A = \{a, b, c, d, e, f\}$ generated by sample 2-tree T

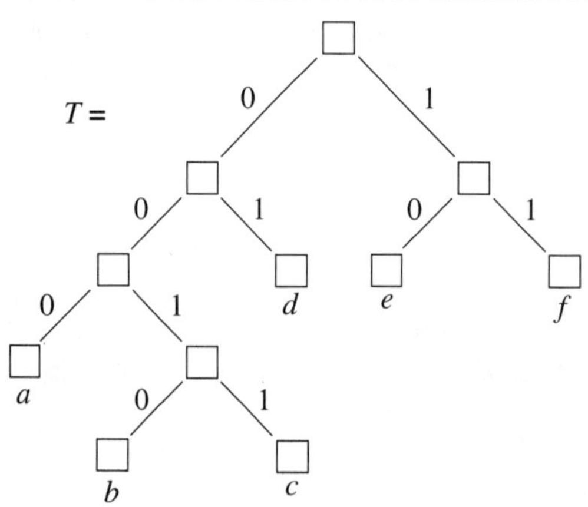

Symbol	Binary encoding
a	000
b	0010
c	0011
d	01
e	10
f	11

be computed from $B(N)$ by concatenating '0' and '1', respectively, to the end of $B(N)$; that is, $B(L) = B(N)$ + '0' and $B(R) = B(N)$ + '1' (where + denotes the operation of concatenation). We now give pseudocode for the algorithm *GenerateCode*, which utilizes this formula to compute the array *Code*, where *Code*[i] is the binary code for symbol a_i, $i = 1, \ldots, n$. *GenerateCode* computes $B(N)$ for each node by performing a preorder traversal of the 2-tree, where the generalized visit operation at node N involves computing $B(L)$ and $B(R)$ from $B(N)$ using the formula described above. We assume that the 2-tree T is implemented using pointers and dynamic variables, where each node contains two pointer fields corresponding to the left and right children and two information fields *BinaryCode* and *SymbolIndex*. Before calling *GenerateCode*, we assume that the value of *SymbolIndex* for each leaf node has been initialized to contain the index i of the alphabet symbol a_i corresponding to the leaf node (*SymbolIndex* is left undefined for the internal nodes of T). The computed value $B(N)$ for each node N is stored in the variable *BinaryCode*. The root node's value of *BinaryCode* is initialized to the null string before calling *GenerateCode*.

```
procedure GenerateCode(Root,Code[1:n]) recursive
Input: Root (a pointer to root of 2-tree T)   {Root→BinaryCode is initialized
           to the null string. The leaf node corresponding to a_i
           has its SymbolIndex field initialized to i, i = 1, . . . , n.}
Output: Code[1:n] (array of binary strings, where Code[i] is the code for
           symbol a_i)
   if Root→LeftChild = nil then        {a leaf is reached}
      Code[Root→SymbolIndex] := Root→BinaryCode
   else
      (Root→LeftChild)→BinaryCode := Root→BinaryCode + '0'
      (Root→RightChild)→BinaryCode := Root→BinaryCode + '1'
      call GenerateCode(Root→LeftChild,Code)
      call GenerateCode(Root→RightChild,Code)
   endif
end GenerateCode
```

The table *Code*[1:n] of binary codes can be used to encode text consisting of strings of symbols from the alphabet A. Going the other direction, encoded text can be decoded using the 2-tree T in the obvious way.

2.3.5 Traversing General Trees

The three standard traversals of a binary tree can be generalized to any tree as follows. To define the preorder, inorder, and postorder traversals of a general T, we assume that an ordering is associated with the subtrees $T_1, T_2, \ldots, T_j$ of the root as indicated by the subscripting (so that T_1 is the left-most subtree, T_2 is the subtree to the immediate right of T_1, and so forth). A similar ordering is assumed for the subtrees of any node in T.

DEFINITION 2.3.2 The *preorder* traversal of T is defined recursively as follows:
If T is not empty, then

1. visit the root node R,
2. perform successively a preorder traversal of $T_1, T_2, \ldots, T_j$.

Again, the postorder traversal is identical to the preorder traversal except that steps 1 and 2 are interchanged. The definition of the inorder traversal of a general T is not as natural as it was for binary trees, since there are many intermediate orders in which to perform a visit. In Figure 2.15 we formulate an inorder traversal that visits a node immediately after traversing the leftmost subtree of the node.

Pseudocode for implementing preorder, inorder, and postorder traversals of general trees, as well as their generalizations to traversals of rooted forests, is developed in the exercises. As with binary trees, each of the standard traversals of general trees has its particular uses. For example, what amounts to a postorder traversal of a forest arises in an algorithm for determining the strongly connected components of the digraph (see Chapter 19). It turns out that a multivisit version of an inorder traversal is very useful in multiway search trees.

Traversals with Multivisits For general trees, visiting a node might occur more than once in the course of a traversal. For example, an inorder traversal might actually take the following form:

inorder: traverse T_1 visit$_1 R$ traverse T_2 visit$_2 R$ ... visit$_{j-1} R$ traverse T_j

where each visit of the node R, visit$_1 R$, visit$_2 R$, ..., visit$_{j-1} R$ results in a (possibly) different operation performed on the information fields in R. We call this type of traversal a *multivisit inorder traversal.* Multivisit preorder and postorder traversals are defined similarly.

The following algorithm *MultiVisitTreeInOrder* implements a multivisit inorder traversal of a general tree. The procedure *Visit* called by *MultiVisitTreeInOrder* has an addition parameter *Count*, which is used to keep track of the number of visits made at a particular node so that the appropriate operations can be performed for each visit of a node.

Figure 2.15 Preorder, inorder, and postorder traversals of general T

preorder:	visit R	traverse T_1	traverse T_2 ...	traverse T_j
inorder:	traverse T_1	visit R	traverse T_2 ...	traverse T_j
postorder:	traverse T_1	traverse T_2	traverse T_j ...	visit R

```
procedure MultiVisitTreeInOrder(Root) recursive

Input:  Root (a pointer to the root node of a tree T)
Output: the preorder traversal of T

  if Root ≠ nil
    if Root→ LeftMostChild = nil then
      call Visit(Root→Info,1)      {single-node tree}
    else
      Count := 1
      CurrentSibling := Root→ LeftMostChild
      while CurrentSibling ≠ nil do
        call MultiVisitTreeInOrder(CurrentSibling)
        call Visit(Root→ Info,Count)
        Count := Count + 1
        CurrentSibling := CurrentSibling → NextSibling
      endwhile
    endif
  endif

end MultiVisitTreeInOrder
```

Multivisit inorder traversals are the basis of searching multiway search trees as defined in Section 2.6.

2.4 Complete and Full Binary Trees

Given a binary tree T, for ease of discussion, we denote the number of nodes, the number of leaf nodes, the number of internal nodes, and the depth by $N = N(T)$, $L = L(T)$, $I = I(T)$, and $D = D(T)$, respectively, Let n_i denote the number of nodes of T at level i, $i = 0, \ldots, D$. Since T is a binary tree, each node of T has at most two children, which yields the following recurrence relation.

$$n_i \leq 2n_{i-1}, \quad i = 1, \ldots, D, \quad \textbf{init. cond. } n_0 = 1 \tag{2.4.1}$$

Thus, we have

$$n_i \leq 2^i, \quad i = 0, \ldots, D. \tag{2.4.2}$$

For a given level $i \in \{1, \ldots, D\}$, we say that T is *full at level* i, if $n_i = 2^i$. We can easily verify that if a binary tree T is full at level i, then it is full at every level j smaller than i. A binary tree T is full if it is full at every level. Equivalently, a binary tree is full if it is full at the last level. Note that a full binary tree exists only for those N satisfying

$$N = 1 + 2 + 2^2 + \cdots + 2^D = 2^{D+1} - 1.$$

We denote the full binary tree on n nodes by T_n, $n = 1, 3, 7, 15, \ldots$. The full binary tree T_{15} is shown in Figure 2.16a. Full binary trees are special cases of complete binary trees. To define the complete binary tree T_n when n is not one less than a power of 2, we let k be the (unique) positive integer such that $2^k - 1 < n < 2^{k+1} - 1$. Then T_n is defined as a canonical subtree of the complete (full) binary tree T_N, $N = 2^{k+1} - 1$ having the same depth and differing from T_N only at the last level. The $q = n - (2^k - 1)$ nodes of T_n on the last level occupy the leftmost q positions of T_N. More precisely, suppose we label the nodes of the T_N as follows. We label the root node 1. Inductively, we label the left and right children of vertex i as $2i$ and $2i + 1$, respectively. Then the complete binary tree T_n on n nodes is the subtree of T_N consisting of those nodes labeled 1,2, . . . , n (see Figure 2.16b).

The labeling of the nodes of the complete binary tree T_n illustrated in Figure 2.16 is particularly useful when the tree is implemented using an array $T[1{:}n]$, where $T[i]$ corresponds to the ith node in the labeling, $i = 1, \ldots, n$. For example, in Chapter 4 we will see how the heap ADT is implemented very naturally using

Figure 2.16 (a) Complete (full) binary tree T_{15}; (b) complete binary tree T_{10}

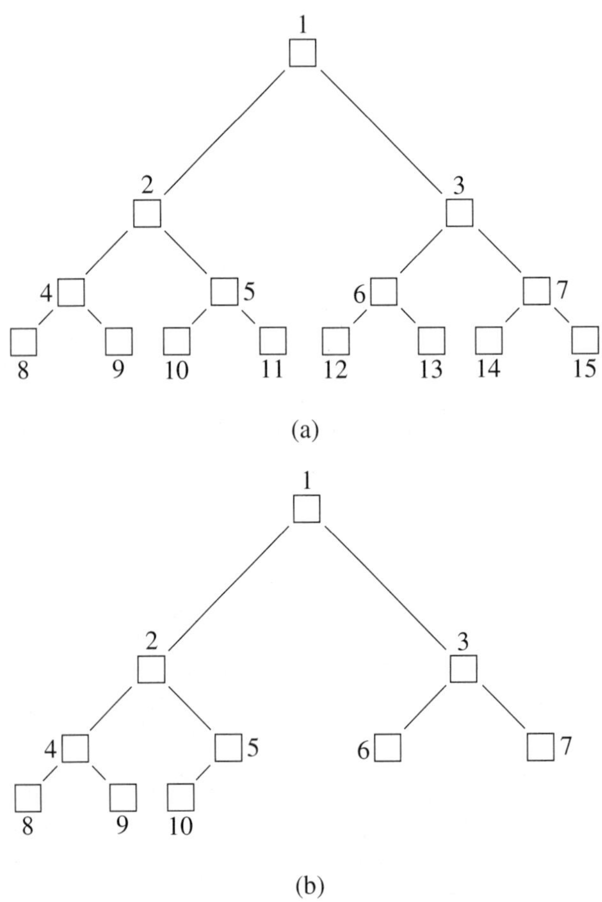

the array implementation of T_n. Pseudocode for algorithms implementing the creation of a heap, and deletion and insertion operations, is greatly facilitated using the fact that the children of the ith node $T[i]$ are $T[2i]$ and $T[2i + 1]$, and the parent of the ith node is $T[\lfloor i/2 \rfloor]$.

The depth of the complete binary tree T_n for a general n is given by (see Exercise 2.17):

$$D(T_n) = \lfloor \log_2 n \rfloor. \tag{2.4.3}$$

Since T_n is as full as possible for a binary tree on n nodes, it is not surprising that T_n has minimum depth for all binary trees on n nodes. The latter fact and other mathematical properties of binary trees are discussed in Chapter 7.

Remark The terminology for complete and full binary trees varies in the literature. For example, sometimes the name *complete* is used for what we have called *full* and the name *left complete* is used for what we have called *complete*.

2.5 Binary Search Trees

Given a totally ordered set S (elements of S will be called *keys* or *identifiers*), the *dictionary problem* is the well-known problem of designing an ADT to maintain a collection of items drawn from S. The classical dictionary problem restricts attention to the *dictionary* operations of inserting a key, deleting a key, and searching for a key, but we also might consider performing additional operations, such as finding the maximum or minimum elements or accessing the keys in sorted order.

In this section we introduce the binary search tree ADT and show how it supports the dictionary operations. In what follows we assume that the entire binary search tree is kept in internal memory, which is a reasonable assumption if the number of keys to be maintained is not too large. For larger sets of keys (such as typically encountered in databases stored in external memory) other ADTs such as B-trees are more appropriate (see Chapter 17). However, internal memory is rapidly becoming larger and cheaper, so that binary trees held in internal memory and containing keys from database records held in external memory are becoming more realistic.

Binary search trees are explicit generalizations of the implicit tree underlying the algorithm *BinarySearch*. The algorithm for inserting an element in a binary search tree is a variation of the searching strategy. Deletion is slightly more complicated, requiring in some cases a replacement of the deleted node by its inorder successor. A useful property of a binary search tree is that an inorder traversal visits the nodes in sorted order.

2.5.1 Definition and Examples of Binary Search Trees

A binary search tree T with respect to keys in the nodes of T can be defined recursively as follows.

DEFINITION 2.5.1 A binary tree T is called a *binary search tree* (with respect to a key field in each node) if T is empty, or if T has root R, and its right and left subtrees RT, LT satisfy the following.

1. Each key in LT is not larger than the key in R.
2. Each key in RT is not smaller than the key in R.
3. Both LT and RT are binary search trees.

Figure 2.17 gives six of the fourteen different binary search trees for the set of four keys: 'Ann', 'Joe', 'Pat', 'Ray'.

Given any set of n distinct keys (identifiers) and an *arbitrary* binary tree T on n nodes, there is a unique assignment of the keys to the nodes in T such that T becomes a binary search tree for the keys (see Exercise 2.20). Hence, the number of binary search trees containing a given set of n distinct keys equals the number $b_n = \frac{1}{n+1}\binom{2n}{n}$ of binary trees on n nodes.

2.5.2 Searching a Binary Search Tree

A simple and efficient algorithm (*BinSrchTreeSearch*) exists for locating information stored in binary search trees. For example, if we wish to determine which node in the tree (if any) has its key equal to a search element X, then we first compare X to the key contained in the root. If X equals this key, we are done. Otherwise, if X is

Figure 2.17 Sample binary search trees for the keys Ann, Joe, Pat, and Ray

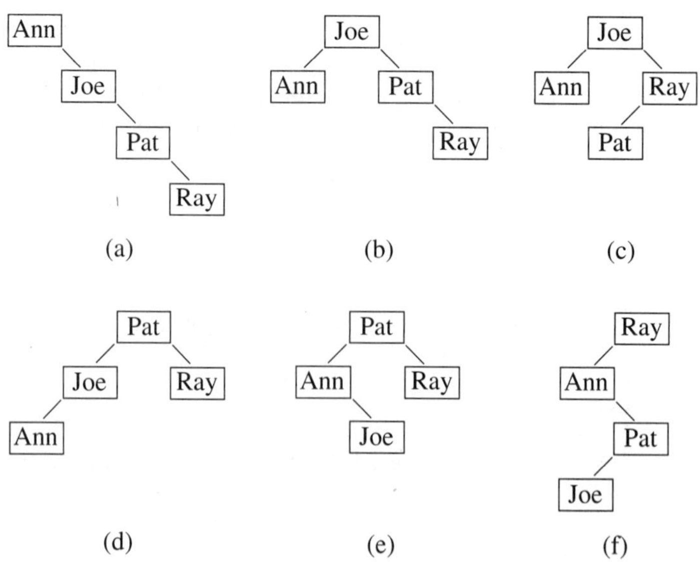

less than the key in the root, then we search the left subtree, else we search the right subtree. The following algorithm *BinSrchTreeSearch* based on this simple process returns a pointer *Location* to a node where a search element *X* is found. If *X* is not found in the tree, then *BinSrchTreeSearch* sets *Location* to a node in the tree where *X* can be inserted as a child and still maintain a search tree. In the pseudocode for *BinSrchTreeSearch*, we assume that the nodes in the tree have the structure as described in Section 2.2.1.

```
procedure BinSrchTreeSearch(Root,X,Location,Found)
Input:  Root (→BinaryTreeNode)      {points at root of a binary search tree}
        X (KeyType)
Output: Location (→BinaryTreeNode) {points at occurrence of X, if any}
        Found (Boolean)       {.false. if X not in tree, .true. otherwise}
  Found := .false.
  Location := nil
  Current := Root
  while Current ≠ nil .and. .not. Found do
    Location := Current
    if X = Current→Key then
      Found := .true.
    else
      if X < Current→Key then
        Current := Current→LeftChild
      else
        Current := Current→RightChild
      endif
    endif
  endwhile
end BinSrchTreeSearch
```

In Figure 2.17 we have not drawn the (implicit) leaf nodes corresponding to unsuccessful searches. The tree in Figure 2.17d is redrawn in Figure 2.18 with leaf

Figure 2.18 Search tree with leaf nodes drawn representing unsuccessful searches

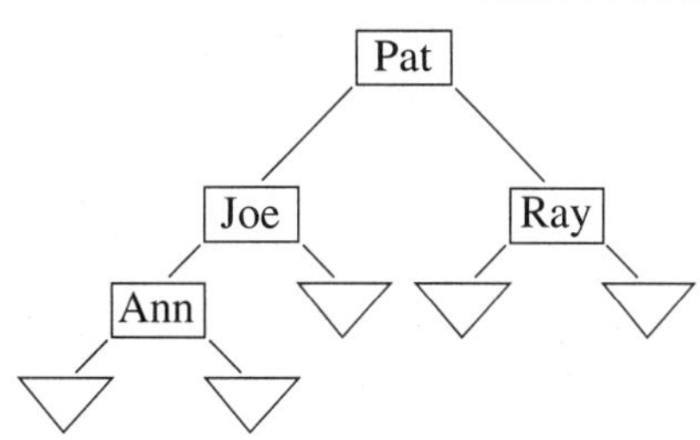

nodes corresponding to unsuccessful searches added. For example, a search for 'Mary' would end up at the right child of 'Joe'. The latter node is where the key 'Mary' can be inserted and still maintain the binary search tree property.

2.5.3 Inserting Nodes into Binary Search Trees

The following algorithm inserts a key *X* into a binary search tree *T* pointed to by *Root*. We assume that duplicate keys are not allowed in the tree (the case where duplicate keys are allowed is developed in the exercises). Hence, *BinSrchTreeInsert* begins by calling *BinSrchTreeSearch* to determine whether or not the key *X* is in the tree *T*. If *X* is not in the tree, then the output parameter *Location* of *BinSrchTreeSearch* points to the node in the tree that is the appropriate parent node for the insertion of *X*. *BinSrchTreeInsert* assumes that *Root* = **nil** when it points to an empty tree.

```
procedure BinSrchTreeInsert(Root,X)
Input:  Root (→BinaryTreeNode)        {points to root of a binary search tree T}
        X (KeyType)
Output: Root (→BinaryTreeNode)        {points to root of the binary search tree
                                          obtained from T by inserting X}
  if Root = nil then                  {create a single-node tree containing X}
     call AllocateBinaryTreeNode(NodePtr)
     NodePtr→Key := X
     Root := NodePtr
  else
     call BinSrchTreeSearch2(Root,X,Location,Found)
     if .not. Found then              {insert X into tree}
        call AllocateBinaryTreeNode(NodePtr)
        NodePtr→Key := X
        if X < Location→Key then
           Location→LeftChild = NodePtr
        else
           Location→RightChild = NodePtr
        endif
     endif
  endif
end BinSrchTreeInsert
```

A binary search tree can be created from a given set of keys by repeated invocations of *BinSrchTreeInsert* with the keys as input parameters. We discuss this further in Chapter 4 in connection with the sorting algorithm *TreeSort*. *TreeSort* is based on the fundamental property that inorder traversals of binary search trees visit the keys in increasing order.

2.5.4 Deleting Nodes from Binary Search Trees

There are three cases to consider when deleting a node v from a binary search tree:

1. v is a leaf node.
2. v is an internal node with a single child.
3. v is an internal node with two children.

In cases 1 and 2, we delete v by "splicing" an appropriate child of v to v's parent. In case 1, the appropriate child of v can be taken as either v's right or left child, since they are both **nil** (the pseudocode takes the right child). In case 2, the appropriate child of v is the single non-nil child of v. The "splicing out" of v is illustrated in Figure 2.19, where we show the appropriate child of v as its right child.

Case 3 is reduced to case 1 or 2 by replacing the values in the information and key fields in the node v by the values in corresponding fields of its inorder successor w, and then deleting w (which has at most one child). In particular, we do not alter the child pointer fields v or its parent. Recall that the inorder successor of any node v having a right child can be found by moving to the right child and then traveling left as far as possible, until a node w having no left child is encountered. Then w is the inorder successor of v. We leave it as an exercise to verify that the resulting tree is a binary search tree.

In Figure 2.20 (page 62) we illustrate the three cases of node deletion. In each case, we assume that the deletion is from the binary search given in Figure 2.20a. We show the value of the key field in each node.

For cases 1 and 2, deleting a node v is done in constant time. In case 3, we must determine the inorder successor of v, which might require depth(T) node accesses and pointer comparisons.

On input to *BinSrchTreeDelete*, the input-output parameters *Node,Parent* are pointers to the node v to be deleted and its parent, respectively. As output parameters, *Node,Parent* remain unchanged if v has at most one child, otherwise *Node,Parent* point to the inorder successor of v and its parent, respectively. If *Parent* = **nil**, then we are deleting the root node.

Figure 2.19 Splicing out node v

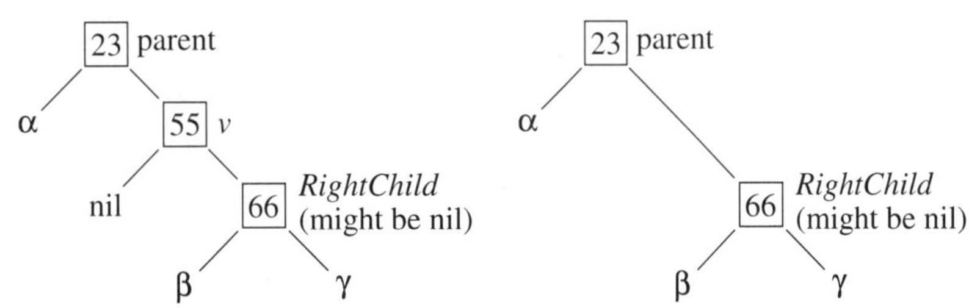

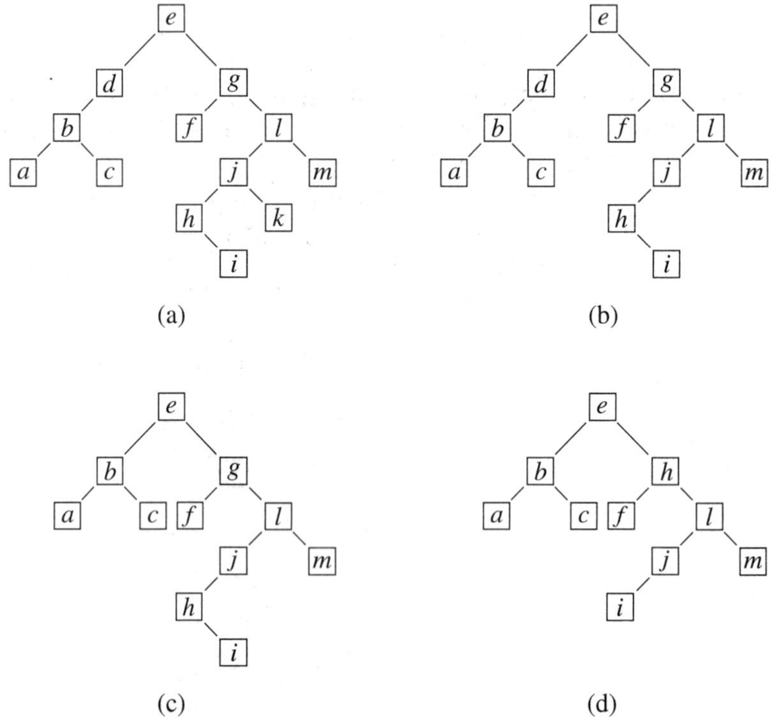

Figure 2.20 Illustrations of the three cases of node deletion from a binary search tree. (a) Binary search tree before deletion of a node; (b) case 1 deletion: delete node *k*; (c) case 2 deletion: delete node *d*; (d) case 3 deletion: delete node *g*

procedure *BinSrchTreeDelete*(*Root*,*Node*,*Parent*)

Input: *Root* (→*BinSrchTreeNode*) {points to root of a binary search tree *T*}
Node (→*BinSrchTreeNode*) {points to node *v* to be deleted}
Parent (→*BinSrchTreeNode*) {points to parent of *v*}
Output: *Root* (→*BinSrchTreeNode*) {points to root of a binary search tree *T′* obtained from *T* by deleting *v*}
Node, *Parent* (→*BinSrchTreeNode*) {*Node*,*Parent* remain unchanged if *v* has at most one child, otherwise *Node*,*Parent* point to the inorder successor of *v* and its parent}

```
if Node→LeftChild ≠ nil .and. Node→RightChild ≠ nil then
    Successor := InOrderSucc(Node)      {v is internal node}
    Node→Info := Successor→Info
    Node→Key := Successor→Key
    Node := Successor
    Parent := Node→Parent
endif
```

```
{Now Node points at node w to be deleted that is not an internal node, so define the
appropriate child x of w for splicing to w's parent. Note that x is nil if w is a leaf.}

  if Node→LeftChild ≠ nil then        {determine appropriate child x of
     AppropChild := Node→LeftChild       w for splicing to parent of w}
  else
     AppropChild := Node→RightChild
  endif
  if Parent = nil then        {make x new root of tree}
     Root := AppropChild
  else                                          {see which child of the parent
     if Parent→LeftChild = Node then             of w should be spliced to x}
        Parent→LeftChild := AppropChild
     else
        Parent→RightChild := AppropChild
     endif
  endif
end BinSrchTreeDelete
```

The input parameter *Node* is typically determined by first calling *BinSrchTreeSearch* to determine where a node having a given key is found. However, the pseudocode for *BinSrchTreeSearch* does not determine *Parent*. We leave it as an exercise to alter *BinSrchTreeSearch* so that the locations of both the search element and its parent are determined.

2.6 Multiway Search Trees

Multiway search trees generalize binary search trees by allowing more than one key to be stored in a given node. Thus, multiway search trees allow multiway branching to occur at a given node, instead of the two-way branching allowed by binary trees. The keys in a given node of a multiway search tree are maintained as an ordered list. During a search of a multiway search tree, when a search element reaches a given node, a search of the keys in the node is performed. (The latter search is usually a linear search, but if the keys are maintained in an array, then binary search might be used if the node contains a sufficiently large number of keys.) Either the search element equals one of the keys, or a branch is made to an appropriate child's subtree (which contains the key if it is in the tree at all), where the search continues. The following recursive definition places the fairly obvious restriction on the keys in a multiway search tree that must hold to support the natural generalization of the search strategy for binary search trees. In the definition, the value of j varies with the particular node in T.

DEFINITION 2.6.1 A *multiway search tree* T is either empty or consists of a root R containing j keys, $key_1 \leq key_2 \leq \cdots \leq key_j$, $j \geq 1$, and a collection of subtrees $T_1, T_2, \ldots, T_{j+1}$ having the following two properties:

1. for any $k_i \in T_i$, $1 \le i \le j + 1$, we have

$$k_1 \le key_1 \le k_2 \le key_2 \le \cdots \le k_j \le key_j \le k_{j+1},$$

2. every T_i is a multiway search tree.

For a given fixed constant m, multiway search trees for which the maximum number of keys in a given node is bounded above by $m - 1$ are called *m-way* search trees. For example, a four-way search tree on 15 nodes is shown in Figure 2.21. Note that a binary search tree is a two-way search tree.

Remark A commonly used multiway search tree is the B-tree discussed in Chapter 17. B-trees are subject to several additional restrictions, including bounds on the minimum and maximum number of keys allowed in each node. B-trees also satisfy a more restrictive condition than (2), namely:

2.′ Either *every* T_i is empty, or *every* T_i is a nonempty multiway search tree.

We adopt the more general condition (2) in our definition of multiway search trees so that multiway search trees generalize the notion of binary search trees.

Given a multiway search tree T and a key k, Figure 2.22 contains a high-level procedure determining a location in T where k is found. In our high-level description, we do not specify how the tree T is implemented. The location of k consists of a pair (v,i), where v is a node of T in which k occurs, and i is a position of the occurrence of k in the ordered list of keys of v. If k does not occur in the tree, we set $i = 0$ and regard v as undefined. We initially perform step 1 in Figure 2.22 with a variable *Root* pointing at the root of T. Before repeating step 1, *Root* is redefined to point to the appropriate subtree being searched.

Figure 2.21 A four-way search tree on 15 nodes

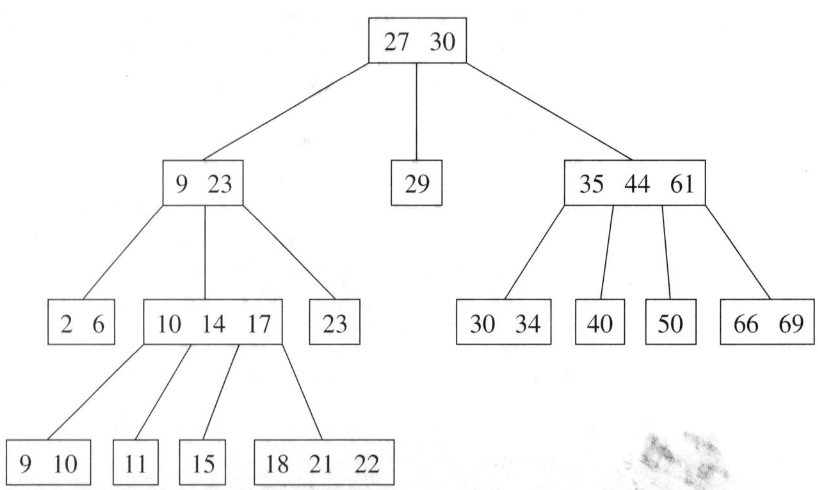

1. If k occurs in position i of the node v pointed to by *Root*, then return (v,i).
2. If we are at a leaf, then return $i = 0$.
3. Redefine *Root* to point at the root of subtree T_i and repeat step 1, where T_i is determined by one of the following three conditions:
   a. $i = 1$ and $k < key_1$,
   b. $1 < i < n$ and $key_{i-1} < k < key_i$,
   c. $i = n$ and $k > key_j$.

Figure 2.22 High-level description of procedure for searching a multiway search tree

2.7 Removing Recursion

Recursion is a very powerful design strategy for algorithms. However, there is usually a fair amount of overhead incurred when actually implementing a recursive function or procedure. This overhead results from the need to save the values of local variables and parameters, as well as a return address, whenever a function or procedure (recursive or not) is invoked. Moreover, there are computer languages in wide use today that do not even support recursion. Implementing recursive functions or procedures in such languages therefore requires that the recursion be removed (simulated). Even when a language supports recursion, it may well be more efficient to simulate the recursion to save the overhead resulting from repeated function or procedure invocations.

2.7.1 Tail Recursion

A recursive procedure or function is *tail recursive* when the last executable statement in the pseudocode is a simple recursive reference. Such a recursive reference (called *tail recursion*) can be removed by (1) redefining parameters or assigning the parameter expressions involved in the recursive call (invocation) to suitable local variables; and (2) making an unconditional branch to the first executable statement in the subprogram. The unconditional branch is typically implemented by introducing a **while** loop that continues to execute until the initial conditions of the recursive version are satisfied. A final statement (or statements) after the **while** loop then returns the appropriate function value or assigns appropriate values to output parameters as determined by the initial conditions.

For example, consider the recursive function *EuclidGCDRec* given in Section 1.3.1. Removing the tail recursion from *EuclidGCDRec* converts it precisely into the nonrecursive version *EuclidGCD* given in Section 1.1.2.

As a second example of tail recursion, consider the procedure *Towers* given in Section 1.3.3. Removing the tail recursion from *Towers* results in the following pseudocode for *Towers*2.

procedure *Towers*2(*First, Auxiliary, Last, n*) **recursive**

Input: *First, Auxiliary, Last* (names of three pegs)
Output: a solution to the Towers of Hanoi puzzle

Continued

```
    while n > 1 do
        call Towers2(First,Last,Auxiliary,n – 1)
        write ('Move disk',n,'from peg',First,'to peg',Last)
        Temp := Auxiliary
        Auxiliary := First
        First := Temp
        n := n – 1
    endwhile
    write ('Move disk',1,'from peg',First,'to peg',Last)
end Towers2
```

While *Towers*2 has removed the tail recursion from *Towers*, *Towers*2 remains recursive. The recursion in *Towers*2 can be removed by applying the general simulation technique using stacks, which is discussed in the next section.

2.7.2 Simulating Recursion Using Stacks

In this section we illustrate the use of stacks to translate a recursive function or procedure into an equivalent nonrecursive version. Roughly speaking, recursive invocations or calls are simulated by pushing all parameters and local variables on a stack, together with a return address for the first executable statement following the recursive reference. Then parameters are recomputed according to the expressions in the recursive invocation, and a branch is made to the beginning of the pseudocode. Returns are simulated by popping the stack and assigning parameters and local variables the appropriate popped values. Then a branch is made to the statement determined by the return address that was popped off the stack (or otherwise determined), and execution is continued.

We begin by considering single-reference recursive functions. Similar to removing tail recursion, the pseudocode for the nonrecursive version generally starts with a **while** loop that continues execution until the parameters satisfy the initial conditions of the original recursion. A local variable, say *RetFun* (*RetFun* is not stacked), is introduced to compute the function value. After completion of the **while** loop, *RetFun* is assigned the appropriate initial condition value, and a loop is entered in which the *Stack* is continually popped, with *RetFun* continually recomputed according to the description of the recursion. When the stack is empty, *RetFun* is recomputed for the last time and returned.

We illustrate the simulation of a single-reference recursive function using stacks with the function *PowersRec*. Recall that even though *PowersRec* is single-reference recursion, there are two recursive invocations in the pseudocode. In particular, the two recursive invocations are the consequent clauses of an **if-then-else** statement, so that *PowersRec* is not tail recursive. Because of the two recursive references, we need a method of determining which recursive reference is being resolved when we simulate the recursion using a stack. We can always do this by generating and pushing an appropriate return address on the stack. However, as our simulation of *PowersRec*

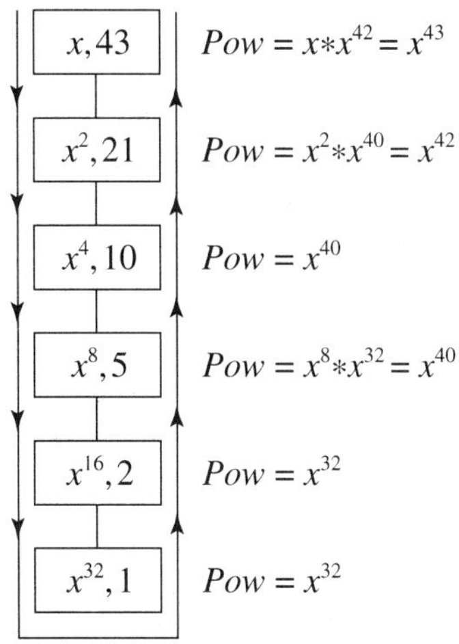

Figure 2.23 Resolution of *PowersRec* with $n = 43$

shows, sometimes the return address can be implied from the value of other variables or parameters, thereby obviating the necessity to maintain an explicit return address.

In Figure 2.23 we illustrate the resolution of the recursive calls generated by *PowersRec* with input parameters x and $n = 43$. The current value of the parameters x and n for each invocation are shown inside the nodes in the path. The value of *Pow* that is returned as we resolve each invocation is shown beside the nodes. (*Pow* plays the role of *RetFun* in our previous simulation discussion.)

Figure 2.24 shows the dynamic trace of the stack simulating the recursive resolution of *PowersRec* when invoked with $n = 43$. We show the status of the stack after each push or pop of the parameters x and n.

Figure 2.24 Dynamic trace of a stack modeling the resolution of *Powers* when invoked with $n = 43$. The value of *Pow* is shown above the stack, the values of the parameters x and n are shown as pushed on the stack, and the current values of the parameters x and n are shown beneath the stack.

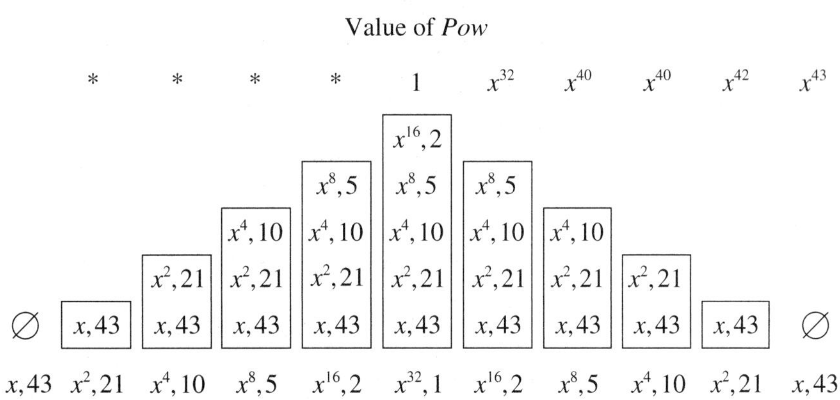

In the pseudocode for *PowersNonRec* that follows, each time we push the stack we store a record containing the two parameters x and n. The actual form that this record takes depends on the particular implementation used for the stack and doesn't concern us here. We also assume that the function *Empty*(*Stack*) returns a Boolean value of **.true.** if *Stack* is empty, otherwise it returns **.false.**

```
function PowersNonRec(x,n) {nonrecursive version}
Input:   x (a real number), n (a nonnegative integer)
Output: x^n (returns the value of x^n)
  dcl Stack       {initialize empty stack}
    while n > 0 do
      call Push(Stack, {x,n})
      x := x*x
      if even(n) then
        n := n/2
      else
        n := (n – 1)/2
      endif
    endwhile
    Pow := 1
    while .not. Empty(Stack) do
      call Pop(Stack, {x,n})
      if odd(n) then
        Pow := x*Pow
      endif
    endwhile
    return(Pow)
end PowersNonRec
```

Even though *PowersRec* is not tail recursion, by observing the dynamic trace of the stack in Figure 2.24, it is straightforward to write pseudocode for an iterative version that eliminates the stack. Assume that *PowersNonRec* is invoked with the parameters x and n. Note that the stack is continually pushed until the parameters x^m,1 are at the top of the stack, where $m = 2^p$, $p = \lfloor \log_2 n \rfloor$. Then the pushing on the stack ceases, and *Pow* is initialized to 1. Thereafter, we continually pop the stack, updating the value of *Pow* by multiplying it by the current value of the parameter x when the current value of the parameter n is odd (otherwise leaving *Pow* unchanged). Thus, the binary representation of n determines exactly how *Pow* changes as we execute the unstacking loop. (This is why the algorithm is referred to as the "binary method.") Working through the binary expression of the original n from lower to higher powers, a 1 corresponds to the current value of the parameter n

being odd, whereas a 0 corresponds to the current value of the parameter n being even. This suggests accumulating the product of those values of the parameter x (powers of the original x), which will be multiplied by *Pow* as we work through the unstacking. Indeed, the correct final value of x^n is then the product of this latter accumulated product and x^m as defined above. The purely iterative pseudocode for *Powers* as given in Section 1.3 is now straightforward.

Stacks can be used to translate multiple-reference recursive functions and procedures into equivalent nonrecursive versions. The basic idea is to use a stack to model the action of resolution of the recursion as we follow the path around the tree of recursive calls or invocations. Each move down the tree corresponds to pushing the current values of parameters, local variables, a return address, and so forth, on the stack. Each move back up the tree corresponds to popping the stack, assigning the popped values to appropriate local variables and parameters, and branching to that part of the code corresponding to the appropriate return address. A completely general canonical description of the stack simulation of recursion that uses a set of formal rules for translating out the recursion in a line-by-line fashion can be found in the references.

2.8 Closing Remarks

In addition to the elementary ADTs discussed in this chapter, more advanced ADTs will be introduced throughout the text. These ADTs include priority queues, heaps, graphs, digraphs, sets, strings, balanced trees, and B-trees.

High-level descriptions of algorithms often hide the information concerning particular implementations of the ADTs they use. Information hiding has several advantages. Most important, information hiding allows the designer of an algorithm to concentrate on the big picture and not get bogged down in low-level implementation details. This, in turn, leads to better modularization of the algorithm. Algorithms utilizing ADTs and information hiding are more easily implemented in a variety of programming environments. Algorithms using ADTs have another aspect of portability; namely, they can be utilized easily by other algorithms.

ADTs thought of as objects with associated operations form a basic building block of programming languages such as C++, ADA, and SmallTalk that utilize the *object-oriented* paradigm. Object-oriented programming (OOP) is a very powerful modern technique that supports a new level of abstraction known as *inheritance*. Inheritance provides a mechanism for defining very general data types, some of whose properties can be inherited by more specialized data types.

References and Suggestions for Further Reading

For an expanded treatment of the standard data structures discussed in this chapter, as well as the more advanced data structures discussed later in the text:

Aho, A. V., J. E. Hopcroft, J. D. Ullman. *Data Structures and Algorithms*. Reading, MA: Addison-Wesley, 1983.

Horowitz, E., and S. Sahni. *Fundamentals of Data Structures in Pascal*, 3rd Ed. San Francisco, CA: Computer Science Press, 1990.

Knuth, D. E. *The Art of Computer Programming, Volume I: Fundamental Algorithms*, 2nd Ed. Reading, MA: Addison-Wesley, 1973.

Knuth, D. E. *The Art of Computer Programming, Volume III: Searching and Sorting*. Reading, MA: Addison-Wesley, 1973.

Tarjan, R. E. *Data Structures and Network Algorithms*. Philadelphia, PA: SIAM, 1983.

Wood, D. *Data Structures, Algorithms, and Performance*. Reading, MA: Addison-Wesley, 1993.

For a formal treatment of removing recursion:

Horowitz E., and S. Sahni. *Fundamentals of Computer Algorithms*. Rockville, MD: Computer Science Press, 1978.

Kruse, R. L. *Data Structures & Program Design*. Englewood Cliffs, NJ: Prentice Hall, 1984.

Exercises

Section 2.1 Linear ADTs

2.1 Given a linked list, create a linked list with the same elements but in the reverse direction.

2.2 **a.** Give pseudocode for the push and pop operations on a stack implemented using an array.

b. Repeat (a) for a linked list implementation.

2.3 A *circular queue* is implemented using a circular array *Queue*[0:*Max*-1] and two pointers (indices) *Front* and *Rear.* We view the elements *Queue*[0], *Queue*[1], . . . , *Queue*[*Max*-1] as positioned around a circle in a counterclockwise fashion. Initially, we have *Front* = *Rear* = 0. For nonempty queues, the variable *Rear* points to the position of the element at the rear of the queue. However, *Front* points one position counterclockwise from the element at the front of the queue. An element (in variable) *x* is enqueued by the two statements

Rear := (*Rear* + 1) **mod** *Max*
Queue[*Rear*] := *x*,

and an element is dequeued and assigned to *x* by the two statements

Front := (*Front* + 1) **mod** *Max*
x := *Queue*[*Front*].

Figure 2.25 shows how the circular array *Queue* with *Max* = 6 is updated as a given sequence of enqueues and dequeues are performed.

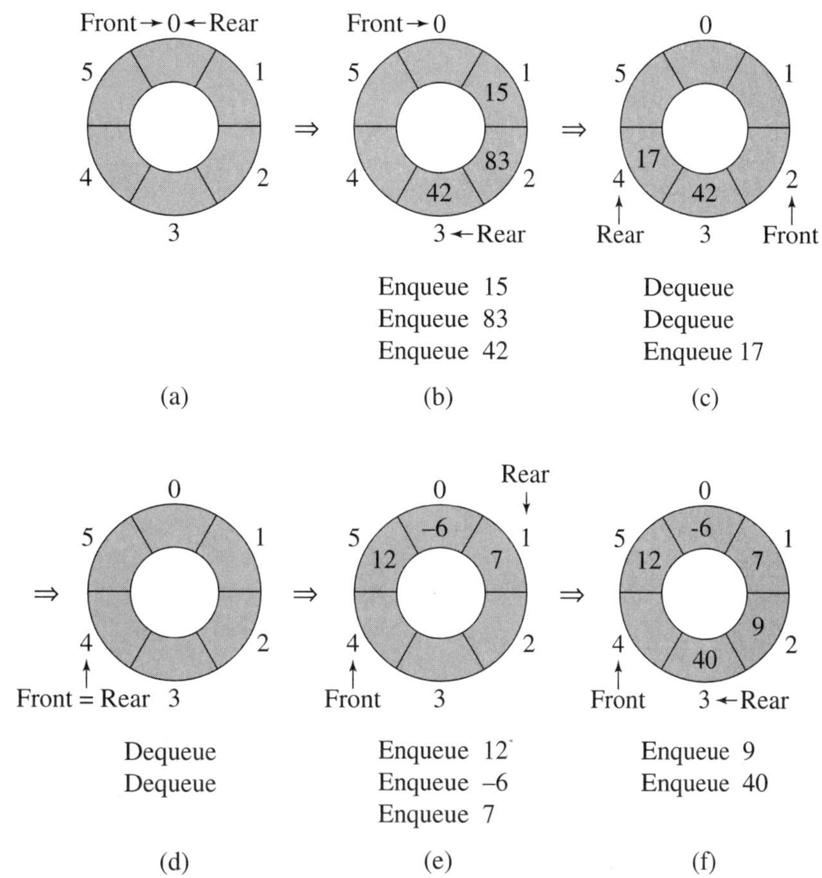

Figure 2.25 Circular array *Queue*[0:5] after a sequence of enqueues and dequeues

If the queue is empty then *Rear* = *Front*. On the other hand, if the queue is full in the sense that the entire array *Queue* is filled, then it is also the case that *Rear* = *Front*. The question then arises as to how to distinguish between these two situations. One way is to introduce a Boolean variable, which tags the queue as being empty or full. However, maintaining such a variable may lead to a noticeable increase in the computing time, since the procedures for implementing a queue are usually called repetitively. Another solution is to consider the queue full when every position of the array *Queue* but one is filled with elements of the queue. (Of course, this wastes a single memory location in the array *Queue*, but this is hardly significant.)

a. Give pseudocode for procedures *EnqueueCirc* and *DequeueCirc* for enqueueing and dequeueing an element using the latter implementation of a circular queue.

b. Starting with an empty circular queue *Queue*[0:8], show the final state of *Queue*[0:8] and the values of *Rear* and *Front* after the following sequence of enqueues and dequeues.

enqueue 23, enqueue 108, enqueue 55, dequeue, enqueue 61, dequeue dequeue, dequeue, enqueue 44, enqueue 21, dequeue

2.4 Give pseudocode for inserting and deleting into a linked list implemented using two arrays $L[1{:}n]$ and $Next[1{:}n]$.

Sections 2.2–2.3 Tree ADTs and Tree Traversal

2.5 Write a program that inputs a tree using its parent representation and converts the tree to its child-sibling representation.

2.6 Consider the child-sibling implementation of a general tree using pointer and dynamic variables.

a. Give pseudocode for a recursive procedure *TreePreOrder* that performs a preorder traversal of the tree whose root node is pointed to by the pointer variable *Root*.

b. In the case where T is a nonempty binary tree, show that the traversal of T generated by *TreePreOrder* coincides with the traversal generated by the algorithm *PreOrder* for binary trees.

c. Further, show that if T is any tree and T' is the binary tree obtained from T via the Knuth transformation, then the traversal of T generated by *TreePreOrder* coincides with the traversal of T' generated by *PreOrder*.

2.7 a. Show that a binary tree T can be reconstructed from its inorder and preorder traversal sequences.

b. Show that part (a) is not true in general for its postorder and preorder traversal sequences.

2.8 Give a recursive procedure for swapping the left and right children of every node in a binary tree.

2.9 Associated with each arithmetic expression E involving binary or unary operations is a binary expression tree T defined recursively as follows. The leaf nodes of T correspond to the operands in E and the internal nodes correspond to the operators in E and are labeled accordingly. We may assume without loss of generality that E is fully bracketed. If $E = (E_1) \odot (E_2)$ then the root node is labeled $\odot$, and the left and right subtrees of the root are expression trees for E_1 and E_2, respectively.

The reverse Polish (the postfix) form of expression E is defined recursively as follows. If F_1 and F_2 are reverse Polish expressions for E_1 and E_2 and $\odot$ is binary operator, then the reverse Polish expression of $E_1 \odot E_2$ is $F_1F_2\odot$. If $\odot$ is a unary operator, then the reverse Polish expression of $\odot E_1$ is $E_1\odot$. For example, the reverse Polish expression for $(A + B)*(-C)$ is $AB + C - *$.

a. Given a string representing a fully bracketed arithmetic expression E involving the binary operators +, –, *, / and the unary operator of minus, write a program that creates an expression tree for E.

b. Show that a postorder traversal of the expression tree for E yields the reverse Polish expression for E.

2.10 Give pseudocode for a version of the inorder-successor function *InOrderSucc* when there is a *Parent* field in each node.

2.11 Give pseudocode for a simple modification of the algorithm *InOrder*3 that creates an inorder threading.

2.12 Give pseudocode for an inorder traversal of an inorder-threaded binary tree.

2.13 Give pseudocode for an inorder traversal of a binary tree that does not use an additional field for the parent pointer (or inorder threading), but instead simulates a parent pointer by dynamically reversing and resetting child pointers during the course of the traversal.

2.14 Show that a binary code allows for unambiguous (unique) decoding if, and only if, it is a binary prefix code.

2.15 Show that if T is any tree and T' is the binary tree obtained from T via the Knuth transformation, then the traversal of T generated by *TreeInOrder* coincides with the traversal of T' generated by *InOrder*.

Section 2.4 **Complete and Full Binary Trees**

2.16 Show that if a binary tree T is full at level i, then it is full at every level j smaller than i.

2.17 Show that the depth of the complete binary tree T_n for a general n is given by

$$D(T_n) = \lfloor \log_2 n \rfloor.$$

Section 2.5 **Binary Search Trees**

2.18 Give pseudocode for a recursive version of *BinSrchTreeInsert*.

2.19 **a.** Give pseudocode for a procedure *CreateBinSrchTree*, which creates a binary search tree by repeated calls to *BinSrchTreeInsert*.

b. Show the binary search tree created by *CreateBinSrchTree* for the keys 22,11,0,72,27,55,23,108,1, inserted in that order.

c. Give three orderings for insertion of the keys in part (b) for which the binary search tree created by *CreateBinSrchTree* is a path; that is, each node has at most one child.

d. Give two orderings for insertion of the keys in part (b) for which the binary search tree created by *CreateBinSrchTree* is complete.

2.20 Given any set of n distinct keys (identifiers) and an *arbitrary* binary tree T on n nodes, show that there is a unique assignment of the keys to the nodes in T such that T becomes a binary search tree for the keys.

2.21 Show that a binary search tree without the (implicit) external leaf nodes added is full at the second-deepest level if, and only if, the associated binary tree with the external leaf nodes added is full at the second-deepest level.

2.22 Give pseudocode for an altered *BinSrchTreeSearch* that determines the locations of both the search element and its parent.

2.23 **a.** Our pseudocode for inserting a key into a binary search tree assumed that the tree did not contain duplicate keys. Give pseudocode for an alternate version that allows duplicate keys to be inserted. When inserting a duplicate key X, assume that you always move to the right child of a node already containing X (the *move-to-the-right rule*).

b. Show that when creating a binary search tree by successively inserting a sequence of keys using the move-to-the-right rule, an inorder traversal of the resulting search tree will visit duplicate keys in the order in which they were inserted. Moreover, when using *BinSrchTreeSearch* for a tree built with the move-to-the-right rule, show that the position in the tree of first key inserted amongst a given set of duplicate keys is found. Thus, all keys having a given value can be found by executing *BinSrchTreeSearch* to find the position of the first one that was inserted and then executing an inorder traversal of the right subtree of this position until a key having a different value is encountered.

Section 2.6 Multiway Search Trees

2.24 Discuss an implementation of an m-way search tree using pointers and dynamic variables.

2.25 For the implementation of the m-way search tree given in the previous exercise, give pseudocode for the high-level searching procedure described in Figure 2.22.

Section 2.7 Removing Recursion

2.26 Using the standard techniques for removing recursion, give pseudocode for a nonrecursive version of *Towers*.

2.27 Give pseudocode for nonrecursive versions of

a. preorder traversal.

b. postorder traversal.

3

Design and Analysis of Sequential Algorithms

You cannot control what you cannot (do not) measure.

—Lord Kelvin

There are two basic considerations in measuring the performance (*complexity*) of an algorithm: its time and space requirements. We are mainly concerned with analyzing the time complexity of an algorithm. Hence, when using the term *complexity*, unless otherwise stated we will mean the time complexity.

We measure the complexity of an algorithm by identifying a basic operation and then counting how many times the algorithm performs that basic operation for an input of size n. However, the number of basic operations performed by an algorithm usually varies for inputs of a given size. Thus, we use the concepts of best-case, worst-case, and average complexity of an algorithm. Occasionally we measure the complexity of an algorithm by considering several different operations performed by the algorithm and amortizing the performance over all these operations.

Determining the complexities of an algorithm exactly is often difficult, and in practice it is usually sufficient to obtain an asymptotic approximation to these complexities. To aid our description of asymptotic behavior, we introduce and utilize the standard notation for order and growth rate of functions.

3.1 The Complexity of Algorithms

When you can measure what you are speaking about and can express it in numbers, you know something about it. But when you cannot measure it, when you cannot express it in numbers, your knowledge is of a meager and unsatisfactory kind.

—Lord Kelvin

To analyze the complexity of an algorithm, we usually identify a basic operation and count how many times an algorithm performs this basic operation. Analysis based on a suitably chosen basic operation yields measurements that are proportional to actual run time behavior exhibited when running the algorithm on various computers, so that the analysis is not dependent on a particular computer. Basic operations for some sample algorithms are shown in Figure 3.1.

We measure the complexity of an algorithm as a function of the *input size n* to the algorithm. For example, when searching or sorting a list, the input size is the number of elements n in the list; when evaluating a polynomial, the input size is either the degree of the polynomial or the number of nonzero coefficients in the polynomial; when multiplying two square n-by-n matrices, the input size is n; when testing for the primality of a integer, the input size is the number of digits of the integer; when traversing a tree, the input size is the number of nodes in the tree; and so forth (see Figure 3.1).

For some algorithms the number of basic operations performed is the same for any input of size n. However, for many algorithms the number of basic operations performed can differ for two inputs of the same size. For the latter algorithms, we talk about *best-case*, *worst-case*, and *average complexity* as functions of the input size n.

The formal definitions of these complexities require the specification of the set of inputs of size n to an algorithm. An *input I* to an algorithm is the data, such as numbers, character strings, and records, on which the operations of the algorithm are performed. Let $\mathcal{I}_n$ denote the set of all inputs of size n to an algorithm. The second column of Figure 3.1 shows what is typically considered to be $\mathcal{I}_n$ for various types of problems and their associated algorithms. For $I \in \mathcal{I}_n$, let $\tau(I)$ denote the number of basic operations that are performed when the algorithm is executed with input I.

DEFINITION 3.1.1 **Best-Case Complexity** The *best-case complexity* of an algorithm is the function $B(n)$ such that $B(n)$ equals the *minimum* value of $\tau(I)$, where I varies over all inputs of size n. That is,

$$B(n) = \min \{ \tau(I) \mid I \in \mathcal{I}_n \}. \tag{3.1.1}$$

The best-case complexity $B(n)$ is typically easy to compute, but is not a particularly useful measure of the performance of the algorithm. In fact, $B(n)$ is usually

Figure 3.1 Algorithm complexity as a function of input size n

Problem	Input of size n	Basic Operation
Searching a list	lists with n elements	comparison
Sorting a list	lists with n elements	comparison
Multiplying two matrices	two n-by-n matrices	multiplication
Prime factorization	n-digit number	division
Evaluating a polynomial	polynomial of degrees n	multiplication
Traversing a tree	tree with n nodes	accessing a node
Towers of Hanoi	n disks	moving a disk

only of use as a lower bound for the average complexity A(n) (as defined presently) of the algorithm. Far more important than B(n) is the worst-case complexity W(n).

DEFINITION 3.1.2 **Worst-Case Complexity** The *worst-case complexity* of an algorithm is the function W(n) such that W(n) equals the *maximum* value of $\tau(I)$, where I varies over all inputs of size n. That is,

$$W(n) = \max \{ \tau(I) \mid I \in \mathcal{T}_n \}. \tag{3.1.2}$$

In practice, a given algorithm is usually run repeatedly with varying inputs of size n. Thus, another important measure of the performance of an algorithm is its *average complexity* A(n). Average complexity A(n) is defined as the *expected* number of basic operations performed. The notion of *expectation* is formally defined in Chapter 9 and depends on the probability distribution imposed on the sample space $\mathcal{T}_n$. For now, we assume that $\mathcal{T}_n$ is a finite set and that each input $I \in \mathcal{T}_n$ has probability $p(I)$ of occurring as the input to the algorithm.

DEFINITION 3.1.3 **Average Complexity** The *average complexity* of an algorithm with finite input set $\mathcal{T}_n$ is defined as

$$A(n) = \sum_{I \in \tau_n} \tau(I)p(I). \tag{3.1.3}$$

Equivalently, if p_i denotes the probability that the algorithm performs exactly i basic operations on an input of size n, so that $p_i = \Sigma_{\tau(I) = i} p(I)$, the formula (3.1.3) becomes

$$A(n) = \sum_{i=1}^{W(n)} ip_i, \tag{3.1.4}$$

where W(n) is given by (3.1.2).

Formula (3.1.4) is obtained immediately from (3.1.3) by gathering up all the terms such that $\tau(I) = i$, $i = 1, \ldots, W(n)$. Formulas (3.1.3) and (3.1.4) involve summations, so that their growth as functions of the input size n is usually hard to estimate directly. Thus, when analyzing the average complexity of a given algorithm, we usually seek closed-form expressions or estimates for A(n).

A *comparison-based* algorithm for searching or sorting a list is based on making comparisons involving list elements and then making decisions based on these comparisons. In particular, comparison-based algorithms make no a priori assumption about the nature of the list elements, other than knowing their relative order. For comparison-based algorithms we can often reduce or transform the sample space to a simpler sample space without affecting the average complexity. For example, the lists ($\sqrt{5}$,1.3,4,2), (108,23,123,55), (Mary,Ann,Pete,Joe), and (30.2,π,111.23,12) of size 4 can all be regarded as having the same ordering (third largest, first largest, fourth largest, second largest) as the permutation (3,1,4,2). In particular, they each require the same number of comparisons when input to a comparison-based sorting algorithm before they are put into increasing order. For this reason, we may reduce our input space for comparison-based sorting algorithms to the set of all permutations on the n integers $1,2, \ldots, n$. This amounts to making two lists in $\mathcal{I}_n$ equivalent if they have the same ordering.

Multiple-Variable Input Size In all of the problems mentioned so far, the input size was a function of a single variable n (see Figure 3.1). However, there are occasions where the input size is most naturally a function of two or more variables. For example, suppose we consider the problem of finding efficient sorting algorithms for lists of size n with repeated elements. The input size is then a function of n and the number m of *distinct* elements in the list, where $1 \leq m \leq n$. In such cases, best-case, worst-case, and average complexities are functions of the two variables n and m.

3.2 Searching Algorithms and Their Analysis

Searching lists is one of the most common tasks performed by computers. If no preconditioning (such as sorting) of the list is assumed, then there is no better algorithm than *LinearSearch*, which is based on a sequential (linear) scan of the list. On the other hand, for sorted lists there are much more efficient searching algorithms such as *BinarySearch*. The analyses of the two algorithms *LinearSearch* and *BinarySearch* nicely illustrate the computations of $B(n)$, $W(n)$, and $A(n)$.

LinearSearch and *BinarySearch* are based on making comparisons between a search element and the key field of the list elements. For simplicity, in our pseudocode for *LinearSearch* and *BinarySearch* we do not actually identify the key field of a list element, but instead reference the entire list element.

3.2.1 Linear Search

We implement *LinearSearch* as a function that returns the (first) position in a list (array of size n) $L[1{:}n]$ where a search element X occurs, or returns 0 if X is not in the list.

```
function LinearSearch (L[1:n],X)

Input:  L[1:n] (an array of list elements), X (a search item)
Output: returns index of first occurrence of X in the list, or 0 if X is not in the list

   for i := 1 to n do
      if X = L[i] then
         return(i)
      endif
   endfor
   return(0)

end LinearSearch
```

The basic operation of *LinearSearch* is the comparison of the search element to a list element. Clearly, *LinearSearch* performs only one comparison when the input X is the first element in the list, so that the best-case complexity is $B(n) = 1$. The most comparisons are performed when X is not in the list, or when X occurs in the last position only. Thus, the worst-case complexity of *LinearSearch* is $W(n) = n$.

To simplify the discussion of the average behavior of *LinearSearch*, we assume that the search element X is in the list $L[1{:}n]$ and is equally likely to be found in any of the n positions. Note that i comparisons are performed when X is found at position i in the list. Thus, the probability that *LinearSearch* performs i comparisons is given by $p_i = 1/n$. Substituting these probabilities into (3.1.4) yields

$$A(n) = \sum_{i=1}^{W(n)} ip_i = \sum_{i=1}^{n} i\frac{1}{n} = \left(\frac{n(n+1)}{2}\right)\frac{1}{n} = \frac{n+1}{2}. \tag{3.2.1}$$

Formula (3.2.1) is intuitively correct, since under our assumptions X is equally likely to be found in either half of the list. To see that (3.2.1) truly reflects average behavior, suppose that we run *LinearSearch* m times (m large) with a fixed list $L[1{:}n]$ and with search element X being randomly chosen as one of the list elements. Let m_i denote the number of runs in which X was found at position i. Then the total number of comparisons performed over the m runs is given by $1m_1 + 2m_2 + \cdots + nm_n$. Dividing this expression by m gives the average number $A_m(n)$ of comparisons over the entire m runs, as follows:

$$A_m(n) = 1\left(\frac{m_1}{m}\right) + 2\left(\frac{m_2}{m}\right) + \cdots + n\left(\frac{m_n}{m}\right). \tag{3.2.2}$$

Note that for large m, the ratios m_i/m occurring in (3.2.2) approach the probability p_i that X occurs in the ith position. Since we have assumed that X is equally likely to be found in any of the n positions, each m_i is approximately equal to m/n. Hence, substituting $m_i = m/n$ into (3.2.2) yields

$$A_m(n) \approx \frac{1 + 2 + \cdots + n}{n} = \frac{n+1}{2} = A(n). \tag{3.2.3}$$

To summarize, *LinearSearch* has best-case, worst-case, and average complexities 1, n, and $(n + 1)/2$, respectively. The best-case complexity is a constant independent of the input size n, whereas the worst-case and average complexities are both linear functions of n. For simplicity, we say that *LinearSearch* has *constant* best-case complexity and *linear* worst-case and average complexities.

We calculated $A(n)$ for *LinearSearch* under the assumption that the search element X is in the list. In Chapter 9 we will obtain a formula for $A(n)$ under the assumption that X is in the list with probability p, $0 \le p \le 1$.

3.2.2 Binary Search

LinearSearch assumes nothing about the order of the elements in the list; in fact, it is an optimal algorithm when no special order is assumed. However, *LinearSearch* is not the algorithm to use when searching ordered lists, at least when direct access to each list element is possible (as with an array implementation of the list). For example, if you are looking up the word *riddle* in a dictionary, and you initially open the dictionary to the page containing the word *middle*, then you know that you only need search for the word in the pages that follow. Similarly, if you were looking up the word *fiddle* instead of *riddle*, then you would only need to search for the word in the preceding pages. This simple observation is the basis of *BinarySearch*.

The idea behind *BinarySearch* is to successively cut in half the range of indices in the list where the search element X might be found. We assume that the list is sorted in nondecreasing order. By comparing X with the element $L[mid]$ in the middle of the list, we can determine whether X might be found in the first half of the list or the second half. We have three possibilities:

$$\begin{array}{ll} X = L[mid], & X \text{ is found}, \\ X < L[mid], & \text{search for } X \text{ in } L[1{:}mid-1], \\ X > L[mid], & \text{search for } X \text{ in } L[mid+1{:}n]. \end{array}$$

This process is then repeated (if necessary) for the relevant "half list." Thus, the number of elements in a sublist where X might be found is being (roughly) cut in half for each repetition. When cutting a sublist $L[low{:}high]$ in half, if the size of the sublist is even, then we take the midpoint index to be the smaller of the two middle indices, so that $mid = \lfloor (low + high)/2 \rfloor$. The following pseudocode implements *BinarySearch* as a function with the same parameters and output as *LinearSearch*, except that the list is assumed to be sorted in nondecreasing order.

```
function BinarySearch (L[1:n],X)

Input:  L[1:n] (an array of list elements, sorted in nondecreasing order)
        X (a search item)
Output: returns the index of an occurrence of X in the list, or 0 if X is
        not in the list

   Found := .false.
   low := 1
   high : = n
   while .not. Found .and. low ≤ high do
      mid := ⌊(low + high)/2⌋
      case
         :X = L[ mid]: Found := .true.
         :X < L[ mid]: high := mid – 1
         :otherwise : low := mid + 1
      endcase
   endwhile
   if Found then
      return(mid)
   else
      return(0)
   endif

end BinarySearch
```

Just as with *LinearSearch*, we use comparison of X to a list element as our basic operation when analyzing *BinarySearch*. Even though *two* comparison operators are used in the **case** statement in *BinarySearch*, it is natural to count them as a single basic operation. This reflects the fact that most assembly languages implement a three-way branch based on such comparisons as a *single* assembler instruction. In Chapter 7 we will discuss a variant of *BinarySearch* that does not test for equality and thereby improves its average and worst-case performance when we count the first two statements in the case structure as two basic operations.

The best-case complexity of *BinarySearch* is 1, which occurs when X is found in the midpoint position $\lfloor (n + 1)/2 \rfloor$ of $L[1{:}n]$. The worst-case complexity is equal to the longest string of midpoints (values of *mid*) ever generated by the algorithm for an input X. In particular, if we assume that $n = 2^k - 1$ for some positive integer k, then such a string is generated by searching for $X = L[1]$. We then compare X successively to the midpoints $2^{k-1}, 2^{k-2}, \ldots, 1$, so that this longest string has length k. To express k in terms of n, we note that $n + 1 = 2^k$, so that we have $k = \log_2(n + 1)$. We leave it as an exercise to verify that for any n, the length of the longest string of midpoints ever generated is $\lceil \log_2(n + 1) \rceil$.

3.3 Finding the Maximum and Minimum Elements in a List

We now consider the problem of finding the maximum (or minimum) value of an element in a list $L[1{:}n]$ of size n. Finding the maximum can be done with a variation of *LinearSearch* where we keep track and update the maximum (or minimum) value encountered as we scan the list.

```
function Max (L[1:n])
Input:  L[1:n] (list of size n)
Output: returns the maximum value occurring in L[1:n]
   MaxValue := L[1]
   for i := 2 to n do
      if L[i] > MaxValue then
         MaxValue := L[i]      {update MaxValue}
      endif
   endfor
   return(MaxValue)
end Max
```

When analyzing the function *Max*, we choose comparison between list elements ($L[i] > MaxValue$) as our basic operation. The only other operation performed by *Max* is the updating of *MaxValue*. However, for any input list, the number of comparisons between list elements clearly dominates the number of updates of *MaxValue*, which justifies our choice of basic operation.

Note that *Max* performs $n - 1$ comparisons for any input list of size n. Thus, the best-case, worst-case, and average complexities of *Max* all equal $n - 1$. In Section 3.5, we use lower bound theory to show that any comparison-based algorithm for finding the maximum value of an element in a list of size n must perform at least $n - 1$ comparisons for any input, so that *Max* is an optimal algorithm.

An analogous algorithm *Min* can be given for finding the minimum value in a list. Sometimes it is useful to determine *both* the maximum and the minimum values in a list $L[1:n]$. An algorithm *MaxMin*1 for solving this problem is to successively invoke *Max* and *Min*. Clearly, *MaxMin*1 has best-case, worst-case, and average complexities all equal to $2n - 2$. The following algorithm based on a single sequential scan through the list results in some improvement to the best-case and average complexities.

```
procedure MaxMin2(L[1:n],MaxValue,MinValue)
Input:   L[1:n] (a list of size n)
Output: MaxValue,MinValue (maximum and minimum values occurring in
           L[1:n])
   MaxValue := L[1]
   MinValue := L[1]
   for i := 2 to n do
      if L[i] > MaxValue then
         MaxValue := L[i]
      else
         if L[i] < MinValue then
            MinValue := L[i]
         endif
      endif
   endfor
end MaxMin2
```

*MaxMin*2 performs a minimum of $n - 1$ comparisons when the list $L[1:n]$ is in strictly increasing order, so that *MaxMin* has best-case complexity $B(n) = n - 1$. *MaxMin*2 performs a maximum of $2(n - 1)$ comparisons when $L[1]$ has the maximum value in the list $L[1:n]$. Thus, *MaxMin*2 has the same worst-case complexity as *MaxMin*1, namely, $W(n) = 2n - 2$.

The average complexity $A(n)$ of *MaxMin*2 lies between $B(n) = n - 1$ and $W(n) = 2n - 2$. Unfortunately, it turns out (see Chapter 9) that the average complexity $A(n)$ of *MaxMin*2 is approximately equal to $2n - \ln n$, which is closer to the worst-case complexity. One wonders if there might be an algorithm with better average complexity; in fact, there is. We now design an algorithm *MaxMin*3 whose worst-case complexity is $\lceil 3n/2 \rceil - 2$, which turns out (see Chapter 10) to be *optimal*.

To facilitate our discussion of the algorithm *MaxMin*3, we introduce the following procedure *M&M*, which solves the max-min problem for a list of size 2.

```
procedure M&M(A,B,MaxValue,MinValue)

Input:  A, B
Output: MaxValue, MinValue (the maximum and minimum of A and B)

  if A ≥ B then
     MaxValue := A
     MinValue := B
  else
     MaxValue := B
     MinValue := A
  endif

end M&M
```

Procedure *MaxMin*3 works by pairing elements (except for one element when n is odd) in the list $L[1:n]$ and computing the maximum and minimum for each pair, yielding $\lceil n/2 \rceil$ potential maxima and $\lceil n/2 \rceil$ potential minima.

```
procedure MaxMin3(L[1:n],MaxValue,MinValue)

Input:  L[1:n] (array)
Output: MaxValue,MinValue (the maximum and minimum values in L[1:n])

  if even(n) then {n is even }
     call M&M(L[1],L[2],MaxValue,MinValue)
     for i := 3 to n – 1 by 2 do
        call M&M(L[i],L[i + 1],b,a)
        if a < MinValue then MinValue := a endif
        if b > MaxValue then MaxValue := b endif
     endfor
  else {n is odd}
     MaxValue := L[1]; MinValue := L[1];
     for i := 2 to n – 1 by 2 do
        call M&M(L[i],L[i + 1],b,a)
        if a < MinValue then MinValue := a endif
        if b > MaxValue then MaxValue := b endif
     endfor
  endif

end MaxMin3
```

For n even, $L[1]$ and $L[2]$ are compared, and then the (first) **for** loop of *MaxMin*3 performs $3(n - 2)/2$ comparisons. For n odd, there is no initial comparison, and the (second) **for** loop performs $3(n - 1)/2$ comparisons. Thus, the best-case, worst-case, and average complexities of *MaxMin*3 for input size n are all equal to $\lceil 3n/2 \rceil - 2$.

3.4 Insertion Sort

We now discuss a simple sorting algorithm called insertion sort. In the next chapter we will discuss a number of more sophisticated sorting algorithms.

3.4.1 Array Implementation of Insertion Sort

Insertion sort sorts a given list $L[1:n]$ by successively inserting the list element $L[i]$ into its proper place in the sorted list $L[1:i-1]$, $i = 2, \ldots, n$. It works similarly to how a card player might insert a newly dealt card into the previously dealt hand that was already put in order. One starts a scan at one end of the hand and stops at a place where the new card can be inserted and still maintain an ordered hand. This scan can start either at the low end of the hand (forward scan), or at the high end of the hand (backward scan). A card player has no reason (other than a personality quirk) to prefer one scan over the other. However, when implementing insertion sort there are several reasons for preferring one scan over the other, depending on the situation.

Given the list $L[1:n]$, clearly the sublist consisting of only the element $L[1]$ is a sorted list. Suppose (after possibly reindexing) we have a list L where the sublist $L[1:i-1]$ is already sorted. We then can obtain a sorted sublist $L[1:i]$ by inserting the element $L[i]$ in its proper position. In a backward scan, we successively compare $L[i]$ with $L[i-1]$, $L[i-2]$, and so forth, until a list element $L[position]$ is found that is not larger than $L[i]$. $L[i]$ can then be inserted at $L[position+1]$. In a forward scan, we successively compare $L[i]$ with $L[1]$, $L[2]$, and so forth, until a list element $L[position]$ is found that is not smaller than $L[i]$. $L[i]$ can then be inserted at $L[position]$. Figure 3.2 demonstrates the action of the backward scan version of insertion sort for a list of size 6.

We now give pseudocode for insertion sort using a backward scan.

```
procedure InsertionSort(L[1:n])
Input:  L[1:n] (a list of size n)
Output: L[1:n] sorted in nondecreasing order
  for i := 2 to n do {insert L[i] in its proper position in L[1:i]}
    Current := L[i]
    position := i – 1
      while position ≥ 1 .and. Current < L[position] do
              {Current must precede L[position]}
        L[position+1] := L[position]   {bump up L[position]}
        position := position – 1
      endwhile
              {position + 1 is now the proper position for Current = L[i]}
      L[position+1] := Current
  endfor
end InsertionSort
```

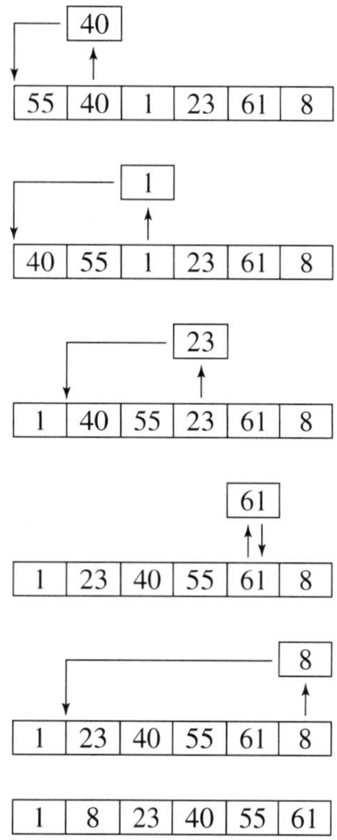

Figure 3.2 Action of insertion sort (backward scan) for a list of size 6

When analyzing *InsertionSort*, we choose comparison between list elements ($Current < L[position]$) as our basic operation. For any input list of size n, the outer loop of *InsertionSort* is executed $n - 1$ times. If the input list is already sorted in nondecreasing order, then the inner loop is iterated only once for each iteration of the outer loop. Hence, the best-case complexity of *InsertionSort* is given by

$$B(n) = n - 1. \tag{3.4.1}$$

The worst-case complexity occurs when the inner loop performs the maximum number of comparisons for each value of the outer loop variable i. For a given i, this occurs when $L[i]$ must be compared to each element $L[i - 1], L[i - 2], \ldots, L[1]$, so that $i - 1$ comparisons are performed in the inner loop. This, in turn, occurs when the list is in strictly decreasing order. Since i varies from 2 to n, we have

$$W(n) = 1 + 2 + \cdots + (n - 1) = \frac{n(n-1)}{2}. \tag{3.4.2}$$

Thus, W(n) for *InsertionSort* is quadratic in the input size n. It turns out (see Chapter 9) that the average complexity A(n) for *InsertionSort* is about half of W(n) and therefore is also quadratic in n. Because of this quadratic complexity, *Insertion-*

Sort is impractical to use for sorting general large lists. However, *InsertionSort* does have five advantages:

1. *InsertionSort* works quickly on small lists. Thus, *InsertionSort* is often used as a threshold sort in conjunction with other sorting algorithms like *MergeSort* or *QuickSort* (see Chapter 4).
2. *InsertionSort* works quickly on large lists that are close to being sorted in the sense that each element is not too far from its proper position (see Proposition 3.4.1). This property of *InsertionSort* makes it useful in connection with other sorts such as *ShellSort* (Chapter 4).
3. *InsertionSort* is amenable to implementation as an on-line sorting algorithm. An *on-line* sorting algorithm is one in which the entire list is not input to the algorithm in advance, but instead elements are added to the list over time. On-line sorting algorithms are required to maintain the dynamic list in sorted order.
4. *InsertionSort* is an in-place sorting algorithm. A sorting algorithm with input parameter $L[1:n]$ is called *in-place* if only a constant amount of memory is used (for temporary variables, loop control variable, sentinels, etc.) in addition to that needed for L.
5. *InsertionSort* is a *stable* sorting algorithm in the sense that it maintains the relative order of repeated elements. More precisely, an algorithm that sorts a list $L[1:n]$ into nondecreasing order is called stable if, given any two elements $L[i]$, $L[j]$, with $i < j$ and $L[i] = L[j]$, the final positions i', j' of $L[i]$, $L[j]$, respectively, satisfy $i' < j'$.

Stable algorithms are useful when we want to sort the elements in a list of records according to primary and secondary keys in the record, but where the sorts on these two keys take place independently. By way of illustration, suppose we have already sorted the records in some list of persons alphabetically according to name. For purposes of bulk mailing, we now wish to sort the list according to the zip code key in each record. However, within a given zip code, we wish to maintain our alphabetical order. Clearly, a stable sorting algorithm is required.

The question arises as to why we use linear scans for finding the correct position to insert the ith list element $L[i]$ into the already sorted list $L[1:i-1]$, when a binary search to find this position would drastically reduce the number of comparisons made by the algorithm. For example, the worst-case complexity of *InsertionSort* would be reduced from $n(n-1)/2$ to approximately $n\log_2 n$. The catch is that this altered version would not reduce the number of array reassignments needed to insert $L[i]$, so that a quadratic number of array reassignments would still be made in the worst case. Thus, even though the altered *InsertionSort* is still comparison-based, the number of comparisons made would no longer be a true measure of the complexity of the algorithm. Another drawback of the binary search version of *InsertionSort* is that some of the advantages of *InsertionSort* listed earlier would be lost. For example, the binary search version of *InsertionSort* would no longer be stable and would not necessarily work fast on large lists that are close to being sorted.

We now prove that *InsertionSort* is efficient for lists that are close to being sorted. Given a list $L[1{:}n]$, we define the *stable final position* of an element in L to be the unique final position of this element after any *stable* sorting algorithm is applied to L. We first prove a lemma about stable sorting algorithms.

LEMMA 3.4.1 Suppose each element in a list $L[1{:}n]$ is no more than k positions from its stable final position. Then for each $i \in \{2, \ldots, n\}$, there are at most $2k - 1$ list elements $L[j]$ such that $j < i$ and $L[i] < L[j]$.

Proof Assume to the contrary that there are at least $2k$ list elements such that $j < i$ and $L[i] < L[j]$. Then there must exist a list element $L[j_0]$ that is strictly greater than $L[i]$, such that $j_0 \le i - 2k$. Let i' and j_0' denote the stable final positions of $L[i]$ and $L[j_0]$, respectively. By hypothesis, every element in the list $L[1{:}n]$ is no more than k positions from its stable final position. In particular, $j_0' \le j_0 + k \le (i - 2k) + k = i - k$, and $i' \ge i - k$. Hence, $j_0' \le i'$, which implies that $L[j_0] \le L[i]$, a contradiction. ■

The following proposition about *InsertionSort* is an easy consequence of Lemma 3.4.1.

PROPOSITION 3.4.2 Given integers n, k, $k \le n$, suppose $L[1{:}n]$ is a list such that every element in the list $L[1{:}n]$ is no more than k positions from its stable final position in the sorted list L. Then *InsertionSort* performs at most $2k(n - 1)$ comparisons when sorting the input list $L[1{:}n]$. ❑

Proposition 3.4.2 shows that for a fixed k, the worst-case complexity of *InsertionSort* for lists satisfying the conditions of the proposition is linear in the input size n. The latter property of *InsertionSort* makes it effective to utilize in conjunction with more efficient sorting algorithms.

An *adjacent-key* comparison-based sorting algorithm is one in which comparisons between list elements are made only between elements that occupy *adjacent* positions in the list. These elements are interchanged if they are out of order. *InsertionSort* is essentially an adjacent-key comparison sort, since the comparison between $L[position]$ and *current* can be thought of as an assignment of *current* to $L[position]$ and then a comparison between $L[position]$ and $L[position-1]$. For reasons of efficiency, we chose not to make the actual assignment until the correct position for insertion was determined. Another well-known adjacent-key sorting algorithm is *BubbleSort* (see Exercise 3.15).

In the next section we show that the worst-case complexity of *any* adjacent-key comparison-based sorting algorithm is at least $n(n - 1)/2$. In view of this result, *InsertionSort* actually has optimal worst-case complexity for an adjacent-key comparison-based sort. If we want to design comparison-based sorting algorithms whose worst-case complexities are smaller than $n(n - 1)/2$, we must look for design strategies that compare nonadjacent (far away) elements in the list. We give examples of algorithms utilizing the latter strategy in the next chapter.

3.4.2 Linked List Insertion Sort

We now implement insertion sort using a linked list rather than an array. To find the correct position to insert an element X, a forward scan of the linked list is performed using the pointer *Place*. This scan first compares X to the first element in the list. If X is not larger than this element, then the new node containing X is inserted at the beginning of the list. Otherwise, the scan continues down the list until the last element in the list smaller than X is pointed to by *Place*. The new node can then be placed immediately after the node containing this latter element.

We illustrate the linked version of insertion sort in Figure 3.3 for the same list of elements used in Figure 3.2. (The pseudocode for this linked version is left as an exercise.) In Figure 3.3, we assume that the elements are placed in nodes and inserted into the sorted linked list immediately after they are input.

3.5 A Glimpse of Lower Bound Theory

Once we have found one or more algorithms solving a given problem, it is important to know whether these algorithms are the best possible. Lower bound theory aids us in answering this question. The purpose of lower bound theory is to obtain (as sharp as possible) lower bounds for the complexities of *any* algorithm that solves the problem. An algorithm whose complexity equals a lower bound that has been established is an optimal algorithm for the problem and the lower bound is sharp. Such a sharp lower bound is the *complexity of the problem*. For example, we don't need to look for a better algorithm for polynomial evaluation than *HornerEval*, since any correct algorithm for evaluating an nth-degree polynomial must perform at least n multiplications.

In this section we introduce lower bound theory by determining sharp lower bounds for a comparison-based searching algorithm to find the maximum value in a list of size n and for sorting a list of size n using an adjacent-key comparison-based sorting algorithm. A lower bound of $n - 1$ for the best-case, worst-case, and average complexities of the first problem is established using a simple counting technique. A lower bound of $n(n - 1)/2$ for the worst-case complexity of the second problem is obtained using simple properties of permutations. More sophisticated strategies for establishing lower bounds, such as comparison trees and adversary arguments, are given in Chapter 10.

3.5.1 Lower Bound for Finding the Maximum

Consider the problem of using a comparison-based algorithm to find the maximum value of an element in a list $L[1{:}n]$ of size n. Clearly, each list element must participate in at least one comparison, so that $\lceil n/2 \rceil$ is trivially a lower bound for the problem. We now use a simple counting argument to show that any comparison-based algorithm for finding the maximum value of an element in a list of size n must perform at least $n - 1$ comparisons for any input. The algorithm *Max* given in Section

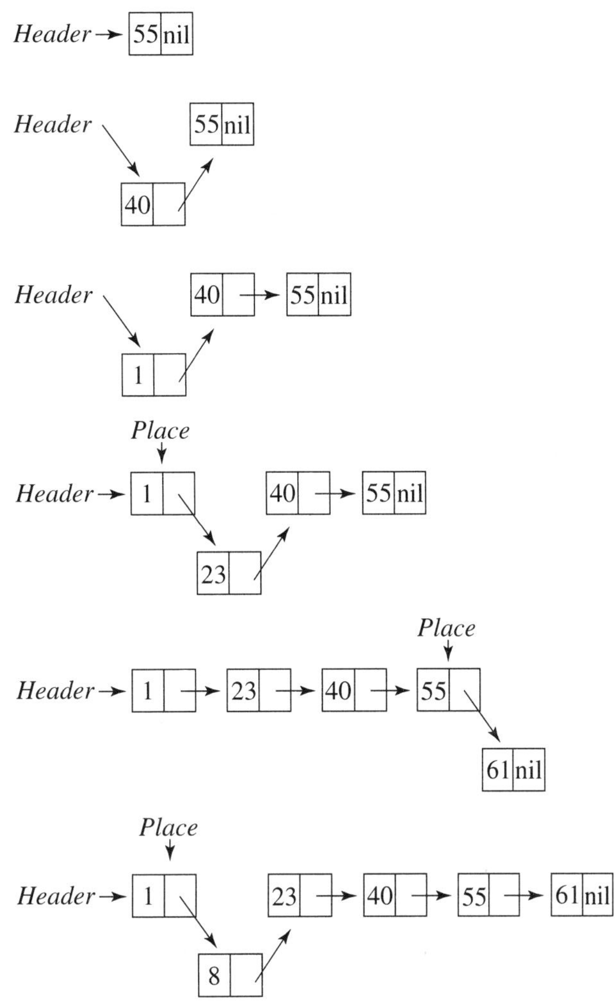

Figure 3.3 Action of the linked version of insertion sort for $L[1:6]$: 55 40 1 23 61 8

3.3 has best-case, worst-case, and average complexities all equal $n - 1$, so that *Max* is an optimal algorithm.

Suppose for convenience that we restrict attention to lists that contain *distinct* elements. When a comparison-based algorithm compares two distinct elements X, Y, we say that X *wins* the comparison if $X > Y$, otherwise X *loses*. Thus, each comparison generates exactly one loss, and this is the unit of work that we associate with the comparison. Now we simply count the number of total losses that must occur if the algorithm is to correctly determine the maximum element. We claim that any comparison-based algorithm must generate at least $n - 1$ units of work. More precisely, we claim that each of the $n - 1$ elements that is *not* the maximum must lose at least one comparison. Indeed, suppose there is an input list $L[1:n]$ that causes the algorithm to terminate with two list elements $L[i]$ and $L[j]$ *both* never having lost a comparison. Assume for definiteness that the algorithm declares $L[i]$ to be the maximum. Now consider the input list $L'[1:n]$ that is identical to L except that $L'[j]$ is increased so that it is larger than $L[i]$ and remains different from any other list element. By the definition of comparison-based algorithms, our algorithm performs *identically* on L and L'. To see this, note that the algorithm could only perform differently when making comparisons involving $L'[j]$. But $L[j]$ won all its comparisons, and since $L'[j] \geq L[j]$, so does $L'[j]$. Thus, the algorithm must again declare $L'[i] = L[i]$ to be the maximum element in L', which is a contradiction.

Thus, we have established that any comparison-based algorithm for finding the maximum element in a list of size n (of distinct elements) must perform at least $n - 1$ comparisons. Since *Max* performs $n - 1$ comparisons for *any* input list of size n, it is an optimal algorithm for the problem, as summarized in the following proposition.

PROPOSITION 3.5.1 The problem of finding the maximum element in a list of size n using a comparison-based algorithm has worst-case, best-case, and average complexities all equal to $n - 1$. Moreover, *Max* is an optimal algorithm for the problem.

3.5.2 Worst-Case Complexity for Adjacent-Key Comparison Sorting

We now consider the problem of sorting a list using an adjacent-key comparison-based algorithm. As discussed earlier, the sorting algorithm *InsertionSort* can be viewed as an adjacent-key comparison sort. Recall that *InsertionSort* has worst-case complexity $W(n) = n(n - 1)/2$. In this section we show that *any* adjacent-key comparison sort performs at least $n(n - 1)/2$ comparisons to sort a list $L[1:n]$ of size n in the worst case. Hence, *InsertionSort* achieves optimal worst-case complexity for an adjacent-key comparison-based sort.

We may assume without loss of generality that the input lists of size n to the algorithm are chosen from the set of all permutations of the set of integers $N_n = \{1, 2, \ldots, n\}$. We associate with each permutation a combinatorial entity called an *inversion* of the permutation and show that the maximum number of inversions in a permutation on n symbols is given by $n(n - 1)/2$. Since interchanging adjacent list elements (keys) removes at most one inversion, the maximum number of inversions then serves as a lower bound for the problem of adjacent-key comparison-based sorting.

PROPOSITION 3.5.2 A lower bound for the worst-case complexity $W(n)$ of any adjacent-key comparison-based sorting algorithm is $n(n - 1)/2$.

Proof Consider input lists of size n to be permutations of $N_n = \{1,2, \ldots, n\}$. Given a permutation π of N_n, an *inversion* of π is a pair $(\pi(a), \pi(b))$, $a, b \in N_n$, where $a < b$ but $\pi(a) > \pi(b)$. In Figure 3.4 we illustrate the inversions corresponding to the six permutations of $\{1,2,3\}$.

$$\begin{pmatrix} 1 & 2 & 3 \\ \pi(1) & \pi(2) & \pi(3) \end{pmatrix}$$

Representation of a permutation π

$\begin{pmatrix} 1 & 2 & 3 \\ 1 & 2 & 3 \end{pmatrix}$	$\begin{pmatrix} 1 & 2 & 3 \\ 1 & 3 & 2 \end{pmatrix}$	$\begin{pmatrix} 1 & 2 & 3 \\ 2 & 1 & 3 \end{pmatrix}$	$\begin{pmatrix} 1 & 2 & 3 \\ 2 & 3 & 1 \end{pmatrix}$	$\begin{pmatrix} 1 & 2 & 3 \\ 3 & 1 & 2 \end{pmatrix}$	$\begin{pmatrix} 1 & 2 & 3 \\ 3 & 2 & 1 \end{pmatrix}$
none	(3,2)	(2,1)	(2,1), (3,1)	(3,1), (3,2)	(3,2), (3,1) (2,1)

Figure 3.4 All permutations of {1,2,3} and their associated inversions

Note that any sorting algorithm must ultimately remove all inversions. Note also that a comparison of list elements utilized by any adjacent-key comparison-based sorting algorithm by definition considers only inversions of the form $(\pi(i),\pi(i+1))$. Removing any such inversion (by interchanging $\pi(i)$ and $\pi(i+1)$) results in a decrease of exactly one inversion (see Exercise 3.16). Thus, the worst-case complexity of any adjacent-key comparison-based sorting algorithm is bounded below by the maximum number of inversions in a permutation of N_n.

Consider the permutation corresponding to a list sorted in decreasing order—that is, the permutation $\pi(i) = n - i + 1$, $i = 1, \ldots, n$. Clearly, every pair $(\pi(i),\pi(j))$, $1 \leq i < j \leq n$ is an inversion of π. The number of such pairs is given by $\binom{n}{2} = n(n-1)/2$ (see Section 1.2), establishing the lower bound given in Proposition 3.5.2. ■

Since *InsertionSort* performs $n(n-1)/2$ comparisons in the worst case (see (3.4.2)), it follows from Proposition 3.5.2 that *InsertionSort* is an optimal worst-case adjacent-key comparison-based algorithm. In the next chapter we develop comparison-based sorting algorithms that compare keys that are far apart and in the worst case perform on the order of $n \log n$ comparisons.

3.6 Asymptotic Behavior of Functions

Suppose $f(n) = 400n + 23$ and $g(n) = 2n^2 - 1$ are the worst-case complexities for two algorithms A and B, respectively, that solve the same problem. Which exhibits better worst-case performance? If $n \leq 200$, then $f(n)$ is greater than $g(n)$, so that algorithm B outperforms A. On the other hand, if $n > 200$, the reverse is true. When analyzing the complexity of an algorithm, it is the behavior of the algorithm as the input size n tends to infinity (asymptotic behavior) that is usually considered to be most important. Thus, algorithm A is considered to have better worst-case behavior.

For purposes of algorithm analysis, the asymptotic behavior (*order*) of a function $f(n)$ not only ignores small values of n, but it also doesn't distinguish between $f(n)$ and $cf(n)$, where c is a positive constant. For example, *InsertionSort* performs $(n^2/2) - (n/2)$

comparisons in the worst case for an input of size n. However, $(n^2/2) - (n/2)$ and n^2 are considered to be asymptotically equivalent measures of the complexity of an algorithm. To see why this is reasonable, note that for large n the quadratic term $n^2/2$ completely dominates the linear term $n/2$ (see Figure 3.5), so that the $n/2$ term can safely be ignored.

Of course, there is a significant difference between n^2 and $n^2/2$, but there are two reasons for ignoring the factor 1/2. The first is that the total number of operations performed by *InsertionSort* (comparisons, array reassignments, loop control updates, and so forth) is bounded by a constant multiple of the number of basic operations (comparisons). The second reason involves the time feasibility of running the algorithm on a particular computer. It is reasonable to assume that it is feasible to perform n^2 operations if, and only if, it is feasible to perform $n^2/2$ operations. By way of illustration, today's fastest computers can perform close to a billion operations a second. Such computers would take more than 3 centuries to perform n^2 operations for an input size of n = 1,000,000,000, but would still take over 1.5 centuries to perform $n^2/2$ operations for this same choice of n. On the other hand, if n = 1,000,000, today's fastest computers can do $n^2/2$ operations in less than 10 minutes and can still do n^2 operations in less than 20 minutes (in many situations both represent a feasible amount of time).

Therefore, we say that the worst-case complexity of *InsertionSort* has order n^2. We now introduce asymptotic notation that formalizes the notion of the order of a function.

3.6.1 Asymptotic Order

Our notion of order is restricted to real-valued functions $f(n):\mathbb{N} \to \mathbb{R}$ defined on the nonnegative integers that are *eventually* positive, so that there exists an integer n_0 (depending on f) such that $f(n) > 0$ for all $n > n_0$. Let $\mathcal{F}$ denote the set of all such functions. Given any function $g(n) \in \mathcal{F}$, we define $\Theta(g(n))$ to be the set of all functions $f(n) \in \mathcal{F}$ that grow at the same rate as $g(n)$ in the sense that they are eventually "squeezed" between two constant multiples of $g(n)$ (see Figure 3.6).

DEFINITION 3.6.1 Given a function $g(n) \in \mathcal{F}$, we define $\Theta(g(n))$ to be the set of all functions $f(n) \in \mathcal{F}$ having the property that there exist positive constants c_1, c_2, and n_0 such that for all $n \geq n_0$,

$$c_1\, g(n) \leq f(n) \leq c_2\, g(n). \tag{3.6.1}$$

If $f(n) \in \Theta(g(n))$, we say that *$f(n)$ is big theta of $g(n)$*.

Figure 3.5 Table comparing linear and quadratic growth

n	$n/2$	$n^2/2$	$(n^2/2) - (n/2)$	n^2
10	5	50	45	100
1,000	500	500,000	499,500	1,000,000
10,000	5,000	50,000,000	49,995,000	100,000,000
100,000	50,000	5,000,000,000	4,999,950,000	10,000,000,000

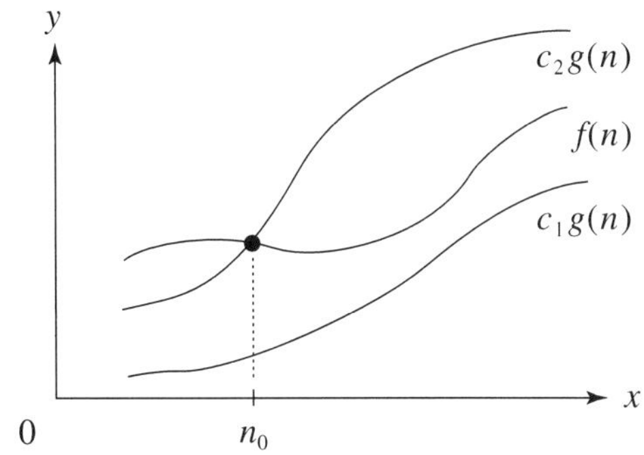

Figure 3.6 $f(n) \in \Theta(g(n))$

Remark

1. When a function $f(n)$ is given by a formula, it is standard to identify the function with the formula for the function. For example, if $f(n)$ is given by the formula $f(n) = n^3 + 2n\log_2 n - 3$, then we simply refer to $f(n)$ as the function $n^3 + 2n\log 2n - 3$. This allows us to say $n^3 + 2n \log_2 n - 3 \in \Theta(n^3)$, $3\sqrt{n} + \log_2 n + 108 \in \Theta(\sqrt{n})$, $864 \in \Theta(1)$, and so forth.
2. It is fairly common in the literature to use the alternate notation $f(n) = \Theta(g(n))$ instead of $f(n) \in \Theta(g(n))$. The same is true for the sets $O(g(n))$ and $\Omega(g(n))$ defined later in Definitions 3.6.3 and 3.6.4.

The following proposition states that Θ determines an *equivalence relation* (see Exercise 3.35) on the set $\mathscr{F}$.

PROPOSITION 3.6.1 Θ determines an equivalence relation on $\mathscr{F}$—that is, for $f(n), g(n) \in \mathscr{F}$,

1. $f(n) \in \Theta(f(n))$ (Θ is reflexive)
2. $f(n) \in \Theta(g(n)) \Rightarrow g(n) \in \Theta(f(n))$ (Θ is symmetric)
3. $f(n) \in \Theta(g(n))$ and $g(n) \in \Theta(h(n)) \Rightarrow f(n) \in \Theta(h(n))$ (Θ is transitive)

The fact that Θ determines an equivalence relation on the set $\mathscr{F}$ shows that the following definition of *order* is well defined. ❑

DEFINITION 3.6.2 Two functions f and g have the *same order* if $f(n) \in \Theta(g(n))$.

Note that the reflexive, symmetric, and transitive properties of Θ can be expressed as follows:

1. f has the same order as itself.
2. If f has the same order as g, then g has the same order as f.
3. If f has the same order as g, and g has the same order as h, then f has the same order as h.

Using these properties of Θ, it follows (see Exercise 3.26) that f and g have the same order if, and only if, $\Theta(f(n)) = \Theta(g(n))$. Thus, $\Theta(f(n))$ can also be denoted by $\Theta(g(n))$ for any function $g(n) \in \Theta(f(n))$. When describing the order of a function $f(n)$, we usually choose a representative in the set (class) $\Theta(f(n))$ having the "simplest" form. For example, we have

$$\Theta((8n^3 + n)^{1/2} + \log_2 n) = \Theta(2n^{3/2} + 6n \ln n) = \Theta(n^{3/2}) = \Theta(n^{3/2} + 23n^{1/3})$$

$$= \Theta(\frac{n^3 + n + 1}{4n^2 + \ln n}).$$

Since $n^{3/2}$ has the simplest form of any function in its Θ-class, we denote this class by $\Theta(n^{3/2})$. Also, we say that any $f \in \Theta(n^{3/2})$ has order $n^{3/2}$.

For purposes of algorithm analysis, we are usually interested in functions whose domain is a subset of the positive integers (input size is usually integral). However, often such a function is given by an expression involving n that is also meaningful when extended over the domain of nonnegative real numbers x. Then we usually sketch the graph of f over the reals, instead of over just the integers (see Figure 3.7). When the extension to the reals of the domain of f yields a differentiable function, we can use various theorems from calculus to help analyze the asymptotic behavior of f.

3.6.2 Illustrations of Orders of Complexity

Many of the algorithms arising in computer science have linear, quadratic, cubic, logarithmic, exponential, or $n\log n$ complexities. We now illustrate the use of asymptotic notation for measuring the complexity of some of the algorithms that we have already discussed. For example, recall that *LinearSearch* has average complexity $A(n) = n/2 + 1/2$ when the search element X is assumed to be in the list. As shown in Figure 3.7, $A(n) = n/2 + 1/2$ is squeezed between $n/2$ and n, so that $A(n) \in \Theta(n)$.

As our second example, recall that the algorithm *InsertionSort* has worst-case complexity $W(n) = n^2/2 - n/2$. As seen in Figure 3.8, $n^2/2 - n/2$ is squeezed between $n^2/3$ and n^2, for $n \geq n_0 = 3$. Thus, $W(n) \in \Theta(n^2)$.

Our third example illustrates logarithmic order. Recall that *BinarySearch* has worst-case complexity $W(n) = \lceil \log_2(n + 1) \rceil$. As seen in Figure 3.9 (page 96), $\lceil \log_2(n + 1) \rceil$ is squeezed between $\log_2 n$ and $2\log_2 n$, for $n \geq n_0 = 2$. Thus, $W(n) \in \Theta(\log_2 n)$.

Logarithms occur frequently in the analysis of algorithms. As was the case in the analysis of *BinarySearch*, the base is usually 2. However, when using asymptotic notation, we drop the base because by (1.2.6), the logarithm to one base is merely a constant multiple of the logarithm to another base.

$$\log_a n = (\log_a b)(\log_b n)$$

($\log_a b$ is a constant that does not depend on n.)

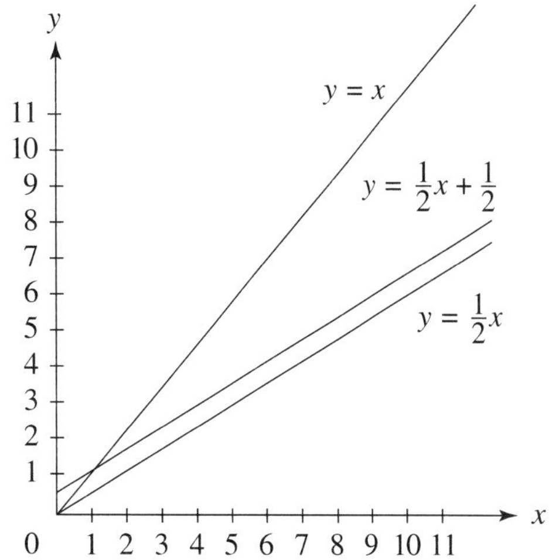

Figure 3.7 Average complexity for *LinearSearch*: $A(n) = n/2 + 1/2 \in \Theta(n)$

Figure 3.8 Worst-case complexity for *InsertionSort*: $W(n) = n^2/2 - n/2 \in \Theta(n^2)$

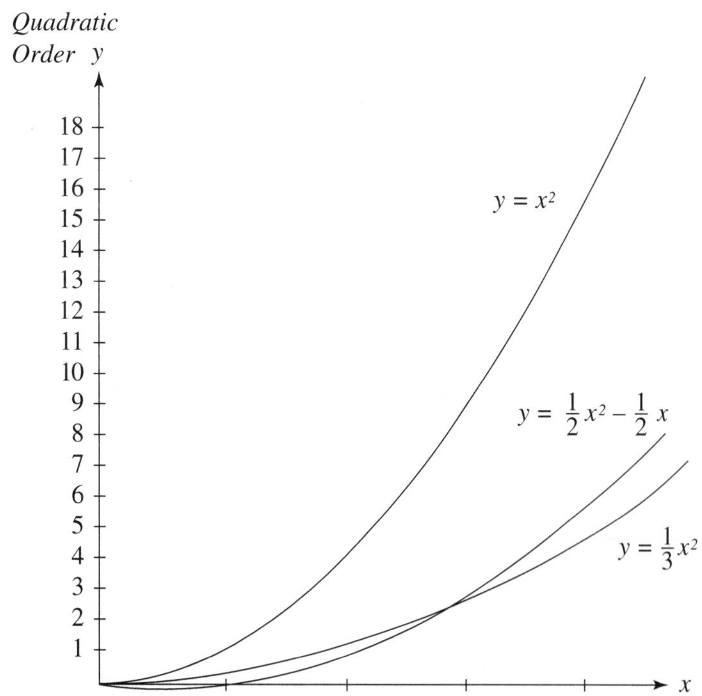

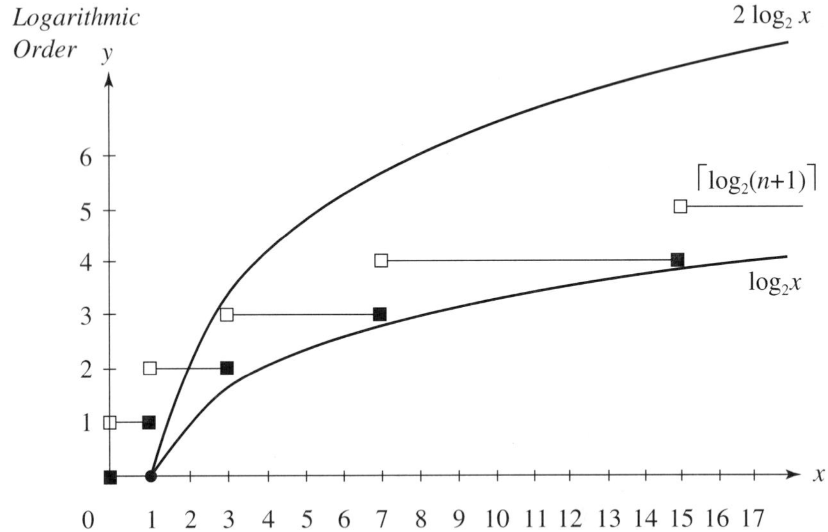

Figure 3.9 Worst-case complexity for *BinarySearch*: $W(n) = \lceil \log_2(n+1) \rceil \in \Theta(\log_2 n)$

Letting $c_1 = c_2 = \log_a b$ and $n_0 = 1$, it follows from Definition 3.6.1 that $\log_a n \in \Theta(\log_b n)$, so that $\Theta(\log_a n) = \Theta(\log_b n)$. Thus, we usually denote this Θ-class simply by $\Theta(\log n)$.

Our fourth example illustrates exponential order. Recall that *Towers* performs $2^n - 1$ moves for n disks. As illustrated in Figure 3.10, $2^n - 1$ is squeezed between $(1/2)2^n$ and 2^n. We say that *Towers* has *exponential* complexity, since $2^n - 1 \in \Theta(2^n)$. More generally, the complexity $t(n)$ of an algorithm is said to be *exponential* if there exist bases a and b, where $b \geq a > 1$, and positive constants c_1, c_2, and n_0 such that

$$c_1 a^n \leq t(n) \leq c_2 b^n, \quad \text{for all } n \geq n_0.$$

3.6.3 The Asymptotic Relations O and Ω

The set $\Theta(g(n))$ consists of those functions $f(n)$ that grow at the same rate as g, meaning that $f(n)$ is eventually bounded above and below by constant multiples of $g(n)$. We obtain a (strictly) larger class of functions, denoted by $O(g(n))$, by only requiring that $f(n)$ is eventually bounded *above* by a constant multiple of $g(n)$ (so that f grows no faster than g). Similarly, we obtain a class $\Omega(g(n))$ containing $\Theta(g(n))$ by only requiring that f is eventually bounded *below* by a constant multiple of $g(n)$ (so that f grows at least as fast as g).

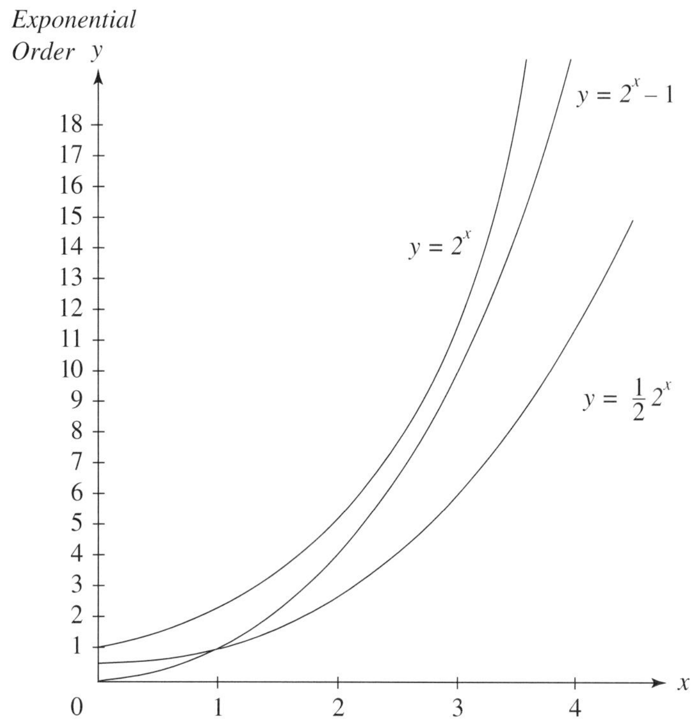

Figure 3.10 Complexity of *Towers*: $2^n - 1 \in \Theta(2^n)$

DEFINITION 3.6.3 Given a function $g(n) \in \mathscr{F}$, we define $O(g(n))$ to be the set of all functions $f(n) \in \mathscr{F}$ having the property that there exist positive constants c and n_0 such that for all $n \geq n_0$,

$$f(n) \leq cg(n). \tag{3.6.2}$$

If $f(n) \in O(g(n))$, we say that *$f(n)$ is big oh of $g(n)$*. (See Figure 3.11, page 98.)

DEFINITION 3.6.4 Given a function $g(n) \in \mathscr{F}$, we define $\Omega(g(n))$ to be the set of all functions $f(n) \in \mathscr{F}$ having the property that there exist positive constants c and n_0 such that for all $n \geq n_0$,

$$cg(n) \leq f(n). \tag{3.6.3}$$

If $f(n) \in \Omega(g(n))$, we say that *$f(n)$ is big omega of $g(n)$*. (See Figure 3.12.)

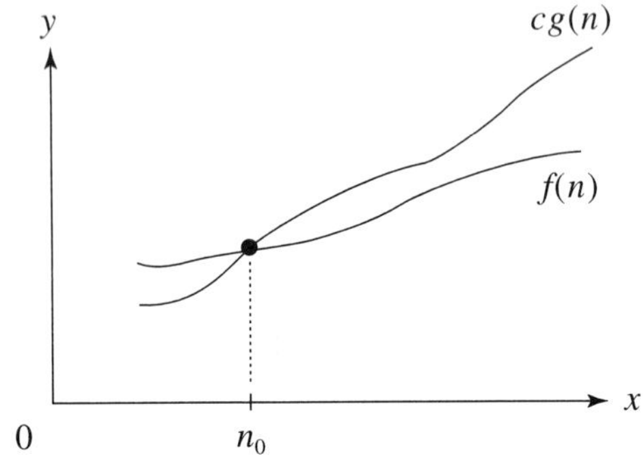

Figure 3.11 $f(n) \in O(g(n))$

Big oh is often used as an upper bound on the performance of a particular algorithm for solving a problem, whereas big omega is often used as a lower bound for the complexity of the problem itself.

It is easy to verify that the set $\Theta(g(n))$ consists precisely of those functions that are in both sets $O(g(n))$ and $\Omega(g(n))$—that is,

$$\Theta(g(n)) = \Omega(g(n)) \cap O(g(n)). \tag{3.6.4}$$

Sample members of the sets $O(n^2)$, $\Theta(n^2)$, and $\Omega(n^2)$ are illustrated in Figure 3.13.

Proposition 3.6.2 states some useful properties of the sets O, Θ, and Ω.

PROPOSITION 3.6.2 The following properties hold for $f(n), g(n) \in \mathscr{F}$.

1. Given any positive constant c, $\Omega(f(n)) = \Omega(cf(n))$, $\Theta(f(n)) = \Theta(cf(n))$, and $O(f(n)) = O(cf(n))$.
2. $f(n) \in \Theta(g(n)) \Leftrightarrow f(n) \in O(g(n))$ and $f(n) \in \Omega(g(n))$.
3. $f(n) \in O(g(n)) \Rightarrow O(f(n)) \subseteq O(g(n))$.
4. $O(f(n)) = O(g(n)) \Leftrightarrow \Omega(f(n)) = \Omega(g(n)) \Leftrightarrow \Theta(f(n)) = \Theta(g(n))$.
5. $f(n) \in O(g(n)) \Leftrightarrow g(n) \in \Omega(f(n))$. ❑

Figure 3.12 $f(n) \in \Omega(g(n))$

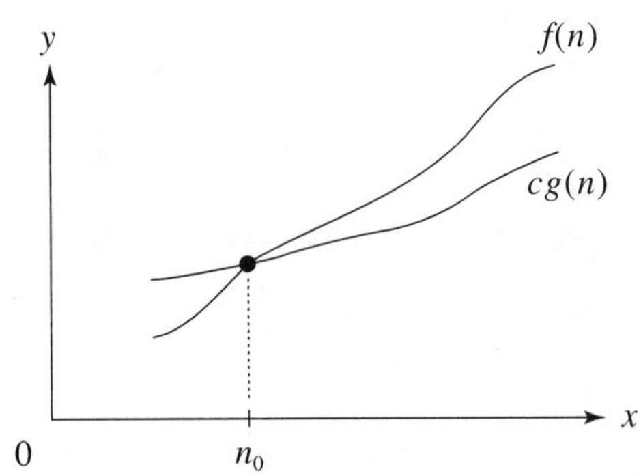

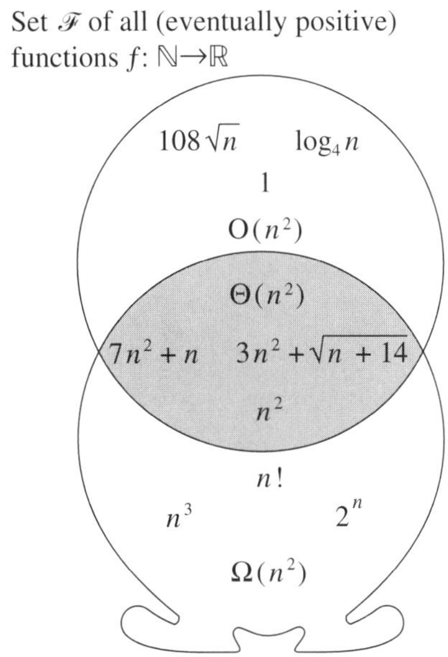

Figure 3.13 Venn diagram for $O(n^2)$, $\Theta(n^2)$, and $\Omega(n^2)$ and sample set members

The proof of Proposition 3.6.2 is left to the exercises.

It is common to use the notation

$$f(n) = g(n) + O(h(n)),$$

when $f(n) - g(n) \in O(h(n))$. Similar notation is used for Θ and Ω. For example, if $f(n) = 3n^3 + 4n^2 + 23n - 108$, then we could write

$$f(n) = 3n^3 + O(n^2).$$

We now give a table (Figure 3.14, page 100) that summarizes the order of the best-case, average, and worst-case complexities for most of the algorithms that we have discussed so far. The table also includes the sorting algorithms *MergeSort*, *HeapSort*, and *QuickSort*, which are discussed in the next chapter.

3.6.4 Hierarchy of Orders

The table in Figure 3.5 indicates that linear complexity is in some sense smaller than quadratic complexity. The question remains, however, whether it might still be true that $n^2 \in O(n)$. Suppose the answer were yes. Then by definition of $O(n)$, there would exist positive constants c and n_0 such that $n^2 \leq cn$, for all $n \geq n_0$. But this implies that $n \leq c$ for all $n \geq n_0$, which is a contradiction. Since $n^2 \notin O(n)$, it follows from property (2) of Proposition 3.6.2 that $n^2 \notin \Theta(n)$, so that n and n^2 have different orders.

Since $n \in O(n^2)$, it follows from property (3) of Proposition 3.6.2 that $O(n)$ is strictly contained in $O(n^2)$—that is, $O(n) \subset O(n^2)$. This makes it natural to say that n

Algorithm	B(n)	A(n)	W(n)
HornerEval	n	n	n
Towers	2^n	2^n	2^n
LinearSearch	1	n	n
BinarySearch	1	$\log n$	$\log n$
Max, Min, MaxMin	n	n	n
InsertionSort	n	n^2	n^2
MergeSort	$n\log n$	$n\log n$	$n\log n$
HeapSort	$n\log n$	$n\log n$	$n\log n$
QuickSort	$n\log n$	$n\log n$	n^2

Figure 3.14 Order of best-case, average, and worst-case complexities

has *smaller order* than n^2 and motivates the following general definition of hierarchy of orders.

DEFINITION 3.6.5 Given functions $f(n)$ and $g(n)$, we say that $f(n)$ has *smaller order than* $g(n)$ if $O(f(n))$ is *strictly* contained in $O(g(n))$—that is, $O(f(n)) \subset O(g(n))$.

Remark In the literature, the notation $O(f(n)) < O(g(n))$ is sometimes used in place of the more precise $O(f(n)) \subset O(g(n))$.

Some of the most frequently encountered functions in the analysis of algorithms are 1, $\log n$, $\sqrt{n}$, n, $n\log n$, n^2, n^3, 2^n, $n2^n$, and $n!$. These functions form an increasing chain in the hierarchy of orders:

$$O(1) \subset O(\log n) \subset O(\sqrt{n}) \subset O(n) \subset O(n \log n)$$
$$\subset O(n^2) \subset O(n^3) \subset O(2^n) \subset O(n2^n) \subset O(n!).$$

This hierarchy is illustrated by the Venn diagram in Figure 3.15.

Even though Figure 3.15 might suggest the contrary, not every two orders are comparable. The orders of two functions $f(n)$ and $g(n)$ are not comparable if $f(n) \notin O(g(n))$ and $g(n) \notin O(f(n))$. For example, consider the functions $f(n) = n^3$ and $g(n) = n^4(n \textbf{ mod } 2) + n^2$. We leave it as an exercise to verify that $O(f(n))$ and $O(g(n))$ are not comparable by showing that neither is a subset of the other. This is illustrated in the Venn diagram in Figure 3.16 (page 102), which includes sample members of $O(f(n))$ and $O(g(n))$. We also leave it as an exercise to find strictly increasing functions whose orders are not comparable.

3.6.5 Establishing Order Relationships

We now discuss some techniques that can be used to establish the order relationship between two functions $f(n)$ and $g(n)$. Sometimes this can be done directly from the definitions of Θ, O, Ω, which we illustrate by showing that a

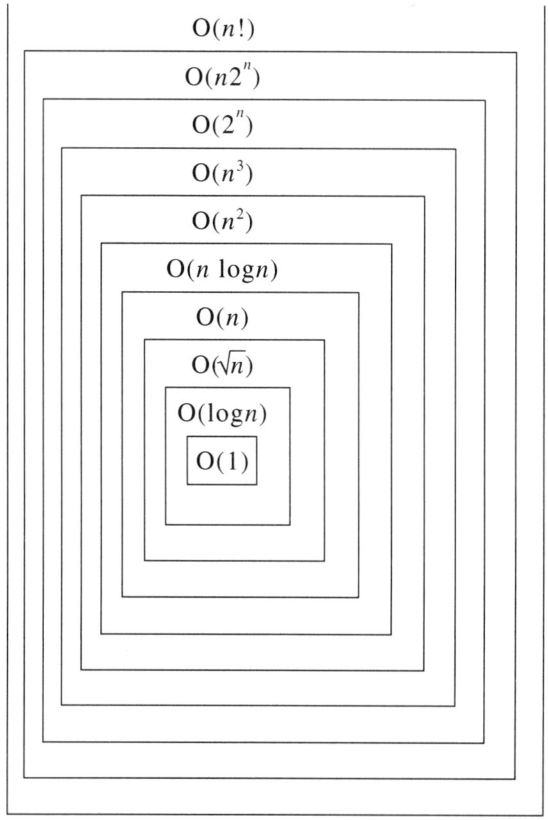

Figure 3.15 An increasing chain in the hierarchy of orders

polynomial of degree k is in $\Theta(n^k)$. The leading coefficient of the polynomial is assumed positive to ensure that the polynomial is eventually positive (and hence belongs to $\mathscr{F}$).

PROPOSITION 3.6.3 Let $P(n) = a_k n^k + a_{k-1}n^{k-1} + \cdots + a_1 n + a_0$ be any polynomial of degree k, $a_k > 0$. Then

$$P(n) \in \Theta(n^k).$$

Proof By property (2) of Proposition 3.6.2, it is sufficient to show that $P(n) \in O(n^k)$ and $P(n) \in \Omega(n^k)$. We first show that $P(n) \in O(n^k)$.

$$\begin{aligned} P(n) &= a_k n^k + a_{k-1}n^{k-1} + \cdots + a_1 n + a_0 \\ &\le a_k n^k + |a_{k-1}|\, n^k + \cdots + |a_1|\, n^k + |a_0|\, n^k \\ &\le (a_k + |a_{k-1}| + \cdots + |a_1| + |a_0|)\, n^k \end{aligned}$$

Thus, if $c = a_k + |a_{k-1}| + \cdots + |a_1| + |a_0|$, then $P(n) \le cn^k$ for all $n \ge n_0 = 1$, so that $P(n) \in O(n^k)$. We now show that $P(n) \in \Omega(n^k)$.

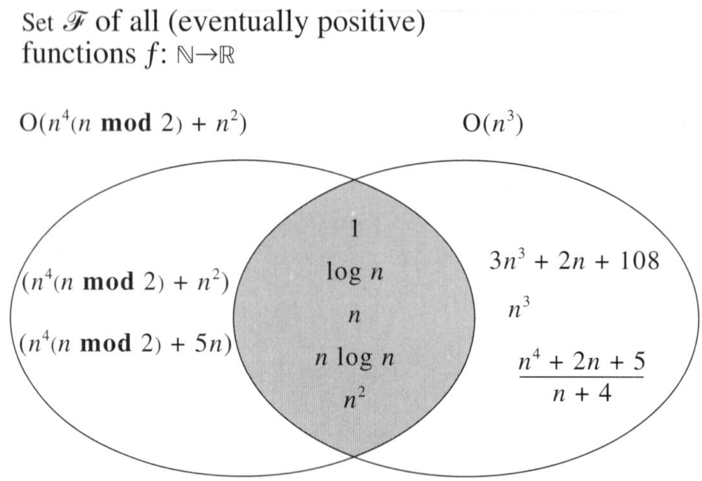

Figure 3.16 Incomparable orders. Sample functions are shown in each region.

$$\begin{aligned}
P(n) &= a_k n^k + a_{k-1} n^{k-1} + \cdots + a_1 n + a_0 \\
&\geq a_k n^k - |a_{k-1}|\, n^{k-1} - \cdots - |a_1|\, n - |a_0| \\
&= \left(\frac{a_k}{2}\right) n^k + \left[\left(\frac{a_k}{2}\right) n^k - |a_{k-1}|\, n^{k-1} - \cdots - |a_1|\, n - |a_0|\right] \\
&\geq \left(\frac{a_k}{2}\right) n^k + \left[\left(\frac{a_k}{2}\right) n^k - (|a_{k-1}|\, n^{k-1} - \cdots - |a_1|\, n^{k-1} - |a_0|\, n^{k-1})\right] \\
&= \left(\frac{a_k}{2}\right) n^k + \left[\left(\frac{a_k}{2}\right) n - (|a_{k-1}| + \cdots + |a_1| + |a_0|)\right] n^{k-1}
\end{aligned}$$

Thus, if $c = a_k/2$ and $n_0 = (|a_{k-1}| + \cdots + |a_1| + |a_0|)(2/a_k)$, the term inside the bracket in the last equation is nonnegative for all $n \geq n_0$, and hence $P(n) \geq cn^k$ for all $n \geq n_0$. This shows that $P(n) \in \Omega(n^k)$, completing the proof of Proposition 3.6.3. ■

Proving that a polynomial of degree k belongs to $\Theta(n^k)$ was not too difficult based directly on the definition of Θ. It would be more difficult, for example, to prove that $2n^{3/2} + 6n \ln n \in \Theta(n^{3/2})$ directly from the definition. The following proposition provides a useful test for comparing the orders of two functions $f, g \in \mathscr{F}$ when the limit of the ratio $f(n)/g(n)$ exists as n tends to infinity.

PROPOSITION 3.6.4 Let $f(n), g(n) \in \mathscr{F}$. If the limit of the ratio $f(n)/g(n)$ exists as n tends to infinity, then f and g are comparable. Moreover, assuming $L = \lim_{n \to \infty} f(n)/g(n)$ exists, then the following results hold.

1. $0 < L < \infty \Rightarrow f(n) \in \Theta(g(n))$. — f and g have the same order.
2. $L = 0 \quad \Rightarrow O(f(n)) \subset O(g(n))$ — f has a *smaller* order than g.
3. $L = \infty \quad \Rightarrow O(g(n)) \subset O(f(n))$. — f has a *larger* order than g.

Proof The definition of $\lim_{n\to\infty} f(n)/g(n) = L$ states that for any given real number $\varepsilon > 0$ there exists an integer N_ε (depending on ε) such that for all $n \geq N_\varepsilon$, $|f(n)/g(n) - L| < \varepsilon$ or, equivalently,

$$(L - \varepsilon)g(n) < f(n) < (L + \varepsilon)g(n), \quad \text{for all } n \geq N_\varepsilon. \tag{3.6.5}$$

***Case* 1: $0 < L < \infty$.** We are free to choose ε to be any positive real number. In particular, we may choose $\varepsilon = L/2$. Setting $c_1 = L/2$, $c_2 = 3L/2$, and $n_0 = N_{L/2}$ and substituting these values into (3.6.5) yields

$$c_1 g(n) < f(n) < c_2 g(n), \quad \text{for all } n \geq n_0. \tag{3.6.6}$$

Hence, $f(n) \in \Theta(g(n))$.

***Case* 2: $L = 0$.** Substituting $L = 0$ into (3.6.5), for *any* positive real number ε we have

$$f(n) < \varepsilon g(n), \quad \text{for all } n \geq N_\varepsilon. \tag{3.6.7}$$

It follows immediately that $f(n) \in O(g(n))$. To prove that $O(f(n)) \subset O(g(n))$, it is necessary to show that $f(n) \notin \Omega(g(n))$. Assume to the contrary that there exist constants c and n_0 such that

$$cg(n) \leq f(n), \quad \text{for all } n \geq n_0. \tag{3.6.8}$$

Substituting $\varepsilon = c$ in (3.6.7) yields $f(n) < cg(n)$ for all $n \geq N_c$. Thus, for any integer n larger than the maximum of n_0 and N_c, $f(n) < cg(n)$ and $cg(n) \leq f(n)$, which is a contradiction. Hence, $f(n) \notin \Omega(g(n))$, which implies that $O(f(n)) \subset O(g(n))$.

***Case* 3: $L = \infty$.** For this case, the proof that $O(g(n)) \subset O(f(n))$ is left as an exercise. ■

Remark A partial converse to case 2 of Proposition 3.6.4 is true: If $O(f(n)) \subset O(g(n))$ and $\lim_{n\to\infty} f(n)/g(n)$ exists, then this limit must be zero. It is easy to construct examples where $O(f(n)) \subset O(g(n))$ but $\lim_{n\to\infty} f(n)/g(n)$ does not exist (see Exercise 3.33). Similar remarks hold in the other two cases of Proposition 3.6.4.

The condition $\lim_{n\to\infty} f(n)/g(n) = 0$ is so useful in mathematics that the following special notation is used to denote the corresponding relation on the set of real-valued functions.

DEFINITION 3.6.6 Given the function $g(n)$, we define $o(g(n))$ to be the set of all functions $f(n)$ having the property that

$$\lim_{n\to\infty} \frac{f(n)}{g(n)} = 0.$$

If $f(n) \in o(g(n))$, we say that $f(n)$ *is little oh of* $g(n)$.

Remark In the mathematical literature, $f(n) \in o(g(n))$ is usually written $f(n) = o(g(n))$.

DEFINITION 3.6.7 Given the function $g(n)$, we define $\sim(g(n))$ to be the set of all functions $f(n)$ having the property that

$$\lim_{n \to \infty} \frac{f(n)}{g(n)} = 1 .$$

If $f(n) \in \sim(g(n))$, then we say that $f(n)$ is *strongly asymptotic* to $g(n)$ and denote this by writing $f(n) \sim g(n)$.

The following proposition follows immediately from Proposition 3.6.4 and the definitions of o and ~.

PROPOSITION 3.6.5 Let $f(n), g(n) \in \mathcal{F}$. Then

1. $f(n) \sim g(n) \Rightarrow f(n) \in \Theta(g(n))$
2. $f(n) \in o(g(n)) \Rightarrow O(f(n)) \subset O(g(n))$.

We illustrate Proposition 3.6.5 by proving the following stronger version of Proposition 3.6.3 concerning the asymptotic behavior of polynomials.

PROPOSITION 3.6.6 Let $P(n) = a_k n^k + \cdots + a_1 n + a_0$ be any polynomial of degree k, $a_k > 0$. Then

$$P(n) \sim a_k n^k.$$

Proof Consider the ratio $(a_k n^k + \cdots + a_0)/a_k n^k$ as n tends to infinity:

$$\lim_{n \to \infty} \frac{(a_k n^k + \cdots + a_0)}{a_k n^k}$$

$$= \lim_{n \to \infty} \left(1 + \frac{a_{k-1}}{a_k n} + \cdots + \frac{a_1}{a_k n^{k-1}} + \frac{a_0}{a_k n^k} \right) = 1 .$$

Proposition 3.6.6 now follows immediately from property (1) of Proposition 3.6.5. ■

Applying Propositions 3.6.4 and 3.6.5 requires calculating $\lim_{n\to\infty} f(n)/g(n)$, which might be difficult to do directly. When calculating $\lim_{n\to\infty} f(n)/g(n)$, a powerful tool from calculus, known as L'Hôpital's Rule, is often helpful. Application of L'Hôpital's Rule requires that $f(n)$ and $g(n)$ be extended to functions $f(x)$ and $g(x)$, respectively, both of which are differentiable for sufficiently large real numbers x.

L'Hôpital's Rule Let $f(x)$ and $g(x)$ be functions that are differentiable for sufficiently large real numbers x. If $\lim_{x\to\infty} f(x) = \infty$ and $\lim_{x\to\infty} g(x) = \infty$, then

$$\lim_{n \to \infty} \frac{f(x)}{g(x)} = \lim_{n \to \infty} \frac{f'(x)}{g'(x)}.$$

We now apply Proposition 3.6.4 to an example that makes use of L'Hôpital's Rule. We show that $2n^{3/2} + 6n\ln n \in \Theta(n^{3/2})$ by considering the ratio $(2n^{3/2} + 6n \ln n)/n^{3/2}$ as n tends to infinity. We have

$$\lim_{x\to\infty} \frac{(2x^{3/2} + 6x\ln x)}{x^{3/2}} = 2 + 6\lim_{x\to\infty} \frac{\ln x}{x^{1/2}}$$

$$= 2 + 6\lim_{x\to\infty} \frac{1/x}{(1/2)\,x^{-1/2}} \qquad \text{(by L'Hôpital's Rule)}$$

$$= 2 + 12\lim_{x\to\infty} \frac{1}{x^{1/2}} = 2 + 0 = 2.$$

As our second illustration, we show that any polynomial has smaller order than any exponential function.

PROPOSITION 3.6.7 For any nonnegative real constants k and a, with $a > 1$,

$$O(n^k) \subset O(a^n).$$

Proof Consider the ratio n^k/a^n. Repeated applications of L'Hôpital's Rule yields

$$\lim_{n\to\infty} \frac{n^k}{a^n} = \lim_{n\to\infty} \frac{kn^{k-1}}{(\ln a)\,a^n}$$

$$= \lim_{n\to\infty} \frac{k(k-1)\,n^{k-2}}{(\ln a)^2 a^n}$$

$$\vdots$$

$$= \lim_{n\to\infty} \frac{k!}{(\ln a)^k a^n} = 0.$$

Hence, $n^k \in o(a^n)$. Proposition 3.6.7 now follows from Proposition 3.6.5. ■

3.6.6 Asymptotic Notation for Functions of Several Variables

Occasionally the input size to an algorithm is a function of two or more variables as opposed to a single variable n. For example, in the case of graph algorithms, the input size is often measured in terms of the number of vertices n and the number of edges m in the graph. In the case of searching or sorting lists, we sometimes measure the input size in terms of the total number n of *list* elements and the number m of *distinct* elements. One such algorithm is *BingoSort* (see Chapter 4), which performs order mn comparisons in the average and worst cases, and $n + m^2$ comparisons in the best case. Adequate description of this behavior clearly requires that the complexity be measured as a function of *both* n and m. Hence, we describe the best-case, worst-case, and average complexities as functions $B(n,m)$, $W(n,m)$, and $A(n,m)$, respectively.

The theory of order for functions of several variables can be developed along the same lines as given for functions of a single variable. For example, given a real-valued function $g(n,m)$ of two nonnegative variables n and m, we say that $g(n,m)$ is *eventually nonnegative* if there exists a positive number n_0 such that $g(n,m) \geq 0$ whenever $n \geq n_0$ and $m \geq n_0$. Similarly, the set $O(g(n,m))$ consists of those eventually nonnegative functions $f(n,m)$ eventually bounded above by a constant multiple of $g(n,m)$. More precisely, $f(n,m) \in O(g(n,m))$ if, and only if, there exist positive integers c and n_0, such that $f(n,m) \leq cg(n,m)$ for all $n,m \geq n_0$. The sets $\Theta(g(n,m))$ and $\Omega(g(n,m))$ are defined similarly. The notions also extend to functions of more than two variables. We do not give a formal treatment of order for functions of two or more variables, but rely on the analogy with the theory developed for the single-variable case.

3.7 The Harmonic Series and Sums of Logarithms

The sum $L(b,n) = \log_b 1 + \log_b 2 + \cdots + \log_b n = \log_b(n!)$ and the harmonic series $H(n) = 1 + 1/2 + \cdots + 1/n$ occur frequently in the analysis of the complexity of algorithms. We now determine the order of these two functions.

We first consider $L(b,n)$. Recall that for any real constants a and $b > 1$, $\log_a n \in \Theta(\log_b n)$, so that we may use *any* base when representing this Θ-class. For convenience, we denote this class simply by $\Theta(\log n)$. Similarly, it is clear that $L(a,n) \in \Theta(L(b,n))$, so that again we may use *any* base when representing this latter Θ-class. Thus, for convenience of notation, we drop the base and simply write

$$L(n) = \log(n!) = \log 1 + \log 2 + \cdots + \log n.$$

PROPOSITION 3.7.1 $\log(n!) \in \Theta(n \log n)$.

Proof Clearly, $L(n) \in O(n\log n)$, since

$$\log 1 + \log 2 + \cdots + \log n \leq \log n + \log n + \cdots + \log n = n \log n.$$

It remains to show that $L(n) \in \Omega(n\log n)$. Let $m = \lfloor n/2 \rfloor$. We have

$$\begin{aligned} L(n) &= (\log 1 + \log 2 + \cdots + \log m) + [\log(m+1) + \log(m+2) + \cdots + \log n] \\ &\geq \log(m+1) + \log(m+2) + \cdots + \log n \\ &\geq \log(m+1) + \log(m+1) + \cdots + \log(m+1) \\ &= (n-m)\log(m+1) \\ &\geq (n/2)\log(n/2) = (n/2)(\log n - \log 2). \end{aligned}$$

Clearly, $(n/2)(\log n - \log 2) \geq (n/2)[\log n - (1/2)\log n]$ for all $n \geq n_0$, where n_0 is sufficiently large (the constant n_0 depends on the base chosen). Thus, $L(n) \geq (1/4)(n \log n)$, so that $L(n) \in \Omega(n \log n)$. ■

PROPOSITION 3.7.2 $H(n) \sim \ln n$. Thus, $H(n) \in \Theta(\log n)$.

Proof By definition, $\int_1^n 1/x \; dx = \ln n$, so that $\ln n$ is the area bounded above by the graph of $1/x$ and bounded below by the interval $1 \leq x \leq n$ on the x axis (see Figure 3.17). We see from Figure 3.17 that this area is larger than the sum of the areas of the rectangles R_i, $i = 1, \ldots, n-1$, where the base of R_i is the interval (of unit

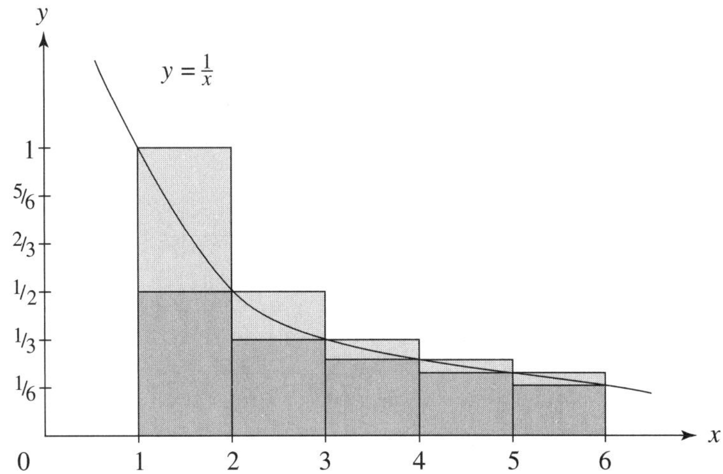

Figure 3.17 $H(n) - 1 \le \int_1^n 1/x\, dx \le H(n) - 1/n$

length) $i \le x \le i + 1$, and the height of R_i equals $1/(i + 1)$. Clearly, this area is smaller than the sum of the areas of the rectangles R_i^*, $i = 1, \ldots, n - 1$, where R_i^* has the same base as R, but has height $1/i$. Thus,

$$H(n) - 1 \le \int_1^n \frac{1}{x}\, dx \le H(n) - \frac{1}{n}. \tag{3.7.1}$$

Proposition 3.7.2 follows after dividing each of the three terms in (3.7.1) by $H(n)$ and taking the limit as n tends to infinity. The two outside limits are 1, so the middle limit must exist and also be 1. Showing that the outside limits are 1 amounts to showing that $\lim_{n\to\infty} H(n) = \infty$ (see Exercise 3.43). ■

3.8 Closing Remarks

For most of the algorithms in this text, our analysis of the complexity consists in identifying a basic operation and then determining the best-case, worst-case, and average complexities with respect to this basic operation. Occasionally, it is important to measure the complexity of an algorithm in terms of more than one operation. This is especially true for algorithms that involve maintaining several data structures, some of which may be altered by insertion or deletion of elements during the performance of the algorithm. *Amortized* analysis of an algorithm involves computing the maximum total number of all operations on the various data structures that are performed during the course of the algorithm for an input of size n. Usually, this amortized total is smaller than the number obtained by simply adding together the worst-case complexities of each individual operation for inputs of size n. Indeed, often for an input where one of the operations achieves its worst-case complexity, another of the operations might perform much less than its worst-case complexity, so that a certain trade-off occurs for any input. When designing algorithms where amortized analysis is appropriate, it then becomes a strategy to exploit this trade-off between the various operations.

Algorithms that have worst-case complexity in $O(n^k)$ are referred to as *polynomial (time)* algorithms. If an algorithm has worst-case complexity whose order is larger than n^k, for every positive integer k, then we say it is *super-polynomial.* Since $O(n^k) \subset O(a^n)$ for any $a > 1$, exponential complexity is super-polynomial.

Running a program implementing an algorithm whose complexity is polynomial is feasible provided that k, n, and the constants involved are not too large. However, a program running an algorithm having exponential worst-case complexity is infeasible to execute for worst-case inputs on any computer, even for relatively small values of n. It turns out that the problem solved by *Towers* is provably exponentially difficult (in fact, *Towers* is an optimal algorithm for its solution). There are many problems that arise naturally in practice that have not been proven to be super-polynomial, but for which no polynomial-time algorithms have been found. An important class of such problems (NP-complete) will be discussed in Chapter 20.

An algorithm's complexity is often described in the literature in terms of belonging to an O-class, even when more exact behavior of the algorithm, such as its Θ-class, might be known. There are several reasons for this, but probably the most important is that the complexity of a particular algorithm for solving a given problem only gives an *upper bound* for the complexity of the problem itself. Thus, the O-class applies to the problem itself, whereas the Θ-class may not.

Of course, when expressing upper bounds for the complexity of an algorithm, we usually use the simplest order representative possible, such as n^2 versus $(n^2/2) - (n/2)$. We are also interested in finding the smallest upper bound possible. For example, if the complexity of an algorithm is $O(n^2)$, then it is also $O(n^3)$. Clearly, in this situation the $O(n^2)$ estimate is more precise. An analogous discussion holds for lower bounds and Ω. Finally, we remark again that when measuring the complexity of an algorithm using O, Ω, or Θ, we are ignoring constants. These constants are either explicitly present in the exact formulas for number of basic operations performed by the algorithm or are implicitly present due do the overhead of such things as function invocations or procedure calls, management of loop control variables, execution of assignment statements, and so forth. Thus, the order of complexity of an algorithm gives us a first approximation to the performance of the algorithm when implemented on a computer.

In practice, the constants sometimes do matter. For example, there are algorithms known for solving certain problems that have linear order, but for which the explicit constants involved are so large that the algorithms are totally impractical to implement. Of course, for sufficiently large input size n, a linear algorithm will outperform, say, a quadratic algorithm, but the latter algorithm might have sufficiently small associated explicit and implicit constants that it outperforms the linear algorithm for any input size n small enough to be of practical interest. The linear algorithm in such cases is then mostly of theoretical interest but nevertheless important: It gives hope that a linear algorithm with reasonably small constants can be found.

References and Suggestions for Further Reading

Baase, S. *Computer Algorithms: Introduction to the Design and Analysis,* 2nd edition. Reading, MA: Addison-Wesley, 1987. A good discussion of measuring the complexity of algorithms in terms of the average and worse-case number of basic operations performed.

de Bruijn, N. G. *Asymptotic Methods in Analysis*. Amsterdam: North Holland, 1961. A good reference for the use of asymptotic methods in mathematics.

Tarjan, R. E. "Amortized Computational Complexity," *SIAM Journal on Applied and Discrete Mathematics*, 14 (November 1985): 862–874. A good reference for the amortized analysis of algorithms.

In the early 1970s, asymptotic notation became widely used in computer science papers relating to the analysis algorithms. The standard forms for asymptotic notation was proposed by Knuth and further refined by Brassard.

Brassard, G. "Crusade for a Better Notation," *SIGACT News*, ACM, 17(1) (1985): 60–64.

Knuth, Donald E. "Big Omicron and Big Omega and Big Theta," *SIGACT News*, vol. 8, no. 2 (April–June 1976): 18–24.

Exercises

Section 3.1 The Complexity of Algorithms

3.1 **a.** Write a procedure that finds both the largest and second-largest elements in a list $L[1{:}n]$ of size n. You should find a more efficient algorithm than simply performing two scans through the list.

b. Determine $B(n)$ and $W(n)$ for the algorithm in (a). (In Chapter 10, we give an optimal algorithm for solving this problem that has worst-case complexity $W(n) = \lceil 3n/2 \rceil - 2$.)

Section 3.2 Searching Algorithms and Their Analysis

3.2 Calculate the average complexity $A(n)$ of *LinearSearch* under the following assumptions about the input list $L[1{:}n]$ and search element X.

a. The probability that X occurs in the list is 2/3.

b. Given that X occurs in the list, X is twice as likely to occur in the first half of the list (positions 1 through $\lfloor n/2 \rfloor$) as in the second half. Further, if X occurs in the first half of the list, it is equally likely to occur in any position in the first half. A similar assumption is made about the second half.

3.3 Trace the action of *BinarySearch*, including listing the values of *Low*, *High*, *Mid* after each iteration, for the list

L: 2 3 5 7 11 13 17 19 23 29 31 37

for each of the following search elements X.

a. $X = 3$

b. $X = 24$

c. $X = 108$

d. $X = 13$

3.4 Show that *BinarySearch* has worst-case complexity $\lceil \log_2(n + 1) \rceil$ for any n.

3.5 Describe a modification of *BinarySearch* that returns an index in the list after which the search element can be inserted and still maintain an ordered list (or zero if the search element is smaller than any element in the list).

3.6 Design a recursive version of *BinarySearch*.

3.7 Design a variant of *BinarySearch* that performs a single check for equality between the search element and a list element.

3.8 Suppose $L[1{:}n]$ is a sorted list of integers. Design an algorithm similar to *BinarySearch* that finds an index i such that $L[i] = i$ or returns 0 if no such index exists.

Section 3.4 **Insertion Sort**

3.9 Design and analyze a forward scan version of *InsertionSort* where the list is implemented using an array. Discuss the drawbacks of the forward scan version as compared with the backward scan version.

3.10 Show that *InsertionSort* is a *stable* sorting algorithm.

3.11 **a.** Design a linked list version of *InsertionSort*.

b. Give the stable (but slightly less efficient) version of the algorithm in (a).

3.12 **a.** Design a recursive version of *InsertionSort*.

b. Design a recursive linked list version of *InsertionSort*.

3.13 The sorting algorithm *SelectionSort* is based on the simple idea of successively selecting the ith largest element in a sublist $L[1{:}n - i + 1]$ and interchanging this element with the element in position $n - i + 1$, $i = 1, \ldots, n - 1$.

a. Give pseudocode for *SelectionSort*.

b. Analyze the complexity of *SelectionSort*.

3.14 Design a recursive version of *SelectionSort* as described in the previous exercise.

3.15 There is another well-known sorting algorithm called *BubbleSort*, which, like *SelectionSort*, results in the ith-largest element being placed in its proper position at the completion of the ith pass. The two algorithms differ in the manner in which this largest element is determined. During the ith pass, *BubbleSort* passes sequentially through the sublist $L[1{:}n - i{+}1]$, comparing adjacent elements in the sublist and interchanging (swapping) them if they are out of order. Since this sequential pass through the sublist starts at the *beginning* of the sublist, after the ith pass we are guaranteed that the ith-largest element will have "bubbled" to the end of the sublist $L[1{:}n - i{+}1]$, its proper place. In *BubbleSort*, this occurs automatically as a consequence of the interchange process, as opposed to *SelectionSort*'s method of actually determining the position in the sublist where the ith-largest element occurred. Note that if on the ith pass no swapping occurred, the list has been sorted. This suggests including a flag to detect this condition.

a. Give the pseudocode for *BubbleSort*.

b. Determine the best-case and worst-case complexities *BubbleSort*.

c. Design and analyze an improvement to *BubbleSort* based on keeping track of the last positions where swaps occur in each pass and where the passes are made alternately in different directions.

Section 3.5 A Glimpse of Lower Bound Theory

3.16 Show that removing an inversion $(\pi(i), \pi(i+1))$ by interchanging $\pi(i)$ and $\pi(i+1)$ results in a decrease of (exactly) one in the total number of inversions in π.

Section 3.6 Asymptotic Behavior of Functions

3.17 Prove each of the following directly from the definition of Θ.

a. $n^2 + 1/n \in \Theta(n^2)$

b. $100{,}000{,}000 \in \Theta(1)$

c. $\log_{10}(n^2) \in \Theta(\log_2 n)$

3.18 Prove each of the following directly from the definitions of O and Ω.

a. $17n^{1/6} \in O(n^{1/5})$

b. $n^2 \in O(n^2/1000)$

c. $n/1000 \in \Omega(n + 1000)$

d. $n\log_2 n \in \Omega(30n + 60)$

e. $O(n!) \subset O((n + 1)!)$

3.19 Repeat Exercise 3.17 using Proposition 3.6.4.

3.20 Repeat Exercise 3.18 using Proposition 3.6.4.

3.21 Using Proposition 3.6.4 and (possibly) L'Hôpital's Rule, prove each of the following.

a. $(n^3 + n)/(2n + \ln n) \in \Theta(n^2)$

b. $O(n^4) \subset O(3^n)$

c. $n2^n + n^9 \in O(e^n)$

d. $n^{1/2} \in \Omega((\log n)^4)$

e. $n^{\log_2 n} \in O(2^n)$

3.22 Using Proposition 3.6.4 prove the following.

$$O(108) \subset O(\ln n) \subset O(n) \subset O(n \ln n) \subset O(n^2) \subset O(n^3) \subset O(2^n) \subset O(3^n)$$

3.23 For any constants k and $b > 1$, show that

$$O(n^k) \subset O(n^{\log_2 n}) \subset O(b^n).$$

3.24 Prove property (1) of Proposition 3.6.2: For any positive constant c,

$$\Omega(f(n)) = \Omega(cf(n)),\ \Theta(f(n)) = \Theta(cf(n)),\ \text{and } O(f(n)) = O(cf(n)).$$

3.25 Prove property (3.6.4):

$$f(n) \in \Theta(g(n)) \Leftrightarrow f(n) \in \Omega(g(n)) \cap O(g(n)).$$

3.26 Show that $f(n) \in \Theta(g(n))$ if, and only if, $\Theta(f(n)) = \Theta(g(n))$.

3.27 Prove properties (3) and (5) of Proposition 3.6.2:

a. If $f(n) \in O(g(n))$, then $O(f(n)) \subseteq O(g(n))$.
b. $f(n) \in O(g(n)) \Leftrightarrow g(n) \in \Omega(f(n))$.

3.28 Prove property (4) of Proposition 3.6.2:

$$O(f(n)) = O(g(n)) \Leftrightarrow \Omega(f(n)) = \Omega(g(n)) \Leftrightarrow \Theta(f(n)) = \Theta(g(n)).$$

3.29 Prove each of the following.

a. $4n^3 + \sqrt{n} \sim n^3 - n^2 + 500$
b. $2^n + 50n^5 + 13n^2 + 42 \sim 2^n$
c. $(6n^2 + 50\sqrt{n})/(3\sqrt{n} + 5 \ln n) \sim 2n^{3/2}$
d. $30n - 50 = o(n^2/23)$
e. $1000 \log_2 n = o(\sqrt{n})$

3.30 Suppose a and c are positive constants.

a. If f is a polynomial, show that $f(n + c) \sim f(n)$ and $f(cn) \in \Theta(f(n))$.
b. If $f(n) = a^n$, show that $f(n + c) \in \Theta(f(n))$.
c. If $f(n) = n!$, show that $O(f(n)) \subset O(f(n + c))$, ($c$ a positive integer).

3.31 Let $P(n)$ be any polynomial of degree k whose leading coefficient is positive, and let a be any real number $a > 1$. Show that $P(n) = o(a^n)$.

3.32 Complete the proof of Proposition 3.6.4 by proving case 3.

3.33 Proposition 3.6.4 states that $\lim_{n\to\infty} f(n)/g(n) = 0$ implies that $O(f(n)) \subset O(g(n))$.

a. a. Show that a partial converse is true: If $O(f(n)) \subset O(g(n))$ and $\lim_{n\to\infty} f(n)/g(n)$ exists, then this limit must be zero.
b. Construct examples where $O(f(n)) \subset O(g(n))$ but $\lim_{n\to\infty} f(n)/g(n)$ does not exist.
c. Give examples similar to (b) for the other two cases of Proposition 3.6.4.

3.34 Prove the following result from the definition of o and ~. If $f(n)$ and $g(n)$ are functions such that $g(n) \in o(f(n))$, then

$$f(n) \pm g(n) \sim f(n).$$

3.35 The notions of Ω, Θ, O can be viewed as binary relations on the set $\mathscr{F}$ of (eventually positive) functions $f: \mathbb{N} \to \mathbb{R}$. *A (binary) relation* R on a set S is any subset of the cartesian product S × S. For $x, y \in$ S, we say that x is *related* to y, written xRy, if the ordered pair $(x, y) \in$ R. Note that Θ determines a relation on the set $\mathscr{F}$ by defining $f\Theta\, g$ to mean $f(n) \in \Theta(g(n))$. In a similar way, Ω and O determine relations on $\mathscr{F}$.

A very important class of relations on a set S are the so-called equivalence relations. Equivalence relations on S correspond precisely to partitions of S into pairwise disjoint subsets (see Exercise 3.36).

DEFINITION A relation R on S is said to be an *equivalence* relation if the following three properties are satisfied.

1. *Reflexive Property:* $x\mathrm{R}x$, $\forall x \in \mathrm{S}$.
2. *Symmetric Property:* $x\mathrm{R}y \Rightarrow y\,\mathrm{R}x$, $\forall x, y \in \mathrm{S}$.
3. *Transitive Property:* $x\mathrm{R}y$ and $y\,\mathrm{R}z \Rightarrow x\mathrm{R}z$, $\forall x, y, z \in \mathrm{S}$.

---

a. Show that the relation Θ is an equivalence relation on $\mathscr{F}$ (that is, verify properties (1), (2), and (3) of Proposition 3.6.1).

b. Show that relations O and Ω are reflexive and transitive but are not symmetric.

3.36 Given an equivalence relation R on a set S, each element $x \in \mathrm{S}$ determines an *equivalence class*, denoted by $[x]$, consisting of all elements y such that $x\mathrm{R}y$. It is easily proved that $[x] = [y]$ iff $x\mathrm{R}y$. A finite or infinite collection of subsets of a set S is said to be a *partition* of S if the sets are pairwise disjoint and their union is S.

a. Given an equivalence relation R on a set S, show that the set of equivalence classes is a partition of the set S.

b. Conversely, given any partition of the set S, show there is a unique equivalence relation on S whose equivalence classes are precisely the subsets of the given partition (define $x\mathrm{R}y$ iff x and y lie in the same subset).

3.37 **a.** Show that the relation ~ is actually an equivalence relation on $\mathscr{F}$.

b. Show that ~ is a stronger relation than Θ in the sense that ~ refines Θ. Given two relations R_1, R_2 on a set S, we say that R_2 *refines* R_1 if $x\mathrm{R}_2y \Rightarrow x\mathrm{R}_1y$. If both R_1 and R_2 are equivalence relations, and R_2 refines R_1, then each R_2-equivalence class is contained in an R_1-equivalence class. We say that the R_2-equivalence classes form a *refinement* of the R_1-equivalence classes.

3.38 Consider the relation K on $\mathscr{F}$ defined as follows:

$$f\,\mathrm{K}g \text{ iff } \lim_{n\to\infty} \frac{f(n)}{g(n)} = L, \quad \text{where } 0 < L < \infty.$$

a. Show that K is an equivalence relation.

b. Show that K refines Θ.

c. Show that K is refined by ~.

3.39 Show that the functions $f(n) = n^3$ and $g(n) = n^4(n \textbf{ mod } 2) + n^2$ are not comparable.

3.40 Find two strictly increasing functions $f,g \in \mathscr{F}$ whose orders are not comparable.

Section 3.7 The Harmonic Series and Sums of Logarithms

3.41 Obtain a formula for the order of $S(n) = \Sigma_{i=1}^{n} (\log i)^2$.

3.42 Obtain an approximation for $\log n! = \Sigma_{i=1}^{n} (\log i)$ by using a technique similar to that used in Section 3.7 for approximating the harmonic series. (*Hint:* Use $\int_1^n \log x \, dx$.)

3.43 Show that $\lim_{n\to\infty} H(n) = \infty$.

3.44 Show that $1 + \left(\frac{1}{2}\right)^k + \left(\frac{1}{3}\right)^k + \cdots + \left(\frac{1}{n}\right)^k \in \Theta(1)$, for all integers $k \geq 2$.

4

Sequential Sorting Algorithms and Their Analysis

It has been estimated that at any given point in time, approximately 20% of the computers in use are executing searching or sorting algorithms. Computer scientists have long been interested in designing efficient sorting algorithms, and highly efficient sequential and parallel sorting algorithms have been devised. The problem of sorting a list has a vast literature, and entire books have been devoted to the subject.

The purpose of this chapter is twofold: to introduce some well-known sorting algorithms and to further illustrate various techniques and features relating to the design and complexity analysis of algorithms. Each sorting algorithm described illustrates a different technique or feature. For example, *ShellSort* is an illustration of how a clever alteration to a straightforward but inefficient algorithm such as *InsertionSort* can lead to a less straightforward but rather more efficient algorithm. *MergeSort* illustrates the divide-and-conquer design strategy and also is an example of an order-optimal sorting algorithm. *QuickSort* is also based on the divide-and-conquer design strategy, and while (unlike *MergeSort*) *QuickSort* has bad worst-case performance, it is often used in practice because of its superior average performance. *TreeSort* is based on an inorder traversal of a binary search tree. Like *QuickSort*, *TreeSort* has bad worst-case performance, but good average behavior. *HeapSort* is another order-optimal sorting algorithm and is based on the important ADT known as a heap. *BingoSort* is an example of a sorting algorithm that is efficient for lists having many repeated elements.

ShellSort, *MergeSort*, *QuickSort*, *TreeSort*, *HeapSort*, and *BingoSort* are each examples of comparison-based sorting algorithms. Our seventh example, *RadixSort*, is not comparison-based, but instead depends on the special nature of the list elements.

All the previous sorts assume that the list is resident in high-speed internal memory (RAM). The last two algorithms we discuss in this chapter are examples of *external sorting* algorithms, where it is assumed that the list is stored in an external device such as in a tape or disk.

4.1 *ShellSort*

As mentioned earlier, *InsertionSort* is an order-optimal adjacent-key sorting algorithm. Thus, if we want to design a more efficient comparison-based sorting algorithm, we must look for design strategies that compare nonadjacent (far away) elements in the list. One such strategy is the basis of *ShellSort*, named after its discoverer, Donald Shell.

Shell's idea was to start out by sorting sublists of a list $L[1:n]$ of size n consisting of equally spaced, but far apart, elements. For example, if the successive elements in the sublists occupy positions k distance apart in the original list, then we have k sublists $S_1 = \{L[1], L[1+k], \dots\}$, $S_2 = \{L[2], L[2+k], \dots\}$, … , $S_k = \{L[k], L[2k], \dots\}$. The process of sorting these k sublists is called a *k-subsort* (note that a 1-subsort corresponds to a sorting of the entire list L). A list in which each of the sublists $S_1, S_2, \dots, S_k$ is sorted is called *k-subsorted* (also referred to in the literature as k-ordered, or k-sorted). Each of the sublists $S_1, S_2, \dots, S_k$ contains at most $\lceil n/k \rceil$ list elements, so for a large k, sorting each sublist is fast using a sort like *InsertionSort*. Although the list L itself is not sorted by the k-subsort, in some sense it is closer to being sorted. *ShellSort* works by performing a succession of k-subsorts with decreasing increments $k_1 > k_2 > \cdots > k_m = 1$.

To accomplish a k-subsort, rather than calling a sorting procedure for each sublist individually, we make a single call to the following variant of *InsertionSort*.

```
procedure InsertionSubsort(L[1:n],k)

Input:  L[1:n] (an array), k (a positive integer increment)
Output: k-subsorted array L[1:n] in nondecreasing order

  for i := k + 1 to n do       {pass = i – k in Figure 4.1}
           {insert L[i] in its proper position in the already sorted sublist
                 L[j], L[j+k] … , L[i–k] of S_j, j = ((i–1) mod k) + 1}
     Current := L[i]
     position := i – k
     while position ≥ 1 .and. Current < L[position] do
           {Current must precede L[position]}
        L[position+k] := L[position]       {bump up L[position]}
        position := position – k
     endwhile
           {position + k is now the proper position for current = L[i]}
     L[position+k] := Current
  endfor

end InsertionSubsort
```

For $i = k + 1, \dots, n$, *InsertionSubsort* parrots the code for *InsertionSort* by comparing $L[i]$ to $L[i-k]$, $L[i-2k]$, … until a proper place is found for its insertion in the already sorted sublist $L[j]$, $L[j+k]$, ... , $L[i-k]$ of S_j, $j = ((i-1)$ **mod** $k) + 1$. The pseudocode for *InsertionSubsort* shows how elegantly the work of sorting the sublists shifts from one sublist to another in a cyclic manner as *pass* varies from 1 to $n - k$ (see Figure 4.1, page 118). The shifting from S_j to S_{j+1} (where $S_{k+1} = S_1$) is an automatic consequence of the action of the pseudocode, and there is no need to explicitly implement the sublists.

The following pseudocode for *ShellSort* merely consists of successively calling *InsertionSubsort* with a given set of diminishing increments.

```
procedure ShellSort(L[1:n],D[1:m])
Input:   L[1:n] (an array of list elements)
         D[1:m] (an array of diminishing increments D[1] > D[2] > ··· > D[m] = 1)
Output: sorted array L[1:n] in nondecreasing order
   for i := 1 to m do
      call InsertionSubsort(L[1:n],D[i])
   endfor
end ShellSort
```

The final stage of *ShellSort* consists of performing *InsertionSort* on the entire list, so that it is unquestionably correct. However, by then the list should be close to being sorted, so by Proposition 3.4.2 *InsertionSort* will work quickly on the list. For example, applying *InsertionSort* to the original list in Figure 4.2 results in 69 comparisons. On the other hand, applying *InsertionSort* as the 1-subsort in the last stage of Figure 4.2 (page 119) results in only 19 comparisons.

The following theorem implies that successive k-subsorting with diminishing k tends to reorder a list closer and closer to being sorted.

THEOREM 4.1.1 For any two integers k and l, suppose we perform an l-subsort on a list L that is already k-subsorted. Then L remains k-subsorted—that is, L becomes *both* k-subsorted *and* l-subsorted. ❑

A proof of Theorem 4.1.1 is left to the exercises.

Even though *ShellSort* is a fairly simple algorithm and has been well-studied and used by many researchers, its analysis is not yet complete. The number of comparisons for a given input of size n performed by *ShellSort* depends on what set of increments is used, and the general mathematical analysis is difficult. For a general n, it is not known what set of increments exhibits optimal behavior.

A number of results are known for special choices of the increments. An order of complexity improvement over *InsertionSort* is already achieved when only two increments d and 1 are used for a suitable d. In fact, it has been shown

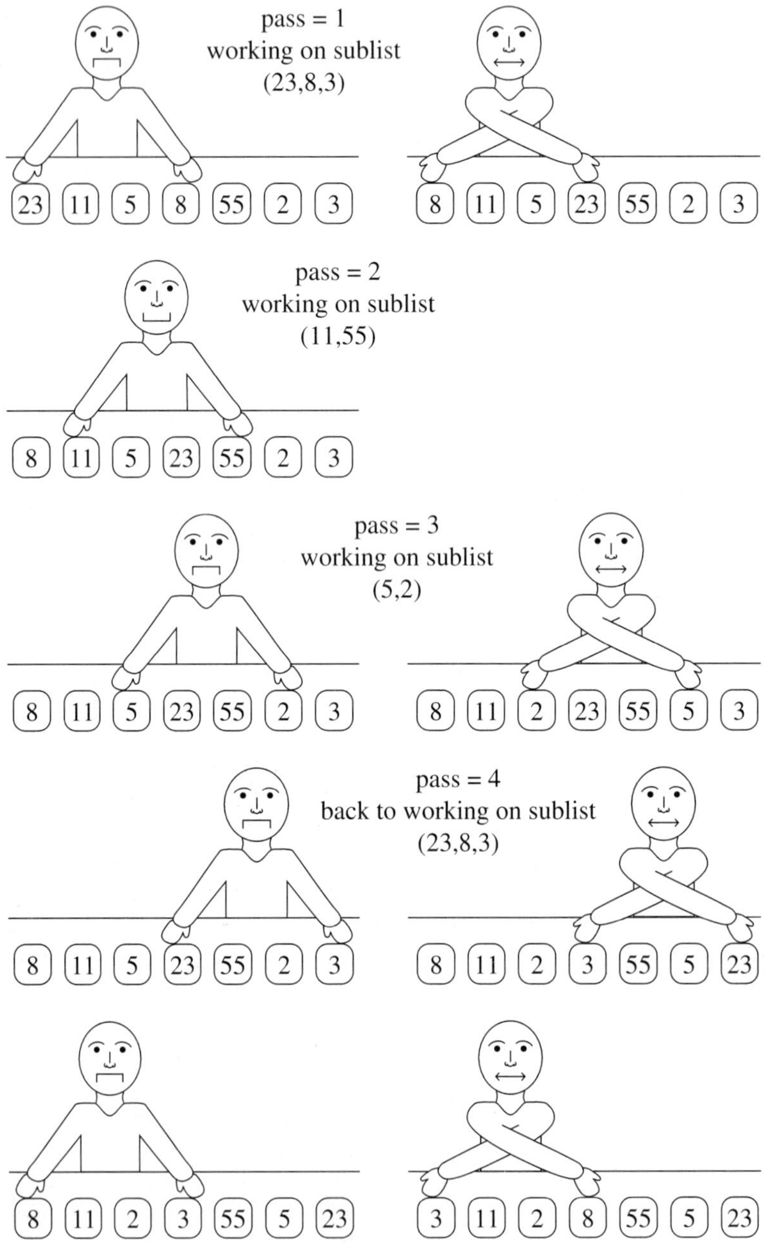

Figure 4.1 Action of *InsertionSubsort* with $k = 3$ on list L: 23,11,5,8,55,2,3

in the two-increment case that the optimal average complexity has order $n^{5/3}$ and is achieved when d is approximately $1.72n^{1/3}$. It has also been shown that when the increments are of the form $2^k - 1$, for all such increments less than n,

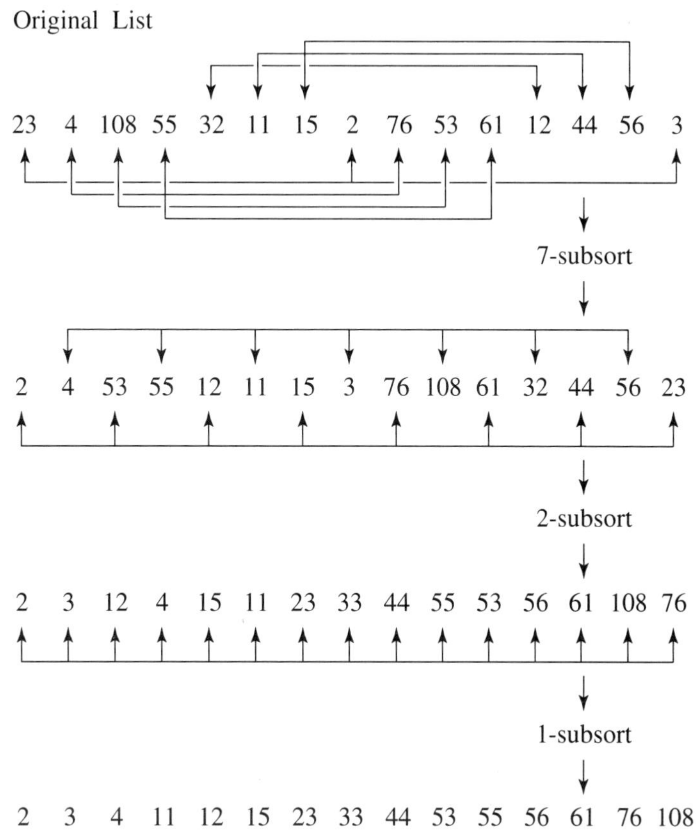

Figure 4.2 Action of *ShellSort* showing results of calls to *InsertionSubsort* with increments 7,2,1

then *ShellSort* exhibits worst-case complexity in $O(n^{3/2})$. Since $2^{2k} - 1 = (2^k - 1) \cdot (2^k + 1)$, a large percentage of these increments divide one another, so we would expect to be able to do even better. Indeed, it has been shown that using increments of the form 2^i3^j for all such increments less than n yields $\Theta(n(\log n)^2)$ worst-case complexity.

4.2 MergeSort

MergeSort is a classical example of an algorithm based on the divide-and-conquer design strategy. Given a list *L*[*low*:*high*], let *x* be any index between *low* and *high*. Let *A*, *B*, *C* denote the sublists *L*[*low*:*x*], *L*[*x*+1:*high*], *L*[*low*:*high*], respectively. The problem of sorting *C* can be solved by first sorting *A* (recursive call), then sorting *B* (another recursive call), and finally calling the procedure *Merge*, which merges the sorted lists *A* and *B* to obtain a sorted list *C*. This leads to the following recursive code. Note in *MergeSort* that we have chosen *x* to be the midpoint of *low* and *high*. The list $L[1{:}n]$ is sorted by calling *MergeSort* and passing 1 to *low* and *n* to *high*.

```
procedure MergeSort(L[1:n],low,high) recursive

Input:  L[1:n] (a list of size n), low,high (indices of L[1:n])
Output: sublist L[low:high] sorted in nondecreasing order

  if low < high then
     mid := ⌊(low + high)/2⌋
     call MergeSort(L[1:n],low,mid)
     call MergeSort(L[1:n],mid+1,high)
     call Merge(L[1:n],low,mid,high)
  endif

end MergeSort
```

We now give the pseudocode for the procedure *Merge* called by *MergeSort*, where an auxiliary array *Temp* is used to aid in the merging process. *Merge* utilizes two pointers *CurPos*1 and *CurPos*2 that refer to the current positions in the already sorted sublists $L[low{:}x]$ and $L[x+1{:}high]$, respectively. *Merge* also utilizes a third pointer *Counter*, which points to the next available position in *Temp*. *CurPos*1, *CurPos*2, and *Counter* are initialized to *low*, $x + 1$, and *low*, respectively. During each iteration of the **while** loop in *Merge*, the elements in the positions *CurPos*1 and *CurPos*2 are compared, and the smaller element is written to the position *Counter* in *Temp*. Then *Counter* and either *CurPos*1 or *CurPos*2, depending on which one points to the element just written to *Temp*, are incremented by one. When one of the sublists has been completely written to *Temp*, the remaining elements in the other sublist are written to *Temp*. Then *Temp* contains a sorting of $L[low{:}high]$, and the algorithm terminates after copying $Temp[low{:}high]$ to $L[low{:}high]$. The pseudocode for *Merge* follows.

```
procedure Merge(L[1:n],low,x,high)

Input:  L[1:n] (a list of size n),
        low,x,high (indices of array L[1:n];   sublists L[low:x], L[x+1:high]
                          are assumed to be sorted in nondecreasing order)
Output: sublist L[low:high] sorted in nondecreasing order

  CurPos1 := low                                  {initialize pointers}
  CurPos2 := x + 1
  Counter := low
  while CurPos1 ≤ x .and. CurPos2 ≤ high do      {while elements remain
    if  L[CurPos1] ≤ L[CurPos2] then               to be written in both
        Temp[Counter] := L[CurPos1]                sublists, merge to Temp}
        CurPos1 := CurPos1 + 1
```

```
    else
        Temp[Counter] := L[CurPos2]
        CurPos2 := CurPos2 + 1
    endif
    Counter := Counter + 1
  endwhile
  if CurPos1 > x then                          {copy remaining elements
   for k := CurPos2 to high do                  in appropriate sublist
        Temp[Counter] := L[k]                   to Temp}
        Counter := Counter + 1
    endfor
  else
    for k := CurPos1 to low do
        Temp[Counter] := L[k]
        Counter := Counter + 1
    endfor
  endif
  for k := low to high do      {copy Temp[low:high] to L[low:high]}
    L[k] := Temp[k]
  endfor
end Merge
```

The tree of recursive calls to *MergeSort* is illustrated in Figure 4.3. *A* node in the tree is labeled by the values *low,mid,high* involved in the call to *Merge*. (In leaf nodes, *mid* is not computed, as indicated by the symbol *.) Initially, *low* = 1 and *high* = 10. Similar to *Fib*1 (see Section 1.3.4), the path around the tree in Figure 4.3 indicates how the recursion resolves. Following this path amounts to a postorder traversal of the tree, where visiting a node corresponds to a call to *Merge*.

Figure 4.3 Recursive calls to *MergeSort* for lists of size 10

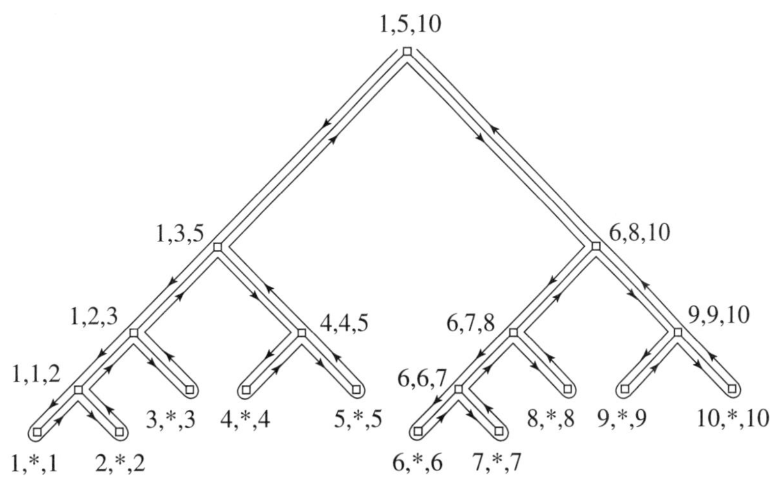

The best-case, worst-case, and average complexities for *MergeSort* all have order $\Theta(n\log n)$. Since $B(n) \leq A(n) \leq W(n)$, it is sufficient to show that $W(n) \in O(n\log n)$ and $B(n) \in \Omega(n\log n)$. We first discuss the worst-case complexity. Note that each call to *Merge* for merging two sublists of sizes m_1 and m_2 performs at most $m_1 + m_2 - 1$ comparisons. Consider the tree of recursive calls to *MergeSort* for a list of size n (see Figure 4.3). The leaf nodes do not generate calls to *Merge*, whereas each internal node generates a single call to *Merge*. At each level of the tree, the total number of comparisons made by all of the calls to *Merge* is at most n. The depth of the tree of recursive calls is $\lceil \log_2 n \rceil$ (see Exercise 4.5). It follows that *MergeSort* performs at most $n\lceil \log_2 n \rceil$ comparisons for any list of size n, so that $W(n) \in O(n\log n)$.

Now consider $B(n)$. Each call to *Merge* for merging two sublists of sizes m_1 and m_2 performs at least $\min\{m_1, m_2\}$ comparisons. We again consider the tree of recursive calls to *MergeSort* for a list of size n. Except for the last two levels, at each level of the tree of recursive calls, the total number of comparisons made by all of the calls to *Merge* is at least $n/2$. It follows that *MergeSort* performs at least $(n/2)(\lceil \log_2 n \rceil - 1)$ comparisons for any list of size n, so that $B(n) \in \Omega(n\log n)$.

Remark An alternate technique for establishing the complexities of *MergeSort* is to solve recurrence relations for $W(n)$ and $B(n)$ (see Exercise 4.6).

By examining Figure 4.3, we can easily come up with a bottom-up version of *MergeSort*. At the bottom level of the tree, we are merging sublists of single adjacent elements in the list. However, as the path around the tree indicates, for a given input list $L[1{:}n]$ *MergeSort* sorts the list $L[1{:}mid]$ before going on to any of the sublists of $L[mid+1, n]$. By contrast, a bottom-up version begins by dividing the list into pairs of adjacent elements, $L[1]{:}L[2]$, $L[3]{:}L[4]$, ... Next, these adjacent pairs are merged yielding the sorted lists $L[1{:}2]$, $L[3{:}4]$, ... The process is repeated by merging the adjacent pairs of sorted two-element sublists, $L[1{:}2]{:}L[3{:}4]$, $L[5{:}6]{:}L[7{:}8]$, ... Continuing this process, we arrive at the root having sorted the entire list $L[1{:}n]$.

Figure 4.4 shows a tree representing this bottom-up merging of adjacent sublists. Each node represents a call to *Merge* with the indicated values of *low,mid,high*. An asterisk * denotes those nodes where a call to *Merge* is not made. Note that the sublists and resulting tree are quite different from the sublists generated by the tree of recursive calls of *MergeSort* given in Figure 4.3. All the calls to *Merge* for a given level are completed before we go up to the next level. The pseudocode for the nonrecursive version of *MergeSort* based on Figure 4.4 is left to the exercises.

Any comparison-based sorting algorithm must perform $\Omega(n\log n)$ comparisons in the worst case to sort a list of size n (see Chapter 10). Thus, *MergeSort* has order-optimal worst-case complexity.

4.3 *QuickSort*

We now discuss the comparison-based sorting algorithm *QuickSort*, discovered by C. A. R. Hoare in 1962. *QuickSort* is often the sorting algorithm of choice because of its superior average behavior. *QuickSort*, like *MergeSort*, is based on dividing a

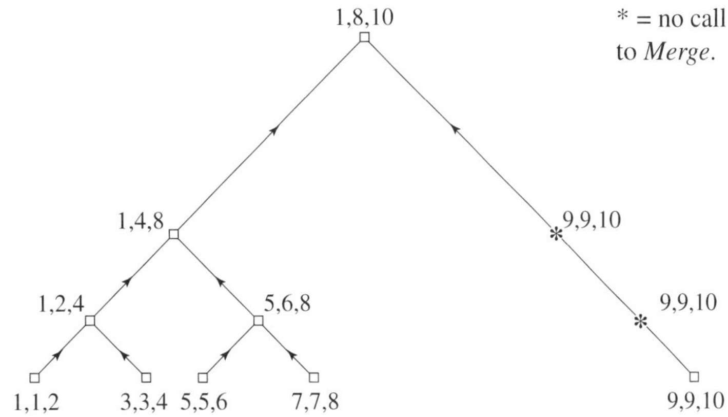

Figure 4.4 Bottom-up nonrecursive *MergeSort* for a list of size 10

list into two sublists (these are actually two sublists of a rearrangement of the original list), and then sorting the sublists recursively. The difference between the two sorting strategies is that *MergeSort* does most of the work in the combine (merge) step, whereas *Quicksort* does most of the work in the divide step.

In *QuickSort* the division into sublists is based on rearranging the list $L[low{:}high]$ with respect to a suitably chosen *pivot element x*. The list $L[low{:}high]$ is rearranged so that every list element in $L[low{:}high]$ preceding x (having smaller index than the index of x) is not larger, and every element following x in $L[low{:}high]$ is not smaller. For example, for the list 23,55,11,17,53,4 and pivot element $x = 23$, such a rearrangement might be 17,11,4,23,55,53. Note that after such a rearrangement, the pivot element x occupies a proper position in a sorting of the list. Thus, if the sublists on either side of x are sorted recursively, then the entire list will be sorted with no need to invoke an algorithm for combining the sorted sublists.

QuickSort sorts $L[low{:}high]$ into nondecreasing order by first calling an algorithm *RearrangeAndPlace* that rearranges the list with respect to pivot element $x = L[low]$ as previously described. *RearrangeAndPlace* assumes that the element $L[high+1]$ is defined and is at least as large as $L[low]$. The output parameter *position* of *RearrangeAndPlace* returns the index where x is placed. *QuickSort* is initially called with $low = 1$ and $high = n$.

procedure *QuickSort*($L[low{:}high]$) **recursive**

Input: $L[low{:}high]$ (an array of list elements) $\{L[high+1]$ is assumed defined and $\geq L[low]\}$

Output: $L[low{:}high]$ sorted in nondecreasing order

```
if high > low then
    call RearrangeAndPlace(L[low:high],position)
    call QuickSort(L[low:position-1])
```

Continued

```
        call QuickSort(L[position+1:high])
    endif
end QuickSort
```

We now describe the algorithm *RearrangeAndPlace. RearrangeAndPlace* is based on the following clever interplay between two moving variables *moveright* and *moveleft*, which contain the indices of elements in *L* and are initialized to *low* + 1 and *high*, respectively.

```
while moveright < moveleft do
    moveright moves to the right (one index at a time) until it assumes the
    index of a list element not smaller than x, then it stops.
    moveleft moves to the left (one index at a time) until it assumes the
    index of a list element not larger than x, then it stops.
    if moveright < moveleft then
            interchange L[moveright] and L[moveleft]
    endif
endwhile
```

To guarantee that *moveright* actually finds an element not smaller than x, we assume that $L[high+1]$ is defined and is not smaller than $L[low]$. We arrange this by introducing a sentinel value $L[n+1] = +\infty$. We leave it as an exercise to check that the condition $L[high+1] \geq L[low]$ is then automatically guaranteed for all subsequent calls to *RearrangeAndPlace* by *QuickSort*. Of course, *RearrangeAndPlace* could be written with explicit checking that *moveright* does not run off the list. However, this checking requires additional comparisons, and we prefer to implement the preconditioning. Figure 4.5 illustrates the movement of *moveright* (*mr*) and *moveleft* (*ml*) for a sample list $L[1:7]$.

The pseudocode for *RearrangeAndPlace* follows.

```
procedure RearrangeAndPlace(L[low:high],position)
Input:  L[low:high] (an array of list elements) {L[high+1] is assumed defined
                                                     and ≥ L[low]}
Output: a rearranged list L[low:high] such that
        L[i] ≤ L[position],        low ≤ i ≤ position,
        L[i] ≥ L[position],        position ≤ i ≤ high
        where, originally, L[low] = L[position]
        position (the position of a proper placement of the original
                element L[low] in the list L[low:high])
  moveright := low
  moveleft := high + 1
  x := L[low]
```

```
    while moveright < moveleft do
       repeat
          moveright := moveright + 1
       until L[moveright] ≥ x
       repeat
          moveleft := moveleft – 1
       until L[moveleft] ≤ x
       if moveright < moveleft then
          call interchange(L[moveright],L[moveleft])
       endif
    endwhile
    position := moveleft
    L[low] := L[position]
    L[position] := x
end RearrangeAndPlace
```

Analysis of *QuickSort* Again, we use list comparisons as our basic operation for analysis, where *QuickSort* is originally called with a list $L[1:n]$ of distinct elements, and $L[n+1] = +\infty$. A call to *RearrangeAndPlace* with $L[low:high]$ performs exactly $high - low + 2$ comparisons, and this forms the basis of our analysis.

The best-case performance of *QuickSort* occurs when all calls to *RearrangeAndPlace* result in having the pivot element placed in the middle of the sublist, so that

Figure 4.5 Action of *RearrangeAndPlace* for a sample list $L[1:7]$

Initially:	index	1	2	3	4	5	6	7	8
	list element	23	9	23	52	15	19	47	$+\infty$
			mr					*ml*	
		Rearrange Step							
				19			23		
1st iteration:		23	9	23	52	15	19	47	$+\infty$
				mr			*ml*		
					15	52			
2nd iteration:		23	9	19	52	15	23	47	$+\infty$
					mr	*ml*			
3rd iteration:		23	9	19	15	52	23	47	$+\infty$
					ml	*mr*			
		Place Step							
		15			23				
After completion of		23	9	19	15	52	23	47	$+\infty$
RearrangeAndPlace:						*position* = 4			

the sublist is divided into two sublists whose sizes differ by at most one. Thus, the tree of recursive calls for best-case performance is full at every level except possibly the last level. In particular, if $n = 2^k - 1$, then the tree of recursive calls is a full tree of depth $k - 1 = \log_2(n + 1) - 1$. At each level except the last level, the total number of comparisons done by all the calls to *RearrangeAndPlace* is $n + 1$. Hence, for $n = 2^k - 1$, $B(n) = (n + 1)\log_2(n + 1) - n - 1$. Interpolating the behavior of $B(n)$ to all n, we conclude that $B(n) \in \Theta(n\log n)$ (Figure 4.6a).

Unfortunately, the worst-case performance of *QuickSort* occurs for a list that is already in order. In this case the recursive calls to *QuickSort* are always with the

Figure 4.6 Tree of recursive calls for *QuickSort:* (a) best-case behavior; (b) worst-case behavior

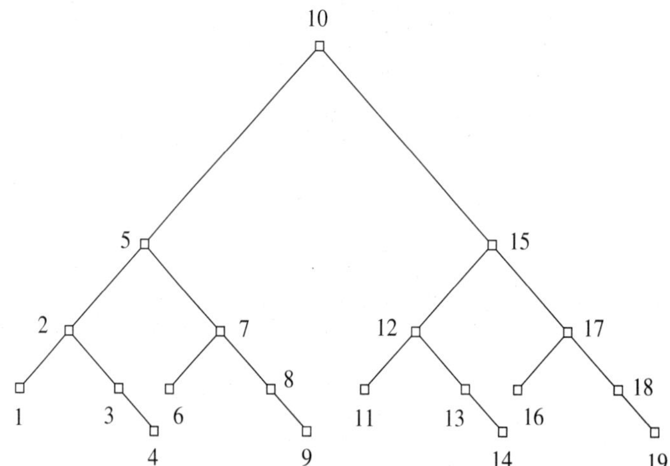

(a) Tree of recursive calls for best-case behavior of *QuickSort* for a list of size 19. Internal nodes are labeled with pivot elements in a call to *RearrangeAndPlace*. Leaf nodes are labeled with single-element sublists that result in immediate returns. For simplicity, we do not show leaf nodes corresponding to calls to *QuickSort* with empty lists.

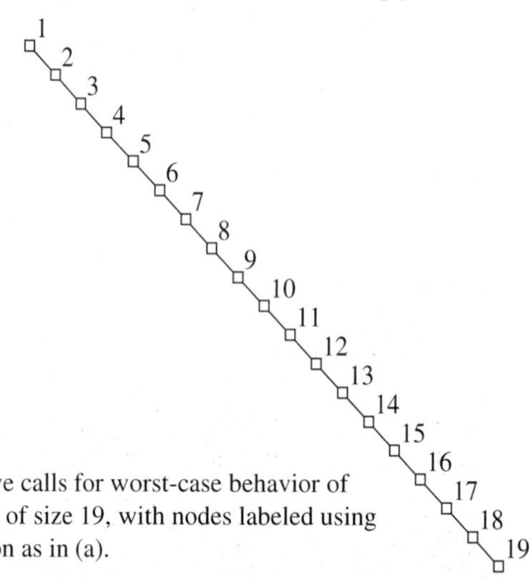

(b) Tree of recursive calls for worst-case behavior of *QuickSort* for a list of size 19, with nodes labeled using the same convention as in (a).

empty list $L[low{:}low-1]$ and the list $L[low+1{:}high]$ (see Figure 4.6b). Thus, the number of comparisons performed is given by $W(n) = (n+1) + n + \cdots + 3 \in \Theta(n^2)$, so that *QuickSort* has quadratic worst-case performance.

Remark For simplicity we chose the pivot element to be the first list element. A common alternative is to take the pivot element to be the median of the three elements in positions *low*, (*low* + *high*)/2, and *high*. Using this *Median of Three Rule* avoids the bad behavior exhibited by lists that are close to being sorted.

In spite of its quadratic worst-case performance, *QuickSort* is often used because it performs so well on average. We will show in Chapter 9 that the average complexity of *QuickSort* belongs to $\Theta(n\log n)$.

4.4 TreeSort

The algorithm *TreeSort* is based on the following interesting observation.

> *An inorder traversal of a binary search tree visits the nodes in nondecreasing order of the keys.*

You should convince yourself of the validity of this observation. (A formal proof of the observation using mathematical induction is given in Chapter 7.) Thus, given a list $L[1{:}n]$, we can sort the list by creating a binary search tree T whose keys are the elements of L, and then performing an inorder traversal of T. In Chapter 2 we developed the algorithm *BinSrchTreeInsert* for inserting a key into a nonempty binary search tree with duplicate keys allowed. Given an input list $L[1{:}n]$, the following procedure *CreateBinSrchTree* creates a binary search tree whose keys are the elements from L.

```
procedure CreateBinSrchTree(L[1:n],Root)
Input:   L[1:n] (a list of size n)
Output: Root (→ BinaryTreeNode), {points to root of the binary search tree
                                   whose keys are the elements from L}
   call AllocateBinaryTreeNode(Root)
   Root→Key := L[1]
   for i := 2 to n do
      call BinSrchTreeInsert(Root,L[i])
   endfor
end CreateBinSrchTree
```

In Figure 4.7 we show the binary search tree created by *CreateBinSrchTree* for a sample input list $L[1{:}16]$.

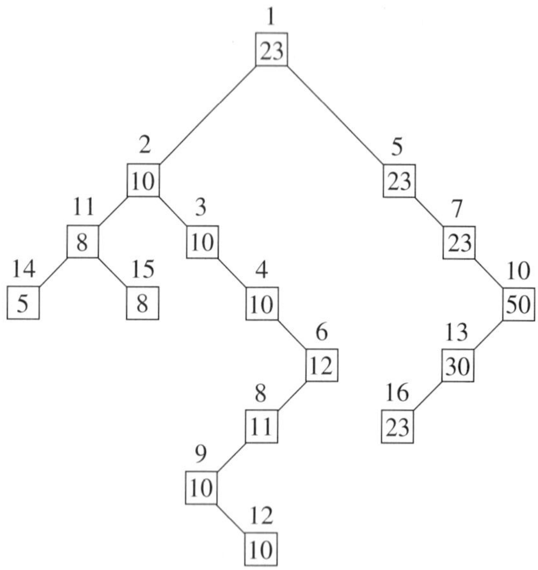

Figure 4.7 The binary search tree created by *CreateBinSrchTree* for input list $L[1:16]$ = (23,10,10,10,23,12,23,11,10,50,8,10,30,5,8,23). We show the list element (key) inside each node and its index in the original list outside each node.

In the case where the list is already sorted, *CreateBinSrchTree* creates a completely skewed tree and has worst-case complexity given by

$$W(n) = 1 + 2 + \cdots + n - 1 = \frac{n(n-1)}{2} \in \Theta(n^2).$$

However, in spite of the quadratic worst-case complexity of *CreateBinSrchTree*, its average complexity belongs to $\Theta(n \log n)$ (see Chapter 9).

In the following pseudocode for *TreeSort*, we assume that the procedure *InOrder* from Chapter 2 is modified slightly to include a global counter that keeps track of the number of nodes visited so far. When the ith node is visited, the procedure *Visit* copies the key in the node to $L[i]$, $i = 1, \ldots, n$.

```
procedure TreeSort(L[1:n])
Input:   L[1:n] (a list of size n)
Output: L[1:n] sorted in nondecreasing order
   call CreateBinSrchTree(L[1:n],Root)
   call InOrder(Root) {when the ith node is visited, the key in the node is
                                   copied to L[i], i = 1, ... , n.}
end TreeSort
```

Since the complexity of the call to *CreateBinSrchTree* dominates the complexity of the call to *InOrder*, *TreeSort* has the same order of complexity as *CreateBinSrchTree*.

Remark Note in Figure 4.7 that when inserting a key for which there already exist keys in the tree having the same value, we always moved to the right when encountering a key with the same value. Of course, the binary search tree property would have also been maintained if we went to the left instead. We choose to use the *move-to-the-right rule* for a good reason: When creating a binary search tree by successively inserting a sequence of keys, an inorder traversal of the resulting search tree will visit duplicate keys in the order in which they were inserted. The latter property ensures that the procedure *TreeSort* is a stable sorting algorithm (see Exercise 4.14).

The following proposition extends to multiway search trees the useful property that an inorder traversal of a binary search tree visits the nodes in sorted order. In our multivisit inorder traversal, if node v contains q keys, then *Visit*(v,i) processes the ith key in the sequential ordering of the keys in the node, $i = 1, \ldots, q$.

PROPOSITION 4.4.1 A multivisit inorder traversal of a multiway search tree visits the nodes in sorted order. ❑

A formal proof of Proposition 4.4.1 requires mathematical induction (see Chapter 7).

4.5 Priority Queues, Heaps, and *HeapSort*

Certainly one of the simplest strategies for sorting a list $L[1:n]$ of size n in nondecreasing order is as follows. Select the largest element in L and interchange it with $L[n]$. Then select the largest element in the sublist $L[1:n-1]$ and interchange it with $L[n-1]$. In general, in the ith step, select the largest element in the sublist $L[1:n-i+1]$ and interchange it with $L[n-i+1]$, $i = 1, \ldots, n-1$. Clearly, this strategy yields a sorted list. The most straightforward strategy for selecting the largest elements is to make a linear scan as in the procedure *Max*. The resulting algorithm is known as *SelectionSort*. Clearly, *SelectionSort* performs $(n-1) + (n-2) + \cdots + 1 = n(n-1)/2$ comparisons for any input list of size n, so that its worst-case, best-case, and average complexities are all in $\Theta(n^2)$.

The inefficiency of *SelectionSort* results from the procedure used to select the largest elements. Our selection procedure amounted to regarding the list as a priority queue, where the priority is the value (key) of the list element. We now discuss a more efficient data structure, called a *heap*, for implementing a priority queue. The algorithm *HeapSort* given later in this section emulates *SelectionSort* except that the priority queue of list elements is maintained as a heap.

4.5.1 Heap Implementation of a Priority Queue

The underlying structure of a heap is a complete binary tree whose root contains the highest priority element.

DEFINITION 4.5.1 A *max-heap* (*min-heap*) H is a complete binary tree whose information fields contain values (keys) having the following *max-heap* (*min-heap*) *property*: Given any node v in H, the value of v is not smaller (not larger) than the value of any node in the subtree having v as a root.

For convenience of notation, throughout the remainder of this section we discuss *max-heaps* and sometimes simply use the word *heap* for *max-heap*. A completely analogous discussion can be given for *min-heaps*. (See Figure 4.8.)

Remark Clearly, the heap property can be equivalently expressed by requiring that the value of each node v is not smaller than the value of its left child or right child (if they exist). This allows us to localize the heap property to each node. Thus, we can talk about the heap property holding or not holding at a given node.

A very convenient implementation of a heap results from storing the n key values in an array $A[1{:}m]$, where $m \geq n$. The convenience of the array implementation results from the ease in which we can reference the children of a node, as well as its parent. Indeed, given a node of index i, its left child has index $2i$ and its right child has index $2i + 1$ (assuming that these children exist). Also, if $i > 1$, the parent of a node of index i is given by $\lfloor i/2 \rfloor$.

Figure 4.9 shows the max-heap given in Figure 4.8 stored in an array $A[1{:}m]$, $m \geq 10$.

Inserting into a Heap We now describe an algorithm for inserting a new element into an existing heap. We suppose that the existing heap occupies positions $A[1], \ldots, A[HeapSize]$ within the array $A[1{:}m]$. The idea is to initially place the new element in position $A[HeapSize+1]$ and let it "rise" along the unique path from

Figure 4.8 (a) Max-heap; (b) min-heap

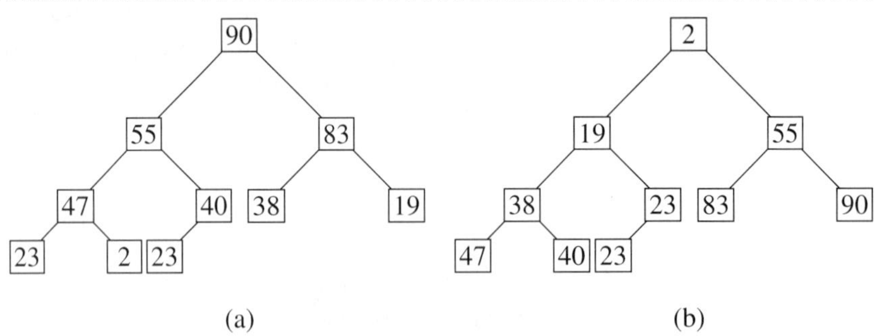

Figure 4.9 Max-heap stored in array $A[1{:}m]$, $m \geq 10$

A											
	Index	1	2	3	4	5	6	7	8	9	10
	Value	90	55	83	47	40	38	19	23	2	23

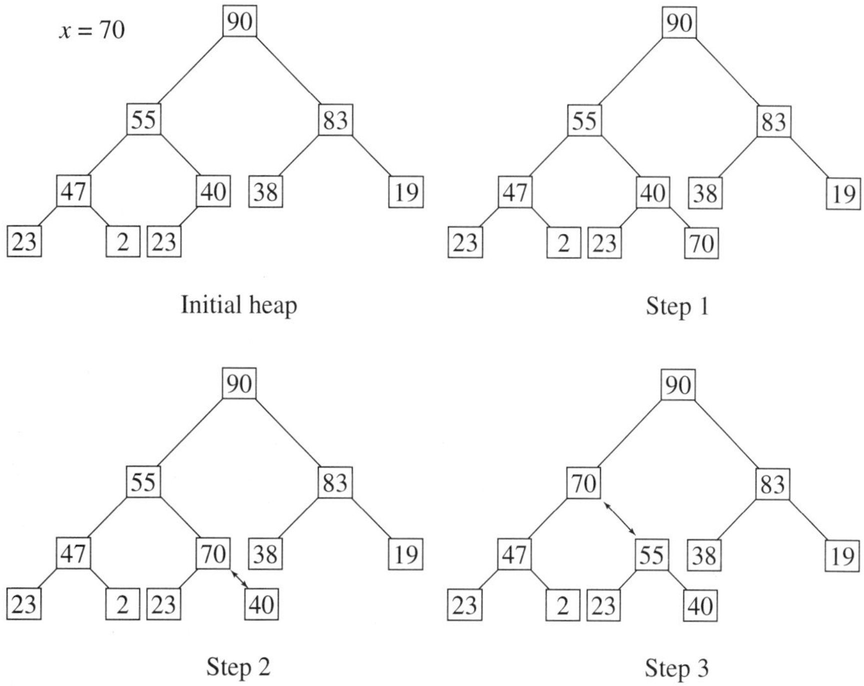

Figure 4.10 Inserting the value $x = 70$ into the existing heap $A[1:10]$

$A[HeapSize+1]$ to the root until it finds a proper position. Letting $i = HeapSize + 1$, we begin by comparing x to the value of the parent node $A[\lfloor i/2 \rfloor]$. If x has a larger value than its parent, we move the parent down to position i and move x up to the (old) position of its parent. The argument is then repeated until x has been moved up the path to a proper position. We illustrate this movement in Figure 4.10, where we insert the value $x = 70$ into the max-heap given in Figure 4.8.

The pseudocode for the insertion operation follows. Note that we do not need to make an assignment of x until a proper place for its insertion into the heap is found.

```
procedure InsertMaxHeap(A[1:m],HeapSize,x)
Input:   A[1:m] (an array containing a heap in positions 1, ... , HeapSize)
         HeapSize (the number of elements in the heap)
         x (a value to be inserted into heap)
Output:  A[1:m] (array altered by addition of x and heap maintained)
         HeapSize (input HeapSize + 1)
   HeapSize := HeapSize + 1
   i := HeapSize
   j := ⌊i/2⌋
   {traverse path to root until proper position for inserting x is found}    Continued
```

```
    while j ≥ 1 .and. A[j] < x do
        A[i] := A[j]            {move parent down one position in path}
        i := j                  {move x up one position in path}
        j := ⌊j/2⌋
    endwhile
    {i is now a proper position for x}
    A[i] := x
end InsertMaxHeap
```

Using comparisons to an element in the heap as our basic operation, the worst-case complexity for *InsertMaxHeap* occurs when the inserted element must rise all the way to the root. Thus, the worst-case complexity of *InsertMaxHeap* belongs to $\Theta(\log n)$.

Removing the Root from a Heap The other basic operation for a priority queue that we need to implement is to remove the highest priority element. For our heap implementation, this element is at the root. The removal of the root is accomplished by elevating the bottom heap element $x = A[HeapSize]$ to the root and then letting x "fall" down an appropriate path in the complete binary tree $A[1{:}HeapSize\text{-}1]$ until it finds a proper position. Initially we compare x to the larger of the two children $A[2]$ and $A[3]$. If x is smaller, the larger child is elevated to the root (made "king of the heap") and x moves down to the position formerly occupied by the elevated child. The argument is then repeated until a proper position for x is found. We illustrate this in Figure 4.11 by removing the root from the max-heap given in Figure 4.10.

The adjustment operation illustrated in steps 1 through 3 in Figure 4.11 can be applied to the subtree rooted at any node having index i in a complete binary tree $A[1{:}n]$, where the subtrees of $A[1{:}n]$ with roots at the two children $2i$ and $2i + 1$ are both heaps. The algorithm *AdjustMaxHeap* shows how to let $A[i]$ follow an appropriate path in the subtree with i as a root so as to make a heap out of this subtree.

```
procedure AdjustMaxHeap(A[1:m],n,i)
Input:  A[1:m] (an array)
        n (consider A[1:n] as complete binary tree, n ≤ m)
        i (index where subtrees of A[1:n] rooted at 2i and 2i + 1 are heaps)
Output: A[1:m] (subtree of A[1:n] rooted at A[i], adjusted so that it
               becomes a heap)
    Temp := A[i]
    {traverse down a path until a proper position for Temp = A[i] is found}
    Found := .false.    {Found signals when a proper position is found}
    j := 2*i            {j is the path finder. At completion of loop, ⌊j/2⌋
                         is a proper position for Temp = A[i]}
```

```
    while j ≤ n .and. .not. Found do
       if j < n .and. A[j] < A[j + 1] then    {then move to right child j + 1}
          j := j + 1              {path finder updated to right child}
       endif
       if Temp ≥ A[j] then
          Found := .true.
       else
          A[⌊j/2⌋] := A[j] {move larger child up one position in path}
          j := 2*j                {move path finder to next possible position for Temp}
       endif
    endwhile
    A[⌊j/2⌋] := Temp
end AdjustMaxHeap
```

Clearly, the worst-case complexity of *AdjustMaxHeap* occurs when the element $A[i]$ must fall all the way down to the bottom of the heap. Thus, when calling *AdjustMaxHeap* with the parameter i, in the worst case we make $\log_2 n - \log_2 i$ comparisons.

Figure 4.11 Removing the root from a heap

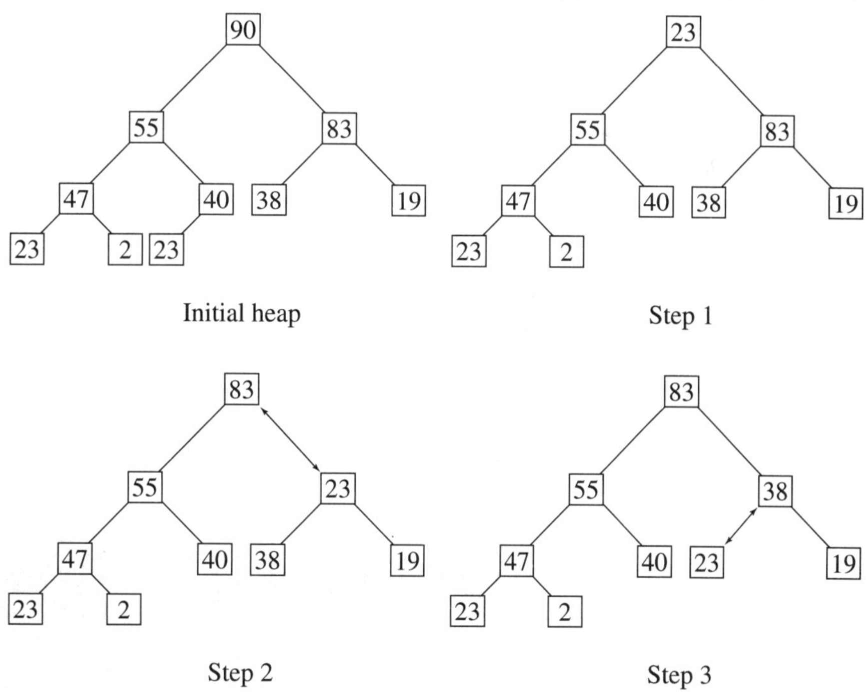

The operation of removing the root from a heap is accomplished by the following pseudocode.

```
procedure RemoveMaxHeap(A[1:m],HeapSize,x)
Input:  A[1:m] (an array containing a heap in positions 1, ... , HeapSize)
        HeapSize (the number of elements in the heap)
Output: A[1:m] (array altered by removal of the root A[1] and heap maintained)
        x (assigned the value of the (original) root A[1])
        HeapSize (input HeapSize - 1)

   x := A[1]
   A[1] := A[HeapSize]
   HeapSize := HeapSize – 1
   call AdjustMaxHeap(A[1:m],HeapSize,1)
end RemoveMaxHeap
```

The worst-case complexity of *RemoveMaxHeap* belongs to $\Theta(\log n)$, so that we have implemented both the insertion of an element and the removal of the highest priority element with worst-case complexity $\Theta(\log n)$.

Creating a Heap Given n values in an array $A[1{:}m]$, the question arises as to how to make $A[1{:}n]$ into a heap. The most obvious way is to simply sequentially invoke *InsertMaxHeap*($A[1{:}m]$,*HeapSize*,x) for $x = A[1], \ldots, A[n]$. In fact, this method is well suited to applications where the heap is maintained dynamically, with new elements being added to an already existing heap.

```
procedure MakeMaxHeap1(A[1:m],n)
Input:  A[1:m] (an array)
        n (values in A[1:n] considered as complete binary tree, n ≤ m)
Output: A[1:m] (A[1:n] made into a heap)

   HeapSize := 0
   for j := 1 to n
      call InsertMaxHeap(A[1:m],HeapSize,A[j])
   endfor
end MakeMaxHeap1
```

Figure 4.12 shows the action of *MakeMaxHeap*1 on the set of elements (10, 5, 20, 30, 25, 7, 40). We show the result after each iteration of the **for** loop.

Since the worst-case complexity of *InsertMaxHeap* belongs to $O(\log n)$, the worst-case complexity $W(n)$ of *MakeMaxHeap*1 belongs to $O(n\log n)$. The worst case occurs when the elements are in increasing order, so that each element must rise to the root when it is inserted. We now show that $W(n)$ belongs to $\Omega(n\log n)$.

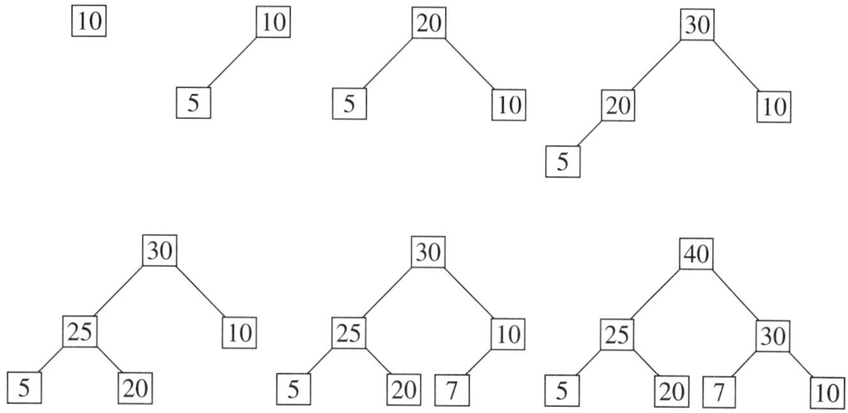

Figure 4.12 Action of *MakeMaxHeap*1 on the set of elements (10, 5, 20, 30, 25, 7, 40)

Let $k = \lfloor \log_2 n \rfloor$ and recall that a complete binary tree on n elements has 2^i nodes at level i for $i = 0, \ldots, k - 1$. Since we make i comparisons for each node at level i, and applying identity (1.2.11), we have

$$W(n) > \sum_{i=0}^{k-1} i2^i = 2^k(k-2) + 2 \in \Omega(n \log n).$$

Hence, $W(n)$ belongs to $\Omega(n \log n) \cap O(n \log n) = \Theta(n \log n)$.

An alternative algorithm *MakeMaxHeap*2 for creating a heap has worst-case complexity in $O(n)$. *MakeMaxHeap*2 works by repeated calls to *AdjustMaxHeap*($A[1{:}m],n,i$). Recall that *AdjustMaxHeap*($A[1{:}m],n,i$) requires that the subtrees in $A[1{:}n]$ with roots $2i$ and $2i + 1$ are already heaps. This will certainly be true if these latter two roots are leaf nodes of $A[1{:}n]$. Thus, we can make a heap by initially calling *AdjustMaxHeap*($A[1{:}m],n,i$) with $i = \lfloor n/2 \rfloor$ and then sequentially calling *AdjustMaxHeap*($A[1{:}m],n,i$) as i is decremented by one each time, until we finally reach the root.

```
procedure MakeMaxHeap2(A[1:m],n)
Input:   A[1:m] (an array)
         n (values in A[1:n] considered as complete binary tree, n ≤ m)
Output:  A[1:m] (A[1:n] made into a heap)
   for i := ⌊n/2⌋ down to 1 do
      call AdjustMaxHeap(A[1:m],n,i)
   endfor
end MakeMaxHeap2
```

The action of *MakeMaxHeap*2 is illustrated in Figure 4.13 on the set (10, 5, 20, 30, 25, 7, 40), which was the same set illustrated in Figure 4.12 for *MakeMaxHeap*1. Note that the heaps created are rather different, but both satisfy the heap property. In Figure 4.13 we show the array before and after each call to *AdjustMaxHeap*. We indicate in each "before" picture the particular subtree that is being made into a heap.

During the execution of the call to *AdjustMaxHeap*($A[1{:}m],n,i$), an element at level j will fall at most $k - j$ levels, where $k = \lfloor \log_2 n \rfloor$. Again, for $j < k$, there are exactly 2^j nodes at level j. We begin calling *AdjustMaxHeap*($A[1{:}m],n,i$) with $i = \lfloor n/2 \rfloor$, which is (the index of) a node at level $k - 1$. Hence, the worst-case complexity of *MakeMaxHeap*2 satisfies

$$W(n) \le \sum_{i=0}^{k-1} 2^i (k-i) = \sum_{i=1}^{k} i2^{k-i} = 2^k \sum_{i=1}^{k} \frac{i}{2^i}.$$

Applying identity (1.2.11) with $x = 1/2$, we obtain

$$W(n) \le 2^k \left[2 - \frac{k+2}{2^k}\right] = 2^{k+1} - (k+2) \le 2n \in O(n).$$

Since *MakeMaxHeap*2 does $\lfloor n/2 \rfloor$ comparisons in the best case (when we have a heap at the outset), we have shown that the best-case, worst-case, and average complexities of *MakeMaxHeap*2 all belong to $\Theta(n)$.

Figure 4.13 Making a max-heap using *MakeMaxHeap*2

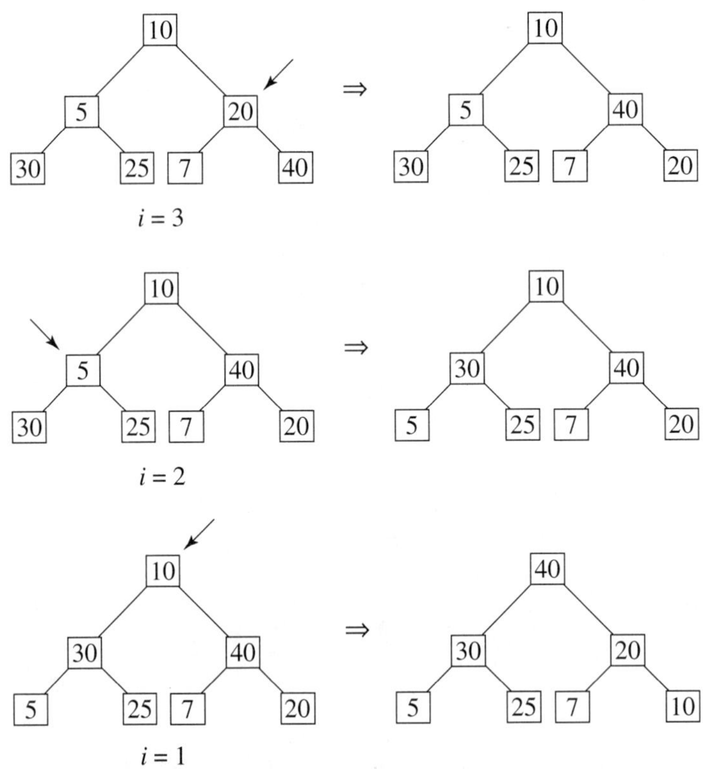

4.5.2 *HeapSort*

An efficient (in-place) sorting algorithm, *HeapSort* utilizes the algorithm *AdjustMaxHeap* as follows. We first invoke *MakeMaxHeap*2 to create the heap $A[1:n]$. Then we interchange the root with the element at the bottom of the heap (position n) and invoke *AdjustMaxHeap*($A[1:m],n-1,1$). The new root is now interchanged with the element at position $n-1$, and the process is continued. The code for *HeapSort* follows.

```
procedure HeapSort(A[1:n])
Input:  A[1:n] (an array)
Output: A[1:n] sorted in nondecreasing order
    call MakeMaxHeap2(A[1:n],n)
    for i := 1 to n – 1 do
        {interchange A[1] with A[n–i+1]}
        call interchange(A[1],A[n–i+1])
        call AdjustMaxHeap(A[1:n],n–i,1)
    endfor
end HeapSort
```

The call to *MakeMaxHeap*2 has best-case and worst-case complexities $\Theta(n)$, whereas the $n-1$ calls to *AdjustMaxHeap* each have $\Theta(1)$ best-case and $\Theta(\log n)$ worst-case complexities. Thus, *HeapSort* has $\Theta(n)$ best-case complexity and $\Theta(n\log n)$ worst-case complexity. *HeapSort* is a comparison-based sorting algorithm, so by lower bound theory (see Chapter 10), it has $\Omega(n\log n)$ average complexity, which implies that *HeapSort* has $\Theta(n\log n)$ average complexity.

4.6 *BingoSort*

In all of the algorithms we have studied so far, the input size was a function of a single variable n. However, as mentioned in Section 3.1, there are situations where the input size is most naturally a function of two or more variables. For example, suppose we consider the problem of finding efficient sorting algorithms for lists of size n with repeated elements. The input size is then a function of n and the number m of *distinct* elements in the list, where $1 \le m \le n$. This situation arises often in practice. For example, we might wish to sort the people in a large mailing list by zip codes for bulk mailing. Then the number of distinct elements m (zip codes) will be rather less than the number of people n. In this section we discuss *BingoSort*, which is an example of a sorting algorithm that works well for input lists with many repeated elements.

We express the complexity of *BingoSort* in terms of the two variables n and m. It turns out that for m small as compared to n, *BingoSort* has average and worst-case complexities in $\Theta(mn)$ and best-case complexity in $\Theta(n + m^2)$. Hence, for $m < \log n$, *BingoSort* performs better than comparison-based sorts such as *MergeSort* and *HeapSort*, which do not work any more efficiently on lists with repeated elements than they do on lists with distinct elements.

BingoSort, which is a variation of *SelectionSort*, works as follows. Each distinct value in the list is considered to be a "bingo" value. Each pass of *BingoSort* corresponds to "calling out" a bingo value. The bingo values are called out in increasing order. During a given pass, all the elements having the current bingo value are placed in their correct positions in the list.

The following pseudocode for *BingoSort* implements this process.

```
procedure BingoSort(L[1:n])
Input:  L[1:n] (an array)
Output: L[1:n] sorted in nondecreasing order
   {first determine the minimum and maximum value for a list element}
   call MaxMin3(L[1:n],MaxValue,MinValue)
   Bingo := MinValue {initialize}
   NextAvail := 1
   NextBingo := MaxValue
   while Bingo < MaxValue do {move up all list elements that equal Bingo}
      StartPos := NextAvail
      for i := StartPos to n do
         if L[i] = Bingo then {BINGO! Interchange L[i] and L[NextAvail]}
            call interchange(L[i],L[NextAvail])
            NextAvail := NextAvail + 1 {update NextAvail}
         else
            if L[i] < NextBingo then NextBingo := L[i] endif
         endif
      endfor
      Bingo := NextBingo {initialize for next pass}
      NextBingo := MaxValue
   endwhile
end BingoSort
```

The action of BingoSort is illustrated in Figure 4.14 for a sample list of size 10.

The best-case complexity $B(n,m)$ of *BingoSort* is achieved by a list where $n - m + 1$ elements have the (same) minimum value, yeilding $B(n,m) \in \Theta(n + m^2)$ (see Exercise 4.23). The worst case occurs for a list where $n - m + 1$ elements have the (same) maximum value, yeilding $W(n,m) \in \Theta(mn)$. When discussing the average complexity, it is natural to assume that each of the m distinct values occurs n/m times. Such a list generates $\Theta(mm)$ comparisons, so that the average complexity $A(n,m)$ of *BingoSort* belongs to $\Theta(mn)$ (see Exercise 4.24).

Pass 1	23	10	15	10	10	23	15	23	23	10
Start position 1	10	23	15	10	10	23	15	23	23	10
Bingo value 10	10	10	15	23	10	23	15	23	23	10
	10	10	10	23	15	23	15	23	23	10
	10	10	10	10	15	23	15	23	23	23
Pass 2										
Start position 5	10	10	10	10	15	23	15	23	23	23
Bingo value 15	10	10	10	10	15	15	23	23	23	23

Figure 4.14 Action of *BingoSort* for list 23,10,15,10,10,23,15,23,23,10 after call to *MaxMin3*

4.7 *RadixSort*

In this section we give an example of a sorting algorithm that is not a comparison-based algorithm, but rather depends on the special nature of the list elements. In *RadixSort*, we assume that the list elements are strings of a given fixed length drawn from a given (ordered) alphabet and that the ordering of the strings is the usual lexicographic ordering (that is commonly used in practice). For example, we might be sorting personnel records based on social security numbers. Then the list elements are character strings of length 9 drawn from the ten character digits 0, 1, ... , 9. Or, we might be sorting the records based on zip codes, so that the list elements are characters strings of length 5 again drawn from 0, 1, ... , 9.

One sorting method that is sometimes used in practice when the sorting is being done by hand is to make a pass through the list, placing the records in "buckets" based on the leading character in the string. For example, suppose we are sorting zip codes. We would use ten buckets $B_0, B_1, \ldots, B_9$. We would then make a linear scan through the list, inserting the record in B_i whenever the zip code has leading digit i, $i = 0, \ldots, 9$. The advantage of this method is that to complete the sort, we now just have to sort the records within each bucket. Sorting in each bucket could then proceed analogously, except we would place the records in sub-buckets based on the second leading digit. The disadvantage of this type of sort is the complication arising from the proliferation of buckets. However, in the case of binary strings, a fairly elegant recursive procedure similar to *QuickSort* can be developed (see Exercise 4.30).

The buckets don't proliferate if instead of looking first at the leading digit, we place the records in buckets based on the last (least significant) digit. We then repeat the process by extracting the list from the buckets in the order $B_0, B_1, \ldots, B_9$ and placing the records in these same ten buckets according to the second to the last digit. After five repetitions, the records will be sorted, as shown in Figure 4.15.

We now give a high-level description of *RadixSort* based on the procedure illustrated in Figure 4.15 (page 140). In our high-level description of *RadixSort*, we assume the built-in function *SubString*(*string*,*i*), which extracts the symbol in position *i* from *string*. We also assume high-level procedures *EnQueue*(*Q*,*X*), which enqueues the element *X* into the queue *Q*, and *DeQueue*(*Q*,*Y*), which dequeues *Q* and assigns the dequeued element to *Y*.

Original list:
45242 45230 45232 97432 74239 12335 43239 40122 98773 41123 61230

Pass 1 Place in buckets based on fifth (least significant) digit	
Bucket 0:	4523**0** 6123**0**
Bucket 2:	4524**2** 4523**2** 9743**2** 4102**2**
Bucket 3:	9877**3** 4112**3**
Bucket 5:	1233**5**
Bucket 9:	7423**9** 4323**9**

Extract list from buckets:
45230 61230 45242 45232 97432 40122 98773 41123 12335 74239 43239

Pass 2 Place in buckets based on fourth digit	
Bucket 2:	401**2**2 411**2**3
Bucket 3:	452**3**0 612**3**0 452**3**2 974**3**2 123**3**5 742**3**9 432**3**9
Bucket 4:	452**4**2
Bucket 7:	987**7**3

Extract list from buckets:
40122 41123 45230 61230 45232 97432 12335 74239 43239 45242 98773

Pass 3 Place in buckets based on third digit	
Bucket 1:	40**1**22 41**1**23
Bucket 2:	45**2**30 61**2**30 45**2**32 74**2**39 43**2**39 45**2**42
Bucket 3:	12**3**35
Bucket 4:	97**4**32
Bucket 7:	98**7**73

Extract list from buckets:
40122 41123 45230 61230 45232 74239 43239 45242 12335 97432 98773

Pass 4 Place in buckets based on second digit	
Bucket 0:	4**0**122
Bucket 1:	4**1**123 6**1**230
Bucket 2:	1**2**335
Bucket 3:	4**3**239
Bucket 4:	7**4**239
Bucket 5:	4**5**230 4**5**232 4**5**242
Bucket 7:	9**7**432
Bucket 8:	9**8**773

Extract list from buckets:
40122 41123 61230 12335 43239 74239 45230 45232 45242 97432 98773

Pass 5 Place in buckets based on first (most significant) digit	
Bucket 1:	**1**2335
Bucket 4:	**4**0122 **4**1123 **4**3239 **4**5230 **4**5232 **4**5242
Bucket 6:	**6**1230
Bucket 7:	**7**4239
Bucket 9:	**9**7432 **9**8773

Extract (sorted) list from buckets:
12335 40122 41123 43239 45230 45232 45242 61230 74239 97432 98773

Figure 4.15 Sorting a sample list of size $n = 11$, where each element is a five-digit numerical character string, so that there are ten buckets and five passes.

```
procedure RadixSort(L,n,k)
Input:  L (a list of size n,L_1,L_2, ... ,L_n, where L_i is a string of
                length k over an alphabet s_0 < s_1 < ··· < s_m)
        n (positive integer)
Output: L (sorted in nondecreasing order)
   for i := k downto 1 do
      {place elements in queues B_0, B_1, ... , B_m based on position i}
      for j := 1 to n do
         if SubString(L_j,i) = s_q then
            call EnQueue(B_q,L_j)
         endif
      endfor
      {reassemble list L by successive dequeueing}
      Ct := 1
      for j := 0 to m do
         while (.not. Empty(B_j)) do
            call DeQueue(B_j,Y)
            L_Ct := Y
            Ct := Ct + 1
         endwhile
      endfor
   endfor
end RadixSort
```

We leave the proof of the correctness of *RadixSort* as an exercise. Clearly, the (best-case, worst-case, and average) complexity of *RadixSort* is in $\Theta(kn)$. A convenient way to implement the queues used in *RadixSort* is to use linked lists. The action of the first pass of *RadixSort* is illustrated in Figure 4.16 (page 142) with this implementation. We assume that the queues are pointed to by the elements in an array *Bucket*[0:9] of pointer variables. We also assume that we have an array *Rear*[0:9] of pointers, where *Rear*[i] points to the end of the queue whose first element is pointed to by *Bucket*[i], $i = 0, \ldots, 9$.

The extracting of the list from the buckets is a simple job of reassigning some pointers, as illustrated in Figure 4.17 (page 143). Note that these pointer reassignments allow us to completely dequeue a given queue in one fell swoop! The reassembled list is a linked list pointed to by *Start*.

In general, *Start* points to the first nonempty queue, that is, *Start* = *Bucket*[i], where i is the smallest integer such that *Bucket*[i] $\neq$ nil. The reassembled linked list is then created by calling the following procedure *Reassemble*. For convenience, we write the pseudocode for *Reassemble* under the assumption that our alphabet is ten numeric characters.

Original list given in Figure 4.15:

45242 45230 45232 97432 74239 12335 43239 41022 98773 41123 61230

Bucket[0] → [45230 |] → [61230 | nil] ← *Rear*[0]

Bucket[1] = nil = *Rear*[1]

Bucket[2] → [45242 |] → [45232 |] → [97432 |] → [40122 | nil] ← *Rear*[2]

Bucket[3] → [98773 |] → [41123 | nil] ← *Rear*[3]

Bucket[4] = nil = *Rear*[4]

Bucket[5] → [12335 | nil] ← *Rear*[5]

Bucket[6] = nil = *Rear*[6]

Bucket[7] = nil = *Rear*[7]

Bucket[8] = nil = *Rear*[8]

Bucket[9] → [74239 |] → [43239 | nil] ← *Rear*[9]

Extract list from buckets:

List → [45230 |] → [61230 |] → [45242 |] → ··· [74239 |] → [43239 | nil]

Figure 4.16 First pass of *RadixSort* for the sample list given in Figure 4.15, where the queues (buckets) are implemented using linked lists

```
procedure Reassemble(Bucket[0:9],Rear[0:9],Start)
Input:  Bucket[0:9] (an array of pointers to queues)
        Rear[0:9] (an array of pointers to end of queues)
Output: Start (a pointer to the reassembled linked list L)
   Ct := 0
   while (Bucket[Ct] = nil) do
      Ct := Ct + 1
   endwhile
   Start := Bucket[Ct]
   SaveCt := Ct
   while (SaveCt < 9) do
      Ct := Ct + 1
      while (Bucket[Ct] = nil .and. Ct ≤ 9) do
         Ct := Ct + 1
      endwhile
      if Ct ≤ 9 then
```

```
      Rear[SaveCt] → Next := Bucket[Ct]
      SaveCt := Ct
    endif
  endwhile
end Reassemble
```

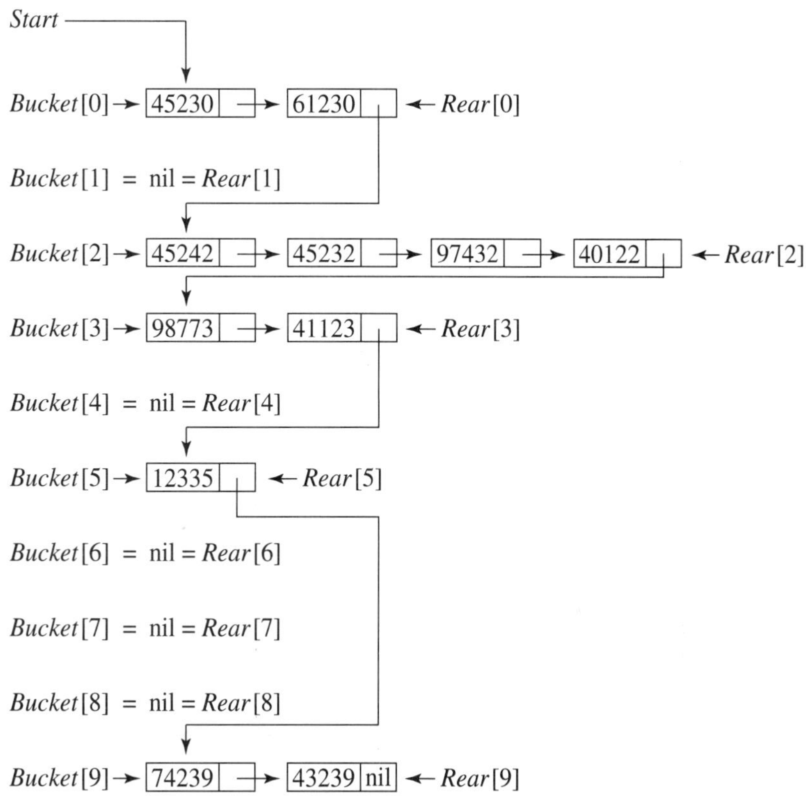

Figure 4.17 Extract list from buckets in the first pass of *RadixSort* for the sample list given in Figure 4.15, where the queues (buckets) and the reassembled list are implemented using linked lists

4.8 External Sorting

In all the sorting algorithms we have examined thus far, we made the assumption that the entire list fits into internal (main) memory. In practice, this may not be the case. Therefore, we now consider the problem of designing *external sorting* algorithms. External sorting algorithms assume that we can only fit a fixed number m of list elements into internal memory, where $m < n$. Thus, the list itself is originally assumed to reside in external memory, such as on a disk or tape.

Our external sorting algorithms assume that the input list originally resides in a single file on disk or tape. However, we allow the sorting algorithms the use of auxiliary external files. We assume that file access is sequential even when the file is on disk. The list elements are normally stored as records, with the sorting done on a given key field in each record. To access a given record in a sequential file, we must have read all the preceding records in the file. This means that if we have already passed a given record of a file and wish to access it again, we must reset the file to the beginning and sequentially read until the given record is located. In our external sorting algorithms, we want to minimize the number of times a file needs to be reset to the beginning. In particular, our external sorting algorithms always sequentially process all the records in a file before resetting the file. When designing a comparison-based external sorting algorithm, we would like to maintain order-optimal $\Theta(n\log n)$ worst-case complexity as measured using the number of comparisons made between keys. At the same time, we would like to minimize the number of times a record is accessed (read or written) in an external file.

One common approach to external sorting is based on a variation of *MergeSort*. To implement this variation, called *MultiMergeSort*, in the most straightforward fashion, we use four external files (either stored on disk or tapes). During the first phase, blocks of size m are brought from the input file into internal memory (the last block might be of size less than m). An efficient internal sorting algorithm is used to sort each block. The sorted blocks are written alternately to two external files (different from the input file). In the external files, a block of sorted records is called a *run*. Thus, after the initial sorting phase, two of the four files contain runs of size m. These two files are then the initial input files for phase 2.

Phase 2 consists of a number of merging passes. The merging of two sublists (runs) in different input files to an external output file is achieved using a slight modification of the algorithm *Merge* described in the Section 4.2. The modified *Merge* compares list elements by reading them from their resident external input files into internal memory. (During the merge, we have at most two elements at a time in internal memory.) The merged list is written one element at a time to a suitable external output file, with the smaller of the two list elements in internal memory written to the output file and another element read in from the appropriate input file.

More precisely, at the start of each merge pass, we have two input files containing runs of a certain size, and the other two files are reset to empty for output. We then merge runs from the input files into runs twice as long and write the longer runs alternately to the two output files. The input-output roles of the four files keep alternating for the different merge passes. The merge passes continue until one of the files contains a single run of size n (that is, contains the sorted list), as illustrated in Figure 4.18.

We now analyze the worst-case complexity $W(n)$ of *MultiMergeSort*. For convenience, we assume that n is a multiple of m. Let $W_1(n)$ and $W_2(n)$ denote the most number of comparisons performed in the first and second phases of *MultiMergeSort*, respectively, so that $W(n) = W_1(n) + W_2(n)$. In the first phase *MultiMergeSort*

Keys of original input list L

File A:
17 6 19 11 10 | 9 2 20 14 18 | 25 7 15 21 16 | 27 22 12 3 8 | 4 23 28 24 26 | 13 0 1 •

Phase 1

Blocks of size $m = 5$ (last block has size 3) are sequentially read from file A to internal memory, sorted using an efficient internal sorting algorithm, and alternately written to files C and D.

File C: 6 10 11 17 19 | 7 15 16 21 25 | 4 23 24 26 28 •
File D: 2 9 14 18 20 | 3 8 12 22 27 | 0 1 13 •

Phase 2–Merge Phase

Pass 1: Runs in input files C and D are merged and written to output files A and B

File A: 2 6 9 10 11 14 17 18 19 20 | 0 1 4 13 23 24 26 28 •
File B: 3 7 8 12 15 16 21 22 25 27 •

Pass 2: Runs in input files A and B are merged and written to output files C and D

File C: 2 3 6 7 8 9 10 11 12 14 15 16 17 18 19 20 21 22 25 27 •
File D: 0 1 4 13 23 24 26 28 •

Pass 3: Runs in input files C and D are merged and written to output file A, which now contains the sorted list L.

File A:
0 1 2 3 4 6 7 8 9 10 11 12 13 14 15 16 17 18 19 20 21 22 23 24 25 26 27 28 •

Figure 4.18 Action of external sorting algorithm *MultiMergeSort* for a sample list L of size $n = 29$, where internal memory has size $m = 5$

sequentially performs internal sorts on n/m lists, each of size m. Using an efficient internal sorting algorithm, each of these lists can be sorted using $\Theta(m\log m)$ comparisons. Thus,

$$W_1(n) \in \Theta\left(\frac{n}{m}(m \log m)\right) = \Theta(n \log m). \tag{4.8.1}$$

Let p denote the number of passes performed during the second (merge) phase of *MultiMergeSort*. Since the initial runs all have size m and the size of the runs is doubled after each pass until a run of size n is obtained, we have $n = 2^p m$ (where for convenience we have assumed n/m is a power of 2). Hence,

$$p = \log_2\left(\frac{n}{m}\right). \tag{4.8.2}$$

Since *Merge* performs at least k and less than $2k$ comparisons to merge two sublists of size k, it follows that the number of comparisons performed during each pass in the merge phase is at least $n/2$ and less than n. Thus, the total number of comparisons performed during the merge phase is at most $pn/2$ and less than pn, so that

$$\left(\frac{n}{2}\right)\log_2\left(\frac{n}{m}\right) \le W_2(n) < n \log_2\left(\frac{n}{m}\right). \tag{4.8.3}$$

Combining (4.8.1) and (4.8.3), we obtain

$$W(n) = W_1(n) + W_2(n) \in \Theta\left(n \log m + n \log\left(\frac{n}{m}\right)\right) = \Theta(n \log n). \qquad (4.8.4)$$

Thus, *MultiMergeSort* has order-optimal worst-case complexity.

We now count the number of times each element of the list L is accessed from an external file. Note that each element is written once to an external file for each time it is read from an external file. Clearly, each element is read from an external file once in phase 1 and p times in phase 2. Hence, the total number of times each element is read from an external file is given by $1 + p = 1 + \log_2(n/m) \in O(\log n)$. Note that this is also the number of times a given file needs to be reset for reading (or writing).

A more sophisticated version of *MultiMergeSort* can be designed where only three external files are used (see Exercise 4.31).

Note that the number of merge passes performed by *MultiMergeSort* in the second phase would be reduced if the initial runs constructed in the first phase were longer than m. We now describe an algorithm *MultiMergeSortQ*, which uses a priority queue maintained in internal memory with the aim of achieving this goal. The priority queue consists of list elements, which have been tagged with a run number in a manner to be described shortly. When comparing two list elements, the list element with the lower run number has the higher priority. If two list elements have the same run number, then the list element with the smaller key has the higher priority.

MultiMergeSortQ begins by reading in the first m list elements (records) from the external input file. These m elements are all given run number 1, since they are all part of the first run (written to the first external output file). A priority queue is constructed out of these m elements. *MultiMergeSortQ* then begins creating runs by an alternating sequence of dequeueing and enqueueing until the input file is exhausted. Elements are enqueued from the input file after tagging on a run number computed as follows. Let x denote the current element to be dequeued and written to the output file as part of run i. Let y denote the element to be read from the input file, which will be enqueued after x has been dequeued. If the key of y is less than the key of x, then y cannot be part of the current run, so its run number is assigned the value $i + 1$. Otherwise, y can be included as part of the current run, and its run number is assigned the value i.

An internal variable *Run* keeps track of the number of the current run being written to an external output file. Before x is written, its run number is compared to *Run*. If the run number of x is (one) greater than *Run*, then we start a new run on the other output file and increment *Run* by one. When the external input file is exhausted, we simply keep dequeueing x in the same fashion until the queue is empty. In Figure 4.19, we illustrate the initial run creating phase of *MultiMergeSortQ* with the same list as used in Figure 4.18. In Figure 4.19, we show the priority of a list element by the pair (run number,key). We show the contents of the priority queue at each stage, but do not show the data structure (such as a heap) that we use to efficiently implement the priority queue.

Original External Input List

17 6 19 11 10 9 2 20 14 18 25 7 15 21 16 27 22 12 3 8 4 23 28 24 26 13 0 1 •

Initial priority queue: (1,17) (1,6) (1,19) (1,11) (1,10)

File *C*	priority queue
6	(1,17) (1,9) (1,19) (1,11) (1,10)
6 9	(1,17) (2,2) (1,19) (1,11) (1,10)
6 9 10	(1,17) (2,2) (1,19) (1,11) (1,20)
6 9 10 11	(1,17) (2,2) (1,19) (1,14) (1,20)
6 9 10 11 14	(1,17) (2,2) (1,19) (1,18) (1,20)
6 9 10 11 14 17	(1,25) (2,2) (1,19) (1,18) (1,20)
6 9 10 11 14 17 18	(1,25) (2,2) (1,19) (2,7) (1,20)
6 9 10 11 14 17 18 19	(1,25) (2,2) (2,15) (2,7) (1,20)
6 9 10 11 14 17 18 19 20	(1,25) (2,2) (2,15) (2,7) (1,21)
6 9 10 11 14 17 18 19 20 21	(1,25) (2,2) (2,15) (2,7) (2,16)
6 9 10 11 14 17 18 19 20 21 25	(1,27) (2,2) (2,15) (2,7) (2,16)
6 9 10 11 14 17 18 19 20 21 25 27	(2,22) (2,2) (2,15) (2,7) (2,16)
File *D*	
2	(2,22) (2,12) (2,15) (2,7) (2,16)
2 7	(2,22) (2,12) (2,15) (3,3) (2,16)
2 7 12	(2,22) (3,8) (2,15) (3,3) (2,16)
2 7 12 15	(2,22) (3,8) (3,4) (3,3) (2,16)
2 7 12 15 16	(2,22) (3,8) (3,4) (3,3) (2,23)
2 7 12 15 16 22	(2,28) (3,8) (3,4) (3,3) (2,23)
2 7 12 15 16 22 23	(2,28) (3,8) (3,4) (3,3) (2,24)
2 7 12 15 16 22 23 24	(2,28) (3,8) (3,4) (3,3) (3,26)
2 7 12 15 16 22 23 24 28	(3,13) (3,8) (3,4) (3,3) (3,26)
File *C*	
6 9 10 11 14 17 18 19 20 21 25 27 3	(3,13) (3,8) (3,4) (4,0) (3,26)
6 9 10 11 14 17 18 19 20 21 25 27 3 4	(3,13) (3,8) (4,1) (4,0) (3,26)
6 9 10 11 14 17 18 19 20 21 25 27 3 4 8	(3,13) (4,1) (4,0) (3,26)
6 9 10 11 14 17 18 19 20 21 25 27 3 4 8 13	(4,1) (4,0) (3,26)
6 9 10 11 14 17 18 19 20 21 25 27 3 4 8 13 26	(4,1) (4,0)
File *D*	
2 7 12 15 16 22 23 24 28 0	(4,1)
2 7 12 15 16 22 23 24 28 0 1	∅

Figure 4.19 Action of the run creation stage of *MultiMergeSortQ*

Note that the first run constructed by *MultiMergeSortQ* for the sample input list *L* of Figure 4.19 has length 12. This first run is more than twice as long as the first run (of length 5) constructed by *MultiMergeSort* for the same input list *L*. Also note that *MultiMergeSortQ* constructs a total of four runs in the first phase, as opposed to the six runs constructed by *MultiMergeSort*. Thus, in the second (merge) phase *MultiMergeSortQ* performs only two passes, as opposed to the three passes performed by *MultiMergeSort*.

We leave the pseudocode of *MultiMergeSort* and *MultiMergeSortQ* as exercises.

4.9 Closing Remarks

In this chapter we have discussed a number of well-known sequential internal and external sorting algorithms, including some whose worst-case and average complexities are in $\Theta(n\log n)$. In Chapter 10 we will establish lower bounds belonging to $\Omega(n\log n)$ for both worst-case and average complexities of any comparison-based sorting algorithm. It follows that *MergeSort* and *HeapSort* exhibit order-optimal worst-case and average complexities. *QuickSort* has $\Theta(n^2)$ worst-case complexity, but it has order-optimal average complexity.

In the next chapter we will introduce parallel algorithms and architectures, and in Chapter 6 we will discuss some parallel sorting algorithms. One strategy for designing a parallel sorting algorithm is to look for inherent parallelism in a sequential algorithm. For example, in Chapter 6 we discuss two classical parallel sorting algorithms that exploit the inherent parallelism in *MergeSort*.

Mergesort was already in use in the earliest electronic computers. In fact, it was one of the stored programs implemented by von Neumann in 1945.

References and Suggestions for Further Reading

Floyd, R. W. "Algorithm 245: Treesort 3," *Communications of the ACM*, 7 (Dec. 1964): 701. An improvement on Williams' algorithm.

Hoare, C. A. R. "Quicksort," *The Computer Journal* (1962): 10–15. The original work on *QuickSort*.

Knuth, D. E. *The Art of Computer Programming, Volume III: Sorting and Searching*. Reading MA: Addison-Wesley, 1973. A comprehensive treatment of sorting algorithms and their history.

Shell, Donald L. "A high-speed sorting procedure," *Communications of the ACM*, vol. 2, no. 7 (July 1959): 30–32. The original paper on *ShellSort*.

Williams, J. W. J. "Algorithm 232: Heapsort," *Communications of the ACM*,7 (June 1964): 347–348. The original work on *HeapSort*.

Exercises

Section 4.1 *ShellSort*

4.1 **a.** Demonstrate the action of *ShellSort* on the list 33,2,56,23,55,78,2,98,61,108,14, 60,56,77,5,3,1 with increments 5,3,1. (See Figure 4.2.)

b. Repeat (a) with increments 4,2,1.

***4.2** Prove Theorem 4.1.1.

4.3 Empirically test the performance of *ShellSort* for various sets of diminishing increments.

Section 4.2 *MergeSort*

4.4 Give the appropriately labeled tree of recursive calls to *MergeSort* for lists of size 18. (See Figure 4.3.)

4.5 Show that the tree of recursive calls for *MergeSort* has depth $\lceil \log_2 n \rceil$. Conclude that *MergeSort* performs at most $n \lceil \log_2 n \rceil$ comparisons for any list of size n, so that $W(n) \in O(n \log n)$.

4.6 **a.** Obtain a recurrence relation for the worst-case complexity $W(n)$ of *MergeSort* when n is a power of two. Solve the recurrence relation by the method of repeated substitution. (*Hint:* See Section 1.3.3 for an example.)

b. Repeat part (a) for the best-case complexity $B(n)$ of *MergeSort.*

4.7 Design a nonrecursive version of *MergeSort* based on Figure 4.4.

4.8 Given a list $L[1{:}n]$, one way of maintaining a sorted order of L is to utilize an auxiliary array $Link[1{:}n]$. The array $Link[1{:}n]$ serves as a linked list determining the next highest element in L, so that the elements of L can be given in nondecreasing order by

$$L[Start], \quad L[Link[Start]], \quad L[Link[Link[Start]]], \ldots .$$

Then, $Link^{n-1}[Start]$ is the index of the largest element in L, and we set $Link[Link^{n-1}[Start]] = Link^{n}[Start] = 0$ to signal the end of the linked list. Design a version of *MergeSort* that utilizes the auxiliary array *Link.*

***4.9** Develop an in-place version of *MergeSort*; that is, a version that does not use an auxiliary array and utilizes only a few additional bookkeeping variables.

Section 4.3 *QuickSort*

4.10 **a.** Give pseudocode for *QuickSort* that incorporates the Median of Three Rule.

b. Show that the worst-case performance of the version of *QuickSort* in part (a) remains quadratic.

4.11 It is useful to analyze how much stacking is generated by the recursion in *QuickSort.*

a. Show that in the worst case (the largest number of unresolved recursive calls active), the size of the stack belongs to $\Omega(n)$.

b. Describe a modification of *QuickSort* for which the size of the stack belongs to $O(\log n)$.

4.12 Give a nonrecursive version of *QuickSort* (utilizing explicit stack operations).

4.13 Give a stable version of *QuickSort* that avoids making extraneous interchanges of elements having identical values.

Section 4.4 *TreeSort*

4.14 Prove that *TreeSort* is a stable sorting algorithm.

4.15 Given a permutation π of a set of keys, the algorithm *BinSearchTreeCreate* creates a binary search tree T_π. Given T_π, show that π can be uniquely determined if, and only if, T_π is a path.

4.16 Show that the best-case and worst-case complexities of *BinSrchTreeCreate* for a list of size n belong to $\Theta(n \log n)$, $\Theta(n^2)$, respectively.

4.17 **a.** Design a sorting algorithm based on a multiway search tree that generalizes *TreeSort*. (*Hint:* Use a multivisit inorder tree traversal.)

b. Discuss how to guarantee that your sorting algorithm in (a) is stable.

Section 4.5 Priority Queues, Heaps, and *HeapSort*

4.18 Demonstrate the action of *MakeMaxHeap*1 on the set of elements

23, 108, 9, 3, 2, 73, 41, 52, 34.

(See Figure 4.12.)

4.19 Demonstrate the action of *MakeMaxHeap*2 (see Figure 4.13) on the same set of elements used in Exercise 4.18.

4.20 Another priority queue operation that is sometimes useful is *ChangePriority*(Q,x,v), which changes the priority value of an element x in the queue Q to v (and maintains a priority queue). Give pseudocode for *ChangePriority* when the priority queue is implemented as a min-heap. *ChangePriority* should have worst-case complexity $O(\log n)$, where n is the number of elements in the min-heap.

4.21 The concept of a complete binary tree extends to the concept of a complete k-ary tree in the obvious way, $k \geq 2$.

a. Show how a complete k-ary tree can be implemented efficiently using an array.

b. Design and analyze a procedure for inserting an element into a k-ary heap implemented using an array.

c. Design and analyze a procedure for deleting an element from a k-ary heap implemented using an array.

d. Design and analyze a k-ary heap sort.

e. Compare the result in (d) to that for an ordinary heap sort ($k = 2$).

Section 4.6 *BingoSort*

4.22 Show that if a list has m distinct elements, then after exactly $m - 1$ passes of *BingoSort* the list is sorted into increasing order and the algorithm terminates.

4.23 **a.** Analyze the best-case complexity $B(n,m)$ of *BingoSort*.

b. Analyze the worst-case complexity $W(n,m)$ of *BingoSort*.

4.24 Show that *BingoSort* performs $\Theta(nm)$ comparisons for an input list of size n having m distinct values, each of which occurs n/m times.

Section 4.7 *RadixSort*

4.25 Demonstrate the action of *RadixSort* for the following sample list, where each element is a 7-digit binary string (see Figure 4.15).

1011101, 0100011, 1010110, 1111101, 0011101, 0000001, 1000000, 1010101

4.26 Demonstrate the first two passes of *RadixSort* for the sample list given in Exercise 4.25, where the buckets are implemented as linked queues (see Figures 4.16 and 4.17).

4.27 Prove the correctness of *RadixSort*.

4.28 Determine whether *RadixSort* is a stable sorting algorithm.

4.29 For binary strings stored in an array, design and analyze a recursive radix sort based on partitioning the list into two sublists according to the most significant digit.

Section 4.8 External Sorting

4.30 **a.** Give pseudocode for *MultiMergeSort*.

b. Give pseudocode for *MultiMergeSortQ*.

4.31 Design a version of *MultiMergeSort* where only three external files are used.

5

Introduction to Parallel Algorithms and Architectures

At current performance levels, our ability to model many important science, engineering and economic problems is still limited .... Parallel computing is the principal source of opportunity to improve computational performance.

—The Federal Performance Computing Program (9/8/89)
Executive Office of the President of the United States
Office of Science and Technology Policy

The above quotation reflects a widely held belief among the scientific community that the supercomputer of the twenty-first century is the massively parallel machine. There are good reasons to look to parallel computing to achieve major improvements in computational performance. The laws of physics place ultimate limitations on the computational speed of individual processors. Thus, the only hope for solving a problem requiring speeds exceeding these limitations is to utilize many processors working on the problem in parallel.

Many problems are most naturally modeled with parallelism. For example, in computational fluid dynamics, weather prediction, and image processing, sampling data uses sensors (real or simulated) placed at a spatial grid of points. Clearly, it would be useful to have a different processor assigned to each sensor in the grid to input and process the data from the sensor and distribute results to other processors. Also, problems in engineering and the physical sciences often involve vector operations. With a processor assigned to each component in the vector, vector operations can take place in each component simultaneously (see Figure 5.1).

Figure 5.1 (a) (Column) vector addition on a serial machine, one row after another; (b) (column) vector addition on a parallel machine, all rows done simultaneously

5.1 Approaches to the Design of Parallel Algorithms

Given the existence of massively parallel machines, a major challenge facing computer scientists today is to design algorithms that exploit the massively parallel paradigm. There are three main approaches to designing parallel algorithms.

1. Modify an existing sequential algorithm exploiting those parts of the algorithm that are naturally parallelizable.
2. Design a completely new parallel algorithm that may have no natural sequential analog.
3. For appropriate problems such as root finding, run the same sequential algorithm on many processors with different seeds until one processor reports "success." That is, all the processors start running a sequential algorithm with different initial conditions, and the first processor to achieve the desired result "wins the race."

Throughout the text we will examine all three strategies. All approaches must take the particular architecture of the parallel machine into consideration.

5.2 Architecture Constraints When Designing a Parallel Algorithm

A number of constraints arise when designing a parallel algorithm that are not present when designing a sequential algorithm. These constraints are imposed by the architecture of the particular parallel computer on which the algorithm is intended to be implemented. The following is a list of the major constraints that must be taken into account.

1. single instruction verses multiple instruction
2. the number and type of processors available
3. shared memory (PRAM) model versus distributed memory (interconnection network) model
4. communication constraints
   a. read and write restrictions for the PRAM model
   b. specification of the graph representing the direct connections that exist between processors in the interconnection network model
5. I/O constraints

We briefly elaborate on each of these constraints to give some notion of the underlying ideas.

5.2.1 Single Instruction versus Multiple Instruction

A single-processor computer can only execute one instruction at a time. However, a parallel computer with p processors $P_1, \ldots, P_p$ can execute p instructions simultaneously, each processor operating on (possibly) different data. One model of parallel computation restricts the processors to each perform the *same* instruction (but with possibly different data) at the same time in a synchronized manner. This common instruction also contains information that can instruct a given processor to remain idle (masked out) during a given step. The operations of each processor are controlled by a front end processor (*central control*) and global clock. During each time interval of the global clock, all the processors synchronously perform the same operation (input or output data, perform computations on data, read from local memories, communicate between processors, and so forth). Since the operations pulse through the system in regular clock intervals, the model is often called *systolic* computation. Parallel computers following this model are called SIMD (single-instruction, multiple-data) machines. Parallel computers that allow *different* instructions to be performed at the same time (on possibly different data) are called MIMD (multiple-instruction, multiple-data) machines. The single-processor computer can be viewed in this terminology as a SISD (single-instruction, single-data) machine. Since the SIMD model is simpler conceptually and easier to implement, we usually assume the SIMD model.

5.2.2 The Number and Type of Processors Available

In practice, computer manufacturers must consider the question of whether to build a parallel computer using, say, tens or hundreds of powerful processors (so-called coarse grain computers) or to build a parallel computer using thousands and thousands of relatively simple processors (fine grain computers). In this text we assume that the processors are powerful enough to execute all the normal instructions of a serial computer.

There are two approaches to designing parallel algorithms with respect to the number of processors available. The first approach is to design an algorithm where the number p of processors utilized by the algorithm is an input parameter. In this approach, p does not depend on the input size n. The second approach, which we adapt in this text, is to allow the number of processors used by the parallel algorithm to grow with the size of the input. Thus, p is not an input parameter to the algorithm, but is a function $p(n)$ of the input size n. By using a *division-of-labor technique* an algorithm designed following the second approach can always be converted into an algorithm suitable for a fixed number of processors. For example, suppose an algorithm has been designed that utilizes $p(n) = n^2$ processors. Then for each parallel step of the original algorithm, in the converted algorithm the work is divided among the p processors into at most n^2/p sequential steps done by each processor.

5.2.3 Shared Memory: PRAMs

In a shared-memory model of a parallel computer, *all* processors have equal access to a single *shared* RAM memory (see Figure 5.2a, page 156). Parallel computers based on this model are referred to as parallel random access machines, or PRAMs, since they are based on parallel random access memory.

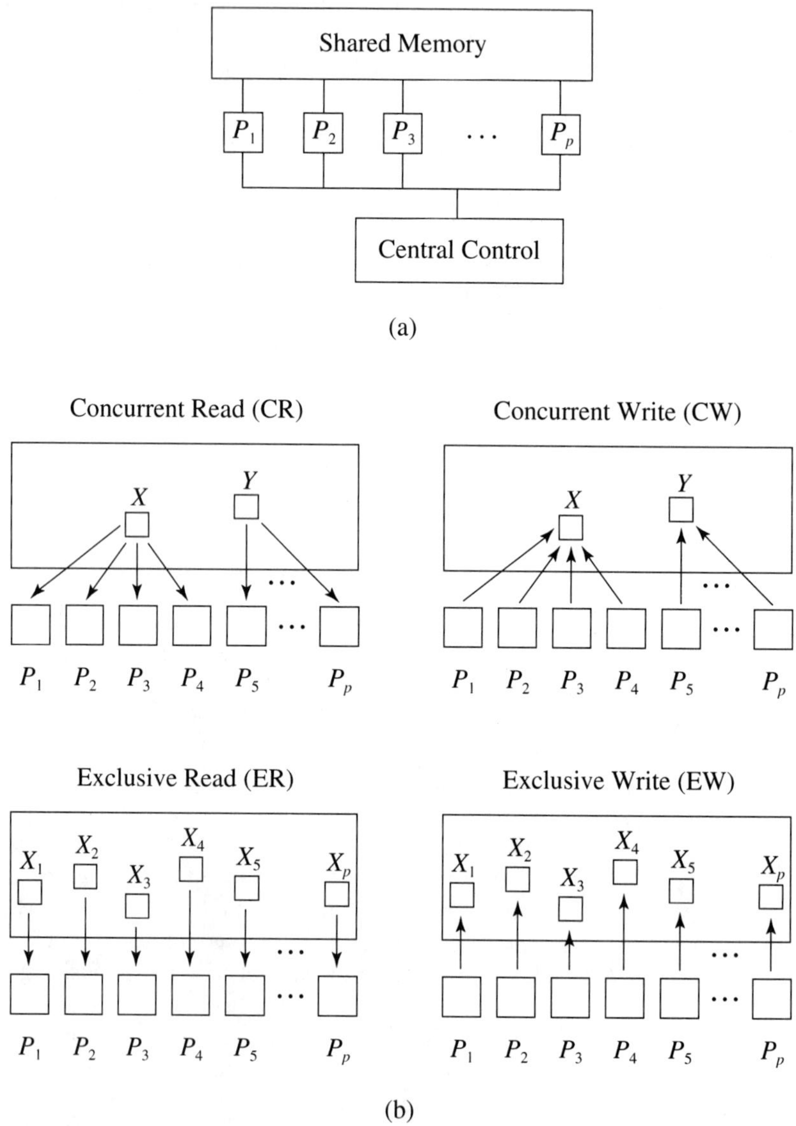

Figure 5.2 (a) A PRAM; (b) read/write possibilities for PRAMs

In the PRAM, each processor has equal access to every memory location in the shared memory. Of course, the possibility of conflict arises if two processors attempt to read or write from the same memory location simultaneously. There are four models (see Figure 5.2b) for dealing with such conflict, EREW (exclusive-read, exclusive-write), CREW (concurrent-read, exclusive-write), ERCW (exclusive-read, concurrent-write), and CRCW (concurrent-read, concurrent-write). The ERCW model is of little theoretical or practical interest and will not be mentioned again.

Algorithms for EREW PRAMs are assumed to be in error if a read or write conflict ever arises. CRCW PRAMs allow multiple processors to read from and write to the same memory location concurrently. Resolution of concurrent writes is done in various ways. A commonly used method only allows concurrent writes when all the processors are attempting to write the same value. Another method, used for numeric data, is to write the sum of all these values. Still other methods involve allowing a randomly chosen processor among contending processors to write its value, or always choosing the processor having smallest index in some ordering of the processors.

The EREW model is the most realistic of the PRAM models to build in practice. Moreover, any algorithm designed for the EREW PRAM also runs without alteration on the other PRAM models. For this reason, when naming an algorithm designed for an EREW PRAM, we simply use the term *PRAM* in the name. Names for algorithms on the other models will explicitly mention the model, such as *CRCW* or *CREW*.

Given the current state of technology, even EREW PRAMs are difficult to build (although there are such efforts currently underway). Nevertheless, PRAMs provide good models for theoretical results and for the initial design of a parallel algorithm unencumbered by processor communication details.

Later in the chapter we will develop pseudocode conventions for parallel algorithms. Now we illustrate some of the main design issues with an informal discussion of a parallel algorithm for searching a list $L[1{:}n]$ of size n for the occurrence of a search element X.

5.2.4 Example: Searching on a PRAM

To keep things as simple as possible, we assume that we have at least n processors $P_1, P_2, \ldots, P_n$ available. A parallel search algorithm comes immediately to mind: In a single parallel step processor P_i compares X to $L[i]$, $1 \le i \le n$. So, what constraints have we assumed?

First, we assumed that each processor accessed (read) the memory location containing X simultaneously. Thus, we have assumed that our PRAM model allows for *concurrent reads* (CR).

Second, we assumed that a successful search can be signaled in a single step. If $L[i] = X$, then we simply have P_i write i to a memory location *Index*. No problem arises if X occurs at most once in the list $L[1{:}n]$. However, if X occurs several times in the list, several processors will attempt to write (different values) to *Index* simultaneously. Thus, we need to assume that our PRAM model allows for *concurrent writes* (CW). Moreover, we must assume a method of resolving the contention between concurrent writes, such as choosing the processor having the smallest index to be the winner. Thus, our solution assumed the CRCW PRAM.

Now suppose we have (the more realistic) EREW PRAM. For each processor to have access to the value of X simultaneously on the EREW PRAM, we need to allocate an auxiliary array *Temp*[1:n] and assign the value of X to each array element *Temp*[i], $1 \leq i \leq n$. Assigning X to each entry in *Temp*[1:n] can be achieved by assigning X to *Temp*[1] and then *broadcasting* X to the other positions in the array as follows. For simplicity, we assume that $n = 2^k$ for some nonnegative integer k. In the first step, P_1 reads X and writes it to *Temp*[1]. In the second step, P_1 reads *Temp*[1] and writes it to *Temp*[2]. In the third step, processors P_1 and P_2 read *Temp*[1] and *Temp*[2], respectively, and write X to *Temp*[3] and *Temp*[4], respectively. The broadcasting process is illustrated in Figure 5.3 with $X = 5$ and $n = 16$. Clearly, after $\log_2 n$ steps, X is stored in all n memory locations of *Temp*[1:n]. Figure 5.3 illustrates the fact that even though the SIMD model restricts the processors to performing the same operation, some of the processors are allowed to remain idle in any given parallel step.

After X has been broadcast throughout *Temp*[1:n], then, in parallel, processor P_i compares $L[i]$ to *Temp*[i] = X and writes i in *Temp*[i] if $L[i] = X$; otherwise it writes the value ∞ in *Temp*[i] (see Figure 5.4). Here, ∞ is a constant equal to the largest value storable in an integer variable and can be computed concurrently by each processor, without the need to access a common memory location.

The array *Temp*[1:n] now contains the results of the search. However, we are still left with the problem of signaling successful search. Note that the value we wish to return is nothing more than the minimum entry in the array *Temp*. Using

Figure 5.3 Broadcasting the value $X = 5$ in $\log_2 n$ steps

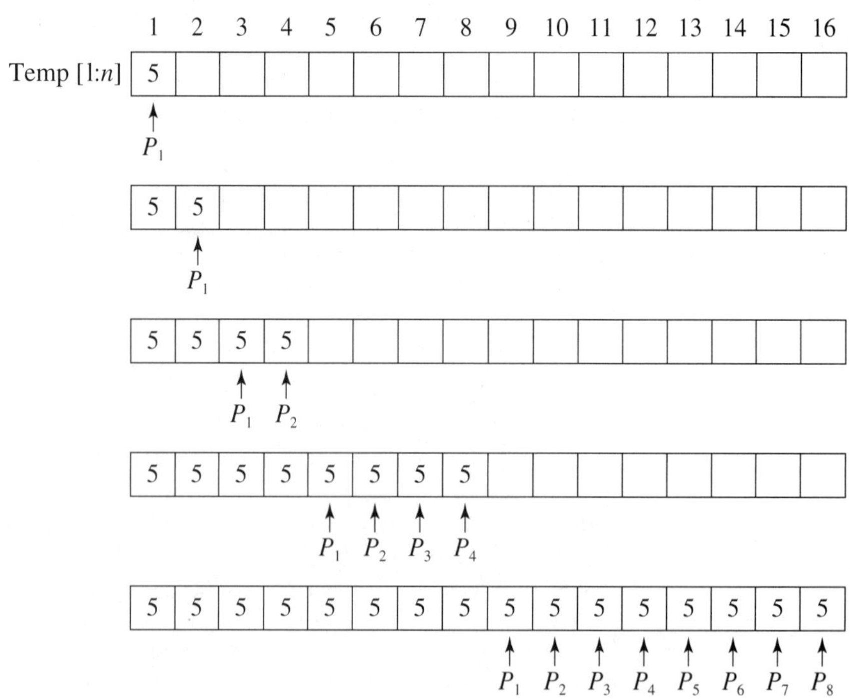

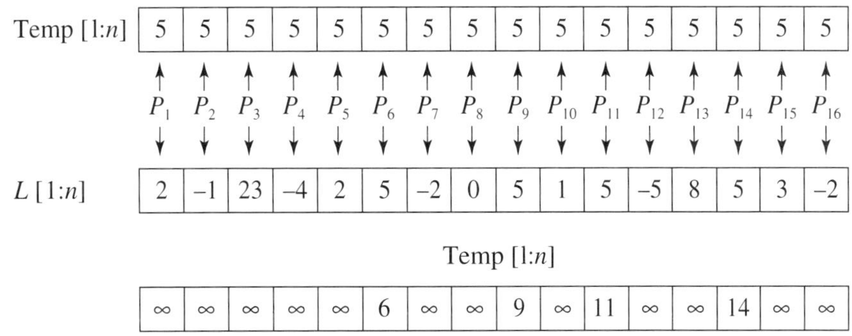

Figure 5.4 A single parallel comparison step between search elements and list elements

a method known as *binary fan-in* (see Figure 5.5), we obtain a straightforward parallel algorithm for determining the minimum of a set of n numbers $x_1, x_2, \ldots, x_n$ on the EREW PRAM with $n/2$ processors. For example, suppose we wish to find the minimum of sixteen numbers using eight processors. In the first parallel step, we divide these numbers into eight pairs and assign these pairs to eight processors for simultaneous computation of their minimums. In the next parallel step, we divide the eight minimums obtained in the first step into four pairs, assign them to four processors and simultaneously compute their minimums. This process is continued as illustrated in Figure 5.5 (page 160). In the last step, one processor contains the desired minimum of all sixteen numbers. Thus, we have reduced the fifteen sequential steps required by a single-processor machine to only four parallel steps, thereby achieving a speedup of 15/4 over the sequential algorithm.

The binary fan-in technique requires $\log_2 n$ parallel steps to find the minimum of an n-element list. The binary fan-in technique is a powerful tool in parallel algorithms and can be used for combining n elements with respect to any associative binary operation such as addition, multiplication, and so forth.

The basic operation for a sequential search algorithm was comparison of a list element to the search element. In the present situation, all such comparisons are made in a single parallel step! Hence, we must choose another basic operation to make a meaningful statement concerning the complexity (number of parallel basic operations) of our search algorithm. We could use either the number of parallel assignments performed when broadcasting the search element or the number of parallel comparison steps in computing the index of a minimum value in *Temp*[1:n] in the final phase. Either choice yields a complexity of $\log_2 n$.

5.2.5 Distributed Memory: Interconnection Networks

The parallel computers being built today more closely follow the lines of the second main model for parallel computers, the interconnection network model. In the interconnection network model, each processor is assumed to have its own dedicated

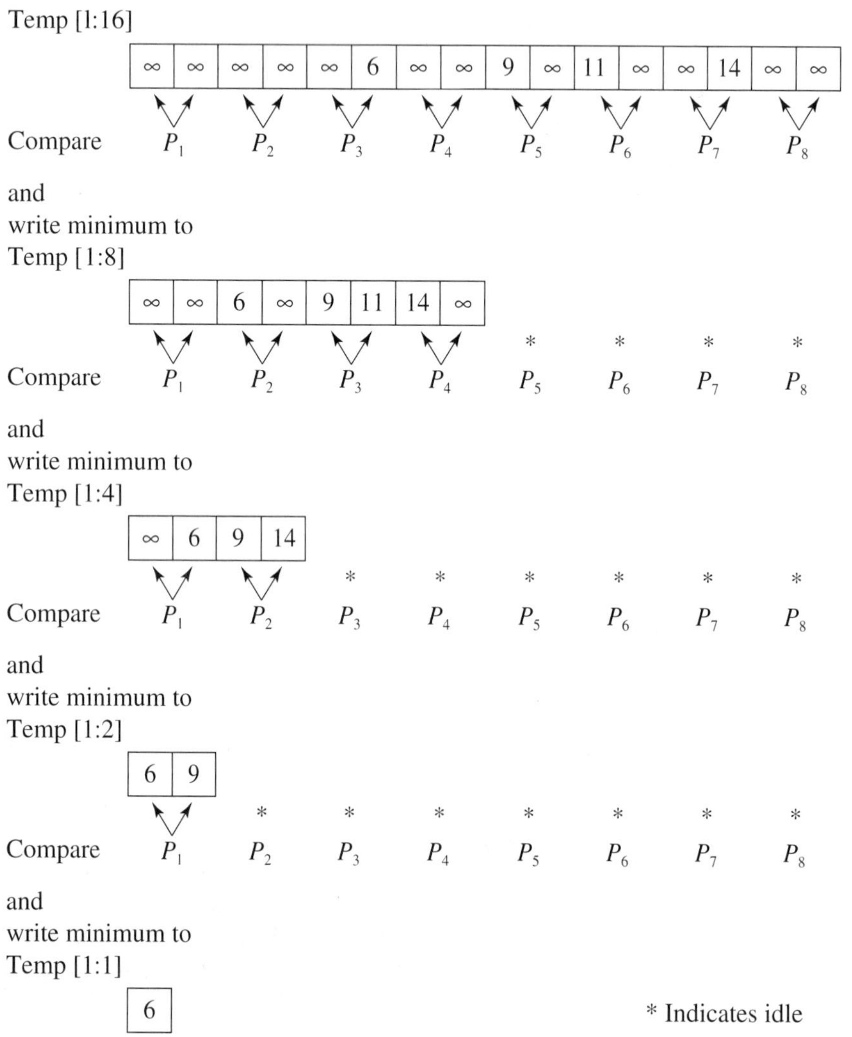

Figure 5.5 Binary fan-in technique for computing the minimum value in *Temp*[1:*n*] in $\log_2 n$ parallel comparison steps

RAM memory, with no common shared memory accessible to each processor. Since we are dealing with a SIMD machine, variables in processor memories each have instantiations in *every* processor, so that they are called *parallel* or *distributed* variables. In other words, if X is a distributed variable, then each processor in the network has a memory location reserved for its own version of X.

In the interconnection network model, we assume that each processor has enough local memory to carry out the task at hand. Nevertheless, our parallel algorithms are usually written so as to only require a constant (independent of the input size) number of distributed variables.

We emphasize that statements involving distributed variables might only be executed in a subset of the processors, with the other processors remaining idle (masked out). If X is a distributed variable, then in our pseudocode for such statements as interprocessor communication, we use the label prefix $\mathbf{P}$:X to indicate the instantiation of X in processor P's local memory. We also assume that central control has a limited amount of local RAM memory for storing such things as loop control variables and processor indices.

In an interconnection network model, information is communicated between processors using messages sent along the network. Such messages pass along routes in the network where each link in a route is between directly connected (*adjacent*) processors. To avoid such things as routing conflicts, unless otherwise specified all of the algorithms in this text are designed so that in each step communication in an interconnection network takes place only between adjacent processors.

To describe how communication takes place between adjacent processors P_i and P_j, suppose the central control instructs processor P_j to assign to the variable Y in its local memory the value of the variable X in the local memory of processor P_i. For our purposes we can view this as being accomplished as follows: Processor P_i reads the value of its local variable X and then sends this value along a link in the interconnection network to processor P_j, which then writes it to its local variable Y (see Figure 5.6). This communication is represented in our pseudocode by $\mathbf{P}_j{:}Y \Leftarrow \mathbf{P}_i{:}X$.

The most commonly used interconnection network models are *meshes* (*arrays*), *trees*, *hypercubes*, and some closely related networks. In this chapter we limit our attention to networks related to meshes and trees. Other networks are introduced in later chapters as they arise naturally in applications.

The simplest interconnection network is the one-dimensional mesh M_p (linear array) with p processors $P_1, \ldots, P_p$, which has P_i directly connected to P_j if, and only if, $|i-j| = 1$ (see Figure 5.7a, page 162). The (square) two-dimensional mesh $M_{q,q}$ with $p = q^2$ processors $P_{i,j}$, $i,j \in \{1, \ldots, q\}$ has $P_{i,j}$ directly connected with $P_{r,s}$ if, and only if, $i = r$ and $|j-s| = 1$ or $j = s$ and $|i-r| = 1$ (see Figure 5.7b). The k-dimensional mesh for $k \geq 3$ is defined similarly.

A *tree* interconnection network is a network whose processors and interconnections correspond to the nodes and edges in a tree. A particularly important exam-

Figure 5.6 $\mathbf{P}_j{:}Y \Leftarrow \mathbf{P}_i{:}X$

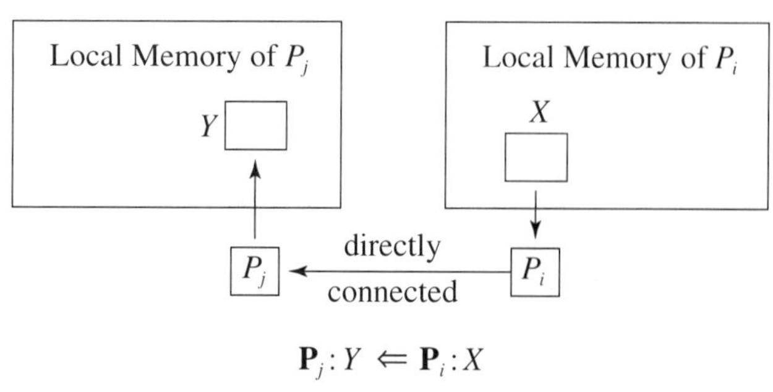

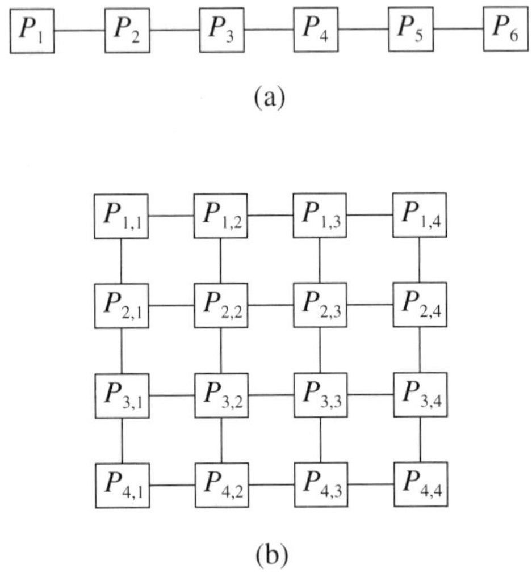

Figure 5.7 (a) One-dimensional mesh M_6; (b) two-dimensional mesh $M_{4,4}$

ple is the *complete binary tree network* on p nodes, $p = 2^k - 1$. We will simply refer to this network as the *processor tree* PT_p. We can conveniently label the processors in PT_p using binary strings that represent a path from the root to a given processor. The root is labeled with the empty string (denoted by ε). Moving to the left adds a zero to the end of the string, whereas moving to the right adds a one (see Figure 5.8).

The extreme version of an interconnection network is the complete graph K_p, where every processor is directly connected to every other processor (see Figure 5.9). Of course, this is a very convenient model for communication, but it is no more practical to build than is the PRAM.

Figure 5.8 Complete binary tree interconnection network PT_{15}

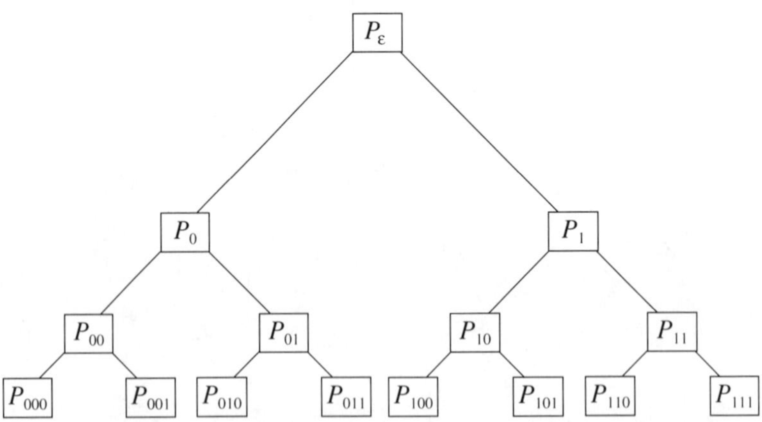

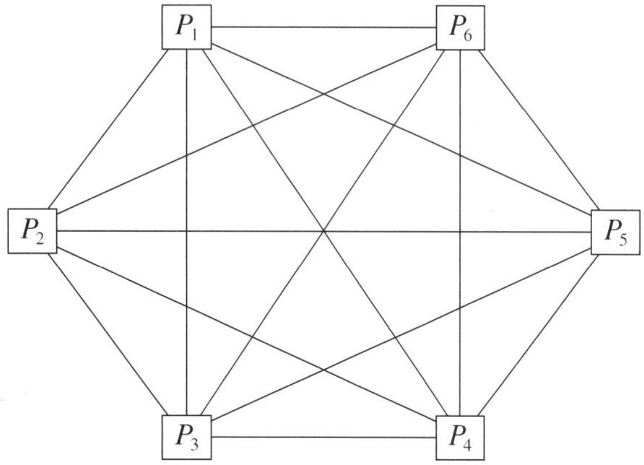

Figure 5.9 The complete graph interconnection network K_6

5.2.6 Three Basic Goodness Measures for Interconnection Networks

Given an interconnection network model M, we now discuss three basic goodness measures of M that are related to communication complexity within the network. For an arbitrary pair of processors P_i, P_j in M, a given algorithm might require that sometime during its execution information initially in the local memories of P_i and P_j be combined (shared) in some way. If $dist(P_i,P_j)$ denotes the number of links in the shortest route connecting P_i and P_j, then the minimum number of communication steps required for information to be shared by P_i and P_j is $dist(P_i,P_j)/2$. This motivates the first goodness measure

$$diameter(M) = \max \{dist(P_i,P_j) : P_i,P_j \in M\}.$$

Clearly, it is desirable to keep the diameter of a network small, because the diameter serves as a lower bound for the communication complexity of algorithms implemented on the network that require information sharing between all the processors. One way to keep the diameter small is to have a large number of links incident with each processor. For example, the complete graph K_p has diameter 1 at the cost of having $p - 1$ links incident with each processor. Although K_p would be the most efficient interconnection network for algorithm design, it is the least feasible to build. With the current state of technology, only interconnection networks with relatively small degree are feasible. Our second goodness measure takes this physical limitation into account. The *degree* of a processor P is defined to be the number of links incident with P. The second goodness measure is then

$$maximum\text{-}degree(M) = \max\{degree(P) : P \in M\}.$$

The third goodness measure is somewhat more subtle than the other two and relates to communication bottlenecks. Consider a bipartition of the processors of M into two sets X and Y. Clearly any route joining a processor in X to a processor in Y

must contain at least one link in the set $C(X,Y)$ of all links joining a processor in X to a processor in Y. If $C(X,Y)$ is small and X and Y are about equal in size, then algorithms requiring many processors in X to communicate with processors in Y will encounter congestion because of the large number of routings that will have to use the small number of links in $C(X,Y)$. The size of the smallest such cut $C(X,Y)$ is called the *bisection width*. More precisely, the third goodness measure of M is given by

$$bisection\text{-}width(M) = \min\{\,|C(X,Y)| : abs(|X| - |Y|) \le 1\}.$$

Unfortunately, there is no interconnection network that optimizes all three goodness measures simultaneously. As the table in Figure 5.10 shows, there are trade-offs between the three goodness measures. In subsequent chapters we discuss other interconnection networks, such as the mesh-of-trees, hypercube, and butterfly networks, which achieve a better balance between the three goodness measures. A verification of the goodness measures in Figure 5.10 is left to the exercises.

5.2.7 Example: Searching on Meshes

To illustrate the difference between shared memory and distributed memory, we again consider the problem of searching for the occurrence of a search element X in a list $L[1{:}n]$, but now on the one-dimensional mesh M_n. The issue now is that the communication constraints place a severe limitation on the efficiency of broadcasting X to each processor's local memory (assuming, for example, that X is initially stored in processor P_1's local memory). We cannot efficiently use the broadcast technique that was described earlier for the EREW PRAM, since it would involve communication between nonadjacent (that is, not directly connected) processors. In fact, it is easy to see that for the one-dimensional mesh, we can do no better than the sequential algorithm *LinearSearch* that requires n comparison steps in the worst case.

Now consider the problem of searching on the two-dimensional mesh $M_{q,q}$. For convenience, we assume that the list has size $n = q^2$. Again, since the search element is a (variable) parameter (as opposed to a constant), it is not possible to initialize the distributed variable $\mathbf{P}{:}X$ with the value of the search element. Thus, we will assume that the search element x is a front-end variable, and initially only $\mathbf{P}_{1,1}{:}X$ is assigned the value x. We assume that the list $L[1{:}n]$ has been input to the distributed variable L in *row-major order*. That is, element $L[k]$ is assigned to processor $P_{i,j}$'s local instantiation of L, where

$$k = (i-1)q + j.$$

Figure 5.10 Values of three basic goodness measures for sample interconnection network models with p processors

Model	Maximum Degree	Diameter	Bisection-Width
K_p	$p-1$	1	$p^2/4$
M_p	2	$p-1$	1
$M_{\sqrt{p},\sqrt{p}}$	4	$2\sqrt{p}-2$	$\sqrt{p} + (p \textbf{ mod } 2)$
PT_p	3	$2\lfloor \log_2 p \rfloor$	1

Each instantiation of the distributed variable *Index* is assigned the row-major value of its associated processor, which, unlike the assignment of the search element to the variable X, can be done in a single parallel step since it is a purely local computation (we assume that each processor $P_{i,j}$ knows its own index i,j). See Figure 5.11.

As with the PRAM, we need to broadcast the value of the search element throughout the distributed variable X. The following simple two-phase procedure broadcasts the value of $x = \mathbf{P}_{1,1}{:}X$ throughout the entire distributed variable $\mathbf{P}_{i,j}{:}X$, $1 \le i,j \le q$ using $2q - 2$ parallel communication steps. In the first phase, x is broadcast

Figure 5.11 Initial states of the distributed variables by *L*, *X*, and *Index*

Initial state of the distributed variables L, X, and *Index* after list L: 2, −1,23,−4, 2,5,−2,0,5,1,5,−5,8,5,3,−2 is input using row-major order, and $\mathbf{P}_{1,1}{:}X$ has been assigned the value 5. The distributive variable *Index* initially is assigned the row-major value of its associated processor (using a local calculation).

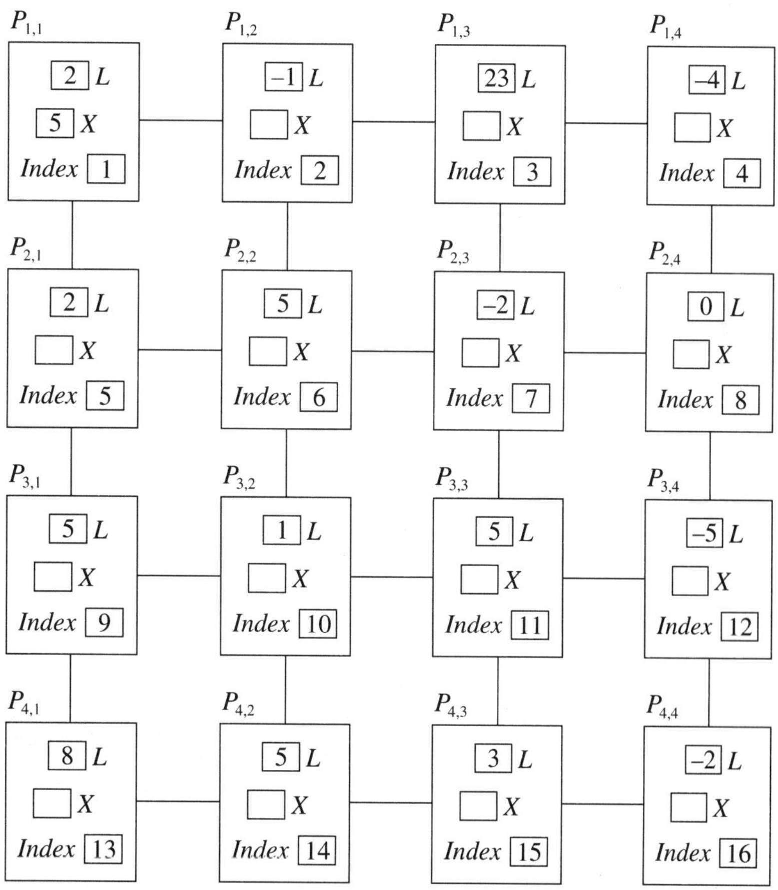

across the first row in $q - 1$ steps, where in the ith step, processor $P_{1,i}$ communicates x to its neighbor $P_{1,i+1}$ on the right, $i = 1, \ldots, q - 1$. In the second phase, x is broadcast down row by row, where in the ith step, $P_{i,j}$ communicates x to its neighbor $P_{i+1,j}$ below, $i = 1, \ldots, q - 1$. We show the action of this broadcasting procedure in Figure 5.12 for the search element $x = 5$.

After the value of the search element has been broadcast to all n processors (so that each $\mathbf{P}_{i,j}{:}X$ contains the value of the search element), in a single parallel comparison step each processor $P_{i,j}$ compares $\mathbf{P}_{i,j}{:}X$ to its list element $\mathbf{P}_{i,j}{:}L$ and writes the value ∞ to $\mathbf{P}_{i,j}{:}Index$ if the search element is not equal to the list element (see Figure 5.13).

After the single parallel comparison step, the distributed *Index* contains the results of the search, but how do we return the result? It will be our convention that whenever a scalar-valued function defined on an interconnection network terminates, the value to be returned by the function resides in a particular processor's local instantiation of a suitable distributed variable. In our example, suppose we specify that $P_{1,1}$ should return the outcome of our search in its instantiation of the distributed variable *Index*. In other words, at the termination of our search algorithm, we want $\mathbf{P}_{1,1}{:}Index$ to hold the smallest index i such that $L[i] = X$, or ∞ if no such index exists. Thus, we want to return the minimum value in *Index*.

To compute the minimum value held in *Index*, we perform what amounts to a reverse-broadcast procedure. In phase 1, we compute column minimums in the obvious way. That is, we first compute the minimums of the elements in the qth row and their immediate neighbors above in the $(q - 1)$st row, leaving the results in the $(q - 1)$st row. Then we repeat the process for the (new) $(q - 1)$st row and the $(q - 2)$nd row, and so forth until the first row contains all the column minimums. In phase 2, we compute the minimum of the column minimums (that is, the minimum of the elements that are now in the first row) proceeding sequentially from right to left. The process is illustrated in Figure 5.14 (page 168). Note that this reverse-broadcast pro-

Figure 5.12 Broadcasting the value 5 throughout the distributed variable X on $M_{4,4}$

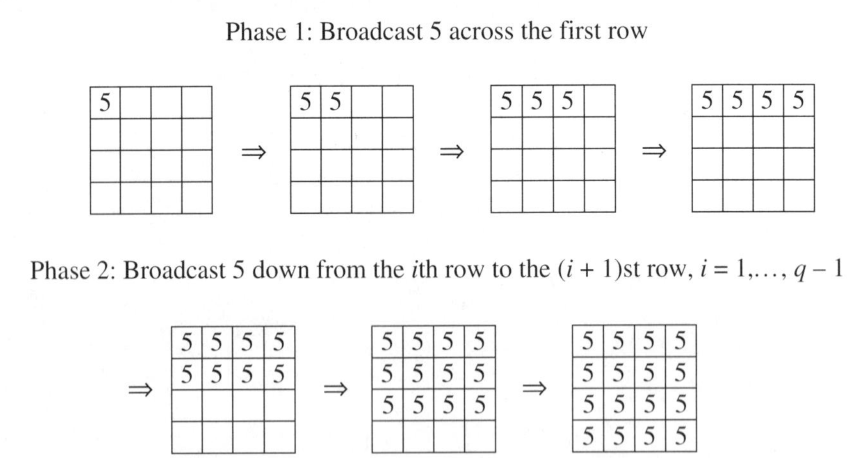

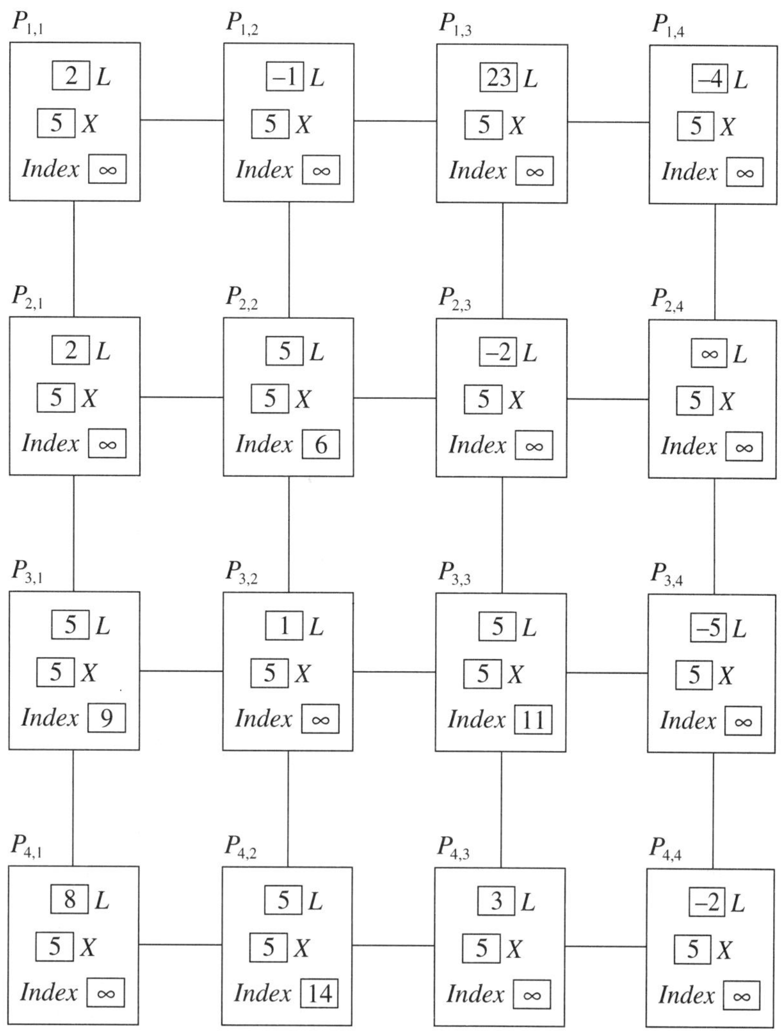

Figure 5.13 State of the distributed variables *L*, *X*, and *Index* after the value 5 has been broadcast throughout *X* and the single parallel comparison step between *X* and *Index* has been performed

cedure works equally well for combining all the elements in *Index* with respect to any associative binary operation such as addition, multiplication, and so forth.

The states of the distributed variables *L*, *X*, and *Index* upon completion of our algorithm are shown in Figure 5.15 (page 169).

Similar to searching on a PRAM, our searching algorithm on $M_{q,q}$ performs a single parallel comparison between the search element and the elements of the list. Hence, the communication complexity is the appropriate measure of the complexity of our algorithm. The number of communication steps for both the broadcast phase and the reverse-broadcast phase is $2q - 2$. Hence, the complexity of our searching

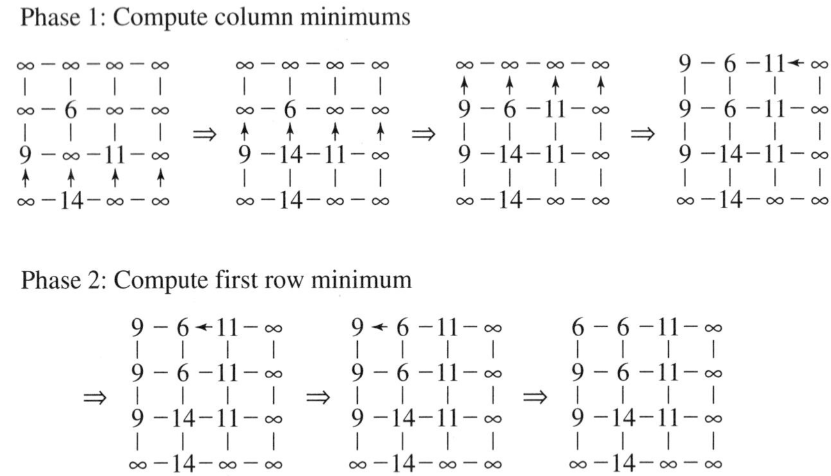

Figure 5.14 Computing the minimum on a two-dimensional mesh. Arrows indicate how the minimums are computed at each stage. The final result is in $P_{1,1}$.

algorithm is $4q - 4 = = 4\sqrt{n} - 4$, so that we achieve a speedup of $n/(4\sqrt{n} - 4)$ over the sequential algorithm *LinearSearch*.

5.2.8 Example: Searching on the Complete Binary Tree

Before describing the searching algorithm, we first note that the complete binary tree network PT_p, $p = 2^k - 1$, readily supports a binary fan-in technique to compute $x_1 \otimes x_2 \otimes \cdots \otimes x_n$, $n = 2^{k-1}$, where $\otimes$ is an associative binary operation such as addition, multiplication, minimum, and so forth. The n operands are input to the leaves, and the computations work their way up the tree, where the parent processor combines the results of the computations performed and communicated by its two children. The final result resides in the root after $\log_2 n$ parallel steps.

Now consider searching on PT_p, where the list $L[1{:}n]$, $n = 2^{k-1}$, is stored in the leaves. For example, if the processors in PT_p are indexed as in Figure 5.8, then the distributed variable L is such that $\mathbf{P}_b{:}L = L[i]$, where b is the $(k-1)$-digit binary representation of $i-1$, i = 1, . . . , n (the other instantiations of L are undefined). Similar to our searching algorithm on the two-dimensional mesh, we assume that the distributed variable *Index* is initialized to ∞. First, the search element x is input to the root local variable $\mathbf{P}_\varepsilon{:}X$, and then x is broadcast throughout the distributed variable X using $2\log_2 n$ parallel communication steps as follows: Starting with the root processor, after receiving the value x, each processor communicates the value x (in 2 steps) to its 2 children. In a single parallel comparison step, the ith leaf processor P_b writes i to $\mathbf{P}_b$:*Index* if $\mathbf{P}_b{:}L = \mathbf{P}_b{:}X$, (otherwise $\mathbf{P}_b$:*Index* remains ∞. Finally, the minimum value in the leaf instantiations of the distributed variable *Index* is sent to the root using the binary fan-in method (see Exercises 5.19 and 5.20)

The algorithm performs 2 $\log_2 n$ communication steps during the broadcast phase and $\log_2 n$ steps for computing the minimum. Hence, the complexity of our searching algorithm is 3 $\log_2 n$, so that we achieve a speedup of $n/(3 \log_2 n)$ over the sequential algorithm *LinearSearch*.

5.2.9 I/O Constraints

Any parallel machine must have some mechanism to read "outside world" data from external input devices into the processors' local memories, as well as to write data from these memories to external output devices. In this text we adopt a high-level

Figure 5.15 State of the distributed variables *L*, *X*, and *Index* upon completion of the searching algorithm

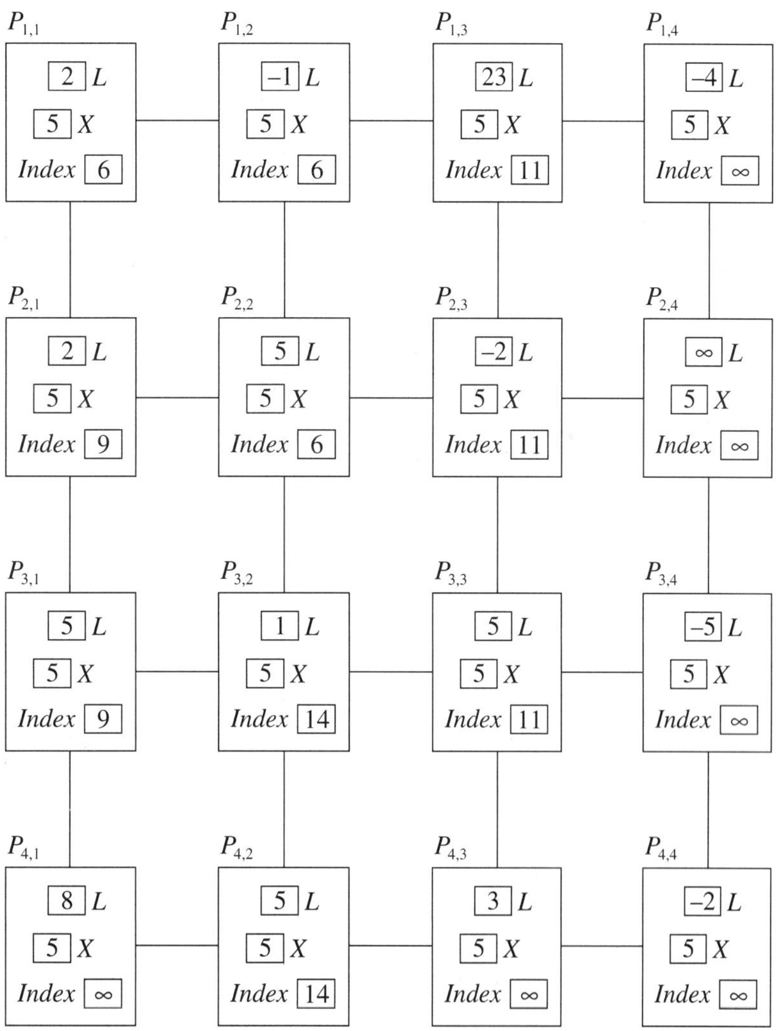

approach where we leave the exact nature of the I/O mechanism unspecified. Similar to serial (single-processor) computers, we simply assume that suitable parallel versions of **read** and **write** statements are available.

In fact, in the case of a PRAM, we almost always assume that central control has already placed input data for algorithms into shared memory, so that little use is made of **read** or **write** statements. Also, in interconnection network models, we often assume that input data for algorithms has already been distributed to the local memory of the relevant processors without specifying the mechanism for accomplishing it. However, we make more use of **read** and **write** statements for interconnection network models than for PRAMs, since it is important in the former models to specify how data gets distributed to the processors' local memories. Also, in interconnection network models, when a procedure is called or a function invoked, we are careful to delineate those processors that have meaningful data supplied to them on input or provide meaningful data on output. For example, if an algorithm for an interconnection network is supposed to sort a list $L[1{:}n]$, we will be explicit about which processors receive the list on input, and we will specify a linear ordering of those processors that contain the sorted list on output. Each local processor manages parameters, local and global variables, in a manner similar to single-processor serial computers.

Figure 5.16 is a very high-level schematic picture and does not address the problem of how parallel computers could actually be built based on this model. As an example of how this might be done in practice, Figure 5.17 shows a schematic picture of the MP-1 parallel computer built by the MasPar Computer Corporation.

5.3 Computing the Dot Product on the EREW PRAM versus the Two-Dimensional Mesh

The simple example of computing the dot product of two vectors serves nicely to further contrast the differences in designing an algorithm for the (EREW) PRAM versus the two-dimensional mesh. The example also serves to illustrate some of our pseudocode conventions.

The dot product of $\mathbf{a} = (a_1, \ldots, a_n)$ and $\mathbf{b} = (b_1, \ldots, b_n)$ is defined by:

$$\mathbf{a} \circ \mathbf{b} = a_1b_1 + \cdots + a_nb_n.$$

We first compute the vector $\mathbf{c} = (c_1, \ldots, c_n)$, where $c_i = a_ib_i$, $i = 1, \ldots, n$, then we compute the sum of the components of $\mathbf{c}$. We develop algorithms *DotProdPram* and *DotProd2DMesh* for this problem, implemented on the PRAM and two-dimensional mesh $M_{q,q}$. In the PRAM model, vectors **a**,**b**,**c**, are stored as the one-dimensional arrays $A[1{:}n]$, $B[1{:}n]$, $C[1{:}n]$ (see Figure 5.18, page 172). In $M_{q,q}$, we store the vectors using the distributed variables A, B, C (see Figure 5.19).

As the first step in computing the dot product **c**, we compute in parallel the vector whose components are the product of the corresponding components of **a** and **b**. In Figure 5.18 we show the pseudocode and memory management for accomplishing

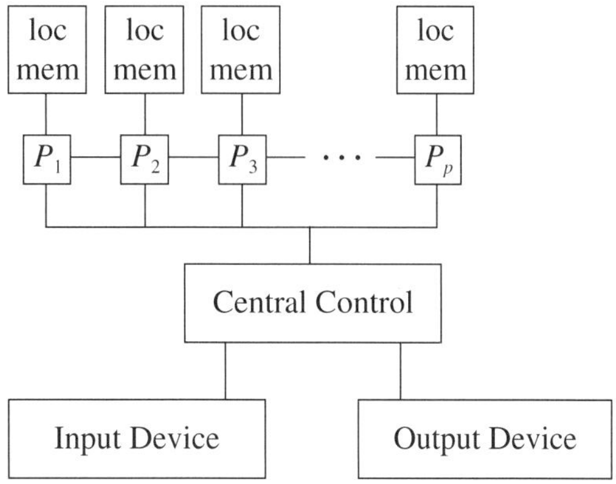

Figure 5.16 Our model of the one-dimensional mesh M_p

Figure 5.17 MasPar Computer Corporation's MP-1 parallel computer (Tom Blank, "The MASPAR MP-1 Architecture," *Proceedings of the IEEE Compcon Spring 90,* IEEE, Feb. 1990, pp. 20–24. Reprinted with permission.)

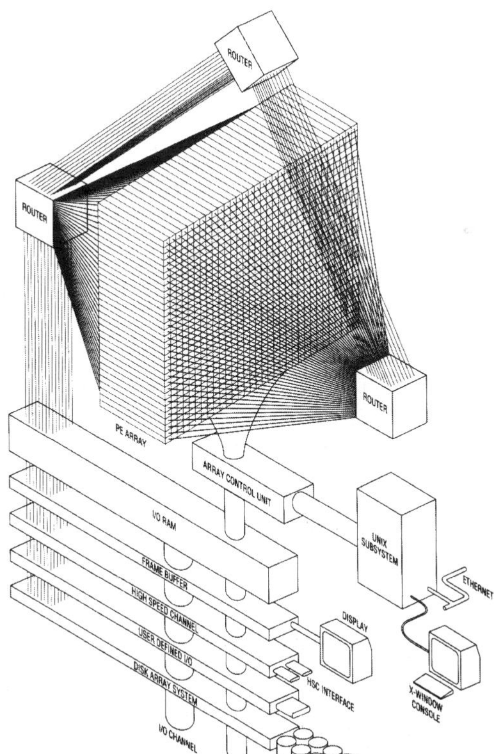

for $1 \le i \le n$ **do in parallel**
  $C[i] := A[i] * B[i]$
end in parallel

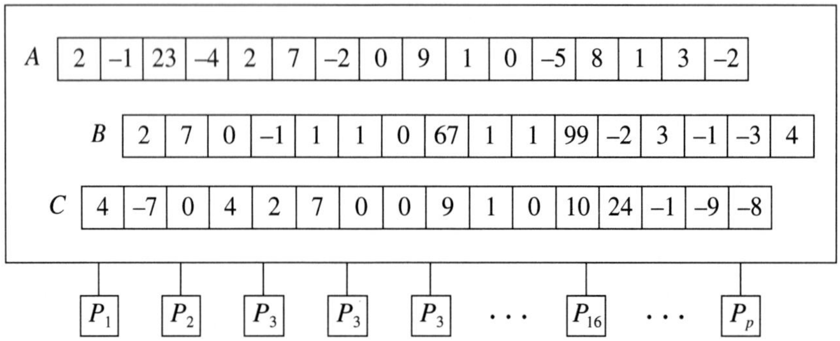

Figure 5.18 Component multiplication on a PRAM

Figure 5.19 Component multiplication on $M_{q,q}$, $q = 4$, $p(n) = n = 16$

for $P_{i,j}$, $1 \le i,j \le q$ **do in parallel**
  $C := A * B$
end in parallel

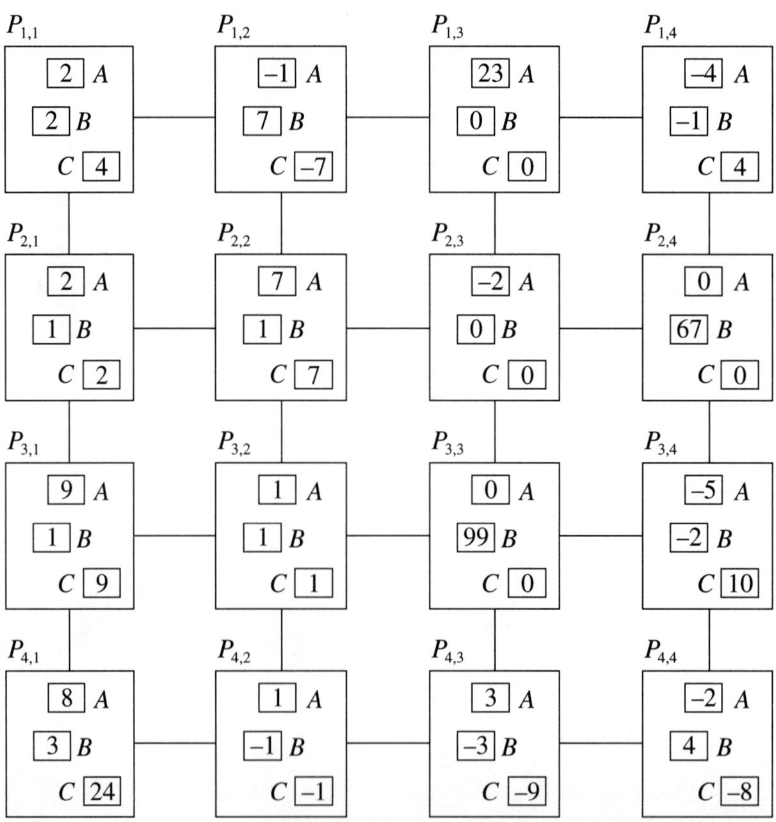

this on the PRAM, with sample vectors **a**,**b** of size 16. In Figure 5.19, we do the same on the two-dimensional mesh $M_{q,q}$. Note the differences between how the vectors are stored in the two models.

In the case of the PRAM, we did not specify which processor actually was assigned the task of computing $A[i]*B[i]$. Indeed, any set of n of the processors could have been assigned the task of computing (in parallel) the n products $A[i]*B[i]$. For PRAMs we don't usually care which of the processors perform an **in parallel** step, just that we have enough processors to do the job.

The final step in the algorithms *DotProdPRAM* and *DotProd2DMesh* is to compute the sum of the values in the array $C[1:n]$ and the distributed variable C, respectively. Functions *SumPRAM* and *Sum2DMesh* for computing these sums can be designed using the binary fan-in and reverse-broadcasting techniques, respectively, that were discussed earlier in connection with computing the minimum value in a list.

5.4 Pseudocode Conventions for PRAMs

In this section we introduce our pseudocode conventions for the PRAM. Similar to sequential algorithms, with both PRAMs and interconnection network models we always make explicit which parameters are input parameters and which are output parameters. Again, to avoid side effects, we assume that all input parameters (that are not also output parameters) are passed *by value*, that is, the corresponding argument is not changed even if the input parameter is altered by the parallel procedure or function. All parameters of a parallel function are assumed to be input parameters.

Since PRAMs have a single shared memory, the pseudocode for PRAMs is an extension of the pseudocode for sequential algorithms. The major new elements are the **in parallel** statement, the **parallelcall** statement, and various opening syntax specifications.

5.4.1 Opening Syntax

The opening syntax for a procedure written for a PRAM has the following form, which is identical to that for a sequential procedure except for the addition of the model specification and the number $p = p(n)$ of processors used.

procedure <Name>(<List of parameters>)

Model: <Model name> **with** p = <function of n> **processors**

Input: <Description of input variables>

Output: <Description of output variables>

dcl: <Declaration of local variables and global variables>

The opening syntax for a function written for the PRAM is similar, except that the **Output** statement then describes the value returned. The name of the procedure usually begins with a reference to the problem being solved, followed by a reference to the model of PRAM. In the procedure name, we simply use the term PRAM for EREW PRAM. However, we use the full term EREW PRAM for <Model Name>. Just as with sequential pseudocode, the **dcl** statement is often omitted, although it is generally included when global variables are used. As an example, the function *SumPRAM* has the following opening syntax.

function *SumPRAM*($X[1:n]$)

Model: EREW PRAM **with** $p = n/2$ **processors**

Input: $X[1:n]$ (array of numbers $x_1, x_2, \ldots, x_n$) $\{n = 2^k\}$

Output: $x_1 + x_2 + \cdots + x_n$

5.4.2 In Parallel Statement

In addition to the usual instructions for serial machines, we have an **in parallel** statement, which allows operations to be performed on more than one component of an array simultaneously.

for <Boolean expression involving array indices> **do in parallel**
<statement 1>
<statement 2>
.
.
.
<statement *k*>
end in parallel

The **for** clause is a Boolean expression involving array indices. Only those array elements whose indices satisfy the Boolean expression participate in the instructions contained in the body of the **in parallel** statement. These instructions can be any statements from the sequential pseudocode that apply to arrays (however, we do have to be careful not to violate the SIMD model, as discussed at the end of this section). Each of these statements are performed simultaneously by a set of active processors. We do not specify in our pseudocode those processors that are active; we only need to be careful that there are enough processors around to do the job. However, in drawing figures to illustrate the action of an algorithm, it is usually helpful to match up processors with operations on array components.

Figure 5.20 illustrates our **in parallel** statement.

Since we have assumed that each processor in a PRAM is as powerful as in a serial machine, the built-in functions and procedures **odd, even, interchange**, **sin**, **read, write**, and so forth that were assumed for serial machines are also assumed capable of execution by each processor in parallel. For example, we can allow parallel statements such as

```
for 1 ≤ i ≤ n do in parallel
   read(A[i],B[i])
   A[i] := A[i] + sin(B[i])
   if A[i] > B[i] then
      call interchange(A[i],B[i])
   endif
   write(A[i],B[i])
end in parallel
```

Figure 5.20

Action of **in parallel** statement for sample arrays

```
for 1 ≤ i ≤ 2k – 1 .and. odd(i) do in parallel
   A[i] := A[i] + B[i]
   B[i] := 2*B[i]
   if i mod 4 = 3 then
      B[i] = B[i] – 6
   endif
end in parallel
```

$A[1:n]$

23	2	108	7	55	1	2	0	17	6	...

$B[1:n]$

55	300	123	11	0	1	2	0	19	66	...

Before execution

$A[1:n]$

78	2	231	7	55	1	4	0	36	6	...

$B[1:n]$

110	300	240	11	0	1	–2	0	38	66	...

After execution

We also assume the existence of the built-in parallel procedures **split**($L[1{:}n]$,$A[1{:}n/2]$,$B[1{:}n/2]$), which splits the list $L[1{:}n]$ into two sublists $A[1{:}n/2] = L[1{:}n/2]$ and $B[1{:}n/2] = L[(n/2) + 1{:}n]$, and **oddevensplit**($L[1{:}n]$,$A[1{:}n/2]$,$B[1{:}n/2]$), which splits the list $L[1{:}n]$ into two sublists $A[1{:}n/2]$ and $B[1{:}n/2]$ consisting of the odd-indexed elements and even-indexed elements of $L[1{:}n]$, respectively.

We also allow parallel procedures and functions to be executed in parallel. For example, in procedure *MatVecProdCREW* discussed in Chapter 11, we compute the product of an $n \times n$ matrix $A[1{:}n,1{:}n]$ and an $n \times 1$ matrix by making a single parallel invocation of the parallel function *DotProdPRAM* as illustrated in the following pseudocode.

```
for 1 ≤ i ≤ n do in parallel
    Prod[i] := DotProdPRAM(A[i,1:n],X[1:n])
end in parallel
```

5.4.3 Parallelcall

In the parallel invocation of *DotProdPRAM*, the n parameter lists

$$(A[i,1{:}n],X[1{:}n]), \quad i = 1, \ldots ,n,$$

are naturally indexed by the variable i. However, many occasions arise, particularly in the presence of recursion, where the different parameters of the parallel invocation to a function or call to a procedure are not so conveniently indexed. For example, in the recursive merging procedure *OddEvenMergePRAM*($L[1{:}n]$) given in the next chapter, simultaneous recursive calls are made with two lists *Odd*$[1{:}n/2]$, *Even*$[1{:}n/2]$. If we renamed these lists $A[1,1{:}n/2]$ and $A[2,1{:}n/2]$, we could do a parallel call to *OddEvenMergePRAM* with aid of A as follows:

```
for 1 ≤ i ≤ 2 do in parallel
    call OddEvenMergePRAM(A[i,1:n/2])
end in parallel
```

However, the renaming (and associated assignments) of *Odd* and *Even* is cumbersome and distracts from the high-level description of the algorithm. Therefore, to accomplish the parallel call we introduce the more compact pseudocode convention

```
parallelcall OddEvenMergePRAM(Odd[1:n/2] | Even[1:n/2]),
```

where we use the symbol | to separate the parameters corresponding to each simultaneous call to *OddEvenMergePRAM*.

5.4.4 SIMD versus MIMD

Since we have adopted the SIMD (single-instruction, multiple-data) model in this text, we must take care not to allow two processors to perform different instructions simultaneously. Thus, **if then else** statements and **case** statements usually need to be altered to maintain the SIMD model. To illustrate this, consider the following example involving an **if then else** statement for a MIMD (multiple-instruction, multiple-data) PRAM and its altered pseudocode for a SIMD PRAM.

```
for 1 ≤ i ≤ n do in parallel
    if i ≤ 10 then
        A[i] := 2*A[i]
    else
        A[i] := A[i] + 1
    endif
end in parallel
```

MIMD PRAM

```
for 1 ≤ i ≤ 10 do in parallel
    A[i] := 2*A[i]
end in parallel
for 11 ≤ i ≤ n do in parallel
    A[i] := A[i] + 1
end in parallel
```

SIMD PRAM

In the MIMD PRAM pseudocode, ten processors would be assigned the task of executing the instruction $A[i]$:= $2*A[i]$, and a different set of $n - 10$ processors would execute the instruction $A[i]$:= $A[i] + 1$, all executions being simultaneous. However, this would not be allowed in the SIMD PRAM, since two different instructions are being performed simultaneously. This necessitates the altered pseudocode for a SIMD PRAM utilizing two **in parallel** statements.

In a similar manner, **case** statements within **in parallel** statements for MIMD PRAMs can be translated for SIMD PRAMs using a sequence of **in parallel** statements, one for each **case** clause.

5.4.5 Example: Pseudocode for Searching on a PRAM

We now give pseudocode for the algorithm discussed in Section 5.2 for searching on an EREW PRAM. The first phase of the algorithm involved assigning the value of the search element to each entry in the array *Temp*[1:n]. At first glance, the following pseudocode seems to do the job.

```
for 1 ≤ i ≤ n do in parallel
    Temp[i] := X
end in parallel
```

However, this statement is not admissible on the EREW PRAM (unless $n = 1$), since it calls for the single memory location X to be read concurrently by n processors. The following pseudocode accomplishes the task using the broadcasting technique described in Section 5.2.

```
procedure BroadcastPRAM(A[1:n],X)
Input:  A[1:n] (an array of size n), {n = 2^k}
        X (a value to be broadcast throughout A[1:n])
Output: A[1:n] (array where A[i] = X, i = 1, ... , n)
    A[1] := X
    for i := 1 to k do
        for 2^(i-1) + 1 ≤ j ≤ 2^i do in parallel
```

Continued

```
      A[j] := A[j – 2^(i – 1)]
    end in parallel
  endfor
end BroadcastPRAM
```

Recall that after X has been broadcast throughout *Temp*[1:*n*], each processor P_i in parallel compares $L[i]$ to $Temp[i] = X$ and writes i back to $Temp[i]$ if $L[i] = X$, otherwise it writes the value ∞ in $Temp[i]$. The comparison step is accomplished by

```
for 1 ≤ i ≤ n do in parallel
  if L[i] = Temp[i] then
    Temp[i] := i
  else
    Temp[i] := ∞
  endif
end in parallel
```

The algorithm terminates by invoking the function *MinPRAM* for computing the minimum value in an array using the binary fan-in technique.

```
function MinPRAM(L[1:n])
Model: EREW PRAM with p = n/2 processors
Input:   L[1:n] (a list of size n, n = 2^k)
Output: the minimum value of a list element in L
  for j := 1 to log2n do
    for 1 ≤ i ≤ n/2^j do in parallel
      if L[2i – 1] > L[2i] then
        L[i] := L[2i]
      else
        L[i] := L[2i-1]
      endif
    end in parallel
  endfor
  return(L[1])
end MinPRAM
```

We now can give pseudocode for the entire algorithm *SearchPRAM*.

```
procedure SearchPRAM(L[1:n],X)
Input:   L[1:n] (a list of size n), {n = 2^k}
         X (a search element)
```

```
Output: the smallest index where X occurs in L, or ∞ if X is not in L
   call Broadcast(Temp[1:n],X)
   for 1 ≤ i ≤ n do in parallel
      if L[i] = Temp[i] then
           Temp[i] := i
      else
           Temp[i] := ∞
      endif
   end in parallel
   return(MinPRAM(Temp[1:n]))
end SearchPRAM
```

5.5 Pseudocode Conventions for Interconnection Network Models

Pseudocode for algorithms designed for interconnection networks is more complex than that for PRAMs, since it must include statements to describe communication between processors, specification of processors handling I/O, central control variables versus distributed variables, and so forth.

5.5.1 Opening Syntax

Each procedure for an interconnection network model begins with opening syntax similar to that for the PRAM. Parameters in the parameter list of a procedure or function on an interconnection network model will now include distributed variables. Therefore, for each distributed variable parameter it is important to describe the set of processors that contain meaningful information on input for that parameter. Thus, our **Input** statement not only describes the high-level nature of the parameter, but also includes a **range** clause specifying the range of indices of processors containing meaningful input data for the parameter. Those parameters for which no range is given in the **Input** statement are so-called *front end* or *central control* variables that are in the local memory of the front end processor (central control). The front end variable parameters are usually used to specify ranges of indices in the **Input** and **Output** statements. The **Output** statement contains similar specifications for all output parameters. We sometimes include a declaration statement **dcl** for local variables. Those local variables that are distributed also have a range specification.

```
procedure <Name>(<List of parameters>)
Model: <Model name> with p = <function of n> processors
Input: <Description of input parameter1> range: <processor range1>
                              .
                              .
                              .
       <Description of input parameteri> range: <processor rangei>
```

Output: <Description of output parameter1> **range:** <processor range1>
.
.
.
<Description of output parameter*j*> **range:** <processor range*j*>
dcl: <variable1> **range:** <processor range1>
.
.
.
<variable*k*> **range:** <processor range*k*>

As an example, the procedure *MatVecProd2DMesh* described in Chapter 11 has the following opening syntax.

procedure *MatVecProd2DMesh*(A,n,X)
Model: two-dimensional mesh $\mathbf{M}_{n,n}$ **with** $p = n^2$ **processors**
Input: A ($\mathbf{P}_{i,j}$:A contains $a_{i,j}$), **range:** $P_{i,j}$, $1 \leq i,j \leq n$
X ($\mathbf{P}_{1,j}$:X contains x_j), **range:** $P_{1,j}$, $1 \leq j \leq n$
Output: X ($\mathbf{P}_{i,1}$:X contains $a_{i1}x_1 + a_{i2}x_2 + \cdots + a_{in}x_n$), **range:** $P_{i,1}$, $1 \leq i \leq n$

All variables occurring in a procedure other than parameters are either additional front end variables used for counters and loop control or distributed variables. The difference between front end variables and distributed variables should always be clear from context. Note that for clarity, the input size n was included as a parameter in *MatVecProd2DMesh*, since the distributed variable A does not include its range as part of its name (as opposed to array parameters for PRAMs). However, in such instances we don't bother including n in the **Input** listing, since its role is made clear from the **range** descriptions.

The opening syntax for a function on an interconnection network is similar to that for functions defined on PRAMs. As an example, we give the opening syntax for the function *Search2DMesh* described later in this section.

function *Search2DMesh*($L,n,X,Index$)
Model: two-dimensional mesh $M_{q,q}$ **with** $p = n = q^2$ **processors**
Input: L (a list of n elements), **range:** $P_{i,j}$, $1 \leq i,j \leq q$
X (a search element), **range:** $P_{1,1}$
Index ($\mathbf{P}_{i,j}$:*Index* contains position (index) in list of $\mathbf{P}_{i,j}$:L),
range: $P_{i,j}$, $1 \leq i,j \leq q$
Output: the smallest index where X occurs in L, or ∞ if
X is not in L

5.5.2 In Parallel Statement

The **in parallel** statement is similar to that for a PRAM, except that the **for** clause contains a Boolean expression involving *processor* indices instead of array indices. The processor index might be a single index as in the one-dimensional mesh, a pair of indices as in the two-dimensional mesh, a bit string in the case of the hypercube, and so forth. It is implicit in all of our **in parallel** statements that a processor recognizes its own index. (In practice, this might be accomplished using a distributed variable *ID* in the read-only local memory of each processor. Then for each processor *P*, **P**:*ID* would contain the index of processor *P*.) Each processor whose index does not satisfy the Boolean expression in the **for** clause is *idle* (masked out). The active processors simultaneously execute the instructions in the body of the **in parallel** statement, while the remaining processors are idle. The general syntax for the **in parallel** statement is as follows.

```
for <processor P>,<Boolean expression involving index of P> do in parallel
   <statement 1>
   <statement 2>
        .
        .
        .
   <statement k>
end in parallel
```

Note that the **in parallel** statement is counted as representing *k* parallel steps. Figure 5.21 (page 182) shows the action of a simple **in parallel** statement that performs three parallel steps on the one-dimensional mesh. Compare the action of this code with the analogous code executed on a PRAM, as illustrated in Figure 5.20.

In Figure 5.21, all the processors with even indices are idle during the execution of the **in parallel** statement. In addition, the processors with indices 1,5,9, ... are idle within the **if then** statement.

For interconnection networks, we consider two main types of functions: those that are simply parallel invocations of sequential functions and those that require communication between processors. The first type of function can be either built-in or user defined. A parallel invocation of a sequential function works within each processor's local memory and does not require communication between processors. For example, consider a parallel invocation of the built-in function **sin** on the one-dimensional mesh M_p. Then, the statement

```
for P_i, 1 ≤ i ≤ n do in parallel
   B := sin(A)
end in parallel
```

where *A* and *B* are distributed variables, assigns **sin**($\mathbf{P}_i$:*A*) to $\mathbf{P}_i$:*B*, i = 1, ... , *n*.

As another example of a parallel invocation of a sequential function, the statement

```
for P_i, 1 ≤ i ≤ 2k − 1 .and. odd(i) do in parallel
   A := A + B
   B := 2*B
   if i mod 4 = 3 then
      B := B − 6
   endif
end in parallel
```

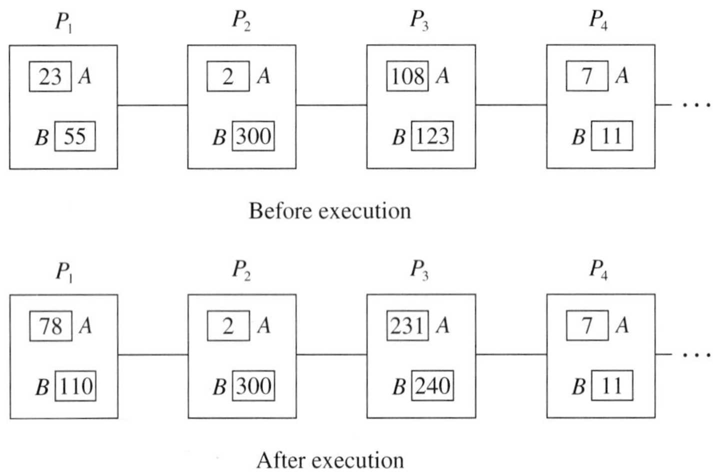

Figure 5.21 Action of **in parallel** statement on the one-dimensional mesh M_p

```
for P_i, 1 ≤ i ≤ n do in parallel
   C := EuclidGCD(A,B)
end in parallel
```

where A, B, and C are distributed variables, assigns $EuclidGCD(\mathbf{P}_i{:}A,\mathbf{P}_i{:}B)$ to $\mathbf{P}_i{:}C$, $i = 1, \ldots, n$. Note that the number of steps performed by $EuclidGCD(\mathbf{P}_i{:}A,\mathbf{P}_i{:}B)$ can vary greatly depending on the value of i. Strictly speaking, this would violate the SIMD model. However, by allowing processors to remain idle until all processors have completed their execution of *EuclidGCD*, we can reconcile it with the SIMD model.

The second type of function, which requires communication between processors, is restricted to functions that return a scalar. When such a function terminates, it is assumed that the value to be returned by the function resides in a particular processor's local instantiation of a suitable distributed variable. For example, suppose the scalar value to be returned by a function implemented on a two-dimensional mesh resides in $\mathbf{P}_{1,1}{:}X$. Then the **return** statement is given by

return($\mathbf{P}_{1,1}{:}X$)

Remark The **return** statement is only used for convenience and must not be interpreted as returning a value to central control that is directly accessible to all the processors. The returned value is resident in a suitable output register of $P_{1,1}$ only. If other processors require this returned value, then it must be broadcast.

5.5.3 Interprocessor Communication Statement

In the interconnection network model, recall (see Figure 5.6) that we introduced the pseudocode $\mathbf{P}_j{:}Y \Leftarrow \mathbf{P}_i{:}X$ for communicating processor P_i*'s* local variable X to an adjacent processor P_j*'s* local variable Y. The communication process $\mathbf{P}_j{:}Y \Leftarrow \mathbf{P}_i{:}X$ can be viewed as consisting of the following steps. Processor P_i fetches the value of X from its local memory and stores it to an output register. Then the contents of this output register is communicated to an input register of the adjacent processor P_j. Finally, P_j writes the value in this input register to the local variable Y. We assume that these steps constitute a single communication step taking unit time.

It is customary to indicate that processors P_i and P_j are adjacent by drawing a single (undirected) edge between them. However, it would perhaps be more realistic to have two directed edges drawn between them to indicate the fact that communication is always directed (see Figure 5.22).

The general form of the interprocessor communication statement is

<target processor:target variable> $\Leftarrow$ <source processor:source variable>

The interprocessor communication statement is usually embedded in the body of an **in parallel** statement, in which case the **for** clause includes a Boolean expression involving source processor indices that determine the set of active source processors. Any active source and corresponding target processor *must be adjacent* in the interconnection network. We illustrate a sample parallel interprocessor communication statement in Figure 5.23 (page 184).

We only allow communication in one direction in a single parallel step, in order to maintain a strict SIMD model. For example, in the one-dimensional mesh, we don't allow some of the active source processors to communicate to the left while others communicate to the right in the same parallel communication step.

If the communication involves the same distributed variable, we use the term *propagate* instead of communicate. Propagating a distributed variable is illustrated in Figure 5.24.

Figure 5.22 Interprocessor communication

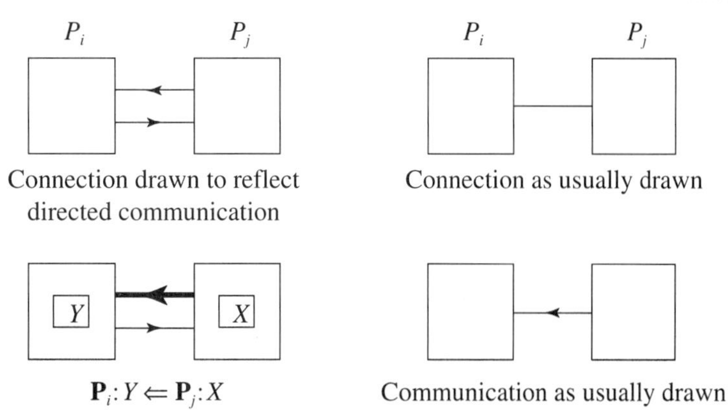

for P_i, $2 \le i \le 2k$ **.and. even**(i) **do in parallel**
  $\mathbf{P}_{i-1}{:}B \Leftarrow \mathbf{P}_i{:}A$ {communicate left from A to B}
end in parallel

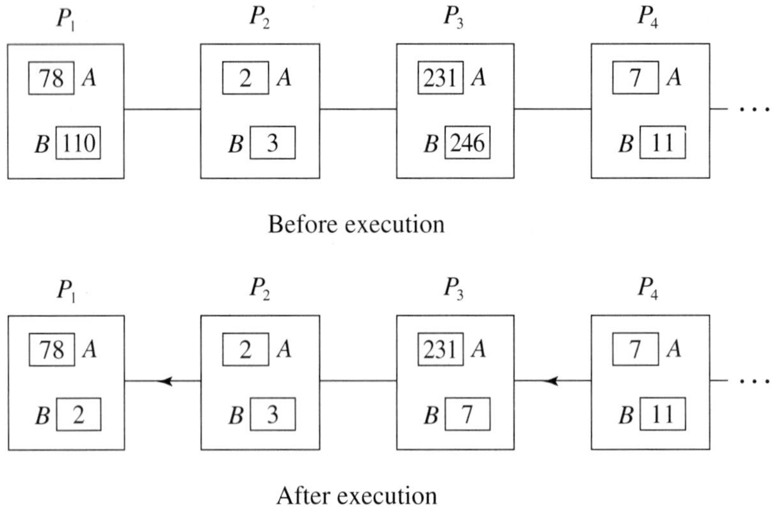

Figure 5.23 Action of parallel communication step

Figure 5.24 Action of parallel communication with processors being both source and target

for P_i, $1 \le i \le n - 1$ **do in parallel**
  $\mathbf{P}_{i+1}{:}A \Leftarrow \mathbf{P}_i{:}A$ {propagate A to the right}
end in parallel

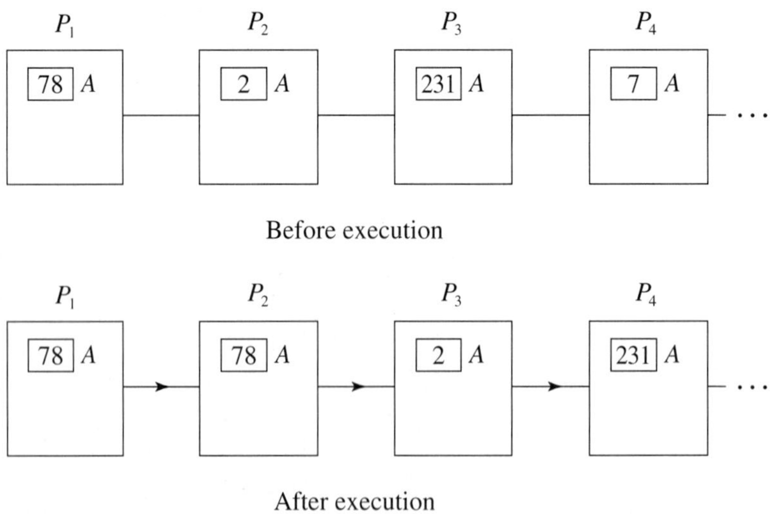

In Figure 5.24 there are processors that are both target and source during a single communication step, which is reasonable in view of the systolic nature of a communication step. At the beginning of a communication step, all active source processors initiate the process of sending (pulsing) data from their local memory to the relevant adjacent processor. At the end (after unit time) of the communication step, this data has been received (and assigned) by the target processor. Thus, it is both a natural and a common feature of many parallel algorithms to have the same processor send information at the beginning of the communication step and receive information at the end of the step. However, we *do not permit* any processor to be the target or source of more than one processor during a single communication step.

When considering complexity issues, interprocessor communication such as illustrated in Figure 5.24 is counted as a single (parallel) communication step. The total number of communication steps performed by a parallel algorithm is called the *communication complexity* of the algorithm. By requiring communication to take place between adjacent processors only, we can usually conclude that the time complexity (as measured by counting the number of parallel basic operations performed) of a parallel algorithm is proportional to the communication complexity.

5.5.4 Read/Write Statements

We assume two parallel I/O statements, **read** for input and **write** for output. Suppose, for example, that we have a list of n values stored sequentially in the input device. These n values might be read into a one-dimensional mesh M_p by the following statement.

for P_i, $1 \leq i \leq n$ **do in parallel**
  read($\mathbf{P}_i$:L)
end in parallel

This **in parallel** statement simultaneously reads the ith value in the external input device into the variable $\mathbf{P}_i$:L in the local memory of processor P_i, $i \in \{1, \ldots, n\}$.

When using **read** or **write** statements in a procedure or function, we include a **dcl** statement for all variables participating in I/O, together with **range** descriptions for such variables in the case of interconnection network models. We also use the terms **External input** and **External output** in the opening syntax.

5.5.5 Example: Pseudocode for Searching on the Two-Dimensional Mesh

We now give the pseudocode for the algorithm discussed in Section 5.2 for searching on a two-dimensional mesh. The first phase of our algorithm, involving broadcasting the value x of the search element throughout the distributed variable X (see Figure 5.12), has the following pseudocode.

```
procedure Broadcast2DMesh(P_{1,1}:X,n)
Model: two-dimensional mesh M_{q,q}, with p = n = q^2 processors
Input:   P_{1,1}:X (element to be broadcast)
Output: P_{1,1}:X is broadcast to each processor in M_{q,q}

  for i := 1 to q – 1 do
      P_{1,i+1}:X ⇐ P_{1,i}:X            {propagate X to the right across first row}
  endfor
  for i := 1 to q – 1 do
    for P_{i,j}, 1 ≤ j ≤ q do in parallel
      P_{i+1,j}:X ⇐ P_{i,j}:X            {propagate X down row by row}
    end in parallel
  endfor
end Broadcast2DMesh
```

After x has been broadcast throughout the distributed variable X, each processor $\mathbf{P}_{i,j}$ in parallel compares $\mathbf{P}_{i,j}:L$ to $\mathbf{P}_{i,j}:X$ and writes the value ∞ in to $\mathbf{P}_{i,j}:Index$ if $\mathbf{P}_{i,j}:L \neq \mathbf{P}_{i,j}:X$. The comparison is accomplished by the following pseudocode.

```
for 1 ≤ i,j ≤ q do in parallel
  if P_{i,j}:L ≠ P_{i,j}:X then
    P_{i,j}:Index := ∞
  endif
end in parallel
```

The algorithm terminates by invoking the function *Min2DMesh* for computing the minimum value in an array using the reverse-broadcast technique.

```
function Min2DMesh(X,n)
Model: two-dimensional mesh M_{q,q} with p = n = q^2 processors
Input:   X (a list of n real numbers x_1,x_2, ... , x_n)      range: P_{i,j}, 1 ≤ i,j ≤ q
Output: min {x_1,x_2, ... , x_n}

  for Row := q – 1 downto 1 do        {compute column minimums}
    for P_{i,j}, i = Row .and. 1 ≤ j ≤ q do in parallel
      {compute (i + 1)st row and ith row minimums in parallel}
      P_{i,j}:Temp ⇐ P_{i+1,j}:X      {communicate up from X to Temp}
      X := min {X,Temp}         {compute min of P_{i,j}:X and P_{i,j}:Temp}
    end in parallel
  endfor
      {compute first row minimum sequentially}
  for Column := q – 1 downto 1 do
```

Continued

```
    for P_{i,j}, i = 1 .and. j = Column do in parallel
        P_{i,j}:Temp ⇐ P_{i,j+1}:X     {communicate left from X to Temp}
        X := min {X,Temp}              {only P_{1,j} is active}
    end in parallel
  endfor
  return(P_{1,1}:X)

end Min2DMesh
```

We now give pseudocode for the algorithm *Search2DMesh*. Note that we input the search element to the front end variable x (as opposed to a distributed variable that has an instantiation in each processor).

```
function Search2DMesh(L,n,x,Index)
Model: two-dimensional mesh M_{q,q} with p = n = q^2 processors
Input: L (a list of n elements),              range: P_{i,j}, 1 ≤ i,j ≤ q
       x (a search element),                  front end variable
       Index (P_{i,j}:Index = (q − 1)i + j),  range: P_{i,j}, 1 ≤ i,j ≤ q
Output: the smallest row-major index where x occurs in L, or ∞ if
       x is not in L

  P_{1,1}:X := x
  call Broadcast2DMesh(Temp[1:n],X)
  for 1 ≤ i,j ≤ q do in parallel
     if P_{i,j}:L ≠ P_{i,j}:X then
        P_{i,j}:Index := ∞
     endif
  end in parallel
  return(Min2DMesh(Index,n))
end Search2DMesh
```

5.6 Performance Measures of Parallel Algorithms

SIMD computers have the property that all the active processors (some may remain idle) perform the *same* operation (on possibly different data) during any parallel step. If this common operation is the basic operation, we refer to the set of these operations performed during a parallel step as a *parallel basic operation*. The best-case, average, and worst-case complexities of a parallel algorithm are defined in terms of the number of parallel basic operations performed. For example, the worst-case complexity $W(n)$ of a parallel algorithm using $p(n)$ processors is defined to be the maximum number of parallel basic operations performed by the algorithm over all inputs of size n. We can define the best-case complexity $B(n)$ and the average complexity $A(n)$ of a parallel algorithm in a similar manner.

For parallel algorithms, the number of parallel basic operations performed usually depends only on the size n of the input, so that the best-case and average complexities are the same as the worst-case complexity.

One measure of the performance of a parallel algorithm results from comparing $W(n)$ to the smallest worst-case complexity $W^*(n)$ over all sequential algorithms for the problem. This leads to the following formal definition of the *speedup* $S(n)$ of the parallel algorithm.

DEFINITION 5.6.1 **Speedup of a Parallel Algorithm** Let $W(n)$ denote the worst-case complexity of a parallel algorithm for solving a given problem, and let $W^*(n)$ denote the smallest worst-case complexity over all known sequential algorithms for solving the same problem. Then, the *speedup* $S(n)$ of the parallel algorithm is defined by

$$S(n) = \frac{W^*(n)}{W(n)}. \tag{5.6.1}$$

The definition of speedup does not explicitly depend on the number of processors used by the algorithm. Thus, the speedup in isolation is not a true measure of the efficiency of the parallel algorithm. Good speedup usually comes with the additional cost of using many processors. Thus, when measuring the efficiency of a parallel algorithm, it is important to consider both the worst-case complexity $W(n)$ and the number of processors $p(n)$ used.

DEFINITION 5.6.2 **Cost of a Parallel Algorithm** Let $W(n)$ denote the worst-case complexity of a parallel algorithm for solving a given problem using $p(n)$ processors. The *cost* $C(n)$ of the parallel algorithm is then defined as

$$C(n) = p(n) \times W(n). \tag{5.6.2}$$

To judge the quality of a parallel algorithm, it is always useful to compare its cost to $W^*(n)$. A parallel algorithm is *cost optimal* if $C(n) = W^*(n)$. A parallel algorithm is considered *cost efficient* if its cost $C(n)$ is within a *polylogarithmic* factor of being cost optimal (a polylogarithmic function is a function belonging to $O(\log^k n)$).

There is a trade-off between using more processors to achieve better speedup and using fewer processors to achieve cost optimality. Indeed, a sequential algorithm having optimal worst-case complexity for a problem is at the same time cost optimal. Note that the cost $C(n)$ equals the total number of basic operations performed by the algorithm only if each *parallel* basic operation consists of $p(n)$ basic operations; that is, only if none of the $p(n)$ processors (utilized by the algorithm) are idle when a parallel basic operation is performed. Thus, if the $p(n)$ processors are active and doing useful work, we can expect the ratio $W^*(n)/C(n)$ to be close to 1 (since the worst-case complexity of a sequential algorithm is measured in terms of the total number of basic operations). The ratio $W^*(n)/C(n)$, which indicates how effectively the processors are used, is known as the *efficiency* $E(n)$:

$$E(n) = \frac{W^*(n)}{C(n)}. \tag{5.6.3}$$

It follows immediately from the definition of speedup that

$$E(n) = \frac{S(n)}{p(n)}. \tag{5.6.4}$$

Note that $E(n) \le 1$, since otherwise a faster sequential algorithm can be obtained from the parallel one. Moreover, a parallel algorithm is cost optimal if, and only if, $E(n) = 1$. Finding cost-optimal algorithms that also exhibit good speedup is usually difficult because of the trade-off mentioned earlier. Indeed, good speedup might come at a cost of using many processors, perhaps forcing more and more of the processors to remain idle as the algorithm progresses (see Figure 5.25 (page 190)).

We illustrate the performance measures by considering the algorithm *MinPRAM*. As shown earlier, *MinPRAM* has complexity

$$W(n) = \log_2 n.$$

Since the best sequential algorithm for finding the maximum of n elements has complexity $W^*(n) = n - 1$, *MinPRAM* has speedup

$$S(n) = \frac{n-1}{\log_2 n}.$$

Since *MinPRAM* utilizes $n/2$ processors, it has cost and efficiency given by

$$C(n) = \left(\frac{n}{2}\right)\log_2 n,$$

$$E(n) = \frac{2(n-1)}{n \log_2 n}.$$

However, *MinPRAM* is not cost optimal; $C(n)$ is greater than $W^*(n)$.

To illustrate the importance of using more than just speedup as a measure of performance, consider the problem of finding the minimum on a CRCW PRAM. Using $\binom{n}{2} = (n^2 - n)/2$ processors and the CRCW PRAM model, we can design an algorithm *MinCRCW* that finds the minimum value in a list $L[1{:}n]$ of size n using a single parallel comparison step! Note, however, that the cost $(n^2 - n)/2$ of *MinCRCW* is even higher than the cost of *MinPRAM*. We use the CRCW PRAM model in which many processors can write concurrently to the same memory location only if they are writing the same value. For purposes of discussion, we denote the $\binom{n}{2} = (n^2 - n)/2$ processors used by *MinCRCW* by $P_{i,j}$, $i,j \in \{1, \ldots, n\}$, $i < j$. In one parallel step a shared

Figure 5.25 (© Sidney Harris. Reprinted with permission. All rights reserved.)

memory array $Win[1{:}n]$ is initialized to the value 0. The array *Win* is used to store the results of "win-loss" comparisons of the elements in *L*. For each pair of numbers $L[i]$ and $L[j]$, $i < j$, $P_{i,j}$ reads $L[i]$ and $L[j]$, compares them, writes a 1 to $Win[i]$ if $L[i] > L[j]$, and writes a 1 to $Win[j]$ otherwise. It is clear that only one index k has the property that the corresponding array element $L[k]$ loses each of the $n - 1$ comparisons involving $L[k]$. Thus, $Win[i] = 1$, $i \neq k$, and $Win[k] = 0$. The value of k is determined in one parallel step by assigning n processors the task of reading the array *Win*.

function *MinCRCW*($L[1{:}n]$)

Model: CRCW PRAM **with** $p = (n^2 - n)/2$ **processors**

Input: $L[1{:}n]$ (a list of size n)
Output: the minimum value of a list element in L
  for $1 \le i \le n$ **do in parallel**
    $Win[i] := 0$

```
    end in parallel
    for 1 ≤ i,j ≤ n .and. i < j do in parallel
        {P_i,j reads and compares L[i] and L[j]}
        if L[i] > L[j] then
            Win[i] := 1    {processors P_i,j concurrently write 1 to Win[i]}
        else
            Win[j] := 1    {processors P_i,j concurrently write 1 to Win[j]}
        endif
    end in parallel
    for 1 ≤ i ≤ n do in parallel
        if Win[i] = 0 then IndexMin := i endif
    end in parallel
    return(L[IndexMin])
end MinCRCW
```

The action of *MinCRCW* is illustrated for a sample list of size 4 in Figure 5.26.

Figure 5.26 Action of *MinCRCW* for the sample input list [*L*: 95, 10, 6, 15]. *MinCRCW* uses six processors: P_{12}, P_{13}, P_{14}, P_{23}, P_{24}, P_{34}.

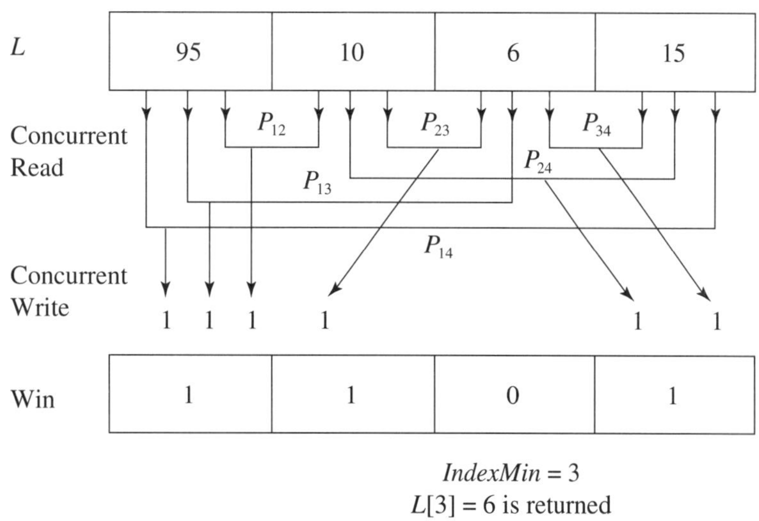

MinCRCW has worst-case complexity

$$W(n) = 1,$$

since it only performs a single parallel comparison step. Thus *MinCRCW* has speedup

$$S(n) = n - 1.$$

The drawback is seen in the large cost and low efficiency measures

$$C(n) = \frac{n^2 - n}{2}, \quad \text{and} \quad E(n) = \frac{2}{n},$$

respectively.

To further illustrate the performance measures of parallel algorithms, consider the problem of computing the dot product. When analyzing the straightforward sequential algorithm *DotProd* for performing the dot product, we would choose multiplication as our basic operation. Choosing multiplication is justified, since it is a more time-consuming operation than addition and the number of multiplications performed by the *DotProd* is one greater than the number of additions. However, in the case of *DotProdPRAM* and *DotProd2DMesh*, a single parallel multiplication step is performed, so that it is appropriate to use addition as the basic operation for these algorithms.

Choosing addition as the basic operation, *DotProdPRAM* has the same complexity $W(n) = \log_2 n$ as *SumPRAM*. Thus, its speedup over *DotProd* is $(n-1)/\log_2 n$ with respect to addition. Since *DotProdPRAM* uses n processors, its cost is $C(n) = n \log_2 n$. With respect to addition, *DotProd2DMesh* has the same complexity $W(n) = 2\sqrt{n} - 2$ and cost $C(n) = n(2\sqrt{n} - 2)$ as *SumProd2DMesh*. Thus, *DotProd2DMesh* achieves a speedup of $S(n) = (n-1)/(2\sqrt{n} - 2)$ over *DotProd*.

Figure 5.27 summarizes the analysis of most of the parallel algorithms discussed in this chapter. Given is the *order* of the number $p(n)$ of processors required, the worst-case complexity $W(n)$ (the average and best-case complexities are the same as the worst-case complexity for all the algorithms we have discussed so far), the speedup $S(n)$, the cost $C(n)$, and the processor efficiency $E(n)$.

Figure 5.27 Order formulae for the performance measures of algorithms discussed in this chapter

Algorithm	Basic Operation	$p(n)$	$W(n)$	$S(n)$	$C(n)$	$E(n)$
SumPRAM	+	n	$\log n$	$n/\log n$	$n \log n$	$1/\log n$
MinPRAM	<	n	$\log n$	$n/\log n$	$n \log n$	$1/\log n$
SearchPRAM	<	n	$\log n$	$n/\log n$	$n \log n$	$1/\log n$
Sum2DMesh	+	n	$\sqrt{n}$	$\sqrt{n}$	$n^{3/2}$	$1/\sqrt{n}$
Min2DMesh	<	n	$\sqrt{n}$	$\sqrt{n}$	$n^{3/2}$	$1/\sqrt{n}$
Search2DMesh	<	n	$\sqrt{n}$	$\sqrt{n}$	$n^{3/2}$	$1/\sqrt{n}$
SumPT	+	n	$\log n$	$n/\log n$	$n \log n$	$1/\log n$
MinPT	<	n	$\log n$	$n/\log n$	$n \log n$	$1/\log n$
SearchPT	<	n	$\log n$	$n/\log n$	$n \log n$	$1/\log n$
MinCRCW	+	n^2	1	n	n^2	$1/n$
DotProdPRAM	+	n	$\log n$	$n/\log n$	$n \log n$	$1/\log n$
DotProd2DMesh	+	n	$\sqrt{n}$	$\sqrt{n}$	$n^{3/2}$	$1/\sqrt{n}$

5.7 Speedup and Amdahl's Law

Utilizing p processors versus a single processor can yield significant speedup for certain problems. Ultimate speedup occurs for problems that involve independent operations requiring little or no communication between the processors. For example, adding two n-dimensional vectors $x = (x_1, x_2, \ldots, x_n)$ and $y = (y_1, y_2, \ldots, y_n)$, $p(n) = n$, can be done in a *single* parallel step: Simultaneously processor P_i adds x_i and y_i for $i \in \{1, \ldots, n\}$ (see Figure 5.1). This is a speedup by a factor of n over the n sequential steps required to do this vector addition on a single processor machine. However, problems yielding such ultimate speedup are rare. Most problems have an inherently sequential component and therefore cannot be completely parallelized. Amdahl's Law expresses an upper bound for the speedup achievable by any parallel algorithm for a given problem in terms of the inherently sequential component of the problem.

Observe that any parallel algorithm can be thought of as a parallelization of the associated sequential algorithm that performs the operations in each parallel operation sequentially. Now suppose that we are given a sequential algorithm that we wish to parallelize and that a fraction f of the basic operations must be performed sequentially for any input (no matter what the input size). If we have at most p processors, then the parallelization of the sequential algorithm has complexity at least $f + (1 - f)/p$ times the complexity of the sequential algorithm. Thus, for any input to the algorithm, the parallelized algorithm achieves a speedup of at most $1/[f + (1 - f)/p]$ over the sequential algorithm. This upper bound on the speedup S achieved for any input by parallelizing a sequential algorithm is known as Amdahl's Law:

$$S \leq \frac{1}{f + (1 - f)/p}.$$

Although Amdahl's Law appears to severely limit the speedup, in practice the fraction f is often dependent on the input size and diminishes as the input size increases.

5.8 Closing Remarks

The wide variety of parallel architectures available presents the fundamental problem of determining the portability of an algorithm written for a specific architecture. In the case of PRAMs, algorithms designed for CRCW and CREW models with p processors can be simulated on the EREW PRAM with p processors at a cost of a multiplicative complexity factor of $\log p$. In the case of interconnection network models, the portability question is usually handled by establishing efficient ways in which to embed one interconnection network model into another. An embedding from an interconnection network model A into another model B yields a canonical translation of any algorithm written for A to one suitable for B. Embeddings between interconnection networks is a very active research area today.

A sequential algorithm is entirely impractical unless it has polynomial complexity $O(n^k)$ for some positive integer k. Sequential algorithms are sometimes called time efficient if they have polynomial worst-case complexity. A parallel algorithm on a PRAM is considered to be time efficient if it has *polylogarithmic* ($O((\log n)^k)$) worst-case complexity. A parallel algorithm for solving a given problem, which can be solved by a sequential algorithm in polynomial time (belongs to the class P), is said to be in the class NC if it has polylogarithmic complexity using a polynomial number of processors. The class name NC is an abbreviation for *N*ick Pippenger's *C*lass. The class NC is an important class from a theoretical point of view, although assuming more than a linear number of processors is impractical for large *n* given the present state of technology. A fundamental unsolved question is whether P = NC (see Chapter 20).

References and Suggestions for Further Reading

Recently, a number of excellent graduate-level textbooks on parallel algorithms have appeared, including:

Akl, S. G. *The Design and Analysis of Parallel Algorithms*. Englewood Cliffs, NJ: Prentice Hall, 1989.

Cosnard, M, and D. Trystram. *Parallel Algorithms and Architectures*. Boston, MA: PWS Publishing Co., 1995.

JaJa, J. *An Introduction to Parallel Algorithms*. Reading, MA: Addison-Wesley, 1992.

Lakshmivarahan, S. and S. K. Dhall. *Analysis and Design of Parallel Algorithms*. New York: McGraw Hill, 1990.

Leighton, F. T. *Parallel Algorithms and Architectures*. San Mateo, CA: Morgan Kaufmann, 1992.

Quinn, M. J. *Designing Efficient Algorithms for Parallel Computers*, 2nd ed. New York: McGraw-Hill, 1994.

Smith, J. R. *The Design and Analysis of Parallel Algorithms*. New York: Oxford Univ. Press, 1993.

Tchueste, M. *Parallel Computation on Regular Arrays*. Manchester, England: Manchester Univ. Press, 1991.

A survey article discussing how parallel algorithms written for the PRAM can be simulated on more realistic machines is:

Harris, T. J. "A survey of PRAM simulation techniques," *ACM Computing Surveys*, 26 (1994): 187–206.

Exercises

Section 5.2 Architecture Constraints When Designing a Parallel Algorithm

5.1 In the column-major ordering of the two-dimensional mesh $M_{q,q}$, the processors are ordered from top to bottom in each column, starting from the first column, going to the second column, and so forth. Obtain a formula for *k*, where processor $P_{i,j}$ occurs *k*th in the column-major ordering.

5.2 **a.** Generalize the concept of row-major and column-major orderings to the three-dimensional mesh $M_{q,q,q}$, and obtain a formula for *l*, where processor $P_{i,j,k}$ occurs *l*th in the respective orderings.

b. Repeat (a) for the k-dimensional mesh $M_{q,q,\ldots,q}$ having q^k processors.

5.3 Argue that the bisection width of K_n is $n^2/4$ (n even).

5.4 Verify that the diameter of the two-dimensional mesh $M_{\sqrt{p},\sqrt{p}}$ is $2\sqrt{p} - 2$.

***5.5** Verify that the bisection width of mesh $M_{\sqrt{p},\sqrt{p}}$ belongs to $O(\sqrt{p})$.

5.6 Verify that PT_p has diameter $2\lfloor \log_2 p \rfloor$ and bisection width 1.

5.7 Consider the 3-dimensional mesh $M_{q,q,q}$ with $n = q^3$ processors.

a. Determine the maximum degree.

b. Determine the diameter.

c. Determine the bisection width.

5.8 Consider the altered two-dimensional mesh $\overline{M}_{q,q}$ where diagonal connections are added between processors.

a. Determine the maximum degree of $\overline{M}_{q,q}$.

b. Determine the diameter of $\overline{M}_{q,q}$.

c. Determine the bisection width of $\overline{M}_{q,q}$.

5.9 **a.** Design a searching algorithm for the three-dimensional mesh $M_{q,q,q}$.

b. Generalize your algorithm in (a) to the k-dimensional mesh.

Section 5.3 Computing the Dot Product on the EREW PRAM versus the Two-Dimensional Mesh

5.10 **a.** Design an algorithm for summing n numbers on the three-dimensional mesh $M_{q,q,q}$, $n = q^3$.

b. Design an algorithm from computing the dot product of two vectors on the three-dimensional mesh $M_{q,q,q}$, $n = q^3$.

c. Generalize the algorithms in (a) and (b) to the k-dimensional mesh.

Sections 5.4 and 5.5 Pseudocode Conventions for PRAMs and Interconnection Network Models

5.11 Show arrays A and B after execution of the following pseudocode.

```
for 1 ≤ i ≤ 2m – 1 .and. odd(i) do in parallel
   A[i] := A[i] – B[i]
   B[i] := 2*B[i]
   if i mod 3 = 0 then
      B[i] := A[i]
   endif
```

end in parallel

$A[1:n]$

7	2	31	4	5	1	4	0	36	3	...

$B[1:n]$

10	40	240	11	0	1	-2	0	-1	66	...

5.12 Show distributive variables A and B after execution of the following pseudocode.

```
for P_i, 1 ≤ i ≤ 2m – 1 .and. odd(i) do in parallel
   A := A – B
   B := 2*B
   if i mod 3 = 0 then
      B := A
   endif
end in parallel
```

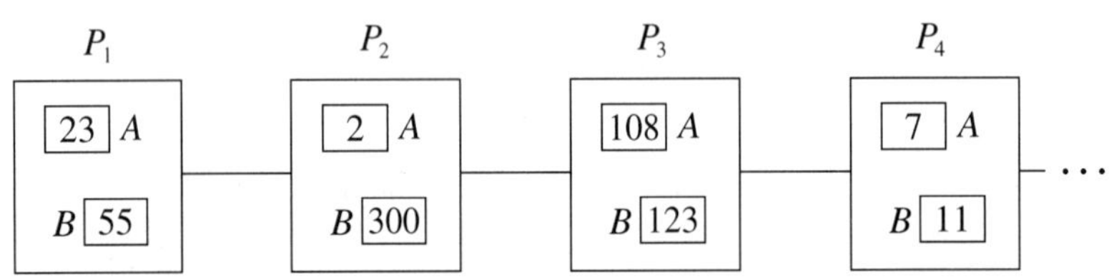

Before execution

5.13 Give pseudocode for an alternative procedure *Broadcast2DMesh2* for broadcasting the value of $P_{1,1}:X$ throughout $M_{q,q}$, based on the following figure.

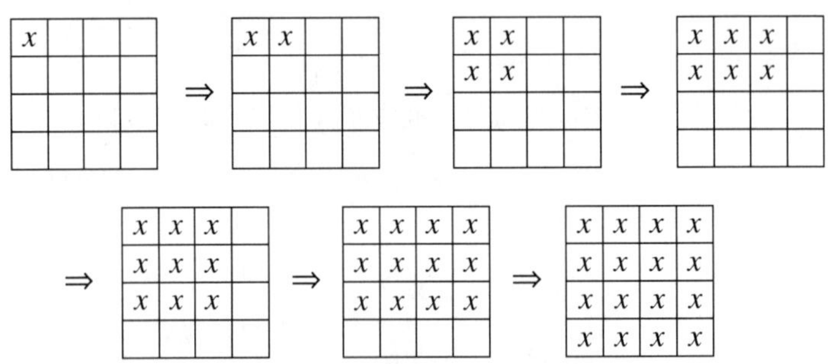

5.14 Give pseudocode for the algorithms designed in Exercise 5.9.

5.15 Give pseudocode for the algorithms designed in Exercise 5.10.

Section 5.6 Performance Measures of Parallel Algorithms

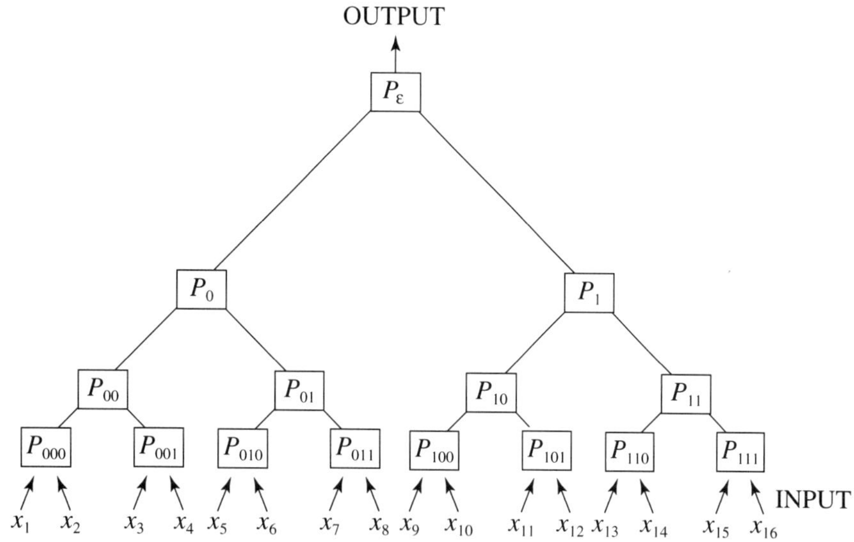

5.16 In the two-dimensional mesh $M_{q,q}$, $n = q^2$, show that $2(\sqrt{n} - 1)$ communication steps are required to relay information from $P_{q,q}$ to $P_{1,1}$.

5.17 Obtain and verify a formula for the number of communication steps required to relay information from $P_{q,q,q}$ to $P_{1,1,1}$ in the three-dimensional mesh $M_{q,q,q}$, $n = q^3$.

5.18 *SumPRAM* assumes that n is a power of two. When $2^{k-1} < n < 2^k$, we can augment the n numbers with $2^k - n$ zeros and invoke *SumPRAM* with the augmented list of 2^k numbers. This yields the desired sum in $k = \lceil \log_2 n \rceil$ parallel steps. Design a slightly more efficient algorithm that does not use this augmentation.

5.19 The processor tree PT_n is ideally suited to execute binary fan-in algorithms, where the n values are initially input in pairs to the leaves, as illustrated above. In this exercise you are asked to incorporate some simple binary string operations needed to describe interprocessor communication in pseudocode.

a. Give a pseudocode for an algorithm *MinPT* for finding the minimum of n numbers on the processor tree PT_n.

b. Suppose now that we have m sets of n numbers, and we wish find the minimum of each set on PT_n. The naive algorithm would be to simply call the algorithm *MinPT* designed in (a) m times. Analyze the complexity, speedup, and cost of this naive algorithm.

c. A more efficient algorithm for finding the minimums of m sets of n numbers results by using pipelining. For example, in the algorithm *MinPT*, as soon as the processors at a given level have computed their minimums and communicated the results up a level, they then remain idle for the remainder of the algorithm. However, they could be working on the other sets of numbers instead of remaining idle. Give pseudocode for this pipelined algorithm for summing m sets of n numbers.

d. Analyze the complexity, speedup, and cost of your algorithm in (c).

5.20 **a.** Give the pseudocode for the algorithm *SearchPT* for searching on the processor tree PT_n containing a list in the leaves, which utilizes the algorithm *MinPT* described in Exercise 5.19 for computing the minimum value stored in the leaves.

b. Analyze the complexity, speedup, and cost of your algorithm in (a).

c. Repeat parts (a) and (b) for a pipelined version of *SearchPT.*

5.21 Suppose that we have m sets of n numbers to sum, and we have a PRAM with $n - 1$ processors (as opposed to the $n/2$ processors used by *SumPRAM*). Design an efficient algorithm using pipelining to compute the m sums. Analyze the complexity, speedup, and cost of your algorithm.

5.22 **a.** Give pseudocode for an algorithm *MaxIndexPRAM* for finding the index of a maximum element in a list $L[1{:}n]$ of size n, where $n = 2^k$ for some nonnegative integer k. Analyze its performance.

b. Demonstrate the action of *MaxIndexPRAM* for the input list $L[1{:}n]$ below, where L is augmented with the appropriate number of ∞'s so that its size is a power of 2.

k	1	2	3	4	5	6	7	8	9	10	11
$L[k]$	18	3	9	2	−11	4	1	−5	20	13	9

5.23 **a.** Give pseudocode for an algorithm *MaxIndex2DMesh* for finding the index of a maximum element in a list $L[1{:}n]$ of size $n = m^2$, where the model is the two-dimensional mesh. Assume that L is stored in row-major order in the mesh. Assume also that processor $P_{i,j}$ can recognize its own indices i,j. Analyze the performance of your algorithm.

b. Demonstrate the action of *MaxIndex2DMesh* for the input list $L[1{:}n]$ below, where L is augmented with the appropriate number of ∞'s so that its size is a perfect square.

k	1	2	3	4	5	6	7
$L[k]$	12	−1	13	−10	2	11	9

5.24 Verify the performance measure formulas given in Figure 5.27.

6

Parallel Sorting

In this chapter we will discuss sorting on PRAMs, interconnection networks, and special-purpose parallel architectures called sorting networks. The design of parallel sorting algorithms is highly dependent on the particular parallel architecture under consideration. For example, using $(n^2 - n)/2$ processors on the CRCW PRAM with concurrent writes resolved by summing, we can sort a list of size n in constant time! On the other hand, the complexity of any sorting algorithm on the EREW PRAM is in $\Omega(\log n)$, and on the one-dimensional mesh is in $\Omega(n)$.

The fastest sequential comparison-based sorting algorithms have worst-case complexity $W^*(n) \in \Theta(n \log n)$ (see Chapter 10). In particular, a parallel sorting algorithm is order cost optimal if its cost $C(n)$ belongs to $\Theta(n \log n)$.

6.1 Sorting on the CRCW and CREW PRAMs

A constant-time parallel sorting algorithm on the CRCW PRAM, *SortCRCW*, is based on a similar idea used to design *MinCRCW* (see Chapter 5). With *SortCRCW*, however, we assume a more powerful (and less realistic) model of resolving concurrent writes that *sums* the values to be concurrently written.

SortCRCW computes a proper position of $L[i]$ in the sorted order of L by determining the number of elements in the list L that are smaller than $L[i]$, $i = 1, \ldots, n$. *SortCRCW* uses $\binom{n}{2} = (n^2 - n)/2$ processors P_{ij} and an auxiliary array $Win[1{:}n]$ in shared memory. $Win[i]$ will contain one less than a proper position of $L[i]$ in the sorted list. The pseudocode for computing the values of $Win[1{:}n]$ is identical to that used by *MinCRCW* for computing its array $Win[1{:}n]$. Recall that in a single comparison step, every pair of list elements $L[i]$ and $L[j]$ are compared, $i < j$. The value 1 is written to $Win[i]$ if $L[i] > L[j]$, otherwise the value 1 is written to $Win[j]$. The difference with *SortCRCW* is that concurrent writes are resolved by summing, so that $Win[i]$ computes the total number of wins of $L[i]$, as opposed to simply whether or not $L[i]$ ever won.

```
procedure SortCRCW(L[1:n])
Model: CRCW PRAM with p = (n^2 – n)/2 processors, write conflicts resolved
                                                  by summing
Input:  L[1:n] (an array of n list elements in shared memory)
Output: L[1:n] (L sorted in nondecreasing order)

  for 1 ≤ i ≤ n do in parallel
     Win[i] := 0
  end in parallel
  for 1 ≤ i,j ≤ n .and. i < j do in parallel
     if L[i] > L[j] then
        processors P_ij concurrently write 1 to Win[i]
     else
        processors P_ij concurrently write 1 to Win[j]
     endif
  end in parallel
  for i := 1 to n do in parallel
     P_i1 reads Win[i] and writes L[i] to position 1 + Win[i] of L
  end in parallel
end SortCRCW
```

In Figure 6.1, the action of *SortCRCW* in computing the array *Win*[1:*n*] is illustrated for a sample list of size 4.

While *SortCRCW* sorts in constant time, it has two major drawbacks: the powerful concurrent write resolution model and the large number of processors used. Indeed, *SortCRCW* is far from cost optimal, since it has cost $C(n) = (n^2 - n)/2 \in \Theta(n^2)$, whereas the optimal cost $W^*(n)$ is in $\Theta(n \log n)$.

Addressing the first drawback, we can transform *SortCRCW* into a $\Theta(\log n)$ algorithm *SortCREW* suitable for the CREW PRAM or a $\Theta(\log n)$ algorithm *SortPRAM* suitable for an EREW PRAM as follows. *SortCREW* uses an auxiliary two-dimensional array *Win*[1:*n*:1:*n*] of n^2 shared-memory locations to record the results of the comparisons of *L*[*i*] with *L*[*j*]. More precisely, P_{ij} writes the results of the comparison *L*[*i*] with *L*[*j*] to *Win*[*i*,*j*] or *Win*[*j*,*i*] instead of to *Win*[*i*] or *Win*[*j*] as in *SortCRCW*. Then the processor P_{i1} computes (one less than) the position of *L*[*i*] by summing the numbers in the *i*th row of *Win*[*i*,*j*]. These sums can be simultaneously computed in $\log_2 n$ parallel steps using *SumPRAM*. Of course, in addition to the $\Theta(n^2)$ processors, *SortCREW* also requires n^2 auxiliary memory locations.

Transforming *SortCRCW* to *SortPRAM* on the EREW PRAM involves the same translation of the concurrent writes. *SortPRAM* translates concurrent reads using replication and broadcasting.

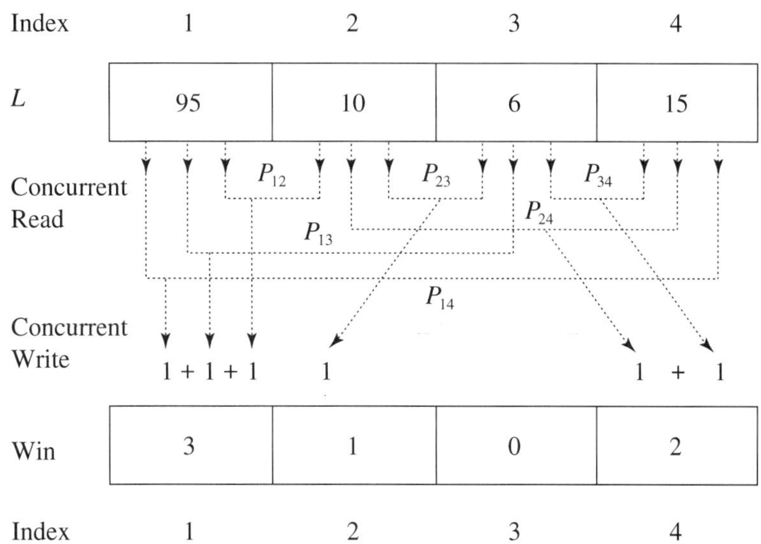

Figure 6.1 Computation of *Win*[1:4] by *SortCRCW* for the sample input list *L* : 95, 10, 6, 15. *SortCRCW* uses 6 processors: P_{12}, P_{13}, P_{14}, P_{23}, P_{24}, P_{34}.

6.2 Odd-Even Merge Sort on the EREW PRAM

SortPRAM is of limited practical interest since it uses a quadratic number of processors. A much more practical algorithm, *OddEvenMergeSortPRAM*, was discovered by K. E. Batcher in the 1960s. *OddEvenMergeSortPRAM* sorts a list of size n on the EREW PRAM using n processors, with complexity $\Theta(\log^2 n)$ and cost $\Theta(n \log^2 n)$. *OddEvenMergeSortPRAM* is within a logarithmic factor of both optimal complexity and optimal cost for a comparison-based parallel algorithm using n processors.

Before describing *OddEvenMergeSortPRAM*, we need to discuss how to exploit the inherent parallelism present in the sequential algorithm *MergeSort*. Recall that *MergeSort* is based on a divide-and-conquer strategy, which involves dividing the input list $L[1:n]$ into two sublists $L[1:n/2]$ and $L[n/2 + 1:n]$, recursively sorting the two sublists, and then merging the resulting sorted sublists. The parallelization *MergeSortPRAM* of *MergeSort* is achieved by performing the two recursive calls concurrently. In the following pseudocode for *MergeSortPRAM*, we call a generic procedure *MergePRAM* for merging the sorted sublists $L[1:n/2]$ and $L[n/2 + 1:n]$, where we assume, for convenience, that n is a power of 2.

procedure *MergeSortPRAM*($L[1:n]$) **recursive**

Model: EREW PRAM {the number of processors needed depends on the implementation of *MergePRAM*}

Input: $L[1:n]$ (a list of size $n = 2^k$)

Continued

Output: $L[1:n]$ (sorted list)

```
if n ≥ 2 then {concurrently make two recursive calls to MergeSortPRAM with
                              input lists L[1:n/2] and L[n/2 + 1:n]}
   parallelcall MergeSortPRAM(L[1:n/2] | L[n/2 + 1:n])
   call MergePRAM(L[1:n])
endif

end MergeSortPRAM
```

Figure 6.2 shows the tree of recursive parallel calls and their resolution. Figure 6.3 shows the action of *MergeSortPRAM* for a sample input list.

We are now left with the problem of designing *MergePRAM*. Our first attempt could be to simply adapt the sequential procedure *Merge* used in the sequential algorithm *MergeSort* using $n/2$ processors. The problem with this approach is the lack of parallelism present in *Merge*. For example, in the resulting parallel algorithm *MergeSortPRAM*, when merging the 2^i sublists of size $n/2^i$ at level i of the tree of recursive calls (see Figure 6.2), only 2^{i-1} processors are active, $i = 1, \ldots, k$. Since *Merge* performs $m - 1$ comparisons in the worst case when merging two sublists of size $m/2$, *MergeSortPRAM* performs $(n/2^{i-1} - 1)$ parallel comparisons in the worst case when concurrently merging the 2^i sublists of size $n/2^i$, $i = 1, \ldots, k$. Thus, the worst-case complexity $W(n)$ of *MergeSortPRAM* is given by

$$\begin{aligned} W(n) &= (n-1) + \left(\frac{n}{2} - 1\right) + \cdots + \left(\frac{n}{2^{k-1}} - 1\right) \\ &= n\left(1 + \frac{1}{2} + \frac{1}{2^2} + \cdots + \frac{1}{2^{k-1}}\right) - k \\ &= n2\left(1 - \left(\frac{1}{2}\right)^k\right) - k \quad \text{(using the geometric series formula)} \\ &= 2n\left(1 - \frac{1}{n}\right) - \log_2 n \\ &= 2n - \log_2 n - 2 \in \Theta(n). \end{aligned}$$

Figure 6.2 Tree of recursive parallel calls to *MergeSortPRAM* for $n = 8$. Each node is labeled with the indices of its associated sublist.

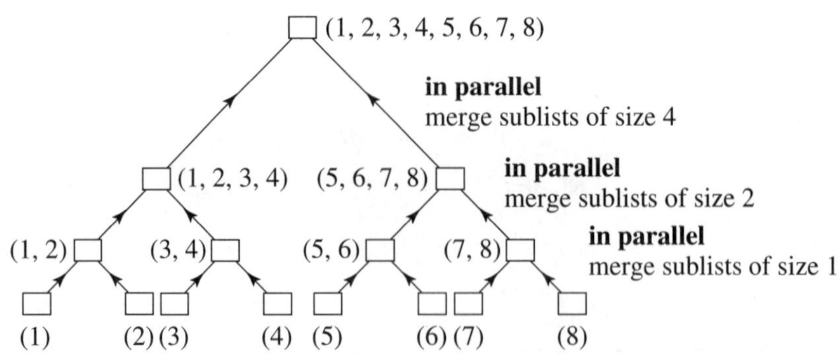

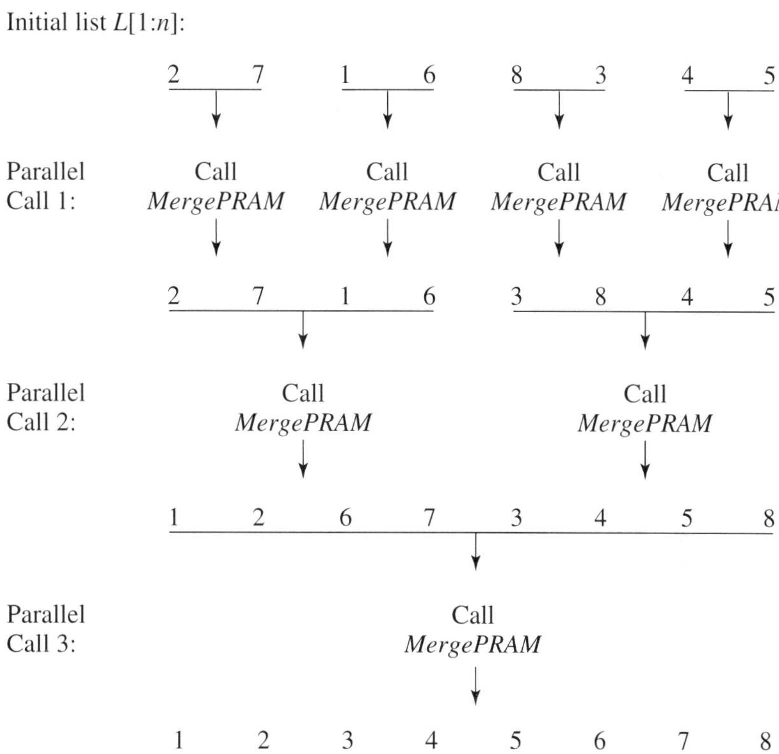

Figure 6.3 Resolution of parallel calls to *MergeSortPRAM* for the sample list $L[1:8] = (2,7,1,6,8,3,4,5)$

Thus, the speedup $S(n)$ of *MergeSortPRAM* over the sequential algorithm *MergeSort* belongs to $\Theta(\log n)$, which is disappointingly low. Further, the cost $C(n) \in \Theta(n^2)$ achieved by *MergeSortPRAM* is high (optimal cost is of the order $\Theta(n \log n)$). To achieve significant speedup and low cost, we must design a sublinear parallel procedure for merging two lists.

More than thirty years ago, Batcher designed two $\Theta(\log n)$ parallel merging procedures that are still widely used today. We now discuss one of these merging procedures: *odd-even merge.* (Batcher's other merging procedure, *bitonic merge*, is discussed in Section 6.5.)

Odd-even merge has some similarity with the sequential algorithm *ShellSort* (see Chapter 4) with increments 2 and 1. Consider a list $L[1:n]$ of size $n = 2^k$, $k \geq 1$, where $L[1:n/2]$ and $L[n/2 + 1:n]$ are sorted. Recall that a 2-subsort of $L[1:n]$ consists of independent sortings of the odd-indexed and even-indexed elements. *OddEvenMergePRAM* is based on the following proposition.

PROPOSITION 6.2.1 For n an even integer, suppose $L[1:n/2]$ and $L[n/2 + 1:n]$ are sorted sublists. Performing a 2-subsort on $L[1:n]$ results in a list in which each element of odd index is at most one position to the right of its proper position and each element of even index is at most one position to the left of its proper position. ❑

The proof of Proposition 6.2.1 is left as an exercise. An *even-odd compare-exchange step* compares $L[2i]$ to $L[2i+1]$ and exchanges them if they are out of order, $i = 1, \ldots, (n/2) - 1$. The following proposition is easily proved from Proposition 6.2.1.

PROPOSITION 6.2.2 For $n = 2^k$, $k \geq 1$, suppose $L[1:n/2]$ and $L[n/2+1:n]$ are sorted sublists. Performing a 2-subsort on $L[1:n]$ followed by an even-odd compare-exchange step results in a sorted list $L[1:n]$. ❑

Figure 6.4 illustrates Proposition 6.2.2 for a sample list $L[1:16]$.

Batcher's odd-even merge proceeds by parallelizing both the 2-subsort step and the even-odd compare-exchange step illustrated in Figure 6.4. For convenience, we assume that n is a power of 2. The 2-subsort is accomplished by first partitioning (pulling apart) $L[1:n]$ into sublists $Odd[1:n/2]$ and $Even[1:n/2]$ of elements having odd and even indices, recursively sorting these sublists, and then simply interleaving the two sorted sublists back together again into $L[1:n]$. Note that the recursion is valid since the arrays *Odd* and *Even* have the same property as the original list in that the first half and the second half of the list is sorted. After the 2-subsort is complete, a single parallel even-odd compare-exchange step yields a sorted list $L[1:n]$.

Pseudocode for the following procedure *OddEvenMergePRAM* is based directly on the preceding discussion. Note the use of the syntax:

parallelcall *OddEvenMergePRAM*(*Odd*[1:*n*/2] | *Even*[1:*n*/2])

Recall from our pseudocode conventions that this statement utilizes n processors, where half of them are used to execute the statement **call** *OddEvenMergePRAM*(*Odd*[1:*n*/2]) and the other half are concurrently used to execute the

Figure 6.4 Given a list $L[1:16]$ where sublists $L[1:8]$ and $L[9:16]$ are sorted, a 2-subsort followed by an even-odd compare-exchange step yields a sorted list.

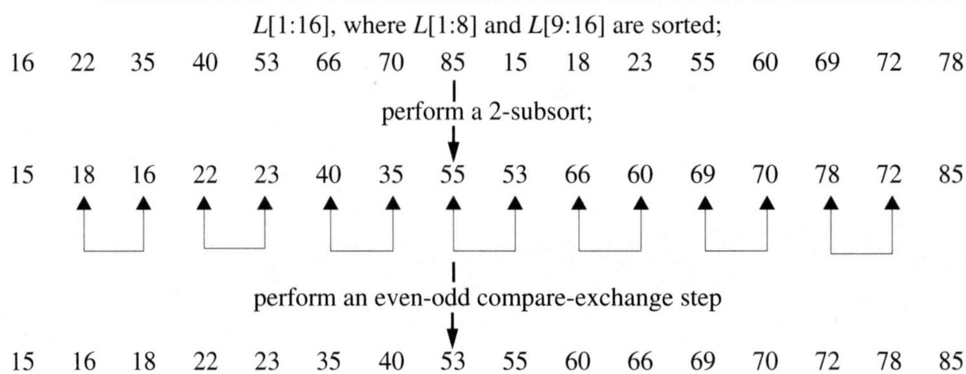

```
procedure OddEvenMergePRAM(L[1:n]) recursive
Model: EREW PRAM with p = n processors
Input:  L[1:n] (list of size n = 2^k, where sublists L[1:n/2] and L[n/2+1:n]
               are sorted)
Output: L[1:n] (sorted list)
    if n = 2 then
      if L[1] > L[2] then
        call interchange(L[1],L[2])
      endif
    else
      {separate list elements of odd and even indices}
    call oddevensplit(L[1:n],Odd[1:n/2],Even[1:n/2])
      {recursively sort list elements of odd and even indices}
    parallelcall OddEvenMergePRAM(Odd[1:n/2] | Even[1:n/2])
      {Interleave computation to achieve a 2-subsort of L[1:n]}
    for 1 ≤ i ≤ n/2 do in parallel
      L[2i–1] := Odd[i]
      L[2i] := Even[i]
    end in parallel
      {finish sort by a parallel even-odd compare and exchange}
    for 1 ≤ i ≤ (n/2) – 1 do in parallel
      if L[2i] > L[2i+1] then
        call interchange(L[2i],L[2i+1])
      endif
    end in parallel
  endif
end OddEvenMergePRAM
```

statement **call** *OddEvenMergePRAM*(*Even*[1:*n*/2]). The high-level action of *OddEvenMergePRAM* is illustrated in Figure 6.5 (page 206) for a sample list *L*[1:*n*] of size 16.

In Figure 6.6 (page 207), we show the complete parallel resolution of the parallel recursive calls to *OddEvenMergePRAM* for a sample list of size 8. Note in Figure 6.6 that for a list of size 8, *OddEvenMergePRAM* performs $3 = \log_2 8$ parallel compare-exchange steps. In general, for a list of size $n = 2^k$, *OddEvenMergePRAM* performs $k = \log_2 n$ parallel compare-exchange steps.

In the general parallel sorting paradigm *MergeSortPRAM*, if *MergePRAM* is chosen to be *OddEvenMergePRAM*, then we obtain the parallel sorting algorithm *OddEvenMergeSortPRAM*. For an input list *L*[1:*n*] of size *n*, *OddEvenMergeSortPRAM* makes a total of $\log_2 n$ parallel calls, each involving concurrent calls to *OddEvenMergePRAM*. Further, each of the input sublists involved in the *i*th parallel call has size 2^{i-1}, $i = 1, \ldots, \log_2 n$. Since *OddEvenMergePRAM* performs $\log_2 2^i = i$ parallel

Original list $L[1:16]$

16 22 35 40 53 66 70 85 | 15 18 23 55 60 69 72 78

After separating elements of odd and even indices:

Odd[1:8]	*Even*[1:8]
16 35 53 70 | 15 23 60 72	22 40 66 85 | 18 55 69 78

After performing **parallelcall** *OddEvenMergePRAM*(*Odd*[1:n/2] | *Even*[1:n/2]):

Odd[1:8]	*Even*[1:8]
15 16 23 35 53 60 70 72	18 22 40 55 66 69 78 85

$L[1:16]$ after performing an interleave computation to achieve a 2-subsort:

15 18 16 22 23 40 35 55 53 66 60 69 70 78 72 85

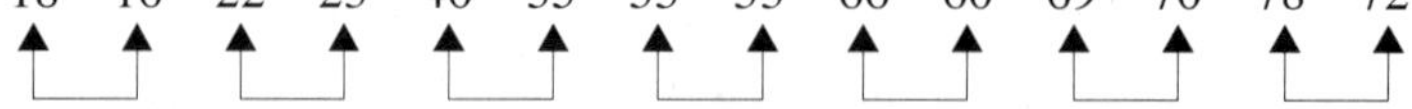

$L[1:16]$ sorted after performing an even-odd compare-exchange step:

15 16 18 22 23 35 40 53 55 60 66 69 70 72 78 85

Figure 6.5 The high-level action of *OddEvenMergePRAM*(*L*[1:16]). The final result of recursive calls to *OddEvenMergePRAM*(*Odd*[1:8]) and *OddEvenMergePRAM*(*Even*[1:8]), is shown, but not the detailed resolution of these recursive calls.

comparisons for an input list of size 2^i, it follows that the total number of parallel comparisons performed by *OddEvenMergeSortPRAM* is given by

$$W(n) = 1 + 2 + \cdots + \log_2 n = \frac{(\log_n n)^2}{2} + \frac{\log_2 n}{2} \in \Theta((\log n)^2).$$

Since *OddEvenMergeSortPRAM* utilizes $p(n) = n$ processors, its cost is given by

$$C(n) = n\left(\frac{(\log_2 n)^2}{2} + \frac{\log_2 n}{2}\right) \in \Theta(n \log^2 n).$$

Hence, as previously stated, *OddEvenMergeSortPRAM* is within a logarithmic factor of being order cost optimal.

6.3 Sorting on the One-Dimensional Mesh

Any comparison-based parallel sorting algorithm on M_p, $p = p(n) = n$, must perform at least $n - 1$ communication steps to properly decide between the relative order of the elements in P_1 and P_n. Thus, we can achieve a speedup of at most log n on the one-dimensional mesh for comparison-based sorting. We now discuss two sorting algorithms on M_p that achieve this speedup: *InsertionSort1DMesh* and *OddEvenSort1DMesh*. As the name suggests, *InsertionSort1DMesh* is a parallel version of *InsertionSort*.

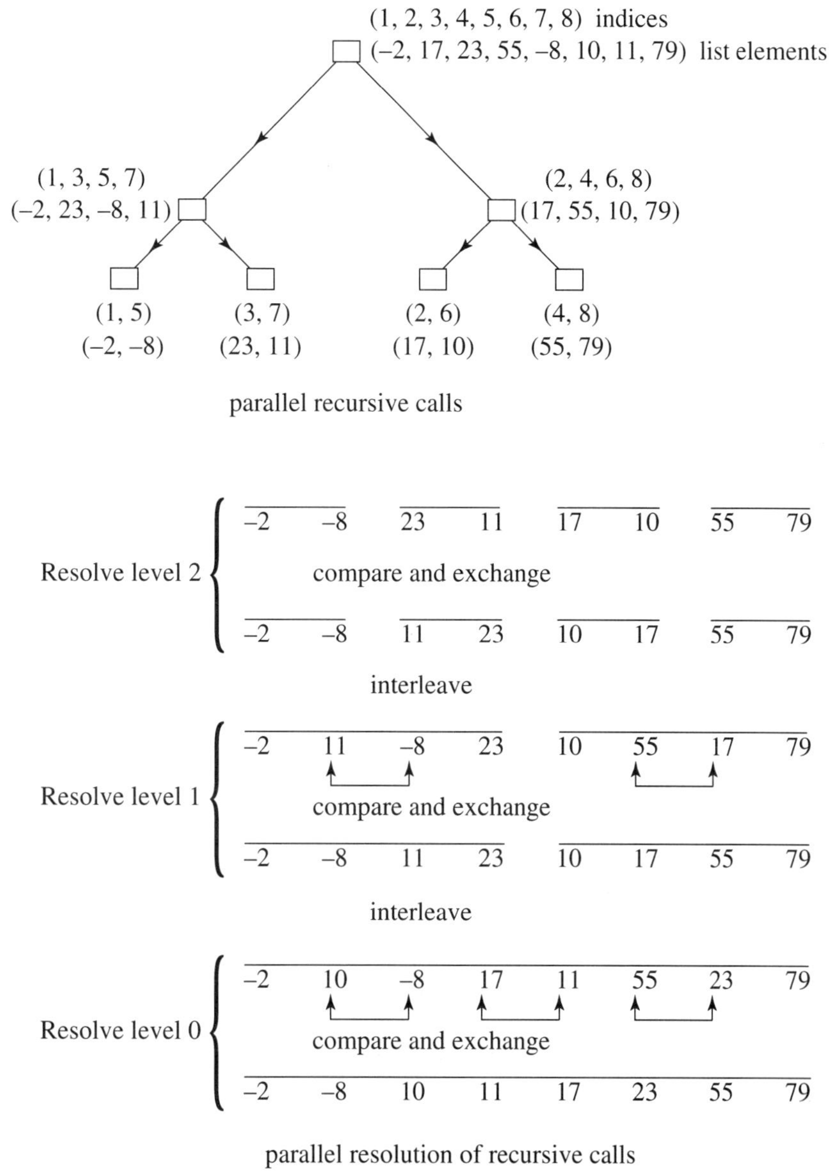

Figure 6.6 Action of *OddEvenMergePRAM* for the list $L[1{:}8] = (-2,17,23,55,-8,10,11,79)$

6.3.1 Parallelized *InsertionSort* on the One-Dimensional Mesh M_p

InsertionSort1DMesh is a parallelization of the forward scan version of *InsertionSort*. Instead of inserting elements one at a time, multiple elements are inserted concurrently in a pipelined fashion. The algorithm *InsertionSort1Mesh* inputs all values in the list to the processor P_1.

The algorithm has two phases. In phase 1, each step consists of inputting the next list element to P_1 and then comparing two values in each of the active processors. The larger of these two values is then communicated to the processor on the right (thereby activating the right-most idle processor). After the n steps in phase 1 have been completed, the next phase consists of $n - 1$ compare-and-communicate steps similar to the steps in phase 1, except at each step one more processor on the left becomes idle.

We use the two distributed variables A and B. In the compare stage, A and B are compared, and an interchange (swap) is made if $A > B$. In the communication step, we propagate the variable B to the right. For convenience of pseudocode, we initialize each A value to the sentinel value $+\infty$ (a value that wins each comparison with a list element) and we assume $p \geq n + 1$. Figure 6.7 illustrates the action of the algorithm for a list of consisting of the five elements 23,2,40,1,9.

The following pseudocode for *InsertionSort1DMesh* is based on Figure 6.7.

```
procedure InsertionSort1DMesh(n)
Model: one-dimensional mesh M_p with p = n + 1 processors
External input: n list elements read into P_1:B one element at a time
Output: sorted list resides in A, range P_i, 1 ≤ i ≤ n
dcl A,B range: P_i, 1 ≤ i ≤ n + 1
  for P_i, 1 ≤ i ≤ n + 1 do in parallel
    A := + ∞
  end in parallel
  {phase one}
  for i := 1 to n do
    read(P_1:B)        {read in ith list element}
    for P_j, 1 ≤ j ≤ i do in parallel
      if A > B then
        call interchange(A,B)
      endif
      P_{j+1}:B ⇐ P_j:B        {propagate B to the right}
    end in parallel
  endfor
  {phase two}
  for i := 2 to n do
    for P_j, i ≤ j ≤ n do in parallel
      if A > B then
        call interchange(A,B)
      endif
```

$\mathbf{P}_{j+1}:B \Leftarrow \mathbf{P}_j:B$ {propagate B to the right}
end in parallel
endfor
{Sort is now complete. Sorted list resides in $P_1:A, \ldots, P_n:A$.}
end *InsertionSort1DMesh*

InsertionSort1DMesh clearly performs $2n - 1$ **in parallel** comparison steps, so that it has complexity $W(n) = 2n - 1$, cost $C(n) = n(2n - 1)$, and speedup $S(n) \in \Theta(\log n)$.

Figure 6.7 Action of *InsertionSort1DMesh* with list 23,2,40,1,9

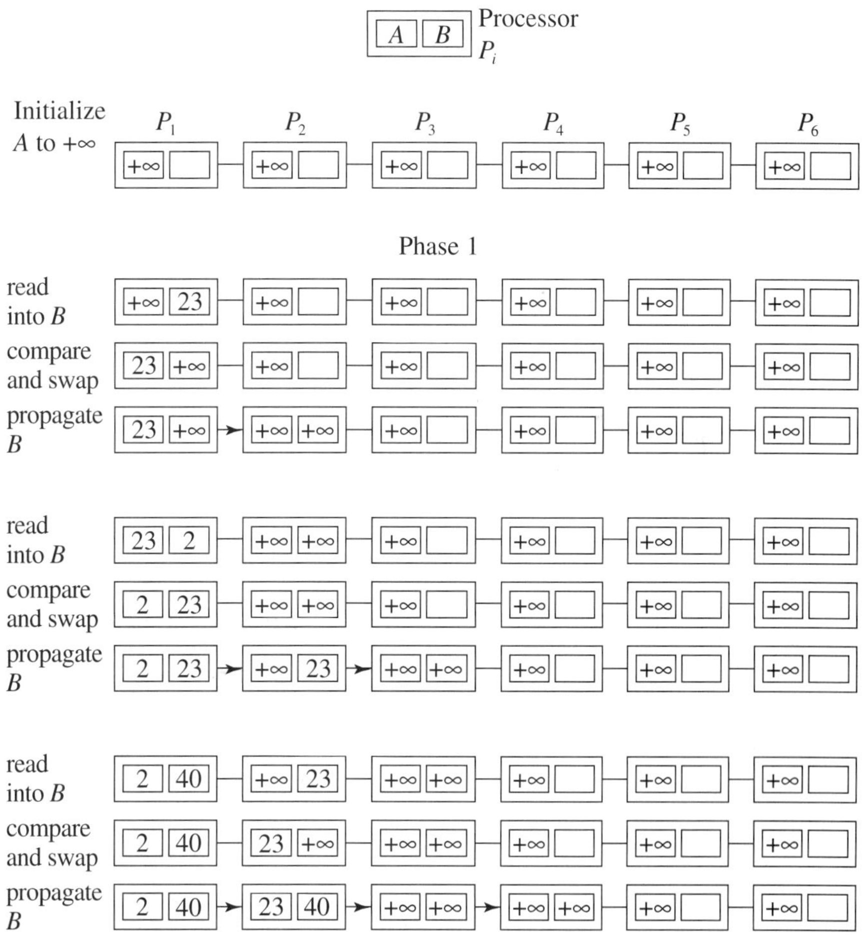

Continued

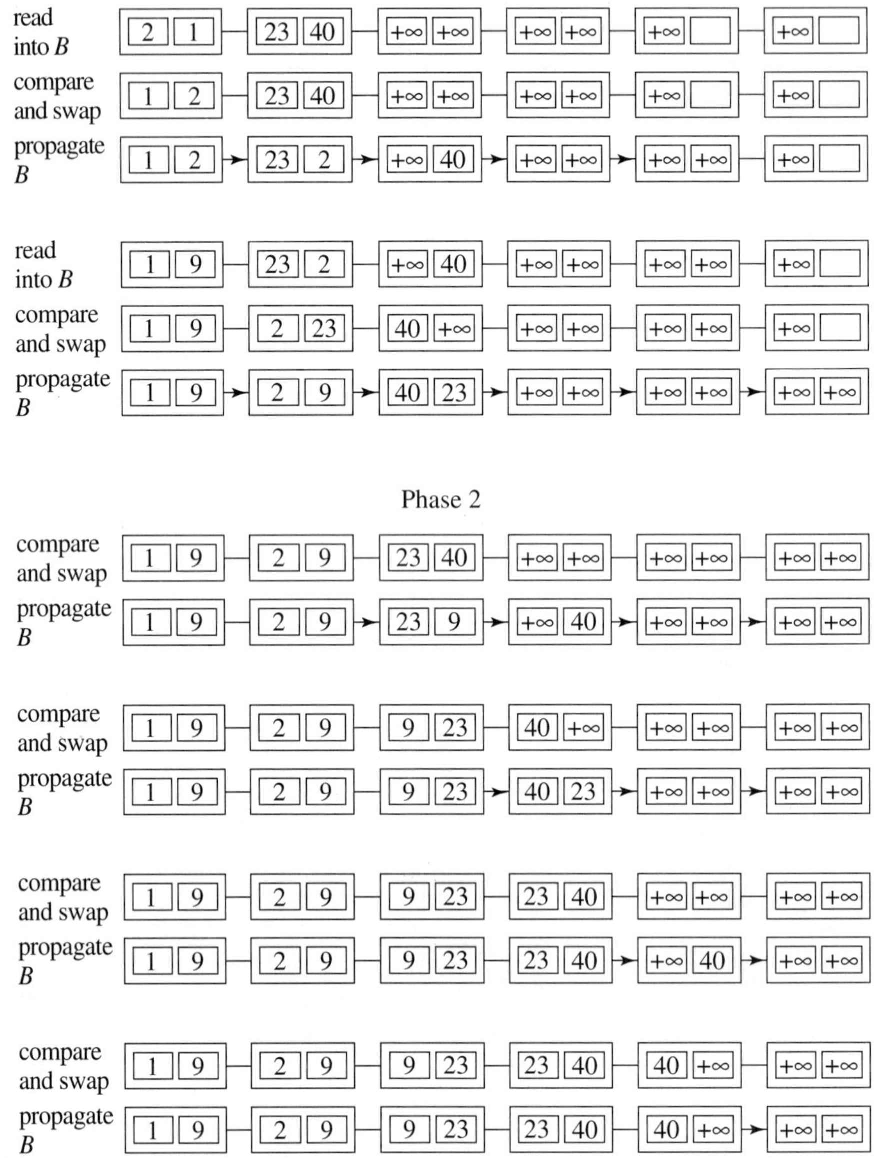

Sorted list now resides in $\mathbf{P}_j{:}A$, $i = 1,\ldots,5$.

Figure 6.7 Continued

6.3.2 Odd-Even Transposition Sort on a One-Dimensional Mesh

The algorithm *OddEvenSort1DMesh* we now describe is not a straightforward parallelization of a standard sequential algorithm. The algorithm requires n iterations, where each iteration alternates between the following two parallel steps (see Figure 6.8 (page 212)):

1. *Odd-even exchange*: For all odd i, $\mathbf{P}_i{:}L$ is compared to $\mathbf{P}_{i+1}{:}L$, and their values are interchanged if $\mathbf{P}_i{:}L > \mathbf{P}_{i+1}{:}L$.
2. *Even-odd exchange*: For all even i, $\mathbf{P}_i{:}L$ is compared to $\mathbf{P}_{i+1}{:}L$, and their values are interchanged if $\mathbf{P}_i{:}L > \mathbf{P}_{i+1}{:}L$.

```
procedure OddEvenSort1DMesh(L,n)
Model: one-dimensional mesh M_p with p = n processors
Input:  L (a list of size n),               range: P_i, 1 ≤ i ≤ n
Output: L sorted in nondecreasing order     range: P_i, 1 ≤ i ≤ n
  for Step := 1 to n do
    if odd(Step) then
      for P_i, 1 ≤ i ≤ n – 1 .and. odd(i) do in parallel {odd-even exchange}
        P_i:Temp ⇐ P_{i+1}:L {communicate left from L to Temp}
        if L > Temp then
          P_{i+1}:L ⇐ P_i:L {propagate L right}
          L := Temp
        endif
      end in parallel
    else {Step is even}
      for P_i, 2 ≤ i ≤ n – 2 .and. even(i) do in parallel {even-odd exchange}
        P_i:Temp ⇐ P_{i+1}:L {communicate left from L to Temp}
        if L > Temp then
          P_{i+1}:L ⇐ P_i:L {propagate L right}
          L := Temp
        endif
      end in parallel
    endif
  endfor
end OddEvenSort1DMesh
```

Processors	P_1	P_2	P_3	P_4	P_5	P_6
List	4	2	0	5	1	3
Step 1	2	4	0	5	1	3
Step 2	2	0	4	1	5	3
Step 3	0	2	1	4	3	5
Step 4	0	1	2	3	4	5
Step 5	0	1	2	3	4	5
Step 6	0	1	2	3	4	5

Figure 6.8 Action of *OddEvenSort1DMesh* for sample list of size 6

A formal proof of the correctness of *OddEvenSort1DMesh* using mathematical induction is given in the next chapter. Clearly, *OddEvenSort1DMesh* has worst-case complexity $W(n) = n$, so that it has cost $C(n) = n^2$ and speedup $S(n) \in \Theta(\log n)$.

Due to the communication complexity of the one-dimensional mesh, *OddEvenSort1DMesh* is actually an optimal comparison-based sorting algorithm on the 1-dimensional mesh. However, its speedup and cost are not very attractive. In the next section we design a sorting algorithm on the two-dimensional mesh that has better speedup and cost.

6.4 Sorting on the Two-Dimensional Mesh

Whenever we implement a sorting algorithm on an interconnection network model, we must specify a linear ordering of the range of processors that contain the sorted list. On the one-dimensional mesh M_n, we used the obvious linear ordering $P_1, P_2, \ldots, P_n$. For the two-dimensional mesh, it would be natural to consider the row-major ordering. However, the algorithm *Sort2DMesh* that we now design requires the snake ordering (see Figure 6.9).

Figure 6.9 Snaking M_{16} through $M_{4,4}$

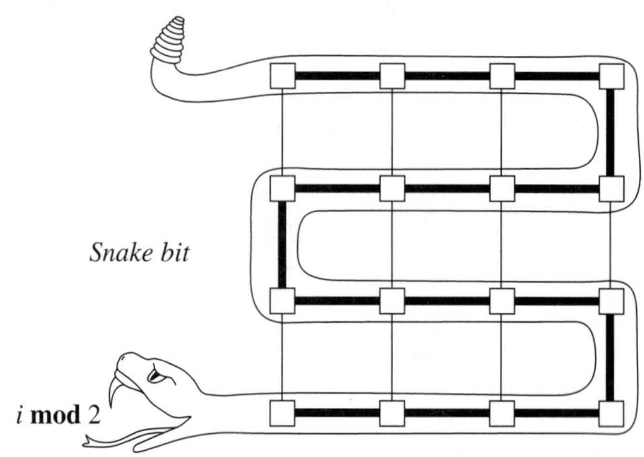

In the snake order, processor $P_{i,j}$ is the kth processor in the order, where k is given by

$$k = \begin{cases} q(i-1)+j & \text{if } i \text{ is odd}, \quad 1 \le i,j \le q, \\ qi - j + 1 & \text{if } i \text{ is even}, \quad 1 \le i,j \le q. \end{cases}$$

In Figure 6.10a, we show a picture of a sample list of size 16 stored in the distributed variable L in snake order. In Figure 6.10b (page 214), the same list is shown after it is sorted.

Figure 6.10 (a) The sample list 23,6,1,5,11,13, 55,19, – 3,12, – 5, – 7,9,55,28, – 2 of size 16 stored in the distributed variable L on the two-dimensional mesh $M_{4,4}$; (b) the sample list stored in the distributed variable L on the two dimensional mesh $M_{4,4}$ after snake ordering

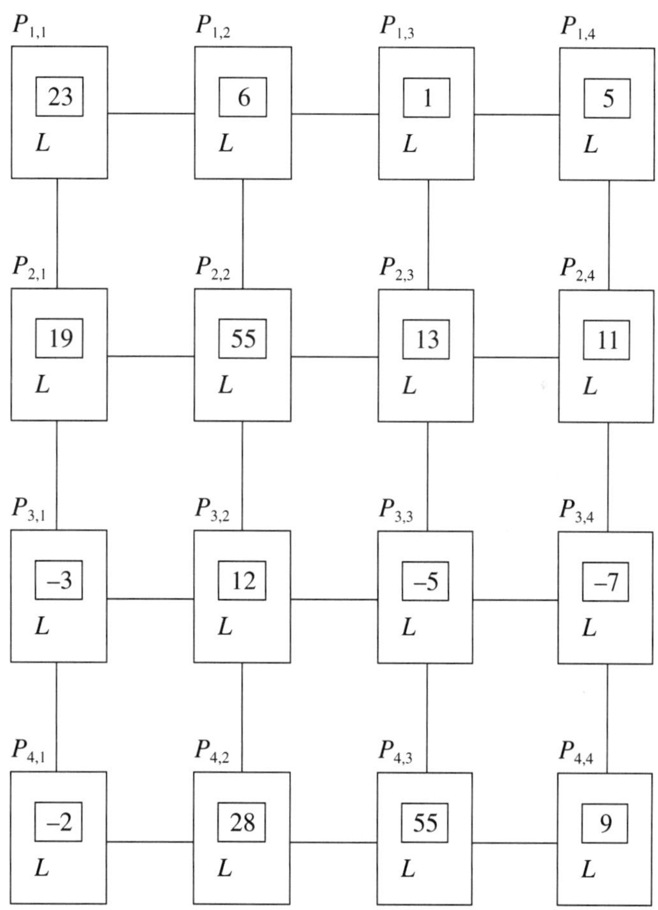

(a)

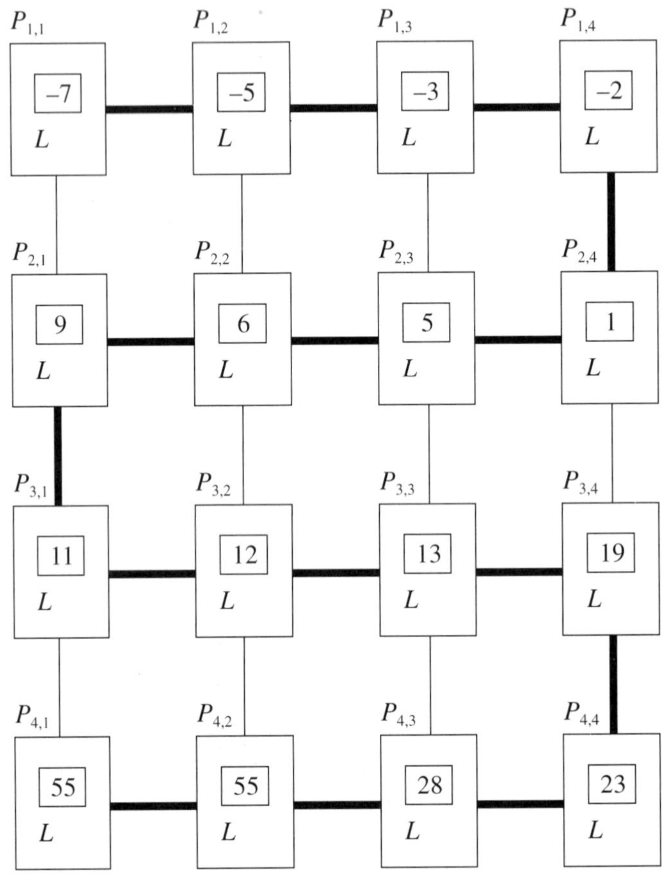

(b)

Figure 6.10 Continued

As usual, when describing our sorting algorithm *Sort2DMesh* on $M_{q,q}$, we assume for convenience that $n = q^2$. Note that each row and each column of $M_{q,q}$ corresponds to the one-dimensional mesh M_q. *Sort2DMesh* involves $\lceil \log_2 n \rceil + 1$ steps, where each odd step is a call to *OddEvenRowSort* and each even step is a call to *OddEvenColumnSort*. As the names suggest, *OddEvenRowSort* concurrently sorts the list elements in row i for all $1 \le i \le q$, and *OddEvenColumnSort* concurrently sorts the list elements in column j for all $1 \le j \le q$. The sorting of row i (and column j) is achieved by emulating the sorting algorithm *OddEvenSort1DMesh* for the one-dimensional mesh M_q. During each row-sorting step of *OddEvenRowSort*, if i is odd then row i is sorted in nondecreasing order, so that $\mathbf{P}_{i,1}{:}L \le \mathbf{P}_{i,2}{:}L, \le \cdots \le \mathbf{P}_{i,q}{:}L$, otherwise row i is sorted in nonincreasing order, so that $\mathbf{P}_{i,1}{:}L \ge \mathbf{P}_{i,2}{:}L, \ge \cdots \ge \mathbf{P}_{i,q}{:}L$. During each column-sorting step of *OddEvenColumnSort*, column j is sorted in nondecreasing order, so that $\mathbf{P}_{1,j}{:}L \le \mathbf{P}_{2,j}{:}L \le \cdots \le \mathbf{P}_{q,j}{:}L$. Figure 6.11 (page 215) shows the action of *Sort2DMesh* for the same sample list used in Figure 6.10.

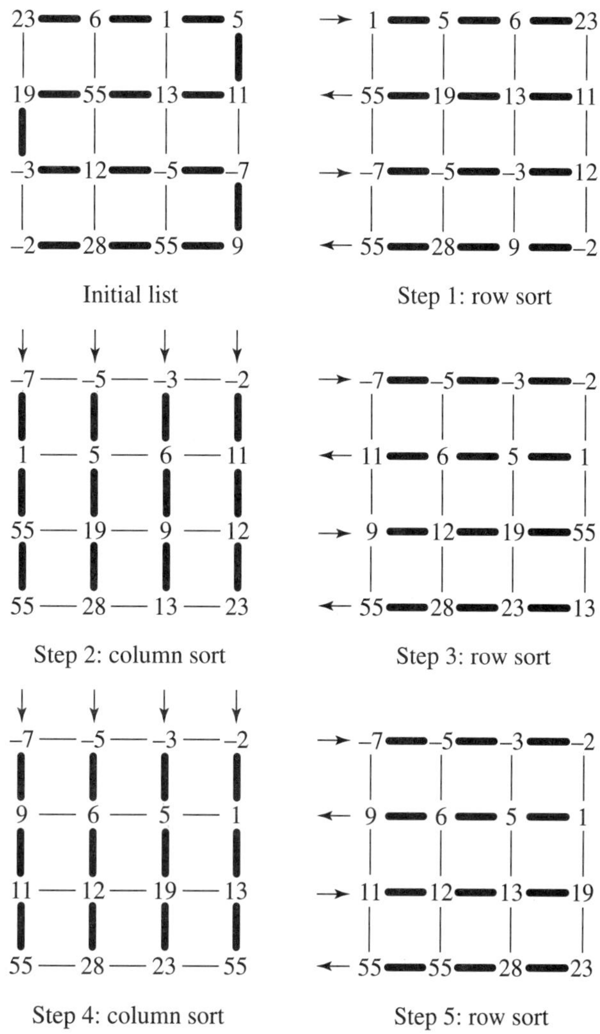

Figure 6.11 Action of *Sort2DMesh* for input list 23,6,1,5,11,13,55,19, – 3,12, – 5, – 7,9,55,28, – 2 stored in the distributed variable *L* on the two-dimensional mesh $M_{4,4}$

Procedure *Sort2DMesh* calls procedures *OddEvenRowSort* and *OddEvenColumnSort*; the pseudocode for these procedures is left as exercises.

procedure *Sort2DMesh(L,n)*

Model: two-dimensional mesh $M_{q,q}$ **with** $p = n - q^2$ **processors**

Input: L (a list of size $n = q^2$) **range:** $P_{i,j}$, $1 \le i,j \le q$

Output: L (sorted in nondecreasing snake order) **range:** $P_{i,j}$, $1 \le i,j \le q$

Continued

```
for Step := 1 to ⌈log2 n⌉ + 1 do
   if odd(Step) then
      call OddEvenRowSort(L,n)
   else
      call OddEvenColumnSort(L,n)
   endif
endfor
end Sort2DMesh
```

Since *OddEvenRowSort* and *OddEvenColumnSort* each perform $q = \sqrt{n}$ parallel comparison steps for an input list of size n, clearly *Sort2DMesh* performs $(\lceil \log_2 n \rceil + 1)\sqrt{n}$ parallel comparison steps. Thus, *Sort2DMesh* has worst-case complexity $W(n) = (\lceil \log_2 n \rceil + 1)\sqrt{n}$ and cost $C(n) = n(\lceil \log_2 n \rceil + 1)\sqrt{n}$. *Sort2DMesh* has speedup $S(n) = n \log_2 n / [(\lceil \log_2 n \rceil + 1)\sqrt{n}]$, which is approximately $\sqrt{n}$. This speedup is much better than the $\log_2 n$ speedup achieved by *OddEven1DMesh*. Further, the cost of *Sort2DMesh* is approximately $n^{3/2}\log_2 n$ versus the n^2 cost of *OddEven1DMesh.*

Remark The snake order utilized in *Sort2DMesh* is not merely an artifact, it is a fundamental design feature on which the correctness of the algorithm is based. For example, at first glance one might conjecture that a sorting algorithm utilizing row-major order would be obtained from *Sort2DMesh* by altering *OddEvenRowSort* to sort each row in increasing order from left to right. However, (see Exercise 6.14), there are lists in $M_{q,q}$ that remain *unaltered* by this modified algorithm, even though the lists are not sorted in row-major order.

*6.5 Sorting Networks

Sorting networks (also called *sorting circuits*) are made up of simple processors called *comparators*, all interconnected by a system of wires. A comparator has two input wires and two output wires, say, *MaxOut* and *MinOut*. When two data elements x,y are input to a comparator, it compares the values of x and y. It then outputs the larger of the two elements in the output wire *MaxOut* and the smaller of the two elements in the output wire *MinOut*. The action of a comparator is illustrated in Figure 6.12a; the simplified drawing shown in Figure 6.12b is sometimes more convenient to use.

In a sorting network for sorting a list of size n, we assume that the list is input and output through n input and output wires, respectively. All other wires in the network connect comparators together. For example, Figure 6.13 illustrates a sorting network for sorting a list of size 4.

When we draw a sorting network for n elements, we show n input wires on the left. A list of size n is input in one step to the n input wires and then moves from left to right through the network, ending in n output wires on the right in sorted order. Comparators are at various levels in the network, and at step i all the comparators at level i are active. The number of levels of a sorting network is called its *depth*; the sorting

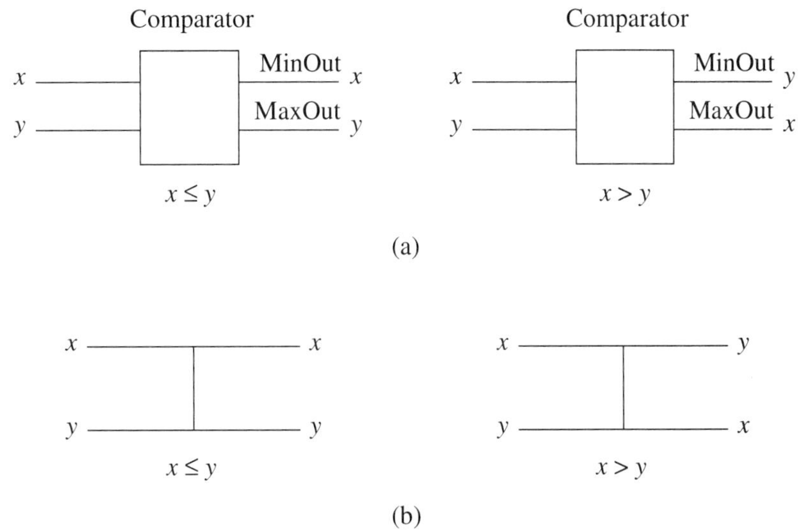

Figure 6.12 (a) The action of a comparator; (b) its simplified drawing

network in Figure 6.13 has depth 3. Note that the depth of a sorting network is the number of parallel comparison steps performed by the network when sorting a list.

6.5.1 Construction of Sorting Networks from Merging Networks

One method of constructing sorting networks is to use merging networks. An (m,n) *merging network* is similar to a sorting network, except that it is assumed that the input to the network is two sorted (in nondecreasing order) lists $a_1 \le a_2 \le \cdots \le a_m$, $b_1 \le b_2 \le \cdots \le b_n$ of size m and n, respectively, where the first list is input to the top set of m wires and the second list is input to the bottom set of n wires. A sample (2,2) merging network is shown in Figure 6.14. (Note that a

Figure 6.13 A sample input to a sorting network for four elements

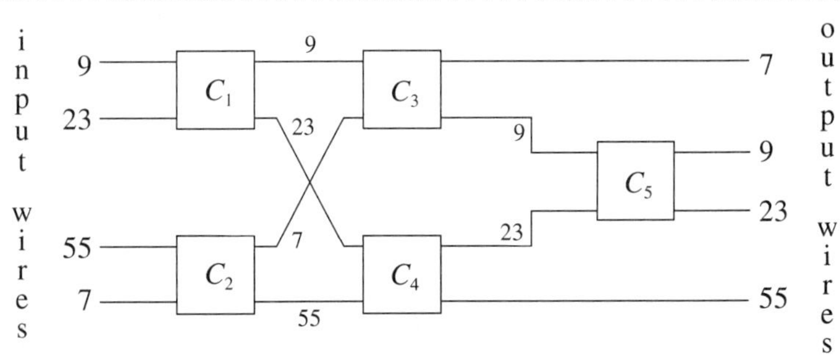

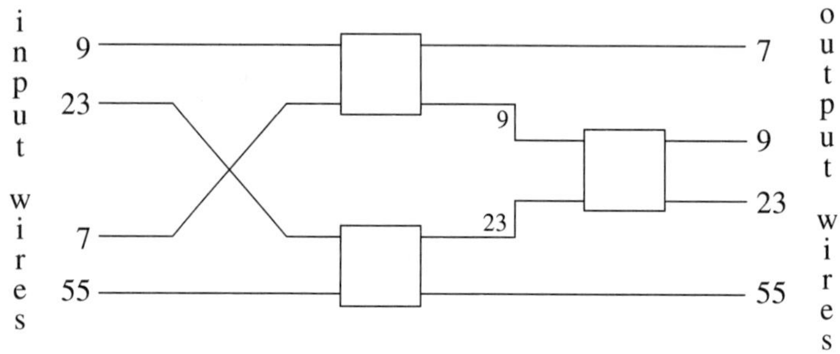

Figure 6.14 A sample input to a (2,2) merging network consisting of that part of the sorting network in Figure 6.13 to the right of comparators C_1, C_2.

(1,1) merging network consists of a single comparator and is also a sorting network for lists of size 2.)

We can construct a sorting network from merging networks in a manner similar to the way the generic procedure *MergeSortPRAM* given in Section 6.2 sorts a list on an EREW PRAM using the generic merging procedure *MergePRAM*. Recall that *MergeSortPRAM* concurrently makes two recursive calls, one to sort the sublists $L[1{:}n/2]$ and the other to sort the sublist $L[n/2 + 1{:}n]$. Then *MergePRAM* is called to merge the sorted sublists $L[1{:}n/2]$ and $L[n/2 + 1{:}n]$. Hence, a sorting network of size n can be built from merging networks using the recursive construction shown in Figure 6.15.

6.5.2 Odd-Even Sorting Network

The sorting network in Figure 6.13 is actually an implementation of Batcher's odd-even merge sort (*OddEvenMergeSortPRAM*) for $n = 4$. We now describe a sorting

Figure 6.15 Recursive construction of a sorting network, *Sorter*[n], from an $\left(\frac{n}{2}, \frac{n}{2}\right)$ merging network, *Merger*[n], for $n > 2$. *Sorter*[2] consists of a single comparator.

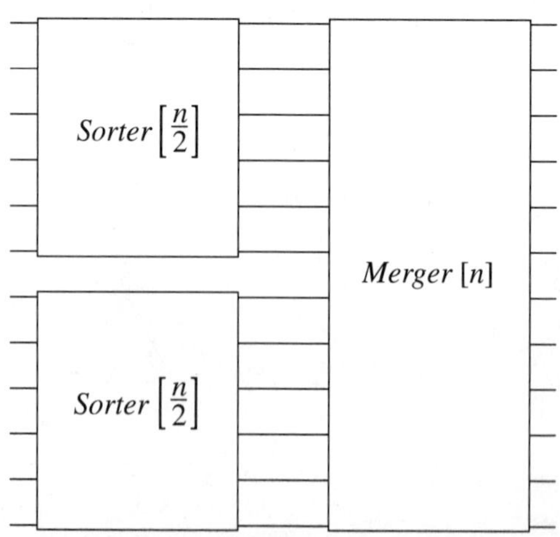

network implementing odd-even merge sort for a general $n = 2^k$. It is sufficient to describe the odd-even merging network corresponding to the merging procedure *OddEvenMergePRAM* utilized by *OddEvenMergeSortPRAM*. The sorting network corresponding to *OddEvenMergeSortPRAM* can then be obtained using the recursive construction shown in Figure 6.15.

Given $a_1 \le a_2 \le \cdots \le a_m$, $b_1 \le b_2 \le \cdots \le b_m$, where m is assumed to be a power of two, recall that odd-even merge first splits two sorted lists into four sorted lists of odd and even indexed elements, respectively.

$$a_1 \le a_3 \le \cdots \le a_{m-1}, \quad b_1 \le b_3 \le \cdots \le b_{m-1},$$
$$a_2 \le a_4 \le \cdots \le a_m, \quad b_2 \le b_4 \le \cdots \le b_m.$$

Then the two lists of odd (even) indexed elements of size $m/2$ are merged recursively. In the next step, the resulting two sorted lists (of size m) of odd and even indexed elements are interleaved, and the merge finishes with an even-odd compare-exchange step. A schematic diagram for the recursive construction of an (m,m) merging network based on odd-even merge is illustrated in Figure 6.16.

Figure 6.16 Recursive construction of odd-even (m,m) merging network for $m = 8$

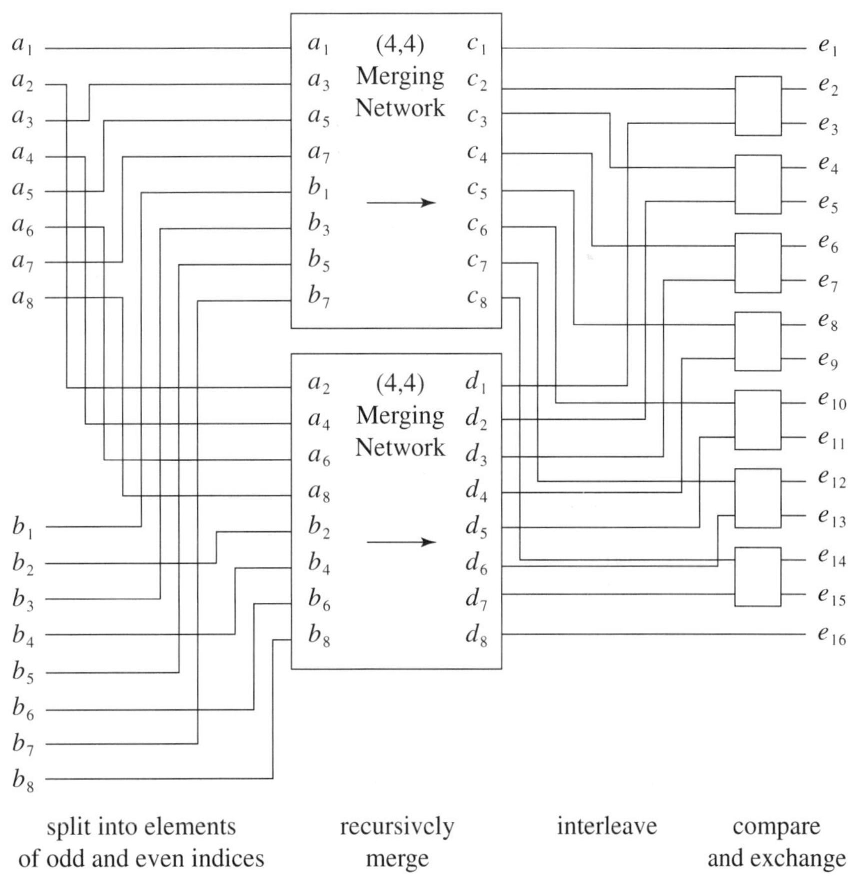

6.5.3 Bitonic Merge Sort

One of the most widely used sorting networks in practice is Batcher's BitonicMerge Sorter. The BitonicMerge Sorter is based on the algorithm *BitonicMergeSortPRAM* obtained from the generic parallel sorting algorithm *MergeSortPRAM* (see Section 6.2) by replacing the generic procedure *MergePRAM* with the merging procedure *BitonicMergePRAM*. *BitonicMergePRAM*, in turn, is based on the procedure *BitonicSortPRAM* for sorting lists consisting of two adjacent sublists (*bitonic lists*), one increasing and the other decreasing. More generally, a *bitonic list* (*sequence*) is a sequence of numbers (or elements of some ordinal type) $x_1, \ldots, x_n$ with the property that

1. there exists an index i, $1 \le i \le n$, such that $x_1 \le x_2 \le \cdots \le x_i$ (increasing) and $x_i \ge x_{i+1} \ge \cdots \ge x_n$ (decreasing) or else
2. there exists a cyclic shift of indices so that condition (1) holds.

For example, the sequence (1,4,6,8,7,5,3,2) is bitonic. The sequence (3,2,1,4,6,8,7,5) is also bitonic since it is a cyclic shift of (1,4,6,8,7,5,3,2).

For $\mathbf{x} = (x_1, \ldots, x_n)$ a bitonic sequence of even length, consider the rearrangement $\mathbf{x}' = (x_1', \ldots, x_n')$ of $\mathbf{x}$ given by

$$x_i' = \min\{x_i, x_{i+n/2}\} \quad \text{and} \quad x'_{i+n/2} = \max\{x_i, x_{i+n/2}\}, \quad i = 1, \ldots, n/2.$$

Note that $\mathbf{x}'$ can be computed from $\mathbf{x}$ by performing a single parallel compare-exchange step: Using $n/2$ processors, element x_i is compared to $x_{i+n/2}$ concurrently for each $i \in \{1, \ldots, n/2\}$, and x_i and $x_{i+n/2}$ are exchanged if x_i is larger than $x_{i+n/2}$. We refer to $\mathbf{x}'$ as the *($i \leftrightarrow n/2 + i$)-rearrangement* of $\mathbf{x}$ (see Figure 6.17).

The following proposition states an important property of bitonic sequences, which forms the basis of bitonic merge sort.

PROPOSITION 6.5.1 Let $\mathbf{x} = (x_1, \ldots, x_n)$ be a bitonic sequence of even length n, and let $\mathbf{x}'$ be the ($i \leftrightarrow n/2 + i$)-rearrangement of $\mathbf{x}$. Then $\mathbf{a} = (x'_1, \ldots, x'_{n/2})$ and $\mathbf{b} = (x'_{n/2+1}, \ldots, x'_n)$ are themselves bitonic sequences. Further, every element in $\mathbf{b}$ is at least as large as any element in $\mathbf{a}$. ❑

Figure 6.17 P_1, P_2, P_3, P_4 performing a compare-exchange step to achieve a ($i \leftrightarrow n/2 + i$)-rearrangement on a list of size 8

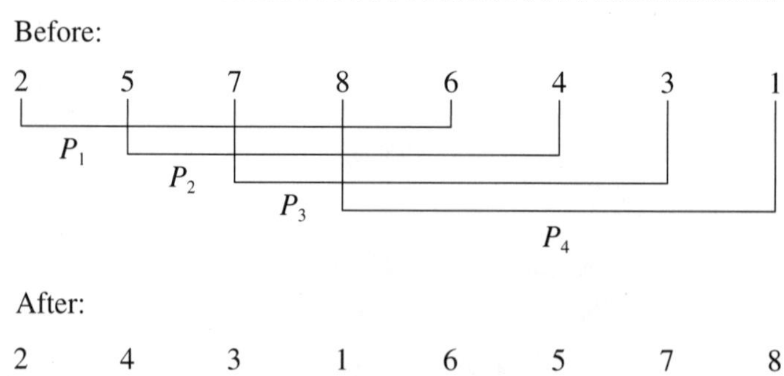

We leave the proof of Proposition 6.5.1 as an exercise. Proposition 6.5.1 is illustrated in Figure 6.18 for a sample bitonic sequence **x** of size $n = 36$.

It is instructive first to give a sequential algorithm *BitonicSort* for sorting a bitonic list. *BitonicSort* assumes as a precondition that the input list $L[1{:}n]$ is a

Figure 6.18 Illustration of Proposition 6.5.1 for a sample bitonic sequence of size 36

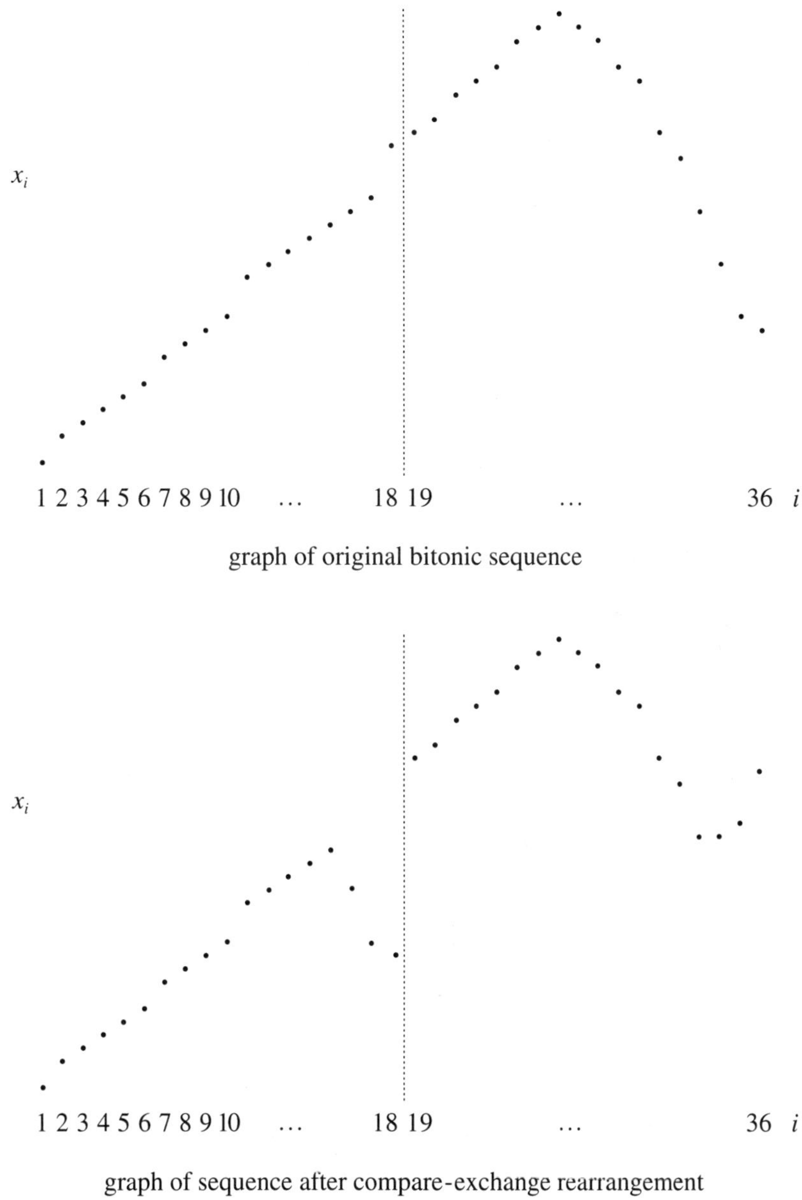

bitonic list of size $n = 2^k$ for some nonnegative integer k. *BitonicSort* is based on the divide-and-conquer paradigm and utilizes Proposition 6.5.1. We first perform a compare-exchange step yielding the $(i \leftrightarrow n/2 + i)$-rearrangement of the input list $L[1:n]$, and then we recursively call *BitonicSort* with the resulting bitonic sublists $L[1:n/2]$ and $L[n/2 + 1,n]$.

```
procedure BitonicSort(L[1:n]) recursive
Input:  L[1:n] (a bitonic list of size n = 2^k)
Output: L[1:n] (sorted list)
   if n ≥ 2 then
      for i := 1 to n/2 do          {perform a (i↔n/2 + i)-rearrangement}
         if L[i] > L[n/2 + i] then
            call interchange(L[i],L[n/2 + i])
         endif
      endfor
      call BitonicSort(L[1:n/2])
      call BitonicSort(L[n/2 + 1:n])
   endif
end BitonicSort
```

Since input list $L[1:n]$ is a bitonic list, by Proposition 6.5.1, we see that after the first **for** loop in the pseudocode for *BitonicSort* is executed, every element in the sublist $L[n/2 + 1:n]$ is at least as great as any element in the sublist $L[1:n/2]$. Therefore, the entire list $L[1:n]$ is sorted after recursively sorting $L[1:n/2]$ and $L[n/2 + 1:n]$.

The tree of recursive calls of *BitonicSort* is illustrated in Figure 6.19 for a sample bitonic list of size 8. In Figure 6.19, each node is labeled with the corresponding bitonic sublist input to *BitonicSort*.

The parallel algorithm *BitonicSortPRAM* for sorting a bitonic list is a straightforward top-down parallelization of the sequential algorithm *BitonicSort*. Starting with

Figure 6.19 Tree of recursive calls for *BitonicSort* for initial bitonic input list L:2,5,7,8,6,4,3,1

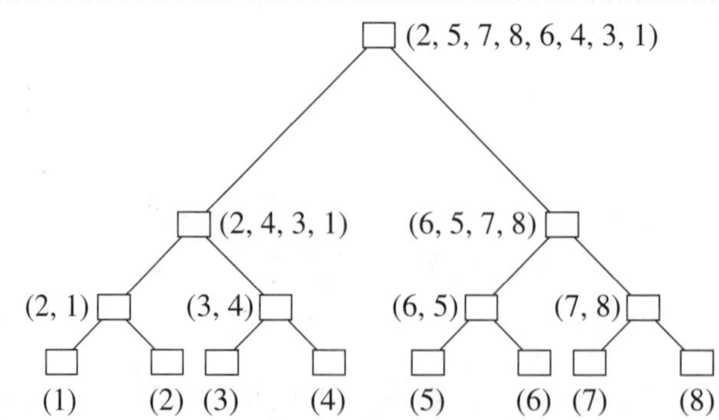

the root node at level 0, the rearrangements of the 2^i sublists at level i of the tree of recursive calls are performed concurrently, $i = 0, \ldots, k - 1$.

```
procedure BitonicSortPRAM(L[1:n]) recursive
Model: EREW PRAM with p = n/2 processors
Input:  L[1:n] (a bitonic list of size n = 2^k, k ≥ 0)
Output: L[1:n] (sorted list)
  if n ≥ 2 then
    for 1 ≤ i ≤ n/2 do in parallel      {perform a (i↔n/2 + i) –
      if L[i] > L[n/2 + i] then          rearrangement of L[1:n]}
        call interchange(L[i],L[n/2 + i])
      endif
    end in parallel
  parallelcall BitonicSortPRAM(L[1:n/2] | L[n/2 + 1:n])
end BitonicSortPRAM
```

The action of *BitonicSortPRAM* is illustrated in Figure 6.20 for the same list given in Figure 6.19.

Since the depth of the recursive calls is $\log_2 n$, *BitonicSortPRAM* has complexity $W(n) = \log_2 n$.

We are now ready to design the merging algorithm *BitonicMergePRAM*. Suppose $L[1:n]$ is a list, where both $L[1:n/2]$ and $L[n/2+1:n]$ are sorted in increasing order. To merge $L[1:n/2]$ and $L[n/2+1:n]$, we could first reverse sublist $L[n/2+1:n]$,

Figure 6.20 Action of *BitonicSortPRAM* for the same sample bitonic input list as in Figure 6.19

Initial bitonic input list L:	x_1	x_2	x_3	x_4	x_5	x_6	x_7	x_8
	2	5	7	8	6	4	3	1
	min	min	min	min	max	max	max	max
Step 1:	(x_1,x_5)	(x_2,x_6)	(x_3,x_7)	(x_4,x_8)	(x_1,x_5)	(x_1,x_5)	(x_2,x_6)	(x_3,x_7)
	2	4	3	1	6	5	7	8
	min	min	max	max	min	min	max	max
Step 2:	(x_1,x_3)	(x_2,x_4)	(x_1,x_3)	(x_2,x_4)	(x_5,x_7)	(x_6,x)	(x_5,x_7)	(x_6,x_8)
	2	1	3	4	6	5	7	8
	min	max	min	max	min	max	min	max
Step 3:	(x_1,x_2)	(x_1,x_2)	(x_3,x_4)	(x_3,x_4)	(x_5,x_6)	(x_5,x_6)	(x_7,x_8)	(x_7,x_8)
	1	2	3	4	5	6	7	8

so that the resulting list is bitonic, and then call *BitonicSortPRAM*. However, we can eliminate the step of reversing $L[n/2+1:n]$ as follows. Define the $(i \leftrightarrow n-i+1)$-rearrangement of $L[1:n]$ to be the list obtained by performing a compare-exchange step, where, in parallel, for $1 \le i \le n/2$, $L[i]$ is compared with $L[n-i+1]$ and exchanged if $L[i]$ is greater. Let $L'[1:n]$ be the list obtained from $L[1:n]$ by reversing the second half of the list, so that $L'[i] = L[i]$ and $L'[n/2+i] = L[n-i+1]$, $i = 1, \ldots, n/2$. By Proposition 6.5.1, the $(i \leftrightarrow n/2+i)$-rearrangement L'' of L' has the property that the sublists $L''[1:n/2]$ and $L''[n/2+1:n]$ are both bitonic and every element in $L''[1:n/2]$ is smaller or equal to any element in $L''[n/2+1:n]$. This fact, together with the observation that the list L'' is the $(i \leftrightarrow n-i+1)$-rearrangement of L, establishes the following proposition.

PROPOSITION 6.5.2 Let $L[1:n]$ be a list, where both $L[1:n/2]$ and $L[n/2+1:n]$ are sorted in increasing order. Then, after performing a compare-exchange step yielding the $(i \leftrightarrow n-i+1)$-rearrangement, $L[1:n]$ has the property that every element in $L[1:n/2]$ is less than or equal to any element in $L[n/2+1:n]$. Further, $L[1:n/2]$ and $L[n/2+1:n]$ are both bitonic lists. ❑

Proposition 6.5.2 immediately yields the following parallel merging algorithm *BitonicMergePRAM*.

```
procedure BitonicMergePRAM(L[1:n])
Model: EREW PRAM with p = n/2 processors
Input:   L[1:n] (a list of size n = 2^k, k ≥ 0, where L[1:n/2] and L[n/2+1:n]
                are sorted in nondecreasing order)
Output: L[1:n] (sorted list)
  if n ≥ 2 then
    for 1 ≤ i ≤ n/2 do in parallel
      if L[i] > L[n-i+1] then
        call interchange(L[i],L[n-i+1])
      endif
    end in parallel
  parallelcall BitonicSortPRAM(L[1:n/2] | L[n/2+1:n])
end BitonicMergePRAM
```

The pseudocode for *BitonicMergeSortPRAM* results from substituting *BitonicMergePRAM* for the generic merging procedure *MergePRAM* in *MergeSortPRAM*.

```
procedure BitonicMergeSortPRAM(L[1:n]) recursive
Model: EREW PRAM with p = n/2 processors
Input:   L[1:n] (a list of size n = 2^k)
Output: L[1:n] (sorted list)
```

```
    if n ≥ 2 then {concurrently make two recursive calls to BitonicMerge-
                    SortPRAM with input lists L[1:n/2] and L[n/2+1:n]}
        parallelcall BitonicMergeSortPRAM(L[1:n/2] | L[n/2+1:n])
        call BitonicMergePRAM(L[1:n])
    endif
end BitonicMergeSortPRAM
```

For an input list $L[1{:}n]$ of size n, *BitonicMergeSortPRAM* makes a total of $\log_2 n$ parallel calls to *BitonicMergePRAM*. Further, each of the 2^i input sublists involved in the ith parallel call has size $n/2^i$, $i = 1, \ldots, \log_2 n$. Since *BitonicMergePRAM* performs $\log_2 2^i = i$ parallel comparisons for an input list of size 2^i, it follows that the total number of parallel comparisons performed by *BitonicMergeSortPRAM* is given by

$$W(n) = 1 + 2 + \cdots + \log_2 n = \frac{(\log_2 n)^2}{2} + \frac{\log_2 n}{2} \in \Theta((\log n)^2).$$

Since *BitonicMergeSortPRAM* utilizes $p(n) = n/2$ processors, its cost is given by

$$C(n) = (n/2)\left(\frac{(\log_2 n)^2}{2} + \frac{\log_2 n}{2}\right) \in \Theta(n \log^2 n).$$

Thus, *BitonicMergeSortPRAM* is within a logarithm factor of being order cost optimal. *BitonicMergeSortPRAM* achieves a speedup whose order is given by

$$S(n) \in \Theta\left(\frac{n}{\log n}\right).$$

6.5.4 Bitonic Merge Sorting Network

BitonicSortPRAM and *BitonicMergePRAM* immediately yield a network *BitonicSorter*[n] for sorting a bitonic list and an $(n/2,n/2)$ merging network *BitonicMerger*[n], respectively. A recursive construction for *BitonicSorter*[n] is illustrated in Figure 6.21.

BitonicMerger[n] is constructed as follows. The first level of the *BitonicMerger*[n] compares elements in positions i and $n - i + 1$, $1 \leq i \leq n/2$. The elements in positions 1 through $n/2$ and positions $n/2 + 1$ through n, respectively, are then sorted by two bitonic sorters of size $n/2$. The general construction of *BitonicMerger*[n] is illustrated in Figure 6.22a, and the action of *BitonicMerger*[8] is illustrated in Figure 6.22*b* for two sample lists of size 4.

The sorting network *BitonicMergeSorter*[n] can now be constructed from the $(n/2,n/2)$ merging network *BitonicMerger*[n] using the recursive construction illustrated in Figure 6.15. The action of *BitonicMergeSorter*[8] is illustrated in Figure 6.23 for a sample list of size 8.

It follows from the analysis of *BitonicMergeSort* that the bitonic merge sorting network has depth

$$\frac{(\log_2 n)^2}{2} + \frac{(\log_2 n)}{2} \in \Theta((\log n)^2).$$

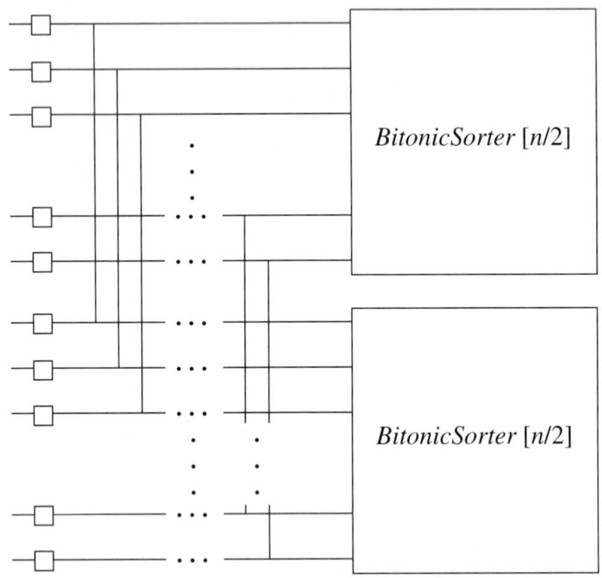

Figure 6.21 Recursive construction of *BitonicSorter*[*n*]

Figure 6.22 (a) *BitonicMerger*[*n*]

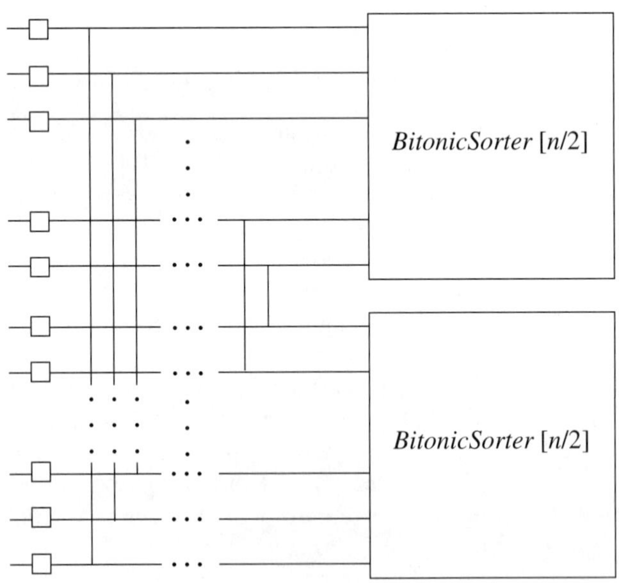

(a)

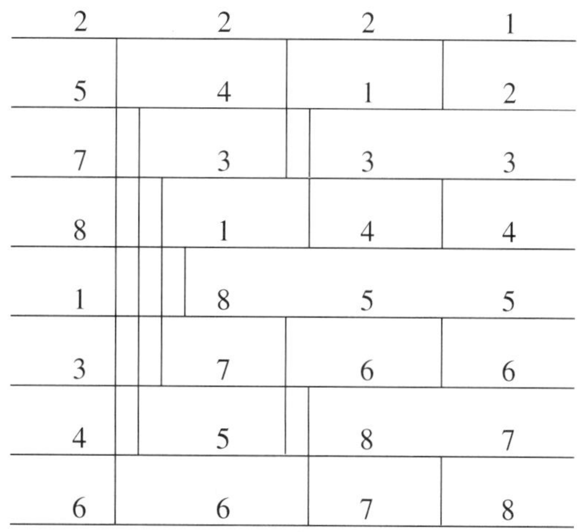

(b)

Figure 6.22 (b) the action of *BitonicMerger*[8] for sample lists 2,5,7,8 and 1,3,4,6

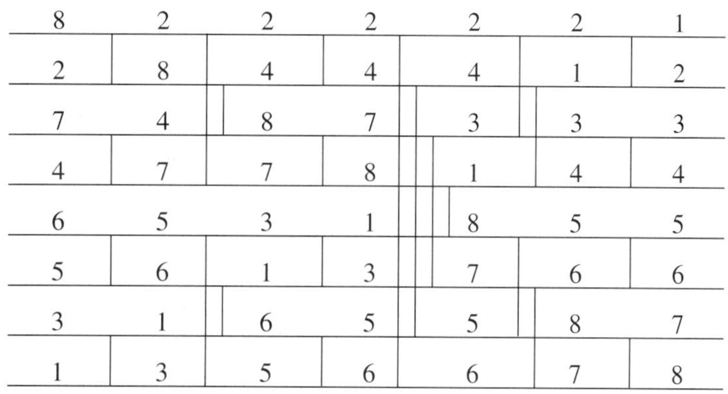

Figure 6.23 Sorting sample list 8,2,7,4,6,5,3,1 on *BitonicMergeSorter*[8]

A given sorting network yields a parallel sorting algorithm whose complexity is its depth. Since the worst-case complexity of any sequential sorting algorithm is in $\Omega(n \log n)$, and since speedup can be at most n using n processors, it follows that the depth of any sorting network is at least $\Omega((n \log n)/n) = \Omega(\log n)$. Thus, the bitonic merge sorting network is within a logarithmic factor of being optimal.

6.6 Closing Remarks

Figure 6.24 summarizes the analysis of some of the parallel algorithms discussed in this chapter. Given is the *order* of the number $p(n)$ of processors required, the worst-

Algorithm	p(n)	W(n)	S(n)	C(n)	E(n)
SortCRCW	n^2	1	$n \log n$	n^2	$(\log n)/n$
SortCREW	n^2	$\log n$	n	$n^2 \log n$	$1/n$
OddEvenMergeSortPRAM	n	$(\log n)^2$	$n/\log n$	$n(\log n)^2$	$1/\log n$
BitonicMergeSortPRAM	n	$(\log n)^2$	$n/\log n$	$n(\log n)^2$	$1/\log n$
InsertionSort1DMesh	n	n	$\log n$	n^2	$(\log n)/n$
OddEvenSort1DMesh	n	n	$\log n$	n^2	$(\log n)/n$
Sort2DMesh	n	$\sqrt{n} \log n$	$\sqrt{n}$	$n^{3/2} \log n$	$1/\sqrt{n}$

Figure 6.24 Order of complexity of some parallel sorting algorithms

case complexity W(n) (the average and best-case complexities are the same as the worst-case complexity for all the algorithms we have discussed so far), the speedup S(n), the cost C(n), and the processor efficiency E(n).

OddEvenMergeSortPRAM and *BitonicMergeSortPRAM* are within a logarithmic factor of being cost optimal. An open problem for about 20 years was to find a depth O(log n) sorting network. (Of course, such a sorting network would yield a cost-optimal parallel sorting algorithm on an EREW PRAM using n processors.) Ajtai, Komlós, and Szemerédi discovered such a network, known as the AKS sorting network, which uses the notion of an *expander graph*. Unfortunately, the constants involved in the AKS sorting network are so large that it is only of theoretical interest. In 1986 a more practical parallel sorting algorithm was discovered by Cole for the EREW PRAM with n processors. However, Cole's sorting algorithm does not yield a sorting network.

References and Suggestions for Further Reading

Akl, S. G. *Parallel Sorting Algorithms*. Orlando, FL: Academic, 1985. An extensive treatment of parallel sorting in general.

Batcher, K. E. "Sorting networks and their applications," Proceedings of the AFIPS 1968 Spring Joint Computer Conference, Atlantic City, NJ, April 30–May 2, 1968, pp. 307–314, AFIPS Press, Montvale, NJ, 1968. Discusses the $\Theta(\log^2 n)$ merging procedures odd-even merge and bitonic merge.

Jaja, J. *An Introduction to Parallel Algorithms*. Reading MA: Addison-Wesley, 1992. A good reference for sorting on PRAMs.

Leighton, F. T. *Parallel Algorithms and Architectures*. San Mateo, CA: Morgan Kaufmann Publishers, 1992. A good reference for sorting algorithms on interconnection networks.

Exercises

Sections 6.1 – 6.2 **Sorting on PRAMs**

6.1 Using broadcasting in the simulation of concurrent reads and binary fan-in in the simulations of concurrent writes, transform *SortCRCW* to an O(log n) algorithm *SortPRAM* suitable for the EREW PRAM.

6.2 Give pseudocode for *SortCREW.*

6.3 Design and give pseudocode for a nonrecursive version of *MergeSortPRAM.*

6.4 Consider the version of *MergeSortPRAM* for sorting a list $L[1{:}n]$, $n = 2^k$, where *MergePRAM* is the sequential procedure *Merge.* An alternate technique for analyzing its complexities B(n) and W(n) is to use recurrence relations.

a. Obtain recurrence relations for B(n) and W(n).

b. Using the technique of repeated substitution (see Section 1.3), solve the recurrence relations in part (a).

6.5 Repeat Exercise 6.4 for *OddEvenMergePRAM.*

6.6 Verify Proposition 6.2.1.

6.7 Verify Proposition 6.2.2.

6.8 Consider the problem of designing a parallel version of *QuickSort* for the PRAM.

a. Analyze the straightforward parallelization *QuickSortPRAM* that simply performs a recursive parallel call to *RearrangeAndPlace.*

b. Design and analyze a more efficient version of *QuickSortPRAM* involving parallelizing *RearrangeAndPlace.*

***6.9** Design and analyze parallel algorithms for the heap operations and *HeapSort* on the PRAM.

Sections 6.3 – 6.4 **Sorting on the Meshes**

6.10 Demonstrate the action of *InsertionSort1DMesh* for the following list.

k	1	2	3	4	5	6	7	8	9	10	11
$L[k]$	18	3	9	2	–11	4	1	–5	20	13	9

6.11 Demonstrate the action of *OddEvenSort1DMesh* for the list given in Exercise 6.10.

6.12 Demonstrate the action of *Sort2DMesh* for the following list (appropriately augmented).

k	1	2	3	4	5	6	7
$L[k]$	12	–1	13	–10	2	11	9

6.13 The median of a list of n elements can be computed by first sorting the elements and then finding the middle element if n is odd and the average of the two middle elements of n is even. Describe a function for computing the median on the two-dimensional mesh, where the sort is accomplished by calling *Sort2DMesh*. Have the median returned by processor P_{11}.

6.14 Show that the alteration of *Sort2DMesh* obtained by utilizing the row-major ordering instead of the snake ordering (that is, by altering *OddEvenRowSort* to sort each row in increasing order from left to right) fails to yield a sorting algorithm. In fact, show that there are lists in $M_{q,q}$ that remain *unaltered* by this modified algorithm, even though the lists are not sorted in row-major order.

6.15 Adapt Batcher's odd-even merge to the two-dimensional mesh.

Section 6.5 Sorting Networks

6.16 Consider the following bitonic sequence **x** of length 12: 23, 55, 94, 108, 123, 88, 53, 21, 12, 2, 7, 11.

List the $(i \leftrightarrow 6 + i)$-rearrangement of **x**.

6.17 Prove Proposition 6.5.1.

6.18 Suppose we say that a bitonic sequence is *increasing-decreasing* if it consists of an increasing sequence followed by a decreasing sequence (that is, no cyclic shift). Does Proposition 6.5.1 hold when "bitonic" is replaced by "increasing-decreasing"?

6.19 Show the action of *BitonicSortPRAM* for the input list 28, 55, 94, 108, 123, 88, 53, 21, 12, 2, 7, 11, 15, 17, 23, 25.

6.20 Design an alternate version of *BitonicMergeSortPRAM* that involves merging an increasing and decreasing list instead of two increasing lists, without simply reversing one of the lists.

6.21 Show the action of *BitonicMerger*[16] for sample lists 2,5,7,8,9,25,66,77 and –3,1, 2,4,18,21,24,108.

6.22 Show the action of *BitonicMergeSorter*[16] for the input list 13,2,5,4,12,16,15,7,9,8,11,10,1,3,6,14.

6.23 Verify the entries in Figure 6.24.

***6.24** Adapt Batcher's bitonic merge to the two-dimensional mesh.

Part Two

Expanding the Design and Analysis of Algorithms Toolkit

7

Induction, Correctness Proofs, and Recurrence Relations

In this chapter we introduce mathematical induction, which is the most commonly used technique for establishing the correctness of an algorithm. Correctness proofs often proceed by establishing loop invariants with the aid of mathematical induction. Mathematical induction is also useful in both the analysis of algorithms and as an algorithm design tool.

In the second part of this chapter we study recursion and recurrence relations in more detail. Recurrence relations often express the complexity of algorithms that are either explicitly or implicitly based on recursion. We have already solved such a recurrence relation in our analysis of *Towers* in Chapter 1. In this chapter we further illustrate the method by solving recurrence relations for the complexities of the previously discussed algorithms *BinarySearch*, *InsertionSort*, and *MergeSort*. We also give a brief discussion of the technique of using generating functions to solve recurrence relations. Generating functions are particularly useful in solving multiple-reference recurrence relations such as the Fibonacci recurrence.

7.1 Mathematical Induction

Often we would like to establish the validity of a proposition or formula concerning the set of positive integers such as:

$$1^2 + 2^2 + \cdots + n^2 = \frac{n\,(n+1)\,(2n+1)}{6} \quad \text{is true for each positive integer } n.$$

This formula can be easily verified for any particular (small) n by directly computing and comparing both sides of the equation. For example, if $n = 5$, we get $1^2 + 2^2 + 3^2 + 4^2 + 5^2 = 1 + 4 + 9 + 16 + 25 = 55 = 5(6)(11)/6$. The question is: How do we verify that the formula is true for *all* positive integers n? More generally, suppose we have a sequence of propositions $P(1)$, $P(2)$, … , $P(n)$, … indexed by the positive integers and we wish to establish the truth of each one of these propositions. For example, the preceding formula is the sequence of propositions

$$P(n)\colon 1^2 + 2^2 + \cdots + n^2 = \frac{n\,(n+1)\,(2n+1)}{6}.$$

Another example might be

$P(n)$: *OddEvenSort1DMesh* sorts a list of size n.

Given such a sequence, we clearly can't prove each proposition one at a time; that would take forever. Thus we use the Principle of Mathematical Induction, a very powerful method by which the truth of such a sequence of propositions $P(1)$, $P(2), \ldots, P(n), \ldots$ can be established.

7.1.1 Principle of Mathematical Induction

We first state the principle of mathematical induction in its usual form. We then state the principle in various alternative (basically equivalent) forms, which are often used in algorithm analysis.

DEFINITION 7.1.1 **Principle of Mathematical Induction** Suppose we have a sequence of propositions $P(1), P(2), \ldots, P(n), \ldots$ for which the following two steps have been established:

Basis step: $P(1)$ is true

Induction (or Implication) step: *if* $P(k)$ is true for any given k, *then* $P(k+1)$ must also be true.

Then $P(n)$ is true for all positive integers n.

The validity of the Principle of Mathematical Induction can be seen as follows. Since $P(1)$ is true, the induction step shows that $P(2)$ is true. But the truth of $P(2)$ in turn implies that $P(3)$ is true, and so forth. The induction step allows this process to continue indefinitely. The truth of $P(n)$ for all n rests ultimately on the following property of the positive integers: Every nonempty subset of the positive integers has a smallest element. The assumption that $P(k)$ is false for some k then leads to a contradiction. Indeed, if the set $\{k\colon P(k) \text{ is false}\}$ is nonempty, then it has a smallest element l. But $l > 1$, so that $P(l-1)$ is true. Letting $n = l - 1$, our induction step implies that $P(n+1) = P(l)$ is true, a contradiction.

As our first illustration of the use of mathematical induction, we now establish that the formula $1^2 + 2^2 + \cdots + n^2 = n(n+1)(2n+1)/6$ is true for all n. We proceed as follows.

Basis step: $1^2 = 1 = 1(1+1)\dfrac{1\,(1+1)\,(2+1)}{6}$ ($P(1)$ is true).

Induction step: Assume that $P(k)$ is true for a given k, so that $1^2 + 2^2 + \cdots + k^2 = k(k+1)(2k+1)/6$. We must show that it would follow that $P(k+1)$ is true, namely, that $1^2 + 2^2 + \cdots + (k+1)^2 = (k+1)(k+2)(2k+3)/6$. We have

$$
\begin{aligned}
1^2 + 2^2 + \cdots + k^2 + (k+1)^2 &= (1^2 + 2^2 + \cdots + k^2) + (k+1)^2 \\
&= k(k+1)\frac{(2k+1)}{6} + (k+1)^2 \text{ (since } P(k) \text{ is assumed true)} \\
&= \frac{(k+1)\,[k\,(2k+1) + 6\,(k+1)\,]}{6} = \frac{(k+1)\,(2k^2 + 7k + 6)}{6} \\
&= \frac{(k+1)\,(k+2)\,(2k+3)}{6},
\end{aligned}
$$

and therefore $P(k + 1)$ is true.

Thus, we have proved the induction step, and by the Principle of Mathematical Induction, the proposition $1^2 + 2^2 + \cdots + n^2 = n(n + 1)(2n + 1)/6$ is true for all positive integers n.

7.1.2 Variations of the Principle of Mathematical Induction

The following three variations of the principle of mathematical induction are frequently encountered in the analysis of algorithms. In practice, a combination of these variants might be used.

1. Often the sequence of propositions starts with an index different from 1, such as 0. Then the basis step starts with this initial index. The induction step remains the same, and the two steps together establish the truth of the propositions $P(n)$ for all n greater than or equal to this initial index.
2. Sometimes the propositions are only finite in number, $P(1), \ldots, P(l)$. Then the induction step is modified to require that $k < l$. Of course, the conclusion then drawn is that $P(1), \ldots, P(l)$ are all true if the basis and induction steps are valid.
3. The Principle of Mathematical Induction can also be stated in the following so-called *strong form*, where the induction step is as follows:

   Induction step (strong form): For any positive integer k, *if* $P(j)$ is true for *all* positive integers $j \le k$, *then* $P(k + 1)$ must also be true.

We illustrate the strong form of induction by proving the following proposition, which is the basis for the algorithm *TreeSort* given in Chapter 4.

PROPOSITION 7.1.1 An inorder traversal of a binary search tree visits the keys in sorted order.

Proof To use induction, we need to find a parameter n associated with binary search trees (depth, number of nodes, and so forth) that facilitates the induction. Here, the natural choice for n turns out to be the number of nodes in the binary search tree.

Basis step: An inorder traversal of a binary search tree on a single node visits a single key, and a list of size 1 is sorted ($P(1)$ is true).

Induction step: Assume that Proposition 7.1.1 is true for *all* binary search trees having at most k nodes. (Equivalently, we assume that $P(j)$ is true for all $1 \le j \le k$.) We now must show that it would follow that Proposition 7.1.1 is true for *any* binary search tree T on $k + 1$ nodes (equivalently, $P(k + 1)$ is true). Let L and R denote the left and right subtrees of the root of T, respectively. Note that an inorder traversal of T performs the three steps in succession:

1. performs an inorder traversal of L,
2. visits the root,
3. performs an inorder traversal of R.

Note that both L and R have at most k nodes. Thus, if L is non-null, then by induction hypothesis an inorder traversal of L visits the keys of L in sorted order. Similarly, if R is non-null, then by induction hypothesis an inorder traversal of R visits the keys of R in sorted order. By the definition of a binary search tree, the root key is not smaller than any key in L and not larger than any key in R. Hence, an inorder traversal of T visits the keys of T in sorted order, which completes the induction step. ■

7.2 Correctness Proofs

A complete analysis of an algorithm should not only describe its complexity, but should also include a verification of its *correctness*. For most of the algorithms we study, the correctness of the algorithm will be fairly obvious from a discussion of the design strategy for solving the problem. Other times the correctness is more subtle, and a rigorous proof using a technique such as induction is required.

We use induction to establish the correctness of an algorithm in two basic ways. The first way is to identify certain *loop invariants* associated with loops in the pseudocode for an algorithm. A loop invariant is a statement about the value of a variable or condition after each iteration of the loop. Establishing these values is typically done by an induction on a variable that controls the iterations of the loop.

The second way is to perform an induction on the input size n. This technique is particularly appropriate for recursive algorithms. Recursive invocations or calls typically involve problem instances of smaller input size than the original invocation or call, and the induction assumption (in the strong form) then applies. We now illustrate each of these two methods in turn.

7.2.1 Loop Invariants

To illustrate the idea of loop invariants, consider the algorithm *EgyptianPowers* given in Chapter 1. For convenience, we repeat the pseudocode for *EgyptianPowers* here.

function *EgyptianPowers*(x,n)

Input: x (a real number), n (a power of two, $n = 2^m$)
Output: x^n

$j := 1$
Product $:= x$

```
   while j < n do
      Product := Product * Product
      j := j + j
   endwhile
   return(Product)
end EgyptianPowers
```

Since j is incremented by at least 1 for each iteration of the **while** loop, clearly the **while** loop terminates after a finite number, say t, of iterations. We now prove by induction that the following two loop invariant conditions hold after the first k passes, $0 \le k \le t$:

1. j has the value 2^k,
2. *Product* has the value $x^j = x^{2^k}$.

Basis step: After $k = 0$ passes (that is, before the first pass), j has the value $1 = 2^0$ and *Product* has the value $x = x^{2^0}$. Thus, the two loop invariants have the stated values.

Induction step: Assume that after k passes, j has the value 2^k, and *Product* has the value x^j, for $k < m$. Then $j = 2^k < 2^m = n$, and another pass through the **while** loop is performed. During each pass of the **while** loop in *EgyptianPowers*, j is doubled, and *Product* is squared. Thus, after $k + 1$ passes, j has the value $2^k + 2^k = 2^{k+1}$ and *Product* has the value $x^{2^k}x^{2^k} = x^{2^{k+1}}$, completing the induction step.

By the principle of mathematical induction, we have shown that after k passes, j has the value 2^k, and *Product* has the value x^j, for $0 \le k \le t$. In particular, when $k = t$, we have $j = 2^t = n$, yielding $t = m$. Moreover, we leave the **while** loop with the value of *Product* being $x^{2^m} = x^n$. Thus, we have verified that after m passes, *EgyptianPowers* returns the value x^n, completing our correctness proof.

For another illustration of using loop invariants for correctness proofs, we consider the implementation *HornerEval* of Horner's Rule given in Chapter 1. Again, for convenience, we repeat the pseudocode for *HornerEval* here.

function *HornerEval*(a[0:n],v)

Input: $a[0{:}n]$ (an array of real numbers), v (a real number)

Output: the value of the polynomial $a_nx^n + a_{n-1}x^{n-1} + \cdots + a_1x + a_0$ at $x = v$

```
   Sum := a[n]
   for i := n – 1 downto 0 do
      Sum := Sum * v + a[i]
   endfor
   return(Sum)
end HornerEval
```

We establish the correctness of *HornerEval* by using induction to show that the following loop invariant condition holds after the kth pass of the loop in *HornerEval*, $k = 0, 1, \ldots, n$:

Sum has the value $a[n]*v^k + a[n-1]*v^{k-1} + \cdots + a[n-k+1]*v + a[n-k]$.

Basis step: Before the loop is entered, *Sum* is assigned the value $a[n]$, so that the claim is true for $k = 0$.

Induction step: Assuming that the claim is true for $k < n$, we show that it is also true for $k + 1$. By assumption, the value of *Sum* after the kth pass is

$$a[n]*v^k + a[n-1]*v^{k-1} + \cdots + a[n-k+1]*v + a[n-k].$$

Now the $(k + 1)$st pass corresponds to the loop variable i having the value $n - k - 1$. Thus, after this pass *Sum* has the value

$$(a[n]*v^k + a[n-1]*v^{k-1} + \cdots + a[n-k+1]*v + a[n-k])*v + a[n-k-1]$$
$$= a[n]*v^{k+1} + a[n-1]*v^k + \cdots + a[n-k+1]*v^2 + a[n-k]*v + a[n-k-1].$$

Thus, our claim is true for $k + 1$. By induction, our claim is therefore proved for all $k = 0, 1, \ldots, n$. The correctness of *HornerEval* now follows from our claim, since the algorithm terminates after the nth pass with the correct evaluation of the polynomial at v.

7.2.2 Induction on Input Size

To illustrate a correctness proof that uses induction on the input size, we consider a variation of the algorithm *BinarySearch* given in Chapter 3. The variation *BinarySearch2* only performs an equality test after the list has been narrowed to a single element list.

```
function BinarySearch2(L[low:high],X) recursive
Input:  L[low:high]       (a list of elements with indices between low and
                          high, sorted in nondecreasing order)
        X (search item)   {We assume that the initial invocation of
                          BinarySearch2 has low = 1 and high = n ≥ 1}
Output: Returns the index of an occurrence of X in the list, or 0 if X is
        not in the list

  if low = high then
     if X = L[low] then
        return(low)
     else
        return(0)
     endif
  else
     mid := ⌊(low + high + 1)/2⌋
     if X < L[mid] then
```

```
        return(BinarySearch2(L[low:mid – 1],X))
      else
        return(BinarySearch2(L[mid:high],X))
      endif
    endif
end BinarySearch2
```

It is easy to verify that if $low < high$, then $low < mid \leq high$. Thus, each recursive invocation narrows the size of the sublist $L[low{:}high]$ by at least one, so that eventually $low = high$ and the algorithm terminates. We complete the correctness proof of *BinarySearch2* using (the strong form of) induction on the size $k = high - low + 1$ of the sublist $L[low{:}high]$ of the original list $L[1{:}n]$.

Basis step: If $k = 1$, then $low = high$, so that the sublist $L[low{:}high]$ consists of the single element $L[low]$, and the algorithm correctly compares X to $L[low]$.

Induction step: Assume *BinarySearch2* works correctly on all sublists $L[low{:}high]$ of size $k = high - low + 1$, for $1 \leq k < n$. Now consider a sublist $L[low{:}high]$ of size $k + 1$. Since $k + 1 > 1$, $low \neq high$, so that mid is assigned the value $\lfloor (low + high + 1)/2 \rfloor$ and X is compared to $L[mid]$. If $X < L[mid]$, then *BinarySearch2* invokes itself recursively with the sublist $L[low{:}mid\text{-}1]$, otherwise *BinarySearch2* invokes itself recursively with the sublist $L[mid{:}high]$. We first consider the case where $X < L[mid]$. Since L is ordered, this means that X does *not* occur in the sublist $L[mid{:}high]$. Thus, if X occurs at all in $L[low{:}high]$, then it must occur in the sublist $L[low{:}mid - 1]$. Since $mid - 1 < high$, the size of the latter sublist is at most k. Our induction hypothesis then implies that *BinarySearch2* correctly returns the index of an occurrence of X in the sublist $L[low{:}mid\text{-}1]$, or returns 0 if X is not in this sublist. Thus, *BinarySearch2* works correctly on $L[low{:}high]$.

It remains to show that *BinarySearch2* works correctly in the case where $X \geq L[mid]$. Again, since L is ordered, if X occurs at all in the sublist $L[low{:}high]$, then it must occur in the sublist $L[mid{:}high]$. Since $mid > low$, the size of the latter sublist is at most k. Thus, our induction hypothesis implies that *BinarySearch2* correctly returns the index of an occurrence of X in the sublist $L[mid{:}high]$, or returns 0 if X is not in this sublist. Thus, *BinarySearch2* works correctly on $L[low{:}high]$, which completes the induction step and proves the correctness of *BinarySearch2*.

7.2.3 Correctness Proofs for Parallel Algorithms: The 0/1 Sorting Lemma

Many simply described parallel sorting algorithms, such as *OddEvenSort1DMesh* are surprisingly difficult to prove for a general input list. However, the powerful 0/1 Sorting Lemma allows us to restrict attention to 0/1 lists when establishing correctness. The 0/1 Sorting Lemma applies to all oblivious comparison-exchange sorting algorithms.

A comparison sorting algorithm is an *oblivious comparison-exchange* sorting algorithm if

1. whenever two list elements in positions i and j, $i < j$, are compared, if they are out of order the smaller element is placed in position i and the larger in position j; and
2. these compare-exchange steps for inputs of size n are oblivious of the input; that is, the set of pairs of list positions involved in a given parallel comparison-exchange step is prespecified, and does not depend on the outcome of any previous steps.

LEMMA 7.2.1 **0/1 Sorting Lemma** Suppose an oblivious comparison-exchange sorting algorithm works correctly on all lists of size n consisting of only 0s and 1s. Then it works correctly on all lists of size n.

Proof Let A be an oblivious comparison-exchange sorting algorithm that correctly sorts 0/1 lists. We prove the lemma by induction on the number k of distinct elements of a list $L[1{:}n]$. Clearly, the lemma is true for $k = 2$ (establishing the base case). Now, assume the lemma is true for all lists of size n having k distinct elements, and consider a list $L[1{:}n]$ having $k + 1$ distinct elements. Let M_1 denote the largest element of $L[1{:}n]$. Let $L'[1{:}n]$ be the list obtained from $L[1{:}n]$ by replacing all occurrences of M_1 with a 1, and all other elements in the list with a 0. Since A is an oblivious comparison-exchange sorting algorithm, it follows that a list element having the value M_1 is moved to exactly the same position in $L[1{:}n]$ as the corresponding 1 is in $L'[1{:}n]$. Thus, A places the largest elements in their correct positions.

Now, let M_2 denote the second largest element of $L[1{:}n]$, and consider the list $L''[1{:}n]$ obtained by replacing every occurrence of M_1 with M_2. By induction hypothesis, A sorts the list $L''[1{:}n]$. Since A is an oblivious comparison-exchange sorting algorithm, it follows that any element x of L smaller than M_2 is placed in the same position in $L[1{:}n]$ as it is in $L''[1{:}n]$. Thus x is placed in its correct position in the list $L[1{:}n]$. We have shown that A places every element different from M_2 in its correct position in the list $L[1{:}n]$. But then every element equal to M_2 necessarily must be placed in its correct position, so that A sorts $L[1{:}n]$. ■

As our first illustration of the 0/1 Sorting Lemma, we consider *OddEvenMergeSortPRAM*. The correctness of *OddEvenMergeSortPRAM* will obviously follow if we prove the correctness of *OddEvenMergePRAM*. We prove the correctness of *OddEvenMergePRAM* by induction on k, where the size of the list is $n = 2^k$. For $k = 2$, *OddEvenMergePRAM* is clearly correct. Assume, then, that *OddEvenMergePRAM* is correct for all 0/1 lists of size $n = 2^k$, and consider a 0/1 list of size 2^{k+1}. *OddEvenMergePRAM* first splits the input list into two sublists each of size 2^k consisting of the odd indexed and even indexed elements, respectively, then calls itself recursively with the latter two sublists. By induction hypothesis, the recursive calls result in sortings of the odd indexed and even indexed elements of the original list. Next, *OddEvenMergePRAM* interleaves the odd indexed list with the even indexed list, resulting in a 2-sorting of the original list. Finally, an even-odd compare-exchange step is performed.

The original list consisted of two halves that are both sorted, so they had the form

$$\underbrace{0,0,0,\ldots,0}_{p \text{ times}},1,1,\ldots,1 \quad \text{and} \quad \underbrace{0,0,0,\ldots,0}_{q \text{ times}},1,1,\ldots,1, \text{ respectively.}$$

Thus, when separating these sublists into the odd indexed elements and the even indexed elements, there are exactly $\lceil p/2 \rceil + \lceil q/2 \rceil$ 0s in the odd indexed list and $\lfloor p/2 \rfloor + \lfloor q/2 \rfloor$ 0s in the even indexed list. There are three cases to consider, depending on the three possible values 0,1,2 of the difference $(\lceil p/2 \rceil + \lceil q/2 \rceil) - (\lfloor p/2 \rfloor + \lfloor q/2 \rfloor)$. It is easy to check that in the first two cases, the interleaved list is already sorted, so that the even-odd exchange step makes no exchanges. In the third case, the interleaved list is of the form

$$\underbrace{0,0,0,\ldots,0}_{\lfloor p/2 \rfloor + \lfloor q/2 \rfloor \text{ times}}\ 0,1,0,1,1,\ldots,1.$$

Hence, the single out-of-place 1 is in an even position, and the even-odd compare-exchange step sorts the list. Thus, by induction, *OddEvenMergePRAM* correctly sorts all 0/1 lists of size $n = 2^k$. But then by the 0/1 Sorting Lemma, *OddEvenMergePRAM* (and hence *OddEvenMergeSortPRAM*) correctly sorts *all* lists of size n.

As our second correctness proof using the 0/1 Sorting Lemma, we consider *OddEvenSort1DMesh*. Recall that *OddEvenSort1DMesh* sorts a list of size n using n iterations, where each iteration is alternately one of the following two parallel steps.

1. *Odd-even exchange*: For all odd i, $\mathbf{P}_i{:}L$ is compared to $\mathbf{P}_{i+1}{:}L$, and their values are interchanged if $\mathbf{P}_i{:}L > \mathbf{P}_{i+1}{:}L$.
2. *Even-odd exchange*: For all even i, $\mathbf{P}_i{:}L$ is compared to $\mathbf{P}_{i+1}{:}L$, and their values are interchanged if $\mathbf{P}_i{:}L > \mathbf{P}_{i+1}{:}L$.

EvenOddSort1DMesh is the same as *OddEvenSort1DMesh*, except that the first exchange is an even-odd exchange (all the odd-even exchanges are replaced with even-odd exchanges and vice versa).

THEOREM 7.2.2 Let L be a list of size n such that $L[i] \in \{0,1\}$, $i = 1, 2, \ldots, n$. Then both *OddEvenSort1DMesh* and *EvenOddSort1DMesh* sort L. Hence, by the 0/1 Sorting Lemma, both *OddEvenSort1DMesh* and *EvenOddSort1DMesh* sort an arbitrary list of size n.

Proof We prove Theorem 7.2.2 by induction on the input size n. The theorem is trivially true for a list of size $n = 1$. Assume that the theorem is true for all lists of size $n = k - 1$, so that both *OddEvenSort1DMesh* and *EvenOddSort1DMesh* sort lists of size $k - 1$ satisfying the hypothesis of the theorem. Now let L be *any* list of size $n = k$ satisfying the hypothesis of the theorem.

We first consider *OddEvenSort1DMesh*. Clearly, during each iteration of *OddEvenSort1DMesh* the 0s either remain fixed or move one position closer to the beginning of the list. Similarly, the 1s either remain fixed or move one position closer to the end of the list. Note that during any iteration of *OddEvenSort1DMesh* the movement of the 1s in the sublist $L[2:n]$ are unaffected, either directly or indirectly, by the movement of $L[1]$ (this is not necessarily true of the 0s). Also note that an odd-even exchange (even-odd exchange) on L induces (with the obvious reindexing of $L[2:k]$) an even-odd exchange (odd-even exchange) on the sublist $L[2:k]$. Thus, the movement of the 1s belonging to $L[2:k]$ during the first $k-1$ iterations of *OddEvenSort1DMesh* applied to L is exactly the same as the movement of the 1s when *EvenOddSort1DMesh* is applied to the sublist $L[2:k]$ (of size $k-1$). It follows from the induction hypothesis that after $k-1$ iterations of *OddEvenSort1DMesh*, all the 1s originally in the sublist $L[2:n]$ have moved to their proper position at the end the list L.

During any iteration of *OddEvenSort1DMesh* the movement of the 0s in the sublist $L[1:k]$ are unaffected, either directly or indirectly, by the movement of $L[k]$. Also, an odd-even exchange (even-odd exchange) on L induces an odd-even exchange (even-odd exchange) on the sublist $L[1:k-1]$. Thus, the movement of the 0s belonging to $L[1:k-1]$ during the first $k-1$ iterations of *OddEvenSort1DMesh* applied to the L is exactly the same as the movement of the 0s when *OddEvenSort1DMesh* is applied to the sublist $L[1:k-1]$. It follows from our induction hypothesis that after $k-1$ iterations of *OddEvenSort1DMesh*, all the 0s originally in the sublist $L[1:k-1]$ have moved to their proper position at the beginning of the list L.

We have just shown that after $k-1$ iterations of *OddEvenSort1DMesh*, all the elements except possibly $L[1]$ and $L[k]$ are in their correct positions. Thus, after $k-1$ iterations of *OddEvenSort1DMesh*, we have either a sorted list or a list with the following properties: For some index j, the elements in positions $1, 2, \ldots, j-1$ are all equal to 0, the element in position j equals 1, the element in position $j+1$ equals 0, and the remaining elements are all equal to 1. We may assume that we have the latter list. Since the element in position j is greater than the element in position $j+1$ after iteration $k-1$, it follows that the elements in positions j and $j+1$ were not compared during iteration $k-1$. Thus, they are compared and interchanged during iteration k, yielding a sorted list. This shows that *OddEvenSort1DMesh* works correctly on L. Similar arguments show that a sorted list is obtained when *EvenOddSort1DMesh* is applied to L. ■

7.3 Mathematical Properties of Binary Trees

The analysis of the complexity of many problems and algorithms discussed in this text depends on various mathematical properties of binary trees. In this section we establish a number of these properties relating to depth, internal path length, and leaf path length.

7.3.1 Lower Bounds for the Depth of Binary Trees

As mentioned in Chapter 2, the following proposition establishes the useful and intuitively clear result that the depth of a binary tree having N nodes is at least the depth of the complete binary tree T_N on N nodes. Recall that, given any binary tree

T, we use the notation $N = N(T)$, $I = I(T)$, $L = L(T)$, and $D = D(T)$ for the number of nodes, number of internal nodes, number of leaf nodes, and the depth of T, respectively.

PROPOSITION 7.3.1 Suppose T is any binary tree having N nodes. Then T has depth at least $\lfloor \log_2 N \rfloor$; that is,

$$D(T) \geq \lfloor \log_2 N \rfloor. \tag{7.3.1}$$

The lower bound $\lfloor \log_2 N \rfloor$ is achieved for the complete binary tree T_N.

Proof Let n_i denote the number of nodes of T at level i, $i = 0, \ldots, D$. Then by (2.4.2), $n_i \leq 2^i$, so that $N \leq 1 + 2 + 2^2 + \cdots + 2^D = 2^{D+1} - 1$. Hence,

$$2^{D+1} \geq N + 1. \tag{7.3.2}$$

Taking logarithms of both sides of (7.3.2) and subtracting one, we have

$$D \geq \log_2(N + 1) - 1. \tag{7.3.3}$$

But D is an integer, so (7.3.3) implies

$$D \geq \lceil \log_2(N + 1) - 1 \rceil. \tag{7.3.4}$$

Since $\lceil \log_2(n + 1) - 1 \rceil = \lfloor \log_2 n \rfloor$ for all integers $n \geq 1$, (7.3.1) follows from (7.3.4). By (2.4.3), T_N has depth $\lfloor \log_2 N \rfloor$, which completes the proof of Proposition 7.3.1. ■

If $N = 2^k - 1$, then T_N is the only binary tree whose depth achieves the lower bound of $\lfloor \log_2 N \rfloor$. For a general N (different from $N = 2^k - 1$), there are many binary trees on N nodes that achieve the lower bound depth of $\lfloor \log_2 N \rfloor$. In Figure 7.1, a tree having 16 nodes and depth 4 ($= \lfloor \log_2 N \rfloor$) is given, which is quite different from the complete tree T_{16}.

Figure 7.1 A noncomplete tree T with $n = 16$ nodes having depth $= \lfloor \log_2 16 \rfloor = 4$

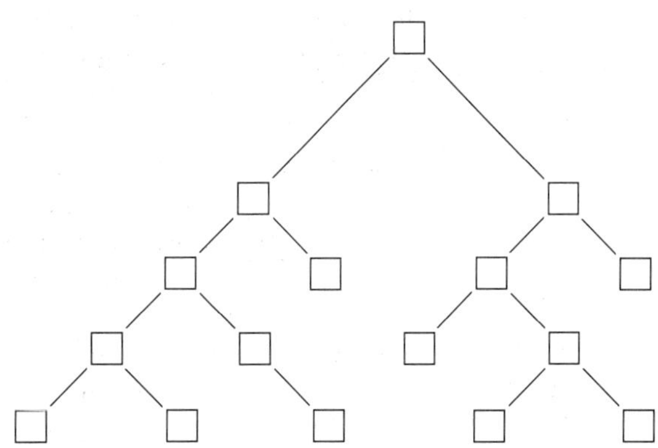

To establish various other mathematical properties of binary trees, we need to introduce the notion of a 2-tree. A *2-tree* is a binary tree such that every node that is not a leaf has exactly two children. Any binary tree T can be canonically mapped to an associated 2-tree $\hat{T}$ having the same number of leaf nodes and no larger depth. The 2-tree $\hat{T}$ is obtained from T as follows: If a node of T has only one child, then we contract the edge joining this node with its child (so that the node is identified with its child). This process is repeated until we end up at the 2-tree $\hat{T}$ (see Figure 7.2). Clearly, $L(T) = L(\hat{T})$, and $D(T) \geq D(\hat{T})$.

The following proposition establishes a simple relationship between the number of internal nodes and the number of leaf nodes of a 2-tree.

PROPOSITION 7.3.2 Suppose T is any 2-tree. Then the number of leaf nodes is one greater than the number of internal nodes of T; that is,

$$I(T) = L(T) - 1. \tag{7.3.5}$$

Equivalently, we have

$$N(T) = 2L(T) - 1. \tag{7.3.6}$$ ❑

Proposition 7.3.2 is easily proved by induction (see Exercise 7.13). Proposition 7.3.2 fails dramatically for binary trees in general, since an entirely skewed binary tree has only one leaf node.

PROPOSITION 7.3.3 Suppose T is any binary tree. Then the depth of T satisfies

$$D(T) \geq \lceil \log_2 L(T) \rceil. \tag{7.3.7}$$

Proof Consider the 2-tree $\hat{T}$ associated with T. By Proposition 7.3.2, $N(\hat{T}) = 2L(\hat{T}) - 1$. Using Proposition 7.3.1 and the fact that $\lfloor \log_2(2n-1) \rfloor = \lceil \log_2 n \rceil$ for any integer $n > 1$, we have

Figure 7.2 A binary tree T and its associated 2-tree $\hat{T}$

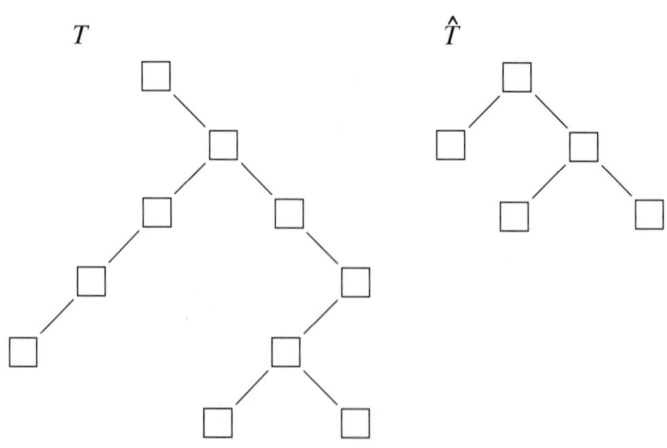

$$D(\hat{T}) \geq \lfloor \log_2(2L(\hat{T}) - 1) \rfloor = \lceil \log_2 L(\hat{T}) \rceil.$$

Since $L(T) = L(\hat{T})$ and $D(T) \geq D(\hat{T})$, it follows that

$$D(T) \geq D(\hat{T}) \geq \lceil \log_2 L(\hat{T}) \rceil = \lceil \log_2 L(T) \rceil.$$ ■

Recall that a binary tree T is called full at level i if T contains the maximum number 2^i nodes at that level. A full binary tree T is full at every level i, $i = 0, \ldots, D(T)$. Proofs of the following two propositions are left to the exercises.

PROPOSITION 7.3.4 Suppose T is a 2-tree. Then T is full at the second-deepest level if, and only if, all the leaf nodes are contained in two levels ($D - 1$ and D). ❑

PROPOSITION 7.3.5 If a 2-tree T is full at the second-deepest level, then the depth $D(T)$ and the number of leaf nodes $L(T)$ are related by

$$D(T) = \lceil \log_2 L(T) \rceil. \tag{7.3.8}$$ ❑

7.3.2 Internal and Leaf Path Lengths of Binary Trees

Inputs to search algorithms that are based implicitly (such as *BinarySearch*) or explicitly on binary trees usually follow paths from the root to an internal or leaf node. Moreover, the number of basic operations performed by the algorithm is usually proportional to the length of such a path. In particular, determining the average complexity of such algorithms requires computing the sum of the lengths of all paths from the root to nodes in the tree, which motivates the following definitions. The *internal path length IPL*(T) of a binary tree T is defined as the sum of the lengths of the paths from the root to the internal nodes as the internal nodes vary over the entire tree. The *leaf path length LPL*(T) is defined similarly. Note that the length of a path from the root to a node at level i is i.

When the binary tree T is a 2-tree, then $LPL(T)$ is precisely $2I$ more than $IPL(T)$. This simple relationship between $IPL(T)$ and $LPL(T)$ for 2-trees is stated as Proposition 7.3.6; its proof is left as an exercise.

PROPOSITION 7.3.6 Given any 2-tree T having I internal nodes,

$$IPL(T) = LPL(T) - 2I. \tag{7.3.9}$$ ❑

The following theorem establishes a useful lower bound for the leaf path length of a binary tree in terms of the number of leaf nodes.

THEOREM 7.3.7 Given any binary tree T with L leaf nodes

$$LPL(T) \geq L\lfloor \log_2 L \rfloor + 2(L - 2^{\lfloor \log_2 L \rfloor}). \tag{7.3.10}$$

Moreover, inequality (7.3.10) is an equality if, and only if, T is a 2-tree that is full at the second-deepest level.

Proof We first verify that inequality (7.3.10) is an equality if T is a 2-tree that is full at the second-deepest level. Suppose that $L = 2^k$, so that $\lfloor \log_2 L \rfloor = \log_2 L = k$. Then T is a full binary tree, and (7.3.10) correctly yields the equality $LPL(T) = L \log_2 L$. Now suppose $2^{k-1} < L < 2^k$ for a suitable $k > 1$. By Proposition 7.3.5, T has depth $D = \lceil \log_2 L \rceil$. Since T is full at level $D - 1 = \lfloor \log_2 L \rfloor$, each of the L paths in T from the root to a leaf reach level $D - 1$. Hence, the total contribution to $LPL(T)$ from reaching level $D - 1$ is $L\lfloor \log_2 L \rfloor$, so that

$$LPL(T) = L\lfloor \log L \rfloor + \text{the number of nodes at level } D.$$

If x denotes the number of nodes at level $D - 1$ having 2 children (non-leaf nodes), then $2x$ equals the number of nodes at level D. The remaining $2^{D-1} - x$ nodes at level $D - 1$ are leaf nodes, so that $L = (2^{D-1} - x) + 2x$. Thus, we have $x = L - 2^{D-1}$, and $LPL(T) = L\lfloor \log_2 L \rfloor + 2(L - 2^{\lfloor \log_2 L \rfloor})$.

We now verify that inequality (7.3.10) holds for any binary tree. Clearly, if T is any binary tree with L leaf nodes, then its associated 2-tree $\hat{T}$ has L leaf nodes and $LPL(\hat{T}) \leq LPL(T)$. Thus, to complete the proof of Theorem 7.3.7, it is sufficient to show that $LPL(T)$ is not minimized for 2-trees T having L leaf nodes that are not full at the second-deepest level.

Suppose T is a 2-tree of depth D having L leaf nodes that is not full at level $D - 1$. We show that there then exists a 2-tree $\tilde{T}$ having L leaf nodes such that $LPL(\tilde{T}) < LPL(T)$. Since level $D - 1$ is not full, there exists a leaf node, say X, at some level $i \leq D - 2$. Let Y be any node at level $D - 1$ that is not a leaf node, and let C_1 and C_2 denote the children of Y. We construct $\tilde{T}$ from T by removing the leaf nodes C_1 and C_2 and adding two new leaf nodes $\tilde{C}_1$ and $\tilde{C}_2$ as the children of X (see Figure 7.3). Let $\mathscr{L}(T)$ and $\mathscr{L}(\tilde{T})$ denote the set of all leaf nodes of T and $\tilde{T}$, respectively. Then,

$$\mathscr{L}(\tilde{T}) = (\mathscr{L}(T) - \{C_1, C_2, X\}) \cup \{\tilde{C}_1, \tilde{C}_2, Y\}.$$

Since C_1, C_2, and X are at levels D, D, and i in T, respectively, and $\tilde{C}_1$, $\tilde{C}_2$, and Y are at levels $i + 1$, $i + 1$, and $D - 1$ in $\tilde{T}$, respectively, it follows that

$$LPL(\tilde{T}) = LPL(T) - (2D + i) + (2(i + 1) + D - 1) = LPL(T) - D + i + 1 < LPL(T).$$

■

COROLLARY 7.3.8 If T is any binary tree having L leaf nodes, then

$$LPL(T) \geq \lceil L \log_2 L \rceil. \tag{7.3.11}$$

Further, if T is a full binary tree, then inequality (7.3.11) is an equality.

Proof Inequality (7.3.11) follows immediately from inequality (7.3.10) of Theorem 7.3.7 by using the inequality

$$x\lfloor \log_2 x \rfloor + 2\left(x - 2^{\lfloor \log_2 x \rfloor}\right) \geq \lceil x \log_2 x \rceil, \quad \text{for any integer } x \geq 1.$$

Moreover, we have already observed in the proof of Theorem 7.3.7 that if T is a full binary tree, then $LPL(T) = L \log_2 L = \lceil L \log_2 L \rceil$. ■

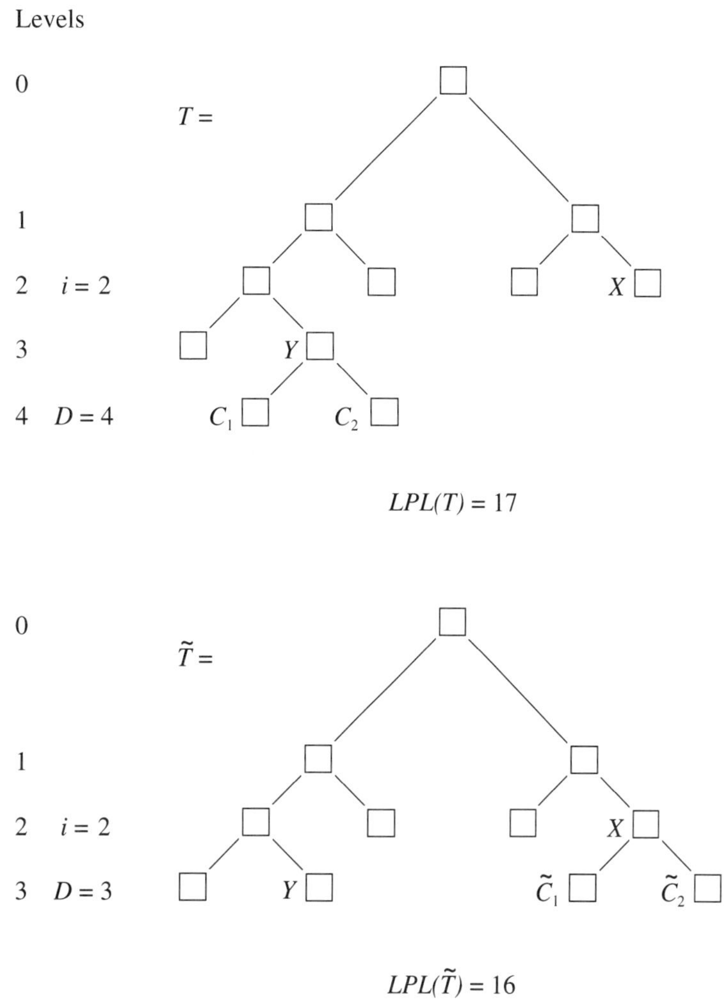

Figure 7.3 Construction of $\tilde{T}$ from T used in the proof of Theorem 7.3.7

7.4 Solving Recurrence Relations for Complexity

Solving recurrence relations for the complexities of algorithms is an important technique in algorithm analysis. In this section we use this technique to analyze recursive versions of *BinarySearch* and *InsertionSort*, as well as *MergeSort*. The recursive versions of these algorithms lead directly to (linear) recurrence relations for their complexities. A *linear recurrence relation* is one of the form

$$t(n) = c_1 t(n-1) + c_2 t(n-2) + \cdots + c_k t(n-k) + f(n),$$
$$\textbf{init. cond. } t(0) = d_0, \ldots, t(k-1) = d_{k-1},$$

for constants $c_1, \ldots, c_k (c_k \neq 0), d_0, \ldots, d_{k-1}$ and some fixed function $f(n)$.

7.4.1 Solving Linear Recurrence Relations Relating the *n*th Term to (*n* – 1)st Term

To further illustrate recurrence relations for the complexities of algorithms, let us reexamine *InsertionSort*. Note that for an input list $L[1:n]$, the action of *InsertionSort* can be viewed as first executing *InsertionSort* with input list $L[1:n-1]$, followed by inserting $L[n]$ into the previously sorted list $L[1:n-1]$. This observation immediately yields the following recurrence relation for $W(n)$.

$$W(n) = W(n-1) + n - 1, \quad \textbf{init. cond. } W(1) = 0 \tag{7.4.1}$$

An explicit formula for $W(n)$ is easily obtained by repeated substitution into (7.4.1). We have

$$\begin{aligned}
W(n) &= W(n-1) + n - 1 \\
&= (W(n-2) + n - 2) + n - 1 \\
&= (W(n-3) + n - 3) + n - 2 + n - 1 \\
&\quad \vdots \\
&= W(1) + 1 + 2 + \cdots + n - 1 \\
&= n^2/2 - n/2.
\end{aligned}$$

Recurrence relation (7.4.1) is a special case of the following more general recurrence relation, which can be solved by a simple repeated substitution (unwinding).

$$t(n) = at(n-1) + f(n), \quad \textbf{init. cond. } t(1) = b, \tag{7.4.2}$$

where a and b are positive constants and $f(n)$ is a given function of n.

Repeated substitution into (7.4.2) yields

$$\begin{aligned}
t(n) &= at(n-1) + f(n) \\
&= a(at(n-2) + f(n-1)) + f(n) = a^2t(n-2) + af(n-1) + f(n) \\
&= a^2(at(n-3) + f(n-2)) + af(n-1) + f(n) \\
&= a^3t(n-3) + a^2f(n-2) + af(n-1) + f(n) \\
&\quad \vdots \\
&= a^{n-1}b + \sum_{i=2}^{n} a^{n-i} f(i).
\end{aligned} \tag{7.4.3}$$

Since (7.4.2) involves a general function $f(n)$, the solution for $t(n)$ given in (7.4.3) is a summation over n terms. For algorithm analysis, it is desirable to obtain a simple closed-form formula for the exact value of this summation, or at least for its order. For many algorithms considered in this text, such closed-form formulas can be derived using the standard formulas for arithmetic series, geometric series, sum of logarithms, harmonic series, and so forth.

7.4.2 Solving Linear Recurrence Relations Relating the *n*th Term to the (*n*/*b*)th Term

Another type of recurrence relation that can be unwound easily is one in which $t(n)$ relates directly to $t(n/b)$ for some constant b. As our first illustration of this type of recurrence relation, we consider the worst-case complexity $W(n)$ of *BinarySearch2*. For an input list generating worst-case behavior, after comparing to the midpoint element, *BinarySearch2* always invokes itself with the larger sublist. It is easy to check that for an input $L[low,high]$ of size n ($n = high - low + 1$), the larger of the two sublists is the sublist $L[mid,high]$ of size $\lceil n/2 \rceil$. Hence, we obtain the following recurrence for $W(n)$.

$$W(n) = W(\lceil n/2 \rceil) + 1, \quad \textbf{init. cond. } W(1) = 1 \tag{7.4.4}$$

The occurrence of the ceiling function $\lceil\ \rceil$ in (7.4.4) makes it cumbersome to solve. However, if we assume that $n = 2^k$, the relation becomes

$$W(n) = W(n/2) + 1, \quad \textbf{init. cond. } W(1) = 1. \tag{7.4.5}$$

The recurrence (7.4.5) can be solved by repeated substitution.

$$\begin{aligned}
W(n) &= W\left(\frac{n}{2}\right) + 1 \\
&= \left(W\left(\frac{(n/2)}{2}\right) + 1\right) + 1 = W\left(\frac{n}{2^2}\right) + 2 \\
&= \left(W\left(\frac{(n/2^2)}{2}\right) + 1\right) + 2 = W\left(\frac{n}{2^3}\right) + 3 \\
&\ \ \vdots \\
&= \left(W\left(\frac{(n/2^{k-1})}{2}\right) + 1\right) + k - 1 = W\left(\frac{n}{2^k}\right) + k = W(1) + k = 1 + k
\end{aligned}$$

Since $k = \log_2 n$, we have

$$W(n) = 1 + \log_2 n, \quad n = 2^k. \tag{7.4.6}$$

We now show how formula (7.4.6) can be used to obtain an approximation for $W(n)$, for general n, which is accurate to within a single comparison. Given any positive integer n, there exists a positive integer j such that

$$2^{j-1} \le n < 2^j. \tag{7.4.7}$$

Clearly, $W(m) \le W(n)$ for $m \le n$. Thus,

$$W(2^{j-1}) \le W(n) \le W(2^j). \tag{7.4.8}$$

By (7.4.6), $W(2^{j-1}) = j$ and $W(2^j) = j + 1$. Substituting these results into (7.4.8) yields

$$j \le W(n) \le j + 1. \tag{7.4.9}$$

By taking base-2 logarithms in (7.4.7), we obtain

$$j - 1 \le \log_2 n < j. \tag{7.4.10}$$

Combining (7.4.9) and (7.4.10) yields

$$\log_2 n < W(n) \le 2 + \log_2 n, \quad \forall\, n \ge 1. \tag{7.4.11}$$

Since W(n) is an integer, (7.4.11) implies

$$1 + \lfloor \log_2 n \rfloor \le W(n) \le 2 + \lfloor \log_2 n \rfloor, \quad \forall\, n \ge 1. \tag{7.4.12}$$

It follows from (7.4.12) that $W(n) \in \Theta(\log n)$.

As a second illustration, we consider the algorithm *MergeSort*, which we analyzed in Chapter 4 using properties of binary trees. We now rederive the $\Theta(n \log n)$ formulas for the best-case and worst-case complexities of *MergeSort* using recurrence relations.

Assume that $n = 2^k$ for some nonnegative integer k, so that $k = \log_2 n$. Since the best-case complexity of *Merge* is $n/2$ for input size n, and since *MergeSort* recursively calls itself with two sublists of size $n/2$, B(n) satisfies the following recurrence relation.

$$B(n) = 2B\left(\frac{n}{2}\right) + \frac{n}{2}, \quad B(1) = 0 \tag{7.4.13}$$

Unwinding recurrence relation (7.4.13), we obtain

$$\begin{aligned} B(n) &= 2B\left(\frac{n}{2}\right) + \frac{n}{2} \\ &= 2\left(2B\left(\frac{n}{2^2}\right) + \frac{n}{2^2}\right) + \frac{n}{2} = 2^2B\left(\frac{n}{2^2}\right) + 2\left(\frac{n}{2}\right) \\ &= 2^2\left(2B\left(\frac{n}{2^3}\right) + \frac{n}{2^3}\right) + 2\left(\frac{n}{2}\right) = 2^3B\left(\frac{n}{2^3}\right) + 3\left(\frac{n}{2}\right) \\ &\quad\vdots \\ &= 2^kB(1) + k\left(\frac{n}{2}\right) = k\left(\frac{n}{2}\right). \end{aligned}$$

Since $k = \log_2 n$, we have

$$B(n) = \frac{n \log_2 n}{2}, \quad \text{for } n = 2^k,\ k = 0,1,2,\ldots. \tag{7.4.14}$$

Using a technique similar to that used to establish (7.4.11) from (7.4.6), we use (7.4.14) to obtain an approximation for B(n), for general n. Given any positive integer n, there exists a positive integer j such that

$$2^{j-1} \le n < 2^j. \tag{7.4.15}$$

Since $B(m) \le B(n)$ for $m \le n$, we have

$$B(2^{j-1}) \le B(n) \le B(2^j). \tag{7.4.16}$$

By (7.4.14), $B(2^{j-1}) = \frac{1}{2}2^{j-1}(j-1)$, and $B(2^j) = \frac{1}{2}2^j j$. Substituting these results into (7.4.15) yields

$$2^{j-2}(j-1) \le B(n) \le 2^{j-1}j. \tag{7.4.17}$$

By (7.4.15), we have

$$j - 1 \leq \log_2 n < j. \tag{7.4.18}$$

Combining (7.4.17) and (7.4.18) yields

$$2^{((\log_2 n) - 2)}((\log_2 n) - 1) < \mathrm{B}(n) \leq 2^{\log_2 n}((\log_2 n) + 1). \tag{7.4.19}$$

Simplifying (7.4.19) results in

$$\frac{1}{4}n((\log_2 n) - 1) < \mathrm{B}(n) \leq n((\log_2 n) + 1) \qquad \forall\, n \geq 1. \tag{7.4.20}$$

It follows from (7.4.20) that $\mathrm{B}(n) \in \Theta(n \log n)$.

We now consider the worst-case complexity of *MergeSort*. Again, we assume that $n = 2^k$ for some nonnegative integer k. Since the worst-case complexity of *Merge* is $n - 1$, $\mathrm{W}(n)$ satisfies the following recurrence relation.

$$\mathrm{W}(n) = 2\mathrm{W}\left(\frac{n}{2}\right) + n - 1, \qquad \mathrm{W}(1) = 0 \tag{7.4.21}$$

Unwinding recurrence relation (7.4.21), we obtain

$$\begin{aligned}
\mathrm{W}(n) &= 2\left(2\mathrm{W}\left(\frac{n}{2^2}\right) + \frac{n}{2} - 1\right) + n - 1 = 2^2\mathrm{W}\left(\frac{n}{2^2}\right) + 2n - (1 + 2) \\
&= 2^2\left(2\mathrm{W}\left(\frac{n}{2^3}\right) + \frac{n}{2^2} - 1\right) + 2n - (1 + 2) = 2^3\mathrm{W}\left(\frac{n}{2^3}\right) + 3n - (1 + 2 + 2^2) \\
&\quad \vdots \\
&= 2^k\mathrm{W}(1) + kn - (1 + 2 + 2^2 + \cdots + 2^{k-1}) = kn - (2^k - 1).
\end{aligned}$$

Since $k = \log_2 n$, we have

$$\mathrm{W}(n) = n \log_2 n - n + 1, \quad \text{for } n = 2^k, \quad k = 0,1,2, \ldots. \tag{7.4.22}$$

Now that we have established a formula for $\mathrm{W}(n)$ when n is a power of two, we can obtain an approximation to $\mathrm{W}(n)$ for a general n using arguments similar to those used for $\mathrm{B}(n)$ (see Exercise 7.24). We conclude that $\mathrm{W}(n) \in \Theta(n \log n)$.

Recurrence relations (7.4.5), (7.4.13), and (7.4.21) are all examples of the following more general recurrence relation that arises often in divide-and-conquer algorithms.

$$t(n) = at\left(\frac{n}{b}\right) + f(n), \quad t(1) = c_1, n = b^k, b > 1, k = 0,1,2, \ldots \tag{7.4.23}$$

Recurrence relation (7.4.23) can be solved by repeated substitution. The recurrence is stated for powers of a base b. However, in applications the function being analyzed is typically Θ-invariant under scaling as defined in the next section, so the order of the function is determined by its behavior at powers of b.

Unwinding (7.4.23), we obtain

$$
\begin{aligned}
t(n) &= at\left(\frac{n}{b}\right) + f(n) \\
&= a\left(at\left(\frac{n}{b^2}\right) + f\left(\frac{n}{b}\right)\right) + f(n) = a^2 t\left(\frac{n}{b^2}\right) + af\left(\frac{n}{b}\right) + f(n) \\
&= \qquad \vdots \\
&= a^k c_1 + a^{k-1} f\left(\frac{n}{b^{k-1}}\right) + a^{k-2} f\left(\frac{n}{b^{k-2}}\right) + \cdots + af\left(\frac{n}{b}\right) + f(n) \\
&= a^k c_1 + \sum_{i=0}^{k-1} a^i f\left(\frac{n}{b^i}\right), \quad \text{where } k = \log_b n. \qquad (7.4.24)
\end{aligned}
$$

Two special cases of (7.4.24) that arise often are when $f(n) = c$ or $f(n) = cn$. Consider the case $f(n) = c$. First suppose that $a \neq 1$. Substituting $f(n) = c$ into (7.4.24), and simplifying using the identities

$$\sum_{i=0}^{k-1} a^i = \frac{a^k - 1}{a - 1} \quad \text{and} \quad a^k = \left(b^{\log_b a}\right)^k = \left(b^k\right)^{\log_b a} = n^{\log_b a}$$

yields:

$$
\begin{aligned}
t(n) &= c_1 a^k + c \sum_{i=0}^{k-1} a^i \\
&= c_1 a^k + c\frac{a^k - 1}{a - 1} \\
&= \left(c_1 + \frac{c}{a-1}\right) a^k - \frac{c}{a-1} \\
&= \left(c_1 + \frac{a}{a-1}\right) n^{\log_b a} - \frac{c}{a-1}. \qquad (7.4.25)
\end{aligned}
$$

We leave it as an exercise to show that when $a = 1$, $t(n) = c_1 + c \log_b n$.

Now consider the case $f(n) = cn$. Substituting $f(n) = cn$ into (7.4.24) and setting $a = b$ immediately yields

$$t(n) = c_1 n + cn \log_b n, \quad n = 1, b, b^2, \ldots.$$

Substituting $f(n) = cn$ in the case $a \neq b$ and simplifying yields

$$
\begin{aligned}
t(n) &= c_1 a^k + cn \sum_{i=0}^{k-1} \left(\frac{a}{b}\right)^i \\
&= c_1 a^k + cn\frac{(a/b)^k - 1}{(a/b) - 1}. \quad \left(\text{using the identity } \sum_{i=0}^{k-1} x^i = \frac{x^k - 1}{x - 1}\right) \\
&= c_1 n^{\log_b a} + c\frac{n^{\log_b a} - n}{(a/b) - 1}. \quad (\text{substituting } a^k = n^{\log_b a} \text{ and } b^k = n).
\end{aligned}
$$

Thus, in the case where $f(n) = cn$, (7.4.24) simplifies to

$$t(n) + \begin{cases} c_1 n + cn \log_b n & a = b, \\ c_1 n^{\log_b a} + c\dfrac{n^{\log_b a} - n}{(a/b) - 1} & a \neq b, \end{cases} \quad n = 1, b, b^2, \ldots. \tag{7.4.26}$$

It must be emphasized that formulas (7.4.25) and (7.4.26) for $t(n)$ only hold for powers of b. If we know that $t(n)$ is a nondecreasing function, then we can determine the order of $t(n)$ in a manner similar to that used for the complexities of *BinarySearch2* and *MergeSort*. In other words, we can interpolate the asymptotic behavior of $t(n)$ for all n knowing its values only for powers of b. We now discuss an important class of functions for which such interpolation of asymptotic behavior is possible.

7.4.3 Interpolating Asymptotic Behavior

Interpolating asymptotic behavior is a useful tool in algorithm analysis, especially when solving recurrence relations. Suppose we have two functions $f(n), g(n) \in \mathscr{F}$, and we know that $f(s(n)) \in \Theta(g(s(n))$, where $s(0) < s(1) < s(2) < \cdots$. For example, in *BinarySearch2* we showed that $W(n)$ is equal to $1 + \log_2 n$ for $n = 1, 2, 2^2, \ldots$ (this result does not hold for all n). Thus, setting $f(n) = W(n)$, $g(n) = \log_2 n$, and $s(k) = 2^k$, we find that $f(s(n)) \in \Theta(g(s(n))$. Using the fact that $W(n)$ is an increasing function, we were able to show that $W(n) \in \Theta(\log n)$. Given general functions $f(n)$ and $g(n)$, where $f(s(n)) \in \Theta(g(s(n))$, we seek conditions on $f(n)$ and $g(n)$ that imply $f(n) \in \Theta(g(n))$.

For our purposes it is sufficient to restrict $s(n)$ to powers of a base b (usually $b = 2$), namely, $s(n) = b^n$, $n \in \mathbb{N}$. For such a sequence $s(n)$ there are simple conditions on $f(n)$ and $g(n)$ that yield our desired interpolation result. The first condition on $f(n)$ and $g(n)$ is that they be *eventually nondecreasing*. A function $h(n)$ is *eventually nondecreasing* if there exists an integer n_0 such that $h(n) \leq h(m)$, whenever $m \geq n \geq n_0$. Most of the time the functions $W(n)$, $B(n)$, and $A(n)$ (measuring the complexity of an algorithm) are eventually nondecreasing, since the number of basic operations performed usually increases as the input size n increases. A second condition that we place on $g(n)$ is that

$$g(cn) \in \Theta(g(n)), \quad \text{for all positive constants (integers) } c. \tag{7.4.27}$$

PROPOSITION 7.4.1 If $g(n) \in \mathscr{F}$ has the property that $g(cn) \in \Theta(g(n))$, then $f(cn) \in \Theta(g(n)) = \Theta(f(n))$ for all $f(n) \in \Theta(g(n))$. ❑

The proof of Proposition 7.4.1 is left as an exercise. Proposition 7.4.1 implies that the class $\Theta(g(n))$ is invariant under any "change of scale" of the input n to any representative of $\Theta(g(n))$. We refer to such a Θ-class as *scale invariant*, and we refer to any function in the class as *Θ-invariant under scaling*. Functions that are Θ-invariant under scaling include logarithm functions and polynomials, but do not include exponential functions. Also, sums and products of functions that are Θ-invariant under scaling are themselves Θ-invariant under scaling.

PROPOSITION 7.4.2 Suppose $g(n) \in \mathscr{F}$ is eventually nondecreasing and Θ-invariant under scaling. Further, suppose $f(n) \in \mathscr{F}$ is eventually nondecreasing and $f(b^n) \in X(g(b^n))$, $n \in \mathbb{N}$, where X is one of the classes Θ, Ω, O, and $b > 1$. Then $f(n) \in X(g(n))$.

Proof First we consider the case $X = O$. Clearly, any positive integer n is either a power of b or lies between two powers of b. That is, given any positive integer n there exists a positive integer $k(n)$ (dependent on n) such that $b^{k(n)-1} \leq n < b^{k(n)}$. Since $f(n)$ is eventually nondecreasing, there exists a positive integer constant N (independent of n) such that

$$f(b^{k(n)-1}) \leq f(n) \leq f(b^{k(n)}), \quad \forall\, n \geq N. \tag{7.4.28}$$

Now $f(b^n) \in O(g(b^n))$. Thus, there exist positive constants d and m_0 such that

$$f(b^m) \leq dg(b^m), \quad \forall\, m \geq m_0. \tag{7.4.29}$$

Since g is Θ-invariant under scaling, it follows that there exist positive constants d' and m_0' such that

$$g(b^m) = g(bb^{m-1}) \leq d'\, g(b^{m-1}), \quad \forall\, m \geq m_0'. \tag{7.4.30}$$

We now choose $m = n(k)$, so that $b^{m-1} \leq n < b^m$. Since g is eventually nondecreasing, it follows that there exists a positive constant N' such that

$$g(b^{m-1}) \leq g(n), \quad \forall\, n \geq N'. \tag{7.4.31}$$

It follows from (7.4.28), (7.4.29), (7.4.30), and (7.4.31) that there exists a positive constant n_0 such that

$$f(n) \leq f(b^m) \leq dg(b^m) \leq dd'g(b^{m-1}) \leq dd'g(n), \quad \forall\, n \geq n_0. \tag{7.4.32}$$

Hence, $f(n) \in O(g(n))$, which proves Proposition 7.4.2 for the case $X = O$. A similar proof (see Exercise 7.31) yields the case $X = \Omega$. The case $X = \Theta$ then follows immediately from the cases $X = O$ and $X = \Theta$. ■

Using the fact that the functions n, $n \log n$, and $n^{\log_a b}$ are Θ-invariant under scaling, the next two propositions follow from (7.4.25) and (7.4.26), respectively.

PROPOSITION 7.4.3 If $t(n)$ is eventually nondecreasing and satisfies the recurrence relation $t(n) = at(n/b) + c$, with initial condition $t(1) = c_1$, where a, b, c, c_1 are positive constants, $b > 1$, then

$$t(n) \in \begin{cases} \Theta(n^{\log_b a}), & a \neq 1, \\ \Theta(\log n), & a = 1. \end{cases} \tag{7.4.33}$$

❑

PROPOSITION 7.4.4 If $t(n)$ is eventually nondecreasing and satisfies the recurrence relation $t(n) = at(n/b) + cn$, with initial condition $t(1) = c_1$, where a, b, c, c_1 are positive constants, $b > 1$, then

$$t(n) \in \begin{cases} \Theta(n) & a < b, \\ \Theta(n \log n) & a = b. \\ \Theta(n^{\log_b a}) & a > b. \end{cases} \tag{7.4.34}$$

❑

7.5 Generating Functions

Multiple-reference recurrence relations, such as fib(n) = fib(n – 1) + fib(n – 2) determining the Fibonacci numbers 0, 1, 1, 2, 3, 5, 8, 13, 21, 34, . . . , are often rather difficult to solve (in the sense of finding an explicit formula for the nth term in the resulting sequence). In this section we discuss the method of *generating functions*, which is of great help in solving these relations.

Suppose we are considering a sequence $a_0, a_1, a_2, \ldots, a_n, \ldots$ of numbers (the sequence might be finite) that has been well defined (by a recurrence relation or otherwise), but where we don't as yet have an explicit formula for the nth term a_n. (For example, a_n might denote the complexity of an algorithm for input of size n.) We denote this sequence by (a_n). The *generating function* for this sequence is defined to be the formal power series

$$g(x) = \sum_{n=0}^{\infty} a_n x^n = a_0 + a_1 x + a_2 x^2 + \cdots + a_n x^n + \cdots. \tag{7.5.1}$$

For example, the generating function for the Fibonacci sequence is the formal power series

$$x + x^2 + 2x^3 + 3x^4 + 5x^5 + 8x^6 + 13x^7 + 21x^8 + \cdots.$$

Note that (7.5.1) is a "formal" sum in the sense that we are usually not concerned with determining the values of x for which the associated power series actually converges. (In fact, determining the radius of convergence of the power series usually requires knowing an explicit formula for a_n, which is the problem at hand!) The generating function for a sequence (a_n) is simply used to "encode" the sequence in a form that is amenable to algebraic manipulation. For example, in the case of the Fibonacci sequence, we will see that formal algebraic manipulations of its generating function lead to the explicit formula

$$\text{fib}(n) = \frac{1}{\sqrt{5}}\left[\left(\frac{1+\sqrt{5}}{2}\right)^n - \left(\frac{1-\sqrt{5}}{2}\right)^n\right]. \tag{7.5.2}$$

We illustrate how generating functions can be used to determine explicit formulas for terms in a sequence by considering again the recurrence relation for the minimum number of disk moves required to solve the Towers of Hanoi problem for n disks. Letting $t(n)$ denote the minimum number of moves required to solve the Towers of Hanoi problem with n disks, we saw in (1.3.4) that the terms in the sequence $(t(n))$ satisfy the recurrence relation

$$t(n) = 2t(n-1) + 1, \quad \textbf{init. cond. } t(0) = 0. \tag{7.5.3}$$

The method of finding an explicit formula for $t(n)$ using generating functions begins by forming the generating function for the sequence $(t(n))$.

$$g(x) = \sum_{n=0}^{\infty} t(n)x^n = x + 3x^2 + 7x^3 + \cdots + a_n x^n + \cdots \tag{7.5.4}$$

Since $t(n) - 2t(n-1) = 1$, $n \geq 1$, this suggests multiplying $g(x)$ by $2x$ and subtracting the resulting formal power series from $g(x)$. We obtain

$$\begin{aligned} g(x) - 2xg(x) &= x + 3x^2 + 7x^3 + \cdots + a_n x^n + \cdots \\ &\quad\ - 2x^2 - 6x^3 - \cdots - 2a_{n-1}x^n - \cdots \\ &= x + x^2 + x^3 + \cdots + x^n + \cdots \\ &= \frac{x}{1-x}, \end{aligned} \tag{7.5.5}$$

where the last expression is obtained from the geometric series

$$\frac{1}{1-x} = 1 + x + x^2 + \cdots + x^n + \cdots. \tag{7.5.6}$$

It follows from (7.5.5) that

$$g(x) = \frac{x}{(1-2x)(1-x)} = \frac{1}{1-2x} - \frac{1}{1-x}. \tag{7.5.7}$$

Substituting $2x$ for x in the geometric series (7.5.6), we obtain

$$\frac{1}{1-2x} = 1 + 2x + 2^2x^2 + \cdots + 2^n x^n + \cdots. \tag{7.5.8}$$

Finally substituting (7.5.6) and (7.5.8) into (7.5.7) yields

$$g(x) = \sum_{n=0}^{\infty} 2^n x^n - \sum_{n=0}^{\infty} x^n = \sum_{n=0}^{\infty} (2^n - 1)x^n. \tag{7.5.9}$$

We now have achieved our goal. That is, by equating coefficients of x^n in (7.5.4) and (7.5.9), we obtain the explicit formula $t(n) = 2^n - 1$. Note that the expression $2^n - 1$ agrees with the formula for $t(n)$ obtained earlier using a simple unwinding of the recurrence (7.5.3).

7.5.1 The Generating Function Paradigm for Solving Recurrence Relations

The way we used a generating function to solve the recurrence relation for the Towers of Hanoi problem is illustrative of a general paradigm. Suppose we are considering a sequence $a_0, a_1, a_2, \ldots, a_n, \ldots$ of numbers (the sequence might be finite) that has been well defined (by a recurrence relation or otherwise), but where we don't as yet have an explicit formula for the nth term a_n. The generating function paradigm for finding such a formula consists of the following three steps.

Step 1. Form the formal power series (generating function)

$$g(x) = \sum_{n=0}^{\infty} a_n x^n = a_0 + a_1 x + a_2 x^2 + \cdots + a_n x^n + \cdots.$$

Step 2. Based on the recurrence relation, use algebraic or other manipulation to find a closed-form expression for $g(x)$.

Step 3. Expand this closed-form expression into a power series with explicit coefficients. Then evaluate the sequence (a_n) using term-by-term comparison.

We now employ the generating functions paradigm to obtain an explicit formula for the nth Fibonacci number fib(n) given by the recurrence relation

$$\text{fib}(n) = \text{fib}(n-1) + \text{fib}(n-2), \quad \textbf{init. cond. } \text{fib}(0) = 0, \text{fib}(1) = 1. \tag{7.5.10}$$

Step 1. Form the generating function.

$$g(x) = \sum_{n=0}^{\infty} \text{fib}(n)\, x^n = 0 + x + x^2 + 2x^3 + \cdots + \text{fib}(n)x^n + \cdots. \tag{7.5.11}$$

Step 2. Formula (7.5.10) suggests forming the series $g(x) - xg(x) - x^2g(x)$. As shown below, the coefficients of all terms x^k, $k \geq 2$, in this series are zero.

$$\begin{array}{rl}
g(x) = & 0 + 1x + 1x^2 + 2x^3 + \cdots + \text{fib}(n)x^n + \cdots \\
-xg(x) = & \quad - 0x - 1x^2 - 1x^3 - \cdots - \text{fib}(n-1)x^n - \cdots \\
-x^2g(x) = & \qquad\quad - 0x^2 - 1x^3 - \cdots - \text{fib}(n-2)x^n - \cdots \\
\hline
(1 - x - x^2)g(x) = & 0 + 1x + 0x^2 + 0x^3 + \cdots + 0x^n + \cdots.
\end{array}$$

Hence,

$$g(x) = \frac{x}{1 - x - x^2}. \tag{7.5.12}$$

Step 3. Using the quadratic formula, we have

$$g(x) = \frac{x}{(1-\alpha x)(1-\beta x)}, \tag{7.5.13}$$

where $\alpha = (1 + \sqrt{5})/2$ and $\beta = (1 - \sqrt{5})/2$. By the method of partial fractions, we obtain

$$g(x) = \frac{x}{(1-\alpha x)(1-\beta x)} = \frac{A}{1-\alpha x} + \frac{B}{1-\beta x}. \tag{7.5.14}$$

We now determine the values of A and B as follows. Summing the two terms on the right-hand side of (7.5.14) and comparing numerators, we obtain

$$x = A(1 - \beta x) + B(1 - \alpha x). \tag{7.5.15}$$

Substituting $x = 1/\alpha$ in (7.5.15) yields $A = \dfrac{1}{\alpha - \beta}$, and substituting $x = 1/\beta$ in (7.5.15) yields $B = \dfrac{-1}{\alpha - \beta}$.

Continuing from (7.5.14), we have

$$\begin{aligned} g(x) &= \frac{A}{1-\alpha x} + \frac{B}{1-\beta x} \\ &= A\sum_{n=0}^{\infty} \alpha^n x^n + B\sum_{n=0}^{\infty} \beta^n x^n \qquad \text{(by (7.5.6))} \\ &= \sum_{n=0}^{\infty} (A\alpha^n + B\beta^n)\, x^n. \end{aligned}$$

Since fib(n) is the coefficient of x^n,

$$\text{fib}(n) = A\alpha^n + B\beta^n = \frac{\alpha^n - \beta^n}{\alpha - \beta}. \qquad (7.5.16)$$

Substituting for α and β in (7.5.16), we obtain

$$\text{fib}(n) = \frac{1}{\sqrt{5}}\left[\left(\frac{1+\sqrt{5}}{2}\right)^n - \left(\frac{1-\sqrt{5}}{2}\right)^n\right], \quad n \ge 0. \qquad (7.5.17)$$

The method we used to solved the recurrence relation (7.5.10) for the Fibonacci numbers can be generalized to solve the general recurrence relation:

$$t(n) = c_1 t(n-1) + c_2 t(n-2), \quad \textbf{init. cond. } t(0) = b_0, \quad t(1) = b_1. \qquad (7.5.18)$$

We first form the generating function $g(x) = \Sigma_{n=0}^{\infty} t(n)x^n$. Then, we form the series

$$\begin{aligned} & g(x) - c_1 x g(x) - c_2 x^2 g(x) \\ &= t(0) + (t(1) - c_1 t(0))x + (t(2) - c_1 t(1) - c_2 t(0))x^2 + \cdots \\ &\qquad + (t(n) - c_1 t(n-1) - c_2 t(n-2))x^n + \cdots \\ &= b_0 + (b_1 - c_1 b_0)x. \qquad \text{(by (7.5.18))} \end{aligned}$$

Hence,

$$g(x) = \frac{b_0 + (b_1 - c_1 b_0)\,x}{1 - c_1 x - c_2 x^2}. \qquad (7.5.19)$$

Formula (7.5.19) gives a closed-form expression for the generating function $g(x)$ as the ratio of a linear polynomial and a quadratic polynomial. An explicit formula for $t(n)$ can be obtained by deriving the power series for $g(x)$ given by (7.5.19) and then equating $t(n)$ to the coefficient of x^n in this power series. How the series is derived depends on the nature of the two roots of the quadratic $1 - c_1 x - c_2 x^2$. If the roots are distinct, then we can obtain the power series expansion of (7.5.19) using the method of partial fractions in a manner similar to that used to solve the

Fibonacci recurrence. We now illustrate the case of repeated roots by considering the following example.

$$t(n) = 4t(n-1) - 4t(n-2), \quad \textbf{init. cond. } t(0) = 1,\ t(1) = 3. \tag{7.5.20}$$

By (7.5.19), we have

$$g(x) = \frac{1-x}{1-4x+4x^2} = \frac{1-x}{(1-2x)^2}. \tag{7.5.21}$$

To obtain a power series expansion for $g(x)$, consider the following formula obtained by taking the derivative of both sides of (7.5.6).

$$\frac{1}{(1-x)^2} = 1 + 2x + 3x^2 + \cdots + (n+1)x^n + \cdots \tag{7.5.22}$$

Substituting $2x$ for x in (7.5.22) yields

$$\frac{1}{(1-2x)^2} = 1 + 2(2)x + 3(2^2)x^2 + \cdots + (n+1)2^n x^n + \cdots. \tag{7.5.23}$$

Multiplying both sides of (7.5.23) by $1 - x$ and collecting coefficients of x^n for each n, we obtain

$$\begin{aligned} g(x) &= (1-x)(1 + 2(2)x + 3(2^2)x^2 + \cdots + (n+1)2^n x^n + \cdots) \\ &= (1 + 2(2)x + 3(2^2)x^2 + \cdots + (n+1)2^n x^n + \cdots) \\ &\quad - (x + 2(2)x^2 + \cdots + n2^{n-1}x^n + \cdots) \\ &= 1 + (2(2) - 1)x + (3(2^2) - 2(2))x^2 + \cdots + ((n+1)2^n - n2^{n-1})x^n + \cdots \\ &= \sum_{n=0}^{\infty} ((n+2)2^{n-1})\, x^n. \end{aligned}$$

Since $t(n)$ by definition is the coefficient of x^n, we obtain

$$t(n) = (n+2)2^{n-1}, \quad n \ge 0. \tag{7.5.24}$$

To check our work, we can substitute $t(n) = (n+2)2^{n-1}$ back into (7.5.20):

$$t(0) = (0+2)2^{0-1} = 1, \quad t(1) = (1+2)2^{1-1} = 3, \quad \text{and}$$
$$4t(n-1) - 4t(n-2) = 4(n+1)2^{n-2} - 4n2^{n-3} = (n+2)2^{n-1} = t(n) \quad \text{for } n \ge 2.$$

The examples that we have discussed in this section illustrate the power of generating functions in solving certain types of recurrence relations. We have only described how generating functions are used to solve linear recurrences for $t(n)$ in terms of $t(n-1)$ and $t(n-2)$. However, this method extends to general linear recurrences of the following form:

$$t(n) = c_1 t(n-1) + \cdots + c_k t(n-k), \quad \textbf{init. cond. } t(0) = b_0, \cdots, t(k-1) = b_{k-1}. \tag{7.5.25}$$

Applying the method used to derive (7.5.19), we obtain the following closed-form formula for the generating function $g(x)$ associated with (7.5.25):

$$g(x) = \frac{b_0 + (b_1 - c_1 b_0)\,x + \cdots + (b_{k-1} - c_1 b_{k-1} - \cdots - c_k b_0)\,x^{k-1}}{1 - c_1 x - \cdots - c_k x^k} \qquad (7.5.26)$$

In the exercises we develop techniques for deriving the power series for $g(x)$ based on (7.5.26).

7.5.2 Obtaining a Formula for the Number of Binary Trees on *N* Nodes

We now use generating functions to prove formula (2.2.1), which states that the number b_n of binary trees on n nodes equals the nth Catalan number; that is,

$$b_n = \frac{1}{n+1}\binom{2n}{n}, \qquad (7.5.27)$$

The number b_n satisfies the following recurrence relation (see Exercise 7.36).

$$b_n = \sum_{i=0}^{n-1} b_i b_{n-i-1} = b_0 b_{n-1} + b_1 b_{n-2} + \cdots + b_{n-1} b_0, \text{ init. cond. } b_0 = 1 \qquad (7.5.28)$$

Recurrence relation (7.5.28) is an example of a *quadractic* recurrence relation because each term in the summand consists of a product of the function b_n evaluated at two of the previous points $0, 1, \ldots, n-1$. Let $g(x)$ be the generating function for the number b_n of binary trees on n nodes; that is,

$$g(x) = \sum_{n=0}^{\infty} b_n x^n = b_0 + b_1 x + b_2 x^2 + \cdots + b_n x^n + \cdots.$$

Clearly,

$$xg(x) = \sum_{n=1}^{\infty} b_{n-1} x^n = b_0 x + b_1 x^2 + \cdots + b_{n-1} x^n + \cdots$$

so that

$$\begin{aligned} g(x)(xg(x)) &= (g(x))^2 x \\ &= \sum_{n=1}^{\infty}\left(\sum_{i=0}^{n-1} b_i b_{n-i-1}\right)x^n \qquad (7.5.29) \\ &= (b_0 b_0)x + (b_0 b_1 + b_1 b_0)x^n + (b_0 b_{n-1} + \cdots + b_{n-1} b_0)x^n + \cdots. \end{aligned}$$

Substituting (7.5.28) into (7.5.29) yields

$$(g(x))^2 x = b_1 x + b_2 x^2 + \cdots + b_{n-1} x^n + \cdots = g(x) - 1. \qquad (7.5.30)$$

Hence,

$$(g(x))^2 x - g(x) + 1 = 0. \qquad (7.5.31)$$

Using the quadratic formula we have

$$g(x) = \frac{1 \pm (1-4x)^{1/2}}{2x}. \qquad (7.5.32)$$

Since $b_0 = 1$, it follows that

$$g(x) = \frac{1 - (1-4x)^{1/2}}{2x}. \qquad (7.5.33)$$

Employing the Binomial Theorem (1.2.21) we obtain

$$\frac{1-(1-4x)^{1/2}}{2x} = -\sum_{n=1}^{\infty} \binom{1/2}{n} \frac{(-4x)^n}{2x}$$

$$= \sum_{n=0}^{\infty} \binom{1/2}{n+1} (-2)^n 2^{n+1} x^n$$

$$= \sum_{n=0}^{\infty} \left[\frac{-(1/2)(-1/2)(-3/2)(-5/2)\cdots(-(2n-1)/2)(-2)^{n+1}2^n}{(n+1)!} \right] x^n$$

$$= \sum_{n=0}^{\infty} \left[\frac{(1)(3)(5)\cdots(2n-1)2^n}{(n+1)!} \right] x^n.$$

Since b_n is the coefficient of x^n in $g(x)$, we have

$$b_n = \frac{(1)(3)(5)\cdots(2n-1)2^n}{(n+1)!} \qquad (7.5.34)$$

The numerator in (7.5.34) satisfies the identity (see Exercise 7.35)

$$(1)(3)(5)\cdots(2n-1)2^n = (n+1)(n+2)\cdots(2n-1)(2n). \qquad (7.5.35)$$

Substituting identity (7.5.35) into the right-hand side of (7.5.34) yields (7.5.27).

7.6 Polynomial Interpolation

Given a polynomial $P(x) = a_{n-1}x^{n-1} + \cdots + a_1x + a_0$, Horner's Rule is an optimal algorithm for evaluating $P(x)$ at a given input value x (see Chapter 1). In this section we look at the related problem of determining the coefficients of $P(x)$ given that we know its evaluation $y_i = P(x_i)$ at a set of n input values x_i, $i = 1, \ldots, n$. Geometrically, we are asking for the unique (*interpolating*) polynomial of degree at most $n-1$ whose graph passes through the n points $(x_1, y_1), (x_2, y_2), \ldots, (x_n, y_n)$. For example, suppose we are given the four points $(-1, 1), (0, 0), (1, 0), (3, 2)$, and we wish to find the polynomial of degree at most 3 that passes through (interpolates) these points (see Figure 7.4, page 262).

The straightforward linear algebraic method for finding the interpolating polynomial $P(x) = a_3x^3 + a_2x^2 + a_1x + a_0$ is simply to solve the following system of four linear equations in the four unknowns a_3, a_2, a_1, a_0 obtained by using the equalities $y_i = P(x_i)$, $i = 1,2,3,4$.

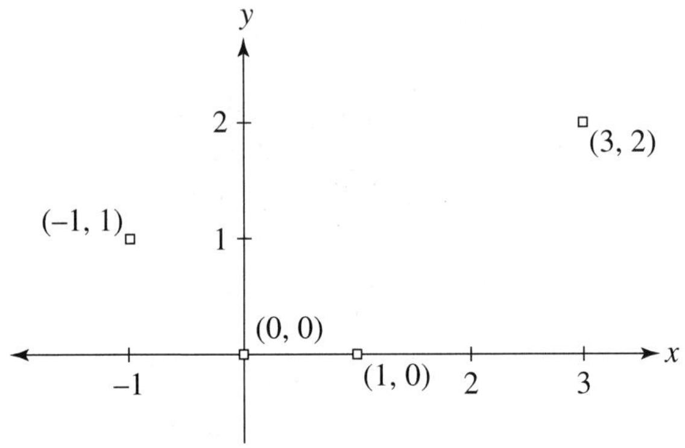

Figure 7.4 Problem: Find polynomial of degree at most 3 that interpolates four specified points.

$$
\begin{aligned}
-a_3 + a_2 - a_1 + a_0 &= -1 \\
a_0 &= 0 \\
a_3 + a_2 + a_1 + a_0 &= 0 \\
27a_3 + 9a_2 + 3a_1 + a_0 &= 2
\end{aligned}
$$

Solving this system of equations yields $a_3 = -1/24$, $a_2 = 1/2$, $a_1 = -11/24$, $a_0 = 0$. Thus, the polynomial of degree at most 3 that interpolates the points $(-1, 1)$, $(0, 0)$, $(1, 0)$, $(3, 2)$ is the cubic $(-1/24)x^3 + (1/2)x^2 + (-11/24)x$, whose graph is given in Figure 7.5.

In general, we can obtain the interpolating polynomial $P(x) = a_{n-1}x^{n-1} + \cdots + a_1x + a_0$ for the n points $(x_1, y_1), (x_2, y_2), \ldots, (x_n, y_n)$ by solving the following system

Figure 7.5 Cubic polynomial interpolation of the four points given in Figure 7.4

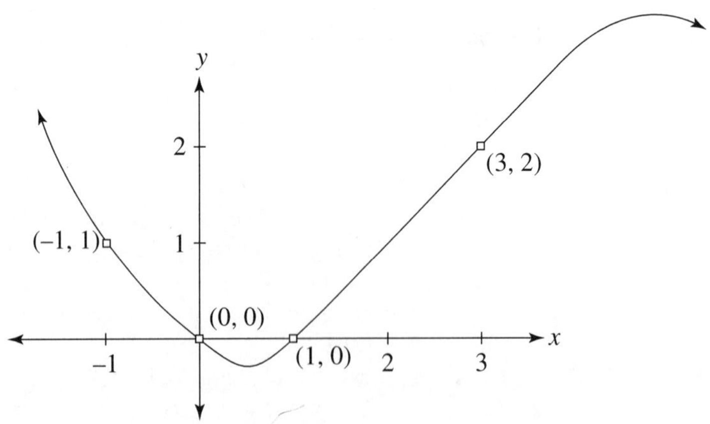

$$y = P(x) = \frac{-1}{24}x^3 + \frac{1}{2}x^2 - \frac{11}{24}x$$

of n linear equations in the n unknowns $a_{n-1}, \ldots, a_1, a_0$ obtained by using the equalities $y_i = P(x_i)$, $i = 1,2, \ldots, n$.

$$\begin{aligned} a_{n-1}x_1^{n-1} + a_{n-2}x_1^{n-2} + \cdots + a_1x_1 + a_0 &= y_1 \\ a_{n-1}x_2^{n-1} + a_{n-2}x_2^{n-2} + \cdots + a_1x_2 + a_0 &= y_2 \\ \vdots \qquad\qquad \vdots \qquad\qquad \vdots \\ a_{n-1}x_n^{n-1} + a_{n-2}x_n^{n-2} + \cdots + a_1x_n + a_0 &= y_n \end{aligned} \tag{7.6.1}$$

The standard linear algebra method for solving equations (7.6.1) is Gaussian elimination (see any linear algebra textbook). An algorithm based on solving (7.6.1) using Gaussian elimination has cubic complexity. Let D denote the determinant associated with the system of linear equations given by (7.6.1):

$$D = \begin{vmatrix} x_1^{n-1} & x_1^{n-2} & \cdots & x_1 & 1 \\ x_2^{n-1} & x_2^{n-2} & \cdots & x_2 & 1 \\ \vdots & \vdots & & \vdots & \\ x_n^{n-1} & x_n^{n-2} & \cdots & x_n & 1 \end{vmatrix} \tag{7.6.2}$$

Determinant D is the famous Vandermonde determinant, and it is well known that D is nonzero. Thus, the system of linear equations given by (7.6.1) has a unique solution, proving that there is a unique polynomial of degree at most $n - 1$ interpolating the set of n points $(x_1, y_1), (x_2, y_2), \ldots, (x_n, y_n)$.

7.6.1 Lagrange Interpolation

The following explicit formula for the interpolating polynomial $P(x) = a_{n-1}x^{n-1} + \cdots + a_1x + a_0$ for the n points $(x_1, y_1), (x_2, y_2), \ldots, (x_n, y_n)$ was given by Lagrange:

$$P(x) = \sum_{1 \le i \le n} \left(\prod_{\substack{i \ne j \\ 1 \le j \le n}} \frac{(x - x_j)}{(x_i - x_j)} \right) y_i. \tag{7.6.3}$$

For example, when $n = 3$, (7.6.3) becomes

$$P(x) = \frac{(x - x_2)(x - x_3)}{(x_1 - x_2)(x_1 - x_3)} y_1 + \frac{(x - x_1)(x - x_3)}{(x_2 - x_1)(x_2 - x_3)} y_2 + \frac{(x - x_1)(x - x_2)}{(x_3 - x_1)(x_3 - x_2)} y_3.$$

A simple substitution into (7.6.3) shows that $P(x_i) = y_i$, so that $P(x)$ is indeed the desired interpolating polynomial. A naive algorithm that simply translates (7.6.3) directly into pseudocode has $\Theta(n^3)$ complexity . However, it is not difficult to improve the naive algorithm to obtain an algorithm (still based on (7.6.3)) having $\Theta(n^2)$ complexity. We leave the pseudocode for the improved algorithm as an exercise.

7.6.2 Newtonian Interpolation

Newtonian interpolation is another method of polynomial interpolation which, like Lagrangian interpolation, is based on an explicit mathematical formula for the interpolating polynomial, but which, unlike Lagrangian interpolation, is a recursive formula. The recursion is based on formula (7.6.4), which expresses the interpolating polynomial $P_k(x)$ for the k points $(x_1, y_1), (x_2, y_2), \ldots, (x_k, y_k)$ in terms of the interpolating polynomial $P_{k-1}(x)$ for the $k - 1$ points $(x_1, y_1), (x_2, y_2), \ldots, (x_{k-1}, y_{k-1})$. Let $Q_k(x) = (x - x_1)(x - x_2) \cdots (x - x_k)$.

THEOREM 7.6.1 The interpolating polynomial $P_k(x)$ satisfies the recurrence relation

$$P_k(x) = (y_k - P_{k-1}(x_k))\left(\frac{Q_{k-1}(x)}{Q_{k-1}(x_k)}\right) + P_{k-1}(x), \quad 2 \le k \le n,$$

$$\textbf{init. cond. } P_1(x) \equiv y_1. \tag{7.6.4}$$

Proof We show by induction on k that $P_k(x)$ given by (7.6.4) is the interpolating polynomial for the k points $(x_1, y_1), (x_2, y_2), \ldots, (x_k, y_k)$. The initial condition of (7.6.4) obviously defines the interpolating polynomial for the single point (x_1, y_1), so our statement is true for $k = 1$. For the inductive step, consider any $k = j$, where $1 \le j < n$, and assume that $P_j(x)$ as defined by (7.6.4) is the interpolating polynomial for the j points $(x_1, y_1), (x_2, y_2), \ldots, (x_j, y_j)$. We show that this implies that $P_{j+1}(x)$ as defined by (7.6.4) is the interpolating polynomial for the $j + 1$ points $(x_1, y_1), (x_2, y_2), \ldots, (x_{j+1}, y_{j+1})$, that is, $P_{j+1}(x_i) = y_i$ for every $i \in \{1, 2, \ldots, j + 1\}$.

Consider first the case where $i < j + 1$. Clearly, $Q_j(x_i) = 0$, and substituting x_i for x in formula (7.6.4) with $k = j + 1$ yields

$$P_{j+1}(x_i) = (y_{j+1} - P_j(x_{j+1}))\left(\frac{Q_j(x_i)}{Q_j(x_{j+1})}\right) + P_j(x_i) = P_j(x_i).$$

But by induction hypothesis, $P_j(x)$ interpolates the point (x_i, y_i). Thus, $P_{j+1}(x_i) = P_j(x_i) = y_i$.

Now consider the case $i = j + 1$. Substituting x_{j+1} for x in formula (7.6.4) with $k = j + 1$ yields

$$\begin{aligned} P_{j+1}(x_{j+1}) &= (y_{j+1} - P_j(x_{j+1}))\left(\frac{Q_j(x_{j+1})}{Q_j(x_{j+1})}\right) + P_j(x_{j+1}) \\ &= (y_{j+1} - P_j(x_{j+1}))(1) + P_j(x_{j+1}) = y_{j+1}. \end{aligned}$$ ■

We illustrate Newtonian interpolation for the set of four points used in our previous example, namely, $(-1, 1)$, $(0, 0)$, $(1, 0)$, and $(3, 2)$. For this example $x_1 = -1$, $x_2 = 0$, $x_3 = 1$, $x_4 = 3$, $y_1 = 1$, $y_2 = 0$, $y_3 = 0$, and $y_4 = 2$.

Step 1: $P_1(x) \equiv y_1 = 1.$

Step 2: $$P_2(x) = \frac{(y_2 - P_1(x_2))(Q_1(x))}{Q_1(x_2)} + P_1(x)$$

$$= \frac{(0-1)(x-(-1))}{(0-(-1))} + 1$$

$$= -x$$

Step 3: $Q_2(x) = (x - x_2)Q_1(x) = (x - 0)(x - (-1)) = x^2 + x$

$$P_3(x) = (y_3 - P_2(x_3))\left(\frac{Q_2(x)}{Q_2(x_3)}\right) + P_2(x)$$

$$= (0 - (-1))\left(\frac{x^2 + x}{2}\right) + (-x)$$

$$= \frac{x^2}{2} - \frac{x}{2}$$

Step 4: $Q_3(x) = (x - x_3)Q_2(x) = (x - 1)(x^2 + x) = x^3 - x$

$$P_4(x) = (y_4 - P_3(x_4))\left(\frac{Q_3(x)}{Q_3(x_4)}\right) + P_3(x)$$

$$= \left(2 - \left(\frac{3^2}{2} - \frac{3}{2}\right)\right)\left(\frac{x^3 - x}{3^3 - 3}\right) + \frac{x^2}{2} - \frac{x}{2}$$

$$= -\frac{x^3 - x}{24} + \frac{x^2}{2} - \frac{x}{2}$$

$$= \left(-\frac{1}{24}\right)x^3 + \left(\frac{1}{2}\right)x^2 + \left(-\frac{11}{24}\right)x$$

Note that our formula for $P_4(x)$ agrees with our earlier calculation using linear algebra techniques. Figure 7.6 (page 266) illustrates the generation of the polynomials $P_1(x)$, $P_2(x)$, $P_3(x)$, $P_4(x)$.

The following algorithm *NewtonInterp* for polynomial interpolation is based on formula (7.6.4). *NewtonInterp* invokes the function *HornerEval*, which was given in Chapter 1. For convenience, we use the notation $P(x)$ for the polynomial instead of its coefficient array $a[0:n]$.

procedure *NewtonInterp*($X[1:n]$,$Y[1:n]$,n,$P(x)$)

Input: $X[1:n]$, $Y[1:n]$ (arrays of real numbers)
n (a positive integer)

Output: $P(x) = a_{n-1}x^{n-1} + \cdots + a_1x + a_0$ (interpolating polynomial)

$P(x)$:= $Y[1]$ {$P(x)$ is initialized to be the constant polynomial y_1}
$Q(x)$:= $x - X[1]$ {$Q(x)$ is initialized to be the linear polynomial $x - x_1$}
for i := 2 **to** n **do**
PEval := *HornerEval*($P(x)$,$X[i]$)

Continued

```
    QEval := HornerEval(Q(x),X[i])
    P(x) := (Y[i] – PEval)(Q(x)/QEval) + P(x)
    if i < n then Q(x) := PolyMult(Q(x),x – X[i]) endif
  endfor
end NewtonInterp
```

We leave it as an exercise to show that the complexity (number of multiplications) of *NewtonInterp* is quadratic.

7.7 Multivariate Recurrence Relations

All of the recurrence relations discussed so far have involved functions of a single variable. In this section we discuss two important recurrence relations that involve functions of two variables. The first is Pascal's recurrence relation for binomial coefficients. The second one involves summing the kth powers of the first n integers.

7.7.1 Binomial Coefficients

Before stating Pascal's recurrence relation, recall from (1.2.17) of Chapter 1 that the binomial coefficient $C(n,k)$ (alternately denoted by $\binom{n}{k}$) satisfies the formula

Figure 7.6 Stages of Newtonian interpolation for the points given in Figure 7.4

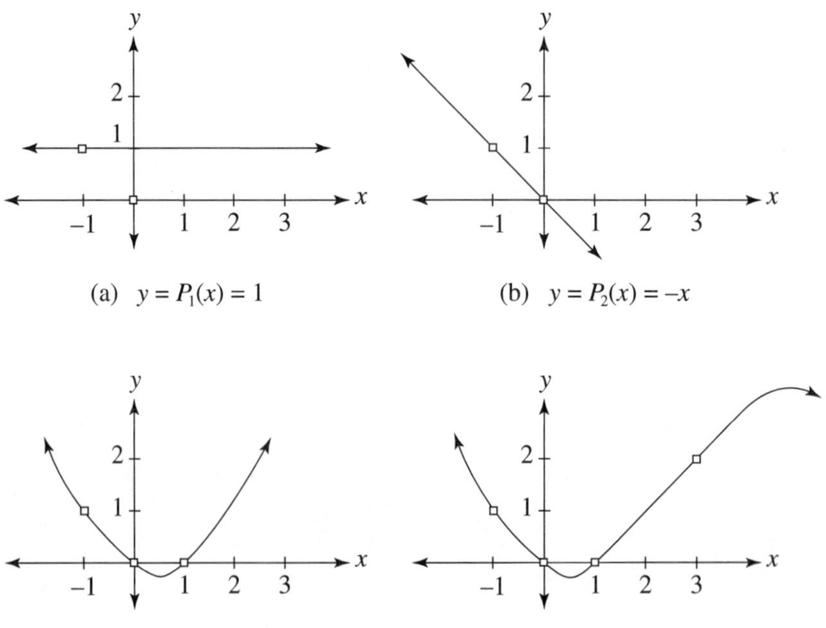

(a) $y = P_1(x) = 1$ (b) $y = P_2(x) = -x$

(c) $y = P_3(x) = \frac{x^2}{2} - \frac{x}{2}$ (d) $y = P_4(x) = \frac{-1}{24}x^3 + \frac{1}{2}x^2 - \frac{11}{24}x$

$$C(n,k) = n(n-1)\cdots(n-k+1)/k!, \quad \text{for } 0 \le k \le n. \tag{7.7.1}$$

Note that for a *fixed* k, $C(n,k) \in \Theta(n^k)$.

A problem arises when attempting to implement an algorithm for $C(n,k)$ based directly on (7.7.1), namely, the rapid growth of the factorial function $k!$. For example, many compilers store integer variables using four bytes of storage, so that the largest integer that can be stored without overflow is $2^{31} - 1 = 2{,}147{,}483{,}648$. This value is already exceeded by 13!. One way around this is to calculate $C(n,k)$ as the product of the k fractions $\frac{n}{1}\frac{(n-1)}{2}\cdots\frac{(n-k+1)}{k}$. This avoids integer overflow, but has the disadvantage of using real arithmetic and introducing round-off errors. For large n, the best way to calculate $C(n,k)$ is based on Pascal's recurrence relation (7.7.2):

$$C(n,k) = C(n-1,k-1) + C(n-1,k), \quad \textbf{init. cond. } C(n,0) = C(n,n) = 1. \tag{7.7.2}$$

To verify (7.7.2), consider a set of n elements $S = \{s_1, s_2, \ldots, s_n\}$. The initial conditions in (7.7.2) are true since the only subset having zero elements is the empty set and the only subset having n elements is S itself. To prove the recurrence relation, consider the element s_1. The subsets of S having k elements fall into two disjoint classes: those that contain s_1 and those that don't. If s_1 is in a subset of size k, then the remaining $k - 1$ elements of the subset must be chosen from the $n - 1$ elements in the subset $S \setminus \{s_1\}$. The number of such choices is $C(n-1,k-1)$. On the other hand, if s_1 is not in a subset of size k, then all k elements of this subset must be chosen from $S \setminus \{s_1\}$. The number of such choices is $C(n-1,k)$. Thus, the total number of ways to choose a subset of k elements is the sum of $C(n-1,k-1)$ and $C(n-1,k)$, which establishes the recurrence relation (7.7.2).

In Figure 7.7, we show the famous *Pascal's Triangle*, which is generated using recurrence relation (7.7.2). Each successive row in Pascal's Triangle is obtained from the row above by using (7.7.2). The initial conditions yield the 1s forming two sides of the (infinite) triangle. The famous philosopher and mathematician Blaise Pascal

Figure 7.7 Pascal's Triangle generated by $C(n,k) = C(n-1,k-1) + C(n-1,k)$, **init. cond.** $C(n,0) = C(n,n) = 1$

	k = 0	1	2	3	4	5	6	7	8	9	. . .
n = 0	1										
1	1	1									
2	1	2	1								
3	1	3	3	1							
4	1	4	6	4	1						
5	1	5	10	10	5	1					
6	1	6	15	20	15	6	1				
7	1	7	21	35	35	21	7	1			
8	1	8	28	56	70	56	28	8	1		
9	1	9	36	84	126	126	84	36	9	1	
⋮	⋮				⋮						⋱

exploited various properties of this triangle in a paper appearing in 1654. However, the triangle itself was already known to the Chinese as early as the eleventh century.

We now give pseudocode for a recursive function *BinomCoeff* that evaluates the binomial coefficients based directly on the recurrence (7.7.2).

```
function BinomCoeff(n,k) recursive
Input:  n,k (nonnegative integers)
Output: (n k) (returns the binomial coefficient C(n,k) = (n k))
   if k = 0 .or. k = n then
      return(1)
   else
      return(BinomCoeff(n – 1,k) + BinomCoeff(n – 1,k – 1))
   endif
end BinomCoeff
```

An examination of the tree of recursive calls for *BinomCoeff* shows that it is very inefficient due to redundant recalculations. In Figure 7.8 the redundant recalculations are illustrated by the tree of recursive calls generated by the resolution of *BinomCoeff* when invoked with $n = 6$ and $k = 3$.

Figure 7.8 Tree of recursive calls of *BinomCoeff* when $n = 6$ and $k = 3$

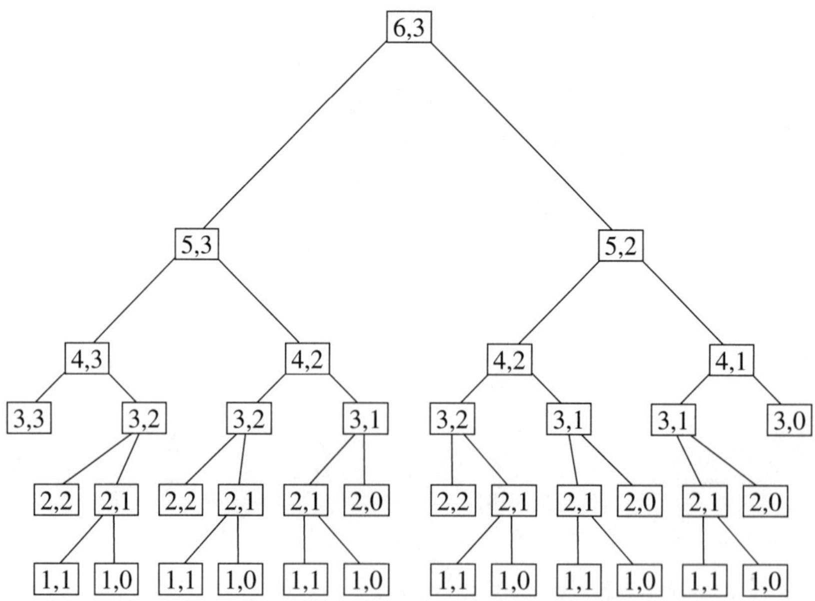

As we discussed in Section 1.3 of Chapter 1, there is a canonical way to remove redundant recalculations from recursive functions such as *Fib*1 and *BinomCoeff*: A table is maintained for storing the calculated values, and recursive invocations are made only after checking the corresponding table entry to see if the value has already been calculated. We rewrite the pseudocode for *BinomCoeff* using this idea of saving values in a table. *BinomCoeff* calls a procedure *InitTable*, which initializes the entries of the array *Table*[0:*n*,0:*k*] to zeros.

```
function BinomCoeff2(n,k)
Input:  n,k (nonnegative integers)
Output: (n k) (returns the binomial coefficient C(n,k) = (n k))
   call InitTable(Table[0:n,0:k])
   return(BinomCoeff2Rec(n,k))

   function BinomCoeff2Rec(n,k) recursive
      if k = 0 .or. k = n then
         return (1)
      else
         if Table(n,k) = 0 then
            Table(n,k) := BinomCoeff2Rec(n – 1,k) + BinomCoeff2Rec(n – 1,k – 1)
         endif
         return(Table(n,k))
      endif
   end BinomCoeff2Rec
end BinomCoeff2
```

Figure 7.9 (page 270) shows the tree of recursive calls generated by invoking *BinomCoeff*2 with $n = 6$ and $k = 3$. Leaf nodes correspond to either using initial conditions or using values from *Table*. We indicate those leaf nodes where *Table* values are used by placing a T underneath the node. Storing previously calculated values of *BinomCoeff*2 results in a significant pruning of the tree in Figure 7.8.

In Figure 7.10 (page 271) we indicate the part of Pascal's Triangle generated when invoking *BinomCoeff*2(n,k) for $n = 9$ and $k = 3$. Except for $C(0,0)$, the coefficients generated by *BinomCoeff*2 are exactly those that lie in the indicated *Pascal's Parallelogram*.

Thus, the procedure *InitTable* only needs to initialize those entries of *Table*[1:*n*,1:*n*] that lie within Pascal's Parallelogram with bottom vertex n,k. An examination of Figure 7.10 leads to the following pseudocode for *InitTable*.

```
procedure InitTable(Table[1:n,1:k])
Input:  Table[1:n,1:k] (two-dimensional array)
Output: Table[1:n,1:k] (Pascal's Parallelogram initialized to 0)
```

Continued

```
    for Row := 1 to n – k do
        for Col := 1 to k do
            Table[Row + Col,Col] := 0
        endfor
    endfor
end InitTable
```

Clearly, *InitTable* performs $(k + 1)(n - k + 1)$ assignments to the two-dimensional array *Table*. To count the number of additions performed by *BinomCoeff*2, we place an addition symbol + between every pair of elements in Pascal's Parallelogram, as illustrated in Figure 7.11. The + symbols determine a k by $n - k$ parallelogram, so that the total number of + symbols is $k(n - k)$. Choosing addition as our basic operation, it follows that the best-case, average, and worst-case complexities of *BinomCoeff*2 all equal $k(n - k)$.

A purely iterative version of *BinomCoeff* can be written that computes $C(n,k)$ by generating the binomial coefficients in Pascal's Parallelogram. Pseudocode for this iterative version is left as an exercise.

Figure 7.9 Tree of recursive calls of *BinomCoeff2* when $n = 6$ and $k = 3$

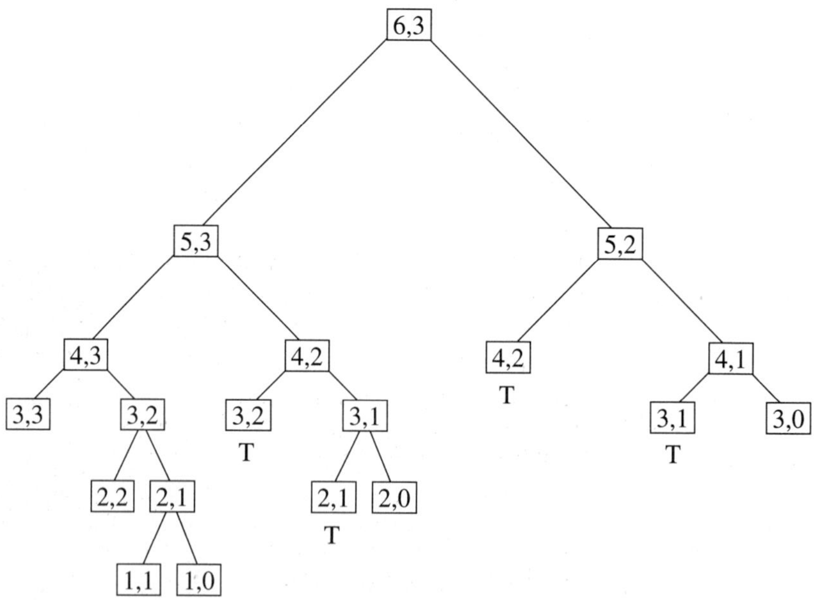

7.7.2 Sums of Powers: $S(n,k) = 1^k + 2^k + \cdots + n^k$

Our second multivariate recurrence relation involves the sum of powers $S(n,k) = 1^k + 2^k + \cdots + n^k$, for k a nonnegative integer. Unlike the recurrence relation for $C(n,k)$, the recurrence relation (7.7.3) that we now give for $S(n,k)$ has n fixed. Recurrence relation (7.7.3) does not yield an explicit formula for $S(n,k)$ for a general k. However, (7.7.3) can be used to show that $S(n,k)$ is a polynomial in n of degree $k + 1$. Thus, for *k fixed*, $S(n,k) \in \Theta(n^{k+1})$. The fact that $S(n,k) \in \Theta(n^{k+1})$ can also be established directly (see Exercise 7.47). It is interesting that recurrence relation (7.7.3) was also known to Pascal, who in fact published it in a slightly more general form (see Exercise 7.50).

	k = 0	1	2	3	4	5	6	7	8	9	. . .
n = 0	1										
1	1	1									
2	1	2	1								
3	1	3	3	1							
4	1	4	6	4	1						
5	1	5	10	10	5	1					
6	1	6	15	20	15	6	1				
7	1	7	21	35	35	21	7	1			
8	1	8	28	56	70	56	28	8	1		
9	1	9	36	84	126	126	84	36	9	1	
⋮	⋮				⋮						⋱

Figure 7.10 Binomial coefficients calculated when invoking *BinomCoeff* (either version) with $n = 9$ and $k = 3$

	k = 0	1	2	3	4	5	6	7	8	9	. . .
n = 0	1										
1	1 +	1									
2	1 +	2 +	1								
3	1 +	3 +	3 +	1							
4	1 +	4 +	6 +	4	1						
5	1 +	5 +	10 +	10	5	1					
6	1 +	6 +	15 +	20	15	6	1				
7	1	7 +	21 +	35	35	21	7	1			
8	1	8	28 +	56	70	56	28	8	1		
9	1	9	36	84	126	126	84	36	9	1	
⋮	⋮				⋮						⋱

Figure 7.11 Additions performed by *BinomCoeff2* with $n = 9$ and $k = 3$

PROPOSITION 7.7.1 Given any nonnegative k, $S(n,k) = 1^k + 2^k + \cdots + n^k$ satisfies the recurrence relation

$$S(n,k) = \frac{1}{k+1}\left[(n+1)^{k+1} - 1 - \left(\binom{k+1}{0}S(n,0) + \binom{k+1}{1}S(n,1) + \cdots + \binom{k+1}{k-1}S(n,k-1)\right)\right] \quad \textbf{init cond. } S(n,0) = n. \tag{7.7.3}$$ ❑

Proof Applying the binomial expansion (1.2.18) with $x = 1$ and y ranging from 1 to n yields

$$
\begin{aligned}
(1+1)^{k+1} &= \binom{k+1}{0}1^0 + \binom{k+1}{1}1^1 + \cdots + \binom{k+1}{k+1}1^{k+1} \\
(1+2)^{k+1} &= \binom{k+1}{0}2^0 + \binom{k+1}{1}2^1 + \cdots + \binom{k+1}{k+1}2^{k+1} \\
&\ \ \vdots \\
(1+n)^{k+1} &= \binom{k+1}{0}n^0 + \binom{k+1}{1}n^1 + \cdots + \binom{k+1}{k+1}n^{k+1}.
\end{aligned}
\tag{7.7.4}
$$

Now, summing all the terms on the left-hand side of each equation in (7.7.4) yields

$$S(n,k+1) + (n+1)^{k+1} - 1,$$

whereas summing all the expressions on the right-hand side yields

$$\binom{k+1}{0}S(n,0) + \binom{k+1}{1}S(n,1) + \cdots + \binom{k+1}{k+1}S(n,k+1).$$

Equating these latter two expressions and simplifying yields (7.7.3). ■

Let us apply (7.7.3) to obtain an explicit formula for $S_1(n)$ and $S_2(n)$. Substituting in (7.7.3) with $k = 1$, we obtain

$$S(n,1) = \frac{1}{2}\left[(n+1)^2 - 1 - \binom{2}{0}S(n,0)\right] = \frac{1}{2}(n^2 + 2n - n) = \frac{1}{2}n(n+1).$$

Substituting in (7.7.3) with $k = 2$, we obtain

$$
\begin{aligned}
S(n,2) &= \frac{1}{3}\left[(n+1)^3 - 1 - n + \left(\binom{3}{0}S(n,0) + \binom{3}{1}S(n,1)\right)\right] \\
&= \frac{1}{3}\left[(n+1)^3 - 1 - \left(n + \binom{3}{2}n(n+1)\right)\right] \\
&= \frac{1}{6}n(n+1)(2n+1).
\end{aligned}
$$

Note that the latter formula agrees with the formula for $S(n,2)$ established in Section 7.1.

The following useful proposition is applied in Chapter 9 to establish the average complexity of the algorithm *ProbeLinSrch* for searching a link-ordered list.

PROPOSITION 7.7.2 $S(n,k) = 1^k + 2^k + \cdots + n^k$ is a polynomial in n of degree $k + 1$ with leading coefficient $1/(k + 1)$. In particular,

$$S(n,k) \sim \frac{n^{k+1}}{k+1}. \tag{7.7.5}$$

Proof The proof proceeds by induction (strong form) on k.

Basis step: $S(n,0) = 1^0 + 2^0 + \cdots + n^0 = n$, a polynomial of degree 0 + 1 with leading coefficient 1/(0 + 1).

Induction step: Given a positive integer k, assume $S(n,j)$ is a polynomial in n with leading coefficient $1/(j + 1)$ for all positive integers $j < k$. We must show that $S(n,k)$ is a polynomial in n of degree $k + 1$ with leading coefficient $1/(k + 1)$. From (7.7.3),

$$S(n,k) = \frac{(n+1)^{k+1}}{k+1} + P_k(n), \tag{7.7.6}$$

where $P_k(n) = \dfrac{-1}{k+1}\left(1 + \binom{k+1}{0} S(n,0) + \binom{k+1}{1} S(n,1) + \cdots + \binom{k+1}{k+1} S(n,k-1)\right)$.

The strong form of our induction step implies that $P_k(n)$ is a polynomial in n of degree k. By the binomial expansion (1.2.18), $(n + 1)^{k+1}$ is a polynomial in n of degree $k + 1$ with leading coefficient 1. Proposition 7.7.2 follows from this latter observation and (7.7.6). ■

7.8 Closing Remarks

Induction and recurrence relations play important roles in both the analysis and design of algorithms. We will see numerous applications of induction and recurrence relations in subsequent chapters. Many additional interesting applications of induction and recurrence relations can be found in the references.

References and Suggestions for Further Reading

Material on mathematical induction and recurrence relations can be found in a host of books on discrete mathematics. Some recent texts include the following.

Albertson, M. O. and J. P. Hutchinson. *Discrete Mathematics with Algorithms*. New York: Wiley, 1988.

Fletcher, P. *Foundations of Discrete Mathematics*. Boston, MA: PWS, 1991.

Graham, R. L., D. E. Knuth, and O. Patashnik. *Concrete Mathematics*. Reading, MA: Addison-Wesley, 1989.

Johnsonbaugh, R. *Discrete Mathematics*. New York: Macmillan, 1990.

Maurer, S. B. *Discrete Algorithmic Mathematics*. Reading MA: Addison-Wesley, 1991.

Gries, D. *The Science of Programming*. New York: Springer-Verlag, 1981. Contains material specifically relating to loop invariants and proving correctness of programs.

Hofri, M. *Analysis of Algorithms, Computational Methods & Mathematical Tools*. New York: Oxford University Press, 1995. A recent book presenting the methods and tools needed to determine the effectiveness of algorithms.

Manber, U. *Introduction to Algorithms: A Creative Approach.* Reading, MA: Addison-Wesley, 1989. A text stressing the use of mathematical induction in the design as well as the analysis of algorithms.

Exercises

Section 7.1 Mathematical Induction

7.1 Prove by induction that

$$1(1!) + 2(2!) + \cdots + n(n!) = (n + 1)! - 1.$$

7.2 Prove by induction that

$$\frac{1}{1 \times 2} + \frac{1}{2 \times 3} + \cdots + \frac{1}{n \times (n+1)} = \frac{n}{n+1}.$$

7.3 Prove by induction that

$$C(n,0) + C(n,1) + C(n,2) + \cdots + C(n,n) = 2^n.$$

7.4 Prove by induction that

$$C(n,0) - C(n,1) + \cdots + (-1)^i C(n,i) + \cdots + (-1)^n C(n,n) = 0.$$

7.5 **a.** For A_1 and A_2 subsets of U, verify DeMorgan's Laws:

$$(A_1 \cup A_2)^c = A_1^c \cap A_2^c,$$

$$(A_1 \cap A_2)^c = A_1^c \cup A_2^c.$$

b. State and prove by induction a generalization of DeMorgan's Laws to n sets.

7.6 Prove by induction that

$$\text{fib}(1) + \text{fib}(2) + \cdots + \text{fib}(n) = \text{fib}(n + 2) - 1.$$

7.7 Prove by induction that for any set of nonnegative real numbers $x_1, x_2, \ldots, x_n$:

$$(x_1 + x_2 + \cdots + x_n)^2 \le n(x_1^2 + x_2^2 + \cdots + x_n^2).$$

7.8 **a.** Prove by induction that

$$\begin{pmatrix} \text{fib}(n) \\ \text{fib}(n+1) \end{pmatrix} = \begin{pmatrix} 0 & 1 \\ 1 & 1 \end{pmatrix}^n \begin{pmatrix} 0 \\ 1 \end{pmatrix}$$

where fib(n) denotes the nth Fibonacci number.

b. Briefly describe how the preceding formula can be employed to design an algorithm for computing fib(n) using only O(log n) multiplications.

7.9 Suppose you have a collection of n lines in the plane such that no two are parallel and no three meet in a point. Find a formula for the number of regions in the plane determined by these lines, and prove your formula using induction.

Section 7.2 Correctness Proofs

7.10 Prove the correctness of *BubbleSort* (see Exercise 3.15).

7.11 Prove the correctness of the procedure *RearrangeAndPlace* (see Chapter 4).

Section 7.3 Mathematical Properties of Binary Trees

7.12 Using induction, prove Proposition 4.4.1.

7.13 Using induction, prove Proposition 7.3.2.

7.14 Using induction, give a direct proof of Proposition 7.3.3 without using the transformation to 2-trees.

7.15 Prove Proposition 7.3.4.

7.16 Prove Proposition 7.3.5.

7.17 Using induction, prove Proposition 7.3.6.

7.18 Complete the verification of Corollary 7.3.8 by establishing the inequality

$$x\lfloor \log_2 x\rfloor + 2(x - 2^{\lfloor \log_2 x\rfloor}) \geq \lceil x \log_2 x\rceil, \quad \text{for any integer } x \geq 1.$$

7.19 Given a 2-tree T having L leaf nodes, let T_{left} and T_{right} denote the left and right subtrees (of the root) of T, respectively.

a. Give a recurrence relation for $LPL(T)$ in terms of $LPL(T_{\text{left}})$, $LPL(T_{\text{right}})$, and L.

b. Solve the recurrence relation in (a) to give an alternate proof of Corollary 7.3.8.

7.20 Show that the implicit search tree for *BinarySearch* is a 2-tree that is full at the second-deepest level.

7.21 A *k-ary tree* T is a rooted tree where every node has at most k children. A *k-tree* T is a k-ary tree where every nonleaf node has exactly k children. In each of the following, state and prove a generalization of the result for binary trees or 2-trees to k-ary trees or k-trees.

a. Proposition 7.3.1

b. Proposition 7.3.2

c. Proposition 7.3.3

d. Proposition 7.3.4

e. Proposition 7.3.6

f. Theorem 7.3.7

Section 7.4 Solving Recurrence Relations for Complexity

7.22 Solve the following recursion relation, where $n = 3^k$ for some nonnegative integer k.

$$t(n) = 2t(n/3) + n, \quad n \geq 1, \quad \textbf{init. cond. } t(0) = 0.$$

7.23 *TriMergeSort*, given below, is a variant of *MergeSort*, where the list is split into three equal parts instead of two. For convenience, assume that $n = 3^k$, for some nonnegative integer k.

```
procedure TriMergeSort(Low,High) recursive
Input:  Low,High (indices of a global array L[1:n], initially Low = 1
                                                   and High = n = 3^k)
Output: sublist L[Low:High] is sorted in nondecreasing order
   if High ≤ Low then return endif
   Third := ⌊(Low + High)/3⌋
   TwoThirds := 2*Third
   call TriMergeSort(Low,Third)
   call TriMergeSort(Third+1,TwoThirds)
   call TriMergeSort(TwoThirds+1,High)
{merge sublists L[Low:Third] and L[Third:TwoThirds] of sizes n/3 and n/3}
   call Merge(Low,Third,TwoThirds)
{merge sublists L[Low:TwoThirds] and L[TwoThirds:High] of sizes 2n/3 and n/3}
   call Merge(Low,TwoThirds,High)
end TriMergeSort
```

a. Give a recurrence relation for the worst-case complexity $W(n)$ of *TriMergeSort* for an input list of size n.

b. Solve the recurrence formula you have given in (a) to obtain an explicit formula for the worst-case complexity $W(n)$ of *TriMergeSort*.

c. Which is more efficient in the worst case, *MergeSort* or *TriMergeSort*? Discuss.

d. Repeat (a), (b), and (c) for the best-case complexity $B(n)$.

7.24 In this section we established a formula for the worst-case complexity $W(n)$ of *MergeSort* when n is a power of two. Find an approximation to $W(n)$ for a general n, showing that $W(n) \in \Theta(n \log n)$.

7.25 Prove Proposition 7.4.1.

7.26 Show that the following functions are Θ-invariant under scaling.

a. n

b. $\log_b n$

c. $n \log_b n$

d. any polynomial

7.27 Show that exponential functions are not Θ-invariant under scaling.

7.28 Show that sums and products of functions that are Θ-invariant under scaling are themselves Θ-invariant under scaling.

7.29 Show that the condition $g(cn) \in \Theta(g(n))$ is equivalent to the condition $g(cn) \in O(g(n))$ for eventually nondecreasing functions.

7.30 Show that if there exists an integer $b \geq 2$ such that $g(bn) \in \Theta(g(n))$, then $g(cn) \in \Theta(g(n))$ for *all* positive integers c.

7.31 Complete the proof of Proposition 7.4.2 for the case $X = \Omega$.

Section 7.5 Generating Functions

7.32 Derive formula (7.5.7) using the method of partial fractions.

7.33 Using *generating functions* solve the recurrence relation

$$t(n) = t(n-1) + 2t(n-2), \quad n \geq 2, \quad \textbf{init. cond. } t(0) = 2,\ t(1) = 1.$$

7.34 Consider the formula (7.5.26)

$$g(x) = \frac{b_0 + (b_1 - c_1 b_0)x + \cdots + (b_{k-1} - c_1 b_{k-1} - \cdots - c_k b_0)x^{k-1}}{1 - c_1 x - \cdots - c_k x^k}$$

where the coefficients of the polynomials in the numerator and denominator are real. Develop a technique for generating the power series for $g(x)$ for the following cases.

a. The roots of the denominator polynomial are real and distinct.

b. The roots are all distinct.

c. The general case.

7.35 Prove by induction (see 7.5.35) that

$$(1)(3)(5) \cdots (2n-1)2^n = (n+1)(n+2) \cdots (2n-1)(2n).$$

7.36 Verify that the number b_n of binary trees on n nodes satisfies the recurrence (7.5.28)

7.37 Prove by induction that

$$\binom{2n}{n} \geq \frac{4^n}{2n+1}, \quad n \geq 1,$$

and conclude that the nth Catalan number $b_n = \frac{1}{n+1}\binom{2n}{n} \in \Omega(4^n/n^2)$.

7.38 Prove by induction that given any set of n distinct keys $k_1 < k_2 < \cdots < k_n$ and any binary tree T on n nodes, there is a unique way to assign the n keys to nodes in T such that T becomes a binary search tree for the keys. Hence, the number of binary search trees containing n distinct keys coincides with the number

$b_n = \frac{1}{n+1}\binom{2n}{n}$ of binary trees on n nodes.

7.39 **a.** Let t_n denote the number of 2-trees with n leaf nodes. Give a recurrence relation for t_n, and then solve the recurrence relation using generating functions to obtain the formula

$$t_n = \frac{1}{n}\binom{2n-2}{n-1}, \quad n \geq 2.$$

b. Derive the formula in (a) by relating the number t_n of 2-trees with n leaf nodes to the number b_{n-1} of binary trees on $n-1$ nodes, and then employ formula (7.5.27) for b_n.

Section 7.6 Interpolation of Polynomials

7.40 Design and analyze a divide-and-conquer algorithm for polynomial evaluation. Compare your algorithm to Horner's method.

7.41 Show that the algorithm for solving linear equations using a direct implementation of Gaussian elimination has cubic complexity.

7.42 Given points $(x_1, y_1), \ldots, (x_n, y_n)$, the Lagrange formula for the polynomial $P(x)$ interpolating these points is given by

$$P(x) = \sum_{1 \leq i \leq n} \left(\prod_{\substack{i \neq j \\ 1 \leq j \leq n}} \frac{(x - x_j)}{(x_i - x_j)} \right) y_i.$$

a. Verify that

$$P(x_i) = y_i, \; i = 1, \ldots, n.$$

b. Show that the algorithm based on simply translating the above formula has $\Theta(n^3)$ complexity.

c. Give pseudocode for an algorithm for implementing the Lagrange formula that has $\Theta(n^2)$ complexity.

7.43 Show that the complexity (number of multiplications) of the procedure *Newton-Interp* is quadratic.

Section 7.7 Multivariate Recurrence Relations

7.44 Using induction, prove the binomial expansion (1.2.18).

7.45 Give pseudocode for a nonrecursive version of *BinomCoeff*.

7.46 Given nonnegative integers n and $k_1, k_2, \ldots, k_j, j \geq 2$, such that $n = k_1 + k_2 + \cdots + k_j$, let $\binom{n}{k_1, k_2, \ldots, k_j}$ denote the number of partitions of a set of size n into j subsets of size $k_1, k_2, \ldots, k_j$. Note in this setting that the binomial coefficient $C(n,k)$, also denoted by $\binom{n}{k}$, is equal to $\binom{n}{k, n-k}$. The numbers $\binom{n}{k_1, k_2, \ldots, k_j}$ are known as *multinomial coefficients* because of the occurrence in the multinomial expansion

$$(x_1 + x_2 + \cdots + x_j)^n = \sum_{\substack{k_1, \ldots, k_j \\ n = k_1 + \cdots + k_j}} \binom{n}{k_1, k_2, \ldots, k_j} x_1^{k_1} x_2^{k_2} \ldots x_j^{k_j}.$$

a. Verify the multinomial expansion.

b. Generalize formula (7.6.1) to obtain the following formula.

$$\binom{n}{k_1, k_2, \ldots, k_j} = \frac{n!}{k_1!\, k_2! \,\ldots\, k_j!}$$

c. Obtain a recurrence relation for the multinomial coefficients.

7.47 Show directly that $S(n,k) \in \Theta(n^{k+1})$ without using recurrence relation (7.7.3).

7.48 Proposition 7.7.2 states that $S(n,k) = 1^k + 2^k + \cdots + n^k$ is a polynomial in n of degree $k + 1$ whose leading coefficient is $1/(k + 1)$. Using induction and Proposition 7.6.1, show that the coefficients of n^k and n^{k-1} in $S(n,k)$ are $-1/2$ and $k/12$, respectively.

7.49 **a.** $S(n,k) = 1^k + 2^k + \cdots + n^k$ is a polynomial in n of degree $k + 1$ whose constant term is zero. Let $B(j,k)$ denote the coefficient of n^j in polynomial $S(n,k)$, $j = 1, \ldots, k + 1$. Using the recurrence relation (7.7.3) for $S(n,k)$, obtain a recurrence relation for $B(j,k)$.

b. Using the recurrence relation for $B(j,k)$, obtained in (a), give pseudocode for the procedure *SumOfPowers*, which outputs the (coefficients of) the polynomial $S(n,k)$.

c. Modify your algorithm to avoid truncation errors when dividing by large integers by storing the coefficients $B(j,k)$ as fractions in lowest form. These fractions can be stored as pairs of integers (*Numer*,*Denom*), where *Numer* is the numerator and *Denom* is the denominator. Obtain the lowest form of the fraction represented by (*Numer*,*Denom*) by employing Euclid's gcd algorithm.

7.50 Generalize the recurrence relation (7.7.3) for $S(n,k)$ to a recurrence relation for the sum of the kth powers of the first n terms in the arbitrary arithmetic progression

$$S(a,d,n,k) = a^k + (a + d)^k + \cdots + (a + (n - 1)d)^k.$$

8

Graphs, Digraphs, and Sets

Many problems are naturally modeled using graphs and digraphs, and data structures implementing the graph or the digraph ADT are commonly used throughout computer science. The subject of graph algorithms is a very active area of research today. Graphs also play an important role in parallel architectures, since they serve as models for interconnection networks. In this chapter we introduce the two basic search strategies known as depth-first search and breadth-first search, and apply these strategies to such problems as topological sorting and finding shortest paths in graphs. We finish the chapter with a discussion of union and find operations for the disjoint sets ADT.

8.1 Graphs and Digraphs

In this section, we give a brief introduction to the theory of graphs and digraphs. We introduce some basic terminology, prove several elementary results, and present two standard representations of graphs and digraphs.

8.1.1 Graphs

Formally, a *graph* $G = (V,E)$ is a set $V = V(G)$ called *vertices* (or *nodes*), together with a set $E = E(G)$ called *edges*, such that an edge is an *unordered* pair of vertices. An edge $\{u,u\}$ is called a *loop*. Unless otherwise stated, we restrict our attention to graphs without loops. For simplicity, we sometimes denote the edge $\{u,v\}$ by uv. Given an edge $e = uv$ in a graph $G = (V,E)$, we refer to vertices u and v as the *end* vertices of e, and we say that *e joins u* and v. Two vertices u and v are *adjacent* if they are the two end vertices of an edge in the graph (that is, $uv \in E$). The set of all vertices adjacent to u is the *neighborhood of* u. A vertex v and an edge e are *incident* if e contains vertex v, that is, if $e = vw$ for some vertex w. Two edges e and f are *adjacent* if they have an end vertex in common.

A graph can be represented pictorially by a drawing in the plane, where the vertices are represented by points in the plane and an edge $\{u,v\} \in E$ is represented by a continuous curve joining u and v (see Figure 8.1). Note from Figure 8.1*b* that such drawings allow the continuous curves to intersect at points (called *crossings*) other than vertices. It turns out that the graph in Figure 8.1b cannot be drawn in the plane without at least one crossing.

Graphs such as the one in Figure 8.1a, which can be drawn in the plane without crossings, are called *planar graphs*. Such a drawing is called an *embedding* of G in the plane. Planar graphs are particularly important in computer science. For example, they are useful in VLSI design and flow diagrams. If a graph is not planar, then we say that it is *nonplanar.*

Two graphs $G = (V,E)$ and $G' = (V',E')$ are *isomorphic* if there exists a bijective mapping $\beta : V \rightarrow V'$ from the vertex set V of G onto the vertex set V' of G' such that adjacency relationships are preserved; that is, $\{u,w\} \in E$ if, and only if, $\{\beta(u),\beta(w)\} \in E'$. The mapping β is called an *isomorphism.* Deciding whether two graphs are isomorphic is, in general, a difficult problem. A graph is *complete* if every pair of distinct vertices is joined by an edge. Clearly, any two complete graphs having the same number of vertices are isomorphic. A complete graph on n vertices is denoted by K_n.

Figure 8.1 (a) A planar graph; and (b) a nonplanar graph

$V = \{1,2,3,4,5,6,7,8,9\}$

$E = \{\{1,2\},\{1,3\},\{2,3\},\{2,4\},\{2,6\},\{3,4\},\{4,5\},\{4,6\},\{7,8\},\{7,9\},\{8,9\}\}$

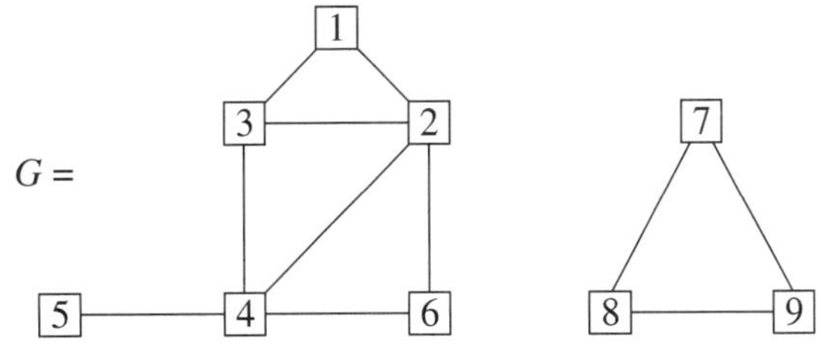

(a)

$V = \{1,2,3,A,B,C\}$

$E = \{\{1,A\},\{1,B\},\{1,C\},\{2,A\},\{2,B\},\{2,C\},\{3,A\},\{3,B\},\{3,C\}\}$

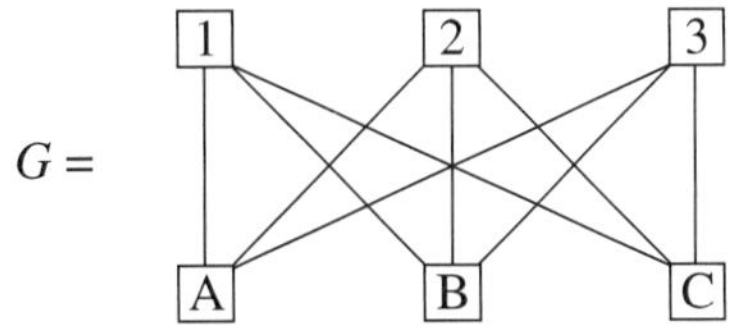

(b)

We denote the number of vertices and edges by $n = n(G)$ and $m = m(G)$, respectively, so that $n = |V|$ and $m = |E|$. The *degree* of a vertex $v \in V$, denoted by $d(v)$, is the number of edges incident with v. Let $\delta = \delta(G)$ and $\Delta = \Delta(G)$ denote the minimum and maximum degrees, respectively, over all the vertices in G, so that $\delta \le d(v) \le \Delta$ for all $v \in V$.

The following useful formula due to Euler relates the number m of edges to the sum of the degrees of the vertices.

PROPOSITION 8.1.1 The sum of the degrees over all the vertices of a graph G equals twice the number of edges; that is,

$$\sum_{v \in V} d(v) = 2m. \tag{8.1.1}$$

Proof Every edge is incident with exactly two vertices. Therefore, when summing the degrees over all the vertices we count each edge exactly twice, once for each of its end vertices. ■

A graph is *r-regular* if every vertex has degree r. A useful corollary of Proposition 8.1.1 is the following result relating the number of vertices and edges of an r-regular graph.

COROLLARY 8.1.2 If G is an r-regular graph with n vertices and m edges, then

$$m = rn/2. \tag{8.1.2}$$ ❑

Another corollary of Proposition 8.1.1 is obtained by taking both sides of (8.1.1) **mod** 2.

COROLLARY 8.1.3 For any graph G, the number of vertices of odd degree is even. ❑

A *path P of length p* ($p \ge 0$) joining vertices u and v is an alternating sequence of $p + 1$ vertices and p edges $u_0e_1u_1e_2 \ldots e_pu_p$ such that $u = u_0$, $v = u_p$, where e_i joins u_{i-1} and u_i, $i = 1,2, \ldots, p$. We call u_0 and u_p the *initial* and *terminal* vertices, respectively, and the remaining vertices in the path the *interior* vertices. Vertices in a path can be repeated, but edges must be distinct. Since the path P is completely defined by the sequence of vertices $u_0u_1 \ldots u_p$, we often use this shorter sequence to denote P. If $u = v$, then the path is called a *closed path* or *circuit*. We call a path *simple* if the interior vertices in the path are all distinct and are different from the initial and terminal vertices. A *simple circuit* or *cycle* is a simple closed path. A path that contains every edge in the graph exactly once is called an *Eulerian path*. A circuit that contains every edge exactly once is called an *Eulerian circuit* or *Eulerian tour*. A simple path that contains every vertex in the graph is called a *Hamiltonian path*. A cycle that contains every vertex in the graph is called a *Hamiltonian cycle*.

Two vertices $u,v \in V$ are *connected* if there exists a path (of length 0 when $u = v$) that joins them. The relation *u is connected to v* is an equivalence relation on V. Each equivalence class C of vertices, together with all incident edges, is called a *(con-*

nected) component of G. When G has only one component, then G is *connected*; otherwise, G is *disconnected.*

The *distance* between u and v, denoted by $d(u,v)$, is the length of a path from u to v that has the shortest (minimum) length among all such paths. By convention, if u and v are not connected, then $d(u,v) = \infty$. The *diameter* of G is the maximum distance between any two vertices. In Figure 8.2 the distance between every pair of vertices and the diameter is given for a sample graph G. We show the distances using a 6-by-6 *distance matrix*, whose *ij*th entry is given by $d(i, j)$.

A *tree* is a connected graph without cycles. A rooted tree is simply a tree with one vertex designated as the root. Every tree is planar, so that it can be drawn in the plane without crossings. The following propositions are not difficult, and their proofs are left as exercises.

PROPOSITION 8.1.4 A graph T is a tree if, and only if, there exists a unique path joining every pair of distinct vertices of T. ❑

PROPOSITION 8.1.5 A connected graph T is a tree if, and only if, the number of edges of T is one less than the number of vertices. ❑

A *subgraph* H of G is a graph such that $V(H) \subseteq V(G)$ and $E(H) \subseteq E(G)$. A subgraph H that is a tree is a *subtree* of G, or simply a *tree* of G. A subgraph H of G is called a *spanning* subgraph if H contains all the vertices of G. When a subgraph H of G is a tree, we use the term *spanning tree* of G. Given a subset U of vertices of G, the subgraph $G[U]$ *induced by* U is the subgraph with vertex set U and edge set consisting of all edges in G having both end vertices in U. In Figure 8.3b we show a

Figure 8.2 Distances between vertices for a sample graph G on six vertices

$G =$

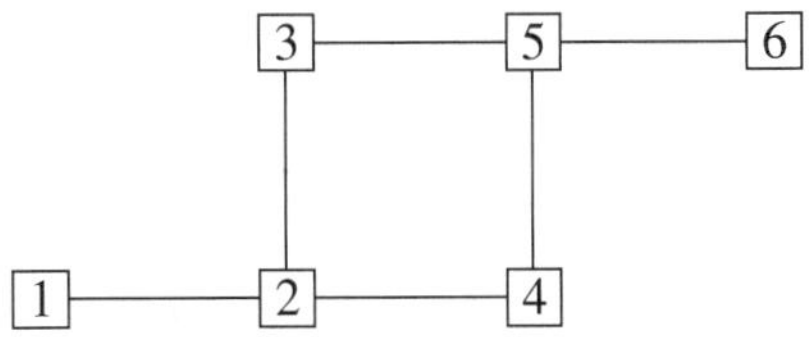

Distance Matrix of G

$$\begin{bmatrix} 0 & 1 & 2 & 2 & 3 & 4 \\ 1 & 0 & 1 & 1 & 2 & 3 \\ 2 & 1 & 0 & 2 & 1 & 2 \\ 2 & 1 & 2 & 0 & 1 & 2 \\ 3 & 2 & 1 & 1 & 0 & 1 \\ 4 & 3 & 2 & 2 & 1 & 0 \end{bmatrix}$$

Diameter $G = 4$

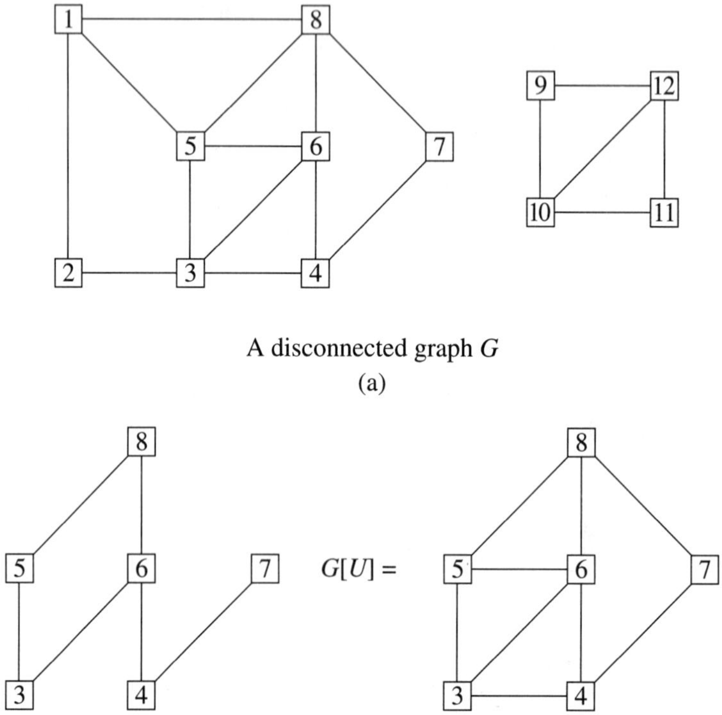

A disconnected graph G

(a)

A subgraph spanning U The subgraph induced by U

(b)

Figure 8.3 (a) The components of G are the subgraphs, $G[A]$ and $G[B]$, induced by the vertex sets $A = \{1,2,3,4,5,6,7,8\}$ and $B = \{9,10,11,12\}$, respectively; (b) a subgraph spanning $U = \{3,4,5,6,7,8\}$ and the induced subgraph $G[U]$

subgraph and an induced subgraph on the same set of vertices U. Note that a component of a graph G is a connected induced subgraph of G that is not contained in a strictly larger connected subgraph (see Figure 8.3a).

A graph is *bipartite* if there exists a bipartition of the vertex set V into two sets X and Y such that every edge has one end in X and the other in Y. The *complete bipartite* graph $K_{i,j}$ is the bipartite graph with $n = i + j$ vertices, i vertices belonging to X and j vertices belonging to Y, such that there is an edge joining every vertex $x \in X$ to every vertex $y \in Y$. The complete bipartite graph $K_{3,3}$ is shown in Figure 8.1b. Two other examples of complete bipartite graphs are given in Figure 8.4.

The graphs K_5 and $K_{3,3}$ are both nonplanar graphs. If you try drawing them in the plane without edge crossings, you will have trouble. In fact, K_5 and $K_{3,3}$ are in some sense the quintessential nonplanar graphs (see Theorem 8.1.6). A *subdivision* S of G is a graph obtained from G by replacing each edge e with a path joining the same two vertices as e (subdividing the edge e, as in Figure 8.5). Observe that G is a sub-

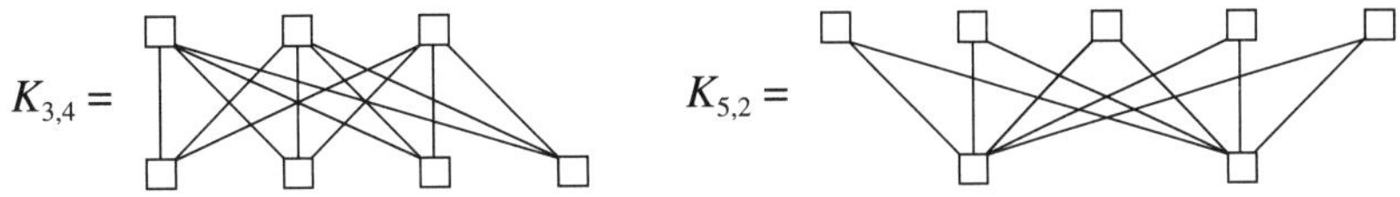

Figure 8.4 Two complete bipartite graphs

division of itself (replace each edge with a path of length 1). Clearly, if G is nonplanar, then every subdivision of G is nonplanar. It is also clear that if G contains a nonplanar graph, then G is nonplanar. It follows that a graph is nonplanar if it contains a subgraph that is a subdivision of K_5 or $K_{3,3}$. The fact that the converse is also true is a surprising and deep result discovered by Kuratowski. We state Kuratowski's theorem without proof.

THEOREM 8.1.6 **Kuratowski's Theorem** A graph G is nonplanar if, and only if, it contains a subgraph that is (isomorphic to) a subdivision of K_5 or $K_{3,3}$. ❑

A *clique of size p* in a graph G is a complete subgraph on p vertices. The *clique number* $\omega(G)$ of a graph G is the size of the largest clique in G. No polynomial algorithm is known to determine the clique number of an arbitrary graph. However, the following result of Turan gives a lower bound on the clique number in terms of the number of vertices n and edges m in G.

THEOREM 8.1.7 **Turan's Theorem** Let $G = (V,E)$ be a graph with n vertices and m edges. Then, the clique number $\omega(G)$ of G satisfies

$$\omega(G) \geq \lceil n^2/(n^2 - 2m) \rceil. \qquad ❑$$

The following equivalent formulation of Theorem 8.1.7 is used in Chapter 10 to establish a lower bound for the problem of finding the maximum value in a list on the EREW PRAM with n processors. Given a graph $G = (V,E)$, a subset $U \subseteq V$ is an *independent* set of vertices if no two vertices in U are joined by an edge in E. The

Figure 8.5 Subdividing a graph

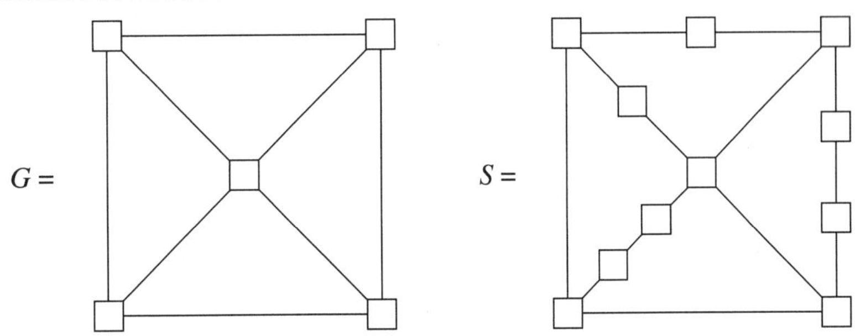

independence number $\beta(G)$ of G is defined to be the size of a largest set of independent vertices in G (see Figure 8.6). The complement $\overline{G}$ of a graph G is the graph with the same vertex set as G, such that $\{u,v\}$ is an edge of $\overline{G}$ if, and only if, $\{u,v\}$ is not an edge of G. Clearly, we have

$$\beta(G) = \omega(\overline{G}). \tag{8.1.3}$$

Further, if m, $\overline{m}$ denote the number of edges of G, $\overline{G}$, respectively, then

$$m + \overline{m} = \binom{n}{2} \frac{n\,(n-1)}{2}. \tag{8.1.4}$$

Using (8.1.3) and (8.1.4), Theorem 8.1.7 can be restated in terms of the independence number.

THEOREM 8.1.7′ Let $G = (V,E)$ be a graph with n vertices and m edges. Then the independence number $\beta(G)$ of G satisfies

$$\beta(G) \geq \lceil n^2/(2m + n) \rceil.$$

❑

For the graph G of Figure 8.6, Turan's Theorem guarantees an independence number of at least $\lceil 6^2/(2 \times 8 + 6) \rceil = \lceil 36/22 \rceil = 2$. Actually, $\beta(G) = 3$, since vertices 1,3,4 form an independent set (there is no independent set of four vertices). Removal of vertex 4 would yield a graph G' with $\beta(G') = 2$, which equals the lower bound of $\lceil 5^2/(2 \times 7 + 5) \rceil$ given by Turan's Theorem.

8.1.2 Graphs as Interconnection Networks

The underlying structure for an interconnection network model for parallel communication is a graph. The nodes of the graph correspond to the processors. Two nodes are joined with an edge (adjacent) whenever the corresponding two processors communicate directly with one another. As discussed in Chapter 5, in the interconnection network model, information is communicated between nonadjacent processors P and Q by relaying the information along a path in the network joining P and Q. The diameter is an upper bound on the number of communication steps needed to relay information between any two processors. Having small diameter is clearly a desirable property of an interconnection network. Another desirable property is having small maximum degree, since such interconnection network models are easier to build than those having large maximum degree.

Figure 8.6 A graph G with $\beta(G) = 3$. For this graph, {1,3,4} is the only set of independent vertices of size 3.

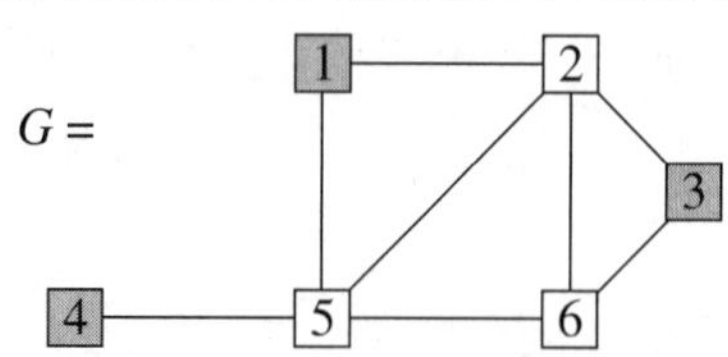

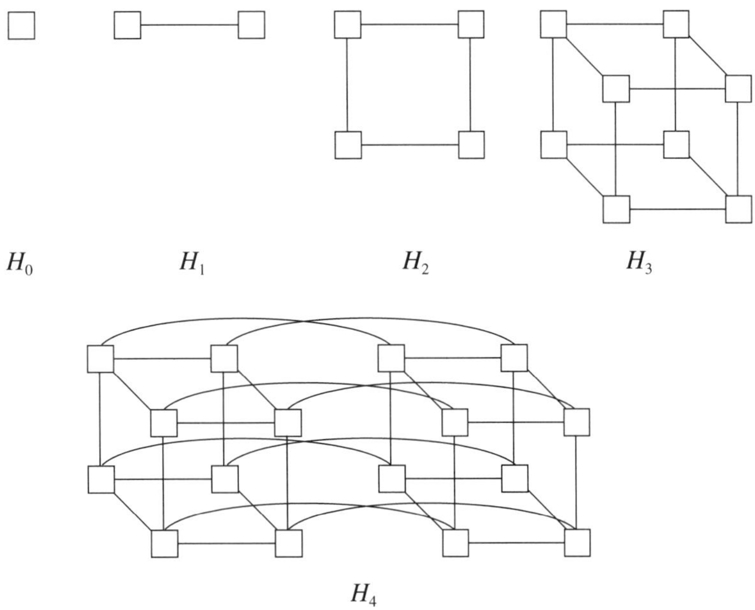

Figure 8.7 Hypercubes H_i, $i = 0,1,2,3,4$

The complete graph interconnection network K_n having n processors has diameter 1 (since the distance between every pair of vertices is 1), but is not practical to build since each vertex has very high degree $(n - 1)$. At the other extreme is the one-dimensional mesh M_p, which has maximum degree 2 and diameter $n - 1$ (the distance between processor P_1 and processor P_p is $n - 1$). The two-dimensional $M_{q,q}$ $(n = q^2)$ has maximum degree 4 but has diameter $2(\sqrt{n} - 1)$, which is still quite large (the maximum distance $2(q - 1)$ occurs between processors $P_{1,1}$ and $P_{q,q}$).

There are some other interconnection networks based on graphs that achieve nice compromises between extreme values of degree and diameter. One such example is the hypercube interconnection network model. The k-dimensional hypercube H_k has 2^k nodes consisting of the set of all 0/1 k-tuples; that is, $V(H_k) = \{(x_1, \ldots, x_k) \mid x_i \in \{0,1\}, i = 1, \ldots, k\}$. Two nodes in $V(H_k)$ are joined with an edge of H_k if, and only if, they differ in exactly one component. We illustrate the hypercubes up to dimension 4 in Figure 8.7. Note that the hypercube of dimension k can be obtained by joining corresponding vertices in two copies of the hypercube of dimension $k - 1$.

The hypercube H_k of dimension k is k-regular and has diameter $k = \log_2 n$ (see Exercise 8.5). Thus, the maximum degree and diameter of H_k are both logarithmic in the number of vertices n.

8.1.3 Digraphs

Graphs with directed edges are natural models for many physical phenomena. For example, the pairings of the players in a (match-play golf, table tennis, etc.)

$V = \{1,2,3,4,5,6,7\}$
$E = \{\{1,2\},\{2,1\},\{1,6\},\{4,2\},\{2,7\},\{3,4\},\{4,5\},\{4,7\},\{7,5\},\{6,5\},\{7,6\}\}$

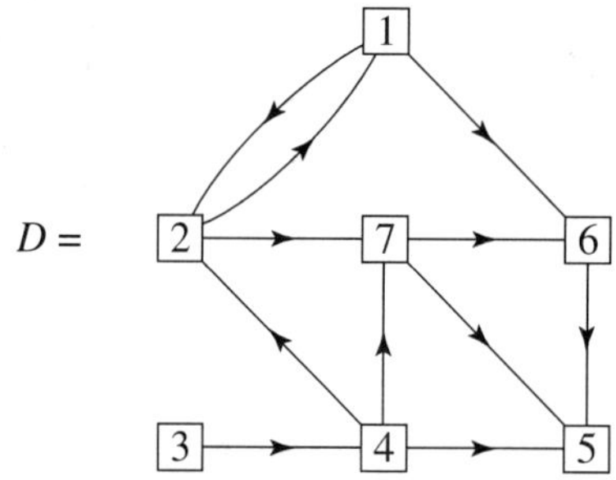

Figure 8.8 Drawing a digraph

tournament can be represented by a graph, where the vertices correspond to the players and where two vertices A and B are joined with an edge if players A and B are to play each other. The winner of the match between each pair of players A and B can be recorded by "orienting" the edge joining A and B, so that it is directed from A to B if A defeats B. The resulting oriented graph is called a *directed graph* or *digraph.*

Formally, a *digraph D* is a set $V = V(D)$ of vertices together with a set $E = E(D)$ called *directed edges* (sometimes called *arcs*) such that a directed edge is an *ordered* pair of vertices. A digraph can be represented by a drawing in the plane in the same way that a graph can, except that we add an arrow to the curve representing each edge to indicate the orientation of that edge (see Figure 8.8).

Since an unordered pair $\{a,b\}$ is combinatorially equivalent to the two ordered pairs (a,b), (b,a), a digraph is actually a generalization of a graph. Associated with any given undirected graph $G = (V,E)$ is the combinatorial equivalent digraph $\hat{G} = (\hat{V},\hat{E})$, where $V = \hat{V}$, and where the ordered pairs (u,v), (v,u) are in $\hat{E}$ if, and only if, the unordered pair $\{u,v\}$ is in E (see Figure 8.9).

For simplicity, we sometimes denote the directed edge (u,v) by uv. We refer to u as the *tail* of e and v as the *head* of v. The *out-degree* of a vertex $v \in V$, denoted by $d_{out}(v)$ is the number of edges having tail v. Similarly, the *in-degree* of a vertex $v \in V$, denoted by $d_{in}(v)$, is the number of edges having head v. It is easily verified that

$$\sum_{v \in V} d_{out}(v) = \sum_{v \in V} d_{in}(v) = m,$$

where m denotes the number of directed edges of D. The *out-neighborhood (in-neighborhood) of vertex u* is the set of all vertices v such that uv (vu) is a directed edge of D.

8.1.4 Implementing Graphs and Digraphs

Two standard implementations of a graph G are the adjacency matrix implementation and the adjacency lists implementation. For convenience, assume that the verti-

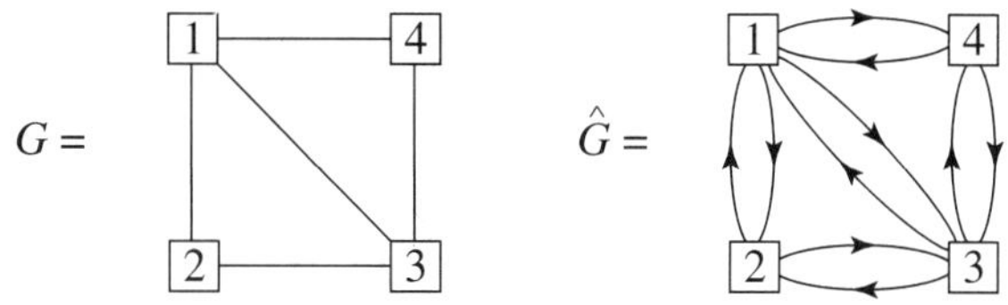

Figure 8.9 Graph G and its combinatorial equivalent digraph $\hat{G}$

ces of G are labeled 1,2, … , n (when referring to a vertex we do not distinguish between the label of the vertex and the vertex itself). The *adjacency matrix* of a graph G is the $n \times n$ symmetric matrix $A = (a_{ij})$ given by

$$a_{ij} = \begin{cases} 1 & \text{vertices } i \text{ and } j \text{ are adjacent in } G. \\ 0 & \text{otherwise,} \end{cases} \qquad i,j \in \{1, \ldots, n\}.$$

A sample graph G and its adjacency matrix are given in Figure 8.10.

Implementing G using its adjacency matrix makes it easy to perform many standard operations on G, such as adding a new edge or deleting an existing edge. Note that the adjacency matrix of G allocates n^2 memory locations no matter how many edges are in the graph. Thus, implementing G using its adjacency matrix is inefficient if the number of edges of G is small relative to the number of vertices. For example, if G is a tree with n vertices, then G has only $n - 1$ edges. On the other hand, if $m \in \Theta(n^2)$, then the adjacency matrix representation is probably an efficient way to implement G.

Given a graph $G = (V,E)$, for $v \in V$, the *adjacency list of* v is the list consisting of all the vertices adjacent to v. The order in which the vertices are listed usually does not matter. The adjacency lists of G are generally implemented as linked lists, where

Figure 8.10 Adjacency matrix of a graph G

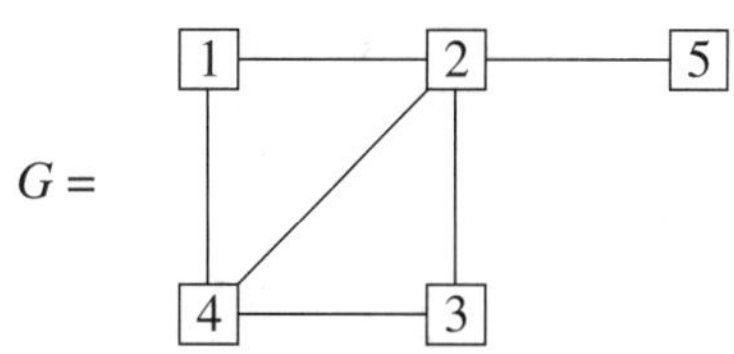

$$A = \begin{bmatrix} 0 & 1 & 0 & 1 & 0 \\ 1 & 0 & 1 & 1 & 1 \\ 0 & 1 & 0 & 1 & 0 \\ 1 & 1 & 1 & 0 & 0 \\ 0 & 1 & 0 & 0 & 0 \end{bmatrix}$$

pointers to the beginning of the linked lists are stored in *header* nodes. The header nodes can themselves be implemented using a linked list, or they can be stored in an array *Header*[1:*n*]. Figures 8.11a and 8.11b illustrate the adjacency lists implementation using linked lists for the graph given in Figure 8.10, where the header nodes are implemented using a linked list and an array, respectively. We do not assume that the nodes in each adjacency list are maintained in increasing order of node labels.

Each node of the adjacency list of vertex i corresponds to an edge $\{i,j\}$ incident with i. In the adjacency lists implementation of a graph, where the header nodes belong to a linked list (see Figure 8.11*a*), it is convenient in practice to maintain a second pointer p in each list node defined as follows: If the list node belongs to the adjacency list of vertex i and corresponds to edge $\{i,j\}$ then p points to the header node of adjacency list j.

Both the adjacency matrix and adjacency lists implementations of graphs generalize naturally to digraphs. The *adjacency matrix* of a digraph D, whose vertices are labeled $1,2, \ldots, n$, is the $n \times n$ matrix $A = (a_{ij})$ defined by

$$a_{ij} = \begin{cases} 1 & \text{if there is an arc from } i \text{ and } j, \\ 0 & \text{otherwise,} \end{cases} \qquad i,j \in \{1, \ldots, n\}.$$

The adjacency matrix of a sample digraph D is given in Figure 8.12.

Figure 8.11 Adjacency lists implementations for the sample graph *G* given in Figure 8.10

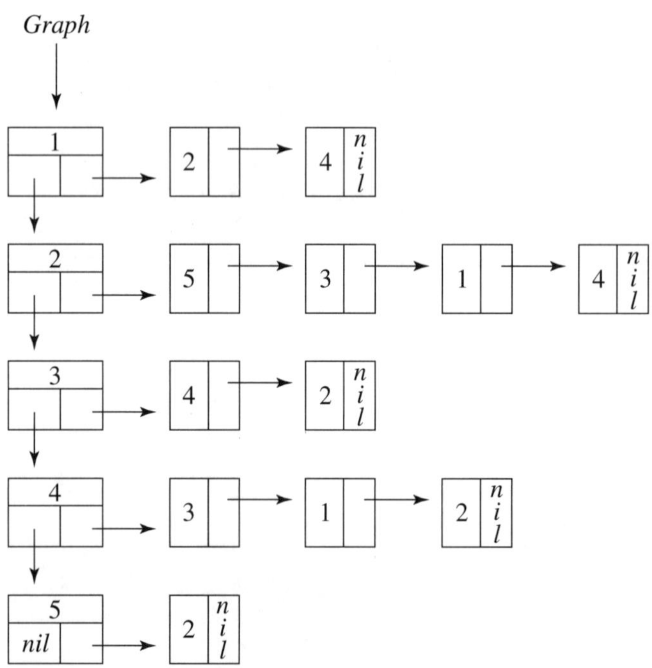

(a) Linked list of header nodes

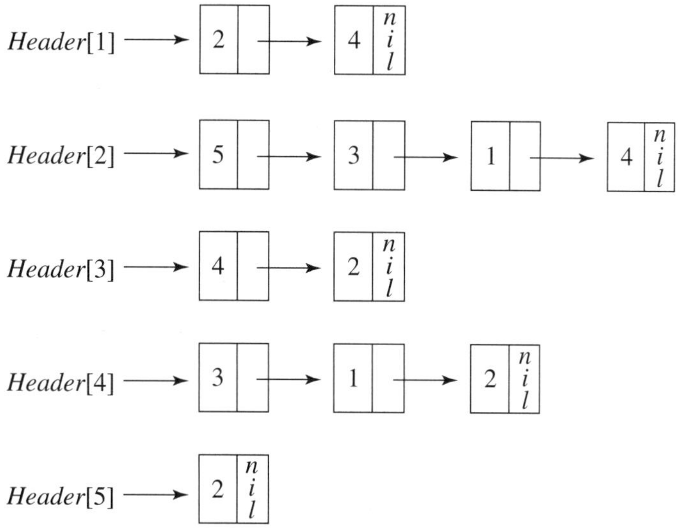

(b) Array of header nodes

Figure 8.11 Continued

Figure 8.12 Adjacency matrix of a digraph D

$D =$

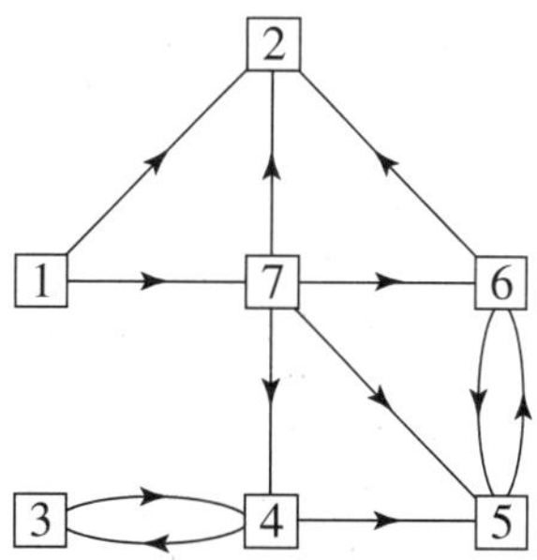

$$A = \begin{bmatrix} 0 & 1 & 0 & 0 & 0 & 0 & 1 \\ 0 & 0 & 0 & 0 & 0 & 0 & 0 \\ 0 & 0 & 0 & 1 & 0 & 0 & 0 \\ 0 & 0 & 1 & 0 & 1 & 0 & 0 \\ 0 & 0 & 0 & 0 & 0 & 1 & 0 \\ 0 & 1 & 0 & 0 & 1 & 0 & 0 \\ 0 & 1 & 0 & 1 & 1 & 1 & 0 \end{bmatrix}$$

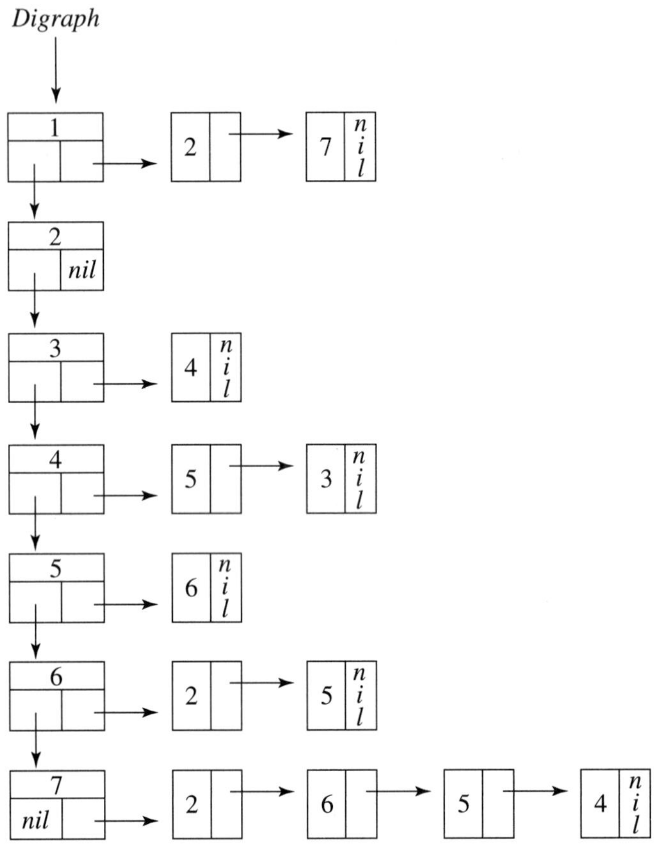

(a) Linked list of header nodes

Figure 8.13 Adjacency list implementations of the sample digraph D given in Figure 8.12

In Figure 8.13a and 8.13b, the adjacency lists implementations using a linked list of header nodes and an array of header nodes, respectively, are illustrated for the sample digraph given in Figure 8.12.

The adjacency matrix and adjacency lists representations of digraphs are generalizations of the adjacency matrix and adjacency lists representations of (undirected) graphs. To see this, consider any undirected graph G and its combinatorially equivalent digraph $\hat{G}$ (see Figure 8.9). Clearly, the adjacency matrix and adjacency lists of G and $\hat{G}$ are identical.

8.2 Search and Traversal of Graphs

The solutions to many important problems require an examination (visit) of the nodes of a graph. Two standard search techniques are *depth-first search* and *breadth-first search*.

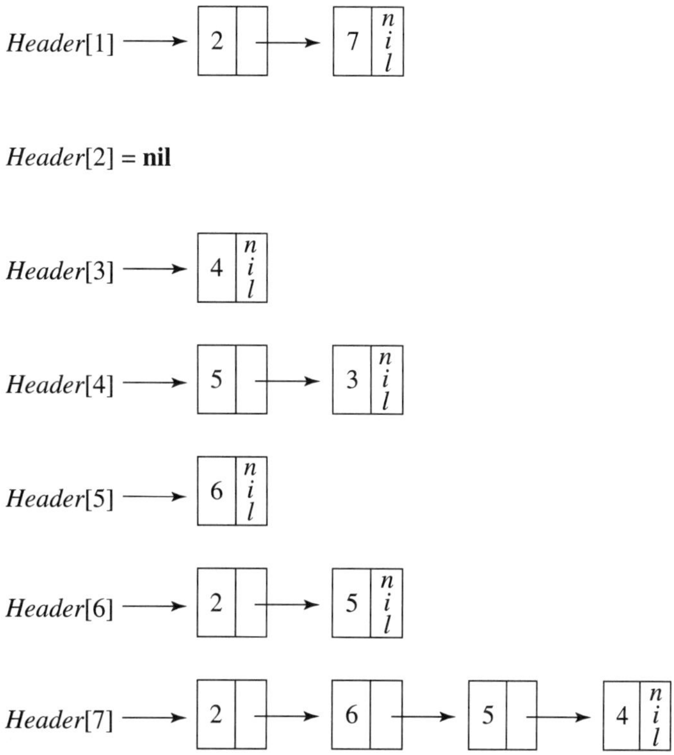

(b) Array of header nodes

Figure 8.13 Continued

Depth-first search and breadth-first search differ in their exploring philosophies: Depth-first search always longs to see what's over the next hill (where pastures might be greener), whereas breadth-first search visits the immediate neighborhood thoroughly before moving on. After visiting a node, breadth-first search explores this node (visits all neighbors of the node that have not already been visited) before moving on. On the other hand, depth-first search immediately moves on to an unvisited neighbor, if one exists, after visiting a node. Whenever depth-first search is at an explored node, it "backtracks" until an unexplored node is encountered and then continues. This backtracking often returns to the same node many times before it is explored.

8.2.1 Depth-First Search

We first give the pseudocode *DFS* for depth-first search. For simplicity, we assume that G has n vertices labeled 1,2, … , n; we use the symbol v to simultaneously denote a vertex and its label. We maintain an auxiliary array $Mark[1{:}n]$ to keep track of the nodes that have been visited.

```
procedure DFS(G,v) recursive
Input:   G (a graph with n vertices and m edges)
         v (a vertex) {The array Mark[1:n] is global and initialized to 0s.}
Output: the depth-first search of G with starting vertex v
   Mark[v] := 1
   call Visit(v)
   for each vertex u adjacent to v do
      if Mark[u] = 0 then call DFS(G,u) endif
   endfor
end DFS
```

The **for** loop in the *DFS* pseudocode did not explicitly describe the order in which the vertices are considered. This order is incidental to the nature of the search, and would, in general, depend on the particular implementation of the graph G (for example, adjacency matrix, adjacency lists, and so forth). Since we have labeled the vertices $1,2, \ldots, n$, we assume that the vertices are accessed by the **for** loop in increasing order of their labels (thereby making the order of visiting the nodes independent of the implementation). The graph in Figure 8.14 illustrates this convention.

For the graph G in Figure 8.14, *DFS* visits all the vertices of G. For a general graph G, *DFS* with starting vertex v visits all the vertices in the (connected) component containing v.

To further illustrate the **for** loop in *DFS*, we show how this loop can be written in the two standard ways to implement a graph. Suppose first that G is implemented using its adjacency matrix $A[1{:}n,1{:}n]$. We assume that the vertex v is labeled i. In this case the **for** loop becomes

```
for j := 1 to n do
   if (A[i,j] = 1) .and. (Mark[j] = 0) then
      call DFS(G,j)
   endif
endfor
```

On the other hand, suppose that G is implemented using adjacency lists. Recall that a typical version of this implementation has an array $Header[1{:}n]$ of header nodes, where $Header[v]$ is a pointer to the adjacency list of v. A node in the adjacency list for v corresponding to vertex u contains a field *Vertex* containing the index (label) of u. It also contains a pointer *NextVertex* to the next vertex in the adjacency list. Under these assumptions, the **for** loop in *DFS* becomes

```
p := Header[v]
while (p ≠ nil) do
   if Mark[p → Vertex] = 0 then
      call DFS(G,p → Vertex)
   endif
   p := p → NextVertex
endwhile
```

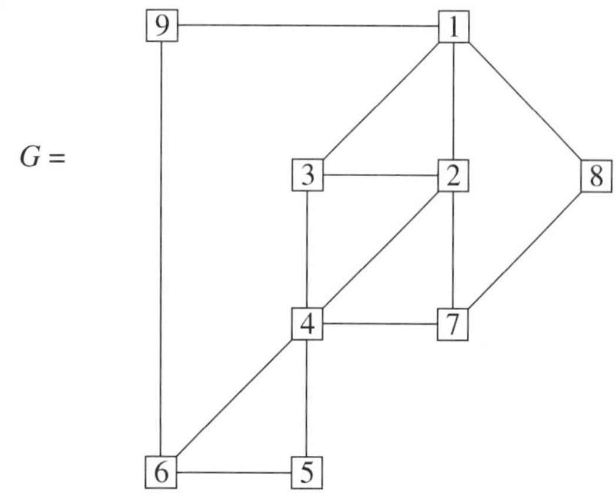

Figure 8.14 *DFS* with $v = 7$ visits vertices in the order 7,2,1,3,4,5,6,9,8

It is useful to write *DFS* as a nonrecursive procedure. For convenience, the nonrecursive version calls a procedure *Next* that determines the next unvisited node *w* adjacent to the node *u* just visited. If no such node *w* exists, then a Boolean parameter *found* is set to be **.false.**.

```
procedure DFS(G,v)
Input:  G (a graph with n vertices and m edges)
        v (a vertex)
Output: the depth-first search of G starting from vertex v

  dcl   Stack          {stack initialized as empty}
        Mark[1:n]      {array initialized to 0s}

  Mark[v] := 1
  call Visit(v)
  u := v
  call Next(u,w,found)
  while found .or. (.not. Empty(Stack))
     if found then          {go deeper}
        call Push(Stack,u)
        Mark[w] := 1
        call Visit(w)
        u := w
     else
        call Pop(Stack,u)         {backtrack}
     endif
     call Next(u,w,found)
  endwhile

end DFS
```

Visit 7
Push 7 *Visit* 2
Push 2 *Visit* 1
Push 1 *Visit* 3
Push 3 *Visit* 4
Push 4 *Visit* 5
Push 5 *Visit* 6
Push 6 *Visit* 9
Pop $(u = 6)$
Pop $(u = 5)$
Pop $(u = 4)$
Pop $(u = 3)$
Pop $(u = 1)$
Push 1 *Visit* 8
Pop $(u = 1)$
Pop $(u = 2)$
Pop $(u = 7)$
Stack is now empty.

$G =$

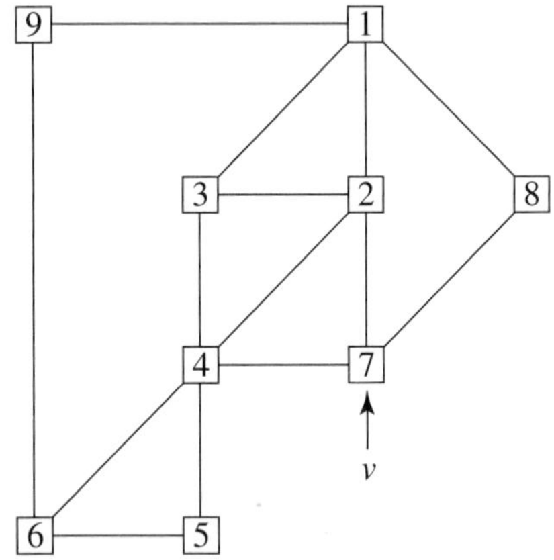

Figure 8.15 *DFS* with $v = 7$ visits vertices in the order 7,2,1,3,4,5,6,9,8

Figure 8.15 illustrates the sequence of pushes, visits, and pops performed by *DFS* for the graph of Figure 8.14.

We analyze the complexity of *DFS* from the point of view of two basic operations: visiting a node and examining a node (by calls to *Next*) to see if it has been marked as visited. The worst-case complexity for both operations occurs when the graph is connected. If G is connected, then it is easily verified that every vertex is visited exactly once, so that we have a total of n node visits. Each edge uw in the graph gives rise to exactly two vertex examinations by *Next*, one with u as input parameter and one with w as input parameter. This shows that the worst-case complexity of *DFS* in terms of the number of vertices examined by *Next* is $2m$. Therefore, the total number of basic operations of the two types performed by *DFS* in the worst case is $n + 2m \in O(n + m)$.

8.2.2 Depth-First Search Tree

Depth-first search with starting vertex v determines a tree, called the *depth-first search tree* (*DFS-tree*) *rooted at* v. During a depth-first search, whenever we move from a vertex u to an adjacent unvisited vertex w, we add the edge uw to the tree. This tree is naturally implemented using the parent representation where u is the parent of w.

The following procedure *DFSTree* for computing the DFS-tree rooted at v is a simple modification of the (nonrecursive) procedure *DFS*. We maintain the parent implementation of the DFS-tree using the array *TreeDFS*[1:n].

```
procedure DFSTree(G,v,TreeDFS[1:n])
Input:  G (graph with n vertices and m edges)
        v (vertex)
Output: TreeDFS[1:n] (parent-array implementing the DFS tree rooted at v)
   dcl  Stack                     {stack initialized as empty}
        Mark[1:n]                 {array initialized to 0s}
   for u := 1 to n do
      TreeDFS[u] := 0
   endfor
   Mark[v] := 1
   call Visit(v)
   u := v
   call Next(u,w,found)
   while found .or. (.not. Empty(Stack))
      if found then               {go deeper}
         TreeDFS[w] := u
         call Push(Stack,u)
         Mark[w] := 1
         call Visit(w)
         u := w
      else
         call Pop(Stack,u)        {backtrack}
      endif
      call Next(u,w,found)
   endwhile
end DFSTree
```

The depth-first search tree rooted at vertex 7 of the graph of Figure 8.15 is illustrated in Figure 8.16a, and the associated array *TreeDFS* is given in Figure 8.16b.

The array *TreeDFS*[1:*n*] gives us the unique path from any given vertex *w* back to *v* in the depth-first search tree rooted at *v*. For example, suppose we wish to find the path in the depth-first search tree of Figure 8.16 from vertex 9 to vertex 7. By simply starting at vertex 9 and following the pointers in the array *TreeDFS*[1:*n*], we obtain the path

9, *TreeDFS*[9] = 6, *TreeDFS*[6] = 5, *TreeDFS*[5] = 4, *TreeDFS*[4] = 3, *TreeDFS*[3] = 1, *TreeDFS*[1] = 2, *TreeDFS*[2] = 7.

8.2.3 Depth-First Traversal

If the graph G is connected, then *DFS* visits *all* the vertices of G so that *DFS* performs a *traversal* of G. For a general graph G, the following simple algorithm based on repeated calls to *DFS* with different starting vertices performs a traversal of G.

```
procedure DFT(G)

Input:  G (a graph with n vertices and m edges)
Output: the depth-first traversal of G

dcl     Mark[1:n]     {an array initialized to 0s}
  for v := 1 to n do
    if Mark[v] = 0 then
      call DFS(G,v)
    endif
  endfor

end DFT
```

If we call the procedure *DFSTree* instead of *DFS*, then a forest of DFS-trees is generated (a DFS-tree rooted at i is generated for each i such that i is not visited when $v = i$). We refer to this forest as the *depth-first traversal forest* (*DFT-forest*) of G. In the following procedure *DFTForest*, the DFT-forest is implemented using an array *ForestDFT*[1:n].

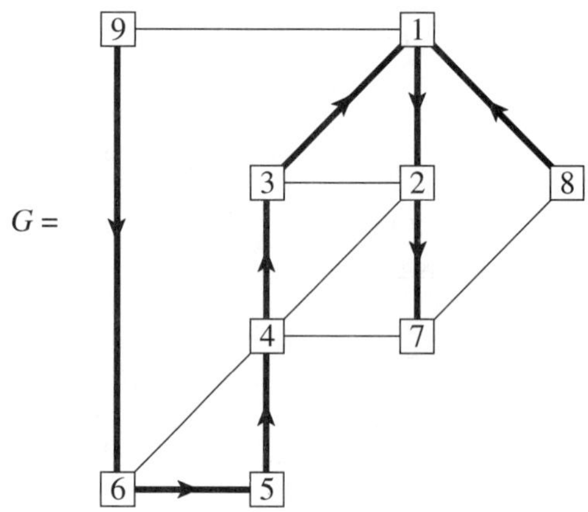

(a)

v	1	2	3	4	5	6	7	8	9
TreeDFS[v]	2	7	1	3	4	5	0	1	6

(b)

Figure 8.16 (a) The DFS-tree rooted at vertex 7 for the graph given in Figure 8.15; (b) the associated array *TreeDFS*

```
procedure DFTForest(G,ForestDFT[1:n])

Input:  G (a graph with n vertices and m edges)
Output: ForestDFT[1:n] (an array implementing the DFT forest of G)
```

```
   dcl Mark[1:n] {array initialized to 0s}
   for v := 1 to n do
      ForestDFT[v] := 0
   endfor
   for v := 1 to n do
      if Mark[v] = 0 then
         call DFSTree(G,v,ForestDFT[1:n])
      endif
   endfor
end DFTForest
```

8.2.4 Breadth-First Search

Recall that the strategy underlying breadth-first search is to visit all unvisited vertices adjacent to a given vertex before moving on. The breadth-first strategy is easily implemented by visiting these adjacent vertices and placing them on a queue. Then a vertex is removed from the queue, and the process repeated. Breadth-first search starts by inserting a vertex *v* onto an initially empty queue, and continues until the queue is empty.

```
procedure BFS(G,v)
Input:  G (a graph with n vertices and m edges)
        v (vertex) {the array Mark[1:n] is global and initialized to 0s}
Output: the breadth-first search of G starting from vertex v
   dcl   Queue        {initialized to be empty}
   call  Enqueue(Queue,v)
   Mark[v] := 1
   call Visit(v)
   while .not. Empty(Queue) do
      Dequeue(Queue,u)
      for each vertex w adjacent to u do
         if Mark[w] = 0 then
            call Enqueue(Queue,w)
            Mark[w] := 1
            call Visit(w)
         endif
      endfor
   endwhile
end BFS
```

Clearly, *BFS* has the same $O(n + m)$ complexity as *DFS*. Figure 8.17 illustrates *BFS* starting at vertex 7 for the same graph as given in Figure 8.15.

Breadth-first search with starting vertex *v* determines a tree spanning the component of the graph *G* containing *v*. As with *DFS*, a minor modification of

Enqueue 7	*Visit* 7
Dequeue	(u = 7)
Enqueue 2	*Visit* 2
Enqueue 4	*Visit* 4
Enqueue 8	*Visit* 8
Dequeue 1	(u = 2)
Enqueue 3	*Visit* 1
Enqueue 6	*Visit* 3
Dequeue	(u = 4)
Enqueue 5	*Visit* 5
Enqueue 6	*Visit* 6
Dequeue	(u = 8)
Dequeue	(u = 1)
Enqueue 9	*Visit* 9
Dequeue	(u = 3)
Dequeue	(u = 5)
Dequeue	(u = 6)
Dequeue	(u = 9)

Queue is now empty.

G =

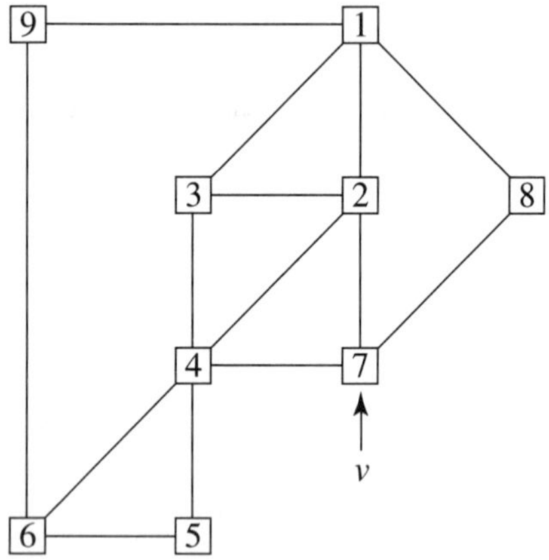

Figure 8.17 *BFS* with v = 7 visits vertices in the order 7,2,4,8,1,3,5,6,9

the pseudocode for *BFS* generates this *breadth-first search tree* (*BFS-tree*) rooted at v.

```
procedure BFSTree(G,v,TreeBFS)
Input:   G (a graph with n vertices and m edges)
         v (vertex) {the array Mark[1:n] is global and initialized to zeros}
Output: TreeBFS[1:n] (array implementing the BFS-tree rooted at vertex v)
  dcl    Queue           {initialized to be empty}
  for    u := 1 to n do
         TreeBFS[u] := 0
  endfor
  call Enqueue(Queue,v)
  Mark[v] := 1
  while .not. Empty(Queue) do
    Dequeue(Queue,u)
    for each vertex w adjacent to u do
      if Mark[w] = 0 then
        call Enqueue(Queue,w)
        Mark[w] := 1
        TreeBFS[w] := u
      endif
    endfor
  endwhile
end BFSTree
```

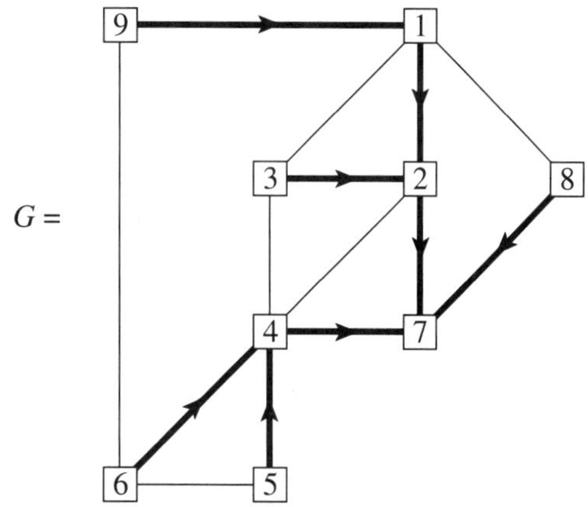

(a)

v	1	2	3	4	5	6	7	8	9
TreeBFS[v]	2	7	2	7	4	4	0	7	1

(b)

Figure 8.18 BFS-tree rooted at vertex 7 and its associated array *TreeBFS*

The breadth-first search tree rooted at vertex 7 and the associated array *TreeBFS* for the graph in Figure 8.17 is illustrated in Figure 8.18.

Just as was the case for depth-first search, breadth-first search can be used to determine whether or not the graph G is connected. Also, a breadth-first *traversal* of an arbitrary graph (connected or not) is obtained from the following algorithm.

```
procedure BFT(G)

Input:  G (a graph with n vertices and m edges)
Output: the breadth-first traversal of G

dcl      Mark[1:n]      {array initialized to 0s}
  for v := 1 to n do
    if Mark[v] = 0 then
      call BFS(G,v)
    endif
  endfor

end BFT
```

8.2.5 Breadth-First Search Tree and Shortest Paths

The breadth-first search tree starting at vertex v is actually a *shortest path tree* in the graph rooted at v; that is, it contains a shortest path from v to every vertex in the same component as v. We leave the proof of this shortest path property as an exercise. The shortest path property is in sharp contrast to the paths generated by depth-first search. Indeed, we have seen that paths generated by depth-first search with starting vertex v are often rather longer than shortest paths to v, a fact well illustrated by the complete graph K_n on n vertices. Given any $v \in V(K_n)$, breadth-first search generates (shortest) paths of length 1 from v to all other vertices. On the other hand, depth-first search applied to K_n generates a depth-first search tree that is a path of length $n - 1$.

Clearly, any algorithm that finds a shortest path tree must visit each vertex and examine each edge. Thus, a lower bound on the complexity of the shortest path problem is $\Omega(n + m)$. Since the algorithm *BFSTree* has worst-case complexity $O(n + m)$, it is an (order) optimal algorithm.

8.3 Searching and Traversing Digraphs

Each search and traversal technique for a graph G generalizes naturally to a digraph D. It is convenient to define both an in-version and an out-version of these searches and traversals (the out-version is equivalent to the in-version in the digraph with the edge orientations reversed). A *directed path from u to v* is a sequence of vertices $u = w_0 w_1 \ldots w_p = v$, such that $w_i w_{i+1}$ is a directed edge of D, $i = 0, \ldots, p - 1$. For convenience, we sometimes refer to a directed path simply as a path. Given a vertex r in a digraph D, an *out-tree T rooted at r* is a minimal subdigraph that contains a directed path from r to any other vertex v in T. An *in-tree rooted at r* is defined analogously. See the example shown in Figure 8.19.

Corresponding to the algorithm *DFS* for graphs we have the two algorithms in-directed depth-first search *DFSIn* and out-directed depth-first search *DFSOut*. Similarly, corresponding to *DFSTree* we have the algorithms *DFSInTree* and *DFSOutTree*,

Figure 8.19 An in-tree and an out-tree of a digraph D

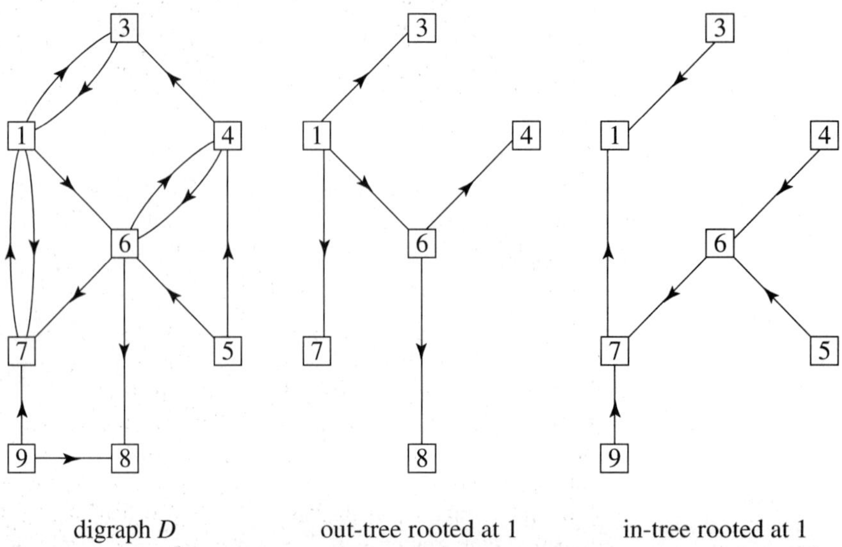

and corresponding to *DFTForest* we have *DFTInForest* and *DFTOutForest*. Analogously, we have algorithms *BFSIn*, *BFSOut* and *BFSInTree*, *BFSOutTree*, respectively, for breadth-first search, and similar algorithms for breadth-first traversals of digraphs.

Figure 8.20a (page 304) illustrates the sequence of pushes, visits, and pops performed by *DFSOut* and *DFSIn* for a sample digraph D starting at vertex 2, and 8.20b gives the DFS out-tree and DFS in-tree rooted at 2. Figure 8.21a (page 305) illustrates the sequence of enqueues, visits, and dequeues performed by *BFSOut* and *BFSIn* for the digraph D of Figure 8.20. Figure 8.21b gives the BFS out-tree and BFS in-tree rooted at 2.

8.4 Topological Sorting

Suppose that there are n tasks to be performed and that certain tasks must be performed before others. For example, if we are building a house, the task of pouring the foundation must precede the task of laying down the first floor. However, another pair of tasks might not need to be done in a particular order, such as painting the kitchen and painting the bathroom. The problem is to obtain a linear ordering of the tasks in such a way that if task u must be done before task v, then u occurs before v in the linear ordering.

The dependencies of the tasks can be naturally modeled using a directed acyclic graph (*dag*) D—that is, a directed graph without any directed cycles. The vertices of D correspond to the tasks, and a directed edge from u to v is in D if, and only if, task u must precede task v. A *topological sorting of* D is a listing of the vertices such that if uv is an edge of D, then u precedes v in the list. A *topological-sort labeling* of D is a labeling of the vertices in D with the labels $1, \dots, n$ such that for any edge uv in D, the label of u is smaller than the label of v.

A topological-sort labeling is obtained by performing an out-directed depth-first traversal, where we keep track of the order in which vertices become explored. A vertex becomes *explored* when the traversal acceses u having visited all vertices in the out-neighborhood of u. The topological-sort labeling is the explored labeling in reverse order. That is, a vertex u has topological-sort label $n - i + 1$ if u is the ith vertex to become explored, $i = 1, \dots, n$.

```
procedure: TopologicalSort(D,TopList[1:n])
Input:  D (dag with vertex set V = {1, ... ,n} and edge set E)
Output: TopList[1:n] (array containing a topologically sorted list of the
                                                         vertices in D)
dcl     Mark[1:n]     {array initialized to 0s}
  Counter := n
  for v := 1 to n do
    if Mark[v] := 0 then
      call DFSOutTopLabel(D,v)
    endif
  endfor
  procedure DFSOutTopLabel(D,v) recursive
    Mark[v] := 1
    for each w ∈ V such that vw ∈ E do
```

(Continued on page 306)

DFSOut starting at 2	
Visit 2	
Push 2	*Visit* 1
Push 1	*Visit* 3
Pop	$(u = 1)$
Push 1	*Visit* 8
Push 8	*Visit* 7
Pop	$(u = 8)$
Pop	$(u = 1)$
Pop	$(u = 2)$
Push 2	*Visit* 4
Pop	$(u = 4)$
Pop	$(u = 2)$
Stack is now empty.	

DFSIn starting at 2	
Visit 2	
Push 2	*Visit* 7
Push 7	*Visit* 4
Push 4	*Visit* 5
Push 5	*Visit* 6
Pop	$(u = 5)$
Pop	$(u = 4)$
Pop	$(u = 7)$
Push 7	*Visit* 8
Push 8	*Visit* 1
Push 1	*Visit* 3
Pop	$(u = 1)$
Push 1	*Visit* 9
Pop	$(u = 1)$
Pop	$(u = 8)$
Pop	$(u = 7)$
Pop	$(u = 2)$
Stack is now empty.	

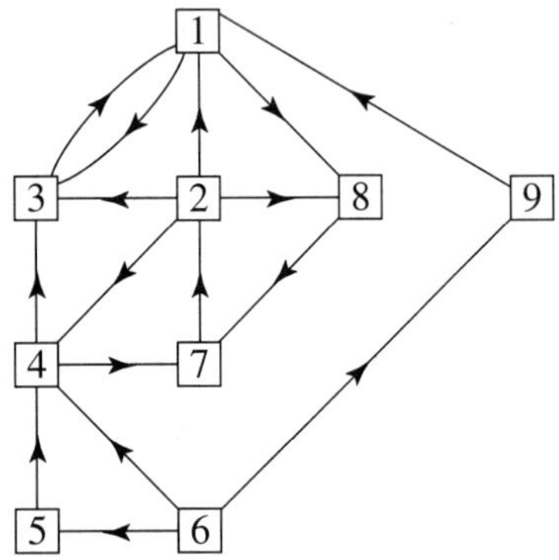

DFSOut starting at 2   *DFSIn* starting at 2   digraph *D*

(a)

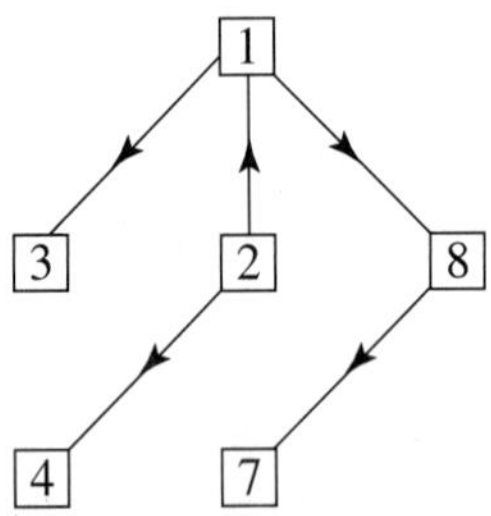

DFS out-tree rooted at 2

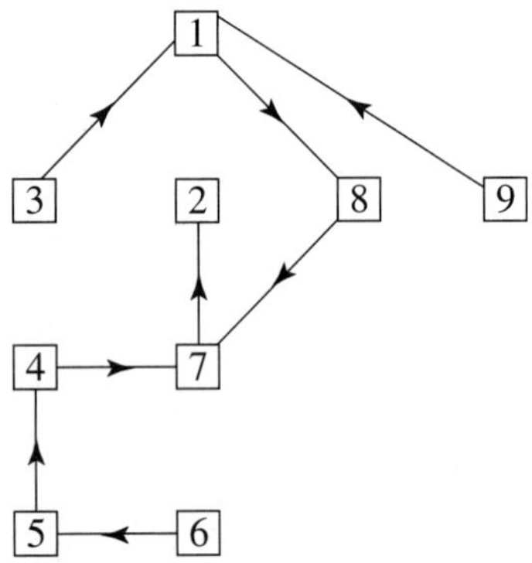

DFS in-tree rooted at 2

v	1	2	3	4	5	6	7	8	9
OutTreeDFS[v]	2	0	1	2	0	0	8	1	0
InTreeDFS[v]	8	0	1	7	4	5	2	7	1

(b)

Figure 8.20 (a) *DFSOut* and *DFSIn* with input *D* and starting vertex $v = 2$ visits vertices in the order 2,1,3,8,7,4 and 2,7,4,5,6,8,1,3,9 respectively; (b) *DFS* out-tree and in-tree rooted at 2 and their associated arrays

Enqueue 2	*Visit* 2
Dequeue	$(u = 2)$
Enqueue 1	*Visit* 1
Enqueue 3	*Visit* 3
Enqueue 4	*Visit* 4
Enqueue 8	*Visit* 8
Dequeue	$(u = 1)$
Dequeue	$(u = 3)$
Dequeue	$(u = 4)$
Enqueue 7	*Visit* 7
Dequeue	$(u = 8)$
Dequeue	$(u = 7)$

Queue is now empty.

BFSOut starting at 2

Enqueue 2	*Visit* 2
Dequeue	$(u = 2)$
Enqueue 7	*Visit* 7
Dequeue	$(u = 7)$
Enqueue 4	*Visit* 4
Enqueue 8	*Visit* 8
Dequeue	$(u = 4)$
Enqueue 5	*Visit* 5
Enqueue 6	*Visit* 6
Dequeue	$(u = 8)$
Enqueue 1	*Visit* 1
Dequeue	$(u = 5)$
Dequeue	$(u = 6)$
Dequeue	$(u = 1)$
Enqueue 3	*Visit* 3
Enqueue 9	*Visit* 9
Dequeue	$(u = 3)$
Dequeue	$(u = 9)$

Queue is now empty.

BFSIn starting at 2

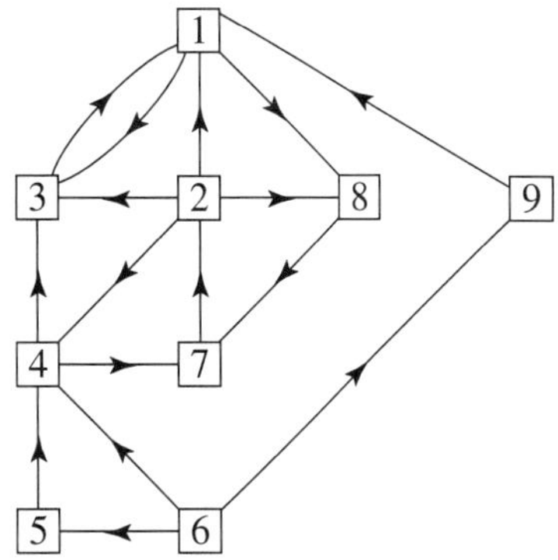

digraph *D*

(a)

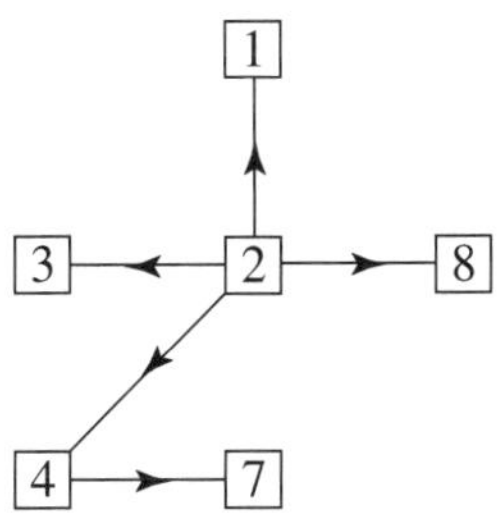

BFS out-tree rooted at 2

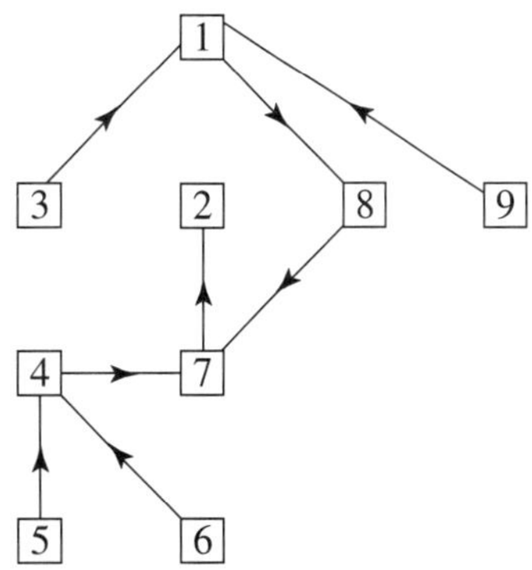

BFS in-tree rooted at 2

v	1	2	3	4	5	6	7	8	9
OutTreeBFS[v]	2	0	2	2	0	0	4	2	0
InTreeBFS[v]	8	0	1	7	4	4	2	7	1

(b)

Figure 8.21 (a) *BFSOut* and *BFSIn* with input *D* and starting vertex $v = 2$ visits vertices in the order 2,1,3,4,8,7, and 2,7,4,8,5,6,1,3,9 respectively; (b) BFS out-tree and in-tree rooted at 2 and their associated arrays

```
        if Mark[w] = 0 then
            call DFSOutTopLabel(D,w)
        endif
      endfor
      TopList[Counter] := v
      Counter := Counter – 1
    end DFSOutTopLabel
end TopologicalSort
```

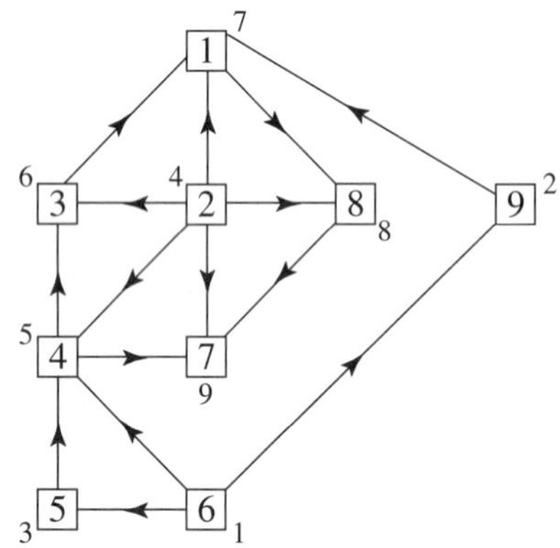

TopList	1	2	3	4	5	6	7	8	9
	6	9	5	2	4	3	1	8	7

Figure 8.22 A dag *D* and the array *TopList*[1:9] output by *TopologicalSort*

The proof that *TopList*[1:*n*] output by the procedure *TopologicalSort* is a topologically sorted list of the vertices of *D* is developed in the exercises. The array *TopList*[1:*n*] output by the procedure *TopologicalSort* is illustrated in Figure 8.22 for a sample dag *D*. In Figure 8.22 we show the position in the list *TopList* outside each vertex.

8.5 Shortest Paths in Weighted Graphs and Digraphs

Given a vertex *r* in a graph or digraph, breadth-first search yields shortest paths from *r* to every vertex reachable from *r*. In many applications, a weight (cost) *c*(*uv*) is associated with each edge *uv*. For example, a graph might model a network of cities, and the weight might be the distance or driving time between adjacent cities. As another example, a digraph might model a network of (direct) airplane flights, and the weight of a flight might be its cost or its flying time. The (*weighted*) *length* of a path is the sum of the weights of its edges. A natural and very important problem is to find a *shortest* (*weighted*) *path* from *r* to a vertex *v*, that is, a path from *r* to *v* of minimum length. Note that a shortest path between *r* and *v* with respect to the weighting *c* is not necessarily a shortest path from *r* to *v* in the unweighted graph or digraph. Given a weighted graph or digraph, the (*weighted*) *distance* from r to v is the length of the shortest path from r to v.

Since a weighted graph is a special case of a weighted digraph, we consider the problem of finding shortest paths in a weighted digraph D. A *shortest path (out) tree rooted at vertex r in D* is an out-tree that contains a shortest path from r to u for every vertex u for which there exists such a path. A *negative* (*directed*) *cycle* is a directed cycle such that the sum of the costs on the edges of the cycle is negative. If there exists a negative cycle, then we can obtain arbitrarily small (weighted) paths by repeatedly traversing the negative cycle. Thus the notion of a shortest path tree is not well defined in this case.

We now discuss an algorithm due to Bellman and Ford, which determines whether there exists a negative cycle in D and computes a shortest path tree rooted at r if no such cycle exists. The procedure *BellmanFord* that we present maintains an array *Pred*[1:*n*], which upon termination of the algorithm stores the parent array implementation of a shortest path tree rooted at r. In addition, we maintain the array *Dist*[1:*n*], which eventually stores the distances from vertex r. The array *Dist*[1:*n*] is initialized to ∞. At any intermediate stage in procedure *BellmanFord*, either *Dist*[v] is ∞, or *Dist*[v] is the length of some path from r to v. Thus, *Dist*[v] is always at least as large as the actual distance from r to v. Further, if *Dist*[v] is different from ∞, then *Pred*[1:*n*] determines a path from r to v having length *Dist*[v]. Starting from v, the sequence of predecessors in this path is given by

$$v, Pred[v], Pred[Pred[v]], \ldots, Pred^k[v] = r.$$

Procedure *BellmanFord* first performs $n - 1$ scans of all the edges of D. During each scan the arrays *Pred* and *Dist* are updated as follows. When an edge uv is scanned, if $Dist[v] > Dist[u] + c(uv)$ then a path from r to v has been found that is shorter than the current path (namely, the current path from r to u together with the edge uv). Hence, *Pred*[v] is updated to the value u and *Dist*[v] is updated to the value $Dist[u] + c(uv)$. Note that the current path from r to u may be updated many times during the same scan. After completion of the first $n - 1$ scans, procedure *BellmanFord* makes a final scan of the edges to check for negative cycles. If there is any edge uv such that $Dist[v] > Dist[u] + c(uv)$, then a negative cycle exists.

```
procedure BellmanFord(D,r,c,Dist[1:n],Pred[1:n],NegativeCycle)
Input:  D (a digraph with vertex set V = {1, ... , n} and edge set E)
        c (a weighting of the edges with real numbers)
        r (a vertex of D)
Output: Pred[1:n] (an array implementing a shortest path tree rooted at r)
        Dist[1:n] (an array of distances from r)
        NegativeCycle (a Boolean variable having value .true. if, and only
                       if, there exists a negative cycle)

  for i := 1 to n do       {initialize Dist[1:n] and Pred[1:n]}
     Dist[i] := ∞
     Pred[i] := ∞
  endfor
```

Continued

```
    Dist[r] := 0
    Pred[r] := 0

    for Pass 1 to n – 1 do          {compute Dist[1:n] and Pred[1:n]}
       for each edge uv ∈ E do      {scan all the edges}
          if Dist[u] + c(uv) < Dist[v] then
             Pred[v] := u
             Dist[v] := Dist[u] + c(uv)
          endif
       endfor
    endfor
    NegativeCycle := .false. {check for negative cycles}
    for each edge uv ∈ E do
       if Dist[v] > Dist[u] + c(uv) then
          NegativeCycle := .true.
       endif
    endfor
end BellmanFord
```

In procedure *BellmanFord* we did not specify the order in which the edges are scanned. In fact, any order will do. In Figure 8.23, during each pass, we scan the edges (u,v) in lexicographical order.

Since procedure *BellmanFord* examines each edge in D n times, it has (best-case, average, and worst-case) complexity belonging to $\Theta(nm)$. The best-case and average performance of *BellmanFord* can be improved by adding a flag to test whether any changes to *Dist* occur during a given pass. If no change occurs in any particular pass, then the algorithm can terminate. For example, for the digraph illustrated in Figure 8.23, only four passes are required.

The correctness of procedure *BellmanFord* involves establishing the loop invariant stated in the following lemma.

LEMMA 8.5.1 After completing k iterations of the *Pass* **for** loop, for each vertex v, $Dist[v]$ is no larger than the minimum length of a path from r to v having at most k edges (where the minimum is ∞ if no such path exists), $k = 0, \ldots, n - 1$.

Proof We prove the lemma by induction on the number k of passes that have been completed. Clearly, the lemma is true before any pass has been completed, so that the basis step is established. Now assume that after k passes have been completed, for each vertex v, $Dist[v]$ is no larger than the minimum length $L(v)$ of a path from r to v having at most k edges. Consider any vertex v for which there is a path from r to v having at most $k + 1$ edges (the lemma is trivially true if there is no such path), and let P be such a path of minimum length. Let uv be the terminal edge of P, and let Q be the subpath of P from r to u. The length of Q is at least $L(u)$. Thus, the length of P is at least $L(u) + c(uv)$. But, by induction assumption, after the kth pass, $Dist[u]$ is at most $L(u)$. Therefore, after the $(k + 1)$st pass, the value of $Dist[v]$ is at most $L(u) + c(uv)$. Thus, we have verified that the loop invariant holds after $k + 1$ passes. ■

Pass	Index	1	2	3	4	5	6	7	8	9		1	2	3	4	5	6	7	8	9
0	*Dist*:	∞	∞	∞	∞	∞	∞	0	∞	∞	*Pred*:	∞	∞	∞	∞	∞	∞	0	∞	∞
1	*Dist*:	∞	3	∞	2	∞	∞	0	∞	∞	*Pred*:	∞	7	∞	7	6	∞	0	∞	∞
2	*Dist*:	1	3	1	2	5	9	0	5	6	*Pred*:	9	7	4	7	6	4	0	2	6
3	*Dist*:	1	3	1	2	5	9	0	–3	6	*Pred*:	9	7	4	7	6	4	0	1	6
4	*Dist*:	1	3	1	2	5	9	0	–3	6	*Pred*:	9	7	4	7	6	4	0	1	6
5	*Dist*:	1	3	1	2	5	9	0	–3	6	*Pred*:	9	7	4	7	6	4	0	1	6
6	*Dist*:	1	3	1	2	5	9	0	–3	6	*Pred*:	9	7	4	7	6	4	0	1	6
7	*Dist*:	1	3	1	2	5	9	0	–3	6	*Pred*:	9	7	4	7	6	4	0	1	6
8	*Dist*:	1	3	1	2	5	9	0	–3	6	*Pred*:	9	7	4	7	6	4	0	1	6

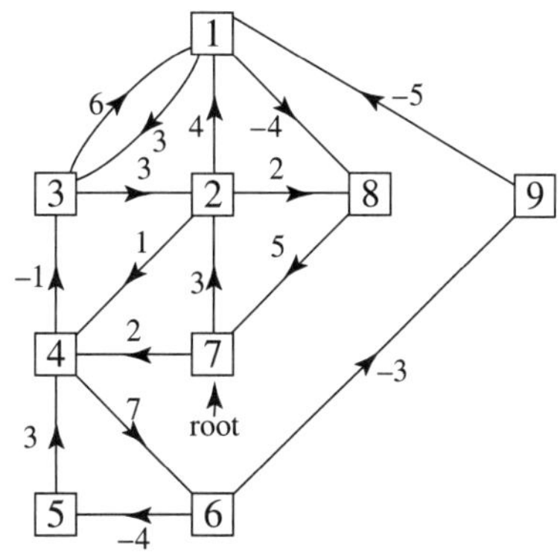

Figure 8.23 Action of procedure *BellmanFord* for a sample weighted digraph and root vertex $r = 7$, where edges (u,v) are scanned in lexicographical order

When there are no negative cycles, a shortest path contains at most $n - 1$ edges. Hence, the correctness of *BellmanFord* when there are no negative cycles follows immediately from Lemma 8.5.1. We leave the correctness proof of *BellmanFord* when there are negative cycles as an exercise.

8.6 Implementing Disjoint Sets

Many algorithms, particularly graph algorithms, involve maintaining a (dynamically changing) equivalence relation on some universal (or base) set $S = \{s_1, s_2, \ldots, s_n\}$. Since an equivalence relation on S determines a partition of S into disjoint subsets, the problem of maintaining equivalence relations reduces to the problem of maintaining disjoint sets. The two main operations associated with the disjoint set ADT are efficiently finding the set containing a given element and forming the union of two sets.

8.6.1 *Union* and *Find* Algorithms

For disjoint sets, an efficient implementation of the union and find operations is based on the parent array representation of a forest. Consider any collection $\mathscr{C}$ of k

Disjoint collection of sets: $\mathscr{C} = \{\{1, 3, 5, 9\}, \{4, 6, 8, 10, 11\}\ \{2, 7\}, \{12\}\}$

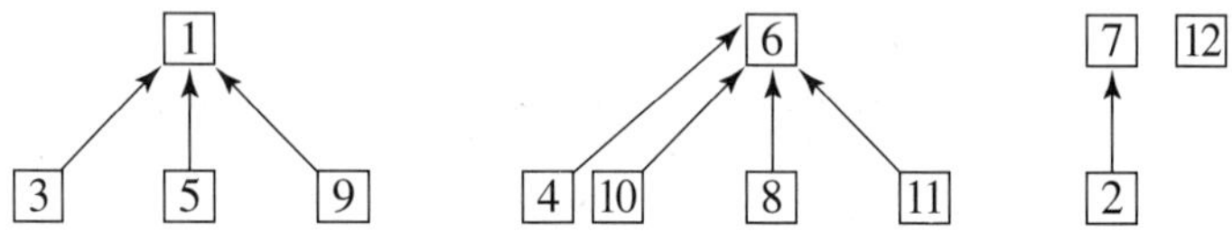

i	1	2	3	4	5	6	7	8	9	10	11	12
$Parent[i]$	0	7	1	6	1	0	0	6	1	6	6	0

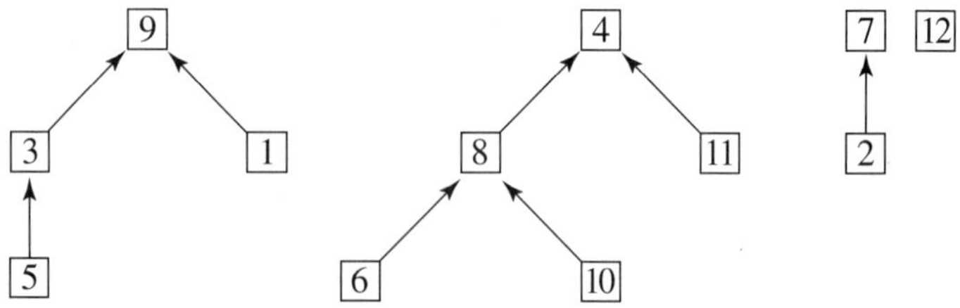

i	1	2	3	4	5	6	7	8	9	10	11	12
$Parent[i]$	9	0	9	0	3	8	2	4	0	8	4	0

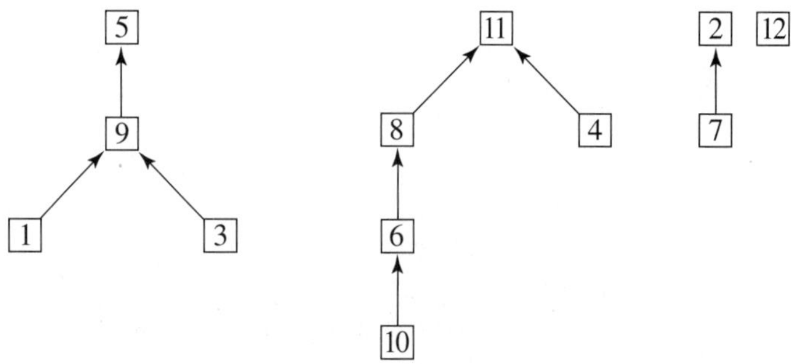

i	1	2	3	4	5	6	7	8	9	10	11	12
$Parent[i]$	9	0	9	11	0	8	2	11	5	6	0	0

Figure 8.24 Three sample forest representations of the collection $\mathscr{C}$

disjoint subsets $A_1, A_2, \ldots, A_k$ of the base set S, where for convenience we assume that $S = \{1,2, \ldots, n\}$. A rooted forest F *represents* the collection $\mathscr{C}$ of disjoint sets if it consists of k rooted trees $T_1, T_2, \ldots, T_k$ such that the vertex set of T_i is A_i, $i = 1,2, \ldots, k$. Clearly, the k sets $A_1, A_2, \ldots, A_k$ are (uniquely) determined by the single array *Parent*[1:n] of the parent-array implementation of F. Further, once the array *Parent*[1:n] is given, the set A_i can be identified by the single element r_i, where r_i is the root of T_i, $i = 1,2, \ldots, k$ (see Figure 8.24).

Given an element $x \in A_1 \cup A_2 \cup \cdots \cup A_k$, the following procedure *Find*1 returns the root r of the tree in forest F containing x.

```
procedure Find1(Parent[1:n],x,r)
Input:   Parent[1:n] (array representing disjoint subsets of S)
         x (an element of S)
Output: r (the root of the tree corresponding to the subset containing x)

   r := x
   while Parent[r] > 0 do
      r := Parent[r]
   endwhile
end Find1
```

The complexity of *Find*1 is measured by the number of iterations of the **while** loop. Clearly, the complexity of *Find*1 depends on the depths of the trees T_i in F. Thus, it is desirable to keep the depths of the trees in the forest relatively small.

Given $\mathscr{C} = \{A_1, A_2, \ldots, A_k\}$, let $\mathscr{C}_{ij}$ denote the collection of $k - 1$ sets obtained from $\mathscr{C}$ by removing sets A_i and A_j and adding the set $A_i \cup A_j$. Consider the two forests F_{ij} and F_{ji}, where F_{ij} is obtained from F by setting $Parent(r_j) = r_i$, and F_{ji} is obtained from F by setting $Parent(r_i) = r_j$. Both forests represent the collection of sets $\mathscr{C}_{ij}$. To help minimize the depth of the tree representing the union of A_i and A_j, we choose the forest F_{ij} if T_i has more vertices than T_j, otherwise we choose the forest F_{ji}. This choice can be made instantly (in constant time) by dynamically keeping track of the number n_i of vertices in each tree T_i. An efficient way of keeping track of n_i without allocating additional memory locations is to (dynamically) store the negative of n_i in $Parent[r_i]$ (instead of the value zero). The root vertices are still distinguishable from the other vertices, since they index those locations of the array $Parent[1{:}n]$ having negative values.

The following pseudocode for the algorithm *Union* computes the parent array implementation of the union of two sets based on the preceding discussion.

```
procedure Union(Parent[1:n],r,s)
Input:   Parent[1:n] (an array representing disjoint sets)
         r,s (roots of the trees representing two disjoint sets A,B)
Output: Parent[1:n] (an array representing disjoint sets after forming A ∪ B)

   sum := Parent[r] + Parent[s]
   if  Parent[r] > Parent[s] then      {tree rooted at s has more vertices than
        Parent[r] := s                   tree rooted at r}
        Parent[s] := sum
   else
        Parent[s] := r
        Parent[r] := sum
   endif
end Union
```

When implementing disjoint sets in a particular algorithm, the initial collection $\mathscr{C}$ of disjoints sets is usually the collection of all singleton subsets of the base set S. Figure 8.25 illustrates a sample sequence of calls to the procedure *Union*, starting with the collection of singleton subsets of the base set $S = \{1,2, \ldots, 9\}$.

Figure 8.25 Representative forests and parent arrays for a sample sequence of calls to procedure *Union*

Initial collection: $\{\{1\},\{2\},\{3\},\{4\},\{5\},\{6\},\{7\},\{8\},\{9\}\}$

i	1	2	3	4	5	6	7	8	9
Parent[i]	–1	–1	–1	–1	–1	–1	–1	–1	–1

Call *Union*[*Parent*[1:9],3,6] $\Rightarrow\{\{1\},\{2\},\{3,6\},\{4\},\{5\},\{7\},\{8\},\{9\}\}$

i	1	2	3	4	5	6	7	8	9
Parent[i]	–1	–1	–2	–1	–1	3	–1	–1	–1

Call *Union*[*Parent*[1:9],3,8] $\Rightarrow\{\{1\},\{2\},\{3,6,8\},\{4\},\{5\},\{7\},\{9\}\}$

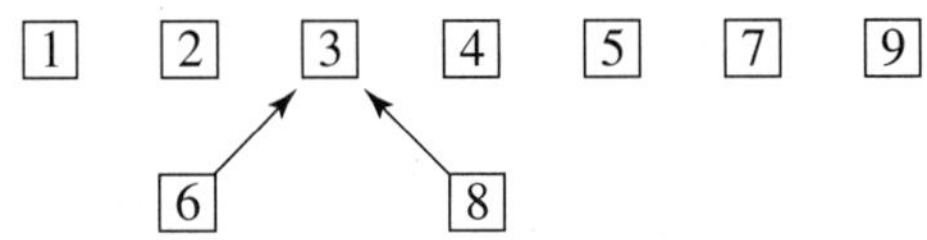

i	1	2	3	4	5	6	7	8	9
Parent[i]	–1	–1	–3	–1	–1	3	–1	3	–1

Call *Union*[*Parent*[1:9],5,9] $\Rightarrow\{\{1\},\{2\},\{3,6,8\},\{4\},\{5,9\},\{7\}\}$

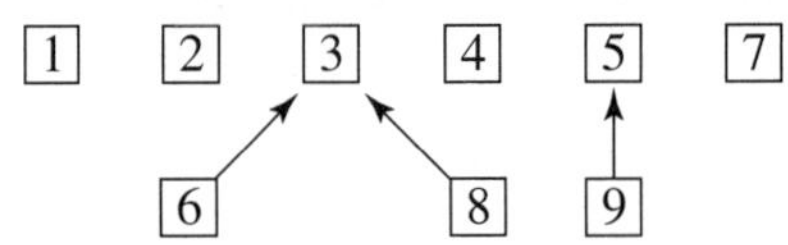

i	1	2	3	4	5	6	7	8	9
Parent[i]	–1	–1	–3	–1	–2	3	–1	3	5

Call *Union*[*Parent*[1:9],1,5] $\Rightarrow\{\{1,5,9\},\{2\},\{3,6,8\},\{4\},\{7\}\}$

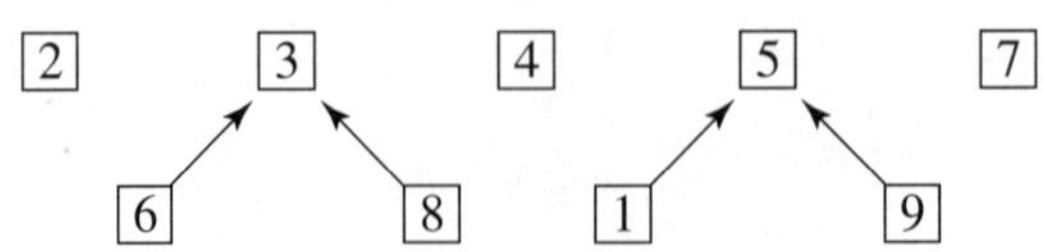

i	1	2	3	4	5	6	7	8	9
Parent[i]	5	–1	–3	–1	–3	3	–1	3	5

Call *Union*[*Parent*[1:9],3,5] ⇒{{1,3,5,6,8,9},{2},{4},{7}}

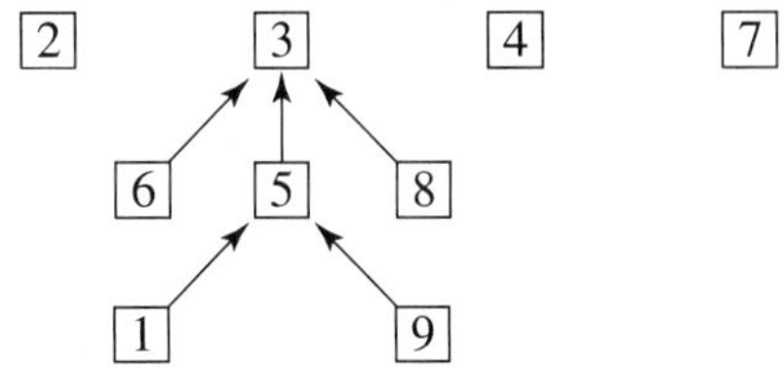

i	1	2	3	4	5	6	7	8	9
Parent[i]	5	–1	–6	–1	3	3	–1	3	5

Call *Union*[*Parent*[1:9],2,7] ⇒{{1,3,5,6,8,9},{2,7},{4}}

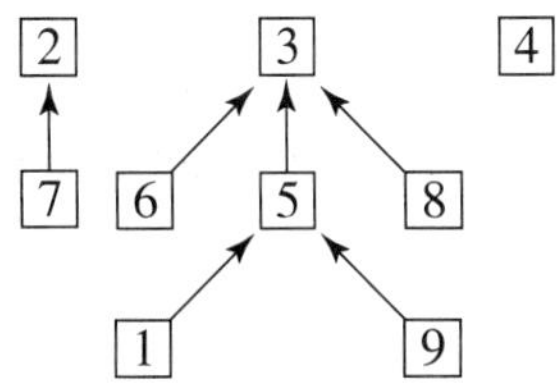

i	1	2	3	4	5	6	7	8	9
Parent[i]	5	–2	–6	–1	3	3	2	3	5

Call *Union*[*Parent*[1:9],2,3] ⇒{{1,2,3,5,6,7,8,9},{4}}

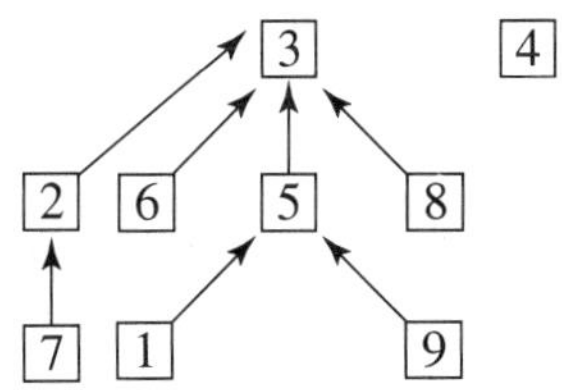

i	1	2	3	4	5	6	7	8	9
Parent[i]	5	3	–8	–1	3	3	2	3	5

Figure 8.25 Continued

PROPOSITION 8.6.1 Let F be any forest resulting from some sequence of calls to *Union*, where the initial input forest consists of isolated vertices. Then the depth of any tree in F having j vertices is at most $\lfloor \log_2 j \rfloor + 1$.

We leave the proof of Proposition 8.6.1 as an exercise. It follows from Proposition 8.6.1 that if we start with a collection $\mathscr{C}$ consisting of singleton sets, and perform a sequence of calls to *Union*, then the computing time of *Find*1 for a given input x is O(log j), where j is the cardinality of the set containing x. Thus, the worst-case complexity in making an intermixed sequence of calls to *Union* and *Find*1, n calls to *Union*, and $m \geq n$ calls to *Find*1, is O(m log n).

8.6.2 Improved *Union* and *Find* Using the Collapsing Rule

Even though the procedures *Union* and *Find*1 are very efficient and might seem the best possible, there is still room for improvement. The improvement comes by modifying *Find*1(*Parent*[1:*n*],*x*,*r*) as follows. After r is obtained, we can traverse the path from x to r again and set *Parent*[v] = r for each vertex v encountered. This "collapsing" of the path basically doubles the computing time of *Find*1. However, it tends to reduce the depth of the trees in the forest so that an overall improvement is achieved in the efficiency of making an intermixed sequence of union and find operations. The following procedure *Find*2 implements the collapsing rule.

```
procedure Find2(Parent[1:n],x,r)
Input:  Parent[1:n] (an array representing disjoint subsets of S)
        x (an element of S)
Output: r (the root of the tree corresponding to subset containing x)

   r := x
   while Parent[r] > 0 do
      r := Parent[r]
   endwhile
   y := x
   while y ≠ r do
      Temp := Parent[y]
      Parent[y] := r
      y := Temp
   endwhile
end Find2
```

Figure 8.26 illustrates the collapsing rule for a call to *Find*2 with $x = 9$, for a sample input forest representing the sets

$$\{\{1,3,4,5,6,8,9,10,11,12,14,15,16\},\{2,7\},\{13\}\}.$$

Tarjan has shown that the worst-case complexity in making an intermixed sequence of calls to *Union* and *Find*2, $n - 1$ calls to *Union* and $m \geq n$ calls to *Find*2, is $O(m\ \alpha(m,n))$, where $\alpha(m,n)$ is an extremely slow-growing function. In fact, it grows so slowly that, for all practical purposes, we can regard $\alpha(m,n)$ as a constant function of m and n. The function $\alpha(m,n)$ is related to a functional inverse of the extremely fast-growing Ackermann's function $A(m,n)$ given by the recurrence relation:

$$A(m,n) = A(m-1, A(m, n-1)),\ m \geq 1 \text{ and } n \geq 2,$$
$$A(0,n) = 2n,\ A(m,0) = 0,\ m \geq 1,\ A(m,1) = 2,\ m \geq 1.$$

Forest representing collection {{1,3,4,5,6,8,9,10,11,12,14,15,16}{2,7},{13}} before call to *Find*2:

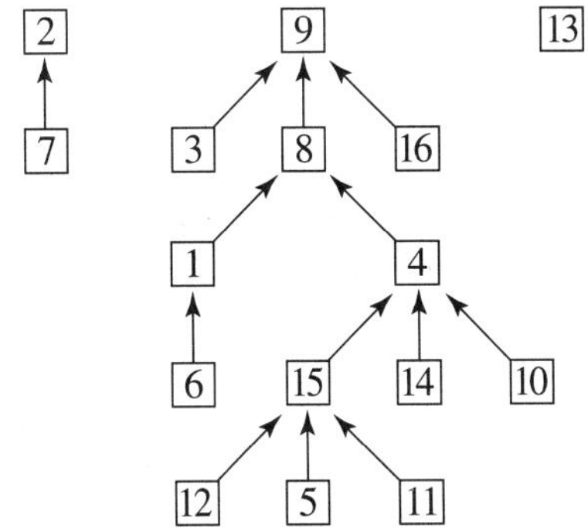

i	1	2	3	4	5	6	7	8	9	10	11	12	13	14	15	16
Parent[i]	8	–2	9	8	15	1	2	9	–13	4	15	15	–1	4	4	1

Forest representing collection {{1,3,4,5,6,8,9,10,11,12,14,15,16}{2,7},{13}} after call to *Find*2 with $x = 15$:

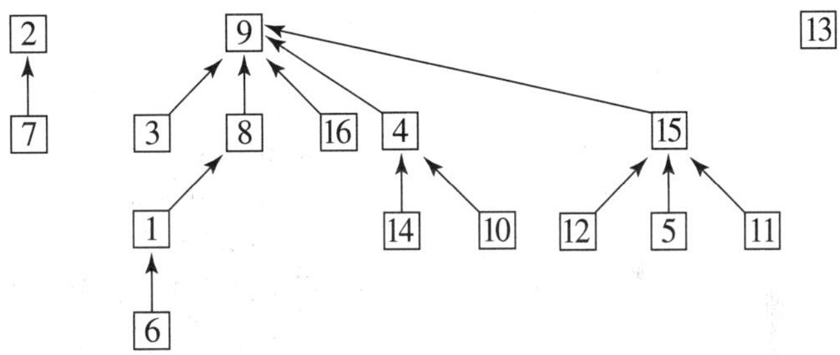

i	1	2	3	4	5	6	7	8	9	10	11	12	13	14	15	16
Parent[i]	8	–2	9	9	15	1	2	9	–13	4	15	15	–1	4	9	1

Figure 8.26 Illustration of collapsing rule for sample call to *Find*2

8.6.3 Example: Computing Components of a Graph On Line

To illustrate the *Union* and *Find* algorithms, we consider the problem of computing the components of a graph $G = (V,E)$, where V is a fixed set and E is given as an initially empty set that grows over time in an *on-line* fashion. More precisely, edges are added to E over time, but we do not know in advance which particular edges will be added. We maintain the vertex sets of the components using an array *Parent*[1:n], which is initialized by setting *Parent*[i] = –1, $i = 1, \ldots, n$. When an edge $\{u,v\}$ is added, we call procedures *Find*2(*Parent*[1:n],u,r) and *Find*2(*Parent*[1:n],v,s) to determine whether u and v are in the same component ($r = s$). If u and v are not in the same component ($r \neq s$), then we call procedure *Union*(*Parent*[1:*n*], *r*,*s*) to merge the two components into the single new component containing u and v.

Suppose the final graph G has m edges. Then, as described earlier, the complexity of our on-line algorithm belongs to $O(m\ \alpha(m,n))$. Also, note that a single pass through the array *Parent*[1:*n*] determines whether or not G is connected.

Pseudocode for the algorithm just described depends on how the edge set E is implemented (adjacency matrix, adjacency lists, and so forth). We leave specific implementation details to the exercises.

8.7 Closing Remarks

In this chapter we have seen how breadth-first search yields shortest paths in (unweighted) graphs and digraphs. Further, we discussed the Bellman-Ford algorithm for finding shortest paths in weighted graphs and digraphs. The study of shortest path algorithms has a long history, and there are many such algorithms to be found in the literature. We will present algorithms for finding shortest paths based on the greedy method and dynamic programming in Chapters 12 and 14, respectively.

Depth-first search has a host of applications. We saw how a labeling generated by a depth-first traversal yielded a topological sort of a dag. Other vertex labelings generated by depth-first search are the basis of many classical graph algorithms (see Chapter 19).

References and Suggestions for Further Reading

There is a vast literature on theory of graphs and numerous textbooks have been written on the subject. Some of the classic books on graph theory include:

Berge, C. *Graphs and Hypergraphs.* Amsterdam: North-Holland Pub. Co., 1973.

Biggs, N. *Graph Theory 1736–1936.* Oxford: Clarendon Press, 1976.

Bollobas, B. *Graph Theory: An Introductory Course.* New York: Springer Verlag, 1979.

Bondy, J. A. and U. S. R. Murty. *Graph Theory with Applications.* New York: North Holland, 1976.

Harary, F. *Graph Theory.* Reading, MA: Addison-Wesley, 1969.

Wilson, R. J. *Introduction to Graph Theory*, 3rd Ed. Harlow, England: Longman, 1985.

Some graph theory books that have appeared more recently include:

Chartrand, G. and L. Lesniak. *Graphs & Digraphs.* Wadsworth & Brooks/Cole Mathematics Series, Monterey, CA, 1986.

West, D. *Introduction to Graph Theory.* Englewood Cliffs, NJ: Prentice Hall, 1995.

References and further readings for graph algorithms are included at the end of Chapter 19.

Exercises

Section 8.1 Graphs and Digraphs

8.1 Show that for a digraph D having m directed edges, we have

$$\sum_{v \in V} d_{out}(v) = \sum_{v \in V} d_{in}(v) = m.$$

8.2 Prove Proposition 8.1.4.

8.3 Prove Proposition 8.1.5.

8.4 Prove by induction that the hypercube H_k of dimension k has

a. 2^k vertices

b. $k2^{k-1}$ edges

8.5 Consider the hypercube H_k of dimension k on $n = 2^k$ vertices.

a. Show that H_k is k-regular.

b. Show that H_k has diameter $k = \log_2 n$.

c. Show that H_k has bisection width $n/2$.

8.6 Give an alternate proof of the result in part (b) of Exercise 8.4 using part (a) of that exercise and Corollary 8.1.2.

8.7 Prove by induction that the k-dimensional hypercube H_k has exactly $C(k,j)2^{k-j}$ different sub-hypercubes of dimension j, $j = 0, \ldots, k$.

8.8 Give an example of each of the following types of graph.

a. A 2-regular graph with diameter 8

b. A 3-regular graph with diameter 5

8.9 Suppose a 6-regular graph G has 20 more edges than vertices. How many vertices and edges does G have?

8.10 Does there exist a 5-regular graph with 44 edges? Explain.

8.11 **a.** Using Euler's degree formula and Proposition 8.1.5 prove Proposition 7.3.2, which states that the number of interior vertices in a 2-tree is one less than the number of leaf nodes.

b. Repeat part (a) for k-ary trees (see Exercise 7.15).

8.12 Design an O(log n) algorithm for summing n numbers on the k-dimensional hypercube H_k, where $n = 2^k$.

8.13 **a.** Prove that a graph G has an Eulerian path from a to b if, and only if, G is connected, every vertex different from a and b has even degree, and a and b have odd degree.

b. Prove that a graph G has an Eulerian cycle if, and only if, every vertex has even degree.

c. State and prove a necessary and sufficient condition for the existence of Eulerian paths and cycles in digraphs.

8.14 Give pseudocode for the standard ADT operations of adding and deleting an edge for a graph G for each of the following implementations.

a. adjacency matrix

b. adjacency lists with array of header nodes

c. adjacency lists with a linked list of header nodes

8.15 In the adjacency lists implementation of a graph, where the header nodes belong to a linked list (see Figure 8.11a), it is convenient in practice to maintain a second pointer p in each list node defined as follows: If the list node belongs to the adjacency list of vertex i and corresponds to edge (i,j), then p points to the header node of adjacency list j. Redo Exercise 8.14c with this enhanced implementation.

Sections 8.2–8.3 Search and Traversal of Graphs and Digraphs

8.16 Consider the following graph.

$G =$

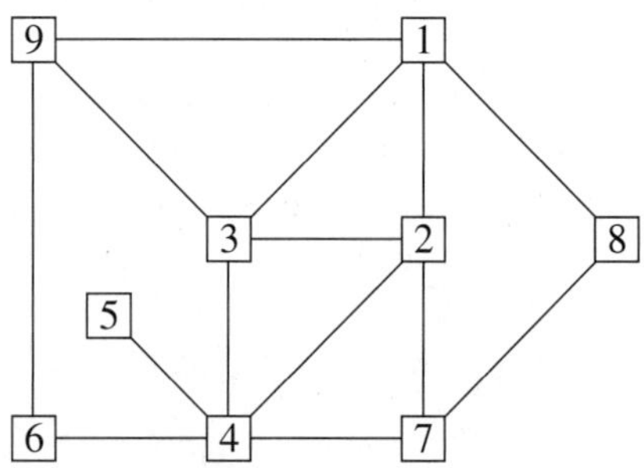

a. Starting at vertex 3, list the vertices in the order in which they are visited by breadth-first search, and give the array *TreeBFS*[1:9] that gives a parent implementation of the breadth-first search spanning tree.

b. Repeat part (a) for depth-first search, giving the *TreeDFS*[1:9].

8.17 **a.** Prove that a depth-first search with starting vertex v of a graph G visits every vertex of G if, and only if, G is connected.

b. More generally, prove that a depth-first search with starting vertex v of a graph G visits every vertex of the component of G containing v.

c. Repeat (a) and (b) for breadth-first search.

8.18 Give complete pseudocode for *DFS* for the following implementations of a graph *G*, where the array *Current*[1:*n*] is used by the procedure *Scan*:

a. adjacency matrix

b. adjacency lists with an array of header nodes

c. adjacency lists with a linked list of header nodes

8.19 Repeat Exercise 8.18 for *BFS*.

8.20 Prove that the breadth-first search out-tree rooted at a vertex *v* is a *shortest path* out-tree rooted at *v* in the digraph; that is, it contains a shortest path from *v* to every vertex accessible from *v*.

8.21 Design an algorithm that determines whether a graph is bipartite and identifies the bipartition of the vertices if it is bipartite for the following implementations of the graph.

a. adjacency matrix

b. adjacency lists

8.22 The *girth* of a graph *G* is the length of the smallest cycle in *G*. Design and analyze an algorithm for determining the girth of *G*.

Section 8.4 Topological Sorting

8.23 Show the topological-sort labeling generated by *TopologicalSort* for the following dag.

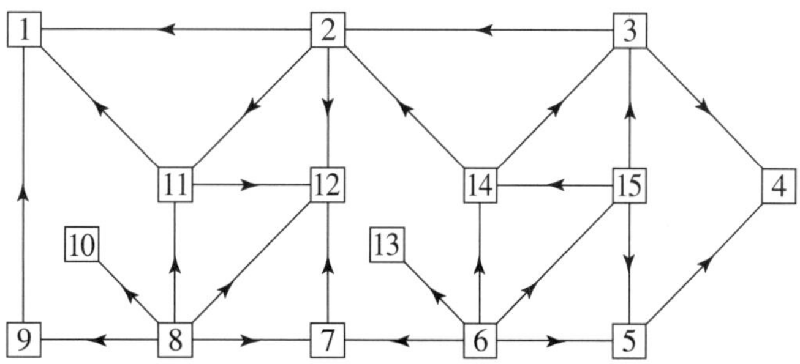

8.24 Consider an (out-directed) depth-first traversal of a directed graph *D*. A *back edge* is an edge *uv* not belonging to the depth-first search forest *F*, such that *u* and *v* belong to the same tree in *F* and *u* is an ancestor of *v*. Show that a directed graph *D* is a dag if, and only if, there are no back edges.

8.25 Use Exercise 8.24 to verify the correctness of the procedure *TopologicalSort*.

8.26 **a.** Design an algorithm for determining whether a digraph *D* is a dag.

b. Modify your algorithm in (a) to output a directed cycle if *D* is not a dag.

Section 8.5 Shortest Paths in Weighted Graphs and Digraphs

8.27 **a.** Redo the action of procedure *BellmanFord* shown in Figure 8.23 when $r = 2$.

b. Redo the action of procedure *BellmanFord* shown in Figure 8.23 when $r = 7$ and where the cost of the edge (3,2) is changed to -2.

8.28 **a.** Design a modification of *BellmanFord* that improves the best-case and average performance by terminating if no changes to *Dist* occur during a pass.

b. Analyze the best-case and worst-case complexity of your algorithm in part (a) in terms of the parameters n and m. Explicitly exhibit input digraphs achieving the best-case and worst-case complexities, respectively.

c. Repeat part (b) in terms of the single parameter n.

8.29 A high-level description of *BellmanFord* is given in Section 8.5. Give a description of *BellmanFord* for the following implementations of the digraph.

a. adjacency matrix

b. adjacency lists

8.30 Show that if a weighted digraph contains no negative cycles, then a shortest path contains at most $n - 1$ edges. Conclude the correctness proof of *BellmanFord* when there are no negative cycles.

8.31 Give the correctness proof of *BellmanFord* when there are negative cycles.

8.32 Modify *BellmanFord* to output a negative cycle, when one exists.

8.33 Consider the version of procedure *BellmannFord*, which checks each pass for any change in *Dist*[1:*n*]. For this version of *BellmanFord*, the order in which the edges are scanned can significantly affect the worst-case performance of the algorithm. Give an example of a weighted digraph, and two orderings *A* and *B* of the edges, such that if ordering *A* is used during each pass, *BellmanFord* performs only two passes, but if ordering *B* is used during each pass, *BellmanFord* performs n passes.

Section 8.6 Implementing Disjoint Sets

8.34 Prove Proposition 8.4.1.

8.35 Suppose we have the following parent implementation of a forest representing a partition of a set of 10 elements:

i	1	2	3	4	5	6	7	8	9	10	11	12	13	14	15	16	17
Parent[i]	8	−3	9	8	15	1	2	9	−13	4	15	5	−1	4	4	1	2

a. Sketch the trees in *F*.

b. Show the state of *Parent*[1:17] after a call to *Union*(*Parent*[1:17],2,9) and sketch the trees in *F*.

c. Given the state of *Parent*[1:17] in part (b), show the state of *Parent*[1:17] after an invocation of *Find2*(*Parent*[1:17],5) and sketch the trees in *F*.

9

Probability and Average Complexity of Algorithms

The definition of the average behavior of an algorithm given in Chapter 3 informally used some concepts from elementary probability theory. In this chapter we formally define such notions as a probability distribution on a sample space, conditional probability, and the expectation of a random variable. These notions allow us to define the average behavior of an algorithm in a general setting. Various properties of expectation are established that are useful tools in average behavior calculations.

9.1 Elementary Probability Theory

Consider the simple illustration of throwing a pair of dice. We assume that each number on a given die occurs with equal likelihood. Many games of chance are based on the outcome (event) of rolling a pair of dice. The set of all possible outcomes (the sample space) resulting from a roll of dice D_1 and D_2 is illustrated in the table given in Figure 9.1.

In most games the critical number (random variable) associated with each event is the sum of the numbers showing on the dice (see Figure 9.2).

Figure 9.1 Sample space of all possible outcomes resulting from rolling two dice

D_1 \ D_2	1	2	3	4	5	6
1	(1,1)	(1,2)	(1,3)	(1,4)	(1,5)	(1,6)
2	(2,1)	(2,2)	(2,3)	(2,4)	(2,5)	(2,6)
3	(3,1)	(3,2)	(3,3)	(3,4)	(3,5)	(3,6)
4	(4,1)	(4,2)	(4,3)	(4,4)	(4,5)	(4,6)
5	(5,1)	(5,2)	(5,3)	(5,4)	(5,5)	(5,6)
6	(6,1)	(6,2)	(6,3)	(6,4)	(6,5)	(6,6)

D_1 \ D_2	1	2	3	4	5	6
1	2	3	4	5	6	7
2	3	4	5	6	7	8
3	4	5	6	7	8	9
4	5	6	7	8	9	10
5	6	7	8	9	10	11
6	7	8	9	10	11	12

Figure 9.2 Random variable X on the sample space given in Figure 9.1 corresponding to the sum of the numbers showing on the two dice

Since die D_1 is fair, on any single roll the probability that D_1 shows the number i is 1/6, $i \in \{1, \ldots, 6\}$. Similar probabilities hold for D_2. In games of chance, we usually wish to calculate the probability p_j that the sum of the two dice is j, for a given value of j. Since the sum of the two dice is at least 2 and at most 12, the latter probability is zero unless $j \in \{2, \ldots, 12\}$.

Suppose, for example, that we wish to calculate p_8. Observe that there are 36 possible outcomes resulting from the roll of the two dice (see Figure 9.1). Out of these 36 possible outcomes, exactly 5 have the property that the two dice sum to 8. Thus, on average we can expect the dice to sum to 8, 5 out of 36 times. We conclude that $p_8 = 5/36$. More generally, $p_j = n_j/36$, where n_j denotes the number of outcomes in which the two dice sum to j. The values of p_j, $j = 2, \ldots, 12$, are shown in the table given in Figure 9.3.

9.1.1 Sample Spaces and Probability Distributions

We now formalize the notions of sample space, event, probability distribution, random variable, and expected value. Consider an experiment such as tossing two dice, flipping a coin, playing the lottery, and so forth. The sample space of the experiment, denoted by S, is the set of all possible outcomes of the experiment. (For example, Figure 9.1 shows the sample space of the experiment of tossing two dice.) A sample space is called *discrete* if it is finite, or countably infinite (that is, in one-to-one correspondence with the positive integers). For our purposes, it suffices to restrict attention to discrete sample spaces.

When analyzing the average complexity of a given algorithm, we consider as our sample space the set $\mathcal{I}_n$ of all possible inputs of size n to the algorithm. For most of

Figure 9.3 Probabilities p_j for the sum j of two dice

j	2	3	4	5	6	7	8	9	10	11	12
p_j	$\frac{1}{36}$	$\frac{2}{36}$	$\frac{3}{36}$	$\frac{4}{36}$	$\frac{5}{36}$	$\frac{6}{36}$	$\frac{5}{36}$	$\frac{4}{36}$	$\frac{3}{36}$	$\frac{2}{36}$	$\frac{1}{36}$

the algorithms that we consider, the sample space $\mathcal{T}_n$ is either finite or can be transformed to a finite sample space without changing the average complexity.

An *event* in the sample space S is a subset of outcomes from S. If E is a singleton subset $\{s\}$, where $s \in S$, then E is a *simple event*. For convenience, we sometimes denote the single element subset $\{s\}$ simply by s. A probability distribution arises by assigning a probability $P(E)$ to each event E in S. We first consider the case where S is finite and any outcome in S has an equal chance of occurring. Then, any particular outcome from S occurs with probability $1/|S|$ (where $|S|$ denotes the cardinality of S). Given an event E in S, the probability that E occurs, denoted by $P(E)$, is the probability that the outcome of the experiment lies in E. Clearly, if E is a simple event, then $P(E) = 1/|S|$. More generally, since E contains $|E|$ outcomes out of a total of $|S|$ possible outcomes, the probability of event E is given by

$$P(E) = \frac{|E|}{|S|}. \tag{9.1.1}$$

Formula (9.1.1) defines the probability $P(E)$ of an event E, under the special assumptions that S is finite and each outcome is equally likely to occur. The mapping P defined by (9.1.1) is called the *uniform probability distribution* on the sample space S. We now give the three basic axioms of probability theory that characterize the probability $P(E)$ of an event occurring in sample spaces S (finite or infinite) where simple events (outcomes) are not necessarily equally likely. For example, if we have loaded dice, then we can't use (9.1.1) to calculate the probability of events resulting from those dice. We need the axioms of probability to handle this more general situation.

Two events E and F are said to be *mutually exclusive* if they have no outcomes in common, that is, if $E \cap F = \varnothing$. More generally, a given sequence $E_1, E_2, \ldots$ of events in S is said to be *mutually exclusive* if $E_i \cap E_j = \varnothing$, for every pair of distinct events E_i and E_j in the sequence. A *probability distribution* P on the sample space S is a mapping from the events in S to the real numbers satisfying the following three axioms.

Axiom 1: $0 \le P(E) \le 1$.

Axiom 2: $P(S) = 1$.

Axiom 3: For any sequence $E_1, E_2, \ldots$ of mutually exclusive events in S,

$$P\left(\bigcup_{i=1}^{\infty} E_i\right) = \sum_{i=1}^{\infty} P(E_i).$$

A probability distribution is called *discrete* if it is defined on a discrete sample space.

PROPOSITION 9.1.1 Suppose P is defined for the simple events in a discrete sample space S and satisfies the following two conditions:

a. $0 \le P(\{s\}) \le 1, \quad$ for $s \in S$, and

b. $\sum_{s \in S} P(\{s\}) = 1$.

Further, suppose for any given event E in S we define $P(E)$ by

$$P(E) = P(\bigcup_{s \in E} \{s\}) = \sum_{s \in E} P(\{s\}). \tag{9.1.2}$$

Then P satisfies the three axioms for a probability distribution on S. ❑

The proof of Proposition 9.1.1 is easy and is left as an exercise. We illustrate Proposition 9.1.1 by considering the experiment of rolling a single loaded die D. The sample space for this experiment is the set $S = \{1,2,3,4,5,6\}$. Assume, for example, that the die is loaded and that the simple events $\{1\}$, $\{2\}$, $\{3\}$, $\{4\}$, $\{5\}$, $\{6\}$ occur with probabilities 1/10, 1/10, 1/10, 1/5, 1/5, 3/10, respectively. (Such an assignment of probabilities to the simple events might be arrived at by rolling the die a large number of times and recording the results.) We then apply formula (9.1.2) to arrive at a probability distribution on S. Now suppose we wish to compute the probability of the occurrence of the event E that the number showing on the die is odd. Applying (9.1.2), we obtain

$$P(E) = P(\{1\}) + P(\{3\}) + P(\{5\}) = \frac{1}{10} + \frac{1}{10} + \frac{1}{5} = \frac{2}{5}.$$

9.1.2 Conditional Probability

Conditional probabilities often play a role when we analyze the average behavior of algorithms. To motivate the notion of conditional probability, consider again the experiment of rolling two fair dice D_1 and D_2. Suppose that you have been informed that the first die came up 2. You now are asked to determine the probability that the sum of the two dice is at most 5. In other words, you are asked to compute the *conditional probability* that the sum of the dice is at most 5, *given* that the first die has come up 2. Let E denote the event that the sum of the two dice is at most 5, and let F denote the event that the first die is a 2. Referring back to Figure 9.1, we see that E contains 10 outcomes, and F contains 6 outcomes, so that $P(E) = 10/36 = 5/18$ and $P(F) = 6/36 = 1/6$. Let $P(E \mid F)$ denote the conditional probability the E occurs given that F has occurred. The condition that D_1 comes up 2 amounts to restricting the sample space to $F = \{(2,i) \mid i \in \{1, \ldots, 6\}\}$. This restricted sample space has 6 simple events. Of these 6 simple events only the 3 events in $E \cap F = \{(2,1), (2,2), (2,3)\}$ yield a sum that is at most 5. Thus,

$$P(E|F) = \frac{|E \cap F|}{|F|} = \frac{3}{6} = \frac{1}{2}. \tag{9.1.3}$$

Since $\frac{|E \cap F|}{|F|} = \frac{|E \cap F|/|S|}{|F|/|S|} = \frac{P(E \cap F)}{P(F)}$, we may express $P(E \mid F)$ in terms of the probabilities of the events $E \cap F$ and F:

$$P(E \mid F) = \frac{P(E \cap F)}{P(F)}. \tag{9.1.4}$$

Substituting $P(E \cap F) = 3/36 = 1/12$ and $P(F) = 1/6$ into (9.1.4) also yields $P(E \mid F) = 1/2$.

Formula (9.1.3) can only be used for sample spaces where simple events are equally likely. However, (9.1.4) can be used as the general definition of the conditional probability $P(E \mid F)$. Formula (9.1.4) is often rewritten as (9.1.5), which expresses $P(E \cap F)$ in terms of $P(E \mid F)$ and $P(F)$:

$$P(E \cap F) = P(E \mid F)P(F). \tag{9.1.5}$$

Events E and F are said to be *independent* if $P(E \mid F) = P(E)$. Equivalently, E and F are independent if

$$P(E \cap F) = P(E)P(F). \tag{9.1.6}$$

The following useful proposition generalizes (9.1.5).

PROPOSITION 9.1.2 Suppose $F_1, F_2, \ldots, F_n$ are mutually exclusive events in S, and $S = \bigcap_{i=1}^{n} F_i$. Let E be any event in S. Then we have

$$P(E) = P(F_1)P(E \mid F_1) + P(F_2)P(E \mid F_2) + \cdots + P(F_n)P(E \mid F_n). \tag{9.1.7}$$ ❑

Formula (9.1.7) is easily proved by noting that $E = (E \cap F_1) \cup (E \cap F_2) \cup \cdots \cup (E \cap F_n)$, and then applying Axiom 3 and (9.1.5).

An important special case of (9.1.7) occurs when $n = 2$, $F_1 = F$, and $F_2 = F^c = S \setminus F$. First note that Axioms 2 and 3 imply that

$$P(F^c) = 1 - P(F). \tag{9.1.8}$$

Using (9.1.7) and (9.1.8), we have

$$P(E) = P(F)P(E \mid F) + (1 - P(F))P(E \mid F^c). \tag{9.1.9}$$

9.1.3 Random Variables and Expectation

A (*discrete*) *random variable* X on a (discrete) sample space S is a mapping from S to the set $\mathbb{R}$ of real numbers. For example, consider the sample space given in Figure 9.1 corresponding to the experiment of rolling two dice. Figure 9.2 illustrates the random variable X that maps the outcome (r_1,r_2) onto $r_1 + r_2$, $r_1,r_2 \in \{1, \ldots, 6\}$. Another example of a random variable on the same sample space is the random variable that maps the outcome (r_1,r_2) onto $\min\{r_1,r_2\}$, $r_1,r_2 \in \{1, \ldots, 6\}$.

Given a random variable X, for x a real number, let

$$P(X = x) \tag{9.1.10}$$

denote the probability of the occurrence of the event $E = \{s \in S \mid X(s) = x\}$. The function $f(x) = P(X = x)$ is called the *probability density function of* X. Consider the sample space S_X consisting of all real numbers x such that $f(x) \neq 0$. It is easily verified that $f(x)$ determines a probability distribution on S_X, which we refer to as the *distribution of the random variable* X.

For example, consider the random variable X illustrated in Figure 9.2 corresponding to the sum of the two dice. The probability density function of X is given by

$$P(X = i) = \frac{6 - |7 - i|}{36} \qquad i = 2, \ldots, 12. \tag{9.1.11}$$

As a second example, consider an experiment (trial) that has two outcomes, "success" or "failure," where the probability of success is p for any given trial. For example, the trial might be flipping a coin, where we consider the coin coming up heads as success. A fair coin would have $p = 1/2$. Suppose now that n independent trials are performed (called *Bernoulli trials*), so that we obtain a sample space consisting of all 2^n possible outcomes of the n trials. Since the trials are independent, any outcome with i successes and $n - i$ failures occurs with probability $p^i(1 - p)^{n-i}$. Given n and p, the *binomial random variable* X is defined to be the number of successes associated with a given outcome. Since there are $\binom{n}{i}$ events with i successes, the probability density function of X is given by

$$P(X = i) = \binom{n}{i} (1 - p)^{n-i} p^i, \quad i = 0, \ldots, n. \tag{9.1.12}$$

The distribution on $\{0,1, \ldots, n\}$ determined by (9.1.12) is called a *binomial distribution*. The binomial distribution for $n = 8$ and $p = 2/5$ is illustrated in Figure 9.4.

As a third example, we now consider another distribution involving Bernoulli trials, but this time for an infinite sample space S. Corresponding to each positive integer i, we obtain a simple event in S by running i trials, where the first $i - 1$ trials result in failure, and the last trial results in success. The probability distribution on S is $P(s) = (1 - p)^{i-1}p$, where s is the simple event consisting of i trials. The random variable X defined on S maps s to the number of trials associated with s. Hence,

$$P(X = i) = (1 - p)^{i-1}p, \quad i = 1,2, \ldots \tag{9.1.13}$$

Figure 9.4 A histogram showing the binomial distribution for $n = 8$ and $p = 2/5$

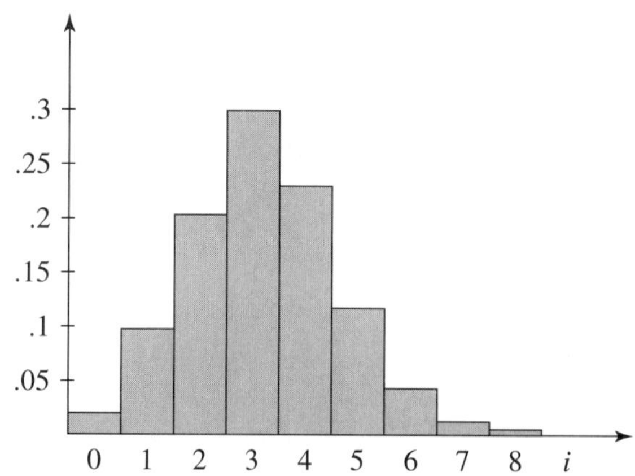

The probability distribution on $\{1,2, \ldots\}$ determined by (9.1.13) is called a *geometric distribution*. The geometric distribution with $p = 2/5$ is illustrated in Figure 9.5.

Random variables occur naturally when analyzing the average behavior of an algorithm. For example, consider the algorithm *LinearSearch*, where elements of the sample space are pairs consisting of a given fixed list $L[1{:}n]$ of size n and a search element X. In Figure 9.6 we use Y to denote the random variable to avoid confusion with the search element X. The random variable Y maps the outcome (L,X) to the number of comparisons performed by *LinearSearch*.

In the following discussion of random variables, we assume that the sample space S is finite. One of the most important concepts in probability theory is that of the expectation $E[X]$ of a random variable X on a (finite) sample space S. The *mean* or *expectation* of X is defined by

$$E[X] = \sum_{s \in S} X(s)P(s). \tag{9.1.14}$$

Collecting terms having a common value of X, (9.1.14) is equivalent to the formula

$$E[X] = \sum_{x} x\ P(X = x), \tag{9.1.15}$$

where the summation is over all real numbers x such that $P(X = x) > 0$. For example, suppose X is the random variable illustrated in Figure 9.2 corresponding to the sum of the two dice. Substituting the values for $P(X = i)$ from (9.1.11) into (9.1.15) yields

$$E[X] = \sum_{i=2}^{12} iP(X = i) = \sum_{i=2}^{12} i\frac{(6 - |7 - i|)}{36} = 7.$$

Figure 9.5 A histogram showing the geometric distribution for $p = 2/5$

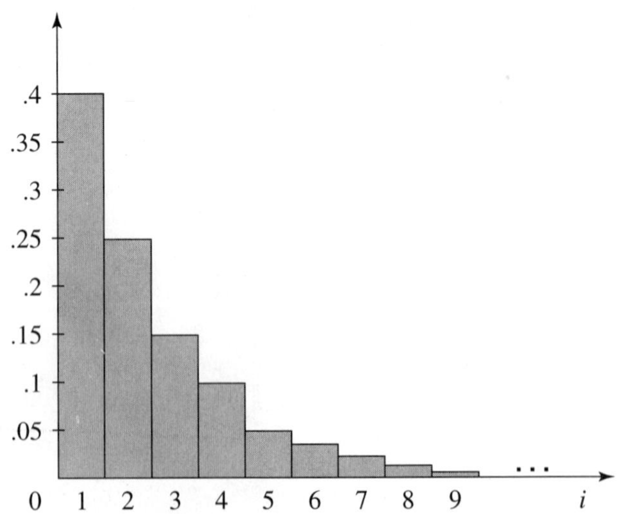

Outcome	Y
$(L,L[1])$	1
$(L,L[2]$	2
.	.
.	.
.	.
$(L,L[n])$	n
$(L,X),\ X \notin L$	n

Figure 9.6 Random variable Y mapping (L,X) onto the number of comparisons performed by *LinearSearch* for input list $L[1:n]$ and search element X

As another example, suppose X is the binomial random variable corresponding to a given n and p. Substituting the values for $P(X = i)$ from (9.1.12) into (9.1.15) yields

$$E[X] = \sum_{i=0}^{n} \binom{n}{i} (1-p)^{n-i} p^i i. \qquad (9.1.16)$$

Applying identity (1.2.19) of Chapter 1 with $x = 1 - p$ and $y = p$ to the right-hand side of (9.1.16) we obtain

$$E[X] = np. \qquad (9.1.17)$$

Formula (9.1.17) is intuitively obvious since we are performing n trials where the success of each trial is p. Another intuitively obvious result is that the expectation of the geometric distribution with probability p for success is given by

$$E[X] = \frac{1}{p}. \qquad (9.1.18)$$

We leave the verification of (9.1.18) as an exercise.

A random variable X on the sample space S is a *constant* if it maps every element in S onto the same real number c (denoted $X \equiv c$). It follows immediately from the definition of expectation that $E[X \equiv c] = c$.

The following two useful propositions are easily proved.

PROPOSITION 9.1.3 Let X be a random variable on S. Then for any constant c,

$$E[cX] = cE[X]. \qquad (9.1.19)\ \square$$

PROPOSITION 9.1.4 If $X_1, X_2, \ldots, X_n$ are n random variables on the sample space S, and $X = X_1 + X_2 + \cdots + X_n$, then

$$E[X] = E[X_1] + E[X_2] + \cdots + E[X_n]. \qquad (9.1.20)\ \square$$

In addition to computing the expectation $E[X]$ of a random variable, it is also useful to measure how much X deviates, on average, from $E[X]$. A small deviation indicates that X assumes, with high probability, only values close to its expectation, whereas a large deviation indicates that X assumes values far off $E[X]$ with relatively high probabilities. The deviation of X from $E[X]$ is formally defined in terms of the *variance*:

$$V[X] = E[(X - E[X])^2]. \tag{9.1.21}$$

The *standard deviation* $\sigma[X]$ of X is defined to be the square root of the variance of X; that is,

$$\sigma[X] = \sqrt{V[X]}. \tag{9.1.22}$$

Note that $E[X]$ can be viewed as a constant random variable on the sample space S, mapping every outcome in S to the same value $c = E[X]$. Another formulation of the variance of X is obtained from (9.1.21) by employing (9.1.19) and (9.1.20):

$$\begin{aligned} V[X] = E[(X - E[X])^2] &= E[(X - c)^2] \\ &= E[X^2 - 2cX + c^2] \\ &= E[X^2] - E[2cX] + E[c^2] \\ &= E[X^2] - 2cE[X] + c^2 \\ &= E[X^2] - c^2. \end{aligned}$$

Hence, we have the following proposition.

PROPOSITION 9.1.5 Let X be a random variable on a sample space S. Then,

$$V[X] = E[X^2] - (E[X])^2. \qquad \square$$

As an illustration of Proposition 9.1.5, we calculate the variance of the geometric distribution. Clearly, the probability density function for the random variable X^2 is given by

$$P(X^2 = i) = \begin{cases} (1-p)^{k-1} & i = k^2 \text{ for some positive integer } k \\ 0 & \text{otherwise.} \end{cases} \tag{9.1.23}$$

Substituting (9.1.23) in formula (9.1.15), we obtain

$$\begin{aligned} E[X^2] &= \sum_{k=1}^{\infty} iP(X^2 = i) \\ &= \sum_{k=1}^{\infty} k^2 (1-p)^{k-1} p \\ &= \frac{2-p}{p^2} \end{aligned}$$

(See Exercise 9.8.) Thus, by Proposition 9.1.5 and formula (9.1.18), we have

$$V[X] = E[X^2] - (E[X])^2$$
$$= \frac{2-p}{p^2} - \frac{1}{p^2}$$
$$= \frac{1-p}{p^2}.$$

9.1.4 Conditional Expectation: Partitioning the Sample Space

Formulas (9.1.14) and (9.1.15) for $E[X]$ are sometimes difficult to apply directly. In such cases there is often a natural partitioning of the sample space S into disjoint sets $F_1, F_2, \ldots, F_m$ that facilitates the evaluation of $E[X]$ using a generalization of formula (9.1.15) and the notion of conditional expectation, which we now define.

Given a sample space S, a random variable X defined on S, and a subset $F \subseteq S$, let P_F be the function defined on F by

$$P_F(s) = P(s \mid F), \quad s \in F. \tag{9.1.24}$$

It is easily verified that P_F defined by (9.1.24) satisfies the three axioms for a probability distribution on F. If X is a random variable defined on S, let X_F denote the random variable defined on F by restricting X to F; that is, $X_F(s) = X(s)$, $s \in F$. We then define the *conditional expectation* $E[X \mid F]$ to be $E[X_F]$. Using (9.1.14) and (9.1.24) we have

$$E[X \mid F] = \sum_{s \in F} X_F(s)\, P_F(s) = \sum_{s \in F} X(s) P(s|F)(s)P(s \mid F). \tag{9.1.25}$$

Equivalently, using (9.1.15), we have

$$E[X \mid F] = \sum_x x\, P(X = x \mid F), \tag{9.1.26}$$

where the sum is taken over all x such that $P(X = x \mid F) > 0$.

To illustrate, consider the sample space S consisting of all possible outcomes of rolling two dice (see Figure 9.2) and let X denote the random variable that maps the outcome (r_1,r_2) onto $r_1 + r_2$, $r_1,r_2 \in \{1, \ldots, 6\}$. Suppose F is the event that die D_1 comes up 2, so that F consists of the six outcome pairs $(2,r_2)$, $r_2 \in \{1, \ldots, 6\}$. Then, since the outcome of die D_2 is independent of the outcome of D_1, we have

$$P_F((2,r_2)) = P((2,r_2) \mid F) = \frac{1}{6}, \quad r_2 \in \{1, \ldots, 6\}. \tag{9.1.27}$$

Substituting (9.1.27) into (9.1.25), we obtain

$$E[X \mid F] = \sum_{s \in F} X(s)P(s \mid F) = \left(\frac{1}{6}\right) \sum_{r_2 \in \{1, \ldots, 6\}} (2 + r_2) \tag{9.1.28}$$
$$= \left(\frac{1}{6}\right)(3 + 4 + 5 + 6 + 7 + 8) = 5.5.$$

The next proposition follows from (9.1.25), (9.1.26), and the fact that $s \in F \Rightarrow P(s) = P(s \cap F) = P(s \mid F)P(F)$.

PROPOSITION 9.1.6 Let X be a discrete random variable on the sample space S. Suppose that we are given a partition $S = \bigcup_{i=1}^{m} F_i$ of S into disjoint subsets F_i, $i = 1, \ldots, m$. Then we have

$$E[X] = \sum_{i=1}^{m} E[X \mid F_i]P(F_i). \tag{9.1.29}$$

Moreover, $E[X \mid F_i] = \sum_x x\ P(X = x \mid F_i)$, where the sum is taken over all x such that $P(X = x \mid F_i) > 0$. ❑

A partition of a sample space S is often described using a naturally arising second random variable Y. The random variable Y determines a partition of S into the mutually exclusive events $F_y = \{s \in S \mid Y(s) = y\}$, where $P(Y = y) > 0$. For x and y real numbers, let $P(X = x \mid Y = y)$ denote the conditional probability $P(G_x \mid F_y)$, where G_x is the event $\{s \in S \mid X(s) = x\}$. The following proposition merely restates Proposition 9.1.6 in the context of the sets G_x and F_y.

PROPOSITION 9.1.7 Let X, Y be discrete random variables on sample space S. Then,

$$E[X] = \sum_y E[X \mid Y = y]\ P(Y = y), \tag{9.1.30}$$

where the summation is taken over all y such that $P(Y = y) > 0$. Moreover,

$$E[X \mid Y = y] = \sum_x x\ P(X = x \mid Y = y), \tag{9.1.31}$$

where the summation is over all x such that $P(X = x) > 0$. ❑

Formula (9.1.30) of Proposition 9.1.7 reduces to formula (9.1.15) when $X = Y$. We illustrate Proposition 9.1.7 for the sample space of all possible outcomes resulting from rolling two dice (see Figure 9.1), where X and Y are the random variables mapping the outcome (r_1,r_2) onto $r_1 + r_2$, and r_1, respectively, $r_1,r_2 \in \{1, \ldots, 6\}$. The conditional probabilities $P\{X = x \mid Y = y\}$, $x \in \{2, \ldots, 12\}$, $y \in \{1, \ldots, 6\}$ are easily calculated using formula (9.1.4) (see Figure 9.7).

The conditional expectations $E[X \mid Y = y]$, $y \in Y$, are calculated by substituting the values for $P(X = x \mid Y = y)$ given in Figure 9.7 into formula (9.1.31) (see Figure 9.8). Note, for example, that the event $Y = 2$ coincides with the event F, whose conditional expectation $E[X \mid F]$ was also calculated in (9.1.28).

Since the dice are fair, $P(Y = y) = 1/6$, $y = 1, \ldots, 6$. Substituting the values for the conditional expectations $E[X \mid Y = y]$, $y = 1, \ldots, 6$, given in Figure 9.8, we obtain

$$E[X] = \sum_y [X \mid Y = y]\ P(Y = y)$$
$$= 4.5\left(\frac{1}{6}\right) + 5.5\left(\frac{1}{6}\right) + 6.5\left(\frac{1}{6}\right) + 7.5\left(\frac{1}{6}\right) + 8.5\left(\frac{1}{6}\right) + 9.5\left(\frac{1}{6}\right) = \frac{42}{6} = 7.$$

Note that $E[X] = 7$ is the same value that we obtained earlier. Of course, the computation of $E[X]$ using conditional expectations in this example was no easier than our previous calculation that proceeded directly from (9.1.15). Our purpose here was merely to illustrate how conditional expectation is applied. However, in practice

x / y	2	3	4	5	6	7	8	9	10	11	12
1	$\frac{1}{6}$	$\frac{1}{6}$	$\frac{1}{6}$	$\frac{1}{6}$	$\frac{1}{6}$	$\frac{1}{6}$	0	0	0	0	0
2	0	$\frac{1}{6}$	$\frac{1}{6}$	$\frac{1}{6}$	$\frac{1}{6}$	$\frac{1}{6}$	$\frac{1}{6}$	0	0	0	0
3	0	0	$\frac{1}{6}$	$\frac{1}{6}$	$\frac{1}{6}$	$\frac{1}{6}$	$\frac{1}{6}$	$\frac{1}{6}$	0	0	0
4	0	0	0	$\frac{1}{6}$	$\frac{1}{6}$	$\frac{1}{6}$	$\frac{1}{6}$	$\frac{1}{6}$	$\frac{1}{6}$	0	0
5	0	0	0	0	$\frac{1}{6}$	$\frac{1}{6}$	$\frac{1}{6}$	$\frac{1}{6}$	$\frac{1}{6}$	$\frac{1}{6}$	0
6	0	0	0	0	0	$\frac{1}{6}$	$\frac{1}{6}$	$\frac{1}{6}$	$\frac{1}{6}$	$\frac{1}{6}$	$\frac{1}{6}$

Figure 9.7 $P\{X = x \mid Y = y\}$ for random variables X, Y on sample space of Figure 9.1 where X and Y map (r_1, r_2) onto $r_1 + r_2$ and r_1, respectively, $r_1, r_2 \in \{1, \ldots, 6\}$

y	1	2	3	4	5	6
$E[X \mid Y = y]$	4.5	5.5	6.5	7.5	8.5	9.5

Figure 9.8 Conditional expectations $E[X \mid Y = y]$, $\{y \in Y\}$ for random variables X, Y of Figure 9.7

it may be difficult to determine the probabilities $P(X = x)$ occurring in the formula (9.1.15) for $E[X]$. Thus, more powerful tools for computing $E[X]$, such as formula (9.1.30), are required.

9.2 Average Complexity Revisited

When analyzing the complexity of an algorithm that is to be used over and over again with varying inputs, it is often not the worst-case complexity that is the critical issue, but rather the *average* complexity. In such a situation one would tend to favor one algorithm over another based on average performance. The notion of *expected value* defined in the previous section allows us to formulate precisely what is meant by the average complexity of an algorithm.

DEFINITION 9.2.1 **Average Complexity** Let $\mathcal{I}_n$ denote the set of all inputs of size n to a given algorithm. For $I \in \mathcal{I}_n$, let $\tau(I)$ denote the number of basic operations performed by the algorithm on input I. Given a probability distribution on $\mathcal{I}_n$, the *average complexity* $A(n)$ of an algorithm is the expected value of τ; that is,

$$A(n) = E[\tau] \tag{9.2.1}$$

]

Average behavior depends on the probability distribution on $\mathcal{T}_n$. Often there is a natural way to assign a probability distribution on $\mathcal{T}_n$. For example, when analyzing comparison-based sorting algorithms it is typical to assume that each of the $n!$ permutations (orderings) of a list of size n is equally likely to be input to the algorithm. The average complexity of any comparison-based sorting algorithm is then the sum of the number of comparisons generated by each of the $n!$ permutations divided by $n!$. In practice, it is not feasible to examine each one of these $n!$ permutations individually, as $n!$ simply grows too fast. Fortunately, there are techniques that allow us to calculate this average without resorting to a permutation-by-permutation analysis.

9.2.1 Techniques for Computing Average Complexity

Computing the average complexity of an algorithm frequently requires deep results from probability theory that are beyond the scope of this text. However, the following relatively elementary techniques suffice for the analysis of the average behavior of most of the algorithms discussed in this text. When computing A(n), one or some combination of the following formulas for A(n) will be most applicable, depending on the particular characteristics of the algorithm being analyzed. Each formulation assumes that $\mathcal{T}_n$ is finite.

FORMULATION I

$$A(n) = E[\tau] = \sum_{I \in \mathcal{T}_n} \tau(I) P(I). \tag{9.2.2}$$

Formula (9.2.2) is simply definition (9.1.14) of $E[\tau]$ with $X = \tau$, $s = I$, and $S = \mathcal{T}_n$. Formulation I is rarely used directly since it is usually too cumbersome to examine each element in $\mathcal{T}_n$ individually.

FORMULATION II Let p_i denote the probability that the algorithm performs exactly i basic operations; that is, $p_i = P(\tau = i)$. Then

$$A(n) = E[\tau] = \sum_{i=1}^{W(n)} i p_i, \tag{9.2.3}$$

Formula (9.2.3) is a special case of (9.1.15) with $X = \tau$ and $x = i$. Note that (9.2.2) and (9.2.3) are the same as the formulas (3.1.3) and (3.1.4), respectively, given in Chapter 3.

FORMULATION III Let q_i denote the probability that the algorithm performs *at least i* basic operations; that is, $q_i = P\{\tau \geq i\}$. Then

$$A(n) = E[\tau] = \sum_{i=1}^{W(n)} q_i . \tag{9.2.4}$$

Formula (9.2.4) follows immediately from (9.2.3) by noting that $q_i = \Sigma_{j=i}^{W(n)} p_j$.

FORMULATION IV Suppose $\tau = \sum_{i=1}^{k} \tau_i$. Then we have

$$A(n) = E[\tau] = \sum_{i=1}^{k} E[\tau_i]. \qquad (9.2.5)$$

Formula (9.2.5) is (9.1.20) with $X = \tau$ and $X_i = \tau_i$.

FORMULATION V Suppose Y is another random variable defined on $\mathcal{T}_n$. Then

$$A(n) = E[\tau] = \sum_{y} E[\tau \mid Y = y] \, P\{Y = y\}, \qquad (9.2.6)$$

where the summation is taken over all y such that $P\{Y = y\} > 0$.

Formula (9.2.6) is (9.1.29) with $X = \tau$.

When determining which of the formulations of A(n) are best suited to a given algorithm, usually one or more of the following three basic techniques are utilized:

1. partitioning the algorithm,
2. partitioning the input space, or
3. recursion.

Partitioning the algorithm refers to a situation where the steps in the algorithm can be partitioned into k stages. If τ_i denotes the random variable on $\mathcal{T}_n$ mapping each input onto the number of basic operations performed in stage i, $i = 1, \ldots, k$, then $\tau = \Sigma_{i=1}^{k} \tau_i$ and Formulation IV applies. The partitioning is useful when each τ_i can be effectively computed, $i = 1, \ldots, k$.

Partitioning the input space is appropriate when the inputs of size n to the algorithm can be partitioned into disjoint sets by utilizing a naturally defined random variable Y and where the quantities $E[\tau \mid Y = y]$, $P\{Y = y\}$ occurring in (9.2.6) are effectively computable. Typically, Y involves mapping an input to some sort of integer constraint, such as the position in a list of size n where a maximum element occurs. Then A(n) is computed using Formulation V.

Often it is possible to find a recurrence relation expressing A(n) in terms of one or more of the values A(m) with $m < n$. Of course, if the algorithm itself is written recursively, then the recurrence relation is usually easy to find. In general, finding a recurrence relation for A(n) often involves utilizing the techniques of partitioning the algorithm or the input space.

9.3 Average Complexity of *LinearSearch*

In Chapter 3 we informally showed that the average complexity of *LinearSearch* is $n/2 + 1/2$, under the assumptions that the search element X is in the list $L[1{:}n]$ and the list consists of distinct elements with each element being equally likely to be X. We will now calculate the average complexity of *LinearSearch* under these latter assumptions, but we will now assume that the search element X occurs in the list with probability p. Using the notation for conditional probability given in the previous section, this assumption becomes

$$P(X = L[i] \mid X \text{ is in the list } L) = \frac{1}{n}, \quad i = 1, \ldots, n.$$

We calculate A(n) using (9.2.3). For $1 \le i \le n - 1$, p_i is equal to the probability that X is the ith list element $L[i]$. Hence, from (9.1.5) we have

$$\begin{aligned} p_i &= P(X = L[i] \mid X \text{ is in the list } L)P(X \text{ is in the list } L) \\ &= \left(\frac{1}{n}\right)p = \frac{p}{n}, \quad i = 1, \ldots, n-1. \end{aligned}$$

LinearSearch performs n comparisons when $X = L[n]$, or when X is not in the list L. Thus, $p_n = p/n + 1 - p$. Substituting these values of p_i into (9.2.3) yields

$$\begin{aligned} A(n) &= 1\left(\frac{p}{n}\right) + 2\left(\frac{p}{n}\right) + \cdots + (n-1)\left(\frac{p}{n}\right) + n\left(\frac{p}{n} + 1 - p\right) \\ &= (1 + 2 + \cdots + n)\left(\frac{p}{n}\right) + n(1-p) \\ &= \left(\frac{n(n+1)}{2}\right)\left(\frac{p}{n}\right) + n(1-p) = \left(1 - \frac{p}{2}\right)n + \frac{p}{2}. \end{aligned} \tag{9.3.1}$$

Remark Note that if we set $p = 1$ in (9.3.1), then we obtain the formula $A(n) = (n + 1)/2$ derived in Chapter 3 for the average complexity of *LinearSearch* when X is assumed to be in the list.

9.3.1 Average Complexity of *LinearSearch* with Repeated Elements

We now determine the average behavior A(n,m) of *LinearSearch* for lists $L[1{:}n]$ of size n having m distinct elements drawn from a given fixed set $S = \{s_1, s_2, \ldots, s_m\}$, where the search element $X \in S$. For purposes of computing A(n,m), we assume that $L[i]$ has an equal probability $1/m$ of being any element in S, $i = 1,2, \ldots, n$. Hence, the probability that X does not occur in position i is $(m - 1)/m$, $i = 1,2, \ldots, n$. It follows that the probability that X is not in the list is $((m - 1)/m)^n$ and the probability that X is in the list is $1 - ((m - 1)/m)^n$.

Clearly, the p_i in formula (9.2.3) ($i = 1,2, \ldots, n - 1$) is the probability that the first occurrence of the search element X is in position i, and p_n is the probability that X does not occur in the first $n - 1$ positions. We have:

$$p_i = \begin{cases} ((m-1)/m)^{i-1}(1/m) & \text{if } 1 \le i \le n-1, \\ ((m-1)/m)^{n-1} & \text{if } i = n. \end{cases} \tag{9.3.2}$$

Substituting (9.3.2) into formula (9.2.3) (suitably interpreted), we obtain

$$A(n,m) = \sum_{i=1}^{W(n)} ip_i = \left(\sum_{i=1}^{n-1} i\left(\frac{m-1}{m}\right)^{i-1}\left(\frac{1}{m}\right)\right) + \left(\frac{m-1}{m}\right)^{n-1}. \tag{9.3.3}$$

We now employ formula (1.2.11) of Chapter 1, which states that

$$\sum_{i=1}^{n-1} ix^{i-1} = \frac{(n-1)x^n - nx^{n-1} + 1}{(1-x)^2}.$$

Substituting $x = (m-1)/m$ into (9.3.3), we obtain

$$A(n,m) = \left[(1/m)\frac{(n-1)((m-1)/m)^n - n((m-1)/m)^{n-1} + 1}{m^{-2}}\right] + \left(\frac{m-1}{m}\right)^{n-1}$$

$$= m\left(1 - \left(\frac{m-1}{m}\right)^n\right) + \left(\frac{m-1}{m}\right)^{n-1}.$$

Note that if we hold m constant, then $((m-1)/m)^n$ tends to zero as n tends to ∞, so that $A(n,m) \sim m$. In particular, we have $A(n,m_{\text{fixed}}) \in \Theta(m_{\text{fixed}}) = \Theta(1)$, and thus $A(n,m_{\text{fixed}})$ has *constant* order. The latter behavior is very different from the linear average behavior $A(n)$ of *LinearSearch* for lists of *distinct* elements.

9.4 Average Complexity of *InsertionSort*

Since *InsertionSort* is a comparison-based algorithm, we can assume without loss of generality that inputs to *InsertionSort* are permutations of $\{1,2, \ldots, n\}$. We also assume that each permutation is equally likely to be the input to *InsertionSort*. Unlike our analysis of *LinearSearch*, we cannot compute $A(n) = E[\tau]$ directly by applying formula (9.2.3). Instead, we partition the algorithm *InsertionSort* into $n-1$ stages. The ith stage consists of inserting the $(i+1)$st element $L[i+1]$ into its proper position in the list $L[1], \ldots, L[i]$, where the latter list has already been sorted by the algorithm. Let τ_i denote the number of comparisons performed in stage i, so that $\tau = \tau_1 + \cdots + \tau_{n-1}$ and, by (9.2.5),

$$A(n) = E[\tau] = E[\tau_1] + E[\tau_2] + \cdots + E[\tau_{n-1}]. \tag{9.4.1}$$

We now calculate $E[\tau_i]$, $i = 1, \ldots, n-1$, using formula (9.1.15). We have

$$E[\tau_i] = \sum_{j=1}^{i} jP(\tau_i = j). \tag{9.4.2}$$

Our assumption of a uniform distribution on the input space implies that any position in $L[1:i+1]$ is equally likely to be the correct position for $L[i+1]$. Thus, the probability that $L[i + 1]$ is the jth largest of the elements in $L[1{:}i + 1]$ is equal to $1/(i + 1)$. If $L[i + 1]$ is the jth largest, where $j \le i$, then exactly j comparisons are performed by *InsertionSort* when placing $L[i + 1]$ in its correct position. If $L[i + 1]$ is the $(i + 1)$st largest (that is, the smallest), then exactly i comparisons are performed by *InsertionSort* when placing $L[i + 1]$ in its correct position. It follows that

$$\begin{aligned} P(\tau_i = j) &= \frac{1}{i+1}, \quad j = 1, \ldots, i-1, \\ P(\tau_i = i) &= \frac{2}{i+1}, \quad i = 1, \ldots, n-1. \end{aligned} \tag{9.4.3}$$

Substituting (9.4.3) into (9.4.2) and simplifying yields

$$E[\tau_i] = \left(\sum_{j=1}^{i} \frac{j}{i+1}\right) + \frac{i}{i+1} = \frac{i}{2} + 1 - \frac{1}{i+1}, \quad i = 1, \ldots, n-1. \tag{9.4.4}$$

Substituting (9.4.4) into (9.4.1), we have

$$\begin{aligned} A(n) &= \sum_{i=1}^{n-1}\left(\frac{i}{2} + 1 - \frac{1}{i+1}\right) \\ &= (n-1)\frac{n}{4} + (n-1) - \left(\frac{1}{2} + \frac{1}{3} + \cdots + \frac{1}{n}\right) \\ &= (n-1)\frac{n}{4} + n - H(n), \end{aligned}$$

where $H(n)$ is the harmonic series $1 + 1/2 + \cdots + 1/n \sim \ln n$. In particular, $A(n) \in \Theta(n^2)$. Note that the average complexity of *InsertionSort* is about half that of its worst-case complexity, since the highest-order terms in the expressions for these complexities are $n^2/4$ and $n^2/2$, respectively.

9.5 Average Complexity of *QuickSort*

As with *InsertionSort*, we assume that input lists $L[1{:}n]$ to *QuickSort* are all permutations of $1,2, \ldots, n$, with each permutation being equally likely. We partition *QuickSort* into two stages, where the first stage is the call to *RearrangeAndPlace* and the second stage is the two recursive calls with input lists consisting of the sublists on either side of the proper placement of the pivot element $L[1]$. Thus, $\tau = \tau_1 + \tau_2$, where τ_1 is the (constant) number $n + 1$ of comparisons performed by *RearrangeAndPlace* and τ_2 is the number of comparisons performed by the recursive calls. Hence,

$$A(n) = E[\tau] = E[\tau_1] + E[\tau_2] = n + 1 + E[\tau_2]. \tag{9.5.1}$$

We compute $E[\tau_2]$ using Formulation V by introducing the random variable Y that maps an input list $L[1{:}n]$ into the proper place for $L[1]$ as determined by a call to

RearrangeAndPlace. The uniform distribution assumption on the input space implies that

$$P(Y = i) = \frac{1}{n}, \quad i = 1, \ldots, n. \tag{9.5.2}$$

If $Y = i$, then the recursive calls to *QuickSort* are with the two sublists $L[1{:}i{-}1]$ and $L[i + 1{:}n]$. Our assumption of a uniform distribution on the input space implies that the expected number of comparisons performed by *QuickSort* on the sublists $L[1{:}i - 1]$ and $L[i + 1{:}n]$ is given by $A(i - 1)$ and $A(n - i)$, respectively. Hence,

$$E[\tau_2 \mid Y = i] = A(i - 1) + A(n - i), \quad i = 1, \ldots, n. \tag{9.5.3}$$

Combining (9.2.6), (9.5.1), (9.5.2), and (9.5.3), we have

$$\begin{aligned} A(n) &= (n + 1) + \sum_{i=1}^{n} E[\tau_2 \mid Y = i]P(Y = i) \\ &= (n + 1) + \sum_{i=1}^{n} (A(i - 1) + A(n - i))\left(\frac{1}{n}\right) \\ &= (n + 1) + \frac{2}{n}(A(0) + A(1) + \cdots + A(n - 1)), \end{aligned} \tag{9.5.4}$$

init. cond. $A(0) = A(1) = 0$.

Recurrence relation (9.5.4) is an example of what is sometimes referred to as a *full history* recurrence relation, since it relates $A(n)$ to *all* of the previous values $A(i)$, $0 \le i \le n - 1$. Fortunately, with some algebraic manipulation, we can transform (9.5.4) into a simpler recurrence relation of the form (7.4.2). The trick is to first observe that

$$nA(n) = n(n + 1) + 2(A(0) + A(1) + \cdots + A(n - 2) + A(n - 1)). \tag{9.5.5}$$

Substituting $n - 1$ for n in (9.5.5), yields

$$(n - 1)A(n - 1) = n(n - 1) + 2(A(0) + A(1) + \cdots + A(n - 2)). \tag{9.5.6}$$

Hence, subtracting (9.5.6) from (9.5.5) we obtain

$$nA(n) - (n - 1)A(n - 1) = 2n + 2A(n - 1). \tag{9.5.7}$$

Rewriting (9.5.7) by moving the term involving $A(n - 1)$ to the right-hand side and dividing both sides by $n(n + 1)$ yields

$$\frac{A(n)}{n + 1} = \frac{A(n - 1)}{n} + \frac{2}{n + 1}. \tag{9.5.8}$$

Letting $t(n) = A(n)/(n + 1)$, (9.5.8) becomes

$$t(n) = t(n - 1) + \frac{2}{n + 1}. \tag{9.5.9}$$

Formula (9.5.9) with the initial condition $t(1) = 0$ is the special case of (7.4.2) with $a = 1$ and $f(n) = 2/(n + 1)$. Therefore, by (7.4.3) we have

$$t(n) = 2\left(\frac{1}{3} + \frac{1}{4} + \cdots + \frac{1}{(n+1)}\right)$$
$$= 2H(n+1) - 3, \tag{9.5.10}$$

where $H(n)$ is the harmonic series. Thus $t(n) \sim 2 \ln n$, so that the average complexity $A(n)$ of *QuickSort* satisfies

$$A(n) \sim 2n \ln n. \tag{9.5.11}$$

In particular, *QuickSort* exhibits $O(n \log n)$ average behavior, which is order optimal for a comparison-based sorting algorithm.

9.6 Average Complexity of *MaxMin2*

We analyze the algorithm *MaxMin*2 given in Chapter 3 as our third example. For convenience, we repeat the pseudocode for *MaxMin*2 here.

```
procedure MaxMin2(L[1:n],MaxValue,MinValue)
Input:  L[1:n] (an array of list elements)
Output: MaxValue,MinValue (the maximum and minimum values occurring in
                           L[1:n])
   MaxValue := L[1]
   MinValue := L[1]
   for i := 2 to n do
      if L[i] > MaxValue then
            MaxValue := L[i]
      else
            if L[i] < MinValue then MinValue := L[i]   endif
      endif
   endfor
end MaxMin2
```

As observed in Chapter 3, the best-case complexity of *MaxMin*2 is $n - 1$, and its worst-case complexity is $2(n - 1)$. Recall that the algorithm *MaxMin*3 has best-case, worst-case, and average complexities all equal to $\lceil 3n/2 \rceil - 2$. Thus, to complete our comparison of *MaxMin*2 and *MaxMin*3, we need to compute the average complexity of *MaxMin*2. Unfortunately, it turns out that the average complexity of *MaxMin*2 is closer to its worst-case complexity than to its best-case complexity.

When analyzing the average complexity of *MaxMin*2, we again assume that the inputs are permutations of 1,2, … , n and that each permutation is equally likely. Observe that $n - 1$ comparisons involving *MaxValue* are performed for any input permutation. An additional comparison involving *MinValue* is performed for each iteration of the loop in which *MaxValue* is not updated. Let D denote the random variable that maps the input permutation to the number of times that *MaxValue* is updated. Then, we have

$$\tau = n - 1 + (n - 1 - D) = 2n - 2 - D. \tag{9.6.1}$$

Thus, using (9.6.1), the average complexity of *MaxMin2* is given by

$$A(n) = E[\tau] = 2n - 2 - E[D]. \tag{9.6.2}$$

We compute the expected number of updates $E[D]$ by partitioning the input space utilizing the random variable M that maps an input permutation $\pi:\{1, \dots, n\} \Rightarrow \{1, \dots, n\}$ to $\pi(n)$. Applying formula (9.1.30), we obtain

$$E[D] = \sum_{i=1}^{n} E[D \mid M = i]\, P(M = i). \tag{9.6.3}$$

The assumption of a uniform distribution on the input space implies that the maximum element is equally likely to occur in any position. Thus, we have

$$P(M = i) = \frac{1}{n}, \quad i = 1, \dots, n. \tag{9.6.4}$$

In analogy with the notation $A(n)$ used for $E[\tau]$, we use the notation $\alpha(n)$ for $E[D]$ to facilitate the expression of a recurrence relation for $E[D]$. Clearly, we have

$$E[D \mid M = n] = \alpha(n - 1) + 1, \tag{9.6.5}$$

since permutations of $\{1, \dots, n\}$ with $\pi(n) = n$ are in one-to-one correspondence with permutations of $\{1, \dots, n - 1\}$, and the maximum is updated for such permutations π exactly one more time than it was updated on $\pi(1), \dots, \pi(n - 1)$. On the other hand, for permutations π such that $\pi(n) = i \neq n$, we have

$$E[D \mid M = i] = \alpha(n - 1), \quad i \in \{1, \dots, n - 1\}, \tag{9.6.6}$$

since the maximum is not updated on $\pi(n)$, and (for a fixed i) such permutations are again in one-to-one correspondence with permutations of $\{1, \dots, n - 1\}$. Hence, combining (9.6.3)–(9.6.6) we have

$$\begin{aligned}\alpha(n) &= \left(\frac{1}{n}\right)(\alpha(n - 1) + 1) + ((n - 1)/n)\alpha(n - 1) \\ &= \alpha(n - 1) + 1/n, \quad \textbf{init. cond. } \alpha(1) = 0.\end{aligned} \tag{9.6.7}$$

Recurrence relation (9.6.7) unwinds directly to yield

$$\alpha(n) = \frac{1}{2} + \frac{1}{3} + \cdots + \frac{1}{n} = H(n) - 1, \tag{9.6.8}$$

where $H(n)$ is the harmonic series. Combining (9.6.2) and (9.6.8), we obtain the following formula for the average complexity $A(n)$ of *MaxMin2*:

$$A(n) = 2n - 2 - \alpha(n) = 2n - H(n) - 1. \tag{9.6.9}$$

Since $H(n)$ is approximately equal to $\ln n$, $A(n)$ is approximately equal to $2n - \ln n - 1$. In particular, $A(n) \sim W(n)$ for *MaxMin2*.

9.7 Average Complexity of *BinarySearch* and *BinSrchTreeSearch*

To compute the average complexity A(n) of the algorithm *BinarySearch* given in Chapter 3, we choose the 3-branch comparison of the **do case** statement as the basic operation. Let p denote the probability that the search element X is on the list. Given that X is on the list $L[1:n]$, we assume that it is equally likely to occur in any of the n positions. Given that the search element is not on the list, we assume that it is equally likely to occur in any of the following $n + 1$ intervals

$$X < L[1],\ L[1] < X < L[2],\ \ldots,\ L[n-1] < X < L[n],\ X > L[n].$$

In Figure 9.9, we show the implicit search tree T for *BinarySearch* with an input list $L[1:8]$ of size 9. Note that T is a 2-tree having n internal nodes (squares) corresponding to the elements of $L[1:n]$ and $n + 1$ leaf nodes (triangles) corresponding to these $n + 1$ intervals. Thus, the probability that X occurs on the list and is equal to any given element $L[i]$ is p/n, $i = 1, \ldots, n$, and the probability that X occurs in any one of the $n + 1$ intervals is $(1 - p)/(n + 1)$.

Given a search element X, *BinarySearch* follows a path in the tree from the root to a node of T corresponding to X, or to a leaf node if X is not in the list. Thus, the average complexity A(n) is given by

$$A(n) = \left(\frac{p}{n}\right)(IPL(T) + n) + \left(\frac{1-p}{n+1}\right)LPL(T). \tag{9.7.1}$$

Employing Proposition 7.3.6 and substituting $LPL(T) - 2n$ for $IPL(T)$ in (9.7.1), we obtain

Figure 9.9 Implicit search tree for *BinarySearch* with an input list $L[1:8]$ of size 9. The index of the list element that is compared to the search element is shown inside each node. Leaf nodes correspond to unsuccessful searches.

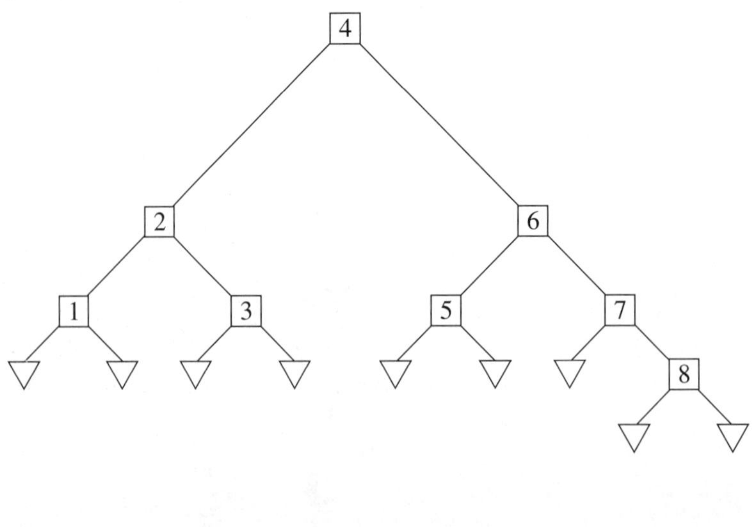

$$A(n) = \left(\frac{p}{n}\right)(LPL(T) - n) + \left(\frac{1-p}{n+1}\right)LPL(T) \tag{9.7.2}$$

$$= \left(\frac{p}{n} + \frac{1-p}{n+1}\right)LPL(T) - p.$$

The implicit search tree for *BinarySearch* is a 2-tree that is full at the second-deepest level (see Exercise 9.23). Hence, by Theorem 7.3.7

$$LPL(T) = \lfloor L \log_2 L \rfloor + 2(L - 2^{\lfloor \log_2 L \rfloor}) \geq \lceil L \log_2 L \rceil = \lceil (n+1)\log_2(n+1) \rceil. \tag{9.7.3}$$

Substituting the expression for $LPL(T)$ given in (9.7.3) into (9.7.2), we obtain

$$A(n) \geq \left(\frac{p}{n} + \frac{1-p}{n+1}\right)\lceil (n+1)\log_2(n+1) \rceil - p. \tag{9.7.4}$$

The lower bound estimate for $A(n)$ in (9.7.4) looks somewhat formidable, but it is easily seen to be asymptotic to the worst-case complexity; that is, $A(n) \sim W(n) = \lceil \log_2(n+1) \rceil$. Furthermore, suppose we restrict attention to successful search, so that $p = 1$ in (9.7.4). Then we have $A(n) \geq (1/n)\lceil (n+1)\log_2(n+1) \rceil - 1$. In other words, for large n the average complexity of *BinarySearch* for successful searching is no more than one less than its worst-case complexity.

Another algorithm whose average behavior can be calculated using leaf path length is *BinSrchTreeSearch* given in Chapter 2. The average complexity $A(n)$ of *BinSrchTreeSearch* depends on the probability distribution on the set of input trees, as well as probabilities associated with the identifiers. The analysis of the average complexity $A(n)$ of *BinSrchTreeSearch* over the set of all binary search trees is difficult. In any case, it is probably more interesting to analyze the average behavior $A(T,n)$ of *BinSrchTreeSearch* for a given fixed tree T and a given set of probabilities associated with the identifiers.

Suppose, then, that T is a fixed input search tree (pointed at by *Root*), whose internal nodes correspond to a fixed set of n identifiers (keys) $K_1, K_2, \ldots, K_n$. Note that T contains $n + 1$ (implicit) leaf nodes, corresponding to the $n + 1$ intervals

$$I_0\colon X < K_1,\ I_1\colon K_1 < X < K_2,\ \ldots,\ I_{n-1}\colon K_{n-1} < X < K_n,\ I_n\colon X > K_n.$$

Given T, the average behavior $A(T,n)$ of *BinSrchTreeSearch* (either version) depends on the probabilities $p_1, p_2, \ldots, p_n$ assigned to the internal nodes, and the probabilities $q_0, q_1, \ldots, q_n$ assigned to the (implicit) leaf nodes of the search tree. Assume now that all the p_i's are equal, and all the q_i's are equal. If p denotes the probability that X is one of the n keys $K_1, \ldots, K_n$, then $p_i = p/n$ for all $i \in \{1, \ldots, n\}$ and $q_j = (1-p)/(n+1)$ for all $j \in \{1, \ldots, n\}$. Under the latter assumptions, it can be shown that

$$A(T,n) = LPL(T)\frac{(1 + p/n)}{n+1} - p. \tag{9.7.5}$$

Applying Theorem 7.3.7 to (9.7.5) yields the following theorem.

THEOREM 9.7.1 Let T be any binary search tree (with external implicit leaf nodes added) for keys $K_1, \ldots, K_n$, where $p_i = p/n$ for each $i \in \{1, \ldots, n\}$, and $q_j = (1-p)/(n+1)$ for each $j \in \{0, \ldots, n\}$. Then,

$$A(T,n) \geq [(n+1)\lfloor \log_2(n+1) \rfloor + 2(n+1-2^{\lfloor \log_2(n+1) \rfloor})]\left(1+\frac{p}{n}\right) - p$$
$$\geq \lceil (n+1)\log_2(n+1) \rceil \left(1+\frac{p}{n}\right) - p \in \Omega(\log n). \quad (9.7.6)$$

Moreover, the first inequality in (9.7.6) is an equality if, and only if, T is a 2-tree that is full at the second-deepest level. ■

An important and natural problem is to determine an optimal search tree T in the sense that T minimizes $A(T,n)$ over all binary search trees containing n identifiers. Theorem 9.7.1 completely characterizes optimal binary search trees for the case where $p_i = p/n$ for each $i \in \{1, \ldots, n\}$ and $q_j = (1-p)/(n+1)$ for each $j \in \{0, \ldots, n\}$. That is, a binary search tree T is optimal for these probabilities if, and only if, it is full at the second-deepest level. The problem of finding an optimal search tree for general probabilities p_i and q_i is more difficult, and will be solved using the technique of dynamic programming in Chapter 14.

Remark Note that a binary search tree without the (implicit) external leaf nodes added is full at the second-deepest level if, and only if, the associated 2-tree with the external leaf nodes added is full at the second-deepest level. Thus, Theorem 9.7.1 could be restated for binary search trees without the external leaf nodes added.

9.8 Searching a Link-Ordered List

Suppose we have a list $L[1:n]$ of records having multiple fields, and we wish to maintain a sorting of L with respect to more than one of these fields (keys). Maintaining multiple sortings is usually done using auxiliary tag arrays or link arrays. For a given key field, a tag array $Tag[1:n]$ has the property that the sorting of L is given by

$$L[Tag[1]], L[Tag[2]], \ldots, L[Tag[n]].$$

In the linked list implementation, we assume that a variable *Head* contains the index of the smallest element in L with respect to the given key field. Then, the link array $Link[1:n]$ has the property that the sorting of L is given by

$$L[Head], L[Link[Head]], L[Link[Link[Head]]], \ldots, L[Link^{n-1}[Head]],$$

where $Link^m$ denotes the m-fold composition of *Link* ($Link^0$ is the identity function). $Link[i] = 0$ indicates the end of the list. In Figure 9.10 we show a sample list $L[1:12]$ and its associated array $Link[1:12]$.

We now consider the problem of searching for an occurrence of a given element in L having a given key value X. The tag array $Tag[1:n]$ facilitates efficient searching using a binary search. Using the link array $Link[1:n]$, we cannot efficiently use a binary search, since there is no way to directly determine the middle list element. Accessing the middle element $L[Link^{(n/2)-1}(Head)]$ requires $(n/2) - 1$ accesses of the array *Link*. In fact, it can be shown that *any* algorithm for searching L requires n accesses to the array *Link* in the worst case. We now describe an algorithm *ProbeLinSrch* for searching L whose average complexity belongs to $O(\sqrt{n})$.

i	1	2	3	4	5	6	7	8	9	10	11	12
$L[i]$	17	11	44	23	7	4	14	55	21	56	9	15
$Link[i]$	9	12	5	8	11	5	1	10	8	0	2	1

$Head = 6$

Figure 9.10 A list $L[1:12]$ maintained in order using an auxiliary array $Link[1:12]$

The algorithm *ProbeLinSrch* utilizes an algorithm *LinkOrdLinSrch* for searching a link-ordered list. *LinkOrdLinSrch* is a variant of the straightforward analog of linear search for link-ordered lists. *LinkOrdLinSrch* is based on the fact that if we utilize the array *Link* to make a search for the occurrence of X in L in sorted order, then we can terminate the search whenever we compare X to a list element that is larger than X. *LinkOrdLinSrch* returns the index of an occurrence of X in the list L, or returns 0 if X is not in list L. For convenience, in the following discussion we identify the record $L[i]$ with its key field.

```
function LinkOrdLinSrch(L[1:n],Link[1:n],Head,X)
Input:  L[1:n] (array of list elements)
        Link[1:n] (array pointing to sorted order of L[1:n])
        Head (index of smallest element in L)
        X (search element)
Output: Returns index of occurrence of X in L, or 0 if X is not in L

   i := Head
   while X > L[i] do
      i := Link[i]
      if i = 0 then return(0) endif
   endwhile
   if X = L[i] then
      return(i)
   else
      return (0)
   endif

end LinkOrdLinSrch
```

To analyze *LinkOrdLinSrch*, note that the number of updates for the index variable i in *LinkOrdLinSrch* is at most one less than the number of comparisons of X with a list element in L. Thus, we choose comparisons of X to list elements as our basic operation. Clearly, the best-case and worst-case complexities of *LinkOrdLinSrch* are $B(n) = 2$ and $W(n) = n + 1$, respectively.

To compute the average complexity of *LinkOrdLinSrch*, we make the usual assumption that each list element is distinct. We also assume that each element is equally likely to be X and that X falls into each of the $n + 1$ "gaps" : $X < L[Head]$,

$L[Head] < X < L[Link[Head]], \ldots, L[Link^{n-2}[Head]] < X < L[Link^{n-1}[Head]], X > L[Link^{n-1}[Head]]$ with equal probability. Let $A_s(n)$ and $A_u(n)$ denote the expected number of comparisons for successful and unsuccessful search, respectively. Then, if p denotes the probability that X occurs on the list, we have

$$A(n) = pA_s(n) + (1 - p)A_u(n). \tag{9.8.1}$$

Formula (9.8.1) can be derived directly from the definition of conditional expectation and is actually a special case of (9.1.29).

To compute $A_s(n)$, note that our assumptions imply that for each i between 1 and n, the probability that X occurs in position i is $1/n$, given that X is in the list. Note also that $i + 1$ comparisons of X to a list element are performed when $X = L[Link^{i-1}[Head]]$, $1 \le i \le n$. Hence,

$$\begin{aligned} A_s(n) &= \frac{2 + 3 + \cdots + n + n + 1}{n} = \frac{1 + 2 + \cdots + n}{n} + 1 \\ &= \frac{n+1}{2} + 1. \end{aligned} \tag{9.8.2}$$

To compute $A_u(n)$, note that our assumptions imply that for each i between 1 and $n + 1$, the probability that X occurs in the ith gap is $1/(n + 1)$, given that X is not in the list. Note also that $i + 1$ comparisons of X to a list element are performed when X is in the ith gap, $1 \le i \le n$, and that n comparisons are made when $X > L[Link^{n-1}[Head]]$. Hence,

$$\begin{aligned} A_u(n) &= \frac{2 + 3 + \cdots + n + 1}{n + 1} + 1 - \frac{1}{n+1} \\ &= \frac{n+2}{2} + 1 - \frac{2}{n+1}. \end{aligned} \tag{9.8.3}$$

Using (9.8.2) and (9.8.3), we have $A_s(n) \sim n/2$ and $A_u(n) \sim n/2$. Hence, it follows from (9.8.1) that $A(n) \sim n/2$.

ProbeLinSrch is a variant of *LinkOrdLinSrch* in which we first scan the list elements $L[1], \ldots, L[\lfloor \sqrt{n} \rfloor]$ for a better starting point than *Head*. Note that the index i is a feasible starting point for our search if $X \ge L[i]$. If there is no feasible starting point in index positions $1, \ldots, \lfloor \sqrt{n} \rfloor$, then *ProbeLinSrch* invokes *LinkOrdLinSrch* with *Probe* = *Head*. Otherwise, *ProbeLinSrch* invokes *LinkOrdLinSrch* with *Probe* equal to the best feasible starting point found.

To illustrate, consider the lists $L[1:12]$ and $Link[1:12]$ shown in Figure 9.10. Given a search element X, we first determine the best feasible starting point *Probe* among $L[1]$, $L[2]$, $L[3] = L[\lfloor \sqrt{12} \rfloor]$. For example, if X = 23,3,46, then *Probe* = 1,6,3, respectively.

function *ProbeLinSrch*($L[1:n]$, $Link[1:n]$, *Head*, X)

Input: $L[1:n]$ (an array of list elements)
$Link[1:n]$ (an array pointing to sorted order of $L[1:n]$)
Head (index of smallest element in L)
X (search element)

```
Output: returns index of occurrence of X in L, or 0 if X is not in L

    Probe := Head
    Max := L[Head]
    {look for the best feasible starting point (if any) between 1 and ⌊√n⌋}
    for i = 1 to ⌊√n⌋ do
      Temp := L[i]
      if Max < Temp .and. Temp ≤ X then
         Probe := i
         Max := Temp
      endif
    endfor
    return(LinkOrdLinSrch(L[1:n],Link[1:n],Probe,X)
end ProbeLinSrch
```

Clearly, the best-case and worst-case complexities of *ProbeLinSrch* are given by $B(n) = \lfloor \sqrt{n} \rfloor + 2$ and $W(n) = n + 1 + \lfloor \sqrt{n} \rfloor$, respectively. To determine the average complexity $A(n)$, it is useful to consider a generalization of *ProbeLinSrch* in which the first k list elements $L[1], \ldots, L[k]$ are probed for a feasible starting point. In the generalized version, k is an additional input parameter to *ProbeLinSrch*, and the ordinary version of *ProbeLinSrch* corresponds to the special case where $k = \lfloor \sqrt{n} \rfloor$. In the following discussion, we continue to refer to the generalized version as simply *ProbeLinSrch*.

We assume that the list elements are distinct and that the search element X is on the list. Further, we assume that X is equally likely to be any of the list elements. Let $x_i = L[Link^{i-1}[Head]]$ denote the ith-smallest list element, $i = 1, \ldots, n$. Let Y denote the random variable that maps the input $(L[1{:}n],X)$ onto m where $X = x_m$. By (9.2.6) we have

$$A(n) = \sum_{m=1}^{n} E[\tau \mid Y = m]P(Y = m) = \frac{1}{n}\sum_{m=1}^{n} E[\tau \mid Y = m]. \tag{9.8.4}$$

To compute $E[\tau \mid Y = m]$, note that *ProbeLinSrch* has three stages: Stage 1 consists of the **for** loop that determines the probe element, stage 2 consists of the **while** loop in *LinkOrdLinSrch*, and stage 3 consists of the final comparison made by *LinkOrdLinSrch*. For $m \in \{1, \ldots, n\}$, Let β_m denote the random variable that maps an input X such that $X = x_m$ onto the number of comparisons performed by *LinkOrdLinSrch* during stage 2 with input X. Then we have

$$E[\tau \mid Y = m] = k + 1 + E[\beta_m], \quad m = 1, \ldots, n. \tag{9.8.5}$$

Given any $m \in \{1, \ldots ,n\}$, we now calculate $E[\beta_m]$ using (9.2.4). Let $q_i = q_i(m)$ denote the probability that *LinkOrdLinSrch* performs at least i comparisons during the second stage. Note that $q_i = 0$ when $i > m$, since the worst-case number of comparisons m occurs when a better start than *Head* was not found among $L[1], \ldots, L[k]$ during stage 1. For $2 \le i \le m$, *LinkOrdLinSrch* performs less than i comparisons precisely when one of the elements $x_{m-i+2}, x_{m-i+3}, \ldots, x_m$ belongs to $L[1], \ldots, L[k]$. Thus, the probability q_i that at least i comparisons are performed is the probability that each of the list elements $L[1], \ldots, L[k]$ belongs to the set of $n - i + 1$ complementary elements to $x_{m-i+2}, x_{m-i+3}, \ldots, x_m$. Since the total number of ways to choose such a sequence $L[1], \ldots, L[k]$ is $(n - i + 1)^{(k)}$ and the total number of ways to choose a sequence $L[1], \ldots, L[k]$ from $x_1, x_2, \ldots, x_n$ is $n^{(k)}$ (where $x^{(k)} = x(x-1)\ldots(x-k+1)$), we have

$$q_i = \frac{(n-i+1)^{(k)}}{n^{(k)}} \le \frac{(n-i+1)^k}{n^k}. \tag{9.8.6}$$

The inequality in (9.8.6) follows from (1.2.15). Using (9.2.4), we have

$$\begin{aligned} E[\beta_m] &= \sum_{i=1}^{m} q_i \le \sum_{i=1}^{m} \frac{(n-i+1)^k}{n^k} \\ &= \left(\frac{1}{n^k}\right) \sum_{i=n-m+1}^{n} i^k \\ &= \left(\frac{1}{n^k}\right)\left[\left(\sum_{i=1}^{n} i^k\right) - \left(\sum_{i=1}^{n-m} i^k\right)\right] \\ &\le \left(\frac{1}{n^k}\right)\left(\sum_{i=1}^{n} i^k\right), \quad m = 1, \ldots, n. \end{aligned} \tag{9.8.7}$$

It was shown in Chapter 7 that $S(n,k) = \sum_{i=1}^{n} i^k$ is a polynomial in n of degree $k + 1$ with leading coefficient $1/(k + 1)$. Hence, (9.8.7) implies

$$E[\beta_m] \le \frac{n}{k+1} + O(1), \quad m = 1, \ldots, n. \tag{9.8.8}$$

Substituting (9.8.8) in (9.8.5), we have

$$E[\tau \mid Y = m] = k + 1 + E[\beta_m] \le k + \frac{n}{k+1} + O(1), \quad m = 1, \ldots, n. \tag{9.8.9}$$

Since the upper bound estimate $k + n/(k + 1) + O(1)$ for $E[\tau \mid Y = m]$ given in (9.8.9) is independent of m, substituting the estimate into (9.8.4) yields

$$A(n) \le \frac{n}{k+1} + k + O(1). \tag{9.8.10}$$

Using calculus, it is easily verified that $n/(k + 1) + k$ achieves a minimum value of $2\sqrt{n} - 1$ at the point $k = \sqrt{n} - 1$, which is approximately $\lfloor \sqrt{n} \rfloor$. Hence, if we use $k = \lfloor \sqrt{n} \rfloor$ in *ProbeLinSrch*, from (9.8.10) we have

$$A(n) \le 2\sqrt{n} + O(1). \tag{9.8.11}$$

Since $A(n) \ge B(n) = \lfloor \sqrt{n} \rfloor + 1$, (9.8.11) implies that $A(n) \in \Theta(\sqrt{n})$.

Remark Choosing a value of k that minimizes $n/(k + 1) + k$ does not automatically guarantee that this k minimizes $A(n)$, since (9.8.10) is an inequality rather than an equality. However, it does give us some intuition as to why $\lfloor \sqrt{n} \rfloor$ is the choice for k used in the design of *ProbeLinSrch*. Moreover, our estimate of $A(n)$ is pretty sharp, since $\sqrt{n} < A(n) < 2\sqrt{n} + O(1)$.

9.9 Closing Remarks

Probability theory is an important basic tool in the analysis of algorithms. The probabilistic analysis of algorithms is a very active area of research today. Another very active research area involves introducing probabilistic techniques into the program logic of the algorithm itself. Algorithms that are not completely deterministic, but rather use random choices as part of the program logic, are called *probabilistic* algorithms (see Chapter 17). These algorithms use probabilistic techniques as a *design strategy* as opposed to merely a tool to analyze the algorithm.

References and Suggestions for Further Reading

Corman, T., C. Leiserson, and R. Rivest. *Introduction to Algorithms*. Cambridge, MA: MIT Press/McGraw-Hill, 1990. A more extended treatment of elementary probability theory as applied to algorithm analysis.

For a more advanced treatment of probabilistic algorithm analysis, see:

Coffman, E. G. Jr. and G. Lueker. *Probabilistic Analysis of Packing and Partitioning Algorithms*. New York: Wiley, 1991.

Hofri, M. *Probabilistic Analysis of Algorithms*. New York: Springer-Verlag, 1987.

Exercises

Section 9.1 Elementary Probability Theory

9.1 Consider the sample space S corresponding to rolling two dice; that is, $S = \{(r_1,r_2) \mid r_1,r_2 \in \{1, \ldots , 6\}\}$. Assume that the first die is fair but the second die is loaded, with probabilities 1/10, 1/10, 1/10, 1/10, 1/10, 1/2 of rolling a 1,2,3,4,5,6, respectively.

a. Give a table showing the probability distribution for rolling these dice (the sample space is shown in Figure 9.1).

b. Compute the probability that at least one of the dice comes up 6.

c. Compute the conditional probability that the sum of the dice is 10 given that the loaded die does not come up 4.

9.2 Consider the random variable $X = r_1 + r_2$ defined on the sample space S given in Exercise 9.1.

a. Compute the density function $f(x) = P(X = x)$ and verify that it is a probability distribution on the sample $S_X = \{2, \dots, 12\}$.

b. Calculate the expectation $E[X]$.

9.3 Repeat Exercise 9.2 for the random variable $X = \max\{r_1, r_2\}$.

9.4 Let S be the sample space consisting of the positive integers. For a fixed p, $0 \le p \le 1$, show that the function $P(i) = (1 - p)^{i-1}p$ is a probability distribution on S.

9.5 Consider the sample space S corresponding to rolling three fair dice; that is, $S = \{(r_1, r_2, r_3) \mid r_1, r_2, r_3 \in \{1, \dots, 6\}\}$. Calculate the expectation $E[X]$ for each of the following random variables X.

a. $X = r_1 + r_2 + r_3$

b. $X = r_1 + r_2$

c. $X = \max\{r_1, r_2, r_3\}$

9.6 Give an alternate derivation of the expectation of a binomial distribution using Proposition 9.1.4. (*Hint:* Let X_i be the random variable that has the value 1 if there was a success in the ith trial, and 0 otherwise.)

9.7 Verify that the expectation of the geometric distribution with probability p of success is $1/p$.

9.8 Verify the following formula, which was used in the derivation of the variance of the geometric distribution with probability p of success.

$$\sum_{k=1}^{\infty} k^2 (1-p)^{k-1} p = \frac{2-p}{p^2}$$

9.9 Calculate the variance of the random variables given in Exercises 9.2 and 9.3.

9.10 Calculate the variance of the random variables given in Exercise 9.5.

9.11 **a.** Two random variables X and Y defined on a sample space S are *independent* if the events $X = x$ and $Y = y$ are independent for every x,y. Show that if $X_1, X_2, \dots, X_n$ are independent random variables, then

$$V(X_1 + X_2 + \cdots + X_n) = V(X_1) + V(X_2) + \cdots + V(X_n).$$

b. Employing the formula of part (a), calculate the variance of the binomial distribution with p and n. (*Hint:* Let X_i be the random variable that has the value 1 if there was a success in the ith trial, and 0 otherwise.)

9.12 Prove Proposition 9.1.1.

9.13 Prove Propositions 9.1.3 and 9.1.4.

9.14 Verify that the probabilities given by (9.1.12) satisfy the three axioms for a probability distribution on $\{0, 1, \dots, n\}$.

9.15 Verify that P_F defined by (9.1.24) satisfies the three axioms for a probability distribution on F.

9.16 Given a (discrete) random variable X, show that the probability density function $f(x) = P(X = x)$ determines a probability distribution on $S_X = \{x : f(x) \neq 0\}$.

9.17 Show that formula (9.1.30) of Proposition 9.1.7 reduces to formula (9.1.15) when $X = Y$.

9.18 Prove the following generalization of (9.1.5) to n events $E_1, E_2, \ldots, E_n$:

$$P(E_1 \cap E_2 \cap \ldots \cap E_n) = P(E_1)P(E_2 \mid E_1)P(E_3 \mid E_1 \cap E_2) \ldots P(E_n \mid E_1 \cap E_2 \cap \ldots \cap E_{n-1}).$$

Section 9.3 Average Complexity of *LinearSearch*

9.19 Calculate the average case complexity A(n) of *LinearSearch* under the following assumptions about the input list $L[1:n]$ and search element X.

(i) The probability that X occurs in the list is 2/3.

(ii) Given that X occurs in the list, X is twice as likely to occur in the first half of the list (positions 1 through $\lfloor n/2 \rfloor$) as in the second half. Further, if X occurs in the first half of the list it is equally likely to occur in any position in the first half. A similar assumption is made about the second half.

Section 9.4 Average Complexity of *InsertionSort*

9.20 **a.** Give pseudocode for a recursive version *InsertionSortRec* of *InsertionSort*.

b. Obtain a recurrence relation for the average complexity A(n) of *InsertionSortRec*.

c. Solve the recurrence relation obtained in (b), and compare it to the formula for A(n) given in Section 9.4.

Section 9.5 Average Complexity of *QuickSort*

9.21 Suppose that we sort a list $L[1:n]$ by first determining an element of maximum value, placing it at position n, and then calling *QuickSort* with the (possibly altered) list $L[1:n-1]$. Analyze the average complexity of the resulting sorting algorithm, and compare it to *QuickSort*.

Section 9.6 and 9.7 Average Complexities of *MaxMin2* and *BinarySearch*

9.22 Show that permutations π: $\{1, \ldots, n\} \rightarrow \{1, \ldots, n\}$ such that $\pi(n) = i \neq n$ are in one-to-one correspondence with permutations of $\{1, \ldots, n-1\}$.

9.23 Show that the implicit search tree for *BinarySearch* is a 2-tree that is full at the second-deepest level.

9.24 Derive formula (9.7.5), which expresses A(T,n) in terms of the leaf path length of T, and use it to prove Theorem 9.7.1.

***9.25** Compute the average complexity A(n) of *BinSrchTreeSearch* over the set of all binary search trees T. Assume that each binary tree search T is equally likely to be the input tree to *BinSrchTreeSearch*. Also assume that X is one of the keys in the search tree and that X has an equal chance of being any of the n keys.

Section 9.8 Searching a Link-Ordered List

9.26 Show that the algorithm for searching a link-ordered list of size n that simulates binary search would perform n index updates in the worst case.

9.27 Generalize Exercise 9.26 by showing that *any* comparison-based algorithm for searching a link-ordered list of size n requires n index updates in the worst case.

9.28 Show that formula (9.8.1) can be derived directly from the definition of conditional expectation and is actually a special case of (9.1.29).

10

Introduction to Lower Bound Theory

Given a problem and a specific algorithm for solving the problem, it is important to determine how close the algorithm comes to exhibiting optimal worst-case or average complexity for the *problem itself*. But how do we determine what optimal behavior *is* for the problem? Of course, the complexity of *any* algorithm that correctly solves the problem gives us *upper bounds* for the complexities of the problem. The object of this chapter is to discuss various techniques for determining good *lower bound* estimates for the worst-case and average complexities of problems such as comparison-based searching and sorting.

10.1 Basic Terminology and Techniques

In Chapter 3 we found sharp lower bound formulas for the worst-case complexity of the problem of adjacent-key comparison-based sorting and for the best-case, worst-case, and average complexities of the problem of finding the maximum value in a list. There are relatively few problems for which sharp lower bound formulas have been established. It is more realistic to search for lower bound formulas for the order of complexity of a problem. An algorithm whose complexity has the same order as a lower bound for the problem is called *order optimal*. Even establishing order optimality is rare and usually only can be done in the context of suitably restricting the problem or type of algorithm allowed.

We will discuss six basic techniques for determining lower bounds: simple counting arguments (based on a principle of minimal work), enumeration arguments, decision (comparison) tree arguments, adversary arguments, information theoretic arguments, and graph theoretic arguments.

10.1.1 Simple Counting Arguments

Simple counting arguments for establishing lower bounds are based on showing that any algorithm for solving a given problem must do a certain amount of minimal work to correctly output the solution for all inputs. This work is measured by associating a numeric quantity with the performance of a basic operation by the algorithm. For example, in

Chapter 3 we used a simple counting argument to show that a comparison-based algorithm for finding a maximum element in an arbitrary list of size n must do at least $n - 1$ comparisons in the best case. Recall that we associated the notion of a loss (one unit of work) with each comparison made by the algorithm, and we simply counted the number of total losses that must occur if the algorithm is to correctly determine the maximum element. We showed that at least $n - 1$ units of work are required for any list of size n.

10.1.2 Enumeration Arguments

Enumeration arguments, which are useful for establishing lower bounds for the average complexity of a problem, are usually more complicated than simple counting arguments. The enumeration method depends on identifying a combinatorial or numeric entity associated with inputs of size n to the problem. Establishing a lower bound for average behavior then depends on showing that the average complexity $A(n)$ of *any* algorithm solving the problem is bounded below by the average number of the combinatorial entities per input. We calculate this average by enumerating the total number of these entities over all inputs of size n and then dividing by the number of inputs (assumed to be finite). For example, consider the problem of sorting a list using an adjacent-key comparison-based algorithm, where for purposes of analysis we assume that the inputs of size n are all permutations of $\{1,2, \ldots , n\}$. Associated with each permutation is a combinatorial entity called an *inversion* of the permutation. We will show in the next section that the average number of inversions in a permutation on n symbols is given by $n(n - 1)/4$. Interchanging adjacent list elements (keys) removes at most one inversion, so that this latter number then serves as a lower bound for the problem of comparison-based adjacent-key sorting.

10.1.3 Decision Trees

A *decision tree* models the hierarchy of all possible branchings resulting from decisions made by an algorithm. The sequence of decisions and resultant branchings performed by the algorithm depends on the input and corresponds to a path from the root to a leaf in the decision tree. Suppose that the decision tree models the branching decisions made on the basis of performing basic operations such as comparing two elements in a list or multiplying two elements in a matrix. Then each internal node of the decision tree corresponds to the performance of one or more basic operations. Leaf nodes correspond to outputs of the algorithm (solutions to the problem). Nodes in the tree that are never reached by any input to the algorithm are pruned. Again, assuming that each internal node corresponds to the performance of at least one basic operation, then it is clear that the depth of the decision tree is a lower bound for the worst-case complexity of the algorithm. The decision tree method for obtaining lower bounds for the worst-case complexity of a problem consists of determining the minimum depth of a decision tree over all algorithms for solving the problem. Similarly, the decision tree method for obtaining lower bounds for the average complexity of a problem consists of determining the minimum leaf path length of a decision tree over all algorithms for solving the problem.

10.1.4 Adversary Arguments

Adversary arguments establish (worst-case) lower bounds by dynamically creating inputs to a given algorithm that force the algorithm to perform many basic operations. Adversary arguments work somewhat like playing a guessing game, where the decisions to be made by the algorithm are regarded as queries to an adversary, and the adversary answers these queries by providing as little information as possible. We only require that the adversary give answers that are consistent and truthful. Clever adversaries force any given algorithm to make many queries before the algorithm has enough information to output the solution. We assume that the algorithm uses a basic operation like comparing two list elements when querying the adversary. The adversary actually constructs a suitable input in response to the queries. Lower bounds for the worst-case complexity $W(n)$ are then obtained by counting the number of basic operations that the adversary forces the algorithm to perform.

10.1.5 Information Theoretic Arguments

Information theoretic arguments establish lower bounds by computing the limitations on information gained by a basic operation and then showing how much information is required before a given problem is solved. The familiar *Rule of Product* from elementary combinatorial theory provides the basis for many information theoretic arguments. This rule is applied as follows: Suppose that a basic operation provides b units of information. Then the upper bound for the cumulative amount of information that can be obtained from performing this basic operation n times in succession is b^n. For example, if the basic operation is a comparison, its outcome is either **.true.** or **.false.**, so that $b = 2$. Sorting algorithms must generate enough information to distinguish among the $n!$ permutations of a list of size n, which implies that the lower bound for the worst-case behavior of comparison-based sorting algorithms is $\lceil \log_2(n!) \rceil \in \Theta(n \log n)$ (see Section 10.5 for a more formal argument).

10.1.6 Graph Theoretic Arguments

Graph theoretic arguments are often used to establish lower bounds for parallel algorithms. For example, on interconnection network models the diameter and the bisection width of the network give lower bounds on the communication complexity (see Chapter 5). As another example, in Section 10. 6 we will obtain a lower bound for finding the maximum element in a list on the CRCW PRAM using Turan's theorem.

10.2 Lower Bound for the Average Complexity of Adjacent-Key Comparison-Based Sorting Algorithms

The worst-case complexity of the problem of sorting a list of size n using an adjacent-key comparison-based algorithm is $n(n - 1)/2$. Since *InsertionSort* performs $n(n - 1)/2$ comparisons in the worst case, it exhibits optimal worst-case performance. Hence, the worst-case complexity of the problem of adjacent-key comparison-based sorting is exactly determined. Proposition 10.3.1 states that a lower bound for the average complexity of the problem of sorting a list of size n using an adjacent-key comparison-based sorting algorithm is $n(n-1)/4$. Thus, *InsertionSort* is also order optimal for the average complexity of this problem.

PROPOSITION 10.2.1 A lower bound for the average complexity $A(n)$ of any adjacent-key comparison-based sorting algorithm is $n(n-1)/4$.

Proof When analyzing the complexity of a comparison-based sorting algorithm, it is sufficient to consider inputs of size n to be permutations of $N_n = \{1,2, \ldots, n\}$. Recall that an inversion of a permutation π of N_n is a pair $(\pi(a),\pi(b))$, $a, b \in N_n$, where $a < b$ but $\pi(a) > \pi(b)$. A comparison of list elements utilized by any adjacent-key comparison-based sorting algorithm by definition only considers inversions of the form $(\pi(i), \pi(i+1))$. Moreover, removing any such inversion (by interchanging $\pi(i)$ and $\pi(i+1)$) results in a decrease of (exactly) one in the total number of inversions. For any input permutation π, any sorting algorithm must ultimately remove all the inversions in π, so that the average complexity of any adjacent-key comparison sorting algorithm is bounded below by the average number of inversions $\iota(n)$ in a permutation of N_n.

We now give a quick proof of the formula $\iota(n) = n(n-1)/4$, from which Proposition 10.2.1 follows. Given any permutation π of N, its *reverse* permutation π_{rev} is defined by $\pi_{\text{rev}}(i) = \pi(n-i+1)$, $i \in N_n$. Now note that given any $x, y \in N_n$, where $x > y$, the pair (x,y) occurs as an inversion in *exactly one* of the permutations π,π_{rev} (see Figure 10.1b). Thus the permutations π,π_{rev} have a total of exactly $n(n-1)/2$ inversions between them. We can sum the number of inversions over all permutations of N_n by summing over each (unordered) pair $\{\pi,\pi_{\text{rev}}\}$, thereby obtaining $(n!/2)(n(n-1)/2)$ inversions altogether. The average is then obtained by dividing this latter number by $n!$, yielding $\iota(n) = n(n-1)/4$. ■

Figure 10.1 Permutations of {1,2,3}

$$\begin{bmatrix}1&2&3\\1&2&3\end{bmatrix} \quad \begin{bmatrix}1&2&3\\1&3&2\end{bmatrix} \quad \begin{bmatrix}1&2&3\\2&1&3\end{bmatrix} \quad \begin{bmatrix}1&2&3\\2&3&1\end{bmatrix} \quad \begin{bmatrix}1&2&3\\3&1&2\end{bmatrix} \quad \begin{bmatrix}1&2&3\\3&2&1\end{bmatrix}$$

none (3,2) (2,1) (2,1),(3,1) (3,1),(3,2) (3,2),(3,1),(2,1)

(a) All permutations of {1,2,3} and their inversions

$$\left\{\begin{bmatrix}1&2&3\\1&2&3\end{bmatrix}, \begin{bmatrix}1&2&3\\3&2&1\end{bmatrix}\right\} \quad \left\{\begin{bmatrix}1&2&3\\1&3&2\end{bmatrix}, \begin{bmatrix}1&2&3\\2&3&1\end{bmatrix}\right\} \quad \left\{\begin{bmatrix}1&2&3\\2&1&3\end{bmatrix}, \begin{bmatrix}1&2&3\\3&1&2\end{bmatrix}\right\}$$

none (3,2),(3,1),(2,1) (3,2) (2,1),(3,1) (2,1) (3,1),(3,2)

(b) Partition of the permutations of {1,2,3} into pairs $\{\pi,\pi_{\text{rev}}\}$, with each pair $(\pi,\pi_{\text{rev}}\}$ yielding a full set of inversions

In Chapter 9 we showed that the average complexity A(n) of *InsertionSort* is given by A(n) = $n^2/4 - 3n/4 - 1 \sim n^2/4$. Thus, by Proposition 10.2.1, *InsertionSort* has order optimal average complexity.

10.3 Decision Trees and Comparison Trees

Decision trees are an important and commonly used method of establishing lower bounds. Decision trees modeling a comparison-based algorithm are referred to as *comparison* trees. An internal node in a comparison tree corresponds to a comparison made by the algorithm using one of the comparison operators <, =, >. For convenience in pseudocode or drawing the comparison tree, we also might use one of the respectively equivalent comparison operators ≥, ≠, ≤. Leaf (external) nodes correspond to outputs of the algorithm. We assume that any nodes in the comparison tree that are never reached by any input to the algorithm are pruned. Each internal node in a (possibly pruned) comparison tree has at most two children, depending on the outcome of a comparison. By convention, we assume that the left child corresponds to the action performed by the algorithm if the comparison is **.true.**. This action is either the next comparison made (internal node), or an output (leaf node). In particular, comparison trees are binary trees, whereas nodes in general decision trees in general may have more than two children. For example, a node corresponding to a **do case** decision may have as many children as the number of cases specified (see Figure 10.3a).

10.3.1 Examples

Figure 10.2 (page 358) shows the comparison tree for *LinearSearch* with $n = 4$. Inside each internal node is the position in the list of the element being compared to the search element.

Figure 10.3a (page 359) shows the *decision* tree that models *BinarySearch* with $n = 5$, whereas Figure 10.3b shows the *comparison* tree that models the same thing. The decision tree in Figure 10.3a is *not* a comparison tree, since a 3-way decision is made based on the **do case** statement in *BinarySearch.* Since we are basing our analysis of the complexity of comparison-based algorithms on 2-way decisions resulting from comparisons, we must use the comparison tree given in Figure 10.3b, rather than the decision tree given in Figure 10.3a. In Figures 10.3a and 10.3b, the current values of *low* and *high* are shown outside each internal node. Inside each internal node is shown the current value of *mid.*

Figure 10.4 (page 360) shows the comparison tree for the procedure *BinarySearch2* introduced in Chapter 7. In Figure 10.4, the current values of *low* and *high*, respectively, are listed outside each internal node, and the current value of *mid* is shown inside each internal node whose children are not leaves. Leaf (output) nodes show the value returned based on the comparison of X to $L[low]$, where *low* is the current value corresponding to the parent node.

Clearly, for any search element X, the action of *BinarySearch2* corresponds to following a path in its comparison tree T from the root to a leaf node. Thus, the

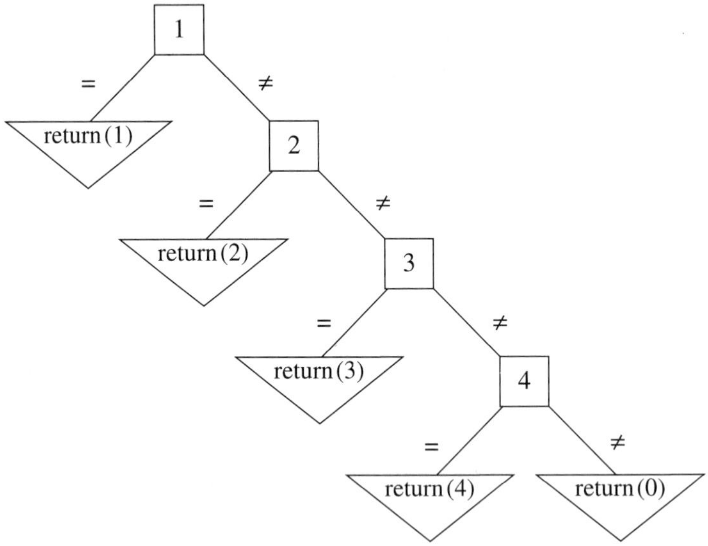

Figure 10.2 Decision tree for *LinearSearch* with $n = 4$

worst-case complexity of *BinarySearch2* equals the depth $D(T)$ of T. Also, the comparison tree T has $2n$ leaf nodes and is a 2-tree that is full at the second-deepest level (see Exercise 10.5). Thus, it follows from Proposition 7.3.5 that

$$W(n) = D(T) = \lceil \log_2 2n \rceil = \lceil \log_2 n \rceil + 1. \tag{10.3.1}$$

Proposition 7.3.4 also implies that all the leaves of T are at levels $D(T) - 1$ and $D(T)$, so that the best-case complexity of *BinarySearch2* is at most one less than its worst-case complexity. Hence, it follows from (10.3.1) that

$$\lceil \log_2 n \rceil \leq B(n) \leq A(n) \leq W(n) = \lceil \log_2 n \rceil + 1. \tag{10.3.2}$$

10.3.2 Lower Bound for the Worst-Case Complexity of Comparison-Based Searching of Lists

Consider now the complexity of the *problem* of searching for the occurrence of a search element X in a list $L[1:n]$. For arbitrary lists $L[1:n]$, the worst-case complexity of the *problem* of searching for X in L using a comparison-based algorithm is n, since X must be compared to *every* list element for a list not containing X.

We now consider the problem of searching an ordered list of size n. A comparison tree T modeling a comparison-based algorithm for this problem must have at least $n + 1$ leaf nodes. Indeed, T must have n leaf nodes corresponding to outputting the n possible occurrences of the search element in the list, and T must also have at least one leaf node outputting 0 when the search element is not in the list. By Proposition 7.2.3, it follows that T has depth at least $\lceil \log_2(n + 1) \rceil$. This establishes the lower bound $\lceil \log_2(n + 1) \rceil$ for comparison-based searching of an ordered list of size n, which we state as a theorem.

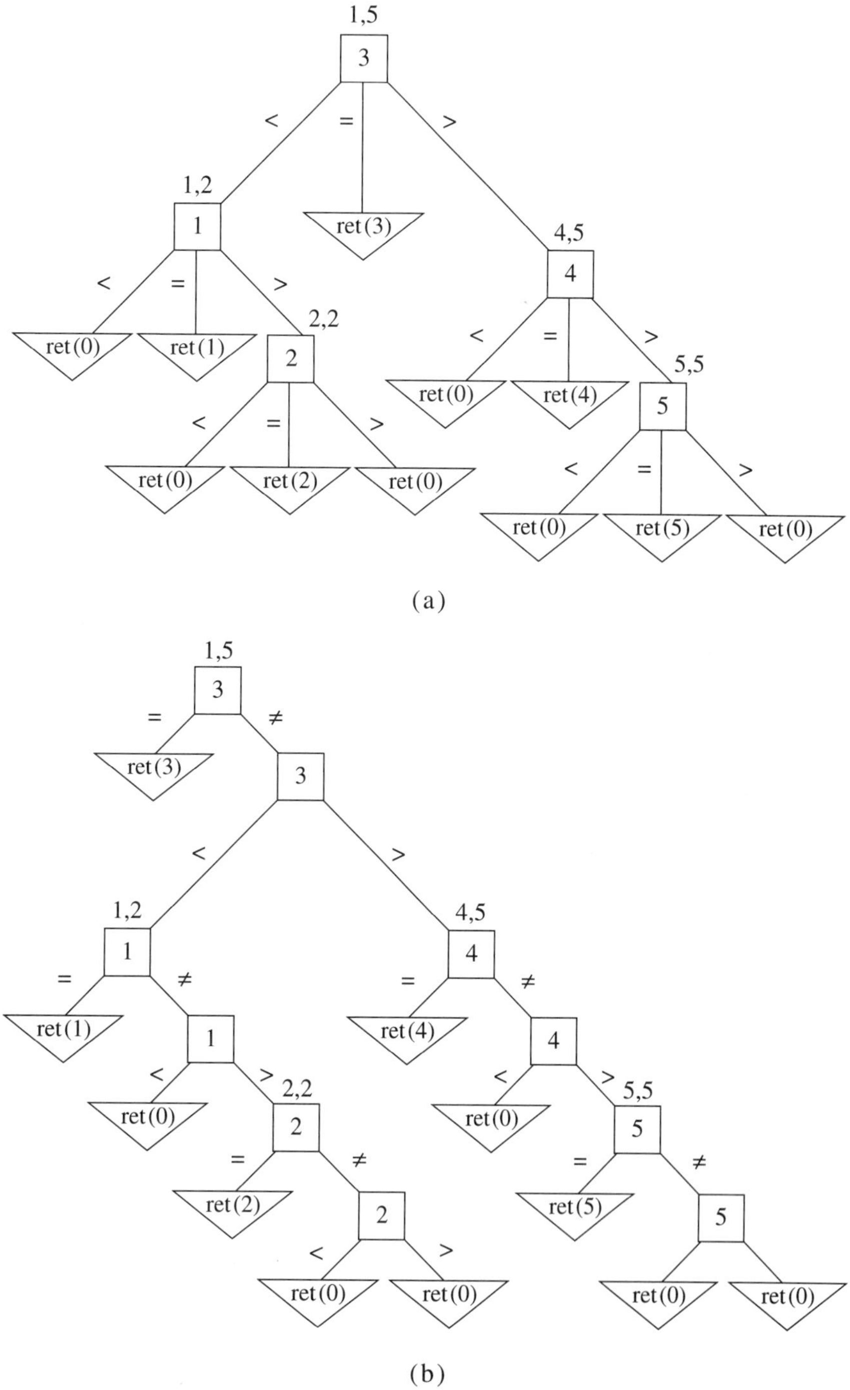

Figure 10.3 (a) The decision tree for *BinarySearch* when $n = 5$; (b) the comparison tree for *BinarySearch* when $n = 5$

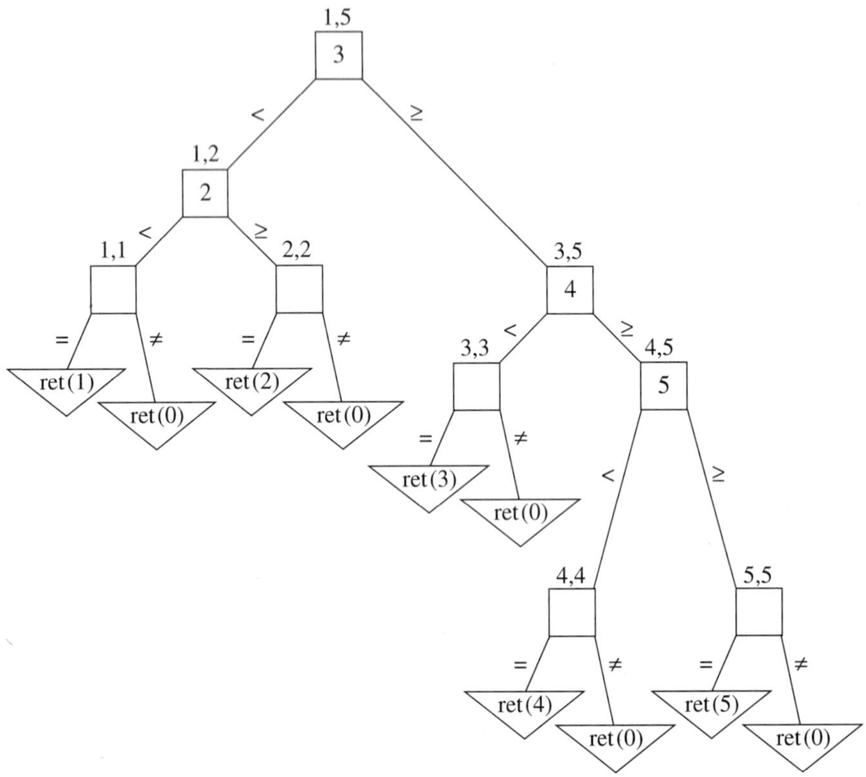

Figure 10.4 Comparison tree *T* for *BinarySearch*2 when $n = 5$

THEOREM 10.3.1 A worst-case lower bound for the problem of searching an ordered list of size n using a comparison-based algorithm is given by

$$\lceil \log_2(n + 1) \rceil. \tag{10.3.3}$$ ■

Since *BinarySearch*2 has $W(n) = \lceil \log_2 n \rceil + 1$, we see from (10.3.3) that *BinarySearch*2 comes within one of being an optimal worst-case algorithm for searching ordered lists. Thus, we have rather precisely determined the worst-case complexity of the problem of searching ordered lists using comparison-based algorithms.

10.3.3 Lower Bound for the Average Complexity of Comparison-Based Searching of Ordered Lists

We now establish a lower bound for the average complexity of searching an ordered list of distinct elements using a comparison-based algorithm. For simplicity, we assume that the search element X is on the list and is equally likely to occur in any position. Consider any algorithm for solving the problem, and let T be its associated comparison tree for a list of size n. Since the search element X can occur in any of the n positions in the list, every index from $\{1, 2, \ldots, n\}$ occurs as the output of some leaf node of T. We now prove that each index occurs as the output of exactly one leaf node. (Note that this property does not hold for all comparison-based algorithms for searching *arbitrary* lists.) Assume to the contrary that there exists an index i such that i is the output of two leaf nodes, say L_1 and L_2. Let N be the root of the smallest subtree of T containing both L_1 and L_2. Now N corresponds to a comparison of the search element X with a list element $L[j]$. The only way X can reach either L_1 or L_2 is for X to be equal to $L[i]$. Since L is ordered, the comparison of X to $L[j]$ corresponding to node N always has the same result, independent of the list. Therefore, X either always goes to the right child of N or to its left, independent of the list. Thus, $X = L[i]$ always ends at L_1 or L_2, so that either L_1 or L_2 can be pruned. Hence, T has exactly n leaf nodes, one for each index from $\{1, \ldots, n\}$.

The number of comparisons made by the algorithm to reach a leaf (output) node equals the length of the path in T from the root to that leaf node. Thus, the average complexity $A(n)$ for a successful search is given by $LPL(T)/n$, where $LPL(T)$ denotes the leaf path length of T defined in Chapter 7. The following theorem then follows from Theorem 7.3.7.

THEOREM 10.3.2 A lower bound for the average complexity of the problem of (successfully) searching an ordered list of size n using a comparison-based algorithm, assuming that each list element has equal probability $1/n$ of being the search element, is

$$\lfloor \log_2 n \rfloor + \frac{2\,(n - 2^{\lfloor \log_2 n \rfloor})}{n} \in \Omega(\log n). \qquad (10.3.4) \quad \square$$

Now the average complexity $A(n)$ of *BinarySearch2* (whether or not the search element is in the list) is no greater than $\lceil \log_2 n \rceil + 1$. Hence, *BinarySearch2* comes within two of having optimal average complexity.

10.3.4 Lower Bound for the Worst-Case Complexity of Comparison-Based Sorting Algorithms

To establish a lower bound for the worst-case complexity of comparison-based sorting algorithms, we can assume that the inputs to the algorithm are the $n!$ permutations of $\{1, 2, \ldots, n\}$. Consider any comparison-based sorting algorithm, and let T be its associated comparison tree. The root node of T corresponds to the first comparison between list elements made by the algorithm. More generally, if we are at an internal node of T corresponding to a comparison of x_i and x_j, $i < j$, the left child of this node corresponds to the action of the algorithm when $x_i < x_j$. The elements are listed in the node $x_i{:}x_j$ with $i < j$. In Figure 10.5, the comparison tree for *ShellSort* is illustrated for $n = 3$ and the two increments

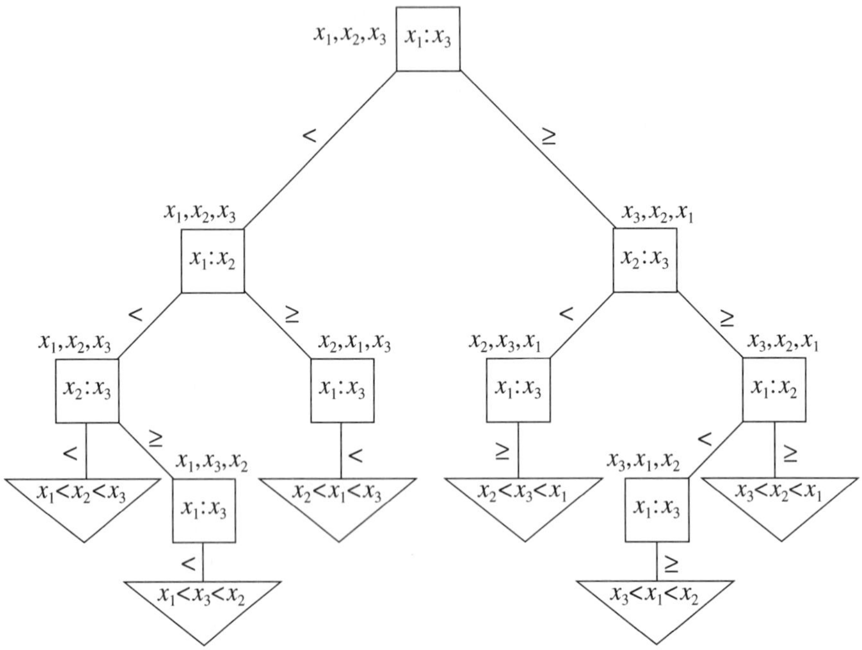

Figure 10.5 Comparison tree for *ShellSort* with distinct elements x_1,x_2,x_3, and with two increments $d_1 = 2$, $d_2 = 1$

$d_1 = 2$, $d_2 = 1$. The current state of the list is shown outside each node. Note that the comparison tree has been pruned to remove nodes that are never reached by inputs to the algorithm.

Note in Figure 10.5 that there are $3! = 6$ leaf nodes corresponding to the 6 possible orderings of a 3-element list x_1,x_2,x_3. In fact, the salient feature of the comparison tree corresponding to *any* comparison-based sorting algorithm is that for an input of size n there must be *exactly* $n!$ leaf nodes. Each leaf node corresponds to one of the $n!$ possible orderings of a list of size n. Indeed, if there were less than $n!$ leaf nodes, then at least two distinct permutations of $\{1,2, \ldots , n\}$ would have to follow identical paths from the root to the same leaf. In other words, the algorithm would perform identically on the two permutations and therefore could not possibly give the right output for both lists. On the other hand, there can't be more than $n!$ leaf nodes, otherwise there would be a leaf node that is never reached by any of the $n!$ permutations, and this node would be pruned. Now by Proposition 7.2.3 any binary tree having $n!$ leaf nodes must have depth at least $\lceil \log_2 n! \rceil$. Thus, any comparison-based sorting algorithm must perform $\lceil \log_2 n! \rceil \in \Omega(n \log n)$ comparisons for some input of size n. We state this in the form of a theorem, where we use the result of Exercise 3.42 to estimate $\lceil \log_2 n! \rceil$.

THEOREM 10.3.3 A lower bound for the worst-case complexity of any comparison-based sorting algorithm is $\lceil \log_2 n! \rceil = n \log_2 n - n/\log_e n + (1/2)\log_2 n + O(1)$. In particular, the worst-case complexity belongs to $\Omega(n \log n)$. ■

Of the sorting algorithms considered so far, it follows from Theorem 10.3.3 that *MergeSort* and *HeapSort* both exhibit order-optimal worst-case performance.

10.3.5 Lower Bound for the Average Complexity of Comparison-Based Sorting Algorithms

We now show that a lower bound for the average complexity of comparison-based sorting algorithms is $\lfloor \log_2 n! \rfloor$. (In particular, this lower bound for average complexity differs by at most one from the lower bound that we established for the worst-case complexity.) Our previous discussion shows that the decision tree for any comparison-based sorting algorithm must have $n!$ leaf nodes. Under our assumption for analyzing average behavior of sorting algorithms, we assume that $\mathcal{T}_n$ consists of all permutations of $\{1,2, \ldots, n\}$ and that each permutation is equally likely. Each permutation must follow some path in the decision tree to a leaf, and our assumption implies that each leaf is equally likely to be a termination point of the algorithm. Since each internal node in the decision tree T corresponds to a comparison performed by the algorithm, it follows that the average number of comparisons performed by the algorithm is at least $LPL(T)/n!$. By Theorem 7.3.7, $LPL(T) \geq n!\lfloor \log_2 n! \rfloor$, so that the average number of comparisons is at least $\lfloor \log_2 n! \rfloor$. Again, we state this in the form of a theorem.

THEOREM 10.3.4 A lower bound for the average complexity of any comparison-based sorting algorithm is $\lfloor \log_2 n! \rfloor$. In particular, the average complexity belongs to $\Omega(n \log n)$. ❑

Of the comparison-based sorting algorithms considered so far, it follows from Theorem 10.3.4 that *MergeSort*, *HeapSort*, and *QuickSort* all have order-optimal average complexity.

10.4 Adversary Arguments

Recall that adversary arguments work by dynamically constructing inputs to an algorithm that makes it work hard. Thus, adversary arguments establish lower bounds for *worst-case* complexity.

10.4.1 Lower Bound for Finding Maximum and Minimum Elements in a List

Our first example of an adversary argument determines a lower bound for worst-case complexity of any comparison-based algorithm for finding both the maximum and minimum elements in a list $L[1:n]$.

THEOREM 10.4.1 Any comparison-based algorithm for finding both the maximum and minimum elements in an arbitrary list of size n must perform at least $\lceil 3n/2 \rceil - 2$ comparisons in the worst case.

Proof We use the simpler word "algorithm" for "comparison-based algorithm" in the proof. Since we are establishing a worst-case lower bound for the number of comparisons that any algorithm performs when determining the maximum and minimum elements for an *arbitrary* list $L[1{:}n]$, we can assume that $L[1{:}n]$ consists of distinct elements. When an algorithm compares two list elements $L[i]$ and $L[j]$, if $L[i] < L[j]$, then we say $L[i]$ lost the comparison and $L[j]$ won it, and vice versa if $L[i] > L[j]$. With any algorithm we associate a *state 4-tuple* modeling the performance of the algorithm as follows. As the algorithm works, we keep track of state 4-tuple $[\#U,\#W,\#L,\#WL]$, where $\#U$ denotes the number of list elements as yet unused in any comparison, $\#W$ denotes the number of elements that have won a comparison and never lost, $\#L$ denotes the number of elements that have lost a comparison and never won, and $\#WL$ denotes the number of elements that have both won a comparison and lost a comparison. Initially, we have the state 4-tuple $[n,0,0,0]$. An argument similar to the proof of Proposition 3.5.1 shows that an algorithm can correctly determine maximum and minimum elements in the list only after transforming this original state 4-tuple into the final state $[0,1,1,n-2]$.

We now design an adversary strategy that forces the algorithm to make at least $\lceil 3n/2 \rceil - 2$ comparisons before this final state is reached. In response to comparisons made by the algorithm, the adversary creates an input list dynamically. As the algorithm progresses, the adversary is allowed to change the values of list elements previously assigned, but such changes must be consistent with all win-loss comparisons that have occurred previously. *The critical property of the final list created by the adversary is that it can be input back into the algorithm and force the following two transition rules*:

1. Whenever $\#U$ changes, $\#WL$ does not, and vice versa.
2. Whenever $\#WL$ changes, it changes by one.

Note that if the transition rules are satisfied by the input list, then Theorem 10.4.1 is established. Indeed, it takes at least $\lceil n/2 \rceil$ comparisons to change $\#U$ from n to 0. By transition rule 1, these comparisons do not alter $\#WL$. Hence, using transition rule 2, at least an additional $n-2$ comparisons are required to change $\#WL$ from 0 to $n-2$. It remains to show that the adversary can produce a list satisfying the transition rules.

At any point in the execution of the algorithm, a list element X can be in one of four states U,W,L,WL, depending on whether X has participated in no comparisons, has won every comparison, has lost every comparison, or has won at least one comparison and lost at least one comparison, respectively. Thus, when the algorithm compares two elements, these elements exhibit one of ten possible (unordered) pair states:

$$\{U,U\},\ \{U,W\},\ \{U,L\},\ \{U,WL\},\ \{W,W\},\ \{W,L\},\ \{W,WL\},\ \{L,L\},\ \{L,WL\},\ \{WL,WL\}.$$

We now show how the adversary creates a list satisfying the transition rules. For ease of discussion, we suppose that X,Y denotes the list elements being compared by the algorithm and that X has state S_1 in the pair state $\{S_1,S_2\}$. When the algorithm compares two list elements X,Y having the pair state $\{U,U\}$, the adversary makes a "whimsical" assignment of values to X and Y (that is, *any* choice will do subject only to the condition of maintaining a list of distinct elements). Similarly, when the algorithm compares two list elements X,Y having the pair state $\{U,WL\}$, the adversary makes a whimsical assignment of a value to X. Clearly, these two pair states cannot lead to a violation of transition rules 1 or 2, since $\#WL$ remains unchanged by a comparison of list elements having either of these states. We now consider those pair states that *can* lead to violations of the transition rules (that is, where $\#WL$ can change).

We first consider transition rule 1. Clearly, the only pair states that can lead to a violation of transition rule 1 are $\{U,W\}$ and $\{U,L\}$. Given, say, a comparison between two list elements X,Y in the pair state $\{U,W\}$, the adversary assigns a value to X that is smaller than Y. This forces the transition $[\#U,\#W,\#L,\#WL] \rightarrow [\#U - 1, \#W,\#L + 1,\#WL]$, which satisfies transition rule 1. Similarly, given a comparison between two list elements X,Y in the pair state $\{U,L\}$, the adversary assigns a value to X that is larger than Y. This completes the adversary strategy with respect to transition rule 1.

Now consider transition rule 2. Clearly, $\{W,L\}$ is the only pair state that can lead to a violation of transition rule 2, since comparisons between list elements having any other pair state can only change $\#WL$ by at *most* one. Suppose, then, that the algorithm makes a comparison between the list elements X,Y having pair state $\{W,L\}$. Then the adversary makes X win, resulting in no change in the state 4-tuple, for otherwise $\#WL$ would change by two. Note, however, that the adversary has to reassign the value of X in case the previous assignments resulted in the relation $X < Y$. If the latter is the case, the adversary merely reassigns a value to X making $X > Y$. This cannot be inconsistent with the results of previous comparisons involving X, since X always won such comparisons (now X is an even bigger winner). Moreover, reassigning the value for X won't affect the previous sequence of comparisons made by the algorithm, since we are restricting attention to comparison-based algorithms. In particular, branching decisions made by such algorithms do not depend on the particular values of list elements, only their relative order. This completes the adversary strategy with respect to transition rule 2.

The remaining pair states, $\{W,W\}$, $\{W,WL\}$, $\{L,L\}$, $\{L,WL\}$, $\{WL,WL\}$, cannot lead to a violation of the transition rules. Thus when the algorithm compares two list elements having any one of these states, the adversary makes no changes to the (already assigned) values of X and Y. Thus, we have shown that the adversary can, in response to comparisons made by the algorithm, create a list dynamically that satisfies the two transition rules. The final list that results can be input back into the algorithm, forcing it to perform $\lceil 3n/2 \rceil - 2$ comparisons. ■

Remark Note that in forcing the transition rules to be satisfied, the adversary doesn't always give away as little information as possible. For example, when the algorithm compares two list elements X,Y having pair state $\{W,WL\}$, the adversary doesn't intervene. Thus, if $X < Y$, the state 4-tuple would undergo the transition $[\#U,\#W,\#L,\#WL] \rightarrow [\#U,\#W-1,\#L,\#WL+1]$. While this does not violate the transition rules, the adversary could reassign a value to X making $X > Y$. Then the state 4-tuple would undergo no change at all. This reflects the fact that a comparison involving the pair state $\{W, WL\}$ is wasted work when $X > Y$. However, for the purposes of the lower bound $\lceil 3n/2 \rceil - 2$, the adversary doesn't care. If the algorithm does even more work, so what? (Of course, if a larger lower bound could be found for *all* comparison-based algorithms, then the adversary *would* care! But, the algorithm *MaxMin3* described in Chapter 3 shows that such a larger lower bound can't be found.)

Since algorithm *MaxMin*3 (see Chapter 3) has worst-case complexity $\lceil 3n/2 \rceil - 2$, by Theorem 10.4.1 it is optimal, and the worst-case complexity of the problem of finding both the maximum and minimum elements in a list of size n is precisely $\lceil 3n/2 \rceil - 2$.

10.4.2 Lower Bound for Finding the Largest and Second-Largest Elements in a List

Our second example of an adversary argument is to find a lower bound for the problem of finding the largest and second-largest elements in a list. Again, we restrict attention to comparison-based algorithms, and we assume that the list contains distinct elements. The basic strategy of the adversary is to force any comparison-based algorithm to have the largest element participate in as many comparisons as possible. The reason for this strategy is clear from the following proposition.

PROPOSITION 10.4.2 Suppose a comparison-based algorithm for finding the largest and second-largest elements in a list of size n compares the largest element to m other elements. Then the algorithm performs at least $m + n - 2$ comparisons altogether.

Proof Suppose for some list $L[1:n]$ of distinct elements that a comparison-based algorithm determines $L[i], L[j]$ to be the largest and second-largest element, respectively, and that it only makes at most $n - 3$ comparisons between elements, neither of which are $L[i]$. Using an argument similar to that used in the proof of Proposition 3.5.1, there must be an element $L[k]$ different from $L[j]$ that never lost a comparison other than with $L[i]$. If we increase the value of $L[k]$ so that $L[j] < L[k] < L[i]$, clearly the algorithm works identically on the altered list, so that it would again declare $L[j]$ to be the second-largest, a contradiction. ■

The question now is, how large can the adversary force the m in Proposition 10.4.2 to be? The next proposition answers this question.

PROPOSITION 10.4.3 Any comparison-based algorithm that determines the largest element in a list of n distinct elements (no matter what else it does) compares the largest element to $\lceil \log_2 n \rceil$ other list elements for a suitable list $L[1:n]$.

Proof Given any comparison-based algorithm that determines the largest element in a list, the adversary will dynamically construct a list that makes this largest element participate in $\lceil \log_2 n \rceil$ comparisons. Whenever a comparison is made between two list elements that have never lost, the adversary chooses the winner to be the one that has won the most previous comparisons (ties are decided arbitrarily). If a comparison is made between an element that has never lost and one that has lost, the adversary chooses the one that has never lost to be the winner. When choosing the winner, the adversary can clearly adjust a previously assigned value (if necessary) to achieve this strategy and remain consistent with previous win-loss comparisons.

We now show that the strategy just described forces the maximum to participate in $\lceil \log_2 n \rceil$ comparisons. We model the strategy as territorial takeover. We assume that initially each list element owns one unit of territory. The adversary adopts a simple "the rich get richer" strategy. Whenever a comparison is made between two list elements that have never lost a comparison, the winner always has at least as much territory as the loser, and then the winner takes over all the territory formerly owned by the loser. The loser from that point on has no territory. The largest element in the list eventually will own all n units of territory. Since each element starts with one unit of territory and since the winner of any comparison can at most double the amount of territory owned, the largest element has to participate in at least $\lceil \log_2 n \rceil$ comparisons. ■

Propositions 10.4.2 and 10.4.3 combine to yield the following theorem.

THEOREM 10.4.4 A comparison-based algorithm for finding the largest and second-largest elements in a list of size n must perform at least $n + \lceil \log_2 n \rceil - 2$ comparisons. ■

The questions arises as to whether there is an optimal algorithm for finding the largest and second-largest elements in a list of size n. The most obvious algorithm is to scan the list for the largest element using $n - 1$ comparisons, and then scan the remaining elements for the second-largest using $n - 2$ comparisons, using $2n - 3$ comparisons altogether. However, $2n - 3$ is almost twice as large as the lower bound given in Theorem 10.4.4. The idea behind a better, in fact optimal, algorithm is to model the algorithm as a match-play golf or table tennis tournament.

We regard the comparison of list elements as a golf or table tennis match between the elements (players). Assume for simplicity that $n = 2^k$. In the first round of the tournament, we pair the n players into $n/2$ matches. In the next round, the $n/2$ winners of the first round are paired into $n/4$ matches, and the $n/4$ winners advance into the third round. After $k = \log_2 n$ rounds, we obtain the winner of the tournament. We draw such a single-elimination tournament in Figure 10.6 (page 368). Note that to arrange for n to be a power of two, we gave Player G a *bye* in the first round.

The algorithm based on this tournament method makes $n/2 + n/4 + \cdots + n/2^k = n(1/2 + 1/4 + \cdots + 1/2^k) = n - 1$ comparisons in determining the largest element. The question remains as to how the algorithm can now proceed to efficiently find the runner-up (second-largest). In actual golf or table tennis tournaments, it is customary to declare the runner-up to be the player who played the winner in the last round (Player

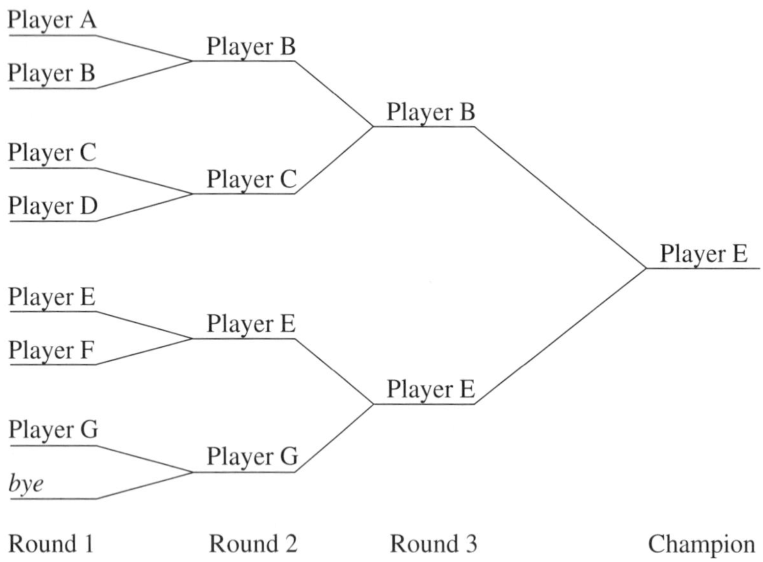

Figure 10.6 A single-elimination tournament

B in Figure 10.6). However, sometimes that runner-up is not really the second-best player in the tournament. (Tournament organizers attempt to minimize the latter problem by avoiding early-round matches between highest-ranked, or so-called *top-seeded*, players. Just as in Figure 10.6, they also use the method of seeding to facilitate tournament *byes*, so that the highest-ranked players don't even have to play in the first round or rounds. *Byes* allow them to assume that the first round is between $n = 2^k$ players.)

Certainly for the tournament method of determining the largest element in a list, we cannot assume that the list element that lost the last comparison made by the algorithm is the second-largest element in the list. However, the key observation here is that the second-largest must have lost a comparison to the largest element in *some* round (not necessarily the last round). Thus, for example, in our tournament shown in Figure 10.6, the candidates for second-best are Player F, Player G, and Player B. More generally, if the algorithm maintains a (dynamic) list of losers for each element, then the second-largest element must be found among the largest element's final list of losers (although a linear number of loser list updates are made, no additional comparisons are incurred). This final list contains $\log_2 n$ elements. A linear scan of these $\log_2 n$ elements (using $\log_2 n - 1$ comparisons) then determines the second-largest element in the list. Hence, the algorithm performs a total of $n + \log_2 n - 2$ comparisons and thus is an optimal algorithm.

10.4.3 Lower Bound for Finding the Median Element in a List

Our third example concerns the problem of determining the median element in a list $L[1:n]$ using a comparison-based algorithm. We show that a lower bound for this problem is $3n/2 - 3/2$, where an adversary argument is used in the second part of the proof. We assume that the list consists of distinct elements. Also, for simplicity we assume that n is odd, so that the median element M has the property that $(n - 1)/2$ elements are smaller (respectively, larger) than M (the case of even n is left to the exercises).

Given a comparison between two list elements X and Y, we say that this comparison is *one-sided* if X and Y are on the same side of the median. Before describing the adversary strategy, we establish a lower bound on the number of *one-sided* comparisons that any comparison-based algorithm for determining the median must perform. By convention, any comparison involving the median is one-sided. We *tag* any one-sided comparison with the comparand that is the farthest from the median (of the two comparands).

PROPOSITION 10.4.5 Any comparison-based algorithm for finding the median element in a list $L[1{:}n]$ performs a one-sided comparison tagged with X for each of the $n - 1$ list elements X different from the median.

Proof Suppose to the contrary that some input list $L[1{:}n]$ to algorithm has the property that there is at least one list element X not involved in any one-sided comparison that is tagged with X. We can assume without loss of generality that X is larger than the median M. Note that if X is compared to any list element Y that is larger than the median, then X is smaller than Y (otherwise X would tag a comparison). Let Z be the largest list element that is smaller than the median M. We now create a new list $L'[1{:}n]$ as follows. We diminish X by a suitable amount yielding X' that lies between Z and M and leave the remaining list elements unaltered. Clearly, all comparisons involving X' have exactly the same outcome as those involving X. Hence, the algorithm works the same on the list L' as it does on L, so that M is also returned as the median element of L'. But this is a contradiction, since the number of elements larger than M in L' is one less than the number of elements larger than M in L. ■

A simple adversary strategy exists that forces any comparison-based algorithm for finding the median to perform at least $(n - 1)/2$ two-sided comparisons. Indeed, for definiteness, we assume that the adversary arbitrarily decides on the value M of the median element (but not its position). We call a comparison between list elements an *assigning comparison* if at least one of the comparands has not previously been involved in any comparison (so that the adversary must assign a value to each such comparand). The adversary ensures that each of the first $(n - 1)/2$ assigning comparisons performed by the algorithm are two-sided by simply assigning values to the previously unassigned comparand(s) that puts the two comparands on opposite sides of the median. This is always possible, since at most $n - 1$ distinct elements can be involved in the first $(n - 1)/2$ assigning comparisons, so that the adversary can put off assigning the median M until after the first $(n - 1)/2$ assigning comparisons have been performed. Thus, any comparison-based algorithm for finding the median can be forced to perform at least $(n - 1)/2$ two-sided comparisons in addition to the $n - 1$ one-sided comparisons guaranteed by Proposition 10.4.5. Hence we have established the following theorem.

THEOREM 10.4.6 A comparison-based algorithm for finding the median element in a list of size n (n odd) must perform at least $3n/2 - 3/2$ comparisons in the worst case. ■

The lower bound in Theorem 10.4.6 is not the best possible. In fact a lower bound of slightly bigger than $2n$ has been established. An upper bound of $n \log_2 n$ for determining the median by a comparison-based algorithm is immediately obtained by sorting the list using an algorithm such as *MergeSort*. In Chapter 13 we will describe a linear comparison-based algorithm for the more general problem of determining the kth-smallest element in a list of size n for any k between 1 and n. Thus, the problem of determining the median element in a list has linear complexity.

10.5 Information Theoretic Arguments

For an example of an information theoretic argument, we reconsider the problem of comparison-based sorting. The information theoretic argument given here yields the same lower bound for the worst-case complexity as the comparison tree argument given earlier. However, it is somewhat more elegant in that it avoids an explicit discussion of a comparison tree.

The result of a given comparison between two list elements yields a single *bit* of information (0 = **.false.**, 1 = **.true.**). Each of the $n!$ permutations of $\{1, 2, \ldots, n\}$ has to be distinguished by a correct algorithm. Thus, a comparison-based algorithm must perform enough comparisons to produce $n!$ cumulative pieces of information. Since each comparison only yields one bit of information, our question amounts to asking what the minimum number of bits of information needed to allow $n!$ different outcomes is. The answer is obtained as follows: m bits of 0-1 information can result in 2^m different 0-1 sequences of length m. Thus, to obtain $n!$ different outcomes, we must have $2^m \geq n!$. Thus, any comparison-based sorting algorithm must make at least $\lceil \log_2 n! \rceil$ comparisons to properly distinguish between the $n!$ inputs of size n, showing (again) that a lower bound for the worst-case complexity of the problem of comparison-based sorting is $\lceil \log_2 n! \rceil \in \Omega(n \log n)$.

10.6 Lower Bounds for Parallel Algorithms

All the lower bounds that we have established for the complexity of problems solved by sequential algorithms made the implicit assumption that a single processor was used to make comparisons, perform arithmetic operations, and so forth. Thus, when we established lower bounds for the complexity of such problems as comparison-based sorting, we implicitly made the assumption that only one comparison could be made at any given step in the algorithm. When determining lower bounds for the complexity of problems solved by parallel algorithms, we must explicitly state the number of processors and the particular model and architecture that we are using. For example, consider the problem of finding the maximum value in a list of size n on the CREW PRAM using a comparison-based algorithm. If there are at most n processors available, then we obtain a lower bound of $\Omega(\log(\log n))$ parallel comparison steps for this problem. On the other hand, we obtain a lower bound $\Omega(\log n)$ for the same problem on the EREW PRAM with an unlimited number of processors.

Similar to sequential algorithms, a lower bound for a problem is sharp if there exists a parallel algorithm on the given architecture and number of processors whose complexity equals the lower bound. In the latter case, we have determined the complexity of the problem itself with respect to the given architecture and number of processors. As with sequential algorithms, we rarely are able to determine the exact complexity of a problem, or even its order.

Clearly, any lower bound for the complexity of a given problem on the EREW PRAM is a lower bound for the complexity of the problem on any particular interconnection network. Further, any lower bound for the complexity of a given problem on a particular PRAM model also is a lower bound for more restrictive PRAM models.

10.6.1 Lower Bound for Finding the Maximum on the EREW PRAM

The first lower bound result that we obtain is for the problem of finding the maximum value in a list on the EREW PRAM.

PROPOSITION 10.6.1 A lower bound for the complexity of the problem of finding the maximum value in a list $L[1{:}n]$, using a comparison-based algorithm on the EREW PRAM with an unlimited number of processors, is

$$\lceil \log_2 n \rceil. \tag{10.6.1}$$

Proof Our proof will use an adversary argument. We assume, without loss of generality, that $L[1{:}n]$ is a list of distinct real numbers. In any parallel comparison step, when two distinct elements X and Y are compared, we say that X *wins* the comparison if $X > Y$, otherwise X *loses*. At each stage of the algorithm, we let W denote the set of elements that have not yet lost any comparison. Initially, $|W| = n$. The algorithm does not terminate until the maximum element is the only element that has not lost a comparison.

A pair of elements from $L[1{:}n]$ must first be read before they can be compared. Since the algorithm is being implemented on an EREW PRAM, no element can be read by more than one processor concurrently. Thus, during any single parallel comparison step, each element of the list is compared to at most one other element. It follows that after the first parallel comparison step, at least half of the elements belong to W, so that $|W| \geq n/2$.

During the second parallel comparison step, for each comparison involving an element in W and an element not in W, the adversary arranges that the element from W wins. The adversary achieves this by replacing input list L with L', where L' is obtained from L by increasing (if necessary) each element of L that lies in W so that (1) the relative order of the elements in W is preserved, and (2) each element of W is strictly larger than any element not in W. It follows that W can be reduced by at most half, so that $|W| \geq n/4$.

In general, using a similar adversary strategy, W is reduced by at most half after each parallel comparison is performed. A simple induction argument shows that $|W| \geq n/2^k$ after the first k parallel comparisons have been performed. Since the algorithm does not terminate before $W = 1$, it follows that the algorithm performs at least $\lceil \log_2 n \rceil$ comparisons. ■

PROPOSITION 10.6.2 The worst-case complexity of the *problem* of finding the maximum value of an element in a list of size n, using a comparison-based algorithm on the EREW PRAM using an unlimited number of processors, is $\lceil \log_2 n \rceil$. Further, the same problem on the EREW PRAM using $\lceil n/2 \rceil$ processors also has complexity $\lceil \log_2 n \rceil$.

Proof The algorithm *MaxPRAM* discussed in Chapter 5, suitably modified to handle input lists of any size n (not necessarily a power of two), has complexity $\lceil \log_2 n \rceil$ on the EREW PRAM using $\lceil n/2 \rceil$ processors. By Proposition 10.6.1, any algorithm for solving the problem on the EREW PRAM performs at least $\lceil \log_2 n \rceil$ parallel comparisons no matter how many processors are used. Hence, the complexities of the problem of finding the maximum value in a list on an EREW PRAM using an unlimited number of processors, or using $\lceil n/2 \rceil$ processors, are both equal to $\lceil \log_2 n \rceil$. ■

10.6.2 Lower Bound for Comparison-Based Sorting on the EREW PRAM

Clearly, any lower bound for the worst-case complexity of the problem of finding the maximum element in a list is also a lower bound for the worst-case complexity of the problem of sorting a list. Hence, by Proposition 10.6.1 a lower bound for the worst-case complexity of the problem of comparison-based sorting of a list of size n on an EREW PRAM having an unlimited number of processors is $\lceil \log_2 n \rceil$. The $\lceil \log_2 n \rceil$ lower bound is order optimal, since the algorithm *SortPRAM* (see Chapter 6) for sorting a list of size n on the EREW PRAM with $n(n-1)/2$ processors has worst-case complexity $O(\log n)$. In fact, the $\lceil \log_2 n \rceil$ lower bound is order optimal even for the EREW PRAM with a linear number of processors, since $O(\log n)$ algorithms for this model have been found.

10.6.3 Lower Bound for Finding the Maximum on the CRCW PRAM

Recall that algorithm *MaxCRCW* finds the maximum element in a list using only a single parallel comparison step (see Chapter 5). Clearly, *MaxCRCW* is an optimal algorithm. However, since *MaxCRCW* utilizes $\Theta(n^2)$ processors, it has cost $\Theta(n^2)$, which is far from the optimal cost of $\Theta(n)$. We now consider the problem of finding the maximum value of an element in a list on a *CRCW PRAM* using only n processors.

We will obtain a worst-case complexity lower bound of $\Theta(\log(\log n))$ for the problem. This lower bound is sharp, since there exists an algorithm *MaxCRCW2* for finding the maximum element in a list of size n on a CRCW PRAM that has worst-case complexity $\Theta(\log(\log n))$ (the design of *MaxCRCW2* is explored in the exercises).

PROPOSITION 10.6.3 The worst-case complexity of the *problem* of finding the maximum element in a list of size n using a comparison-based algorithm on a CRCW PRAM with n processors belongs to

$$\Theta(\log(\log n)).$$

Proof The existence of the algorithm *MaxCRCW2* establishes $O(\log(\log n))$ as an upper bound for the problem. To establish $\Omega(\log(\log n))$ as a lower bound to the problem, we employ the graph theoretic result of Turan on independent sets given in Chapter 8. Recall that Theorem 8.1.7' states that the size $\beta(G)$ of the largest vertex independent set in a graph $G = (V,E)$ satisfies

$$\beta(G) \geq |V|^2/(2|E| + |V|). \tag{10.6.2}$$

We use an adversary strategy to establish the lower bound for the problem, where we make the same assumptions as in the proof of Proposition 10.6.1. We let W_i denote the set of elements in $L[1{:}n]$ that have won every comparison after the ith parallel comparison step has been performed. We obtain a recurrence relation for $t(i) = |W_i|$ as follows. Initially, W_0 consists of all the elements in $L[1{:}n]$, so that $t(0) = n$. Since the maximum is determined only if $|W_i| = 1$, any parallel algorithm for solving the problem cannot terminate until $|W_i| = 1$. As in the proof of Proposition 10.6.1, at any step i, the adversary can arrange that all the elements in W_i are greater than any element not in W_i. Therefore, an element from W_i can lose a comparison during the $(i + 1)$st parallel comparison step only if it is compared to another element from W_i. Consider the graph G_i having vertex set $V(G_i) = W_i$, whose edge set $E(G_i)$ consists of all pairs $\{u,v\}$ such that $u,v \in W_i$, and u and v are compared in the $(i + 1)$st parallel comparison step. Since only n processors are available, we have $|E(G_i)| \leq n$.

Now consider an independent set of vertices U_i of maximum size in G_i. Since U_i is an independent set, no pair of vertices in U_i is involved in a comparison during the $(i + 1)$st comparison step. Thus, by appropriately altering L so that all the elements of L in U_i are strictly greater than any element not in U_i, the adversary arranges that all the elements in U_i win their comparisons in the $(i + 1)$st parallel comparison step, so that $|W_{i+1}| \geq |U_i|$. Using (10.6.2), $|U_i| \geq |W_i|^2 (2n + |W_i|)$, so that $t(i) = |W_i|$ satisfies the recurrence relation

$$t(i + 1) \geq \frac{(t(i))^2}{2n + t(i)}, \quad \textbf{init. cond. } t(0) = n. \tag{10.6.3}$$

It follows from the definition of $t(i)$ that $t(i) \leq n$ for all i, so that

$$t(i + 1) \geq \frac{(t(i))^2}{3n}, \quad \textbf{init. cond. } t(0) = n. \tag{10.6.4}$$

Iterating the inequality (10.6.4) yields

$$t(i) \geq \left(\frac{1}{3}\right)^{2^i} 3n. \tag{10.6.5}$$

Since the algorithm cannot terminate before $t(i) = |W_i| = 1$, it follows from (10.6.5) that in the worst case the algorithm must perform $W(n) \geq i$ iterations, where i is large enough so that $(1/3)^{2^i} 3n \leq 1$. The latter inequality holds only if $n \leq 3^{2^i - 1}$, which implies that

$$W(n) \geq i \geq \log_2(1 + \log_3 n) \in \Omega(\log(\log n)). \tag{10.6.6} \blacksquare$$

10.7 Closing Remarks

In this chapter we have only given a small sampling of the problems for which lower bounds have been successfully determined. However, lower bound theory is a difficult subject in general, and we only know precise, or order-precise, lower bounds in relatively few instances. Indeed, the most important problem in theoretical computer science today is to settle the lower bound question for the wide class of problems known as the *NP-complete* problems. For example, the decision versions of many of the interesting problems in graph theory belong to this class (graph coloring, independent set, Hamiltonian paths, and so forth). The only known algorithms for solving these problems have superpolynomial complexity in the worst case. On the other hand, the only lower bounds known for the worst-case complexity of these problems are typically polynomial with relatively small degree. Thus, the gap between lower bounds and upper bounds for these problems is large indeed. Most computer scientists believe that NP-complete problems are truly computationally difficult in the worst case. In other words, they believe that worst-case lower bounds for these problems really are superpolynomial in nature. However, reducing the upper bounds for worst-case complexity by finding an appropriate algorithm, or raising the lower bound through some mathematical argument, remains elusive in spite of years of effort by some of the best theoreticians in the field. We discuss these matters in more detail in Chapter 20.

References and Suggestions for Further Reading

Some algorithm texts that include discussion of lower-bound theory.

Aho, A. V., J. E. Hopcroft, and J. D. Ullman. *The Design and Analysis of Computer Algorithms*. Reading, MA: Addison-Wesley, 1976.

Baase, S. *Computer Algorithms: Introduction to the Design and Analysis*, 2nd edition. Reading, MA: Addison-Wesley, 1987.

Borodin, A. B., and J. I. Munro. *The Computational Complexity of Algebraic and Numeric Problems*. New York: American Elsevier, 1975.

Horowitz, E. and S. Sahni. *Fundamentals of Computer Algorithms*. Rockville, MD: Computer Science Press, 1978.

Garey, M., and D. Johnson. *Computers and Intractability: A Guide to the Theory of NP-Completeness*. New York: W.H. Freeman & Co., 1979. The classical reference for the theory of NP-completeness.

Exercises

Sections 10.1 – 10.2 Enumeration Techniques in Lower Bound Theory

10.1 What is the maximum number of inversions that can be removed from a permutation of size n by the interchange of a pair of (nonadjacent) elements? Exhibit a permutation and pair of elements whose interchange removes the maximum number of inversions.

10.2 A sloppy chef has delivered a stack of pancakes to a waiter, where the stack is all mixed up with respect to the size of the pancakes. The waiter wishes to rearrange the pancakes in order of their size, with the largest on the bottom. He begins by selecting a pile of pancakes on the top of the stack and simply flips the pile over. He then keeps repeating this pancake-flipping operation until the pancakes are sorted.

a. Show by induction that there always exists a set of flips that sorts any given stack of pancakes.

b. Describe a pancake-flipping algorithm that performs at most $2n$ flips in the worst case for n pancakes.

c. Show that any pancake-flipping algorithm must perform at least n flips in the worst case for n pancakes.

d. Suppose now that the pancakes have a burnt side, and the waiter wishes to have the pancakes stacked in order with the unburnt side up. Redo parts (a), (b), and (c) with for this new problem.

10.3 Obtain and verify a lower bound for the worst-case complexity of adjacent-key comparison sorting restricted to lists where every element is within k of its correct position in the list.

10.4 We have seen that *InsertionSort2* has optimal worst-case complexity for adjacent key comparison sorts. Is it also optimal when the lists are restricted as in the previous exercise?

Section 10.3 Decision Trees and Comparison Trees

10.5 Show that the comparison tree for *BinarySearch2* for a list of size n is a 2-tree with $2n$ leaf nodes, which is full at the second-lowest level.

10.6 **a.** Design an algorithm *TernarySearch* that is analogous to *BinarySearch*, except that it is based on dividing the list into three sublists instead of two.

b. Draw the comparison tree for *TernarySearch* for a list of size 10.

c. Does *TernarySearch* have order-optimal worst-case complexity?

d. Does *TernarySearch* have order-optimal average complexity?

10.7 Repeat Exercise 10.6 for the analog *TernarySearch*2 of *BinarySearch*2.

10.8 Construct an example of a comparison-based algorithm for searching an arbitrary list $L[1{:}n]$ such that an index in L can occur more than once as an output leaf of the associated comparison tree.

10.9 Suppose $L[1{:}n]$ is a sorted list of integers. Exercise 3.8 asked for an algorithm that finds an index i such that $L[i] = i$, or returns 0 if no such index exists. Show that a lower bound for the complexity of any algorithm for solving this problem is $\Omega(\log n)$.

10.10 Stirling's approximation formula for $n!$ states that

$$n! \sim \sqrt{2\pi n}\,(n/e)^n e^{1/(12n)}.$$

Using this formula show that

$$\lceil \log_2 n! \rceil = n\log_2 n - \frac{n}{\ln 2} + \left(\frac{1}{2}\right)\log_2 n + O(1).$$

10.11 Given a list of integers $L[1{:}n]$, where adjacent elements in the list differ by at most one, and $L[1] < L[n]$.

a. Design and analyze an algorithm that finds an index j such that $L[j] = X$ for a given search element X, where $L[1] \le X \le L[n]$.

b. Determine the lower bound for the worst-case complexity of the problem in part (a). Is your algorithm in part (a) optimal?

Sections 10.4 – 10.5 Adversary and Information Theoretic Arguments

10.12 Complete the proof of Proposition 10.4.2 by showing that if the algorithm makes at most $n - 3$ comparisons between elements that are not the largest element $L[i]$ and second-largest element $L[j]$, then there exists an element different from both $L[i]$ and $L[j]$ that never lost a comparison other than with $L[i]$.

10.13 Give pseudocode for the tournament-based algorithm described in Section 10.4.2 for finding the largest and second-largest element in a list.

10.14 Extend Theorem 10.4.6 to the case where n is even.

10.15 Show that *Merge* is an optimal sequential algorithm in the worst-case for merging two ordered sublists of size $n/2$.

Section 10.6 Lower Bound Theory for Parallel Algorithms

10.16 Design a parallel algorithm *MaxCRCW*2 for finding the maximum element in a list of size n on a CRCW PRAM with n processors that has worst-case complexity belonging to $\Theta(\log(\log n))$. (*Hint*: First make $n/2$ comparisons so that $n/2$ winners are determined. Then in successive iterations divide up the winners into appropriately sized groups and use *MaxCRCW* on these groups.)

11

Parallel Prefix, Matrix Multiplication, and Pointer Jumping

In this chapter we discuss three widely used tools in parallel algorithm design. The first tool, referred to as *parallel scan* or *parallel prefix*, is an efficient method for computing the prefix computations: $x_1 \oplus x_2 \oplus \cdots \oplus x_i$, $i = 1, \ldots, n$, where $\oplus$ is any associative operation on a set X. Parallel prefix has many important applications, including carry-lookahead addition, polynomial evaluation, various circuit design problems, solving linear recurrences, scheduling problems, and a variety of graph theoretic problems. Prefix computations are so often encountered that they have come to be regarded as primitive operations in parallel algorithms, and have been implemented in the instruction set of some existing parallel machines.

The second tool, efficient parallel algorithms for computing a matrix-vector product or a matrix-matrix product, arises often in scientific computations. In addition to their role in scientific computations, matrix algorithms are a useful algorithmic design tool, particularly in the design of graph algorithms. We will discuss various algorithms for computing matrix-vector and matrix-matrix products on PRAMs, meshes, and meshes-of-trees.

The third tool, pointer jumping, is an efficient method of determining the root node of the tree containing a given node in a forest of trees implemented with parent pointers. The pointer-jumping technique has many important applications in the theory of parallel algorithms.

11.1 Parallel Prefix Computations

Suppose we have a set X and a binary operation $\oplus$ defined on X. We assume that $\oplus$ is associative, that is $(x \oplus y) \oplus z = x \oplus (y \oplus z)$ for all $x,y,z \in X$. Examples of such an operation include addition, multiplication, minimum of two elements, concatenation of strings, and so forth. For convenience, we will simply refer to $\oplus$ as addition. We also assume the existence of an element 0 (called the identity element) having the property that $0 \oplus x = x \oplus 0 = x$ for all $x \in X$. Given elements $x_1, x_2, \ldots, x_n$ in X, let S_{ij}

denote the sum $x_i \oplus x_{i+1} \oplus \cdots \oplus x_j$, $i \le j$. The sums $S_{1j} = x_1 \oplus x_2 \oplus \cdots \oplus x_j$, $j = 1, \ldots, n$ are called *prefix sums*. Although the problem of computing all the prefix sums might seem inherently sequential, in fact, it is highly parallelizable.

11.1.1 Parallel Prefix Computations on a PRAM

The parallel binary fan-in method of computing $x_1 \oplus x_2 \oplus \cdots \oplus x_n$ (as used by *MinPRAM* to compute the minimum of $x_1, x_2, \ldots, x_n$) performs the following *binary fan-in sums*: $x_1 \oplus x_2$, $x_3 \oplus x_4$, $\ldots$, $x_{n-1} \oplus x_n$, are computed in the first parallel step; $x_1 \oplus x_2 \oplus x_3 \oplus x_4$, $x_5 \oplus x_6 \oplus x_7 \oplus x_8$, $\ldots$, $x_{n-3} \oplus x_{n-2} \oplus x_{n-1} \oplus x_n$, are computed in the second parallel step; and so forth. Thus the only prefix sums that are computed along the way are $x_1 \oplus x_2 \oplus \cdots \oplus x_{2^k}$, $k = 1, \ldots, \log_2 n$.

We now show how to compute all the prefix sums on the EREW PRAM using a *recursive doubling* scheme. We assume that the elements $x_1, x_2, \ldots, x_n$ reside in the array $X[0{:}n]$, where $X[i] = x_i$, $1 \le i \le n$, and $X[0] = 0$. In the first parallel step, P_i reads $X[i-1]$ and $X[i]$, computes $X[i-1] \oplus X[i]$, and assigns the result to $Prefix[i]$, $1 \le i \le n$. (Note that we are not performing concurrent reads, since P_i first reads $X[i-1]$ and then reads $X[i]$, $1 \le i \le n$.) Then $Prefix[1]$ and $Prefix[2]$ contain S_{11} and S_{12}, respectively. In the next parallel step, P_i reads $Prefix[i-2]$ and $Prefix[i]$, computes $Prefix[i-2] \oplus Prefix[i]$, and assigns the result to $Prefix[i]$, $3 \le i \le n$. Then $Prefix[k]$ contains S_{1k}, $k = 1,2,3,4$ (so that we have doubled the number of prefix sums computed). In general, for $2 \le j \le m = \log_2 n$, and $2^{j-1} + 1 \le i \le n$, at the jth parallel step, P_i reads $Prefix[i - 2^{j-1}]$ and $Prefix[i]$, computes $Prefix[i - 2^{j-1}] \oplus Prefix[i]$, and assigns the result to $Prefix[i]$. At the conclusion of the jth step, $Prefix[k]$ contains S_{1k}, $k = 1, \ldots, 2^j$, so that all the prefix sums are computed after $m = \log_2 n$ steps (see Figure 11.1).

The following pseudocode for *PrefixPRAM* is based directly on the preceding discussion.

```
procedure PrefixPRAM(X[1:n],Prefix[1:n])
Model: EREW PRAM with p = n processors
Input:  X[0:n] (an array of elements x1, x2, ..., xn) {X[0] = 0, n = 2^m}
Output: Prefix[1:n] (Prefix[i] = x1 ⊕ ··· ⊕ xi, i = 1, ..., n)
   for 1 ≤ i ≤ n do in parallel
      Prefix[i] := X[i – 1] ⊕ X[i]
   end in parallel
   k := 2
   while k < n do
      for k + 1 ≤ i ≤ n do in parallel
         Prefix[i] := Prefix[i – k] ⊕ Prefix[i]
      end in parallel
         k := k + k
   endwhile
end PrefixPRAM
```

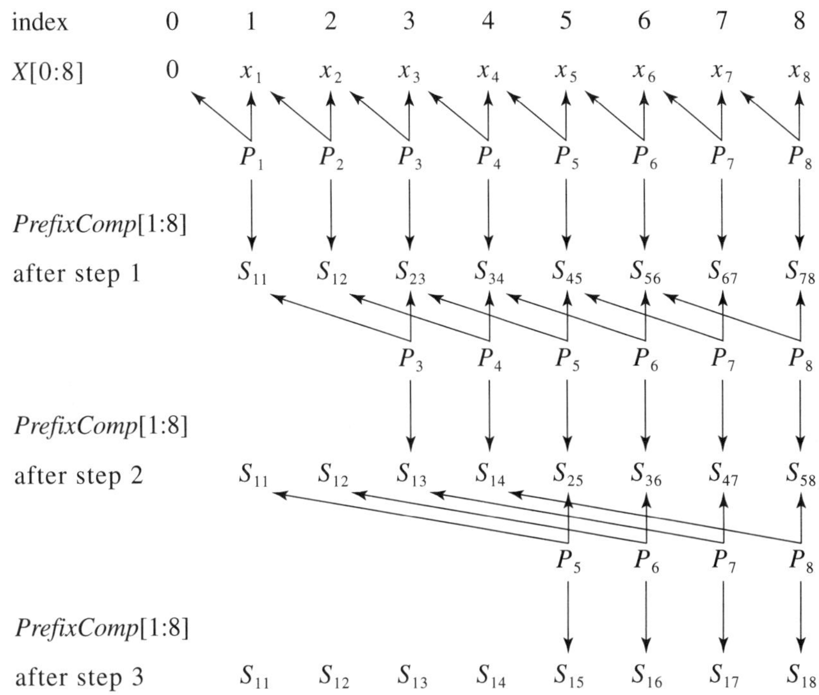

Figure 11.1 Action of *PrefixPRAM* with $n = 8$

The proof of the correctness of *PrefixPRAM* is left as an exercise.

11.1.2 Parallel Prefix Computations on the Complete Binary Tree

We now design a parallel algorithm for performing prefix computations on the processor tree PT_p, $p = 2^j - 1$, having p processors. Algorithms designed for PT_p are useful since there are efficient embeddings of PT_p in other standard interconnection networks such as the hypercube. Hence, the parallel algorithm for performing prefix computations on T_p can be efficiently ported to these other networks.

The algorithm *PrefixPT* begins by assuming that the n operands $x_1, x_2, \ldots, x_n$ are input to the leaves of the complete binary tree PT_{2n-1}. *PrefixPT* works in four phases. In phase 1, binary fan-in computations are performed starting at the leaves and working up to the processors P_0 and P_1 at level one. Phase 1 takes $(\log_2 n) - 1$ communication steps and $(\log_2 n) - 1$ binary operation steps. Phase 1 is illustrated in Figure 11.2a (page 380) for the eight operands $\{x_1, \ldots, x_8\} = \{a,b,c,d,e,f,g,h\}$.

During phase 2, for each pair of operands x_i, x_{i+1} in leaf nodes having the same parent, we replace the operand x_{i+1} in the right child by $x_i \oplus x_{i+1}$, $i = 1,3, \ldots, n-1$. Phase 2 takes two communication steps and a single binary operation step. During Phase 3, each right child that is not a leaf node replaces its binary fan-in computation with that of its (left child) sibling, and the sibling replaces its binary fan-in computation with the identity element. Thus, phase 3 is done in two communication

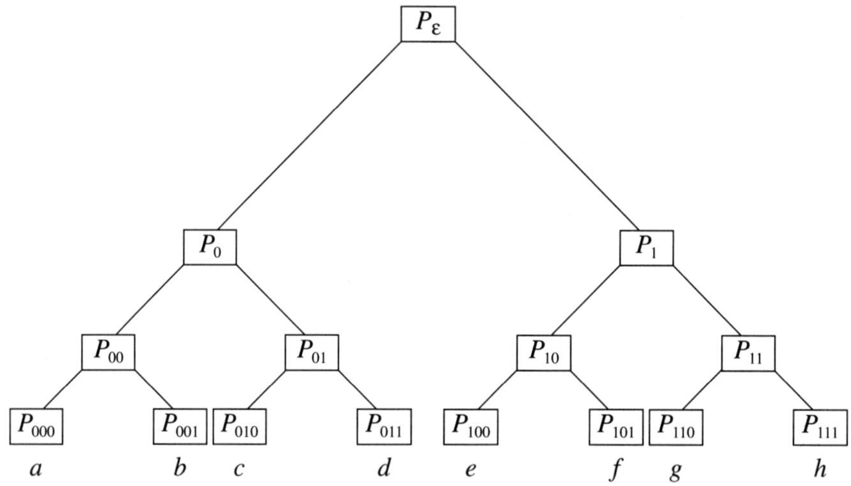

(a) Phase 1: Input the numbers in the leaves of PT_{2n-1}.

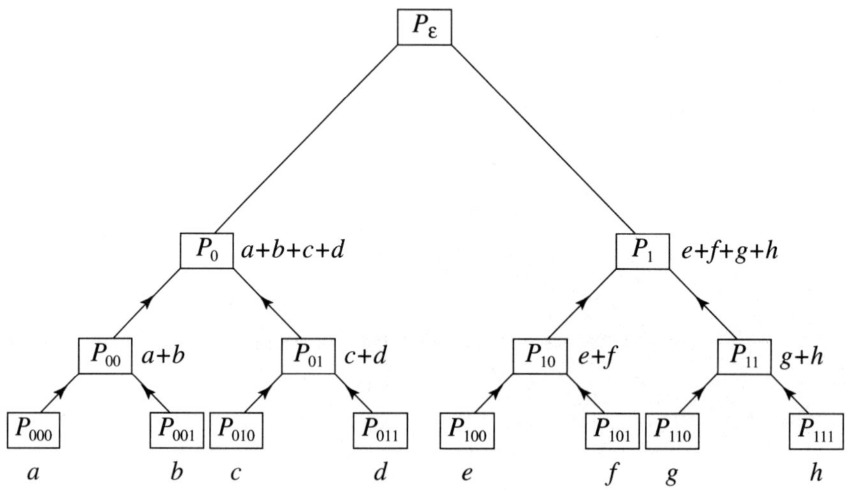

(b) Compute binary fan-in sums.

Figure 11.2 Action of *PrefixPT* for addition with input numbers *a,b,c,d,e,f,g,h*

steps and one assignment step. The action of phases 2 and 3 are illustrated in Figure 11.2c for the addition operator.

In phase 4, binary fan-out computations are performed as follows. Starting with the processors at level one and working our way down level by level to the leaves, a given processor communicates its element to both its children, and then each child adds the parent value to its value.

Note from Figure 11.2c, summing the numbers in the processors in the path from a node at level one to the ith leaf node yields the ith prefix sum, $i = 1,2, \ldots, n$. Phase

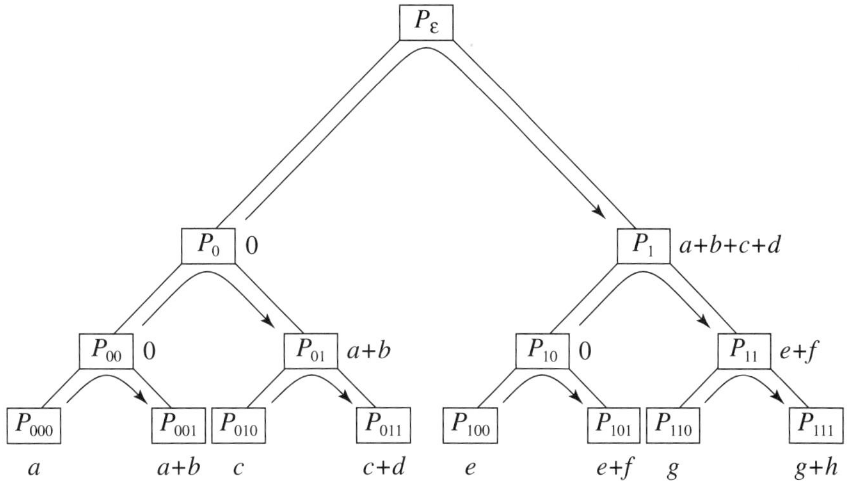

(c) Phase 2: For leaves, add sums in siblings and leave resulting sum in right child sibling. Phase 3: For non-root, non-leaf, left children, transfer binary fan-in sum to sibling then zero out own sum.

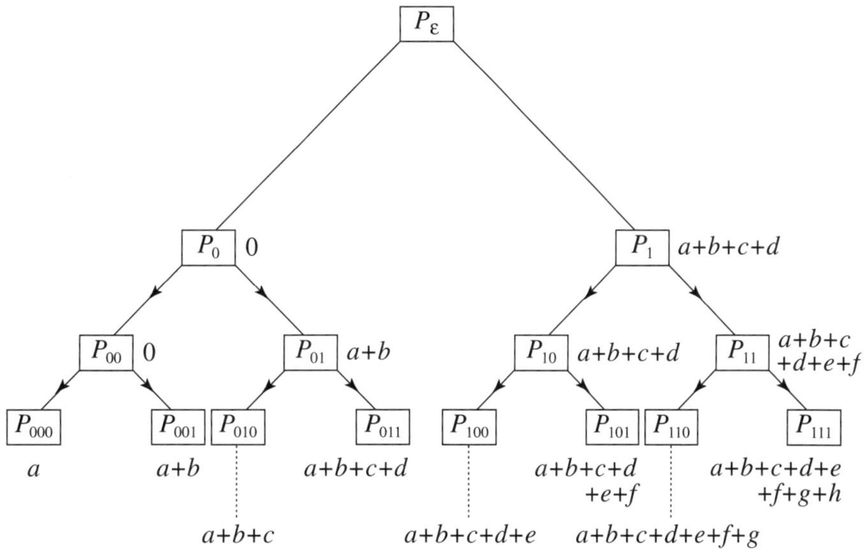

(d) Phase 4: Compute binary fan-out sums. Parallel prefix sums now reside in leaves.

Figure 11.2 Continued

4 consists of $\log_2 n - 1$ steps, where in step i each processor at level $i + 1$ combines its value with the value from its parent. Since each such step requires two communication steps and one parallel binary operation step, phase 4 takes $2 \log_2 n - 2$ communication

steps and $\log_2 n$ binary operation steps. At the end of phase 4, all the prefix computations reside in the leaf nodes. In particular, the prefix computation $x_1 \oplus x_2 \oplus \cdots \oplus x_i$ resides in the local memory of the ith leaf processor, $i = 1,2, \ldots, n$. Phase 4 is illustrated in Figure 11.2d.

The correctness of *PrefixPT* can be verified using induction and is left as an exercise.

11.1.3 Parallel Prefix Computations on the Two-Dimensional Mesh

We now design *Prefix2DMesh* for performing prefix computations on the two-dimensional mesh $M_{q,q}$, $n = q^2$. We assume that the elements $x_1, x_2 \ldots, x_n$ are stored in row-major order in the distributed variable *Prefix*.

Prefix2DMesh works in three phases. Phase 1 consists of $q - 1$ parallel steps, where in the jth step column j of *Prefix* is added to column $j + 1$. This is accomplished by communicating $\mathbf{P}_{ij}$:*Prefix* to processor $P_{i,j+1}$ and adding it to $\mathbf{P}_{i,j+1}$: *Prefix*, $j = 1, \ldots, q - 1$. Since the elements $x_1, x_2, \ldots, x_n$ are stored in row-major order in the distributed variable *Prefix*, after the completion of phase 1, $\mathbf{P}_{i,j}$:*Prefix* has the value $S_{(i-1)q+1,(i-1)q+j}$ (see Figure 11.3a).

Phase 2 consists of $q - 1$ steps, where in the ith step $\mathbf{P}_{iq}$:*Prefix* is communicated to processor $P_{i+1,q}$ and is then added to $\mathbf{P}_{i+1,q}$:*Prefix*, $i = 1, \ldots, q - 1$ (see Figure 11.3b). At the completion of phase 2, the value of $\mathbf{P}_{iq}$:*Prefix* is the desired prefix sum $S_{1,iq}$.

In phase 3, we add the value $\mathbf{P}_{i-1,q}$:*Prefix*, to $\mathbf{P}_{ij}$ thereby obtaining the desired prefix sum $S_{1,(i-1)q+j}$ in $\mathbf{P}_{ij}$:*Prefix*, $2 \le i \le q$ (see Figure 11.3c). Phase 3 can be accomplished in q parallel steps.

11.1.4 Carry-Lookahead Addition and Prefix Computations

The familiar grade school algorithm for adding two numbers adds digits one at a time from right to left (least significant digit to the most significant digit), propagating the appropriate carry from each digit position to its immediate neighbor on the left. For binary integers $x = x_{n-1} \ldots x_0$ and $y = y_{n-1} \ldots y_0$, their sum $z = z_n \ldots z_0$ is given by

$$z_i = (x_i + y_i + c_i) \textbf{ mod } 2, \quad 0 \le i \le n-1,\ c_0 = 0, \tag{11.1.1}$$
$$z_n = c_n,$$

where c_i is the carry, $i = 1, \ldots, n$. Clearly, if all the carries c_i have been computed, then we can compute z on the EREW PRAM with n processors in two parallel addition steps. We now show how computing the set of carries is equivalent to a certain prefix computation problem.

If both x_{i-1} and y_{i-1} are 0, then c_i is 0, and we say that the digit position i is in a *stop carry state*, denoted by s. If both x_{i-1} and y_{i-1} are 1, then c_i is 1, and we say that the digit position i is in a *generate carry state*, denoted by r. If $x_{i-1} \ne y_{i-1}$, then c_i equals the previous carry c_{i-1}, and we say that the digit position i is in a *propagate carry state*, denoted by p.

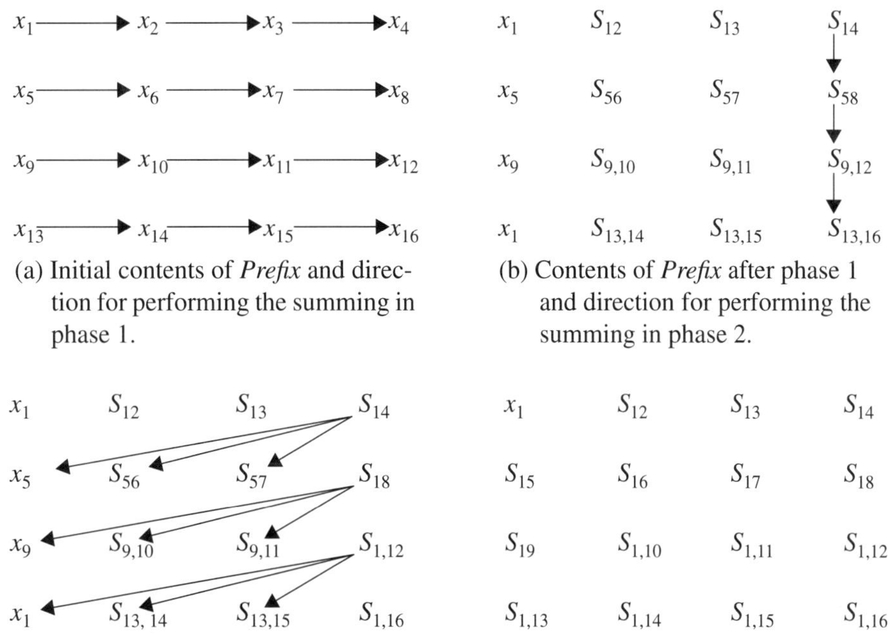

(a) Initial contents of *Prefix* and direction for performing the summing in phase 1.

(b) Contents of *Prefix* after phase 1 and direction for performing the summing in phase 2.

(c) Contents of *Prefix* after phase 2. Arrow from processor $P_{i-1,\,q}$ to P_{ij} indicates that $\mathbf{P}_{ij}$:*Prefix* should be replaced with $\mathbf{P}_{ij}$:*Prefix* $\oplus$ $\mathbf{P}_{i-1,\,q}$:*Prefix* in phase 3. This can be accomplished in q parallel steps.

(d) Contents of *Prefix* upon completion.

Figure 11.3 Action of *Prefix2DMesh* for $n = 16$

The carry states σ_{i-1} and σ_i of adjacent digit positions $i - 1$ and i, respectively, can be combined using the following binary operation $\oplus$. The definition of $\oplus$ is motivated by Figure 11.4 (page 384), where the computation of $\sigma_{i-1} \oplus \sigma_i$ is shown for the nine possible values of (σ_{i-1}, σ_i). (Note that, although the ith digit position is to the left of the $(i - 1)$st digit position, when performing the computation $\sigma_{i-1} \oplus \sigma_i$, we have taken the carry state of the $(i - 1)$st digit position to be on the left.) The ripple effect illustrated in Figure 11.4 suggests defining the binary operation $\oplus$ on $\{s,r,p\}$ given in Figure 11.5.

The following proposition shows that the binary operation $\oplus$ on $\{s,r,p\}$ is amenable to parallel prefix computations.

PROPOSITION 11.1.1 The binary operation $\oplus$ on $\{s,r,p\}$ is associative. ❑

We leave the proof of Proposition 11.1.1 as an exercise. The following proposition shows how prefix computations can be used to compute all the carries c_i, $1 \le i \le n$.

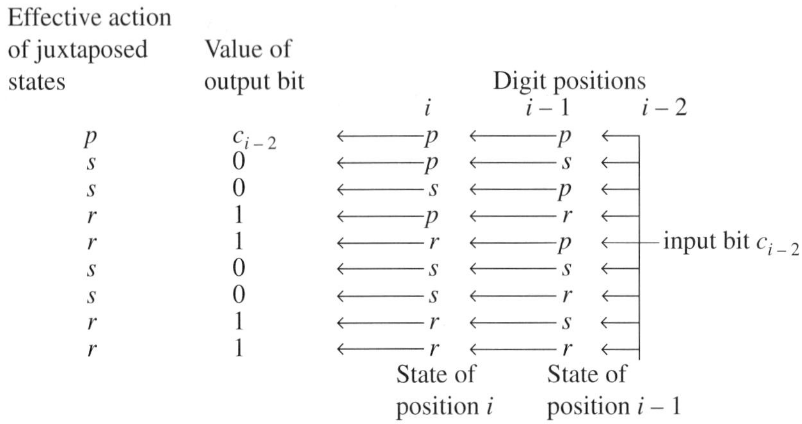

Figure 11.4 Effect of how carry digit c_{i-2} ripples through digit positions $i-1$, i for the nine possible states s,r,p of these two positions

PROPOSITION 11.1.2 Given n-digit binary integers x and y, let σ_i be the carry status of the ith digit of $x + y$, $0 \le i \le n$, so that $\sigma_0 = s$. Then $S_{0i} = \sigma_0 \oplus \sigma_1 \oplus \cdots \oplus \sigma_i$ is either r or s for each $i = 0, \ldots n$. Further

$$c_i = \begin{cases} 0, & S_{0i} = s. \\ 1, & \text{otherwise.} \end{cases}$$ ❑

The proof of Proposition 11.1.2 is left as an exercise. We illustrate Proposition 11.1.2 in Figure 11.6.

Propositions 11.1.1 and 11.1.2 yield an efficient algorithm for computing the sum of two n-digit binary integers x and y on the EREW PRAM with n processors. In the first parallel step, we compute the carry states σ_i, $1 \le i \le n$. We then use *PrefixPRAM* to compute the prefixes $S_{0i} = \sigma_0 \oplus \sigma_1 \oplus \cdots \oplus \sigma_i$, thereby determining all the carries c_i, $1 \le i \le n$. Finally, we compute $x + y$ (using 11.1.1) in two parallel addition steps. The complexity of this algorithm for computing $x + y$ on the EREW PRAM with n processors is in $\Theta(\log n)$.

Using *Prefix2DMesh*, we can compute $x + y$ on the two-dimensional mesh with n processors in time $O(\sqrt{n})$.

Figure 11.5 Defining the binary operation ⊕ on $\{s,r,p\}$

⊕	p	s	r
p	p	s	r
s	s	s	s
r	r	r	r

i	8	7	6	5	4	3	2	1	0	
x_i		1	0	1	0	1	0	0	1	
y_i		0	1	1	0	1	0	1	1	
σ_i	p	p	r	s	r	s	p	r	s	
S_{0i}	r	r	r	s	r	s	r	r	s	
c_i	1	1	1	0	1	0	1	1	0	
z_i	1	0	0	0	1	0	1	0	0	$z_i = (x_i + y_i + c_i)$ **mod** 2

Figure 11.6 Using prefix computations and Proposition 11.1.2 to compute the binary sum of $x = 10101001$ and $y = 01101011$

11.2 Parallel Matrix-Vector Product

Given an $n \times n$ matrix $A = (a_{ij})_{n \times n}$ and the column vector $X = (x_1, x_2, \ldots, x_n)^t$, the matrix-vector product AX is the column vector $B = (b_1, b_2, \ldots, b_n)^t$ defined by

$$b_i = \sum_{j=1}^{n} a_{ij} x_j, \quad i = 1, \ldots, n. \tag{11.2.1}$$

In this section we discuss efficient algorithms for computing AX on PRAMs, Meshes, and Meshes-of-Trees. With the exception of the algorithm *MatVecProd1DMesh*, all of these algorithms perform a *single* parallel multiplication. Therefore, in analyzing their performances, we choose addition as the basic operation. We compute speedup by comparing their complexities to the n^2 worst-case complexity of *MatVecProd*, where *MatVecProd* is the sequential algorithm based directly on definition (11.2.1).

11.2.1 Matrix-Vector Product on a PRAM

Suppose the matrix $A = (a_{ij})_{n \times n}$ and the column vector $X = (x_1, x_2, \ldots, x_n)^t$ are stored as arrays $A[1{:}n,1{:}n]$ and $X[1{:}n]$ in shared memory on a CREW PRAM with n^2 processors. Since the ith component b_i of $B = AX$ is simply the dot product of the ith row $A[i,1{:}n]$ of A and the vector X, each b_i can be computed simultaneously by invoking *DotProdPRAM* with arguments $A[i,1{:}n]$ and $X[1{:}n]$.

procedure *MatVecProdCREW*($A[1{:}n,1{:}n]$,$X[1{:}n]$,*Prod*$[1{:}n]$)

Model: CREW PRAM **with** $p = n^2$ **processors**

Input: $A[1{:}n,1{:}n]$ (an array of real numbers $(a_{ij})_{n \times n}$)
$X[1{:}n]$ (an array of real numbers $x_1, x_2, \ldots, x_n$)

Continued

Output: *Prod*[1:*n*] (matrix-vector product, where *Prod*[*i*] =
$a_{i1}x_1 + a_{i2}x_2 + \cdots + a_{in}x_n$, $i = 1, \ldots, n$)

for $1 \le i \le n$ **do in parallel**
  Prod[*i*] := *DotProdPRAM*(*A*[*i*,1:*n*],*X*[1:*n*])
end in parallel
end *MatVecProdCREW*

Clearly, *MatVecProdCREW* has the same complexity as *DotProdPRAM*, so that it performs $W(n) = \log_2 n$ parallel additions. Hence, *MatVecProdCREW* has a speedup of $n^2/\log_2 n$. However, *MatVecProdCREW* has the drawback that it uses n^2 processors, so that its cost is $C(n) = n^2\log_2 n$.

MatVecProdCREW is easily modified to obtain an algorithm *MatVecProdPRAM* suitable for the EREW PRAM. We merely replicate the vector *X*[1:*n*] in $\log_2 n$ steps to an $n \times n$ array *XX*[1:*n*,1:*n*], each row of *XX* being a copy of *X*[1:*n*]. The algorithm *MatVecProdPRAM* is then completed by replacing the body of the **in parallel** of *MatVecProdCREW* by the statement:

Prod[*i*] := *DotProdPRAM*(*A*[*i*,1:*n*],*XX*[*i*,1:*n*]).

11.2.2 Matrix-Vector Product on the One-Dimensional Mesh

Given the one-dimensional mesh M_p, with $p(n) = n$ processors, we now describe an optimal parallel algorithm *MatVecProd1DMesh* for computing the matrix-vector product *AX*. Since the matrix *A* has n^2 elements, and since we do not want to use more than a constant (independent of *n*) number of distributed variables, we cannot have the matrix $A = (a_{ij})_{n \times n}$ resident in memory. Instead, we input *A* and $X = (x_1, \ldots, x_n)^t$ from an external device. The algorithm *MatVecProd1DMesh* holds *one* value of the matrix $(a_{ij})_{n \times n}$ and vector $(x_1, \ldots, x_n)^t$ in any processor's memory at each stage in the algorithm. The algorithm controls input and interprocessor communication so that a given processor obtains the relevant data for computing a product at exactly the right time, no later and no earlier.

For convenience, we will assume that the elements of matrix $(a_{ij})_{n \times n}$ and vector $(x_1, \ldots, x_n)^t$ are input to distributed variables *A* and *X*, respectively. The algorithm *MatVecProd1DMesh* consists of two phases, where $(a_{ij})_{n \times n}$ and $(x_1, \ldots, x_n)^t$ are input in a staggered fashion as illustrated in Figure 11.7. Phase 1 consists of *n* steps, where in the *i*th step x_i is input to $\mathbf{P}_1$:*X*. At the same time the previously input values $x_{i-1}, \ldots, x_1$ are communicated down the line to the processors $P_2, \ldots, P_i$, respectively. The *i*th row of the matrix $(a_{ij})_{n \times n}$ is input one component at a time to processor P_i, with $\mathbf{P}_i$:*A* receiving a_{ij} at stage $i + j - 1$. Upon receiving values of *A* and *X*, processor P_i multiplies these values and adds the result to $\mathbf{P}_i$:*Prod*, where *Prod* is a distributed variable initialized to 0's. Phase 1 is competed once the entire vector $(x_1, \ldots, x_n)^t$ has been input. Phase 2 consists of performing the same type of steps as phase 1, except that components of the vector $(x_1, \ldots, x_n)^t$ are no longer input to P_1.

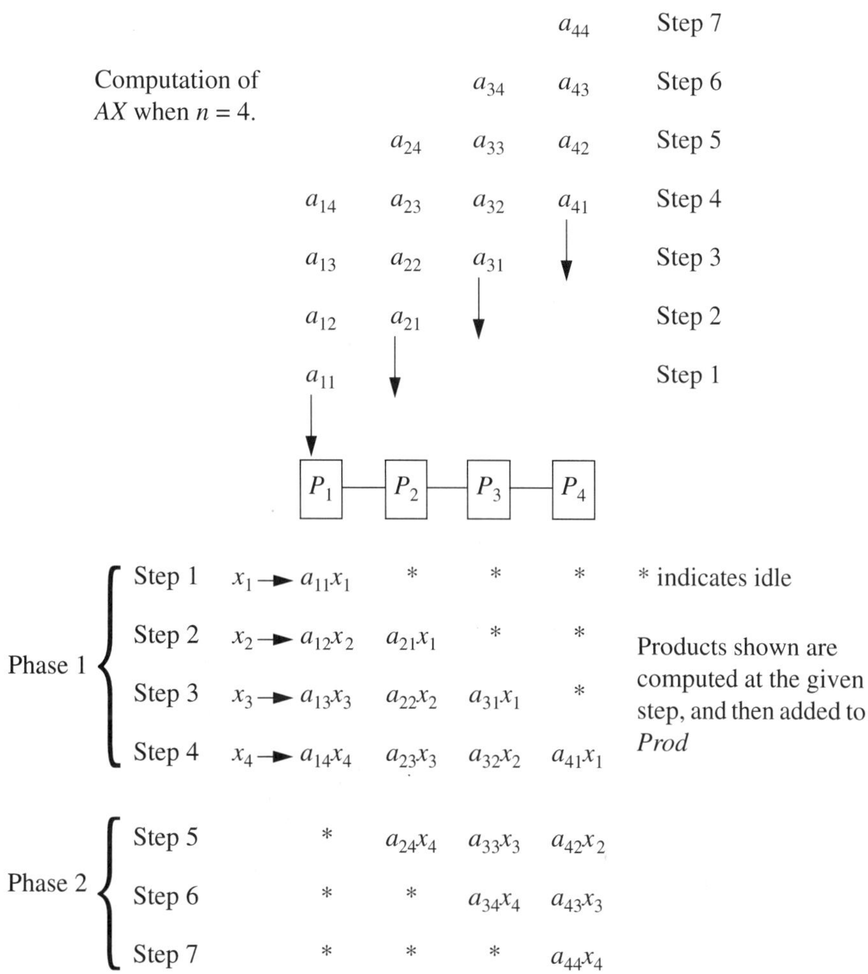

Figure 11.7 Action of *MatVecProd1DMesh* when $n = 4$

The pseudocode for *MatVecProd1DMesh* is fairly easy to write based on Figure 11.7.

procedure *MatVecProd1DMesh(Prod)*
Model: M_p **with** $p = n$ **processors**

External Input: A (a_{ij} input to $\mathbf{P}_i$:A at step $i + j - 1$), **range:** P_i, $1 \le i \le n$
X (x_i input to $\mathbf{P}_1$:X at step i), **range:** P_i, $1 \le i \le n$
Output: *Prod* ($\mathbf{P}_i$:$Prod_i = a_{i1}x_1 + a_{i2}x_2 + \cdots + a_{in}x_n$) **range:** P_i, $1 \le i \le n$

Continued

```
for 1 ≤ i ≤ n do in parallel {initialize Prod}
   P_i:Prod := 0
end in parallel
                                          {Phase 1}
for j := 1 to n do
   for P_i, 1 ≤ i ≤ j do in parallel
      if i < j then
         P_{i+1}:X ⇐ P_i:X                {propagate X right}
      endif
      read(P_1:X)   {P_1:X = x_j}
      read(P_i:A)   {P_i:A = a_{i,j-i+1}}
      Prod := Prod + A*X
   end in parallel
endfor
                                          {Phase 2}
for j := 2 to n do
   for P_i, j - 1 ≤ i ≤ n - 1 do in parallel
      P_{i+1}:X ⇐ P_i:X                   {propagate X right}
      read(P_i:A)                         {input a_{i,n+j-i} to P_i:A}
      Prod := Prod + A*X
   end in parallel
endfor
end MatVecProd1DMesh
```

Choosing multiplication as the basic operation, *MatVecProd1DMesh* has complexity $W(n) = 2n - 1$. Hence *MatVecProd1DMesh* has speedup $S(n) = n^2/(2n - 1)$ and cost $C(n) = n(2n - 1)$.

11.2.3 Matrix-Vector Product on the Two-Dimensional Mesh

We now design an algorithm *MatVecProd2DMesh* for computing the matrix-vector product AX on the two-dimensional mesh $M_{n,n}$. It is natural to assume that the distributed variable A contains the matrix A, where $\mathbf{P}_{i,j}{:}A$ contains a_{ij}. We also assume on input that the vector X is contained in the distributed variable X and that $\mathbf{P}_{1,j}{:}X$ contains x_j. The first step of the procedure *MatVecProd2DMesh* is to broadcast the values of X in the first row to all the other rows. Then we compute the product $A*X$ and store the result in X. Finally, we sum the X values across each row, leaving the final result in the first column of X. The action of *MatVecProd2DMesh* is illustrated in Figure 11.8 for $n = 3$ and a sample A and X.

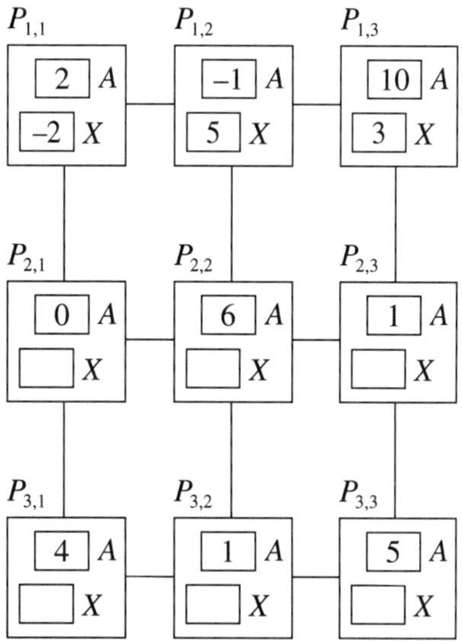

(a) Initial values of distributed variable A and X

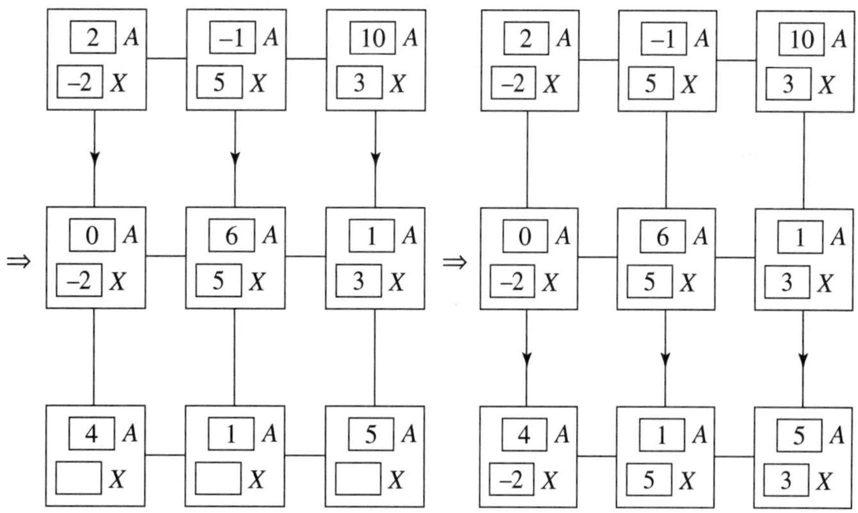

(b) Broadcast values of X in first row to rows 2 and 3.

Figure 11.8 Action of *MatVecProd2DMesh* for sample 3×3 matrix A and vector X of size $n = 3$

Continued

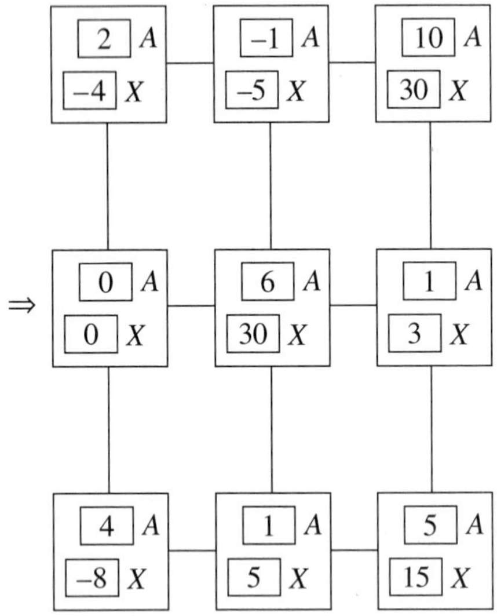

(c) **for** $P_{i,j}$, $1 \le i,j \le q$ **do in parallel**
$X := A * X$
end in parallel

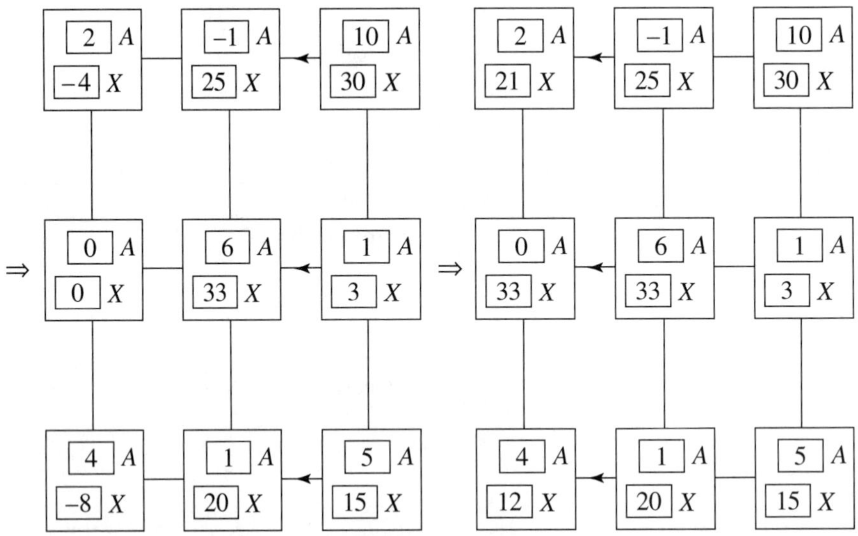

(d) Add third column's X values to those in second column, then add second column's X values to first column. Dot product resides in first column's X.

Figure 11.8 Continued

```
procedure MatVecProd2DMesh(A,n,X)
Model: two-dimensional mesh M_{n,n} with p = n^2 processors
Input:  A (P_{i,j}:A contains a_{ij}), range: P_{i,j}, 1 ≤ i,j ≤ n
        X (P_{1,j}:X contains x_j), range: P_{1,j}, 1 ≤ j ≤ n
Output: X (P_{i,1}:X contains a_{i1}x_1 + a_{i2}x_2 + ··· + a_{in}x_n), range: P_{i,1}, 1 ≤ i ≤ n

  for i := 1 to n – 1 do
          {broadcast X from ith row to (i + 1)st row}
    for P_{i,j}, 1 ≤ j ≤ n do in parallel
      P_{i+1,j}:X ⇐ P_{i,j}:X   {propagate X down}
    end in parallel
  endfor
          {compute a_{ij}x_j in parallel}
  for P_{i,j}, 1 ≤ i,j ≤ n do in parallel
    X := A * X
  end in parallel
          {sum across rows in parallel}
  for j := n down to 2 do
    for P_{i,j}, 1 ≤ i ≤ n do in parallel
      P_{i,j-1}:Temp ⇐ P_{i,j}:X   {communicate left from X to Temp}
      X := X + Temp
    end in parallel
  endfor

end MatVecProd2DMesh
```

Clearly, *MatVecProd2DMesh* performs $W(n) = n - 1$ parallel additions. Thus, *MatVecProd2DMesh* has speedup $S(n) = n^2/(n - 1)$ and cost $C(n) = n^2(n - 1)$.

11.2.4 Matrix-Vector Product on the Two-Dimensional Mesh-of-Trees

The linear worst-case complexity of *MatVecProd2DMesh* is optimal for the two-dimensional mesh architecture because of the need to communicate information from the last column in the mesh to the first column. A related architecture called the *two-dimensional mesh-of-trees* $MOT_{n,n}$ supports computing the matrix-vector product in a logarithmic number of steps. $MOT_{n,n}$ is constructed from the two-dimensional mesh $M_{n,n}$, where n is a power of 2, by removing the edges of $M_{n,n}$ and adding complete binary trees (with processors at each internal node of the tree) above each row and column (see Figure 11.9). $MOT_{n,n}$ has a total of $3n^2 - 2n$ processors. The leaf processors of $MOT_{n,n}$ correspond to the processors of $M_{n,n}$. We denote the leaf processor of $MOT_{n,n}$ corresponding to processor P_{ij} of $M_{n,n}$ by L_{ij}. We let RT_i denote the ith row tree, that is, the tree whose leaves lie along the ith row. Similarly, we let CT_j denote the jth column tree. Note that leaf processor P_{ij} lies in both row tree i and column tree j.

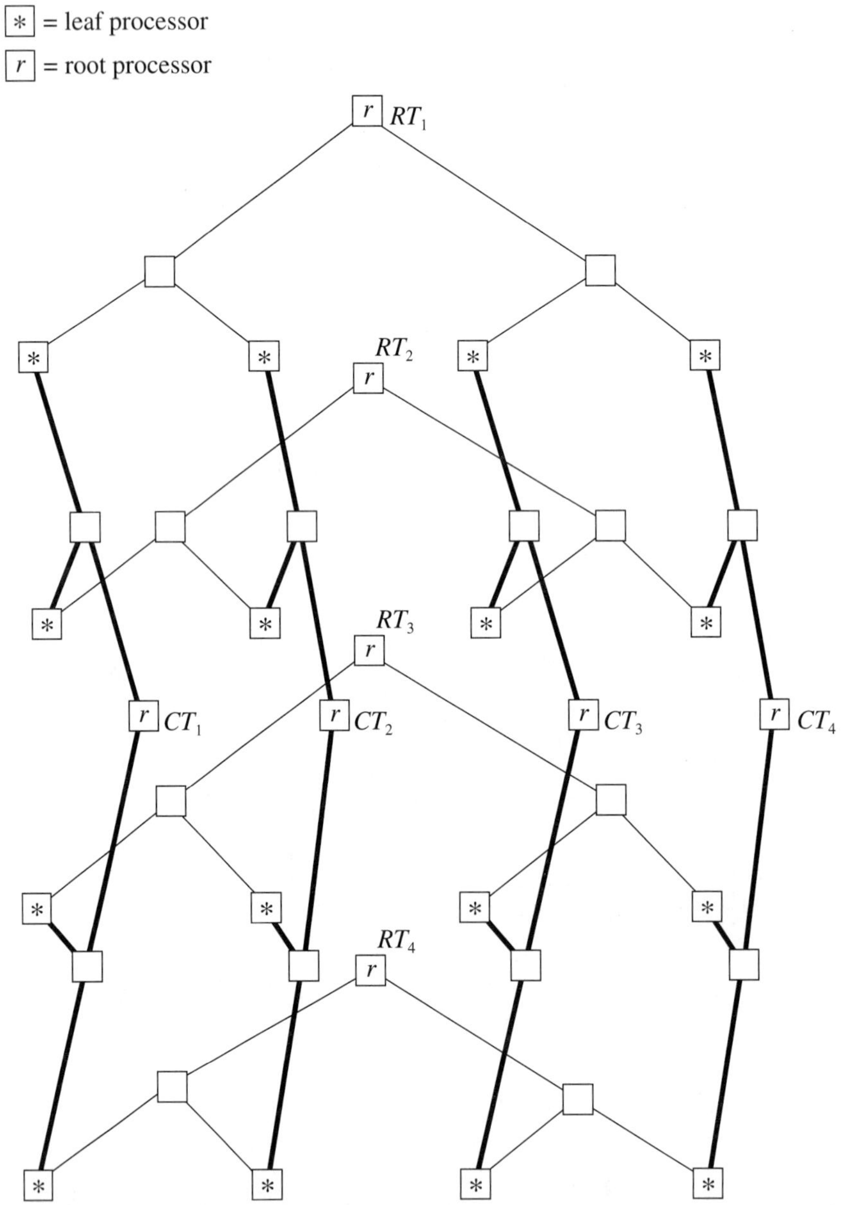

Figure 11.9 Two-dimensional mesh-of-trees $MOT_{4,4}$

In our algorithm *MatVecProd2DMOT* we input the adjacency matrix $A = (a_{ij})$ to the n^2 leaf processors, storing a_{ij} in leaf processors L_{ij}, $i,j = 1, \ldots, n$. We input the vector X to the roots of the column trees, storing x_j in the root of column tree CT_j,

$j = 1, \ldots, n$. The algorithm *MatVecProd2DMOT* consists of three phases. In the first phase, in parallel for $1 \leq j \leq n$ the value x_j in the root of column tree CT_j is broadcast in $\log_2 n$ steps to the n leaf processors $L_{1j}, \ldots, L_{nj}$. Thus, after phase 1, processor L_{ij} contains the values a_{ij} and x_j. In the second phase, in parallel for $1 \leq i,j \leq n$ we compute the product $a_{ij}x_j$. In the final phase, in parallel for $1 \leq i \leq n$ we sum the values in leaf processors $L_{i1}, \ldots, L_{in}$ using the binary fan-in technique on the ith row tree RT_i, so that after $\log_2 n$ steps the value $b_i = \Sigma_{j=1}^{n} a_{ij} x_j$ resides at the root. Hence, after the three phases are completed, the vector AX is stored in the roots of the row trees.

Clearly, *MatVecProd2DMOT* performs $W(n) = \log_2 n$ parallel additions. Thus, *MatVecProd2DMOT* has speedup $S(n) = n^2/\log_2 n$ and cost $C(n) = n^2 \log_2 n$.

11.3 Parallel Matrix Multiplication

Given a $p \times q$ matrix $A = (a_{ij})$ and a $q \times r$ matrix $B = (b_{ij})$, the product AB is defined to be the $p \times r$ matrix $C = (c_{ij})$, where

$$c_{ij} = \sum_{k=1}^{q} a_{ik} b_{kj} \qquad 1 \leq i \leq p,\ 1 \leq j \leq r \tag{11.3.1}$$

The sequential algorithm *MatMult*, based directly on the definition (11.3.1), performs pqr multiplications.

In this section we consider the problem of designing parallel algorithms for computing the matrix product $C = AB$ of square matrices A and B ($p = q = r = n$) on PRAMs and on the three-dimensional mesh-of-trees. The algorithms we design perform a *single* parallel multiplication, so that we choose addition as the basic operation when analyzing their performance. We compute speedup by comparing their complexities to the n^3 worst-case complexity of *MatMult*.

11.3.1 Matrix Multiplication on the PRAM

If B_j denotes the jth column vector of B, then the jth column vector of AB is simply the matrix-vector product AB_j, $j = 1, \ldots, n$. Hence, the following algorithm computes the product AB on the CREW PRAM with n^3 processors.

procedure *MatMultCREW*($A[1{:}n,1{:}n]$,$B[1{:}n,1{:}n]$,$C[1{:}n,1{:}n]$)
Model: CREW PRAM **with** $p = n^3$ **processors**
Input: $A[1{:}n,1{:}n]$ (an array of real numbers $(a_{ij})_{n \times n}$)
$B[1{:}n,1{:}n]$ (an array of real numbers $(b_{ij})_{n \times n}$)
Output: $C[1{:}n,1{:}n]$ (matrix product of A and B, where $Prod[i] =$
$a_{i1}x_1 + a_{i2}x_2 + \cdots + a_{in}x_n$, $i = 1, \ldots, n$)

Continued

```
    for 1 ≤ i,j ≤ n do in parallel
        C[i,j] := DotProdPRAM(A[i,1:n],B[1:n,j])
    end in parallel
end MatMultCREW
```

Clearly, *MatMultCREW* has the same complexity $W(n) = \log_2 n$ as *DotProdPRAM*. Hence, *MatMultCREW* has speedup $S(n) = n^3/\log_2 n$ and cost $C(n) = n^3\log_2 n$.

MatMultCREW is easily modified so that it is suitable for the EREW PRAM.

11.3.2 Matrix Multiplication on the Three-Dimensional Mesh-of-Trees

For the two-dimensional mesh-of-trees, various difficulties arise in managing the computations of *AB*. However, the three-dimensional mesh-of-trees $MOT_{n,n,n}$ is ideally suited for implementing matrix multiplication. The mesh-of-trees $MOT_{n,n,n}$ is constructed from the three-dimensional mesh $M_{n,n,n}$ by replacing each of the n^2 one-dimensional submeshes in each of the three coordinate directions with a complete binary tree. More precisely, for each 3-tuple (i,j,k), $i, j, k \in \{1,2, \ldots, n\}$, we have a leaf processor L_{ijk}. For each dimension $d \in \{1,2,3\}$ and each pair (i,j), $i,j \in \{1,2, \ldots, n\}$, we add a binary tree $T_{ij}^{(d)}$, called the (i,j) *dimension-d tree* such that

$$T_{ij}^{(1)} \text{ has leaf processors } \{L_{xij}, x \in \{1,2, \ldots, n\},$$

$$T_{ij}^{(2)} \text{ has leaf processors } \{L_{ixj}, x \in \{1,2, \ldots, n\},$$

$$T_{ij}^{(3)} \text{ has leaf processors } \{L_{ijx}, x \in \{1,2, \ldots, n\}.$$

We now describe an efficient algorithm *MatMult3DMOT* for multiplying two $n \times n$ matrices $A = (a_{ij})$ and $B = (b_{ij})$ on $MOT_{n,n,n}$. *MatMult3DMOT* consists of the following four phases. In each phase the computations are performed in parallel for the appropriate range of active processors.

Phase 1: Input a_{ij} and b_{ij} to the roots of $T_{ij}^{(1)}$ and $T_{ji}^{(2)}$, respectively.

Phase 2: Broadcast input values to the leaves, so that the leaves of $T_{ij}^{(1)}$ all have the value a_{ij}, and the leaves of $T_{ji}^{(2)}$ all have the value b_{ij}.

Phase 3: After phase 2 is completed, the leaf processor L_{jik} has both the value a_{ik} and the value b_{kj}. In a single parallel step, compute the product $a_{ik}b_{kj}$.

Phase 4: Sum the leaves of tree $T_{ji}^{(3)}$ so that resultant sum is stored in the root of $T_{ji}^{(3)}$. Hence, the root contains the value $c_{ij} = \sum_{k=1}^{n} a_{ik}b_{kj}$.

After phase 4 has been completed the product matrix $C = AB$ resides in the roots of the dimension-3 trees, with c_{ij} stored in the root of tree $T_{ji}^{(3)}$.

Clearly, *MatMult3DMOT* performs $W(n) = \log_2 n$ parallel additions. Therefore, *MatMult3DMOT* has speedup $S(n) = n^3/\log_2 n$ and cost $C(n) = 2n^3\log_2 n$.

11.3.3 Powers of a Matrix

Scientific computations often involve computing the powers of an $n \times n$ matrix A. We can compute the kth power A^k by using the binary method discussed in Chapter 1 and making O(log k) calls to an appropriate matrix multiplication algorithm for the given PRAM or interconnection network under consideration. In particular, using *MatMult3DMesh* to perform matrix products, we obtain an O((log n)(log k)) algorithm for computing A^k on the three-dimensional mesh-of-trees.

11.4 Pointer Jumping

Starting at any node in a linked list of nodes, we can move down the list using a sequential algorithm that uses the pointer at each node to move to the next node until the end is reached. After at most $n - 1$ pointer updates, we will reach the end of the list. Starting from any node, there is a parallel algorithm on the EREW PRAM that can reach the end of the list in $\log_2 n$ parallel steps using the important technique of *pointer jumping*. Each node traces a "path" to the end of the list in parallel, but at each step, the path jumps over twice as many nodes as in the previous step (until it reaches the end). The pointer-jumping technique has many important applications in the theory of parallel algorithms.

11.4.1 Finding Roots in a Forest

The technique of pointer jumping generalizes immediately from linked lists to trees (or, more generally, forests) maintained using the parent implementation. (Given any node in a forest maintained using parent pointers, a path from the node to the root of the tree containing it is simply a linked list.) To illustrate, we consider a rooted forest F on n nodes, where we denote the nodes by $1, \ldots, n$. We suppose that F is implemented using the parent array implementation *Parent*[1:n] (see Chapter 2). To facilitate pointer jumping, we will modify the implementation given in Chapter 2 slightly by setting *Parent*[j] = j when j is a root node rather than setting *Parent*[j] = 0 (see Figure 11.10, page 396).

Given a node i in the forest, where $R[i]$ denotes the root of the tree in F containing the node i, $i = 1, \ldots, n$, we now consider the problem of designing a parallel algorithm for computing the entire array $R[1{:}n]$ on the CREW PRAM. Before presenting the parallel algorithm, we first consider the problem in a sequential setting. For a particular node i in F, clearly the optimal sequential algorithm for determining $R[i]$ is to follow the path up the tree from i to $R[i]$, as accomplished by the following pseudocode.

```
Current := i
while Current ≠ Parent[Current] do
    Current := Parent[Current]
endwhile
R[i] := Current
```

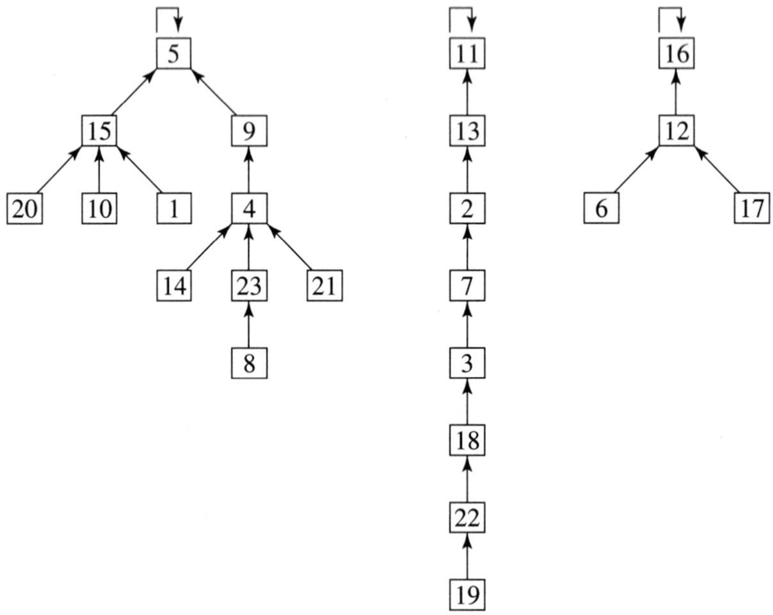

i	1	2	3	4	5	6	7	8	9	10	11	12	13	14	15	16	17	18	19	20	21	22	23
Parent[i]	15	13	7	9	5	12	2	23	5	15	11	16	11	4	5	16	12	3	22	15	4	18	4

Figure 11.10 Parent array implementation of a forest

If i is at level l_i in the tree, then this algorithm requires $l_i + 1$ assignments to *Current*. In particular, when F consists of a single tree and the tree is a path with i as a leaf, then the algorithm performs $W(n) = n$ assignments to *Current*.

To compute the entire array $R[1{:}n]$, the most straightforward sequential algorithm that does the job simply repeats the previous pseudocode for each i between 1 and n. Unfortunately, the worst-case behavior of this algorithm belongs to $\Theta(n^2)$. A better sequential algorithm results from the observation that if j is any node along the path from i to $R[i]$, then $R[j] = R[i]$. Hence, we can stop our journey up the path from i to $R[i]$ whenever we encounter a node such that $R[j]$ has already been determined. Also, after determining $R[i]$, we can retrace the part of the path just traveled and assign $R[k] = R[i]$ for each node encountered. It turns out that the resulting algorithm has linear worst-case complexity. We leave the design and analysis of the improved sequential algorithm based on these observations to the exercises.

The key idea utilized in the design of an efficient parallel algorithm for computing $R[1{:}n]$ on the CREW PRAM is the notion of *pointer jumping* mentioned earlier. At each step we double the number of levels we jump until we reach the root. The jumping is accomplished by replacing *Parent*[i] with *Parent*[*Parent*[i]] exactly $\log_2 n$ times (since *Parent*[i] $= i$, once we reach a root node we remain at the root). The algorithm is shown next and is illustrated in Figure 11.11.

```
procedure ForestRootsCREW(Parent[1:n],R[1:n])
Model: CREW PRAM with p = n processors
Input:  Parent[1:n] (parent array of a forest F)
Output: R[1:n] (array such that R[i] is root of tree of F containing i,
                i = 1, . . . , n)
   for 1 ≤ i ≤ n do in parallel
      for j := 1 to log₂n do
         Parent[i] := Parent[Parent[i]]
      endfor
   end in parallel
   R[i] := Parent[i]
end ForestRootsCREW
```

Figure 11.11 Pointer jumping performed by *ForestRootsCREW* for a sample path of length 16 from a node to the root of a tree in a forest

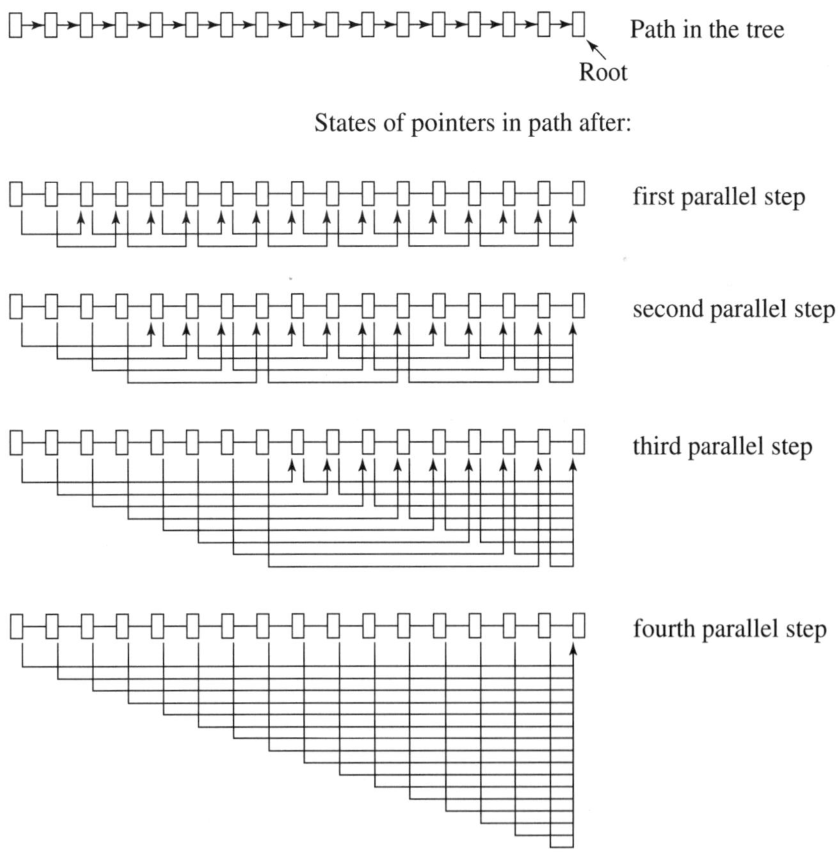

Clearly, *ForestRootsCREW* has complexity $W(n) \in \Theta(\log n)$, speedup $S(n) \in \Theta(n/\log n)$, and cost $C(n) \in \Theta(n \log n)$.

Procedure *ForestRootsCREW* can be adapted to the EREW PRAM at the expense of a logarithmic multiplicative increase in complexity. However, if all the trees in the forest F are paths, then *ForestRootsCREW* can be applied to the EREW PRAM without any slowdown.

11.4.2 List Ranking

Suppose we have a linked list and we want to determine the distance from each element in the list to the end of the list. Determining these distances is called the *list-ranking* problem. An (optimal) $\Theta(n)$ sequential algorithm for list ranking is easy to write. Using pointer jumping, we show how to obtain an $O(\log n)$ algorithm for list ranking on the EREW PRAM.

For definiteness, we suppose that the linked list is maintained using the array $L[1{:}n]$ and the array $Next[1{:}n]$, where the variable *Start* points to the index in *Next* corresponding to the beginning of the linked list. We assume that the end of the linked list is signaled by $Next[i] = 0$. For example, the list 23,7,8,0,12,9,5,3,11,2 might be maintained as the linked list shown in Figure 11.12. The list-ranking problem for the linked list in Figure 11.12 is to compute the array $Distance[1{:}10]$ given the arrays $L[1{:}10]$ and $Next[1{:}10]$.

The following recursive algorithm *ListRank* is clearly an optimal sequential algorithm for the list-ranking problem. *ListRank* is initially invoked with $i = Start$, and during the resolution of this initial call it fills the global array $Distance[1{:}n]$ with the appropriate distances.

```
procedure ListRank(L[1:n],Next[1:n],i) recursive
Input:  L[1:n] (an array of list elements)
        Next[1:n] (an array giving linked list implementation of L)
        i (index of elements in L, i = Start on initial call)
Output: Distance[i] (distance from L[i] to end of list)
dcl Distance[1:n] global
   if Next[i] = 0 then
      Distance[i] := 0
   else
      call ListRank(Next[i])
      Distance[i] := Distance[Next[i]] + 1
   endif
end ListRank
```

Using pointer jumping on the EREW PRAM, we obtain an $O(\log n)$ algorithm for list ranking as follows. We initialize $Distance[1{:}n]$ with 1s, except for that index

Start = 3

Index	1	2	3	4	5	6	7	8	9	10
Next	4	5	7	10	6	1	2	0	8	9
L	9	8	23	5	0	12	7	2	11	3
Distance	4	7	9	3	6	5	8	0	1	2
index	3	7	2	5	6	1	4	10	9	8

23 → 7 → 8 → 0 → 12 → 9 → 5 → 3 → 11 → 2

Figure 11.12 A sample linked list of ten elements using arrays *L*[1:10], *Next*[1:10], together with the array *Distance*[1:10] giving the distance of each element to the end of the list

i such that $Next[i] = 0$, where we set $Distance[i] = 0$. After initialization, we trivially have the following property of the array *Distance*[1:*n*] and the linked sublists determined by *Next*[1:*n*].

PROPERTY A For each index i, $1 \le i \le n$, if we add together the values of Distance[j] for all indices j in the linked sublist headed by i, then we obtain the correct distance from i to the end of the (original) linked list. ❑

The idea behind our parallel algorithm for list ranking is to use pointer jumping to alter the array *Next* so that the length of the sublists is cut in half with each step, while at the same time updating the array *Distance* so that Property A remains valid for the new sublists. After $\lceil \log_2 n \rceil$ parallel pointer-jumping steps, all the linked sublists maintained by *Next* consist of singletons. Since Property A has been maintained for these sublists, *Distance*[1:*n*] necessarily has all the correct distances.

Each parallel pointer-jumping step has the form:

```
for 1 ≤ i ≤ n do in parallel
   if Next[i] ≠ 0 then
      Distance[i] := Distance[i] + Distance [Next[i]]          (11.4.1)
      Next[i] := Next[Next[i]]
   endif
end in parallel
```

During step (11.4.1), each element i, $1 \le i \le n$, adds its successor's value *Distance*[*Next*[i]] to its own value *Distance*[i] before jumping around it by changing *Next*[i]. Hence, the sum of the values *Distance*[j] for all indices j in the linked sublist headed by i will clearly not change in the new linked sublist headed by i resulting from performing (11.4.1). In other words, while (11.4.1) alters the linked sublist headed by i, it leaves the sum of the values *Distance*[j] in the sublist invariant. (When embedding (11.4.1) in a **while** loop, such an invariant property is called a *loop invariant*, as discussed in Chapter 7.) In particular, if *Next*[1:*n*] and *Distance*[1:*n*] satisfy Property A before performing (11.4.1), then they will also satisfy Property A after performing (11.4.1).

The pointer-jumping steps are illustrated in Figure 11.13 for the same list as given in Figure 11.12.

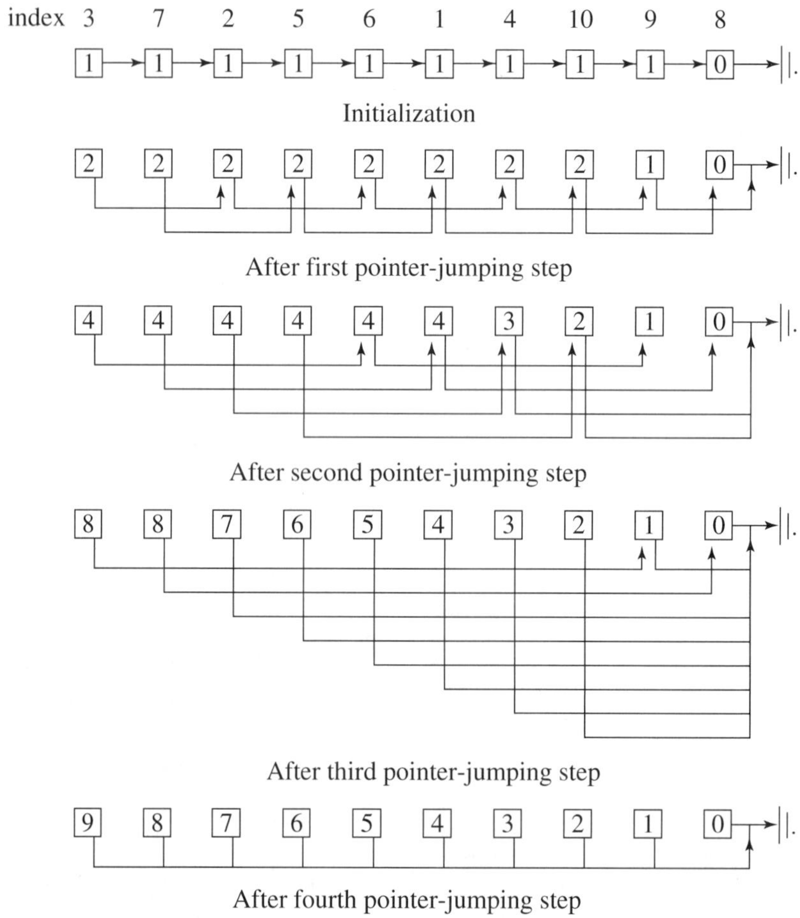

Figure 11.13 Action of pointer-jumping steps for the sample list given in Figure 11.12. After each pointer-jumping step, the state of the array *Next*[1:10] is indicated by the pointers, and the values of *Distance*[1:10] are shown inside each node in the linked sublists.

The pseudocode for *ListRankPRAM* follows immediately from the previous discussion.

procedure *ListRankPRAM*(*L*[1:*n*],*Next*[1:*n*],*Distance*[1:*n*],*Start*)

Model: EREW PRAM **with** $p = n$ **processors**
Input: *L*[1:*n*] (an array of list elements)
Next[1:*n*] (an array giving linked list implementation of *L*)
Start (the index in *L* of first element in linked list)

```
Output: Distance[1:n] (an array such that Distance[i] is the distance of L[i]
                                          to end of linked list, i = 1, . . . , n)

  for 1 ≤ i ≤ n do in parallel   {initialization}
    if Next[i] = 0 then
      Distance[i] := 0
    else
      Distance[i] := 1
    endif
  end in parallel
  while Next[Start] ≠ 0 do
    for 1 ≤ i ≤ n do in parallel
      if Next[i] ≠ 0 then
        Distance[i] := Distance[i] + Distance[Next[i]]
        Next[i] := Next[Next[i]]
      endif
    end in parallel
  endwhile

end ListRankPRAM
```

Since *ListRankPRAM* performs $\lceil \log_2 n \rceil$ iterations of the **while** loop, it has complexity $W(n) = \lceil \log_2 n \rceil$, speedup $S(n) = n/\lceil \log_2 n \rceil$, and cost $C(n) = n\lceil \log_2 n \rceil$.

Remark ListRankPRAM can be altered to a version not using the input parameter *Start*. Moreover, if the number of list elements is known, then this altered version obviously determines the first element in the linked list.

11.5 Closing Remarks

Figure 11.14 (page 402) summarizes the analysis of most of the parallel algorithms discussed in this chapter. Given is the *order* of the number $p(n)$ of processors required, the worst-case complexity $W(n)$ (the average and best-case complexities are the same as the worst-case complexity for all the algorithms we have discussed so far), the speedup $S(n)$, the cost $C(n)$, and the processor efficiency $E(n)$. For those algorithms involving matrix-vector and matrix-matrix products, speedup is calculated with respect to the straightforward sequential algorithms based directly on the definition of these products and having complexity n^2 and n^3, respectively.

We saw in Section 11.3 that the three-dimensional mesh-of-trees $MOT_{n,n,n}$ is particularly well suited for computing matrix products. Since there are efficient embeddings of $MOT_{n,n,n}$ into other standard interconnection networks such as the hypercube, algorithms designed for $MOT_{n,n,n}$ can be adapted to efficient algorithms on such networks. Further information on embeddings can be found in the references.

Algorithm	Basic operation	$p(n)$	$W(n)$	$S(n)$	$C(n)$	$E(n)$
PrefixCompPRAM	$\oplus$	n	$\log n$	$n/\log n$	$n \log n$	$1/\log n$
PrefixCompBinTree	$\oplus$	n	$\log n$	$n/\log n$	$n \log n$	$1/\log n$
PrefixComp2DMesh	$\oplus$	n	$\sqrt{n}$	$\sqrt{n}$	$n^{3/2}$	$1/\sqrt{n}$
MatVecProdCREW	+	n^2	$\log n$	$n^2/\log n$	$n^2\log n$	$1/\log n$
MatVecProd1DMesh	*	n	n	n	n^2	1
MatVecProd2DMesh	+	n^2	n	n	n^3	$1/n$
MatVecProd2DMOT	+	n^2	$\log n$	$n^2/\log n$	$n^2\log n$	$1/\log n$
MatMultCREW	+	n^3	$\log n$	$n^3/\log n$	$n^3\log n$	$1/\log n$
MatMult3DMOT	+	n^3	$\log n$	$n^3/\log n$	$n^3\log n$	$1/\log n$
ForestRootsCREW	ptr update	n	$\log n$	$n/\log n$	$n \log n$	$1/\log n$
ListRankPRAM	ptr update	n	$\log n$	$n/\log n$	$n \log n$	$1/\log n$

Figure 11.14 Order of the complexity measures of algorithms discussed in this chapter

References and Suggestions for Further Reading

A book entirely devoted to applications of prefix computations in parallel computing is:

Lakshmivarahan, S., and S. K. Dhall. *Parallel Computing Using the Prefix Problem*. New York: Oxford University Press, 1994.

Two recent books covering the topics discussed in this chapter, as well as material on network embeddings include:

Leighton, F. T. *Parallel Algorithms and Architectures*. San Mateo, CA: Morgan Kaufmann Publishers, 1992.

Quinn, M. J. *Parallel Computing: Theory and Practice*. New York: McGraw-Hill Inc., 1994.

Exercises

Section 11.1 Parallel Prefix Computations

11.1 Give pseudocode for an algorithm *PrefixPRAM* on the EREW PRAM that performs the n prefix computations $x_1 \oplus x_2 \oplus x_3 \oplus \cdots \oplus x_i$, $i = 1,2, \ldots, n$. Analyze the complexity, speedup, and cost of your algorithm.

11.2 Give pseudocode for parallel prefix computations on the processor tree PT_n.

11.3 Give an inductive proof of the correctness of *PrefixPT*.

11.4 Prove Lemma 11.1.1.

11.5 Prove Lemma 11.1.2.

11.6 Design and analyze an algorithm for parallel prefix computations on a general binary tree (where the operands are input to leaves).

11.7 Design and analyze an algorithm for parallel prefix computations on the complete ternary tree.

11.8 Consider the problem of computing the prefix sums of sets of numbers. For example, for the sets {1,45,3}, {2,5}, {1,8,90,6}, {10,20,30}, the prefix sums are

$$1,\ 46,\ 48;\ 2,\ 7;\ 1,\ 9,\ 99,\ 105;\ 10,\ 30,\ 60.$$

We refer to these prefix sums as the *segmented* prefix sums. Clearly, these segmented prefix sums can be computed by four separate calls to parallel prefix. However, it is interesting that they can be computed by a single call to parallel prefix. More generally, for any binary associative operation $\oplus$, we can compute the *segmented (parallel) prefix* as a special case of parallel prefix. This is accomplished by introducing the symbol $\mid$ and extending the set S of operands to the set $\overline{S}$ consisting of all operands of the form x and $x\mid$, where $x \in S$. We extend the binary operation $\oplus$ to the binary operation $\overline{\oplus}$ on the augmented set of symbols $\overline{S}$ shown in Figure 11.15.

a. Verify that the binary operation $\overline{\oplus}$ given by the table in Figure 11.15 is associative.

b. Show that performing the parallel prefix for the operation $\overline{\oplus}$, where $\oplus = +$, on the list

$$1,45,3, \mid 2,5, \mid 1,8,90,6, \mid 10,20,30$$

yields the segmented prefix sums.

c. Design an algorithm for computing segmented prefix using parallel prefix.

11.9 Design and analyze a parallel pancake-flipping algorithm (see Exercise 10.2 of Chapter 10).

11.10 Suppose the elements in an array $A[1{:}n]$ have two colors, red and blue. The *packing problem* is to rearrange the elements of the array so that all the red elements precede all the blue elements, while at the same time preserving the order of the elements in each color class.

a. Show how parallel prefix can be used to solve the packing problem.

Figure 11.15 Defining the binary operation $\overline{\oplus}$ on $\overline{S}$

$\oplus$	y	$\mid y$
x	$x \oplus y$	$\mid y$
$\mid x$	$\mid (x \oplus y)$	$\mid y$

b. Using part (a), design and analyze a parallelization of *RadixSort* for lists of binary elements.

c. Solve the packing problem for k colors and use the result to design a parallelization of *RadixSort* for a general base k. Analyze your algorithm.

11.11 Design and analyze a parallelization of the version of *RadixSort* described in Exercise 4.30 of Chapter 4 (which is based on proceeding from the most significant digit first). (*Hint:* Use the solution to the packing problem discussed in the previous exercise.)

Section 11.2 Parallel Matrix-Vector Product

11.12 *MatVecProdCREW* utilized the CREW model of a PRAM as opposed to the EREW model. Design and analyze a version for the EREW PRAM.

11.13 Give the pseudocode for *MatVecProd1DMesh2* for computing the matrix-vector product AX on the one-dimensional mesh with n processors, which assumes $A[1:n]$ and $X[1:n]$ are distributed input parameters, where $\mathbf{P}_i$:$A[1:n]$ contains the ith row of the matrix A and $\mathbf{P}_i$:X contains the vector $X = (x_1, \ldots, x_n)^t$. (Note that there is an instantiation of the arrays $A[1:n]$ and $X[1:n]$ in each processor's local memory, so that a total of $2n^2$ memory locations are used.) Return the matrix-vector product AX in the distributed output parameter *Prod*.

Section 11.3 Parallel Matrix Multiplication

11.14 Show that $\mathrm{MOT}_{n,n}$ has a total of $3n^2 - 2n$ processors.

11.15 Modify *MatMultCREW* so that it is suitable for the EREW PRAM.

11.16 Assuming that the ijth entries of the $n \times n$ matrices A,B,C are stored in processors P_{ij}'s local memory on the two-dimensional mesh, various difficulties arise in managing the computation of $C = AB$.

a. Discuss these difficulties.

b. Design an algorithm for computing C on the *wrap-around* two-dimensional mesh obtained from the ordinary mesh $\mathrm{M}_{q,q}$ by joining processors $P_{i,1}$ and $P_{i,q}$ and processors $P_{1,i}$ and $P_{q,i}$, $i = 1, \ldots, q$.

11.17 Modify each of the algorithms in this section to apply to the product of nonsquare matrices.

Section 11.4 Pointer Jumping

11.18 Given a parent array representing a forest, give a linear worst-case sequential algorithm that determines an array $R[1:n]$, where $R[i]$ is the root of the tree containing element i, $i = 1, \ldots, n$.

11.19 Write an algorithm for the EREW PRAM to find the maximum element in a linked list.

11.20 Write a list-ranking algorithm *ListRankPRAM* that does not assume that the variable *Start* pointing to the beginning of the linked list is available.

11.21 Suppose we are maintaining a linked list using two arrays. Design an algorithm on the EREW PRAM that performs prefix computations on the linked list so that the *i*th prefix computation resides in the *i*th node of the linked list.

11.22 Design and analyze an O(log n) algorithm on the CREW PRAM with n processors for determining the depth of each node in a binary tree with n nodes.

***11.23** Repeat Exercise 11.22 for the EREW PRAM.

***11.24** Given a pair of real numbers a and b, the closed interval $[a, b]$ consists of all the real numbers that lie between a and b, inclusive; that is,

$$[a, b] = \{x \in \mathbb{R} \mid a \leq x \leq b\}.$$

Now, suppose we are given a set of n closed intervals $I_1 = [a_1, b_1], \ldots, I_n = [a_n, b_n]$. We say that I_i and I_j have *strict intersection* if I_i and I_j have nonempty intersection, but neither interval is contained in the other. Describe a polylogarithmic parallel algorithm on the CREW PRAM with n processors to determine whether any pair of intervals in the set $I_1, \ldots, I_n$ have strict intersection. (*Hint:* Utilize sorting and pointer jumping.)

Part Three

Major Design Strategies

12

The Greedy Method

The *greedy method* for solving optimization problems follows the philosophy of greedily maximizing (minimizing) short-term gain and hoping for the best without regard to long-term consequences. Algorithms based on the greedy method are usually very simple, easy to code, and efficient. Unfortunately, just as in real life, in the theory of algorithms the greedy method applied to a problem often leads to less than optimal results. However, there are important problems where the greedy method does yield optimal results, such as finding a shortest path between two vertices in a weighted digraph or determining a minimum spanning tree in a weighted graph.

12.1 General Description

Since the subject is greed, it seems appropriate to illustrate the greedy method with a problem about money: making change (see Figure 12.1, page 410). Suppose that we have just purchased a sleeve of golf balls, and the salesperson wishes to give back (exact) change using the *fewest* number of coins. We assume that there is a sufficient amount of pennies (1¢), nickels (5¢), dimes (10¢), quarters (25¢), and half-dollars (50¢) to make change in any manner whatsoever.

A greedy algorithm for making change uses as many coins of the largest denomination as possible, then uses as many coins of the next largest denomination, and so forth. For example, if the change was 74 cents, the greedy solution uses one half-dollar, two dimes, and four pennies, for a total of seven coins. It is easy to see that seven coins is the fewest number of coins that makes 74 cents change. For any amount of change, this greedy method uses the fewest coins for coins of the above denominations. However, when other denominations of coins are used, this greedy method of making change does not always yield the fewest coins. For example, suppose we use the same denomination coins as before, except that we do not have any nickels. For 30 cents change, the greedy method yields one quarter and five pennies, whereas three dimes does the job.

"No one here knows how many tens to give you for a hundred, sir. The computer is down."

Figure 12.1 Making change

The coin-changing example serves as a nice illustration of the general type of problem that might be amenable to the greedy method. These problems generally have the goal of maximizing or minimizing an associated *objective function* over all feasible solutions. In most cases this objective function is a real-valued function defined on the subsets of a given base set S. For example, coin changing is a minimization problem, where the base set S is the set of denominations, a feasible solution is a set of coins making correct change, and the objective function is the number of coins used.

The greedy method might be applied to any problem whose solution can be viewed as the result of building up solutions to the problem via a sequence of partial solutions in the following manner. Partial solutions are built up by choosing elements $x_1, x_2, \ldots$ from a set R, where R is initially equal to some base set S. Whenever the greedy method makes a choice x_i, then x_i is either added to the current partial solution or its addition is found to be infeasible and is rejected forever. In either case, x_i is removed from the set R.

In the following general paradigm for the greedy method, choices are made by invoking a suitably defined function *GreedySelect*(*R*). For example, *GreedySelect*(*R*) might be an element that optimizes (minimizes or maximizes) the change in the value of the objective function or some closely related function. When there are several choices that are equivalent, we assume *GreedySelect* chooses in some arbitrary way from the equivalent choices.

```
procedure Greedy(S,Solution)
Input:   S (base set)        {it is assumed that there is an associated objective
                               function f defined on (possibly ordered) subsets of S}
Output: Solution             (an ordered subset of S that potentially optimizes the
                              objective function f, or a message that Greedy doesn't
                              even produce a solution (optimal or not))
   PartialSolution := ∅      {initialize the partial solution to be empty}
   R := S
   while PartialSolution is not a solution .and. R ≠ ∅ do
      x := GreedySelect(R)
      R := R \ {x}
      if PartialSolution ∪ {x} is feasible then
         PartialSolution := PartialSolution ∪ {x}
      endif
   endwhile
   if PartialSolution is a solution then
      Solution := PartialSolution
   else
      write("Greedy fails to produce a solution")
   endif
end Greedy
```

In most applications of the greedy method, an associated weighting w of the elements of the base set S is given, and *GreedySelect*(*R*) merely chooses the smallest weight or largest weight element in R. For convenience of discussion, assume that *GreedySelect* chooses the smallest weight element. The choice of a smallest element is facilitated by preconditioning the base set by sorting with respect to w or by maintaining the elements in the base set in a priority queue where w determines the priority value of the elements.

Even though the greedy method paradigm can be applied in a given situation, and even though it might output a feasible solution, it still might not output an *optimal* solution (recall the problem of making 30 cents change without nickels). Thus, proofs of optimality should always accompany solutions generated by greedy algorithms. In spite of its somewhat limited applicability, the greedy method solves a number of classic optimization problems. Moreover, when the greedy method does the job optimally, it generally leads to elegant and efficient code for solving the problem.

12.2 An Optimal Sequential Storage Order

Suppose we are given a set of objects $b_1, b_2, \ldots, b_n$ arranged sequentially in some order $b_{\pi(1)}, b_{\pi(2)}, \ldots, b_{\pi(n)}$ (where π is some permutation of $1,2, \ldots, n$), such as files stored on a sequential access tape. We assume that there is a weight c_i associated with object b_i (for example, the length of a file) that represents the individual cost of processing b_i. We also assume that to process object $b_{\pi(i)}$, we must first process the objects $b_{\pi(1)}, b_{\pi(2)}, \ldots, b_{\pi(i-1)}$ that precede $b_{\pi(i)}$ in the sequential arrangement. For example, to read the ith file on a sequential tape (assuming that the tape is rewound to the beginning), we must first read the preceding $i - 1$ files on the tape. Thus the total cost of processing $b_{\pi(i)}$ is $c_{\pi(1)} + c_{\pi(2)} + \cdots + c_{\pi(i)}$. In Figure 12.2, we illustrate this example with six files $b_1, b_2, \ldots, b_6$ of lengths 15,6,1,22,3,10, respectively. A sample storage order of these files on a tape is given in Figure 12.2b. If we wish to read the file of length 22, we first have to read the files of length 1,3,15 that precede this file on the tape. Thus, reading the file of length 22 results in reading files whose total length is $1 + 3 + 15 + 22 = 41$.

The *optimal sequential storage* problem for processing objects $b_1,b_2, \ldots, b_n$ is to find an arrangement (permutation) of the objects $b_{\pi(1)}, b_{\pi(2)}, \ldots, b_{\pi(n)}$ that minimizes the average time it takes to process an object. Let p_i denote the probability that the object b_i is to be processed, $i = 1, \ldots, n$. Given these probabilities and a permutation π, clearly the average total cost of processing an object is given by

$$AveCost(\pi) = p_{\pi(1)}(c_{\pi(1)}) + p_{\pi(2)}(c_{\pi(1)} + c_{\pi(2)}) + \cdots + p_{\pi(n)}(c_{\pi(1)} + \cdots + c_{\pi(n)}). \quad (12.2.1)$$

We first consider the special case where each object is equally likely to be processed, so that $p_i = 1/n$, $i = 1, \ldots, n$. Then (12.2.1) becomes

$$AveCost(\pi) = \frac{1}{n}[(c_{\pi(1)}) + (c_{\pi(1)} + c_{\pi(2)}) + \cdots + (c_{\pi(1)} + c_{\pi(2)} + \cdots + c_{\pi(n)})]$$

$$= \frac{1}{n}\sum_{i=1}^{n} (n - i + 1)\, c_{\pi(i)}. \quad (12.2.2)$$

Figure 12.2 (a) Files of lengths 15,6,1,22,3,10; (b) a sample storage order of files on a sequential access tape corresponding to the permutation π

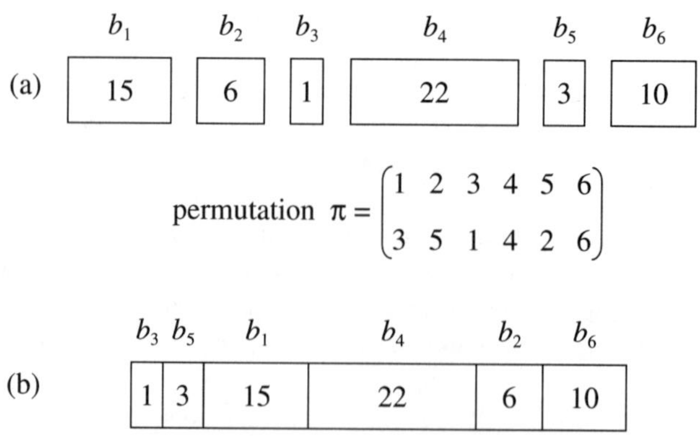

Looking at (12.2.2), we see that the coefficients of $c_{\pi(i)}$, $i = 1, \ldots, n$, decrease as i increases. Thus, it is intuitively obvious that *AveCost*(π) is minimized when $c_{\pi(1)} \leq c_{\pi(2)} \leq \cdots \leq c_{\pi(n)}$, so that the greedy strategy we use is to select the objects in increasing order of their costs.

Now consider the problem of minimizing *AveCost*(π) as given by (12.2.1) for a general set of probabilities p_i, $i = 1, \ldots, n$. Various greedy strategies come to mind. For example, we might use the same greedy strategy given previously, which gives the optimal solution when all the probabilities are equal. However, it is not hard to find examples where this strategy yields suboptimal results. We might also arrange the objects in nondecreasing order of probabilities (placing the objects most likely to be processed near the beginning), but, again, it is not hard to find examples where this strategy fails. Alas, is there a greedy strategy that works? Yes, we leave it as an exercise to show that arranging the objects in nondecreasing order of the ratios c_i/p_i yields an optimal solution.

12.3 The Knapsack Problem

The *Knapsack problem* involves n objects $b_1, b_2, \ldots, b_n$ and a knapsack of capacity C. We assume that each object b_i has a given positive weight (or volume) w_i, and a given positive value v_i (or profit, cost, and so forth) $i = 1, \ldots, n$. We place the objects or fractions of objects in the knapsack, taking care that the capacity of the knapsack is not exceeded. If a fraction f_i of object b_i is placed in the knapsack ($0 \leq f_i \leq 1$), then it contributes $f_i v_i$ to the total value in the knapsack. The Knapsack problem is to place objects or fractions of objects in the knapsack, without exceeding the capacity, so that total value of the objects in the knapsack is maximized. A more formal statement of the Knapsack problem is

$$\text{maximize } \sum_{i=1}^{n} f_i v_i, \tag{12.3.1}$$

$$\text{subject to the constraints: } \sum_{i=1}^{n} f_i w_i \leq C,$$

$$0 \leq f_i \leq 1, \quad i = 1, \ldots, n.$$

As with the sequential storage order problem, several possible greedy methods for solving the Knapsack problem come to mind. For example, we might put as much as possible of the object with highest value into the knapsack, then as much as possible of the object of second-highest value into the knapsack, and so forth, until the knapsack is filled to capacity. Another greedy strategy would be to follow this same scheme, but where the objects are placed in the knapsack according to nonincreasing weights. But we can easily find examples where neither of these strategies yield optimal solutions. We will show that the strategy of putting the objects into the knapsack according to nonincreasing *ratios* (*densities*) v_i/w_i until the knapsack is full yields an optimal solution.

The following interpretation of the Knapsack problem makes it clear that the ratio v_i/w_i is the critical issue. Suppose the (greedy) owner of a gourmet coffee shop

Type	Quantity Available in Pounds	Total Cost in Dollars for That Quantity
Colombian	8	32
Jamaican	2	32
Java	4	25
Bicentennial	1	10
Mountain	2	9
Roast	6	18
Dark	10	55
Special	50	100

Figure 12.3 Coffee shop inventory

wishes to mix a 10-pound bag of coffee using various types of coffee beans in such a way as to produce the coffee blend of maximum cost. The weights of the objects in the Knapsack problem correspond to the quantity in pounds that are available of each type of coffee bean. The value of each quantity of coffee bean is the total cost of that quantity in dollars. In Figure 12.3 we list the types and amounts of coffee beans that are available.

Intuitively, we obtain the most expensive 10-pound bag of coffee if we first use as much as possible of the bean whose cost/pound ratio is the largest, then as much as possible of the bean whose cost/pound ratio is the second-largest, and so on until we reach 10 pounds. Figure 12.4 lists the coffee beans in decreasing order of their cost/pound ratios and the quantity of each type of bean used as determined by this greedy method.

Generalizing the greedy strategy for making expensive coffee leads to a greedy algorithm for the Knapsack problem. The algorithm *Knapsack* assumes as a precondition that the objects have been sorted in nonincreasing order of the ratios v_i/w_i, $i = 1, \ldots, n$.

Figure 12.4 Greedy solution for 10-pound bag of coffee

Type	Cost/Pound	Fraction of Available Quantity Chosen	Quantity Chosen
Jamaican	16	1	2
Bicentennial	10	1	1
Java	6.25	1	4
Dark	5.5	.3	3
Mountain	4.5	0	0
Colombian	4	0	0
Roast	3	0	0
Special	2	0	0

```
Procedure Knapsack(V[1:n], W[1:n], C, F[1:n])
Input:  V[1:n] (an array of positive values)
        W[1:n] (an array of positive weights)
        {it is assumed that V[1]/W[1] ≥ V[2]/W[2] ≥ ··· ≥ V[n]/W[n]}
        C (a positive capacity)
Output: F[1:n] (array of nonnegative fractions)

  for i := 1 to n do
     F[i] := 0
  endfor
  RemainCap = C
  i := 1
  if W[1] ≤ C then
     Fits := .true.
  else
     Fits := .false.
  endif
  while Fits .and. i ≤ n do
     F[i] := 1
     RemainCap := RemainCap – W[i]
     i := i + 1
     if W[i] ≤ RemainCap then
        Fits := .true.
     else
        Fits := .false.
     endif
  endwhile
  if i ≤ n then F[i] := RemainCap/W[i] endif
end Knapsack
```

Clearly, procedure *Knapsack* generates an optimal solution when $\Sigma_{i=1}^{n} w_i \leq C$, since all the objects are then placed in the knapsack. Using a proof by contradiction, we now show that *Knapsack* generates an optimal solution when $\Sigma_{i=1}^{n} w_i > C$. Suppose to the contrary that the solution $F = (f_1, f_2, \ldots, f_n)$ generated by *Knapsack* is suboptimal. Now consider an optimal solution $Y = (y_1, y_2, \ldots, y_n)$, and let j denote the first index i such that $f_i \neq y_i$ so that $1 = f_i = y_i$, $1 \leq i \leq j - 1$. We choose Y such that j is maximized over all optimal solutions. Note that $\Sigma_{i=1}^{n} y_i w_i = C$ (otherwise we could increase one of the y_i's and trivially increase the total value of the solution). Because of the greedy strategy used by *Knapsack*, we have $f_j > y_j$.

By altering Y, we now construct an optimal solution $Z = (z_1, z_2, \ldots, z_n)$ that agrees with F in one more initial position. We set $z_i = f_i$, $1 \leq i \leq j$, and for $i > j$ the value of z_i is obtained by suitably decreasing y_i so that $\Sigma_{i=1}^{n} z_i w_i = C$. It follows that:

$$(z_j - y_j)\, w_j = \sum_{i=j+1}^{n} (y_i - z_i)\, w_i. \tag{12.3.2}$$

Comparing the total values of the solutions Z and Y, we obtain

$$\begin{aligned}
\sum_{i=1}^{n} z_i v_i - \sum_{i=1}^{n} y_i v_i &= (z_j - y_j)\, v_j - \sum_{i=j+1}^{n} (y_i - z_i)\, v_i. \\
&= (z_j - y_j) w_j v_j / w_j - \sum_{i=j+1}^{n} (y_i - z_i) w_i v_i / w_i \\
&\geq \left[(z_j - y_j) w_j - \sum_{i=j+1}^{n} (y_i - z_i) w_i \right] v_j / w_j \\
&\qquad \text{(since } v_1/w_1 \geq v_2/w_2 \geq \cdots \geq v_n/w_n) \\
&= 0 \quad \text{(since the number inside the square brackets is 0 by (12.3.2)).}
\end{aligned}$$

Thus, the value of Z is not less than the value of Y, so that Z is also an optimal solution. But Z agrees with F in the first j initial positions, which contradicts the assumption that Y is an optimal solution that agrees with F in the most initial positions.

Clearly, *Knapsack* has linear complexity in the worst case. However, *Knapsack* assumes the precondition that the objects are ordered so that $v_1/w_1 \geq v_2/w_2 \geq \cdots \geq v_n/w_n$. The preconditioning can be achieved in time $O(n \log n)$.

A natural variation of the Knapsack problem, called the *0/1 Knapsack problem*, is the same as the Knapsack problem except that fractional objects are not allowed. More precisely, the 0/1 Knapsack problem has the following formulation:

$$\text{maximize} \sum_{i=1}^{n} f_i v_i \tag{12.3.3}$$

$$\text{subject to the constraints: } \sum_{i=1}^{n} f_i w_i \leq C$$

$$f_i \in \{0,1\}, \quad i = 1, \ldots, n.$$

The greedy strategy of placing the objects in the knapsack in nonincreasing order of the ratios v_i/w_i, $i = 1, \ldots, n$, rejecting objects when their addition would overflow the knapsack, may yield a suboptimal solution to the 0/1 Knapsack problem. While the Knapsack problem is easily solved, there is no known (worst-case) polynomial algorithm for solving the 0/1 Knapsack problem (see Chapter 20).

12.4 Huffman Codes

In Chapter 2 we showed how to generate prefix codes for a given alphabet $A = \{a_1, a_2, \ldots, a_n\}$ on n symbols. Recall that a binary prefix code is generated for each 2-tree whose n leaf nodes contain the symbols from A. We refer to the binary string representing a symbol as a *coded symbol*. Now suppose we are encoding text composed of symbols from A. Suppose also that the frequency of the occurrence of symbol a_i in such text is given by f_i, $i = 1, \ldots, n$. An important problem in data compression is to find an *optimal* binary prefix code, that is, a binary prefix code that minimizes the expected length of a *coded symbol*. Clearly, an optimal binary prefix code also minimizes the expected length of text generated using the symbols from A. In this section we solve the problem of finding an optimal prefix code using the greedy algorithm due to Huffman.

The problem of finding optimal binary prefix codes is equivalent to the following problem for 2-trees. Each unambiguous binary code can be generated by traversing a suitable 2-tree T (see Chapter 2). We let λ_i denote the length of the path in T from the root to (the leaf node corresponding to) a_i, $i = 1, \ldots, n$. The expected length of the coded symbols determined by T is closely related to a generalization of the leaf path length called the *weighted leaf path length* $WLPL(T)$ *of* T, defined by

$$WLPL(T) = \sum_{i=1}^{n} \lambda_i f_i . \tag{12.4.1}$$

Note that $WLPL(T)$ becomes $LPL(T)$ when all the f_i's equal 1. For example, given frequencies 9, 8, 5, 3, 15, 2 for the symbols a, b, c, d, e, f, respectively, the weighted leaf path length for the binary tree T in Figure 2.14 on page 52 is given by

$$WLPL(T) = (3)(9) + (4)(8) + (4)(5) + (2)(3) + (2)(15) + (2)(2) = 119. \tag{12.4.2}$$

Since symbol a_i occurs with frequency f_i the probability p_i, $i = 1, \ldots, n$ that symbol a_i occurs in a given text position is

$$p_i = \frac{f_i}{\sum_{j=1}^{n} f_j}, \quad i = 1, \ldots, n.$$

Each binary digit in the binary string encoding a_i corresponds to an edge of the path in T from the root node to the leaf node containing a_i. Hence, the length of the binary string encoding a_i (in the binary code determined by T) equals λ_i, $i = 1, \ldots, n$, so that the expected length of a coded symbol is given by $WLPL(T)/(\Sigma_{i=1}^{n} f_i)$. Thus, the problem of finding an optimal binary prefix code is equivalent to finding a 2-tree having minimum weighted leaf path length.

Given the set of frequencies 9, 8, 5, 3, 15, 2 for the symbols a, b, c, d, e, f, respectively, the 2-tree in Figure 2.14, which has weighted leaf path length equal to 119, is not optimal. The binary tree given in Figure 12.5, which has weighted leaf path length equal to 99, turns out to be an optimal 2-tree with respect to these frequencies.

We now describe the greedy algorithm due to Huffman for computing an optimal binary prefix code with respect to a given set of frequencies f_i, $i = 1, \ldots, n$. The procedure *HuffmanCode* computes a 2-tree T minimizing $WLPL(T)$ by constructing a sequence of forests as follows. At each stage in the algorithm, the root of each tree in the forest is assigned a frequency. The initial forest F consists of the n single node trees corresponding to the n elements of A. At each stage *HuffmanCode* finds two

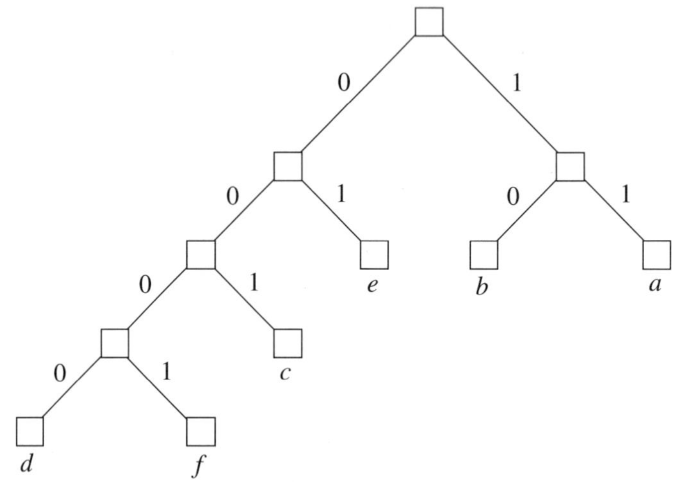

Symbol	Frequency	Encoding
a	9	11
b	8	10
c	5	001
d	3	0000
e	15	01
f	2	0001

$$WLPL(T) = (2)(9) + (2)(8) + (3)(5) + (4)(3) + (2)(15) + (4)(2) = 99$$

Figure 12.5 A tree having minimum weighted leaf path length for alphabet $A = \{a,b,c,d,e,f\}$ with respect to the given frequencies

trees T_1 and T_2 whose roots R_1 and R_2 have the smallest and second-smallest frequencies over all the trees in the current forest. A new (internal) node R is then added to the forest, together with two edges joining R to R_1 and R to R_2, so that a new tree is created with R as the root and T_1 and T_2 as the left and right subtrees of R. The frequency of the new root vertex R is taken to be the sum of the frequencies of the old root vertices R_1 and R_2. The Huffman tree is constructed after $n - 1$ stages, involving the addition of $n - 1$ internal nodes.

The procedure *HuffmanCode* calls the procedure *GenerateCode* (see Chapter 2) to generate the *Huffman code* (the optimal binary prefix code associated with the Huffman tree). We utilize high-level parameter descriptions in our calls to procedures implementing the various priority queue operations. For example, a call to *CreatePriorityQueue* with parameters *Leaf*[1:n] and Q creates a priority queue Q of pointers to the leaf nodes. A call to *RemovePriorityQueue* with parameters Q and L removes the element at the front of the queue Q and assigns it to L.

Procedure *HuffmanCode*(*A*[1:n],*Freq*[1:n],*Code*[1:n])

Input: *Freq*[1:n] (an array of nonnegative frequencies: *Freq*[i] $= f_i$)

```
Output: Code[1:n] (an array of binary strings for Huffman code: Code[i] is the
                   binary string encoding symbol a_i, i = 1, . . . , n)
  for i := 1 to n do   {initialize leaf nodes}
     call AllocateHuffmanNode(P)
     P → SymbolIndex := i
     P → Frequency := Freq[i]
     P → LeftChild := nil
     P → RightChild := nil
     Leaf[i] := P
  endfor
  call CreatePriorityQueue(Leaf[1:n],Q)   {create priority queue of pointers to
                                          leaf nodes with Frequency as the key}
  for i := 1 to n – 1 do
     call RemovePriorityQueue(Q,L)   {L,R point to root nodes of smallest and
     call RemovePriorityQueue(Q,R)   second-smallest frequency, respectively}
     call AllocateHuffmanNode(Root)
     Root → LeftChild := L
     Root → RightChild := R
     Root → Frequency := (L → Frequency) + (R → Frequency)
     call InsertPriorityQueue(Q,Root)
  endfor
  Root → BinaryString := ''   {BinaryString of root initialized to null string}
  call GenerateCode(Root,Code[1:n])
end HuffmanCode
```

The action of *HuffmanCode* is illustrated in Figure 12.6 (page 420) for the same sample alphabet and set of frequencies as in Figure 12.5.

An efficient way to implement the priority queue Q is to use a min-heap (see Chapter 4). Then *CreatePriorityQueue* (*MakeMinHeap*), *RemovePriorityQueue* (*RemoveMinHeap*), and *InsertPriorityQueue* (*InsertMinHeap*) have worst-case complexities belonging to $O(n)$, $O(\log n)$, and $O(\log n)$, respectively. Hence, the worst-case complexity of *HuffmanCode* belongs to $O(n \log n)$.

The correctness of the algorithm *HuffmanCode* depends on the following proposition, whose proof is left to the exercises.

PROPOSITION 12.4.1 Given a set of frequencies f_i, $i = 1, \ldots, n$, the Huffman tree T constructed by *HuffmanCode* has minimal weighted $WLPL(T)$ over all 2-trees whose edges are weighted by these frequencies. ❑

Finding a binary tree of minimum weighted leaf path length has other important applications besides Huffman codes. For example, the problem of finding optimal merge patterns is solved by constructing a binary tree having minimum weighted leaf path length, where the leaf nodes of the tree correspond to files or sublists and the weight of a leaf is the length of its associated file or sublist.

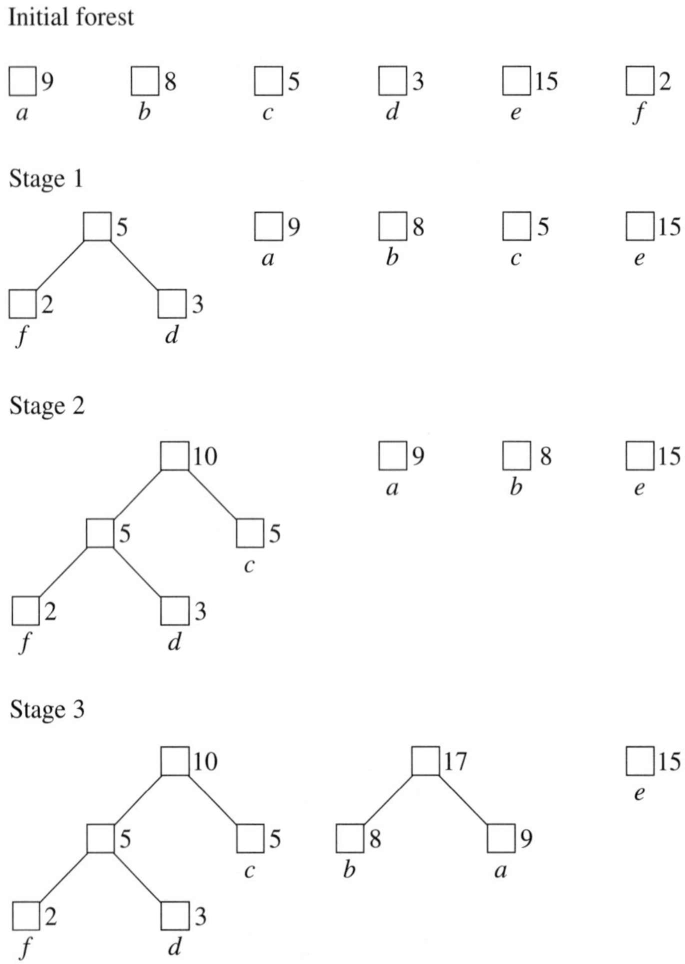

Figure 12.6 Action of *HuffmanCode* for code for alphabet $A = \{a,b,c,d,e,f\}$ with the same set of frequencies as in Figure 12.5

12.5 Minimum Spanning Tree

In this section we consider the problem of finding a spanning tree T of minimum weight in a graph G with respect to a weighting w of the edges. The *weight* of T, denoted *weight*(T), is the sum of the w-weights of its edges. We call such a tree a *minimum spanning tree*. The problem of finding a minimum spanning tree in a weighted graph has a variety of applications. For example, the problem arises when designing any physical network connecting n nodes, where the connections between nodes are subject to feasibility and cost constraints. Examples of such physical networks include telecommunication networks, transportation networks, energy pipelines, VLSI chips, and so forth. In all these examples, the weight of a minimum spanning tree provides a lower bound on the cost of building the network.

Stage 4

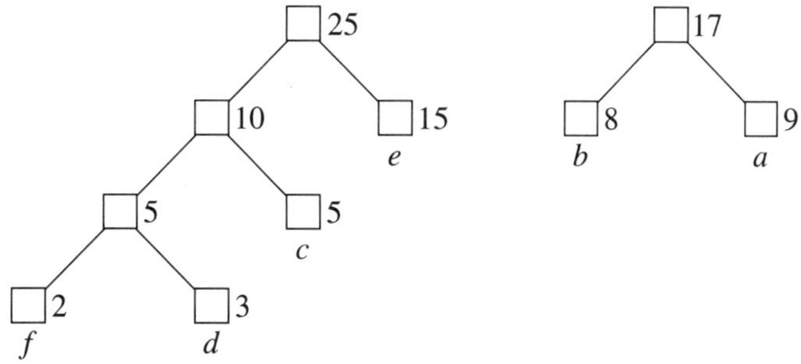

Stage 5: Huffman tree

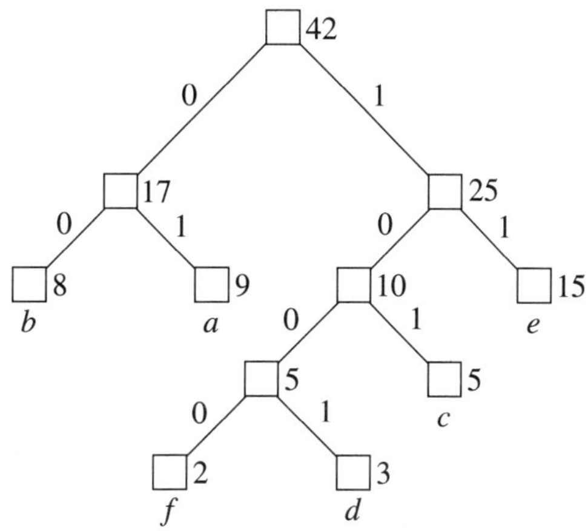

Symbol	Frequency	Encoding
a	9	01
b	8	00
c	5	101
d	3	1001
e	15	11
f	2	1000

$$WLPL(T) = (2)(9) + (2)(8) + (3)(5) + (4)(3) + (2)(15) + (4)(2) = 99$$

Figure 12.6 Continued

In Figure 12.7 (page 422), all the spanning trees of a weighted graph on four vertices are enumerated. The minimum spanning tree is then obtained by inspection.

The number of spanning trees, even for relatively small graphs, is usually enormous, making an enumerative brute-force search infeasible. Indeed, Cayley proved that the complete graph K_n on n vertices contains n^{n-2} spanning trees!

Fortunately, efficient algorithms based on the greedy method do exist for finding minimum spanning trees. We discuss two of the more famous ones, namely, Kruskal's and Prim's algorithms. These algorithms are similar in the sense that their selection functions always choose an edge of minimum weight among the remaining edges. However, Prim's algorithm selection function only considers edges incident to edges already included in the tree. In particular, the partial solutions built by Prim's algorithm are trees, whereas the partial solutions built by Kruskal's algorithm are forests.

12.5.1 Kruskal's Algorithm

Given a connected graph $G = (V,E)$, with $|V| = n$ and $|E| = m$, and a weight function w defined on the edge set E, Kruskal's algorithm finds a minimum spanning tree T of G by constructing a sequence of n forests $F_0, F_1, \ldots, F_{n-1}$, where F_0 is the empty forest and F_i is obtained from F_{i-1} by adding a single edge e. The edge e is chosen so that it has minimum weight among all the edges not belonging to F_i and doesn't form a cycle when added to F_i. Kruskal's algorithm is illustrated for a sample graph in Figure 12.8.

Remark For convenience, we assume that the input graph *G* to Kruskal's algorithm is connected. However, Kruskal's algorithm is easily modified to accept *any* graph *G* (connected or disconnected) and to output a minimum (weight) forest, each of whose trees spans a (connected) component of *G*.

Figure 12.7 Enumeration of the spanning trees of *G*. Minimum spanning tree is T_2.

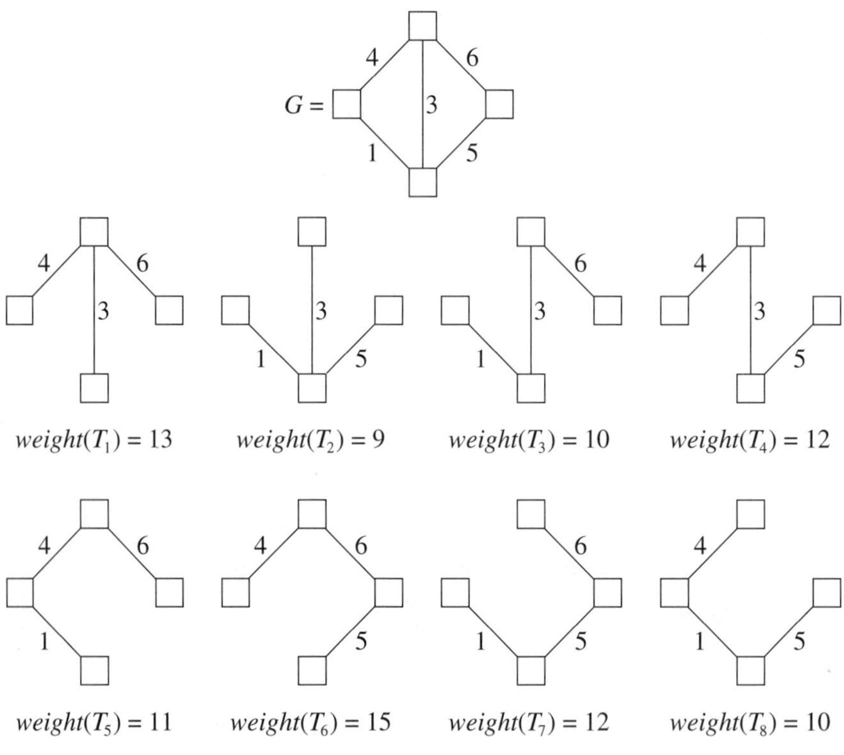

Using a proof by contradiction, we now show that the spanning tree K generated by Kruskal's algorithm is a minimum spanning tree. For convenience, we will assume that the edge weights are distinct. (Our proof can be slightly modified to hold for nondistinct edge weights.) Let the edges of K be denoted by $x_1, x_2, \ldots, x_{n-1}$, listed in increasing order of their weights. Assume that K is not a minimum spanning tree, and let T be a minimum spanning tree with edges $y_1, y_2, \ldots, y_{n-1}$, listed in increasing order of their weights. Let j denote the first index i such that $x_i \neq y_i$, so that $x_i = y_i$, $1 \leq i \leq j - 1$. Since x_j was chosen by the greedy strategy, we have $w(x_j) < w(y_j)$.

We now construct a spanning tree T' whose weight is smaller than T. Consider the subgraph H obtained by adding the edge x_j to T. It is easily verified that H contains a unique cycle C and that this cycle contains x_j. Since a tree contains no cycles, K does not contain the cycle C. Since $y_i = x_i$, $i = 1, 2, \ldots, j - 1$, it follows that C must contain the edge y_k for some index $k \geq j$. Let T' be the spanning tree obtained from H by deleting the edge y_k. Since $w(y_k) \geq w(y_j) > w(x_j)$, we have

$$weight(T') = weight(T) + w(x_j) - w(y_k) < weight(T),$$

contradicting our assumption that T is a minimum spanning tree.

Figure 12.8 Stages in Kruskal's algorithm

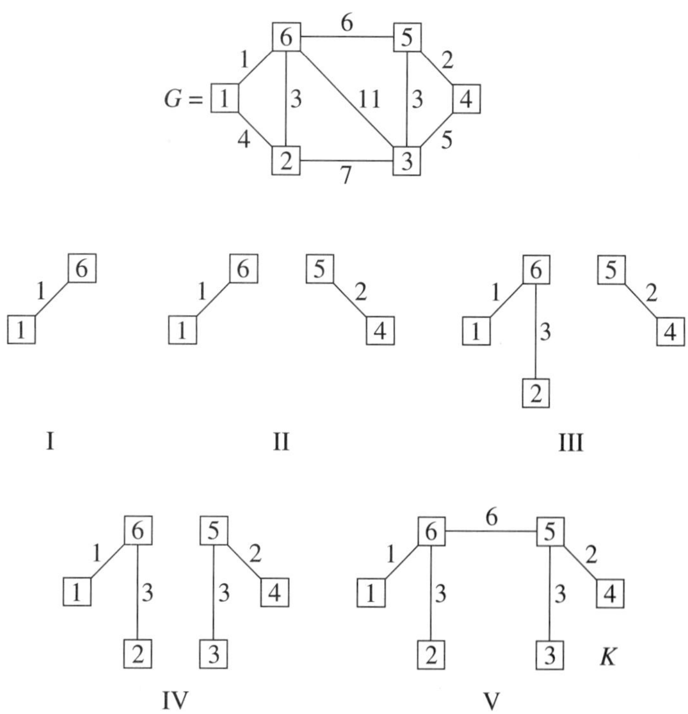

The key implementation detail in Kruskal's algorithm is the detection of whether the addition of a given edge e to the existing forest F creates a cycle. Observe that a cycle is formed by the edge $e = uv$ if, and only if, u and v belong to the same component (tree) in F. Checking whether u and v belong to the same component can be done efficiently using the union and find algorithms for disjoint sets discussed in Chapter 8, where the collection of disjoint sets in the current context is the collection of vertex sets of each tree in forest F.

In the following pseudocode for procedure *Kruskal*, the forest F is maintained using its set of edges.

```
procedure Kruskal(G,w,MCSTree)
Input:  G  (a connected graph with vertex set V of cardinality n
              and edge set E = {e_i = u_i v_i | i ∈ {1, . . . , m}})
        w  (a weight function on E)
Output: MCSTree {a minimum spanning tree}
   Sort the edges in nondecreasing order of weights w(e_1) ≤ · · · ≤ w(e_m)
   Forest := ∅
   Size := 0
   for i := 1 to n do {initialize a disjoint collection of single vertices}
      Parent[i] := – 1
   endfor
   j := 0
   while Size ≤ n – 1 .and. j < m do
      j := j + 1
      call Find2(Parent[1:n],u_j,r)   {e_j = u_j v_j}
      call Find2(Parent[1:n],v_j,s)
      if r ≠ s then {add edge e_j to Forest}
         Forest := Forest ∪ {e_j}
         Size := Size + 1
         call Union(Parent[1:n],r,s) {combine sets containing u_j and v_j}
      endif
   endwhile
   MCSTree := Forest
end Kruskal
```

Since the procedure *Kruskal* does $n - 1$ unions and at most m finds, it follows from the results given in Chapter 8 that checking for cycles has a worst-case complexity that is almost linear in m. In *Kruskal*, we assumed as a precondition that the edges were sorted in nondecreasing order of their weights. Since presorting the edges can be done in time $O(m \log m)$, the overall order of the worst-case complexity is $\Theta(m \log m)$.

Instead of sorting the edges, we could more efficiently select the next minimum weight edge by utilizing a priority queue of edges of the graph, implemented, for example, with a min-heap. *CreatePriorityQueue* (*MakeMinHeap*) and *RemovePriorityQueue* (*RemoveMinHeap*) then have worst-case complexities belonging to O(m) and O(log m), respectively. Hence, the selection process remains O(m log m) in the worst case. However, improvement over *Kruskal* occurs for inputs where the tree is completed early (that is, not many of the selected edges form cycles). For example, in the best case, the first $n - 1$ edges selected form a tree, yielding best-case complexity belonging to O(m).

12.5.2 Prim's Algorithm

Prim's algorithm for generating a minimum spanning tree differs from Kruskal's algorithm in that it maintains a tree at each stage instead of a forest. The initial tree T_0 may be taken to be any single vertex r of the graph G. Prim's algorithm builds a sequence of n trees (rooted at r) $T_0, T_1, \ldots, T_{n-1}$, where T_{i+1} is obtained from T_i by adding a single edge e_{i+1}, $i = 0, \ldots, n - 2$. The edge e_{i+1} is selected (greedily) to be a minimum weight edge among all edges having *exactly* one vertex in T_i. The set of all edges in G having exactly one vertex in T_i is denoted by $Cut(T_i)$. With this notation, at the ith step Prim's algorithm chooses an edge e_i of minimum weight among all the edges in $Cut(T_i)$ (see Figure 12.9).

A proof that Prim's algorithm yields the minimum spanning tree is left to the exercises. Prim's algorithm is illustrated in Figure 12.10 (page 427) using the same sample graph as in Figure 12.8.

As with Kruskal's algorithm, the complexity of Prim's algorithm is dependent on specific implementation details. A straightforward implementation of Prim's algorithm

Figure 12.9 Sample tree T_6 of Prim's algorithm

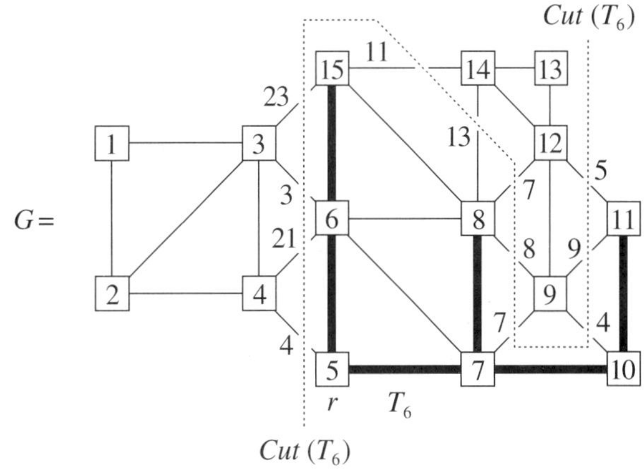

Sample tree T_6 of Prim's algorithm with $V(T_6) = \{5,6,7,8,10,11,15\}$ and $r = 5$.
$Cut(T) = \{\{4,5\},\{4,6\},\{3,6\},\{3,15\},\{14,15\},\{8,14\},\{8,12\},$
$\{8,9\},\{7,9\},\{9,10\},\{9,11\},\{11,12\}\}$.
At stage 7 Prim's algorithm adds edge $e_7 = \{3,6\}$ to T_6,
since $\{3,6\}$ has minimum weight 3 in $Cut(T_6)$.

based on explicitly implementing $Cut(T)$ at each stage is inefficient. A more efficient implementation of Prim's algorithm is based on the following observations. For convenience, we set $w(uv) = \infty$ (or some sufficiently large number) whenever u and v are nonadjacent vertices in G. For each vertex $v \in V(G) - V(T)$, we let $\eta_T(v)$ denote a vertex u of T that is "nearest" to v; that is, $w(uv)$ equals the minimum of $w(xv)$ over all $x \in V(T)$. For example, in Figure 12.9 with $T = T_6$, $V(G) - V(T)$ consists of the vertices $v = 3,4,9,12,14$, with $\eta_T(v) = 6,5,10,11,15$, respectively. Clearly, the following equality holds:

$$\min \{w(\{v,\eta_T(v)\}) \mid v \in V(G) - V(T)\} = \min \{w(e) \mid e \in Cut(T)\}. \qquad (12.5.1)$$

Thus, when implementing Prim's algorithm, we can maintain η_T at each stage rather than explicitly implementing $Cut(T)$.

When implementing Prim's algorithm, the tree T is dynamically maintained using an array $Parent[1{:}n]$. The final state of $Parent[1{:}n]$ yields the parent array implementation of a minimum spanning tree T. Guided by (12.5.1), at any stage of the algorithm a subarray of $Parent[1{:}n]$ gives the parent array of the current tree T, whereas another subarray of $Parent[1{:}n]$ maintains the nearest weight vertices $\eta_T(v)$. More precisely, $Parent[v]$ has the following dynamic interpretation. If $v \in V(T)$, then $Parent[v]$ is the parent of vertex v in T. If $v \notin V(T)$ but is adjacent to at least one vertex of T, then $Parent[v] = \eta_T(v)$. Otherwise, $Parent[v]$ is undefined. Initially, $Parent[r] = 0$.

In the following pseudocode for procedure *Prim*, we also maintain a local array $Nearest[1{:}n]$, which is initialized to ∞'s, except for $Nearest[r]$ which is set to 0. At each stage, if $Parent[v]$ is defined, then $Nearest[v] = w(\{v, Parent[v]\})$, otherwise $Nearest[v] = \infty$. The next vertex u to be added to the tree T is the one such that $Nearest[u]$ is minimized over all $u \notin V(T)$. The arrays $Parent[1{:}n]$ and $Nearest[1{:}n]$ are examined for possible alteration at each stage for each vertex v not belonging to T that is adjacent to u. If $w(uv) < Nearest[v]$, then we set $Parent[v] = u$ and $Nearest[v] = w(uv)$. The procedure *Prim* also maintains a Boolean array *InTheTree* to keep track of the vertices not in T.

```
procedure Prim(G,w,Parent[1:n])
Input:  G (a connected graph with vertex set V and edge set E)
        w (a weight function on E)
Output: Parent[1:n] (parent array of a minimum spanning tree}
  for v := 1 to n do
     Nearest[v] := ∞
     InTheTree[v] := .false.
  endfor
  Parent[r] := 0                     {root the tree at an arbitrary vertex r}
  Nearest[r] := 0
  for Stage := 1 to n – 1 do
     Select vertex u that minimizes Nearest[u] over all u such that
                                                      InTheTree[u] = .false.
     InTheTree[u] := .true.                  {add u to T}
```

```
        for each vertex v such that uv ∈ E do        {update Nearest[v] and
           if .not. InTheTree[v] then                 Parent[v] for all v ∉ V(T)
              if w(uv) < Nearest[v] then              that are adjacent to u}
                 Nearest[v] := w(uv)
                 Parent[v] := u
              endif
           endif
        endfor
     endfor
end Prim
```

The action of procedure *Prim* is illustrated in Figure 12.11 (page 428) for the same graph given in Figure 12.10.

Figure 12.10 Tree *T* and *Cut*(*T*) at each stage in Prim's algorithm with $r = 1$

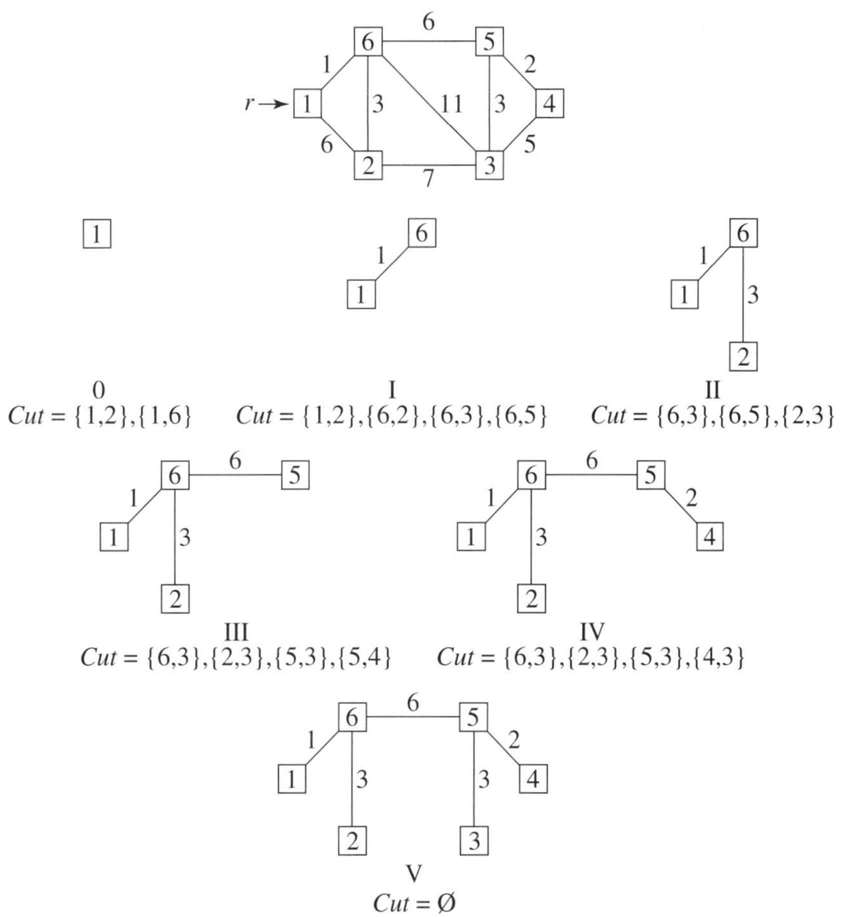

$G =$ 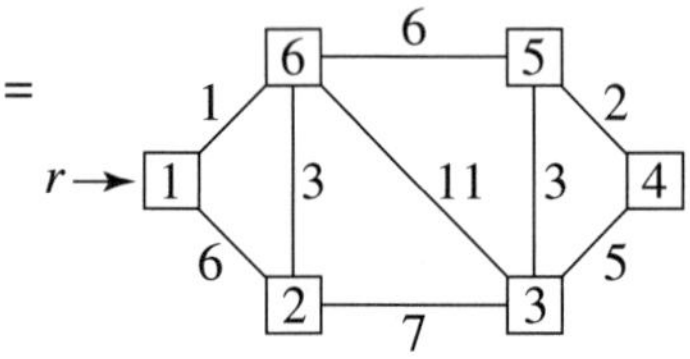

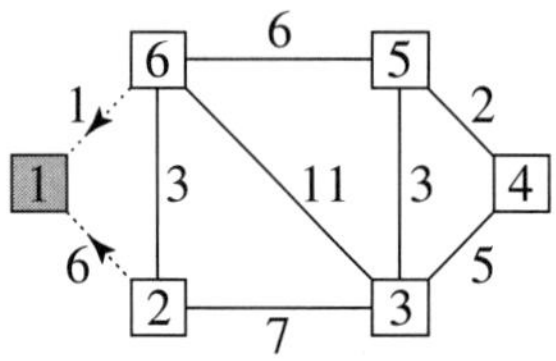

i	1	2	3	4	5	6
Nearest[i]	0	6	∞	∞	∞	1
Parent[i]	0	1	–	–	–	1
InTheTree[i]	T	F	F	F	F	F

Weight = 0

Stage 1

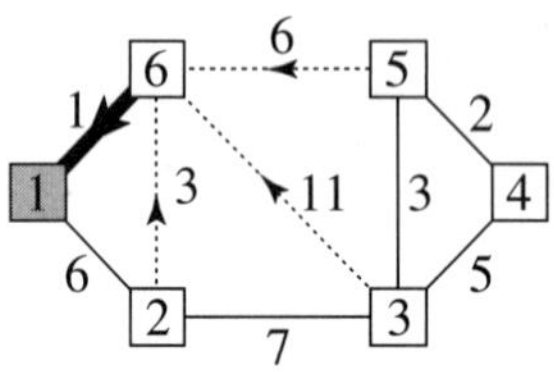

i	1	2	3	4	5	6
Nearest[i]	0	3	11	∞	6	1
Parent[i]	0	6	6	–	6	1
InTheTree[i]	T	F	F	F	F	T

Weight = 1

Stage 2

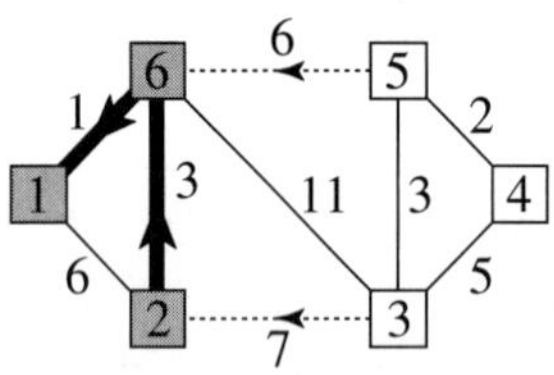

i	1	2	3	4	5	6
Nearest[i]	0	3	7	∞	6	1
Parent[i]	0	6	2	–	6	1
InTheTree[i]	T	T	F	F	F	T

Weight = 4

Stage 3

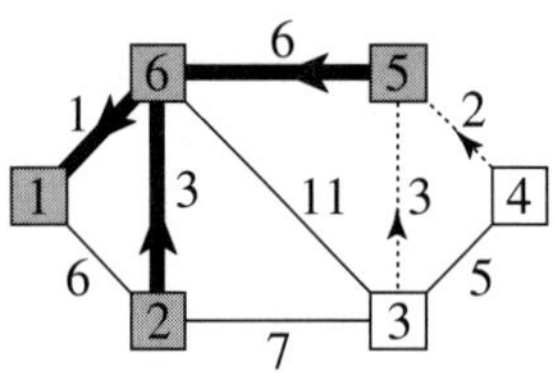

i	1	2	3	4	5	6
Nearest[i]	0	3	3	2	6	1
Parent[i]	0	6	5	5	6	1
InTheTree[i]	T	T	F	F	T	T

Weight = 10

Stage 4

Figure 12.11 Action of *Prim* for the same graph as given in Figure 12.10

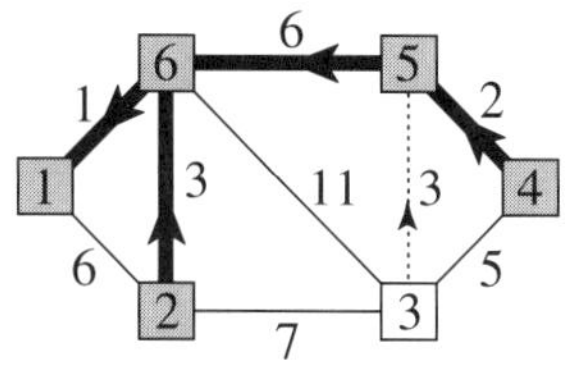

i	1	2	3	4	5	6
Nearest[i]	0	3	3	2	6	1
Parent[i]	0	6	5	5	6	1
InTheTree[i]	T	T	F	T	T	T

Weight = 12

Stage 5

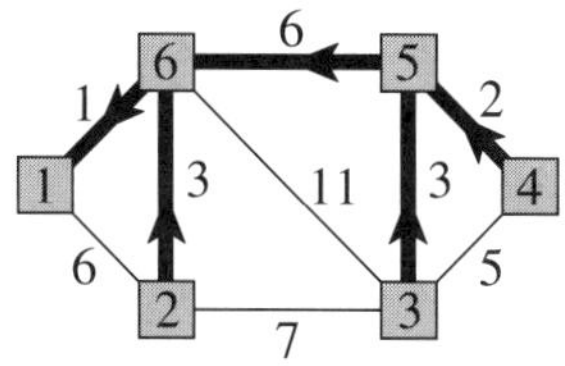

After stage 5, array *Parent* is now complete, and implements a minimum spanning tree with weight 15.

Figure 12.11 Continued

Remark Note that the procedure *Prim* terminates after only $n-1$ stages, even though there are n vertices in the final minimum spanning tree. The reason is simple: After stage $n-1$ has been completed, *InTheTree*[1:*n*] is **.false.** for only one vertex *w*. Thus, another iteration of the **for** loop controlled by *Stage* would result in no change to the arrays *Nearest*[1:*n*] and *Parent*[1:*n*]. In other words, the last vertex and edge in the minimum spanning tree come in for free.

Clearly, the procedure *Prim* has $\Theta(n^2)$ complexity, where we choose the comparison used in updating *Nearest*[1:*n*] in the inner **for** loop as the basic operation. The question arises as to whether *Prim* is more or less efficient than the procedure *Kruskal*, which has $\Theta(m \log m)$ worst-case complexity. The answer depends on how many edges G has in comparison to the number of vertices. When the ratio m/n is large, then *Prim* is more efficient, whereas when m/n is small, *Kruskal* is better. For example, when the number of edges has the same order as the number of vertices (such as for graphs whose vertex degrees are bounded above by some fixed constant), *Kruskal* has $\Theta(n \log n)$ worst-case complexity, which is significantly better than the $\Theta(n^2)$ worst-case complexity of *Prim*. At the other extreme, when the number of edges is quadratic in the number of vertices (such as for the complete graph K_n, so that $m = n(n-1)/2$), *Kruskal* has worst-case complexity $\Theta(n^2 \log n)$, as compared to the $\Theta(n^2)$ worst-case complexity of *Prim*.

Note that in procedure *Prim*, the operations performed on the array *Nearest*[*u*] are essentially those associated with a priority queue. Thus, *Prim* could be rewritten (procedure *Prim*2) using a priority queue Q of the vertices not in the current tree T. The priority of a vertex u is the weight of the edge joining u to its nearest neighbor $\eta_T(u)$ in T. After removing a vertex u from the priority queue and placing it in T, then we must reduce the priorities of each vertex v adjacent to u that is now "nearer" to the tree. If the priority queue Q is implemented as a min-heap, then both the operation of removing an element from Q and the operation of changing the priority value of an element in Q have O(log n) complexity. Since there are n removals of an element from Q and the number of times that the priority of an vertex v is lowered is no more than its degree, it follows that the worst-case complexity of procedure *Prim*2 belongs to $O(n \log n + 2m \log n) = O(m \log n)$. We leave the pseudocode for procedure *Prim*2 as an exercise.

12.6 Shortest Paths in Graphs and Digraphs

In Chapter 8 we presented the Bellman-Ford algorithm for finding shortest paths in weighted graphs and digraphs. By performing successive scans of the edges, procedure *BellmanFord* constructed a shortest path tree rooted at a given vertex r, that is, a tree containing a shortest path in G from r to every vertex v for which such a path exists. We now discuss an algorithm due to Dijkstra that grows a shortest path tree using the greedy method. Although Dijkstra's algorithm has superior worst-case performance to *BellmanFord*, it has the restriction that the edge weights must be nonnegative.

For undirected graphs Dijkstra's algorithm follows a similar strategy to Prim's algorithm, differing only in the way the next edge is selected. At each stage in Dijkstra's algorithm, instead of selecting an edge of minimum weight in $Cut(T)$, we select an edge $uv \in Cut(T)$, $u \in V(T)$, so that the path from r to v in the augmented tree $T \cup \{uv\}$ is shortest; that is, such that $Dist[v] + w(uv)$ is minimized over all edges $uv \in Cut(T)$, where $Dist[v]$ is the length of a path from r to v in T. The following pseudocode is a high-level description of Dijkstra's algorithm.

```
procedure Dijkstra(G,a,w,T)
Input:  G (a graph with vertex set V = {1, . . . , n} and edge set E)
        r (a vertex of G)
        w (a nonnegative weighting of the edges)
Output: T (a shortest path tree rooted at r)
dcl Dist[1:n] (an array initialized to ∞)

   T is initialized to be the tree consisting of the single vertex r
   Dist[r] := 0

   while Cut(T) ≠ ∅ do
      select an edge uv ∈ Cut(T) such that Dist[u] + w(uv) is minimized
      add vertex v and edge uv to T
   endwhile

end Dijkstra
```

Figure 12.12 shows the tree T at each stage of Dijkstra's algorithm for a sample weighted graph G. In this figure the label on the outside of each vertex v already included in the growing tree T is the (weighted) length of the path from v to a in T.

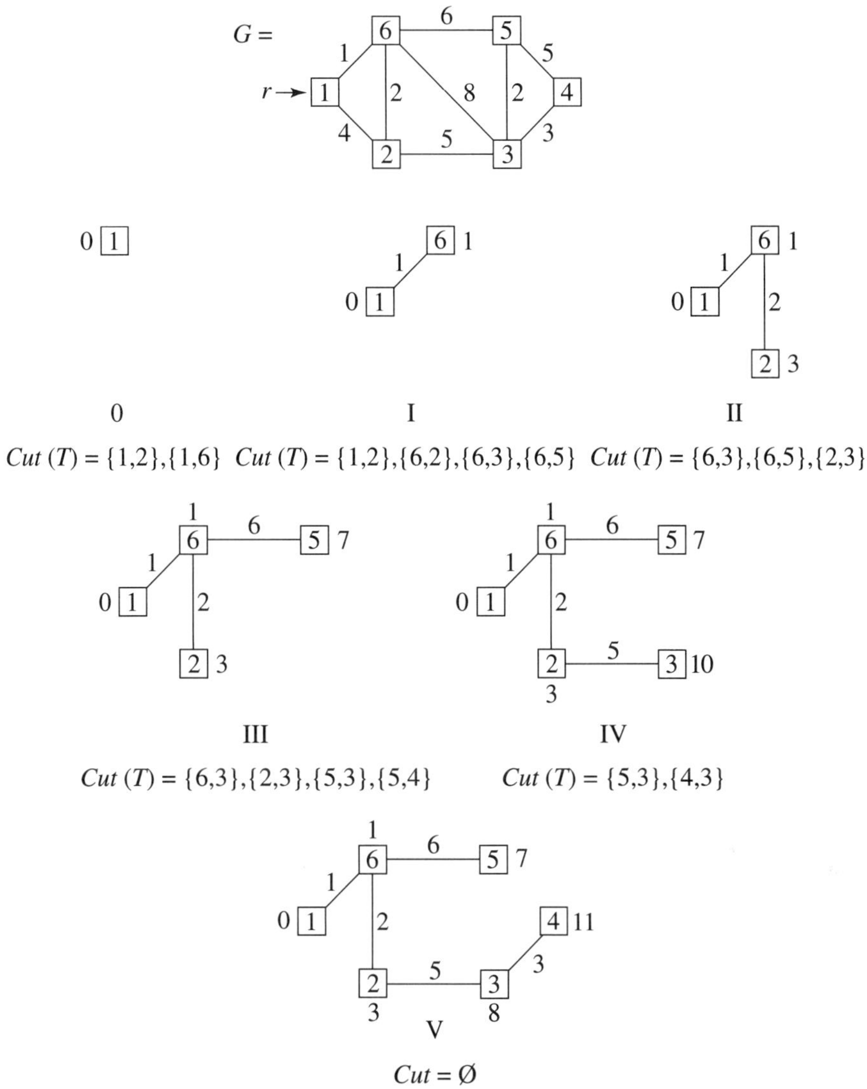

Figure 12.12 Tree *T* and *Cut*(*T*) at each stage in Dijkstra's algorithm

Let T_i be the tree generated by the procedure *Dijkstra* after i iterations of the **while** loop, $i = 0,1, \ldots, t$, where t is the number of iterations of the loop ($t \le n - 1$). For $v \in V(T_i)$, let $P_i^{(v)}$ denote the path from r to v in T_i, $i = 0,1, \ldots, t$. We prove correctness of procedure *Dijkstra* by establishing the following loop invariant: For each vertex $v \in V(T_i)$, $P_i^{(v)}$ is a shortest path in G from r to v, $i = 0,1, \ldots, t$. Clearly, the loop invariant is true for $i = 0$, since T_0 consists of the single vertex r.

Assume that the loop invariant holds for T_i. The tree T_{i+1} is obtained from T_i by adding a single edge uv that minimizes $Dist[u] + w(uv)$ over all edges $uv \in Cut(T_i)$, where $Dist[u]$ is the length of $P_u^{(i)}$. By induction assumption, for all $x \in V(T_i)$, $P_x^{(i+1)} = P_x^{(i)}$ is a shortest path in G from r to x. It remains to show that $P_v^{(i+1)}$ is a shortest path in G from r to v. Consider any path Q in G from r to v. Since the deletion of the edges in $Cut(T_i)$ disconnects G with vertices r and v lying in different components, P must contain edge xy from $Cut(T_i)$, $x \in V(T_i)$, such that the subpath Q_x of Q from r to x lies entirely in T_i. Then we have

$$\begin{aligned} Length(Q) &\geq Length(Q_x) + w(xy) && \text{(since the weight function is nonnegative)} \\ &\geq Length(P_x^{(i)}) + w(xy) && \text{(since } P_x^{(i)} \text{ is a shortest path in } G \text{ from } r \text{ to } x \text{ in } G\text{)} \\ &\geq Length(P_u^{(i)}) + w(uv) && \text{(by the greedy choice } Dijkstra \text{ makes)} \\ &= Length(P_u^{(i+1)}) \end{aligned}$$

Since Q was chosen to be an arbitrary path in G from r to v, it follows that $P_u^{(i+1)}$ is a shortest path G from r to v. This completes the induction step and the correctness proof of procedure *Dijkstra*.

As with Kruskal's and Prim's algorithms, the complexity of Dijkstra's algorithm is dependent on specific implementation details. We now implement a procedure *Dijkstra* that is similar to the procedure *Prim* in that the tree T is dynamically grown using an array $Parent[1{:}n]$. We also maintain a Boolean array *InTheTree* to keep track of the vertices not in T. Again, rather than having to explicitly maintain the sets $Cut(T_i)$, we can efficiently select the next edge uv to be added to the tree by maintaining an array $Dist[1{:}n]$, where $Dist[v]$ is the distance from r to v in T and $Dist[v] = \infty$ if v is not in T. Upon completion of *Dijkstra*, the distance from the root r to all the vertices of G is contained in the array $Dist[1{:}n]$. At each intermediate stage $Dist[v]$ has the following dynamic interpretation. If $v \in V(T)$, then $Dist[v]$ is the length of a path in T from r to v. If $v \notin V(T)$ and v is adjacent to a vertex of T, then $Dist[v]$ is the minimum value of $Dist[u] + w(uv)$ over all $uv \in Cut(T)$. If $v \notin V(T)$ and v is not adjacent to any vertex of T, then $Dist[v] = \infty$. $Dist[v]$ is initialized to ∞ for each vertex $v \neq r$ and $Dist[r]$ is initialized to 0. Each time a vertex u is added to the tree T, we need only update $Parent[v]$ and $Dist[v]$ for those vertices v that are adjacent to u and do not belong to the tree T. If $Dist[v] > Dist[u] + w[uv]$, then we set $Parent[v] = u$ and $Dist[v] = Dist[u] + w[uv]$.

```
procedure Dijkstra(G,w,r,Parent[1:n],Dist)
Input:   G (a connected graph with vertex set V and edge set E)
         r (a vertex of G)
         w (a weight function on E)
Output:  Parent[1:n] (parent array of a shortest path spanning tree)
         Dist[1:n] (array of lengths of shortest paths from a)

  for v := 1 to n do   {initialize Dist[1:n]}
     Dist[v] := ∞
```

```
      InTheTree[v] := .false.
   endfor
   Dist[r] := 0
   Parent[r] := 0
   for Stage := 1 to n – 1 do
      Select vertex u that minimizes Dist[u] over all u such that
                                                    InTheTree[u] = .false.
      InTheTree[u] := .true.
      for each vertex v such that uv ∈ E do   {update Dist and Parent arrays}
         if .not. InTheTree[v] then
            if Dist[u] + w(uv) < Dist[v] then
               Dist[v] := Dist[u] + w(uv)
               Parent[v] := u
            endif
         endif
      endfor
   endfor
end Dijkstra
```

Remark

1. As with procedure Prim, procedure Dijkstra terminates after only n – 1 stages, even though there are n vertices in the final shortest path spanning tree. Another iteration of the for loop controlled by Stage would result in no change to the arrays Dist[1:n] and Parent[1:n].
2. In spite of the similarity between Prim's algorithm and Dijkstra's algorithm and the fact that they both generate spanning trees, they solve different problems. Simple examples show that a shortest path spanning tree can fail to be a minimum spanning tree (see Exercise 12.18).

The shortest path algorithms presented in this section for (undirected) graphs generalize naturally to digraphs. In fact, the code for the procedure *Dijkstra* applies essentially unchanged to digraphs. We merely have to be careful in the case of digraphs to note that *uv* corresponds to a directed edge $uv = (u,v)$ from u to v (as opposed to the undirected edge $uv = \{u,v\}$). We illustrate the action of Dijkstra's algorithm for a sample digraph in Figure 12.13 (page 434), where the current value of *Dist*[*v*] is shown outside each vertex *v*.

As with procedure *Prim*, the worst-case complexity of procedure *Dijkstra* belongs to $\Theta(n^2)$ and belongs to $\Theta(m \log n)$ if a priority queue is used. For either version, the worst-case complexity has smaller order than the $\Theta(mn)$ worst-case complexity of procedure *BellmanFord.*

12.7 Parallelizing Greedy Algorithms

Two main approaches can be taken when designing parallel algorithms based on the greedy method. The first is to generalize the paradigm *Greedy* itself to a parallel

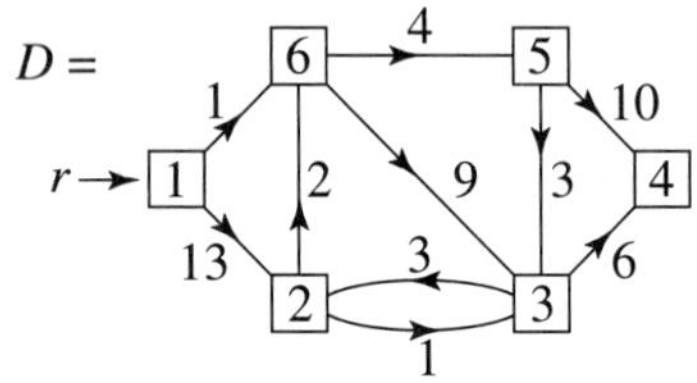

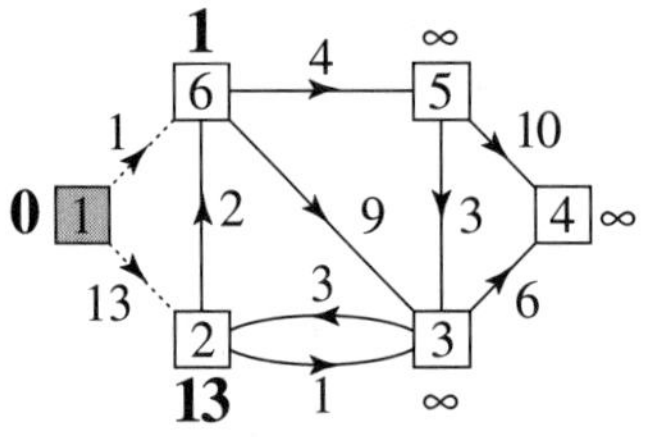

i	1	2	3	4	5	6
Dist[*i*]	0	13	∞	∞	∞	1
Parent[*i*]	0	1	–	–	–	1
InTheTree[*i*]	T	F	F	F	F	F

Stage 1

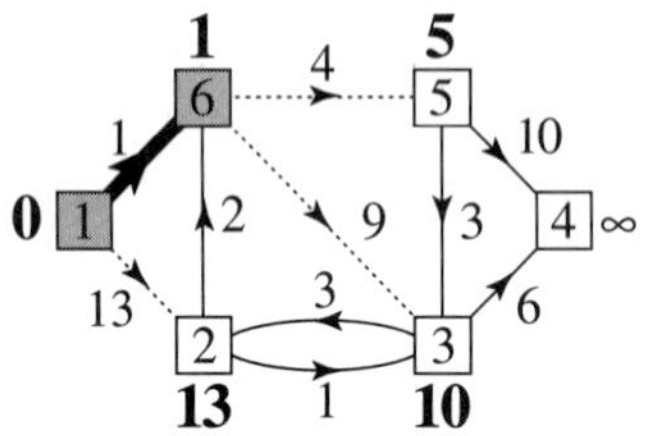

i	1	2	3	4	5	6
Dist[*i*]	0	13	10	∞	5	1
Parent[*i*]	0	1	6	–	6	1
InTheTree[*i*]	T	F	F	F	F	T

Stage 2

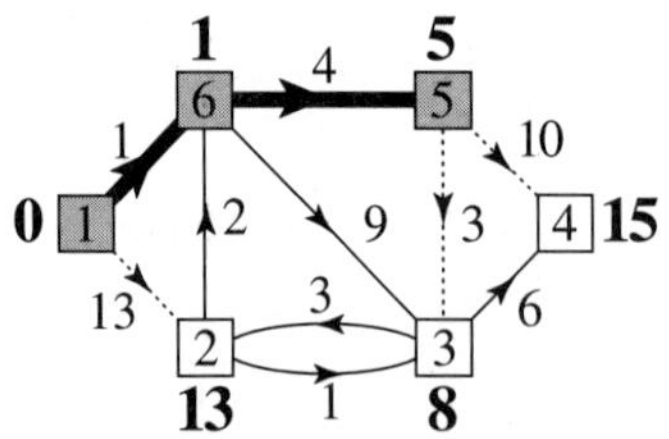

i	1	2	3	4	5	6
Dist[*i*]	0	13	8	15	5	1
Parent[*i*]	0	1	5	5	6	1
InTheTree[*i*]	T	F	F	F	T	T

Stage 3

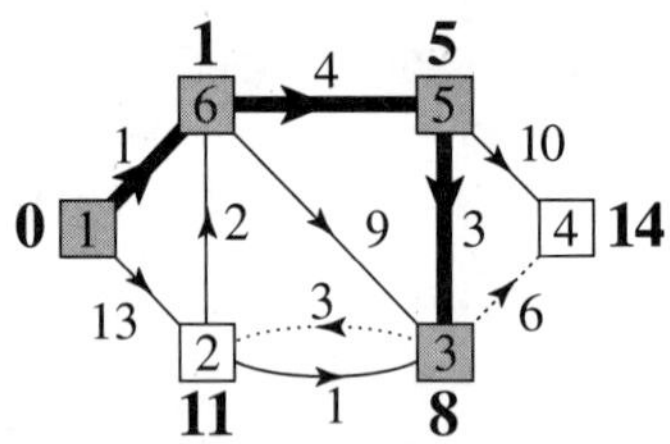

i	1	2	3	4	5	6
Dist[*i*]	0	11	8	14	5	1
Parent[*i*]	0	3	5	3	6	1
InTheTree[*i*]	T	F	T	F	T	T

Stage 4

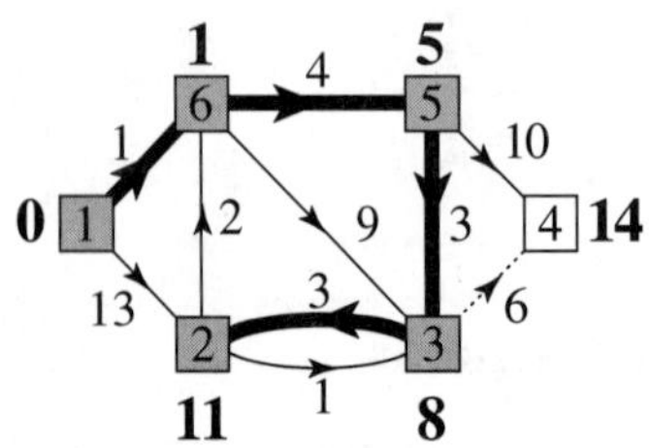

i	1	2	3	4	5	6
Dist[*i*]	0	11	8	14	5	1
Parent[*i*]	0	3	5	3	6	1
InTheTree[*i*]	T	T	T	F	T	T

Figure 12.13 Action of *Dijkstra* for a sample digraph *D*

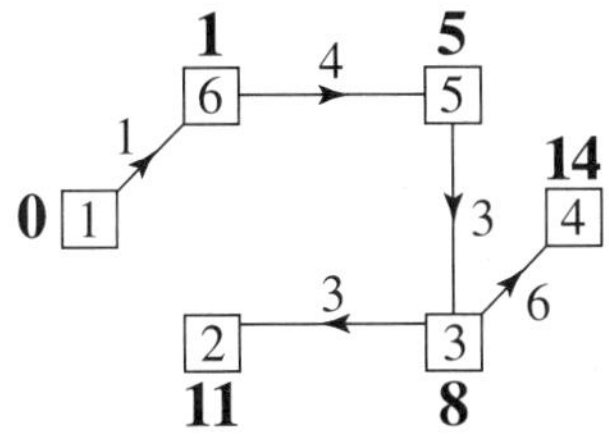

Resulting shortest path spanning out-tree rooted at vertex $r = 1$

Figure 12.13 Continued

paradigm *ParallelGreedy*, which invokes a parallel version of *GreedySelect*. For example, consider the Coin-Changing problem, which seeks to make change using the fewest number of coins of the given denominations. Suppose we add the (paper money) denominations of \$100, \$50, \$20, \$10, \$5, and \$1 to the coin denominations 50¢, 25¢, 10¢, 5¢, and 1¢. Given an amount *Change*, recall that the sequential greedy method would use as many \$100 bills as possible, then use as many \$50 bills as possible, and so forth. For example, if *Change* = \$397.62, then the greedy solution yields three \$100 bills, one \$50 bill, two \$20 bills, one \$5 bill, two \$1 bills, one half-dollar coin, two dimes, and two pennies. A parallel greedy select using two processors works independently on the paper change and the coin change. More precisely, one processor works on making paper money change for $PaperChange = \lfloor Change \rfloor$, and the other processor works on making coin change for $CoinChange = Change - PaperChange$. We leave it as an exercise to verify that the two-processor greedy method makes change using the fewest total number of bills and coins. A more sophisticated example of *ParallelGreedy* is given in Subsection 12.7.3.

In Subsections 12.7.1 and 12.7.2, we follow the second main approach: parallelize existing greedy sequential algorithms. We have seen several examples of sequential algorithms based on the greedy method, where the base set was sorted with respect to a given weighting. For example, in the optimal sequential storage order problem, the solution was shown to consist of storing the objects in nondecreasing order of length. The first step in parallelizing a sequential greedy algorithm is to use an efficient (sublinear) parallel algorithm to perform the sorting of the base set.

Building up a solution sequentially through a sequence of greedy choices (invocations of *GreedySelect*) would obviously lead to an algorithm having at least $\Theta(n)$ (linear) worst-case complexity. To do better, we must look for ways to build up multiple partial solutions in parallel. We illustrate the latter technique by considering again the Knapsack problem.

12.7.1 Knapsack in Parallel

Recall that the Knapsack problem is solved by first sorting the given n objects $b_1, b_2, \ldots, b_n$ in nondecreasing order of their weight per value ratios. Assuming that the objects do not all fit in at once, the knapsack is filled to capacity by placing the first

$k-1$ objects in the knapsack and then a fraction $f = (C - (w_1 + \cdots + w_{k-1}))/w_k$ of the kth object, where k is the largest index i such that $w_1 + \cdots + w_{i-1}$ is no greater than the capacity C of the knapsack. The sequential algorithm computes the prefix sums

$$w_1, w_1 + w_2, w_1 + \cdots + w_{i-1},$$

until the knapsack capacity is exceeded.

The procedure *PrefixSumsPRAM*, which computes the partial weight sums $w_1 + \cdots + w_i$, $i = 1, \ldots, n$, yields a straightforward parallelization of the procedure *Knapsack* on the EREW PRAM. *PrefixSumsPRAM* is the version of *PrefixCompPRAM* where the binary operation is addition (see Chapter 11). In the following pseudocode for *KnapsackPRAM*, the ith partial sum $w_1 + \cdots + w_i$ is stored in the ith position of the array *WeightSums*[1:n]. The solution array F[1:n] can be immediately computed from the array *WeightSums*[1:n].

```
Procedure KnapsackPRAM(Value[1:n], Weight[1:n], Capacity, F[1:n])
Model: EREW PRAM with n processors
Input:  Value[1:n] (an array of positive values)
        Weight[1:n] (an array of positive weights)
        Capacity (a positive capacity)
        precondition: Value[1]/ Weight[1] ≥ Value[2]/Weight[2] ≥ ··· ≥
                                                Value[n]/ Weight[n]
Output: F[1:n] (array of nonnegative fractions)
  dcl WeightSums[0:n] (array of nonnegative integers)
  WeightSums[0] := 0
  call PrefixSumsPRAM(Weight[1:n],WeightSums[1:n])
  for 1 ≤ i ≤ n do in parallel
    if PrefixSums[i – 1] > Capacity then
      F[i] := 0
    else
      if PrefixSums[i] ≤ Capacity then
        F[i] := 1
      else
        F[i] := (Capacity – WeightSums[i – 1])/ Weight[i]
      endif
    endif
  end in parallel
end KnapsackPRAM
```

Since *PrefixSumsPRAM* has $\Theta(\log n)$ complexity, clearly *KnapsackPRAM* also has (best-case, average, and worst-case) $\Theta(\log n)$ complexity. We leave the pseudocode and analysis of the two-dimensional mesh version of *Knapsack* to the exercises.

12.7.2 Parallelizing Prim's and Dijkstra's Algorithms

We now consider a relatively straightforward parallelization of the procedure *Prim*. The **for** loop that initializes *Nearest*[1:*n*], *Parent*[1:*n*], and *InTheTree*[1:*n*] can be performed in a single parallel step. We still must add $n - 1$ vertices and edges to the tree one at a time, so that the outer **for** loop in *Prim* is left unaltered. However, the processes of choosing the *i*th vertex (and edge) and updating the arrays *Nearest* and *Parent* can both be parallelized. Recall that choosing the *i*th vertex amounts to finding the vertex u not in the tree already generated such that *Nearest*[u] is minimized. On the EREW PRAM, the problem of finding u has $\log n$ order of complexity, as opposed to the linear complexity of a sequential scan. The updating of *Nearest*[1:*n*] and *Parent*[1:*n*] can be done in a single parallel step.

To simplify the problem of finding the vertex u to be added to tree at the *i*th stage, vertices v already selected to be added to the tree in previous stages have the value *Nearest*[v] reset to ∞. Then the index of the next vertex u to be added can be obtained by a call to the procedure *IndexMinPRAM* with input *Nearest*[1:*n*].

```
procedure PrimPRAM(G,w,Parent[1:n])

Model: EREW PRAM with n processors
Input:  G (a connected graph with vertex set V and edge set E)
        w (a weight function on E)
Output: Parent[1:n] (parent array of a minimum spanning tree)

  for v := 1 to n do in parallel
     Nearest[v] := ∞
     InTheTree[v] := .false.
  end in parallel
  Parent[r] := 0   {root the tree at an arbitrary vertex r}
  Nearest[r] := 0
  for Stage := 1 to n do   {Select vertex u that minimizes Nearest[u].
                            Note that Nearest[v] = ∞ for v ∈ V(T)}
     u := IndexMinPRAM(Nearest[1:n])
     InTheTree[u] := .true.               {add u to T}
     Nearest[u] := ∞
     call Broadcast(U[1:n],u)             {then U[v] = u, 1 ≤ v ≤ n}
     for v := 1 to n do in parallel       {update Nearest[v] and Parent[v] for
        if .not. InTheTree[v] then         all v ∉ V(T) that are adjacent to u}
           if w({U[v],v}) < Nearest[v] then
              Nearest[v] := w({U[v],v})
              Parent[v] := U[v]
           endif
        endif
     end in parallel
  endfor

end PrimPRAM
```

Since function *IndexMinPRAM* and procedure *Broadcast* both have complexity belonging to O(log n), procedure *PrimPRAM* has complexity belonging to $\Theta(n \log n)$. Thus, *PrimPRAM* has cost

$$C(n) \in O(n^2 \log n),$$

which is within a logarithmic factor of the order-optimal cost $\Theta(n^2)$. Further, *PrimPRAM* has speedup $S(n) = n/\log n$ over *Prim2*.

The procedure *Dijkstra2* can be parallelized in exactly the same way as *Prim2*. We leave the pseudocode for *DijkstraPRAM* as an exercise.

12.7.3 A Polylogarithmic Parallel Algorithm for the Minimum Spanning Tree Problem

Kruskal's greedy algorithm for finding a minimum spanning tree is based on choosing the globally smallest weight edge among the remaining edges and adding that edge to the forest unless a cycle is created. We now design a parallel greedy algorithm *MSTCREW* suitable for implementation on the CREW PRAM with $n^2/2$ processors having polylogarithmic complexity. Each stage of *MSTCREW* involves choosing in parallel the smallest weight edge at each vertex. Without loss of generality, we can assume that all edge weights are distinct. With distinct edge weights, the minimum spanning tree is unique. The key idea behind *MSTCREW* is the following proposition, whose proof is left as an exercise.

PROPOSITION 12.7.1 Consider a graph $G = (V,E)$ with distinct edge weights, and let U be any subset of V. The minimum weight edge with one vertex in U and the other in $V - U$ is in the minimum spanning tree. ❑

For $v \in V$, we let $\mu(v)$ denote the edge of minimum weight incident with vertex v. Setting $U = \{v\}$ in Proposition 12.7.1, we obtain the following corollary.

COROLLARY 12.7.2 Consider a graph $G = (V,E)$ with distinct edge weights. For each vertex v, the edge $\mu(v)$ of minimum weight incident with vertex v is in the minimum spanning tree. ❑

For convenience, we take $V = \{1,2, \ldots, n\}$ as the vertex set V of G, and set $\mu(V) = \{\mu(i): i \in V\}$. We will assume that G is implemented using its *weighted* adjacency matrix $A = (a_{ij})$; that is, $a_{ij} = w(\{i,j\})$ if $\{i,j\} \in E$, and $a_{ij} = \infty$ otherwise. We can greedily choose edge $\mu(i)$ at each vertex $i \in V$ in (approximately) $\log_2 n$ parallel steps by allocating $n/2$ processors to vertex i and using the binary fan-in technique. Note that if $\{a,b\}$ is the minimum weight edge in the graph G, then $\{a,b\} \in \mu(V)$ and $\mu(a) = \mu(b)$.

First suppose that for the remaining $n - 2$ vertices, no two vertices have the same minimum-weight incident edge, so that $\mu(V)$ contains $n - 1$ edges. Since the minimum spanning tree T has $n - 1$ edges and by Corollary 12.7.2 the $n - 1$ edges of $\mu(V)$ all belong to T, it follows that $\mu(V)$ is precisely the edge set of T.

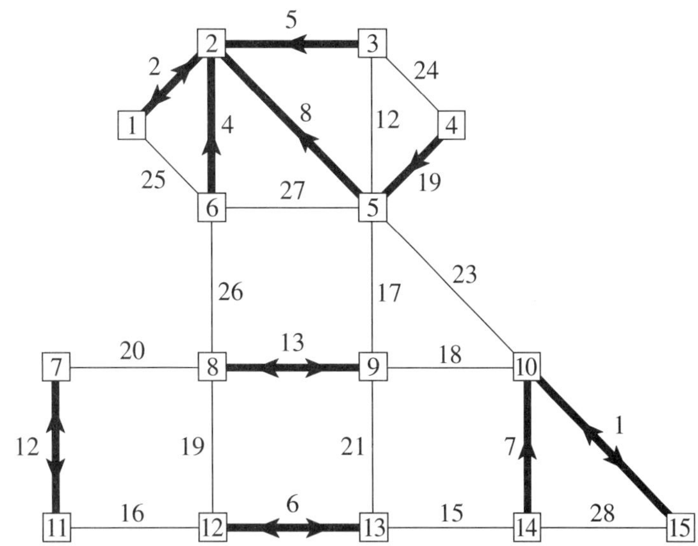

Figure 12.14 A sample weighted graph and the forest F_1 determined by the mapping μ. Edges marked with arrows must belong to minimum spanning tree.

Now suppose that $\mu(V)$ contains fewer than $n - 1$ edges. Then, $\mu(V)$ determines a spanning forest F_1 (the subscript on the forests generated by the algorithm will indicate the iteration during which they are created), where each tree in F_1 contains a unique edge $\{i,j\}$ such that $\mu(i) = \mu(j)$. To see this, for each $i \in V$, mark edge $\mu(i)$ with an arrow having tail i (see Figure 12.14). Then the oriented edges determine a forest F_1, where each tree in F_1 contains exactly one doubly oriented edge.

We now define an equivalence relation on the set of vertices V, where two vertices are equivalent if they belong to the same tree in F_1. We call the equivalence classes *supervertices*. We let S_1 denote the set of all supervertices. Since each tree in F_1 contains at least one edge, F_1 contains at most $n/2$ trees. Thus, S_1 has at most half as many supervertices as G has vertices. For $s \in S_1$, $\mu(s)$ denotes the minimum weight edge having exactly one vertex in s, and $\mu(S_1)$ denotes the set $\{\mu(s) : s \in S_1\}$. By Proposition 12.7.1, the edges in $\mu(S_1)$ belong to the minimum spanning tree. The mapping μ determines a forest F_2 among the supervertices (see Figure 12.15, page 440).

If $\mu(V) \cup \mu(S_1)$ contains $n - 1$ edges, then it must be the edge set of the minimum spanning tree. Otherwise, consider the equivalence relation on the set of vertices V such that two vertices are equivalent if they belong to supervertices that belong to the same tree in F_2, and let S_2 denote the set of all equivalence classes. Then the equivalence classes of S_2 constitute a new set of supervertices and a new set of edges $\mu(S_2)$ that belong to the minimum spanning tree (see Figure 12.16, page 441).

If the forest F_3 determined by S_2 and μ is not a tree, then the process is repeated. We thereby generate larger and larger supervertices, until after at most $\log_2 n$ steps we obtain a supervertex consisting of the entire vertex set V. We let $S_0, S_1, S_2, \ldots,$

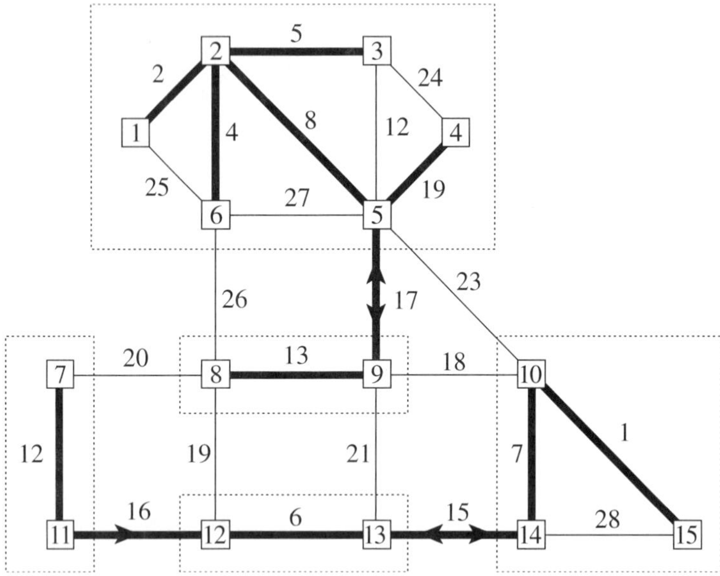

Figure 12.15 Forest F_2 among supervertices $s \in S_1$ determined by the mapping μ. Edges that must be in minimum spanning tree are marked in bold and edges in $\mu(S_1)$ are marked with arrows.

S_k denote the sets of supervertices generated, where $S_0 = V$ and $S_k = \{V\}$. Since S_i has at most half as many supervertices as S_{i-1}, $i = 1, \ldots, k$, it follows that $k \leq \log_2 n$. Further, by Proposition 12.7.1, $\mu(S_0) \cup \cdots \cup \mu(S_{k-1})$ are the edges of the minimum spanning tree. The high-level pseudocode for the algorithm *MST-PRAM* follows.

```
procedure MSTCREW(G,w,MSTree)
Model: CREW PRAM with n^2/2 processors
Input:  G (a connected graph with vertex set V = {1, . . . , n} and edge set E)
        w (a weighting of E using distinct weights)
Output: MSTree (edges of a minimum spanning tree)
  MSTree := ∅          {initialize MSTree to be empty}
  S := V               {initialize S to be the set of n vertices of G}
  while |S| > 1 do
    for all s ∈ S do in parallel
      Compute μ(s)
    end in parallel
    Compute forest F determined by μ(S) = {μ(s) : s ∈ S}
    MSTree := MSTree ∪ μ(S)
```

Compute the new set S of supervertices determined by F
endwhile
end *MSTCREW*

To examine one key implementation detail for *MSTCREW*, consider the computation of $\mu(S)$ during iteration i of the **while loop**. As described earlier, during the first iteration $\mu(S)$ (where $S = V$) can be computed by concurrently using binary fan-in at each vertex. During the ith iteration, $i > 2$, $\mu(S)$ can be computed in approximately 2 $\log_2 n$ steps by two successive applications of the binary fan-in technique.

Another key implementation detail involves the method of maintaining the supervertices and can be solved as follows. Each supervertex s can be identified by a particular vertex $L(s)$, called the *leader*, that serves as a representative of s. At a given stage of the algorithm, the *leader* of a vertex v, denoted by $L(v)$, is the leader of the current supervertex containing v. Initially, $L(v) = v$ for every vertex $v \in V$. During each iteration of the **while** loop in *MSTCREW*, the leaders are updated as follows. Let S_{i-1} denote the set of supervertices at beginning of the *ith* iteration, $i = 1, \ldots, k$, where k denotes the total number of iterations. Consider any supervertex s from S_i. An argument similar to one given earlier shows that the tree of forest F corresponding to s contains a unique (ordered) pair of supervertices (a,b), $a, b \in S_{i-1}$, such that $\mu(a) = \mu(b)$ and $L(a) < L(b)$. We define the leader $L(s)$ of s to be the leader

Figure 12.16 Forest (tree) F_3 is now two supervertices and an edge joining vertex 9 and 10. The algorithm has now generated the indicated minimum spanning tree.

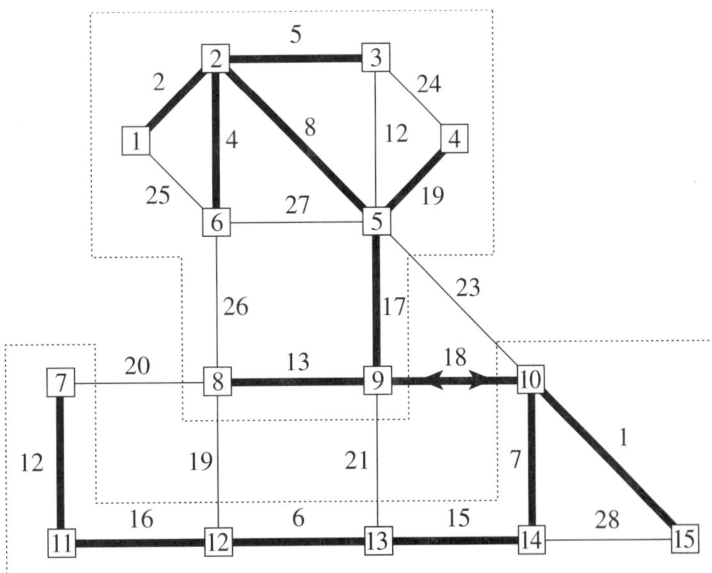

$L(a)$ of a. Thus, for each vertex $v \in V$, the leader $L(v)$ of v is updated to equal $L(s)$, where s is the supervertex from S_i containing v. The leader $L(v)$ of each vertex can be updated in (approximately) $\log_2 n$ steps using pointer jumping.

We have shown that each iteration of the **while** loop in *MSTCREW* can be done in $O(\log n)$ parallel steps. Since the **while** loop performs at most $\log_2 n$ iterations, it follows that *MSTCREW* has complexity $W(n) \in O(\log^2 n)$ and cost $C(n) \in O(n^2 \log^2 n)$. Further, *MSTCREW* has speedup $S(n) \in \Omega(n^2/\log^2 n)$ over *Prim2*.

The procedure *MSTCREW* is readily adapted to implementations on the EREW PRAM and the two-dimensional mesh-of-trees $MOT_{n,n}$. These adaptations are explored in the exercises.

12.8 Closing Remarks

The problem of finding a minimum spanning tree in an undirected graph G generalizes naturally to the problem of finding a minimum spanning in-tree (or out-tree) rooted at a given vertex r in a digraph D. Prim's algorithm extends to digraphs by replacing *Cut*(T) with *CutIn*(T) (or *CutOut*(T)) consisting of all edges of *Cut*(T) having head in T (or having tail in T). Unfortunately, even though this greedy method yields a spanning in-tree (or out-tree) rooted at r, it doesn't always yield one of minimum weight. Algorithms for constructing minimum weight in-trees (out-trees) can be found in the references.

An improvement to Prim's algorithm has been found that uses a Fibonacci heap and has $\Theta(m + n \log n)$ worst-case complexity. Recently, probabilistic $\Theta(m)$ algorithms have been found for the minimum spanning tree problem.

A slight modification of algorithm *MSTCREW* can be used to find connected components in an unweighted graph G. *MSTCREW* can be naturally adapted to the two-dimensional mesh-of-trees $MOT_{n,n}$.

References and Suggestions for Further Reading

It is interesting that the first minimum spanning algorithm, due to O. Boruvka in a 1926 paper, is one of the most naturally parallelizable of all the various algorithms for the problem. The following two excellent references contain the history of the minimum spanning tree problem.

Graham, R. L. and P. Hell. "On the history of the minimum spanning tree problem," *Annals of the History of Computing*, 7 (1985): 43–57.

Tarjan, R. E. *Data Structures and Network Algorithms*. Society for Industrial and Applied Mathematics, 1983.

Gordon, M. and M. Minoux (trans. by S. Vajda). *Graphs and Algorithms*. New York: Wiley, 1984. This text contains algorithms for constructing minimum weight spanning in-trees in digraphs.

The following references discuss generalizations of the minimum spanning algorithms to matroids.

Papadimitriou, C. H. and K. Steiglitz. *Combinatorial Optimization: Algorithms and Complexity*. Englewood Cliffs, NJ: Prentice-Hall, 1982.

Thulasiraman, K. and M. N. S. Swamy. *Graphs: Theory and Algorithms*. New York: Wiley, 1992.

Exercises

Section 12.1 General Description

12.1 Given the standard U.S. coins, pennies (1¢), nickels (5¢), dimes (10¢), quarters (25¢), and half-dollars (50¢), show that the greedy method of making change always yields a solution using the fewest coins.

12.2 For a given integer base $b > 2$, suppose you have coins of denominations $b^0, b^1, \dots, b^{n-1}$. Show that the greedy method of making change for these denominations always yields a solution using the fewest coins.

Section 12.2 An Optimal Sequential Storage Order

12.3 Design and analyze a greedy algorithm for the optimal sequential storage problem when m tapes are available.

12.4 For the sequential storage order problem, consider the general case of minimizing $AveCost(\pi)$ as given by (12.2.1), where p_i denotes the probability that the object to be processed is b_i, $i = 1, \dots, n$.

a. Show that the greedy strategy used when all the probabilities are equal (that is, sorting the objects by nondecreasing order of costs c_i) fails to give an optimal solution for a general set of probabilities p_i.

b. Show that the greedy strategy of arranging the objects in nondecreasing order of probabilities also fails to give optimal solutions in general.

c. Show that the greedy strategy of arranging the objects in nondecreasing order of the ratios c_i/p_i yields an optimal solution.

Section 12.3 The Knapsack Problem

12.5 For the Knapsack problem with weights $w_1, w_2, \dots, w_n$ and profits $p_1, p_2, \dots, p_n$, give a single example where both the greedy solution based on placing the objects in the knapsack in the order $w_1 \le w_2 \le \cdots \le w_n$ as well as the greedy solution based on placing the objects in the knapsack in the order $p_1 \le p_2 \le \cdots \le p_n$ yield suboptimal solutions.

12.6 **a.** Give an example where the greedy method described in Section 12.3 for the 0/1 Knapsack problem yields a suboptimal solution.

b. Show that the greedy method yields arbitrarily bad solutions; that is, given any $\varepsilon > 0$, find an example where the ratio of the greedy solution to the optimal solution is less than ε.

Section 12.4 Huffman Codes

12.7 Prove Proposition 12.4.1.

12.8 Trace the action of *HuffmanCode* for the letters *a,b,c,d,e,f,g,h* occurring with frequencies 10, 7, 3, 5, 9, 2, 3, 2.

Section 12.5 Minimum Spanning Tree

12.9 Consider the following weighted graph *G*.

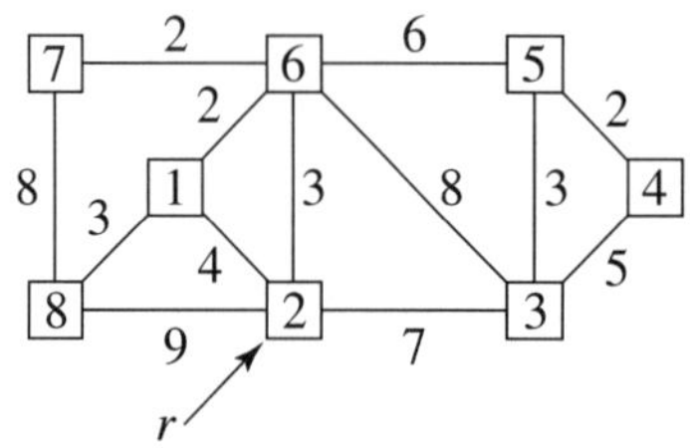

a. Trace the action of *Kruskal* for *G*.

b. Trace the action of *Prim* for *G*, with $r = 2$.

12.10 Give pseudocode for a version of the algorithm *Kruskal*2 that maintains the weighted edges of the graph in a priority queue.

12.11 Prove the following two propositions used in the proof of the correctness of Kruskal's algorithm.

a. Adding an edge to a tree creates a unique cycle.

b. Removing an edge from a cycle in a connected graph still leaves the resulting graph connected.

12.12 Modify *Kruskal* to accept *any* graph *G* (connected or disconnected) and to output a minimum (weight) forest, each of whose trees is a minimum spanning tree for a component of *G*.

12.13 Prove that Prim's algorithm yields a minimum spanning tree.

12.14 Analyze the complexity of the variant of *Prim* based on an explicit implementing of *Cut*(*T*) at each stage of the algorithm. Compare its complexity to that of *Prim*.

12.15 Give pseudocode for *Prim*2 using a priority queue.

12.16 The problem of finding a minimum spanning tree in an undirected graph *G* generalizes naturally to the problem of finding a minimum spanning in-tree (or out-tree) rooted at a given vertex *r* in a strongly connected digraph *D*. Prim's algorithm directly generalizes to digraphs, with the modification that *Cut*(*T*) is replaced by *CutIn*(*T*) (or *CutOut*(*T*)) consisting of all edges of *Cut*(*T*) having head in *T* (or having tail in *T*). Show that this greedy method doesn't always yield a spanning in-tree (or out-tree) rooted at *r*.

Section 12.6 Shortest Paths in Graphs and Digraphs

12.17 Trace the action of *Dijkstra* for the following digraph with initial vertex $r = 3$.

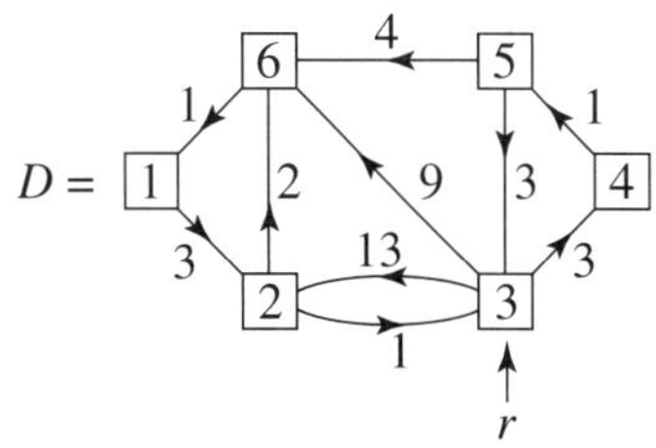

12.18 **a.** Trace the action of *Dijkstra* for the graph G of Figure 12.11 with $r = 1$.

b. Note that the shortest path tree generated in part (a) happens to coincide with the minimum spanning tree. Give an example of a weighted graph G with distinct weights and a vertex r such that the shortest spanning tree T generated by *Dijkstra* is not a minimum spanning tree. In fact, find an example of a graph on n vertices where the shortest path tree has only about half the edges in common with the minimum spanning tree.

12.19 Trace the action of *Dijkstra* for the graph G of Exercise 12.9, with initial vertex $r = 2$.

12.20 Given a weighted graph G and vertices a,b, consider the following greedy strategy that attempts to grow a shortest path from a to b. Choose an edge e having smallest weight among the remaining edges incident with the terminal vertex u of the path P previously generated, such that e contains no vertex of P other than u. These greedy choices continue until either b is reached or no choice for e exists. Give an example where the path so grown

a. never reaches b,

b. reaches b, but is not a shortest path from a to b.

12.21 Prove that the tree generated by *Dijkstra* spans all vertices in the component of the graph G containing a.

12.22 Give pseudocode for *Dijkstra*2 using a priority queue.

12.23 Show that Dijkstra's algorithm can fail in the case where some of the weights are negative, even though no negative cycles exist. (In particular, why doesn't adding a sufficiently large positive constant to each edge always work?)

Section 12.7 Parallelizing Greedy Algorithms

12.24 Consider the parallel greedy select, which using two processors works independently on the paper change and the coin change. Verify that this two-processor greedy method makes change using the fewest total number of bills and coins for U.S. currency.

12.25 Design and analyze an EREW PRAM version of *Knapsack*.

12.26 Design and analyze a two-dimensional mesh version of *Knapsack*.

12.27 Show how the problem of finding a minimum spanning tree in a weighted graph with repeated edge weights can be transformed into an equivalent problem with distinct edge weights.

12.28 Show that the minimum spanning tree is unique when the edge weights are distinct.

12.29 Prove Proposition 12.7.1.

12.30 Show that the μ mapping defined in the discussion of *MSTCREW* determines a spanning forest where each tree in the forest contains a unique pair of supervertices (a,b) such that $\mu(a) = \mu(b)$.

12.31 One key implementation detail for *MSTCREW* is the computation of $\mu(S)$ each iteration of its **while** loop. Show how $\mu(S)$ can be computed in approximately $2 \log_2 n$ steps by two successive applications of the binary fan-in technique.

12.32 Consider the updating protocol for the leader $L(v)$ of each vertex v in each iteration of *MSTCREW* discussed in Subsection 12.7.3. Show how the updating can be computed in (approximately) $\log_2 n$ steps using pointer jumping.

12.33 Give pseudocode for a parallel version of *Dijkstra*2.

12.34 Trace the action of *MSTCREW* for the following weighted graph.

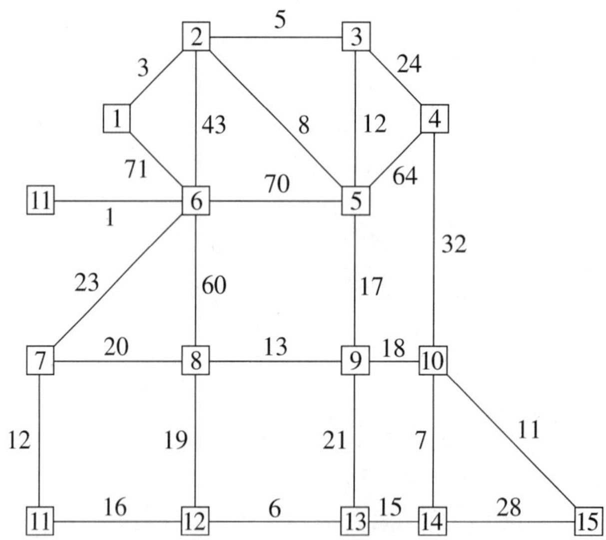

12.35 Design and analyze a version of *MSTCREW* for the EREW PRAM.

12.36 Give an algorithm on the CREW PRAM with $n^2/2$ processors that compute the components of a graph in time $O(\log^2 n)$. (*Hint:* Choose an appropriate weighting, where all weights are distinct, and then use a modification of *MSTCREW*.)

***12.37** Design and analyze a version of *MSTCREW* for the two-dimensional mesh-of-trees $\text{MOT}_{n,n}$.

13

Divide and Conquer

The *divide-and-conquer* paradigm is one of the most powerful design strategies available in the theory of algorithms. The paradigm can be described in general terms as follows. A problem input (instance) is divided according to some criteria into a set of smaller inputs to the same problem. The problem is then solved for each of these smaller inputs, either recursively by further division into smaller inputs or by invoking an ad hoc or a priori solution. Finally, the solution for the original input is obtained by expressing it in some form as a combination of the solution for these smaller inputs. Ad hoc solutions are often invoked when the input size is smaller than some preassigned *threshold* value. Examples of a priori solutions (solutions known in advance) include sorting single element lists or multiplying single-digit binary numbers.

13.1 The Divide-and-Conquer Paradigm

The divide-and-conquer design strategy can be formalized as follows. Let *Known* denote the set of inputs to the problem whose solutions are known a priori or by ad hoc methods. The procedure *Divide_and_Conquer* calls two procedures *Divide* and *Combine*. *Divide* has input parameter I and output parameters $I_1, \ldots, I_m$, where m may depend on I. The inputs $I_1, \ldots, I_m$ must be smaller or simpler inputs to the problem than I, but are not required to always be a *division* of I into subinputs. *Combine* has input parameters $J_1, \ldots, J_m$ and output parameter J. *Combine* is the procedure for obtaining a solution J to the problem with input I by combining the recursively obtained solutions $J_1, \ldots, J_m$ to the problem with inputs $I_1, \ldots, I_m$. For convenience, we formulate *Divide_and_Conquer* as a procedure.

procedure *Divide_and_Conquer*(I,J) **recursive**

Input: I (an input to the given problem)
Output: J (a solution to the given problem corresponding to input I)

```
if I ∈ Known then
    assign the a priori or ad hoc solution for I to J
else
    call Divide(I,I_1, ..., I_m)   {m may depend on the input I}
```

Continued

```
      for i := 1 to m do
         call Divide_and_Conquer(I_i,J_i)
      endfor
      call Combine(J_1, . . . , J_m,J)
   endif
end Divide_and_Conquer
```

Often the work done by an algorithm based on *Divide_and_Conquer* resides solely in one of the two procedures *Divide* or *Combine*, but not both. For example, in the algorithm *QuickSort*, the divide step (call to *RearrangeAndPlace*) is the heart of the algorithm and the combine step requires no work at all. On the other hand, in the algorithm *MergeSort*, the divide step is trivial and the combine step (call to *Merge*) is the heart of the algorithm. Sometimes, both the divide and combine steps are difficult (see Figure 13.1).

For most divide-and-conquer algorithms, the number m of subproblems is a constant (independent of any particular input I). Divide-and-conquer algorithms where m is a constant equal to one are referred to as *simplifications*. The binary search algorithm is an example of a simplification.

One useful technique to improve the efficiency of a divide-and-conquer algorithm is to employ a known ad hoc algorithm when the input size is smaller than some *threshold*. Of course, the ad hoc algorithm must be more efficient than the given divide-and-conquer algorithm for sufficiently small inputs.

For example, consider the sorting algorithm *MergeSort*, which has $\Theta(n \log n)$ average complexity. The sorting algorithm *InsertionSort*, which has $\Theta(n^2)$ average complexity, is much less efficient than *MergeSort* for large values of n. However, due to the constants involved, *InsertionSort* is more efficient than *MergeSort* for small values of n. Thus, we can improve the performance of *MergeSort* by calling *InsertionSort* for input lists of size not larger than a suitable threshold. Finding the optimal value for a threshold is often done empirically in practice. Empirical studies are needed since the best choice of a threshold depends on the constants associated

Figure 13.1 An example of a hard divide–hard combine divide-and-conquer algorithm (Reprinted by permission of Johnny Hart and Creators Syndicate, Inc.)

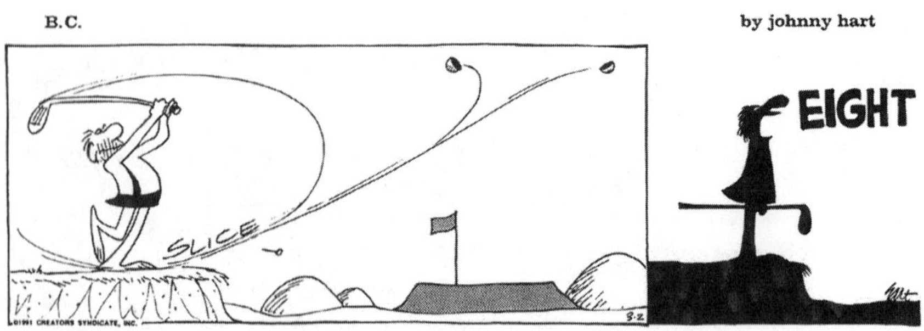

with the implementation on a particular computer. For example, empirical studies have shown that for *MergeSort*, calling *InsertionSort* with a threshold of around $n = 16$ is usually optimal.

13.2 Selecting the *k*th-Smallest Element

Our first illustration of the divide-and-conquer paradigm in this chapter is an instance of simplification. Given a list $L[1{:}n]$, we consider the problem of selecting the kth-smallest element in L. When $k = 1$ or $k = n$, this problem coincides with the problem of finding the minimum or maximum element in L, respectively. Linear algorithms exist for these cases. For a general k, the problem of finding the kth-smallest element can be solved in time $O(n \log n)$ by sorting the list. We will design a linear algorithm for the general problem by utilizing the procedure *RearrangeAndPlace* that formed the basis of *QuickSort*, together with a strategy analogous to binary search.

13.2.1 The Algorithm *Select*

For convenience, we assume that the list $L[1{:}n]$ consists of distinct elements. Suppose that an initial call to procedure *RearrangeAndPlace* with $L[1{:}n]$ returns the value $j = position$, so that in the rearranged list, $L[i] < L[j]$ for $1 \le i < j$, $L[i] > L[j]$ for $j < i \le n$, and the pivot element $L[j]$ has the value of the original $L[1]$. In particular, the pivot element is the jth-smallest element in L. Hence, if $j = k$, then the kth-smallest element has been found. On the other hand, if $k < j$, then since each element in $L[1{:}j-1]$ is smaller than each element in $L[j+1{:}n]$, the kth-smallest element in the list $L[1{:}n]$ coincides with the kth-smallest element in $L[1{:}j-1]$. Similarly, if $k > j$, then the kth-smallest element in $L[1{:}n]$ coincides with the $(k-j)$th-smallest element in the list $L[j+1{:}n]$. Hence, by calling *RearrangeAndPlace* we have either determined the kth-smallest element or a sublist (of size at least one less than the size of the original list) that contains the kth-smallest element.

Whenever $k \ne j$, another call to *RearrangeAndPlace* either determines the kth-smallest element or narrows the length of a sublist containing the kth-smallest element by at least one. After at most $n-1$ calls to *RearrangeAndPlace*, the kth-smallest element will be determined. The recursive procedure *Select* given below formalizes the preceding discussion. The input list to *Select* is the sublist $L[low{:}high]$ of the list $L[1{:}n]$. Procedure *Select* computes the $(k - low + 1)$th-smallest element in the list $L[low{:}high]$, and is initially called with $low = 1$ and $high = n$.

procedure *Select*($L[low,high],k,x$)
Input: $L[low{:}high]$ (array of list elements) {initially called with $low = 1$ and $high = n$}
k (a positive integer such that $low \le k \le high$)
Output: x (the $(k - low + 1)$th-smallest element in the list $L[low{:}high]$) **Continued**

```
    call RearrangeAndPlace(L[low:high],position)
    case
       :k = position: x := L[position]
       :k < position: call Select(L[low:position – 1],k,x)
       :k > position: call Select(L[position + 1:high],k,x)
    endcase
end Select
```

In Figure 13.2 we illustrate the action of *Select* for a sample list of size 7.

We now analyze *Select* for a list of size n, where we assume that $low = 1$ and $high = n$. We will use the list comparisons generated by calls to procedure *RearrangeAndPlace* as our basic operation. Let $W(n,k)$ and $A(n,k)$ denote the worst-case and average complexities of *Select* restricted to inputs where the parameter k is fixed. Hence,

$$W(n) = \max\,\{W(n,k) \mid 1 \le k \le n\}, \qquad (13.2.1)$$

Figure 13.2 Action of *Select* with $k = 5$

Original List	index	1	2	3	4	5	6	7	8
	list element	22	9	23	52	15	19	47	$+\infty$

call *RearrangeAndPlace*(*L*[1:7],*position*)

Output List	index	1	2	3	④	5	6	7	8
	list element	15	9	19	22	52	23	47	$+\infty$

Output value of *position* = 4

call *RearrangeAndPlace*(*L*[5:7],*position*)

Output List	index	1	2	3	4	5	6	⑦	8
	list element	15	9	19	22	47	23	52	$+\infty$

Output value of *position* = 7

call *RearrangeAndPlace*(*L*[5:6],*position*)

Output List	index	1	2	3	4	5	⑥	7	8
	list element	15	9	19	22	23	47	52	$+\infty$

Output value of *position* = 6

call *RearrangeAndPlace*(*L*[5:5],*position*)

Output List	index	1	2	3	4	⑤	6	7	8
	list element	15	9	19	22	23	47	52	$+\infty$

Output value of *position* = 5

$x = 23$

and assuming each k is equally likely,

$$A(n) = \frac{\sum_{k=1}^{n} A(n, k)}{n}. \qquad (13.2.2)$$

Since procedure *RearrangeAndPlace* has complexity $n + 1$ for a list of size n, and since there exists an input list whose size is only narrowed by one with each call to *RearrangeAndPlace*, we have

$$W(n,k) = (n + 1) + n + (n - 1) + \cdots + 2 = \frac{(n + 1)(n + 2)}{2} - 1 \in \Theta(n^2). \qquad (13.2.3)$$

Thus, *Select* has the same quadratic worst-case behavior as *Quicksort*.

When analyzing the average performance of *Select*, as usual we assume that the inputs to *Select* are all permutations of $1, \ldots, n$, and that each permutation is equally likely. We now show by (the strong form of) induction on n that

$$A(n,k) \le 4n \quad \text{for all } n \text{ and all } k \in \{1, \ldots, n\}. \qquad (13.2.4)$$

Since $A(1,1) = 2$, (13.2.4) is true for $n = 1$ and $k = 1$. Assuming that (13.2.4) holds for all pairs m,k with $k \in \{1, \ldots, m\}$ and $m < n$, we show that (13.2.2) also holds for n and all $k \in \{1, \ldots, n\}$. Let k be any integer such that $1 \le k \le n$. Our assumption that all permutations of $1,2, \ldots, n$ are equally likely to be input to *Select* implies that after executing the call to *RearrangeAndPlace* with $L[1{:}n]$, each value $1, \ldots, n$ is equally likely to be the output value of the parameter *position*. There are three cases to consider, depending on whether *position* is less than, equal to, or greater than k.

For convenience, let $i = position$. If $k = i$, then *Select* terminates after a single call to *RearrangeAndPlace*, so that $n + 1$ comparisons are performed. If $k < i$, then *Select* calls itself recursively with the list $L[1{:}i - 1]$, so that the average number of comparisons done in this case is given by

$$n + 1 + A(i - 1,k) \le n + 1 + 4(i - 1) \quad \text{(by induction hypothesis)}. \qquad (13.2.5)$$

Finally, if $k > i$, then *Select* calls itself recursively with the list $L[i + 1{:}n]$, so that the average number of comparisons performed in this case is given by

$$n + 1 + A(n - i,k - i) \le n + 1 + 4(n - i) \quad \text{(by induction hypothesis)}. \qquad (13.2.6)$$

Using inequalities (13.2.5) and (13.2.6) we obtain

$$A(n,k) = \frac{1}{n}\left[n + 1 + \sum_{i=1}^{k-1}(n + 1 + A(n - i, k - i)) + \sum_{i=k+1}^{n}(n + 1 + A(i - 1, k))\right]$$

$$= \frac{1}{n}\left[n(n + 1) + \sum_{i=1}^{k-1} A(n - i, k - i) + \sum_{i=k+1}^{n} A(i - 1,k)\right]$$

$$\le n + 1 + \frac{4}{n}\left[\sum_{i=1}^{k-1}(n - i) + \sum_{i=k+1}^{n}(i - 1)\right]. \qquad (13.2.7)$$

As a function of k, the right-hand side of the inequality in (13.2.7) is maximized when $k = \lceil (n+1)/2 \rceil$. Now assume that n is odd (the proof when n is even is similar). Then using (13.2.7), and rewriting the term inside the last square bracket, yields

$$\begin{aligned}
A(n,k) &\le A\left(n, \frac{n+1}{2}\right) \le n + 1 + \frac{4}{n}\left[2\left(\sum_{i=1}^{n-1} i - \sum_{i=1}^{(n-1)/2} i\right)\right] \\
&= n + 1 + \frac{4}{n}\left[n(n-1) - \left(\frac{n-1}{2}\right)\left(\left(\frac{n-1}{2}\right) + 1\right)\right] \\
&\le n + 1 + \frac{4}{n}\left[n(n-1) - \frac{(n-1)^2}{4}\right] = n + 1 + \frac{4}{n}\left[(n-1)\left(n - \frac{n-1}{4}\right)\right] \\
&= n + 1 + \frac{4}{n}\left[(n-1)\left(\frac{3n+1}{4}\right)\right] = n + 1 + \frac{(n-1)(3n+1)}{n} \le 4n.
\end{aligned}$$

This completes the inductive proof of (13.2.4).

It follows immediately from (13.2.2) and (13.2.4) that $A(n) \le 4n$. Since *Select* performs at least one call to *RearrangeAndPlace* for any input, both $A(n,k)$ and $A(n)$ are at least $n + 1$, so that we have

$$n + 1 \le A(n), A(n,k) \le 4n, \quad \text{for all } n \text{ and all } k \in \{1, \ldots, n\}. \tag{13.2.8}$$

Hence, $A(n), A(n,k) \in \Theta(n)$.

Remark We have shown $A(n,k) \le A(n,(n+1)/2) \le 4n$ and concluded that $A(n) \le 4n$. However, a similar induction argument can be used to show that $A(n,1) \le 2(n+1)$, which is a reduction by about half of our $4n$ upper bound. Since $A(n)$ is the average of $A(n,k)$ over all $k \in \{1, \ldots, n\}$, we might expect to obtain a better upper bound for $A(n)$. In fact, it can be shown that $A(n) \le 3n$ (the verification of this stronger inequality is left as an exercise).

13.2.2 A Version of *Select* with Linear Worst-Case Complexity

We have just seen that *Select* has linear average complexity but quadratic worst-case complexity. By using a more clever choice for the pivot element for *RearrangeAndPlace*, we can achieve a version of *Select* that has *linear* worst-case complexity. The quadratic worst-case behavior of *Select* occurs when the pivot element always happens to be an endpoint of the subarray being rearranged. On the other hand, if the pivot element always happens to be close to the median, then the list is partitioned into sublists of nearly equal size, and the number of comparisons performed would be approximately

$$(n+1)\left(1 + \frac{1}{2} + \frac{1}{4} + \frac{1}{8} + \cdots\right) = 2(n+1) \in \Theta(n)$$

Thus, if we knew the median at each stage of *Select*, we could use it as the pivot element in the call to *RearrangeAndPlace* to compute the kth-smallest element in linear time. However, the problem of computing the exact median (for n odd) is actually the special case $k = (n+1)/2$ of the problem under consideration.

We now describe a technique known as the *median of medians* method for computing an approximation to the median. This approximation, called the *pseudomedian*, is defined as follows. Assume, for convenience, that n is an odd multiple of 5, so that $n = 5q$ for some odd number q. Divide the list $L[1{:}n]$ into q sublists $L_1, L_2, \ldots, L_q$, each of size 5, where

$$L_i = L[5(i-1)+1{:}5i], \quad i = 1, \ldots, q.$$

We now sort each of the lists $L_1, L_2, \ldots, L_q$ and then rearrange them according to increasing medians to obtain the lists $L_1', L_2', \ldots, L_q'$, so that

$$L_i' = (a_i, b_i, c_i, d_i, e_i),\ a_i < b_i < c_i < d_i < e_i, \qquad i = 1, 2, \ldots, q,$$

where $c_1 < c_2 < \cdots < c_q$. Note that c_i is the median of the list L_i', $i = 1, 2, \ldots, q$, so that for a suitable permutation π of $1, \ldots, q$, $C_{\pi(i)}$ is the median of L_i, $i = 1, \ldots, q$. The *pseudomedian*, which we denote by *pm*, is the median $c_{(q+1)/2}$ of the medians $c_1, c_2, \ldots, c_q$ (see Figure 13.3).

Figure 13.3 The pseudomedian is *pm* = 48 for the list $L[1{:}35]$; the actual median is 43.

$$L[1{:}35] = \overset{L_1}{87,23,91,25,66} \mid \overset{L_2}{12,55,8,77,67} \mid \overset{L_3}{88,32,11,44,57} \mid \overset{L_4}{75,21,6,89,80}$$

$$\overset{L_5}{43,84,34,61,48} \mid \overset{L_6}{14,10,37,27,29} \mid \overset{L_7}{36,16,15,78,83}$$

Dividing $L[1{:}35]$ into $q = 7$ sublists of size 5

(a)

	a_i	b_i	c_i	d_i	e_i	
L_1'	10	14	27	29	37	L_6
L_2'	15	16	36	78	83	L_7
L_3'	11	32	44	57	88	L_3
L_4'	34	43	**48**	61	84	L_5
L_5'	8	12	55	67	77	L_2
L_6'	23	25	66	87	91	L_1
L_7'	6	21	75	80	89	L_4

The sublists L_i', $i = 1, \ldots, 7$

(b)

PROPOSITION 13.2.1 The number of elements of $L[1:n]$ greater than (smaller than) the pseudomedian *pm* is at most

$$\frac{7}{10}n - \frac{3}{2}.$$

Proof By the definition of *pm*, each of the following $3(q + 1)/2$ elements is smaller than or equal to *pm*:

$$a_1, b_1, c_1, a_2, b_2, c_2, \ldots, a_{(q+1)/2}, b_{(q+1)/2}, c_{(q+1)/2}.$$

Thus, the number of elements in $L[1:n]$ that are larger than *pm* is at most

$$n - 3\left(\frac{q+1}{2}\right) = n - 3\left(\frac{n/5+1}{2}\right) = \frac{7}{10}n - \frac{3}{2}.$$

By a similar argument, there are at most $\frac{7}{10}n - \frac{3}{2}$ elements smaller than *pm*. ■

Proposition 13.2.1 is the basis for a $\Theta(n)$ worst-case algorithm *Select*2 for computing the *k*th-smallest element in $L[1:n]$. The algorithm *Select*2 works similar to *Select*, except that we compute the pseudomedian and use it as the pivot element in the call to *RearrangeAndPlace*2. *RearrangeAndPlace*2 is a modified version of *RearrangeAndPlace* where we add an input parameter *Pivot* and rearrange the list about *Pivot* (see Exercise 13.13).

The median of 5 elements can be computed using only 6 comparisons. Thus we can compute the medians $c_{\pi(1)}, c_{\pi(2)}, \ldots, c_{\pi(q)}$ of $L_1, L_2, \ldots, L_q$ by performing $6q = \frac{6}{5}n$ comparisons. The pseudomedian *pm* can then be computed by a recursive call to *Select*2 with the input list $c_{\pi(1)}, c_{\pi(2)}, \ldots, c_{\pi(q)}$ and $k = (q + 1)/2$.

In the following pseudocode for the recursive procedure *Select*2, the procedure *MedianOfFive* is called with input parameter $L[low:high]$ and output parameter $M[1:q]$, where $M[i] = c_{\pi(i)}$, $i = 1, \ldots, q$. The initial call to *Select*2 is with $low = 1$ and $high = n$.

```
procedure Select2(L[low:high],k,x) recursive
Input:   L[low:high] (an array of list elements) {initially called with low = 1
                                                   and high = n}
         k (a positive integer such that low ≤ k ≤ high)
Output: x (the (k – low – 1)th-smallest element in the list L[low:high])

   if high – low + 1 ≤ 5 then
      use ad hoc method to find kth-smallest element x
   endif

   q := (high – low + 1)/5

   call MedianOfFive(L[low:high],M[1:q])
   call Select2(M[1:q],(q + 1)/2),Pivot)   {so that PseudoMedian = Pivot}
   call RearrangeAndPlace2(L[Low:High],Pivot,position)
      case
         :k = position: x := L[position]
```

```
        :k < position: call Select(L[low:position – 1],k,x)
        :k > position: call Select(L[position + 1:high],k,x)
    endcase
end Select2
```

It remains to analyze the worst-case complexity W(n) of *Select*2. We assume that *Select*2 is called with a list $L[1{:}n]$ of distinct elements. The call to *MedianOfFive* generates at most $\frac{6}{5}n$ comparisons. The call to *RearrangeAndPlace* generates $n + 1$ comparisons. The recursive call to *Select*2 with input $M[1{:}q]$ results in at most $W(q) = W(n/5)$ comparisons. Since, as discussed earlier, the list used in the recursive call has size at most $\frac{7}{10}n - \frac{3}{2}$, the recursive call to *Select*2 generates at most $W(\frac{7}{10}n - \frac{3}{2})$ comparisons. Thus we obtain the recurrence relation

$$W(n) = W\left(\frac{1}{5}n\right) + W\left(\frac{7}{10}n - \frac{3}{2}\right) + \frac{11}{5}n + 1, \quad \textbf{init. cond. } W(5) = 6. \qquad (13.2.9)$$

Using (13.2.9), a straightforward induction shows that $W(n) \le 22n$. Thus, *Select*2 has linear worst-case complexity.

13.3 Symbolic Algebraic Operations on Polynomials

Algebraic manipulation of polynomials is an essential tool in many applications, and therefore the need arises to design efficient algorithms to carry out the basic arithmetic operations on polynomials, such as addition and multiplication, as efficiently as possible. Given a polynomial $P(x) = a_{m-1}x^{m-1} + \cdots + a_1x + a_0$, it is important to differentiate between the pointwise and the symbolic representation of $P(x)$. The *pointwise representation* of $P(x)$ is simply a function *PtWiseMult*(*P*,*Q*)that maps an input point x to the output point $P(x)$. Thus, given two polynomials $P(x)$ and $Q(x)$, their pointwise product is the function that maps an input point x to the output point $P(x) \times Q(x)$.

The *symbolic representation* of a polynomial $P(x) = a_{m-1}x^{m-1} + \cdots + a_1x + a_0$ is its coefficient array $[a_0, a_1, \ldots, a_{m-1}]$. Thus given two polynomials $P(x) = a_{m-1}x^{m-1} + \cdots + a_1x + a_0$ and $Q(x) = b_{n-1}x^{n-1} + \cdots + b_1x + b_0$, their symbolic product is the coefficient array $[c_0, c_1, \ldots, c_{m+n-2}]$ of the product polynomial $P(x)Q(x)$ given by

$$c_k = \sum_{i+j=k} a_i b_j \qquad 0 \le i \le m-1,\ 0 \le j \le n-1,\ k = 0, \ldots, m+n-2. \qquad (13.3.1)$$

For example, the symbolic product of $P(x) = 3x^2 + 2x - 5$ having coefficient array $[-5, 2, 3]$ and $Q(x) = x^3 - x + 4$ having coefficient array $[4, -1, 0, 1]$ is the polynomial $3x^5 + 2x^4 - 8x^3 + 10x^2 + 13x - 20$ having coefficient array $[-20, 13, 10, -8, 2, 3]$.

The symbolic representation of the product is particularly important for determining various properties of the product polynomial. For example, the function *PtWiseMult* does not lend itself to taking the derivative of the product polynomial, whereas it is a simple matter to compute the symbolic representation of the derivative of a polynomial that is represented symbolically. Similar comments hold for other operations on polynomials such as root finding, integration, and so forth. Throughout the remainder of this chapter, when discussing algebraic operations on polynomials we implicitly assume that we are performing these operations symbolically.

13.3.1 Multiplication of Polynomials of the Same Input Size

An algorithm *DirectPolyMult* based on a straightforward calculation of (13.3.1) has complexity $\Theta(mn)$ (where we choose multiplication of coefficients as our basic operation). We now describe a more efficient algorithm for polynomial multiplication based on the divide-and-conquer paradigm. We first assume that $m = n$. Setting $d = \lceil n/2 \rceil$, we divide the set of coefficients of the polynomials in half, with the higher-order coefficients $a_{n-1}, a_{n-2}, \ldots, a_d$ in one set and the lower-order coefficients of $a_{d-1}, a_{d-2}, \ldots, a_0$ in the other. Setting $P_1(x) = a_{d-1}x^{d-1} + a_{d-2}x^{d-2} + \cdots + a_1x + a_0$ and $P_2(x) = a_{n-1}x^{n-d-1} + a_{n-2}x^{n-d-2} + \cdots + a_{d+1}x + a_d$ we obtain

$$P(x) = x^d P_2(x) + P_1(x).$$

A similar division of the set of coefficients of $Q(x)$ yields polynomials $Q_1(x)$ and $Q_2(x)$ having input size of at most d such that

$$Q(x) = x^d Q_2(x) + Q_1(x).$$

A straightforward application of the distributive law yields

$$P(x)Q(x) = x^{2d}P_2(x)Q_2(x) + x^d(P_1(x)Q_2(x) + P_2(x)Q_1(x)) + P_1(x)Q_1(x). \quad (13.3.2)$$

Note that polynomials $P_1(x)$ and $Q_1(x)$ both have input size d, and polynomials $P_2(x)$ and $Q_2(x)$ both have input size either d or $d - 1$. In the case when $P_2(x)$ and $Q_2(x)$ both have input size $d - 1$, we add a leading coefficient of zero to each so that they have input size d. Thus, the problem of multiplying $P(x)$ and $Q(x)$ has been reduced to the problem of taking four products of polynomials of input sizes $d = \lceil n/2 \rceil$ together with two multiplications by powers of x and three additions. It turns out that the resulting divide-and-conquer algorithm based on (13.3.2) still has quadratic complexity. However, there is a clever way of combining the split polynomials that uses only *three* polynomial multiplications instead of four, based on the following simple identity:

$$P(x)Q(x) = x^{2d}P_2(x)Q_2(x) + x^d((P_1(x) + P_2(x))(Q_1(x) + Q_2(x)) - P_1(x)Q_1(x) - P_2(x)Q_2(x)) + P_1(x)Q_1(x). \quad (13.3.3)$$

Formula (13.3.3) yields the following divide-and-conquer algorithm *PolyMult*1 for polynomial multiplication. *PolyMult*1 calls a (split pea) procedure *Split*($P(x)$,$P_1(x)$,$P_2(x)$), which inputs a polynomial $P(x)$ (of input size n) and outputs the two polynomials $P_1(x)$ and $P_2(x)$. We write *PolyMult*1 as a high-level recursive function whose inputs are the polynomials $P(x)$ and $Q(x)$ and whose output is the polynomial $P(x)Q(x)$. We will not be explicit about how the coefficients of the polynomials are maintained (linked lists, arrays, and so forth). We will also assume that we have well-defined procedures for multiplying a polynomial by x^i. For example, if the polynomial is maintained by an array of its coefficients, then this amounts to shifting indices by i and replacing the first i entries with zeros. We abuse the notation slightly by writing $x^iP(x)$ to mean the result of calling such a procedure. Also, for convenience we simply use the symbol + to denote the addition of two polynomials.

```
function PolyMult1(P,Q,n) recursive
Input:  P(x) = a_{n-1}x^{n-1} + ... + a_1x + a_0, Q(x) = b_{n-1}x^{n-1} + ... + b_1x + b_0
                                                              (polynomials)
        n (a positive integer)
Output: P(x)Q(x) (the product polynomial)
   if n = 1 then
      return(a_0b_0)
   else
      d := ⌈n/2⌉
      call Split(P(x),P_1(x),P_2(x))
      call Split(Q(x),Q_1(x),Q_2(x))
      R(x) := PolyMult1(P_2(x),Q_2(x),d)
      S(x) := PolyMult1(P_1(x) + P_2(x),Q_1(x) + Q_2(x),d)
      T(x) := PolyMult1(P_1(x),Q_1(x),d)
      return(x^{2d}R(x) + x^d(S(x) − R(x) − T(x)) + T(x))
   endif
end PolyMult1
```

When analyzing *PolyMult*1, we choose coefficient multiplication as our basic operation, and thus ignore the two multiplications by powers of x. However, the procedure referred to earlier for multiplying a polynomial by x^i has linear complexity and does not affect the order of complexity of *PolyMult*1. We also assume that n is a power of two, since we can interpolate asymptotic behavior using Θ-scalability (see Section 7.4.3). Since *PolyMult*1 invokes itself three times with n replaced by $d = n/2$, the number of coefficient multiplications $\mathrm{T}(n)$ performed by *PolyMult*1 satisfies the following recurrence relation

$$\mathrm{T}(n) = 3\mathrm{T}\left(\frac{n}{2}\right), \quad n > 1, \quad \mathrm{T}(1) = 1. \tag{13.3.4}$$

It follows from (13.3.4) and Proposition (7.4.4) that $\mathrm{T}(n) \in \Theta(n^{\log_2 3})$. Since $\log_2 3$ is approximately 1.59, we now have an algorithm for polynomial multiplication whose $\Theta(n^{\log_2 3})$ complexity is a significant improvement over the $\Theta(n^2)$ complexity of *DirectPolyMult*. Later in this chapter we develop an even faster polynomial multiplication algorithm utilizing the powerful tool known as the Fast Fourier Transform.

13.3.2 Multiplication of Polynomials of Different Input Sizes

In practice, we often encounter the problem of multiplying two polynomials $P(x)$ and $Q(x)$ of different input sizes m and n, respectively. If $m < n$, then we could merely augment $P(x)$ with $n - m$ leading zeros, but this would be quite inefficient if n is significantly larger than m. It is better to partition $Q(x)$ into blocks of size m. For convenience, we assume n is a multiple of m, that is, $n = km$ for some positive integer k. We let $Q_i(x)$ be the polynomial of degree m given by

$$Q_i(x) = b_{im-1}x^{m-1} + b_{im-2}x^{m-2} + \cdots + b_{(i-1)m+1}x + b_{(i-1)m}, \quad i \in \{1, \ldots, k\}.$$

Clearly,

$$Q(x) = Q_k(x)x^{m(k-1)} + Q_{k-1}(x)x^{m(k-2)} + \cdots + Q_2(x)x^m + Q_1(x).$$

It follows immediately from the distributive law that

$$P(x)Q(x) = P(x)Q_k(x)x^{m(k-1)} + P(x)Q_{k-1}(x)x^{m(k-2)} + \cdots + P(x)Q_1(x).$$

Applying the above ideas, we obtain the following algorithm *PolyMult2* for multiplying two polynomials $P(x)$ and $Q(x)$. *PolyMult2* is efficient even if the degree $m - 1$ of $P(x)$ is much less than the degree $n - 1$ of $Q(x)$. If n is not a multiple of m, we compute the largest integer k such that $n > km$ and augment $P(x)$ with $n - km$ leading zeros.

function *PolyMult2*$(P(x),Q(x),m,n)$

Input: $P(x) = a_{m-1}x^{m-1} + \cdots + a_1x + a_0$, $Q(x) = b_{n-1}x^{n-1} + \cdots + b_1x + b_0$ (polynomials)
n, m (positive integers) $\{n = km$ for some integer $k\}$

Output: $P(x)Q(x)$ (the product polynomial)

ProdPoly(x) := 0 {initialize all coefficients of *ProdPoly*(x) to be 0}
for i := 1 **to** k **do**
  $Q_i(x) = b_{im-1}x^{m-1} + b_{im-2}x^{m-2} + \cdots + b_{(i-1)m}$
endfor
for i := 1 **to** k **do**
  ProdPoly(x) := *ProdPoly*$(x) + x^{(i-1)m}$*PolyMult*1$(P(x),Q_i(x),m)$
endfor
return(*ProdPoly*(x))

end *PolyMult2*

The complexity of *PolyMult*1 for multiplying two polynomials of degree $m - 1$ is $\Theta(m^{\log_2 3})$. Since *PolyMult2* invokes *PolyMult*1 a total of $k = n/m$ times, each time with input polynomials of degree $m - 1$, it follows that the complexity of *PolyMult2* is

$$\Theta(km^{\log_2 3}) = \Theta\left(\frac{nm^{\log_2 3}}{m}\right) = \Theta\left(nm^{\log_2 (3/2)}\right).$$

13.4 Multiplication of Large Integers

Computers typically assign a fixed number of bits for storing integer variables. Arithmetic operations such as addition and multiplication of integers are often carried out by moving the integer operands into *fixed-length* registers and then invoking arithmetic operations built into the hardware. However, there are important applications, such as those that occur in cryptography, when the number of digits is too large to be handled in this way directly by the hardware. In such cases, it is necessary to perform these operations by storing the integers using an appropriate data structure (such as an array or a linked list) and then writing algorithms for carrying out the arithmetic operations. When analyzing the complexity of such algorithms, the size n of an integer A is taken to be the number of digits of A. Any base $b \geq 2$ can be chosen for representing an integer. We denote the ith least significant digit of U (base b) by u_i, $i = 0,1, \ldots, n-1$; that is,

$$U = \sum_{i=0}^{n-1} u_i b^i . \tag{13.4.1}$$

The classic grade-school algorithm for multiplying two integers U and V of size n clearly involves n^2 multiplications of digits. Viewing the right-hand side of (13.4.1) as a polynomial in b leads us to a divide-and-conquer strategy for computing UV based on a formula analogous to (13.3.3). Note that although addition and multiplication of large integers are similar to multiplication of polynomials, there is the additional task of handling carry digits. The design details of the resulting $\Theta(n^{\log_2 3})$ algorithm *MultInt* are left to the exercises.

13.5 Multiplication of Matrices

Given two $n \times n$ matrices $A = (a_{ij})$ and $B = (b_{ij})$, $1 \leq i, j \leq n$, recall that the product AB is defined to be the $n \times n$ matrix $C = (c_{ij})$, where

$$c_{ij} = \sum_{k=1}^{n} a_{ik} b_{kj} . \tag{13.5.1}$$

The straightforward algorithm based on this definition clearly performs n^3 (scalar) multiplications. Around 1970 Strassen devised a divide-and-conquer algorithm for matrix multiplication of complexity $O(n^{\log_2 7})$ using certain algebraic identities for multiplying 2×2 matrices.

The classic method of multiplying 2×2 matrices performs 8 multiplications as follows:

$$AB = \begin{bmatrix} a_{11} & a_{12} \\ a_{21} & a_{22} \end{bmatrix} \begin{bmatrix} b_{11} & b_{12} \\ b_{21} & b_{22} \end{bmatrix} = \begin{bmatrix} a_{11}b_{11} + a_{12}b_{21} & a_{11}b_{12} + a_{12}b_{22} \\ a_{21}b_{11} + a_{22}b_{21} & a_{21}b_{12} + a_{22}b_{22} \end{bmatrix} . \tag{13.5.2}$$

Strassen discovered a way to carry out the same matrix product AB using only the following seven multiplications:

$$\begin{aligned}
m_1 &= (a_{11} + a_{22})(b_{11} + b_{22}) \\
m_2 &= (a_{21} + a_{22})b_{11} \\
m_3 &= a_{11}(b_{12} - b_{22}) \\
m_4 &= a_{22}(b_{21} - b_{11}) \\
m_5 &= (a_{11} + a_{12})(b_{22}) \\
m_6 &= (a_{21} - a_{11})(b_{11} + b_{12}) \\
m_7 &= (a_{12} - a_{22})(b_{21} + b_{22})
\end{aligned} \tag{13.5.3}$$

The matrix product AB is then given by

$$AB = \begin{bmatrix} m_1 + m_4 - m_5 + m_7 & m_3 + m_5 \\ m_2 + m_4 & m_1 + m_3 - m_2 + m_6 \end{bmatrix}. \tag{13.5.4}$$

Consider now the case of two $n \times n$ matrices, where for convenience we assume that $n = 2^k$. We first partition the matrices A and B into four $(n/2) \times (n/2)$ submatrices, as shown.

$$A = \left[\begin{array}{c|c} A_{11} & A_{12} \\ \hline A_{21} & A_{22} \end{array}\right], \quad B = \left[\begin{array}{c|c} B_{11} & B_{12} \\ \hline B_{21} & B_{22} \end{array}\right] \tag{13.5.5}$$

The product AB can be expressed in terms of eight matrix products as follows.

$$AB = \left[\begin{array}{c|c} A_{11}B_{11} + A_{12}B_{21} & A_{11}B_{12} + A_{12}B_{22} \\ \hline A_{21}B_{11} + A_{22}B_{21} & A_{21}B_{12} + A_{22}B_{22} \end{array}\right] \tag{13.5.6}$$

Thus, in complete analogy with the 2×2 case, we can carry out the matrix product AB using only the following seven matrix multiplications.

$$\begin{aligned}
M_1 &= (A_{11} + A_{22})(B_{11} + B_{22}) \\
M_2 &= (A_{21} + A_{22})B_{11} \\
M_3 &= A_{11}(B_{12} - B_{22}) \\
M_4 &= A_{22}(B_{21} - B_{11}) \\
M_5 &= (A_{11} + A_{12})(B_{22}) \\
M_6 &= (A_{21} - A_{11})(B_{11} + B_{12}) \\
M_7 &= (A_{12} - A_{22})(B_{21} + B_{22})
\end{aligned} \tag{13.5.7}$$

As in the case of 2×2 matrices, the matrix product AB is then given by

$$AB = \begin{bmatrix} M_1 + M_4 - M_5 + M_7 & M_3 + M_5 \\ M_2 + M_4 & M_1 + M_3 - M_2 + M_6 \end{bmatrix}. \tag{13.5.8}$$

Equations (13.5.7) and (13.5.8) immediately yield a divide-and-conquer algorithm (Strassen) based on expressing the product of two $n \times n$ matrices in terms of seven products of $(n/2) \times (n/2)$ matrices. The complexity of Strassen's algorithm clearly satisfies the recurrence relation

$$\mathrm{T}(n) = 7\mathrm{T}\left(\frac{n}{2}\right), \quad n > 1, \quad \mathrm{T}(1) = 1. \tag{13.5.9}$$

By Proposition 7.4.4, we have $\mathrm{T}(n) \in \Theta(n^{\log_2 7})$. Since $\log_2 7$ is approximately 2.81, we now have an algorithm for matrix multiplication whose $\Theta(n^{\log_2 7})$ complexity is a significant improvement over the $\Theta(n^3)$ complexity of the classical algorithm for matrix multiplication.

Note that Strassen's identities (13.5.3) and (13.5.4) involve a total of 18 additions (or subtractions). Winograd discovered the following set of identities, which leads to a method of multiplying 2×2 matrices using only 15 additions or subtractions, but still doing only seven multiplications.

$$\begin{aligned}
m_1 &= (a_{21} + a_{22} - a_{11})(b_{22} - b_{12} + b_{11}) \\
m_2 &= a_{11}b_{11} \\
m_3 &= a_{12}b_{21} \\
m_4 &= (a_{11} - a_{21})(b_{22} - b_{12}) \\
m_5 &= (a_{21} + a_{22})(b_{12} - b_{11}) \\
m_6 &= (a_{12} - a_{21} + a_{11} - a_{22})b_{22} \\
m_7 &= a_{22}\,(b_{11} - b_{22} - b_{12} + b_{21})
\end{aligned} \tag{13.5.10}$$

The matrix product AB is then given by

$$AB = \begin{bmatrix} m_2 + m_3 & m_1 + m_2 + m_5 + m_6 \\ m_1 + m_2 + m_4 - m_7 & m_1 + m_2 + m_4 + m_5 \end{bmatrix}. \tag{13.5.11}$$

Although (13.5.10) and (13.5.11) involve a total of 24 additions or subtractions, it can be verified that a total of only 15 distinct additions or subtractions are needed. Thus, we obtain an improvement of three less additions or subtractions over Strassen's identities, but still use only seven multiplications.

13.6 The Discrete Fourier Transform

One of the most celebrated divide-and-conquer algorithms is the Fast Fourier Transform (FFT) implementing the Discrete Fourier Transform. The Fast Fourier Transform algorithm has a wide range of important applications to such fields as signal and image processing, computerized axial tomography (CAT) scans, weather prediction and statistics, and coding theory. In this section we describe the Fast Fourier Transform and show how it can be used to obtain an $O(n \log n)$ algorithm for polynomial multiplication.

The Discrete Fourier Transform DFT_ω of a polynomial of degree n is defined relative to a *primitive nth root of unity* ω. A primitive nth root of unity (over the field of complex numbers) is a complex number ω such that $\omega^n = 1$ and $\omega^k \neq 1$ for $0 < k < n$ (see Section 13.9).

DEFINITION 13.6.1 Given a primitive *n*th root of unity ω, the Discrete Fourier Transform DFT_ω transforms a polynomial $P(x) = a_{n-1}x^{n-1} + \cdots + a_1x + a_0$ with *real* or *complex* coefficients into the polynomial $b_{n-1}x^{n-1} + \cdots + b_1x^1 + b_0$, where $b_i = P(\omega^i)$, $i = 0,1, \ldots, n-1$. That is,

$$DFT_\omega(P(x)) = P(\omega^{n-1})x^{n-1} + \cdots + P(\omega)x + P(1). \tag{13.6.1}$$

For convenience, we (equivalently) regard DFT_ω as transforming the coefficient array $\mathbf{a} = [a_0, a_1, \ldots, a_{n-1}]$ of the polynomial $P(x)$ into the coefficient array $\mathbf{b} = [b_0, b_1, \ldots, b_{n-1}]$ of the polynomial $DFT_\omega(P(x))$. Then **a** and **b** are related by the following matrix equation.

$$\begin{bmatrix} b_0 \\ b_1 \\ \\ \\ \\ b_{n-1} \end{bmatrix} = \begin{bmatrix} 1 & 1 & 1 & \ldots & 1 \\ 1 & \omega & \omega^2 & \ldots & \omega^{n-1} \\ . & . & . & \ldots & . \\ . & . & . & \ldots & . \\ . & . & . & \ldots & . \\ 1 & \omega^{n-1} & (\omega^{n-1})^2 & \ldots & (\omega^{n-1})^{n-1} \end{bmatrix} \begin{bmatrix} a_0 \\ a_1 \\ \\ \\ \\ a_{n-1} \end{bmatrix} \tag{13.6.2}$$

We denote the $n \times n$ matrix in (13.6.2) by $F_\omega = (f_{ij})_{n \times n}$, so that

$$f_{ij} = \omega^{(i-1)(j-1)}, \quad i,j = 1, \ldots, n. \tag{13.6.3}$$

13.6.1 The Fast Fourier Transform

A straightforward way to compute $DFT_\omega(P(x))$ is simply to evaluate $P(x)$ at the points $1, \omega, \ldots, \omega^{n-1}$ using Horner's rule, yielding an $O(n^2)$ algorithm. In fact, for evaluating $P(x)$ at any old set of n points, repeated applications of Horner's rule is the best that we can do. However, the n points $1, \omega, \ldots, \omega^{n-1}$ are very special beyond the mere fact that they are powers of a given number ω. The special nature of these points can be utilized to design an $O(n \log n)$ algorithm called the *Fast Fourier Transform* (FFT) for computing $DFT_\omega(P(x))$. Indeed, we now motivate the choice of points $1, \omega, \ldots, \omega^{n-1}$ by showing how a divide-and-conquer algorithm to evaluate $P(x)$ at n points might proceed if the n points have certain simple relationships with one another.

Suppose we choose the n points so that half of the points are the negatives of the other half, say,

$$z_0, z_1, \ldots, z_{n/2-1}, \quad -z_0, -z_1, \ldots, -z_{n/2-1}. \tag{13.6.4}$$

Evaluating a polynomial $P(x) = a_{n-1}x^{n-1} + \cdots + a_1x + a_0$ at such a set of points requires basically half the number of multiplications required for an arbitrary set of n points. To see this, we merely have to write $P(x)$ as the sum of two polynomials of degree $n/2 - 1$ by gathering up the even and odd powers of x.

$$P(x) = Even(x^2) + xOdd(x^2) \tag{13.6.5}$$

where $Even(x) = a_{n-2}x^{n/2-1} + \cdots + a_2x + a_0$ and $Odd(x) = a_{n-1}x^{n/2-1} + \cdots + a_3x + a_1$. Now if we replace x by $-x$ in (13.6.5), we obtain

$$P(-x) = Even(x^2) - xOdd(x^2). \tag{13.6.6}$$

Thus, to evaluate $P(x)$ at the n points given by (13.6.4), we need only evaluate the two polynomials *Even* and *Odd* at the $n/2$ points $(z_0)^2, (z_1)^2, \ldots, (z_{n/2-1})^2$, and then perform the $n/2$ additions, $n/2$ subtractions, and n multiplications as described in (13.6.5) and (13.6.6). Since we have divided our evaluation problem into two problems of size $n/2$, together with a combine step of $O(n)$ complexity, this looks like a promising start to a divide-and-conquer strategy that satisfies the familiar recurrence relation $T(n) = 2T(n/2) + O(n)$, and therefore yields an algorithm having $O(n \log n)$ complexity.

The same strategy can be applied to evaluate *Even* and *Odd* at $(z_0)^2, (z_1)^2, \ldots, (z_{n/2-1})^2$ if we choose $z_0, z_1, \ldots, z_{n/2-1}$ to satisfy the relation

$$(z_{n/4+j})^2 = -(z_j)^2, \quad j = 0, \ldots, n/4 - 1. \tag{13.6.7}$$

Now (13.6.7) presents a *real* problem (not an *imaginary* problem, see Figure 13.4). More precisely while we cannot find *real* (nonzero) numbers satisfying (13.6.7), there is no problem finding *complex* numbers that do the job. We merely have to take

$$z_{n/4+j} = iz_j, \quad j = 0, \ldots, \frac{n}{4} - 1, \quad i = \sqrt{-1}. \tag{13.6.8}$$

If we carry the divide-and-conquer strategy to the next step, we are looking at the problem of the evaluation of degree $n/8 - 1$ polynomials at the $n/8$ points $(z_0)^8, (z_1)^8, \ldots, (z_{n/8-1})^8$. As before, the trick allowing us to use formulas (13.6.5) and (13.6.6) works if we choose $z_{n/8+j}$, $j = 0, \ldots, n/8 - 1$, so that

$$(z_{n/8+j})^4 = -(z_j)^4, \quad j = 0, \ldots, \frac{n}{8} - 1. \tag{13.6.9}$$

Figure 13.4 Taking a number that leads to numbers satisfying (13.6.7) (FRANK & ERNEST reprinted by permission of Newspaper Enterprise Association, Inc.)

z_0	z_1	z_2	z_3	z_4	z_5	z_6	z_7
1	$\sqrt{i}$	i	$i\sqrt{i}$	-1	$-\sqrt{i}$	$-i$	$-i\sqrt{i}$

Figure 13.5 Table of special points for evaluating polynomials of degree ≤ 7

Formula (13.7.8) is, in turn, satisfied if we set

$$z_{n/8+j} = \sqrt{i}\,(z_j), \quad j = 0, \ldots, \frac{n}{8} - 1. \tag{13.6.10}$$

To see precisely where we are headed in the case of a general $n = 2^k$, we consider the case $n = 8$. The three steps just carried out for $n = 8$ determine the specially chosen eight points for rapid evaluation of polynomials of degree seven or less. Figure 13.5 shows a table containing the eight points that have been chosen subject to (13.6.4), (13.6.7), and (13.6.9). For simplicity, we have added the additional condition that $z_0 = 1$.

Note that $\sqrt{i}$ ($= e^{2\pi i/8} = (1 + i)/\sqrt{2}$) is a complex number ω such that $\omega^8 = 1$, whereas $\omega^j \neq 1$ for $0 < j < 8$. Hence, the points $z_0, z_1, z_2, z_3, z_4, z_5, z_6, z_7$ in Figure 13.5 are all distinct and are of the form $1, \omega, \omega^2, \omega^3, \omega^4, \omega^5, \omega^6, \omega^7$, where ω is a primitive eighth root of unity.

Returning to the case of a general $n = 2^k$, if we continue choosing points by generalizing the constraints expressed by (13.6.7) and (13.6.9) in the obvious way, after $\log_2 n$ steps we will arrive at the problem of evaluating constant polynomials at $(z_0)^n$, $\omega(z_0)^n$, where ω is a primitive nth root of unity. Again, for simplicity, we take z_0 to be 1 and ω to be the complex number $e^{2\pi i/n}$. We can verify that the points that we have then generated by this divide-and-conquer strategy are $1, \omega, \omega^2, \omega^3, \ldots, \omega^{n-1}$. Following our previous discussion, an $O(n \log n)$ recursive algorithm *FFTRec* for evaluating $P(1), P(\omega), \ldots, P(\omega^{n-1})$ is then easily obtained.

When writing the pseudocode for the algorithm *FFTRec*, it is convenient to use the coefficient array $\mathbf{a} = [a_0, a_1, \ldots, a_{n-1}]$ to represent the polynomial $P(x) = a_{n-1}x^{n-1} + \cdots + a_1x + a_0$. *FFTRec* evaluates $P(1), P(\omega), \ldots, P(\omega^{n-1})$ by first splitting $P(x)$ into even- and odd-degree terms $Even(x) = a_{n-2}x^{n/2-1} + \cdots + a_2x + a_0$ and $Odd(x) = a_{n-1}x^{n/2-1} + \cdots + a_3x + a_1$. Then *FFTRec* recursively evaluates $Even(1), Even(\omega^2), \ldots, Even((\omega^2)^{(n/2)-1})$ and $Odd(1), Odd(\omega^2), \ldots, Odd((\omega^2)^{(n/2)-1})$. Finally, *FFTRec* uses (13.6.5) and (13.6.6) to evaluate $P(1), P(\omega), \ldots, P(\omega^{n-1})$.

For *FFTRec* to be a valid recursive procedure, we must verify that it is recursively invoked with the proper type of input. It is immediate that when n is a power of two, so is $n/2$. Moreover, it is easy to verify that when ω is a primitive nth root of unity and n is a power of two, then ω^2 is a primitive $(n/2)$th root of unity. Thus, $n/2$ and ω^2 are appropriate input parameters for a recursive invocation of *FFTRec*.

procedure *FFTRec*($a[0{:}n-1], n, \omega, b[0{:}n-1]$) **recursive**

Input: $a[0{:}n-1]$ (an array of coefficients of the polynomial $P(x) = a_{n-1}x^{n-1} + \cdots + a_1x + a_0$)
n (a power of two) $\{n = 2^k\}$
ω (a primitive nth root of unity)

```
Output: b[0:n – 1] (an array of values b[i] = P(ω^i), i = 0, . . . , n – 1)

  if n = 1 then
    b[0] := a_0
  else
    {divide into even-indexed and odd-indexed coefficients}
    for i := 0 to n/2 – 1 do
      Even[i] := a[2i]          {[a_0,a_2, . . . , a_{n-2}]}
      Odd[i] := a[2i + 1]       {[a_1,a_3, . . . , a_{n-1}]}
    endfor
    call FFTRec(Even[0:n/2 – 1],n/2,ω^2,e[0:n/2 – 1])
    call FFTRec(Odd[0:n/2 – 1],n/2,ω^2,d[0:n/2 – 1])
    {now combine according to (13.6.5) and (13.6.6)}
    for i := 0 to n/2 – 1 do
      b[i] := e[i] + ω^i d[i]
      b[i + n/2] := e[i] – ω^i d[i]
    endfor
  endif

end FFTRec
```

Figure 13.6 illustrates the tree of recursive calls to *FFTRec* resulting from an initial call with $n = 8$. As usual, the left (right) child of a node corresponds to the first (second) recursive call in the code. Each node in the tree lists the relevant **input** parameters for the given invocation. That is, we first list the input polynomial by listing its coefficient array, and we then list the current values of n and associated nth root of unity for the given node.

Figure 13.6 Tree of recursive calls to *FFTRec* with $n = 8$

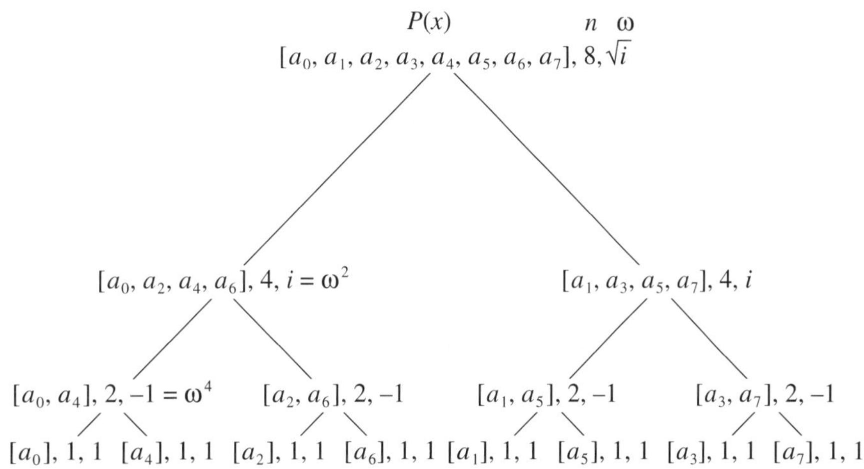

Resolution of the tree of recursive calls made by *FFTRec* with $n = 8$ is shown in Figure 13.7.

A nonrecursive version *FFT* of *FFTRec* can be designed based on the bottom-up resolution of the tree T of recursive calls shown in Figure 13.7. Upon examining Figure 13.7, it is not immediately obvious for a general $n = 2^k$ which coefficient a_j corresponds to the ith leaf node, $i = 0, \ldots, n-1$, in the tree of recursive calls for *FFTRec*. Fortunately, there is a simple procedure for computing the permutation $j = \pi_k(i)$ such that a_j is the coefficient corresponding to the ith leaf node of T.

PROPOSITION 13.6.1 Suppose $n = 2^k$, and $\pi_k: \{0, \ldots, n-1\} \Rightarrow \{0, \ldots, n-1\}$ is the permutation such that a_j is the coefficient corresponding to the ith leaf node of T in the tree of recursive calls of *FFTRec* with input coefficient array $a[0{:}n-1]$. Then the k-digit binary representation of $\pi_k(i)$ is obtained from the k-digit binary representation of i by simply reversing the digits, where leading zeros are included (if necessary).

Figure 13.7 Resolutions of the recursive calls made by *FFTRec* with $n = 8$

$$
\begin{aligned}
DFT_{\sqrt{i}}\,[a_0, a_1, a_2, a_3, a_4, a_5, a_6, a_7]
= [&a_0 + a_4 + a_2 + a_6 + 1(a_1 + a_5 + a_3 + a_7),\\
&a_0 - a_4 + i(a_2 - a_6) + \sqrt{i}(a_1 - a_5 + i(a_3 - a_7)),\\
&a_0 + a_4 - a_2 - a_6 + i(a_1 + a_5 - a_3 - a_7),\\
&a_0 - a_4 - i(a_2 - a_6) + i\sqrt{i}(a_1 - a_5 - i(a_3 - a_7)),\\
&a_0 + a_4 + a_2 + a_6 - 1(a_1 + a_5 + a_3 + a_7),\\
&a_0 - a_4 + i(a_2 - a_6) - \sqrt{i}(a_1 - a_5 + i(a_3 - a_7)),\\
&a_0 + a_4 - a_2 - a_6 - i(a_1 + a_5 - a_3 - a_7),\\
&a_0 - a_4 - i(a_2 - a_6) - i\sqrt{i}(a_1 - a_5 - i(a_3 - a_7))]
\end{aligned}
$$

$$
\begin{aligned}
DFT_i\,[a_0, a_2, a_4, a_6] &= [a_0 + a_4 + a_2 + a_6,\ a_0 - a_4 + i(a_2 - a_6),\ a_0 + a_4 - a_2 - a_6,\ a_0 - a_4 - i(a_2 - a_6)]\\
DFT_i\,[a_1, a_3, a_5, a_7] &= [a_1 + a_5 + a_3 + a_7,\ a_1 - a_5 + i(a_3 - a_7),\ a_1 + a_5 - a_3 - a_7,\ a_1 - a_5 - i(a_3 - a_7)]
\end{aligned}
$$

$$
\begin{aligned}
DFT_{-1}[a_0, a_4] &= [a_0 + a_4, a_0 - a_4) \qquad DFT_{-1}[a_2, a_6] = [a_2 + a_6, a_2 - a_6)\\
DFT_{-1}[a_1, a_5] &= [a_1 + a_5, a_1 - a_5) \qquad DFT_{-1}[a_3, a_7] = [a_3 + a_7, a_3 - a_7)
\end{aligned}
$$

$$[a_0] \quad [a_4] \quad [a_2] \quad [a_6] \quad [a_1] \quad [a_5] \quad [a_3] \quad [a_7]$$

To illustrate Proposition 13.6.1, suppose $k = 4$ ($n = 16$) and $i = 3$. The 4-digit binary representation of 3 is 0011, so that $\pi_k(3) = 1100$ in binary (in decimal, $\pi_k(3) = 12$).

Proof of Proposition 13.6.1 We prove Proposition 13.6.1 using induction on k.

Basis step: $k = 1$. Trivial.

Induction step: Assume that Proposition 13.6.1 is true for k. For $k + 1 = \log_2 n$, note that the leaves of the left subtree of the tree of recursive calls with input coefficients $a_0, a_1, \ldots, a_{2^{k+1}-1}$, and nth root of unity ω, consists of the coefficients $a_0, a_2, \ldots, a_{2^{k+1}-2}$. Moreover, this left subtree corresponds exactly to the tree of recursive calls of *FFTRec* with input coefficients $\tilde{a}_0, \tilde{a}_1, \ldots, \tilde{a}_{2^k-1}$, and $(n/2)$th root of unity ω^2, where

$$\tilde{a}_i = a_{2i}, \quad i = 0, \ldots, 2^k - 1. \tag{13.6.11}$$

It follows from (13.6.11) that

$$\pi_{k+1}(i) = 2\pi_k(i) \quad i = 0, \ldots, 2^k - 1. \tag{13.6.12}$$

By induction hypothesis, for input coefficients $\tilde{a}_i$, $i = 0, \ldots, 2^k - 1$, to *FFTRec*, the k-digit binary representation of $\pi_k(i)$ is obtained from the k-digit binary representation of i by simply reversing the digits. Using (13.6.12), the $(k + 1)$-digit binary representation of $\pi_{k+1}(i)$ is obtained from $\pi_k(i)$ by adding a zero on the right. Since the $(k + 1)$-digit binary representation of i is obtained by from the k-digit binary representation of i by adding a leading zero, we see that Proposition 13.6.1 holds for $\pi_{k+1}(i)$, $i = 0, \ldots, 2^k - 1$. The proof that Proposition 13.6.1 holds for $\pi_{k+1}(i)$, $i = 2^k, \ldots, 2^{k+1} - 1$ is similar (using the right subtree) and is left as an exercise. ■

The pseudocode for a procedure *ReverseBinPerm*($R[0{:}n - 1]$), which computes $R[i] = \pi_k(i)$, $i = 0, \ldots, n - 1$, is easy to write. An iterative version of *FFT* begins by loading an array $b[0{:}n - 1]$ with the values $b[i] = a_{\pi_k(i)}, = a[R[i]]$, $i = 0, \ldots, n - 1$. The values $b[i]$ correspond to the leaves in the tree T of recursive calls to *FFTRec,* and the iterative version of *FFT* proceeds by resolving these calls level-by-level in a bottom-up fashion. With the aid of Figure 13.7, you might want to proceed directly to the pseudocode for *FFT* given at the end of this subsection. However, it is instructive to describe the polynomial evaluations that take place at each level. The leaves are at level k in T and correspond to evaluating the constant polynomials $P_{0,i}(x) = a_{\pi_k(i)}$, $i = 0, \ldots, n - 1$, at $(\omega^n)^0 = 1$. At level $k - j$, $j \in \{1, \ldots, k\}$, the iterative version of *FFT* evaluates $n/2^j$ polynomials $P_{j,s}$, $s = 0, 2^j, \ldots, n - 2^j$ of degree $2^j - 1$ at the points $1, \omega^{n/2^j}, \ldots, (\omega^{n/2^j})^{2^{j-1}-1}, -1, -\omega^{n/2^j}, \ldots, -(\omega^{n/2^j})^{2^{j-1}-1}$. In this notation, the splitting of the polynomial $P_{j,s}$ into even and odd powers corresponds precisely to the two polynomials $P_{j-1,s}$ and $P_{j-1,s+2^{j-1}}$, respectively, at the next lower level in T. Hence, (13.6.5) and (13.6.6) become (see Figure 13.8)

$$P_{j,s}(x) = P_{j-1,s}(x^2) + xP_{j-1,s+2^{j-1}}(x^2), \tag{13.6.13}$$

$$P_{j,s}(-x) = P_{j-1,s}(x^2) - xP_{j-1,s+2^{j-1}}(x^2), \tag{13.6.14}$$

$$j = 1, \ldots, k, \quad s = 0, 2^j, \ldots, n - 2^j, \quad x = 1, \omega^{n/2^j}, \ldots, (\omega^{n/2^j})^{2^{j-1}-1}.$$

The iterative version of *FFT* proceeds by computing (13.6.13) and (13.6.14) with three nested **for-do** loops, the outer loop controlled by $j = 1, \ldots, k$, the middle loop controlled by $s = 0, 2^j, \ldots, n - 2^j$, and the innermost loop controlled by $x = 1, \omega^{n/2^j}, \ldots, (\omega^{n/2^j})^{2^{j-1}-1}$. The calculations at level $k - j$ begin with $b[0{:}n-1]$ containing the previously calculated values.

$$b[s+i] = P_{j-1,s}((\omega^{n/2^{j-1}})^i), \qquad (13.6.15)$$
$$s = 0, 2^{j-1}, \ldots, n - 2^{j-1}, \quad i = 0, 1, \ldots, 2^{j-1} - 1$$

The values corresponding to level $k - j$ are then calculated by first using an auxiliary array $Temp[0{:}n-1]$ to receive the results of calculations in the right-hand sides of (13.6.13) and (13.6.14).

$$Temp[s+i] = b[s+i] + (\omega^{n/2^j})^i\, b[s+i+2^{j-1}], \qquad (13.6.16)$$

$$Temp[s+i+2^{j-1}] = b[s+i] - (\omega^{n/2^j})^i\, b[s+i+2^{j-1}], \qquad (13.6.17)$$

$$j = 1, \ldots, k, \quad s = 0, 2^j, \ldots, n - 2^j, \quad i = 0, 1, \ldots, 2^j - 1$$

Formula (13.6.17) follows from (13.6.14) and the fact that $-b[s+i] = b[s+i+2^{j-1}]$. Then the results in $Temp[0{:}n-1]$ are copied back to $b[0{:}n-1]$.

The level-by-level bottom-up calculations are illustrated in Figure 13.8, where we show using arrows that the new value of $b[s+i]$ is determined from the previously calculated values of $b[s+i]$ and $b[s+i+2^{j-1}]$. Due to space limitations, we only show the initial values of $b[0{:}7]$ and its values after one iteration. However, all the values at each level can be determined by examining Figure 13.7.

The pseudocode on page 470 for the nonrecursive version *FFT* is based on the bottom-up strategy described above.

Figure 13.8 Level-by-level *FFT* computations for $n = 8$

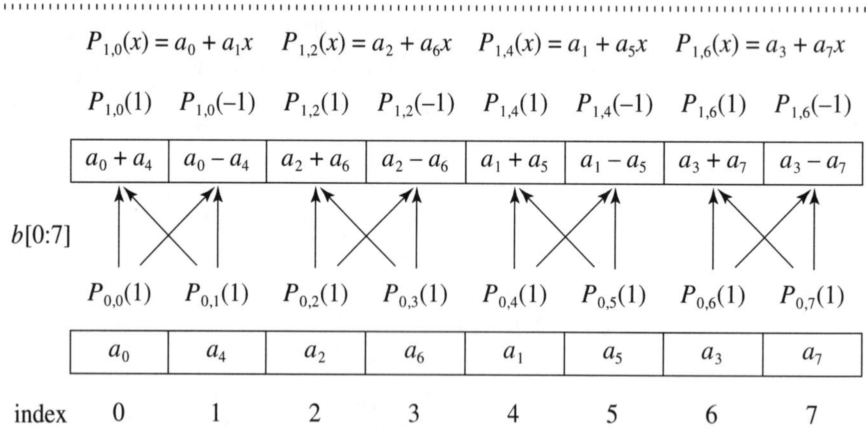

(a) *FFT* calculations performed and stored in the array $b[0{:}7]$ at level 2 ($j = 1$)

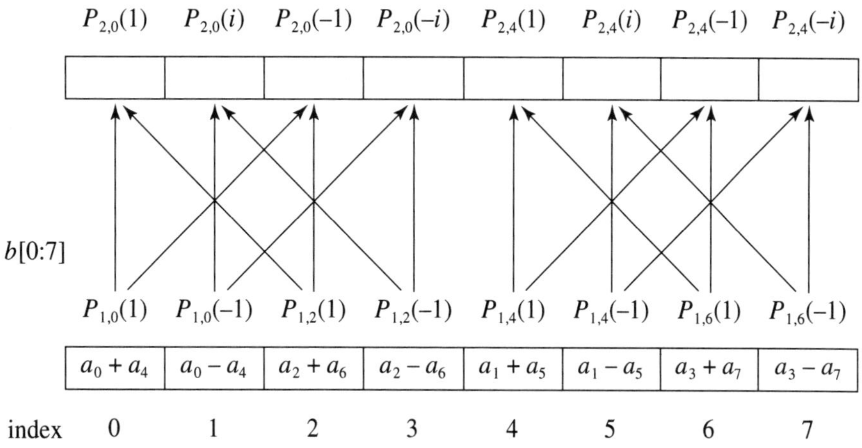

(b) *FFT* calculations performed and stored in the array $b[0:7]$ at level 1 ($j = 2$)

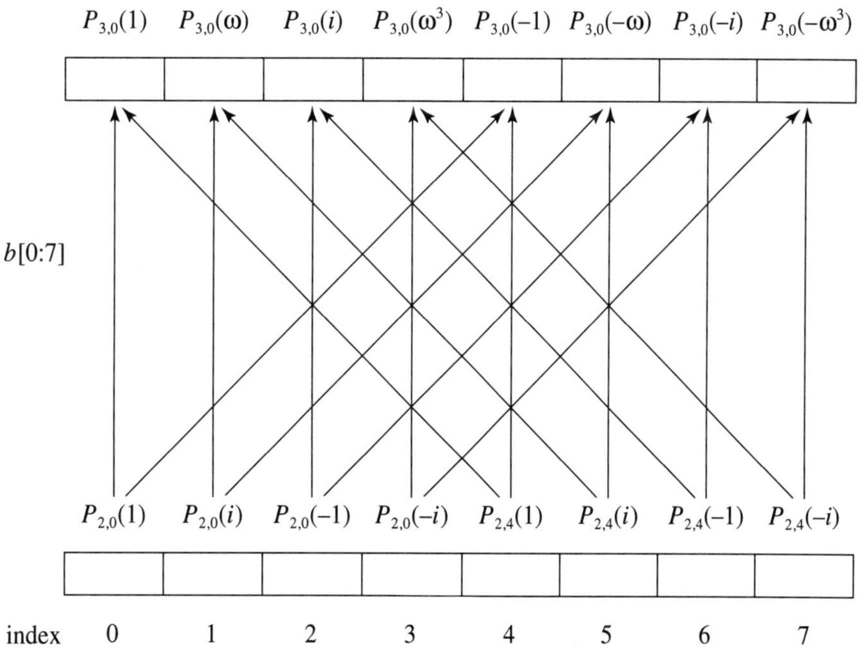

(c) *FFT* calculations performed in the array $b[0:7]$ at level 0 ($j = 3$)

Figure 13.8 Continued

```
procedure FFT(a[0:n – 1],n,ω,b[0:n – 1])
Input:  a[0:n – 1] (an array of coefficients of the polynomial P(x) =
                              a_{n-1}x^{n-1} + · · · + a_1x + a_0)
        n (a positive integer)   {n = 2^k}
        ω(a primitive nth root of unity)
Output: b[0:n – 1] (an array of values b[i] = P(ω^i), i = 0, . . . , n – 1)

  call ReverseBinPerm(R[0:n – 1])
  for i := 0 to n – 1
     b[i] := a[R[i]]
  endfor
  for j := 1 to k do
     for s := 0 to n – 2^j by 2^j do
        for i := 0 to 2^{j-1} – 1
           Temp[s + i] := b[s + i] + (ω^{n/2^j})^i b[s + i + 2^{j-1}]
           Temp[s + i + 2^{j-1}] := b[s + i] – (ω^{n/2^j})^i b[s + i + 2^{j-1}]
        endfor
     endfor
     for i := 0 to n – 1 do
        b[i] := Temp[i]
     endfor
  endfor

end FFT
```

13.6.2 Transformation Techniques

In secondary school mathematics, you were introduced to the idea of transforming problems into equivalent problems that are simpler or easier to solve. For example, it is often convenient to deal with large numbers by transforming them using logarithms to a certain base, say, base 10. The problem of multiplying two large numbers a and b is then transformed into the simpler but equivalent problem of adding their logarithms and then using the inverse transformation (exponentiation) to compute the given product. The idea behind this transformation can be nicely captured by the commutative diagram shown in Figure 13.9.

The diagram given in Figure 13.9 is called commutative since the product ab can be computed either directly by following the top arrow or indirectly by following the three arrows along the sides and bottom of the diagram. The mathematical expression of this commutativity is simply the equation

$$ab = 10^{\log_{10}a + \log_{10}b}$$

The utility of this transformation resides in the fact that addition can be carried out more quickly than multiplication. It is precisely to make the above transformations useful in practice that extensive tables of logarithms and antilogs have been generated.

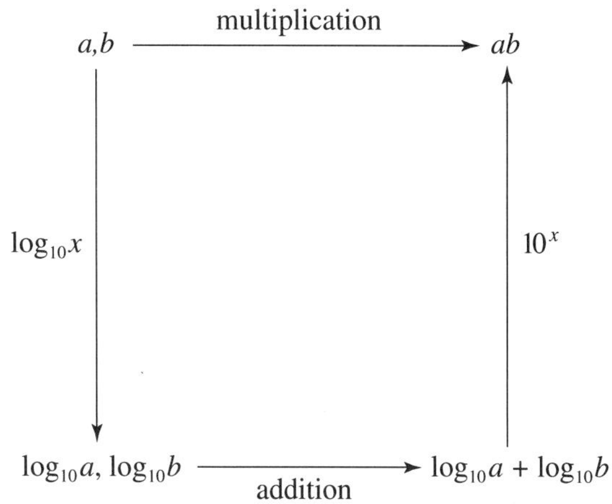

Figure 13.9 Commutative diagram for transforming multiplication via logarithms

The general idea behind transforming a given problem can be captured by the generic commutative diagram shown in Figure 13.10. In the next subsection we apply Figure 13.10 in connection with the Fast Fourier Transform to obtain an efficient algorithm for multiplying polynomials.

13.6.3 The Inverse Fourier Transform and Fast Polynomial Multiplication

We have mentioned that the *FFT* can be used to design an O(n log n) algorithm for polynomial multiplication. Of course, *FFT* worked on polynomials whose input size

Figure 13.10 Generic commutative diagram for transforming problems

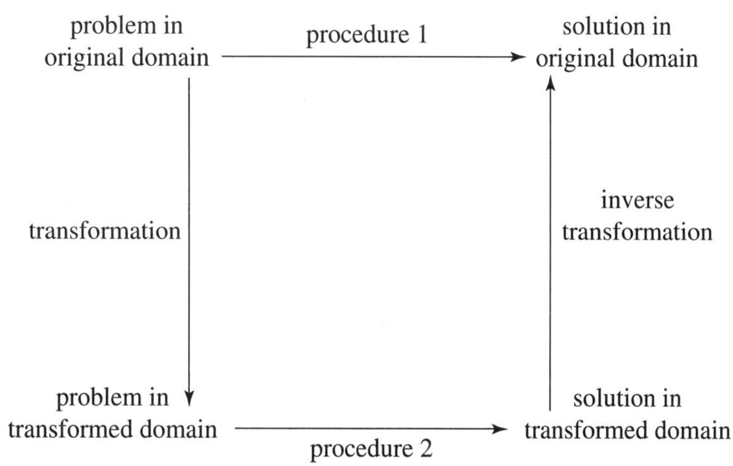

is a power of two. The product of two polynomials both of input size m is a polynomial of input size $n = 2m - 1$. Thus, for convenience in discussing polynomial multiplication, we assume that n is a power of two and that we are multiplying two polynomials of input size m, where $n = 2m - 1$. In practice, when multiplying any two polynomials we can arrange for these conditions by adding leading terms with zero coefficients, if necessary.

Figure 13.11 shows how the Fourier Transform can be utilized to obtain the product polynomial $P(x)Q(x)$. We first apply the Fourier Transform to find the coefficient arrays $[P(1), P(\omega), \ldots, P(\omega^{n-1})]$ and $[Q(1), Q(\omega), \ldots, Q(\omega^{n-1})]$ of both $P(x)$ and $Q(x)$. Then we simply compute the coefficient array of the product polynomial $P(x)Q(x)$ by forming the n scalar products $P(1)Q(1), P(\omega)Q(\omega), \ldots, P(\omega^{n-1})Q(\omega^{n-1})$. Thus by following the left side and bottom of the commutative diagram in Figure 13.11 we have actually computed the Fourier Transform of the product polynomial $P(x)Q(x)$. Hence, to recover $P(x)Q(x)$ we need to be able to perform the inverse to the Fourier Transform. The inverse certainly exists, since it amounts to finding the polynomial interpolating the n points $(1, P(1)Q(1)), (\omega, P(\omega)Q(\omega)), \ldots, (\omega^{n-1}, P(\omega^{n-1})Q(\omega^{n-1}))$.

The left side of the diagram in Figure 13.11 can be computed using *FFT* with $O(n \log n)$ complexity. The bottom of the diagram can be computed with $O(n)$ complexity. However, the best interpolation algorithms that we have seen (see Chapter 7) have complexity $O(n^2)$. Hence, to arrive at an $O(n \log n)$ algorithm for polynomial multiplication we need to look for an $O(n \log n)$ algorithm to compute the inverse Discrete Fourier Transform. Fortunately, the inverse Discrete Fourier Transform can be computed by using another application of *FFT*. This is because of the remarkable fact that the inverse of the Discrete Fourier Transform DFT_ω is (except for the multiplicative factor $1/n$) simply another Discrete Fourier Transform, namely $DFT_{\omega^{-1}}$:

Figure 13.11 Commutative diagram for computing product of two polynomials using DFT_ω

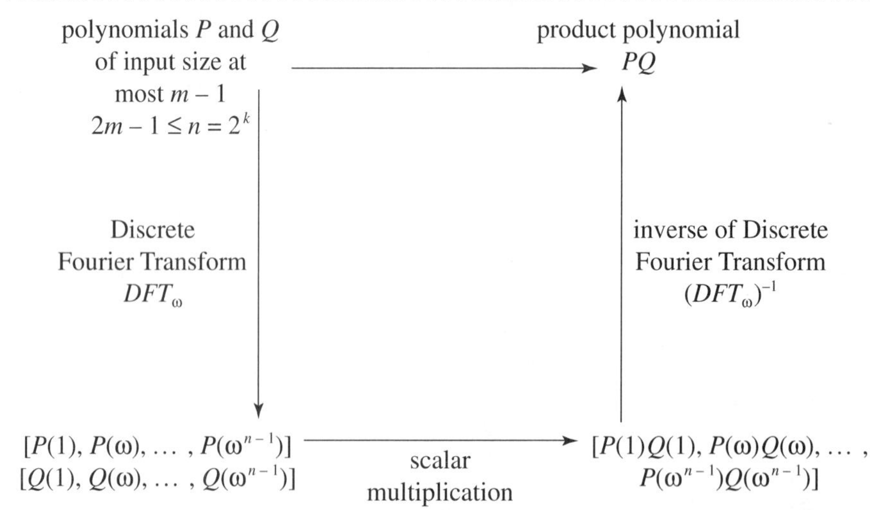

$$(DFT_{\omega})^{-1} = \left(\frac{1}{n}\right) DFT_{\omega^{-1}}. \tag{13.6.18}$$

To verify (13.6.18), it is useful to utilize the array and matrix notation discussed earlier for the Discrete Fourier Transform DFT_{ω}. Then (13.6.18) is equivalent to verifying that the inverse of the matrix A_{ω} defined by (13.6.2) is given by the formula

$$F_{\omega}^{-1} = \left(\frac{1}{n}\right) F_{\omega^{-1}}. \tag{13.6.19}$$

Equivalently, we must verify that

$$F_{\omega}^{-1} = \frac{1}{n}\begin{bmatrix} 1 & 1 & 1 & \cdots & 1 \\ 1 & \omega^{-1} & \omega^{-2} & \cdots & \omega^{-(n-1)} \\ \cdot & \cdot & \cdot & \cdots & \cdot \\ \cdot & \cdot & \cdot & \cdots & \cdot \\ \cdot & \cdot & \cdot & \cdots & \cdot \\ 1 & \omega^{-(n-1)} & (\omega^{-2})^{n-1} & \cdots & (\omega^{-(n-1)})^{n-1} \end{bmatrix} \tag{13.6.20}$$

We leave the proof of (13.6.20) to the exercises.

The commutative diagram in Figure 13.12 summarizes the application of *FFT* for $O(n \log n)$ polynomial multiplication. In Figure 13.12, we denote the application of *FFT* with input parameter ω by FFT_{ω}.

We illustrate the actions of the commutative diagram in Figure 13.12 to compute the product of the polynomials $P(x) = x^3 - x + 2$ and $Q(x) = 2x^2 - 1$. The product polynomial has degree 5, so we take $n = 8$. Then, $\omega = \sqrt{i} = (1 + i)/\sqrt{2}$. Using *FFT* yields the evaluations

Figure 13.12 Commutative diagram summarizing $O(n \log n)$ polynomial multiplication using FFT_{ω}

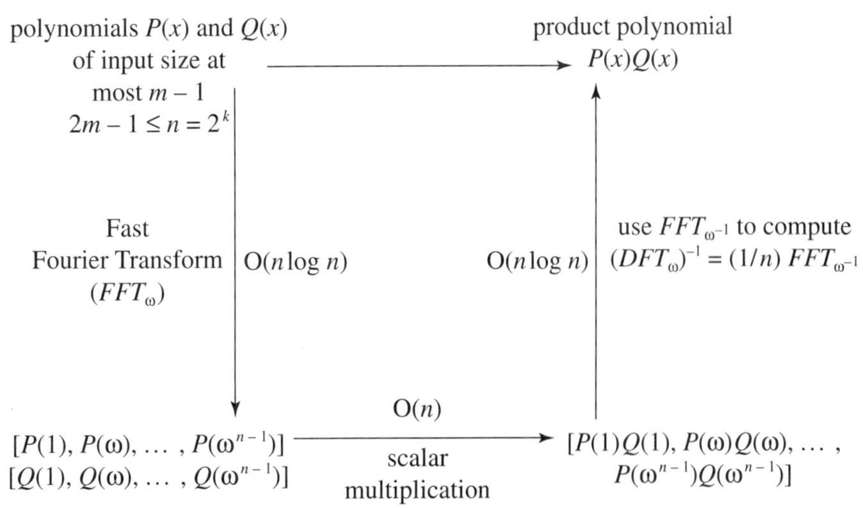

$$[P(1), P(\omega), \ldots, P(\omega^7)] = [2, 2 - \omega + \omega^3, 2 - 2i, 2 + \omega - \omega^3, 2, 2 + \omega - \omega^3, 2 + 2i, 2 - \omega + \omega^3]$$

$$[Q(1), Q(\omega), \ldots, Q(\omega^7)] = [1, 2i - 1, -3, -2i - 1, 1, 2i - 1, -3, -2i - 1].$$

Next we form the pointwise products $P(i)Q(i)$, $i = 1, \omega, \ldots, \omega^7$ and acquire the coefficient array for $DFT_\omega(P(x)Q(x))$ given by

$$[2, 4i - 3\omega^3 - \omega - 2, 6i - 6, -4i - \omega^3 - 3\omega - 2, 2, 4i + 3 + \omega - 2, -6i - 6, -4i + \omega^3 + 3\omega - 2]. \quad (13.6.21)$$

To complete the illustration of computing the product polynomial $P(x)Q(x)$ using the commutative diagram, we use *FFT* to compute $(DFT_\omega)^{-1} = (1/8)\ DFT_{\omega^{-1}}$ for the polynomial (coefficient array) given by (13.6.21). This yields the coefficient array $[-2, 1, 4, -3, 0, 2, 0, 0]$ of the product polynomial

$$P(x)Q(x) = 2x^5 - 3x^3 + 4x^2 + x - 2.$$

13.7 *Divide_and_Conquer* in Parallel

The sequential divide-and-conquer paradigm clearly has an inherent parallel component. Each level of the recursion divides the problem(s) at the previous level into subproblems of a similar type, and these subproblems can be solved in parallel assuming that a sufficient number of processors are available. In the sequential procedure *Divide_and_Conquer*, each of the subproblems is solved using a sequence of recursive calls. In the parallel procedure *Parallel_Divide_and_Conquer*, each of the subproblems is solved by a single parallel recursive call.

```
procedure Parallel_Divide_and_Conquer(I,J) recursive
Input:  I (an input to the given problem)
Output: J (a solution to the given problem corresponding to input I)
   if I ∈ Known then assign the a priori or ad hoc solution for I to J
   else
      call Divide(I,I_1, . . . , I_m) {m may depend on the input I}
         parallelcall Parallel_Divide_and_Conquer(I_1,J_1 | I_2,J_2 | . . . | I_m,J_m)
      call Combine(J_1, . . . , J_m, J)
   endif
end Divide_and_Conquer
```

Clearly, some speedup over the sequential version is achieved by *Parallel_Divide_and_Conquer*. However, in a specific application of *Parallel_Divide_and_Conquer*, unless efficient parallel algorithms for implementing *Divide* and *Combine* are designed, the overall speedup achieved is usually not significant. For example, consider the algorithm *MergeSort*. If

Parallel_Divide_and_Conquer is applied to *MergeSort*, then the resulting parallel version of *MergeSort* has complexity $W(n) = \Theta(n)$ and cost $C(n) = \Theta(n^2)$, both of which are far from optimal. However, as we saw in Chapter 6, replacing *Merge* with the more efficient parallel merging algorithm *OddEvenMergePRAM* resulted in complexity $W(n) = \Theta((\log n)^2)$ and cost $C(n) = \Theta(n(\log n)^2)$, which are both within a logarithmic factor of being optimal.

13.7.1 The Fast Fourier Transform on a PRAM

For the Fast Fourier Transform algorithm both the divide and combine steps can be efficiently parallelized. Applying the paradigm *Parallel_Divide_and_Conquer* directly to *FFTRec* yields the following implementation of the Fast Fourier Transform on a PRAM.

```
procedure FFTRecPRAM(a[0:n – 1],n,ω ,b[0:n – 1]) recursive

Model:  EREW PRAM with p = n processors
Input:  a[0:n – 1] (an array of coefficients of the polynomial P(x) =
                        a_{n-1}x^{n-1} + ··· + a_1x + a_0)
        n (a power of two)  {n = 2^k}
        ω (a primitive nth root of unity)
Output: b[0:n – 1] (an array of values b[i] = P(ω^i), i = 0, . . . , n – 1)

  if n = 1 then
     b[0] := a_0
  else
     {divide into even-indexed and odd-indexed coefficients}
     for 0 ≤ i ≤ n/2 – 1 do in parallel
        Even[i] := a[2i]     {[a_0,a_2, . . . , a_{n-2}]}
        Odd[i] := a[2i + 1]  {[a_1,a_3, . . . , a_{n-1}]}
     end in parallel
     parallelcall FFTRecPRAM(Even[0:n/2 – 1],n/2,ω^2,e[0:n/2 – 1]) |
                                 Odd[0:n/2 – 1],n/2,ω^2,d[0:n/2 – 1])
     for 0 ≤ i ≤ n/2 – 1 do in parallel
        b[i] := e[i] + ω^i d[i]
        b[i + n/2] := e[i] – ω^i d[i]
     end in parallel
  endif

end FFTRecPRAM
```

We can easily verify that *FFTRecPRAM* has complexity $W(n) \in \Theta(\log n)$. Thus, *FFTRecPRAM* has cost $C(n) \in \Theta(n \log n)$, which is optimal.

Note that a $\Theta(\log n)$ algorithm for computing DFT_ω can be based directly on its definition: each of the n polynomial evaluations $f(1), f(\omega), \ldots, f(\omega^{n-1})$ can be computed separately in O(log n) parallel steps on the PRAM. The drawback is that n^2 processors must be utilized, so that the cost C(n) is in $\Theta(n^2 \log n)$.

13.7.2 The Fast Fourier Transform on the Butterfly Network

The *butterfly* interconnection network is particularly well-suited for implementation of divide-and-conquer algorithms, such as the Fast Fourier Transform and odd-even merge sort, where the divide step involves partitioning an array into even indexed and odd indexed elements. The butterfly network B_k of dimension k contains $n = 2^k$ processors, $P_{j,0}, P_{j,1}, \ldots, P_{j,n-1}$ at level j, $j = 0, \ldots, k$, so that B_k contains $n(\log_2 n + 1)$ processors altogether. Only processors from adjacent levels are directly connected. For $0 \le j < n$, processor $P_{j,q}$ is directly connected to processors $P_{j+1,q}$ and $P_{j+1,r}$, where the binary representations of q and r differ only in the jth most significant bit. The butterfly network B_3 is shown in Figure 13.13.

The butterfly network is doubly recursive in nature. By deleting the processors at the top level, the butterfly network of dimension k splits into two butterfly networks of dimension $k - 1$. The same phenomenon occurs by deleting the processors at the bottom level.

Figure 13.13 The butterfly network B_3

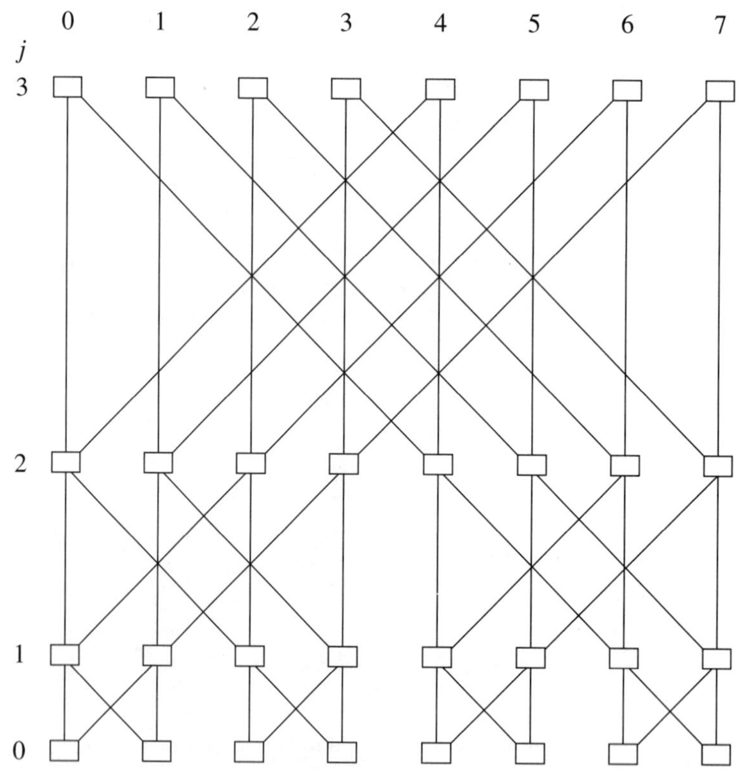

The connection between processors of B_k on adjacent levels are such that moving up a level in the bottom-up resolution of recursion based on splitting arrays into even indexed and odd indexed elements can be accomplished using a single parallel communication step. For example, the bottom-up resolution of Figure 13.8 yields a Fast Fourier Transform algorithm *FFTButterfly* for the butterfly interconnection network. Clearly the complexity of *FFTButterfly* is the same as *FFTRecPRAM*. However, *FFTButterfly* uses $n(\log_2 n + 1)$ processors, so that it is off by a logarithmic factor n from being cost optimal.

For any given k, the degree of each node in the top and bottom levels of the butterfly network B_k is two, and all other nodes have degree four. The small degree property of the butterfly network makes it attractive to build. Moreover, the communication complexity of the butterfly network is small, since the maximum distance between any two nodes in B_k is $2k = 2(\log_2 n + 1)$.

As with the PRAM, DFT_ω can be computed on an interconnection network directly from its definition as a matrix-vector product. In particular, DTF_ω can be computed in $O(\log n)$ time on the two-dimensional mesh of trees. Again, the drawback is in the cost, since we would use $\Theta(n^2)$ processors.

13.7.3 Inverting Triangular Matrices

A *lower triangular* matrix $L = (l_{ij})$ is a matrix such that $l_{ij} = 0$ for all pairs ij such that $i > j$. Such a matrix is invertible if, and only if, the product of the diagonal elements is nonzero. The problem of inverting a lower triangular matrix arises often in applications. In fact, in the next subsection we show how inverting lower triangular matrices can be used to invert general matrices.

An $O(n^3)$ sequential algorithm for finding the inverse of a lower triangular matrix is easily designed. However, it is not as straightforward to obtain a parallel polylogarithmic algorithm for inverting lower triangular matrices. Using a divide-and-conquer technique, we now present an $\Theta(\log^2 n)$ algorithm for this problem on a CREW PRAM with $\Theta(n^3)$ processors.

For convenience, we will assume that $n = 2^k$, $k \geq 0$. For $n \geq 2$, we rewrite L as a 2×2 block matrix.

$$L = \left[\begin{array}{c|c} L_1 & 0 \\ \hline A & L_2 \end{array}\right].$$

We can verify that

$$L^{-1} = \left[\begin{array}{c|c} L_1^{-1} & 0 \\ \hline -L_2^{-1} A\, L_1^{-1} & L_2^{-1} \end{array}\right]. \qquad (13.7.1)$$

The above formula yields a divide-and-conquer algorithm *TriangMatInvert-CREW*, which involves recursively inverting two lower triangular matrices of size $n/2$ and performing two parallel matrix multiplications. We can implement this algorithm on the CREW PRAM with $\Theta(n^3)$ processors by using the matrix multiplication algorithm *MatMultCREW* described in Chapter 11. Since *MatMultCREW* has complexity in $\Theta(\log n)$, we have the following recurrence relation for the complexity W(n) of *TriangMatInvertCREW*.

$$W(n) = W(n/2) + \Theta(\log n)$$

Thus, we have

$$\begin{aligned} W(n) &= W(n/2) + \Theta(\log n) \\ &= \Theta\left(\log n + \log \frac{n}{2} + \cdots + \log \frac{n}{2^k}\right) \\ &= \Theta(\log n + (\log n) - 1 + \cdots + (\log n) - k) \\ &= \Theta\left(k \log n - k\frac{(k+1)}{2}\right) \\ &= \Theta(\log^2 n). \end{aligned}$$

Since *TriangMatInvertCREW* utilizes $\Theta(n^3)$ processors, it has cost $C(n) \in \Theta(n^3 \log^2 n)$. We leave it to the exercises to design versions of *TriangMatInvertCREW* for the EREW PRAM and the three-dimensional mesh of trees.

13.7.4 Inverting General Matrices

We now show how the problem of inverting a general matrix can be reduced to the problem of inverting a lower triangular matrix.

The *characteristic polynomial* $Char_A(x)$ of an $n \times n$ matrix A is given by

$$Char_A(x) = \det(xI - A) = x^n + c_1 x^{n-1} + \cdots + c_{n-1}x + c_n, \tag{13.7.2}$$

where I is the $n \times n$ identity matrix. By the classical Cayley-Hamilton Theorem, the matrix A is a zero of its own characteristic polynomial; that is,

$$A^n + c_1 A^{n-1} + \cdots + c_{n-1}A + c_n I = 0. \tag{13.7.3}$$

Setting $x = 0$ in (13.7.2), we immediately obtain $c_n = \det A$. Hence, $c_n \neq 0$ if, and only if, A is invertible. Moreover, using (13.7.3) we have

$$A^{-1} = \left(\frac{-1}{c_n}\right)(A^{n-1} + c_1 A^{n-2} + \cdots + c_{n-1}I). \tag{13.7.4}$$

Thus, to compute A^{-1} it is sufficient to compute the coefficients $c_1, \ldots, c_n$ of the characteristic polynomial and the powers $A^2, \ldots, A^{n-1}$, and substitute their values into formula (13.7.4). In Chapter 11, we gave an algorithm for computing the kth power of A in time $\Theta(\log^2 n)$ on the CREW PRAM with $\Theta(n^3)$ processors. Thus, we can compute all the powers $A^2, \ldots, A^{n-1}$ in time $\Theta(\log^2 n)$ on the CREW PRAM with $\Theta(n^4)$ processors.

The coefficients $c_1, \ldots, c_n$ satisfy the following formula, due to Leverier.

THEOREM 13.7.1 **Leverier Theorem** Given the matrix $n \times n$ matrix A, let s_i denote the *trace* of A^i, that is, the sum of the diagonal entries of A^i, $i = 1, \ldots, n$. Then we have

$$\begin{bmatrix} 1 & 0 & 0 & 0 & 0 & \ldots & 0 \\ s_1 & 2 & 0 & 0 & 0 & \ldots & 0 \\ s_2 & s_1 & 3 & 0 & 0 & \ldots & 0 \\ s_3 & s_2 & s_1 & 4 & 0 & \ldots & 0 \\ \cdot & & & & & & \\ \cdot & & & & & & \\ \cdot & & & & & & \\ s_{n-1} & s_{n-2} & \ldots & & & s_1 & n \end{bmatrix} \begin{bmatrix} c_1 \\ c_2 \\ c_3 \\ c_4 \\ \cdot \\ \cdot \\ \cdot \\ c_n \end{bmatrix} = - \begin{bmatrix} s_1 \\ s_2 \\ s_3 \\ s_4 \\ \cdot \\ \cdot \\ \cdot \\ s_n \end{bmatrix},$$

where $c_1, \ldots, c_n$ are the coefficients of the characteristic polynomial of A. ❑

Once the matrix A^i has been computed, the trace of A^i can be computed in $O(\log_2 n)$ parallel addition steps. By Leverier's Theorem we can compute the c_i's from the s_i's by inverting a lower triangular matrix using *TriangMatInvertCREW.* The resulting parallel algorithm *MatInvertCREW* uses $O(n^4)$ processors, has complexity $W(n) \in O(\log^2 n)$, and cost $C(n) \in O(n^4 \log^2 n)$.

13.8 Appendix: Complex Numbers and the *n*th Roots of Unity

Often when solving problems whose inputs and outputs are real numbers, it is useful in intermediate stages to utilize the complex numbers. For example, in this chapter we developed a fast algorithm for computing the symbolic product of two input polynomials with *real* coefficients (so that the output product polynomial also has *real* coefficients) by using the Fast Fourier Transform (FFT) to evaluate the polynomials at the set of *n*th roots of unity. We now quickly review the necessary theory of complex numbers required by the FFT and similar algorithms.

The use of counting numbers is part of the earliest written historical record. However, the use of negative numbers occurs much later. There are understandable reasons for the relatively late appearance of negative numbers, but their introduction apparently caused little controversy. This was not the case, however, when numbers such as $\sqrt{-1}$ were introduced (see Figure 13.4). Mathematicians wanted to solve simple equations such as $x^2 + 1 = 0$, which had no solutions in the real numbers. Thus, the so-called *complex numbers* of the form $a + ib$ were introduced, where $i = \sqrt{-1}$ stood for a number whose square was -1. To make the introduction of i more acceptable, it was called an *imaginary number.* More generally, ib is referred to as the *imaginary part* of the complex number $a + ib$ and a is called the *real part.*

The early apologies made for $i = \sqrt{-1}$ are somewhat amusing nowadays, since the complex numbers $a + ib$ can be formally introduced simply as pairs (a, b) of real numbers (so that $a + ib$ corresponds to (a, b)), together with arithmetic operations of addition, subtraction, multiplication (denoted by juxtaposition), and reciprocation defined by

DEFINITIONS

I. $(a, b) \pm (c, d) = (a \pm c, b \pm d)$,

II. $(a, b)(c, d) = (ac - bd, ad + bc)$,

III. $(a, b)^{-1} = (a/(a^2 + b^2), -b/(a^2 + b^2))$.

The right-hand sides of I, II, and III use the usual arithmetic operations on real numbers, and the left-hand sides define the corresponding operation (using the same symbols) on the complex numbers. Using III, division is defined as $(a, b)/(c, d) = (a, b)(c, d)^{-1}$. The definitions II and III seem somewhat unmotivated, but they arise naturally as explained shortly. Note that the real number a can be identified with the complex number as $(a, 0)$. The rules I, II, and III when restricted to the pairs $(a, 0)$ as a varies over the real numbers coincide with the ordinary arithmetic rules for real numbers. Thus we have naturally extended the real numbers to the larger set of complex numbers. Moreover, $(0, 1)^2 = (-1, 0)$, so that we have indeed found $\sqrt{-1}$, which we denote by i. Of course, $-i = (0, -1)$ also squares to -1, so that we now have two complex numbers whose square is -1, whereas the reals have none.

Note that under our identification of b with $(b, 0)$ (and using i to denote $(0, 1)$), we can think of ib as the product of $(0, 1)$ with $(b, 0)$, that is, ib corresponds to $(0, b)$. In the same way, the addition $(a, 0) + (0, b)$ can be written as $a + ib$, and the multiplication $(c, 0)(a, b) = (ca, cb)$ can be viewed as $c(a + ib) = ca + icb$. With similar identifications, we see how definition II arises naturally from the distributive and commutative laws.

$$\begin{aligned}(a, b)(c, d) \text{ corresponds to } (a+ ib)(c + id) &= ac + iad + ibc + i^2bd \\ &= ac - bd + i(ad + bc) \\ \text{which corresponds to } (ac - bd, ad + bc).\end{aligned}$$

Definition III for reciprocals takes its motivation from

$$\frac{1}{a + ib}\,\frac{a - ib}{a - ib} = \frac{a - ib}{a^2 + b^2}.$$

Just as we consider the real numbers as corresponding to points on a straight line, we think of the complex numbers as corresponding to points in a plane. Thus, we identify the complex numbers $z = x + iy$ with points (two-dimensional vectors) (x, y) in the xy plane, with the reals x identified with the points $(x, 0)$ on the x axis. The *modulus of* $z = x + iy$, denoted by $|z|$, is defined to be the distance from (x, y) to the origin $(0, 0)$, that is,

$$|z| = (x^2 + y^2)^{1/2}.$$

Note that the modulus of a point on the x axis corresponds to the absolute value of the point.

Using the analogy with polar coordinates in the plane, given the complex number $z = x + iy$, it is useful to write z as

$$z = |z|\left(\frac{z}{|z|}\right), \tag{13.8.1}$$

where $\frac{z}{|z|}$ is a point on the unit circle. Hence, $\frac{z}{|z|}$ can be written as $\cos(\theta) + i\sin(\theta)$ for an appropriate angle θ between the vector determined by z and the x axis (see Figure 13.14), so that (13.8.1) becomes

$$z = |z|\,(\cos(\theta) + i\sin(\theta)). \tag{13.8.2}$$

In the representation (13.8.2), the angle θ is referred to as an *argument* of z. If $0 \le \theta < 2\pi$, then θ is called the *principal argument* of z.

Using the exponential function e^z, we rewrite (13.8.2) as follows. In analogy with the Taylor's expansion for e^x for real x, the Taylor's expansion for e^z for complex z is given by

$$e^z = 1 + z + \frac{z^2}{2!} + \cdots + \frac{z^n}{n!} + \cdots. \tag{13.8.3}$$

Substituting $z = i\theta$ into (13.8.2) yields

$$\begin{aligned} e^{i\theta} &= 1 + i\theta - \frac{\theta^2}{2!} - \frac{i\theta^3}{3!} + \frac{\theta^4}{4!} + \frac{i\theta^5}{5!} - \frac{\theta^6}{6!} - \frac{i\theta^7}{7!} + \cdots \\ &= \cos(\theta) + i\sin(\theta). \end{aligned} \tag{13.8.4}$$

Setting $r = |z|$ and substituting (13.8.4) into (13.8.2) yields

$$z = re^{i\theta}, \quad r = |z|, \frac{z}{|z|} = \cos(\theta) + i\sin(\theta). \tag{13.8.5}$$

Figure 13.14 Modulus $|z|$ and principle argument θ of a sample complex number z

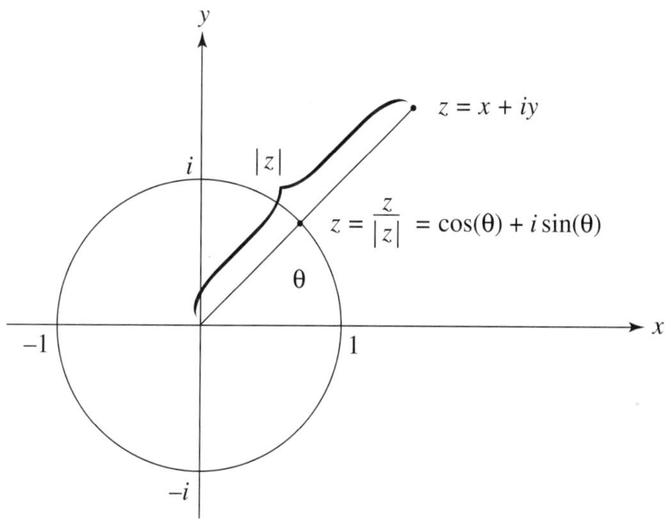

Remark Substituting $\theta = \pi$ into (13.8.4) yields Euler's formula

$$e^{i\pi} + 1 = 0,$$

a remarkable formula indeed, since it relates the fundamental mathematical constants e, i, π, 0, and 1 together in such a wonderfully simple way.

The expression (13.8.5) for z yields a nice geometric interpretation (see Figure 13.15) for the product of two complex numbers $z_1 = r_1e^{i\theta_1}$, $z_2 = r_2e^{i\theta_2}$:

$$\begin{aligned} z_1z_2 &= (r_1e^{i\theta_1})(r_1e^{i\theta_2}) \\ &= r_1r_2e^{i(\theta_1+\theta_2)}. \end{aligned} \tag{13.8.6}$$

From the generalization of (13.8.6) to the product of n complex numbers, we have the following multiplication rule for obtaining the modulus and argument of the product of n complex numbers.

Multiplication Rule for *n* Complex Numbers

Multiply the moduli and add the arguments.

An important special case of the multiplication rule is a formula for the nth power of a complex number $z = re^{i\theta}$, known as DeMoivre's formula:

$$(re^{i\theta})^n = r^ne^{in\theta}. \tag{13.8.7}$$

Given the complex number $z \neq 0$, (13.8.7) yields a formula for the n different complex numbers whose nth powers equal z (that is, the nth roots of z):

$$z^{1/n} = \{r^{1/n}(e^{i((\theta+2\pi k)/n)}) : z = re^{i\theta}, \quad 0 \le \theta < 2\pi, \quad k = 0,1,\ldots,n-1\} \tag{13.8.8}$$

It is easily verified that the nth roots of z defined by (13.8.8) are all distinct. When $z = 1$, (13.8.8) yields the set of nth *roots of unity.* Note that these points all lie on the unit

Figure 13.15 Geometry of the multiplication of complex numbers

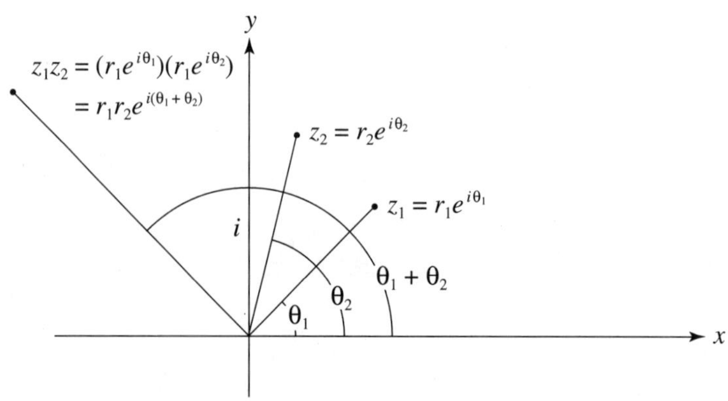

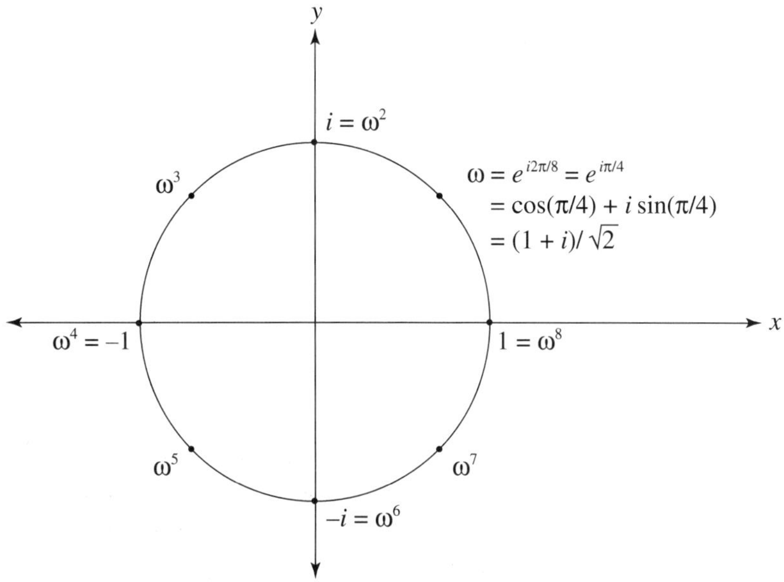

Figure 13.16 Eighth roots of unity

circle and are equally spaced, starting at 1 (see Figure 13.16). Moreover, they are all powers of the single nth root $e^{2\pi/n}$, which is a primitive nth root of unity. More generally, a primitive nth root of unity ω is a complex number such that $\omega^n = 1$, but $\omega^k \neq 1$ for $k = 1, \ldots, n - 1$.

The following useful propositions are easily verified.

PROPOSITION 13.8.1 Let ω be a primitive nth root of unity, and let k be an integer such that $0 < k < n$. Then

$$1 + \omega^k + (\omega^k)^2 + \cdots + (\omega^k)^{n-1} = 0. \quad \square$$

PROPOSITION 13.8.2 If ω is a primitive nth root of unity, then ω^2 is a primitive $(n/2)$th root of unity. $\square$

PROPOSITION 13.8.3 If ω is a primitive nth root of unity and k is a positive integer, then $\omega^k = 1$ if, and only if, n divides k. $\square$

13.9 Closing Remarks

Since Strassen's result for matrix multiplication appeared in 1970, researchers have been trying to improve the basic method. Identities are sought that perform fewer multiplications when multiplying matrices of a certain fixed size m. Then this reduction is used as the basis of a divide-and-conquer algorithm using decomposition into m^2 blocks of size n/m, where we assume for convenience that $n = m^k$

for some positive integer k. It has been shown that for 2×2 matrices, at least seven multiplications are *required*. Thus, divide-and-conquer algorithms, which use as their starting point a method of multiplying two $m \times m$ matrices using less than the number of multiplications used by Strassen's method, require that m be larger than 2.

Nearly ten years after Strassen discovered his identities, Pan found a way to multiply two 70×70 matrices that involves only 143,640 multiplications (compared to over 150,000 multiplications used by Strassen's method), yielding an algorithm that performs $O(n^{2.795})$ multiplications. Improvements over Pan's algorithm have been discovered, and the best result currently known (due to Coppersmith and Winograd) can multiply two $n \times n$ matrices using $O(n^{2.376})$ multiplications. However, all of these methods require n to be quite large before improvements over Strassen's method are significant. Moreover, they are very complicated due to the large number of identities required to achieve the savings in the number of multiplications. Hence, the currently known order of complexity improvements over Strassen's algorithm are mostly of theoretical rather than of practical interest.

It is interesting that the complexity of parallel algorithms for various basic matrix operations are interrelated in a simple way. Up to polynomial factors in the size of the matrices and the number of processors used and up to constant factor changes in the running times of the algorithms, the amount of parallel computing time needed to perform each of the following calculations is the same.

1. Computing the nth power of a matrix
2. Inverting a lower triangular matrix
3. Inverting a general matrix
4. Computing a determinant
5. Computing the characteristic polynomial

The interdependence of the above operations is explored in the exercises.

References and Suggestions for Further Reading

Brigham, E. O. *The Fast Fourier Transform.* Englewood Cliffs, NJ: Prentice-Hall, 1974. A survey of various formulations of the Fast Fourier Transform algorithm.

Cooley, J. W. and J. W. Tukey. "An algorithm for the machine calculation of complex Fourier series," *Mathematical Computation* 19 (1965): 297–301. The Fast Fourier Transform has a long history; the basic ideas were known as long as a century ago. However, this paper is credited with first showing how the Fast Fourier Transform can be effectively implemented on a computer.

For references to matrix multiplication algorithms, see:

Gonnet, G. H. *Handbook of Algorithms and Data Structures.* Reading, MA: Addison-Wesley, 1984.

Wilf, H. S. *Algorithms and Complexity.* Englewood Cliffs, NJ: Prentice-Hall, 1986.

Exercises

Section 13.1 The Divide-and-Conquer Paradigm

13.1 Design and analyze a divide-and-conquer algorithm for finding the maximum element in a list $L[1{:}n]$.

13.2 Design and analyze a divide-and-conquer algorithm for finding the maximum and minimum elements in a list $L[1{:}n]$.

13.3 Suppose we have a list $L[1{:}n]$ representing the results of an election, so that $L[i]$ is the candidate voted for by person i, $i = 1, \ldots, n$. Design a linear algorithm to determine whether a candidate got a majority of the votes, that is, whether there exists a list element that occurs more than $n/2$ times in the list. Your algorithm should output a majority winner if one exists.

13.4 Design a version of *QuickSort* that uses a threshold. Do some empirical testing to determine a good threshold value.

Section 13.2 Selecting the *k*th-Smallest Element

13.5 **a.** Show that for a list of size n, *Select* makes at most $n - 1$ calls to *RearrangeAndPlace* when determining the kth-smallest element.

b. For a given n and k, find a list exhibiting worst-case behavior, that is, for which the size of the sublist containing the kth-smallest element is only reduced by one with each call to *RearrangeAndPlace*.

13.6 Design and give pseudocode for an iterative version of procedure *Select*.

13.7 Show that the median of five elements can be computed directly using six comparisons.

13.8 Consider the function $W(n)$ defined by the recurrence relation (13.2.9). Using induction, show that $W(n) \le 22n$.

13.9 Design and give pseudocode for an iterative version of the procedure *Select2*.

13.10 Design a version of *Select2* that works for lists $L[1{:}n]$ where the elements are not necessarily distinct and still has linear worst-case complexity.

13.11 Design a version of *Select2* that rearranges the list elements by directly comparing them to *PseudoMedian*.

13.12 Design a $O(\log k)$ divide-and-conquer algorithm for finding the kth-smallest (largest) element in a list $L[1{:}2n]$, where the sublists $L[1{:}n]$ and $L[n + 1,2n]$ are in nondecreasing order.

13.13 Design an algorithm *RearrangeAndPlace2* ($L[low{:}high]$, *Pivot*, *position*) that rearranges the input list about the element $Pivot \in L[low{:}high]$ so that each element to the left of *Pivot* is no larger than *Pivot*, and each element to the right is no smaller than *Pivot*. The final position of *Pivot* in the rearranged list should be returned in the parameter *position*.

Section 13.3 Symbolic Algebraic Operations on Polynomials

13.14 Give pseudocode and analyze the procedure *DirectPolyMult*, which are based directly on (13.3.1).

13.15 Repeat Exercise 13.14 when the polynomials are implemented by storing the coefficients and the associated powers of the polynomials in a linked list.

13.16 Give pseudocode for a version of *PolyMult*1 when the polynomials are implemented by storing the coefficients and the associated powers of the polynomials in a linked list.

13.17 When analyzing *PolyMult*1, we choose coefficient multiplication as our basic operation. This ignores the two multiplications by powers of x. Show how the procedure for multiplying a polynomial by x^i can be implemented with linear complexity. Taking this operation into account, verify that the order of complexity of *PolyMult*1 remains unchanged.

13.18 Show that the divide-and-conquer algorithm for multiplying two polynomials based on the recurrence relation (13.3.2) has quadratic complexity.

Section 13.4 Multiplication of Large Integers

13.19 Design and analyze an algorithm *MultInt* for multiplying large integers.

13.20 **a.** Design an algorithm for adding two large integers implemented using arrays.

b. Repeat part (a) for linked list implementations.

13.21 Give pseudocode for the algorithm *MultInt* you designed in Exercise 13.19 for the following two implementations of large integers.

a. arrays

b. linked lists

Section 13.5 Multiplication of Matrices

13.22 Verify formula (13.5.4).

13.23 Verify formula (13.5.11).

13.24 Verify that the evaluation of formulas (13.5.10) and (13.5.11) requires 15 distinct additions or subtractions, which is three less than what is performed when using Strassen's identities.

13.25 Give pseudocode for the procedure *Strassen*, which implements Strassen's matrix multiplication algorithm. Assume that the input matrices A and B are both $n \times n$ matrices, where n is a power of 2.

13.26 Demonstrate the action of the procedure *Strassen* in Exercise 13.25 for the following input matrices.

$$A = \begin{bmatrix} 2 & 0 & 1 & 1 \\ 3 & 0 & 0 & 4 \\ 3 & -4 & 5 & 7 \\ 0 & 1 & 2 & 3 \end{bmatrix} \qquad B = \begin{bmatrix} 0 & 2 & 1 & 1 \\ 0 & 1 & 0 & -1 \\ 3 & 0 & 5 & 0 \\ 6 & 1 & 4 & 0 \end{bmatrix}$$

Section 13.6 The Discrete Fourier Transform

13.27 Verify formula (13.6.2).

13.28 Show that when ω is a primitive nth root of unity and n is a power of two, then ω^2 is a primitive $(n/2)$th root of unity.

13.29 Give the tree of recursive calls to *FFTRec* for $n = 16$.

13.30 For $i = 1, \ldots, n = 2_k$, consider the permutation $\pi_k(i) = j$ such that a_j is the coefficient corresponding to the ith leaf node of the tree of recursive calls to *FFTRec*. Complete the inductive proof that the k-digit binary representation of $\pi_k(i)$ is obtained from the k-digit binary representation of i by simply reversing the digits, where leading zeros are included (if necessary).

13.31 Design and analyze the procedure *ReverseBinPerm*($R[0{:}n - 1]$), which computes $R[i] = \pi_k(i)$, $i = 0, \ldots, n - 1$.

13.32 Given the polynomials $P(x) = x^3 - x + 2$, $Q(x) = 2x^2 - 1$,

a. Show that going along the left-hand side of the commutative diagram in Figure 13.12 and using *FFT* yields the evaluations:

$$[P(1), P(\omega), \ldots, P(\omega^7)] = [2, 2 - \omega + \omega^3, 2 - 2i, 2 + \omega - \omega^3, 2, 2 + \omega - \omega^3, 2 + 2i, 2 - \omega + \omega^3]$$

$$[Q(1), Q(\omega), \ldots, Q(\omega^7)] = [1, 2i - 1, -3, -2i - 1, 1, 2i - 1, -3, -2i - 1].$$

b. Verify that the polynomial $P(x)Q(x)$ results from going along the right-hand side of the commutative diagram in Figure 13.12 by using *FFT* to compute $(DFT_\omega)^{-1}) = (1/8)DFT_{\omega^{-1}}$.

Section 13.7 *Divide_and_Conquer* in Parallel

13.33 Give pseudocode for the nonrecursive procedure *FFTPRAM* that is based on a direct parallelization of the nonrecursive procedure *FFT*.

13.34 Give pseudocode for the procedure *FFTButterfly*.

13.35 Determine the diameter, maximum degree, and bisection width of the butterfly network B_k.

13.36 **a.** Give pseudocode for an algorithm on the EREW PRAM implementing DFT_ω directly from its definition.

b. Repeat part (a) for the two-dimensional mesh of trees.

13.37 Design a version of odd-even merge sort for the butterfly network.

13.38 Design an $O(n^3)$ sequential algorithm for finding the inverse of a lower triangular matrix.

13.39 Verify formula (13.7.1).

13.40 Design an $O(\log^2 n)$ algorithm for inverting a lower triangular matrix on the EREW PRAM with $O(n^3)$ processors.

13.41 Design an $O(\log^2 n)$ algorithm for inverting a lower triangular matrix on the three-dimensional mesh of trees with $O(n^3)$ processors.

13.42 Give pseudocode for *MatInvertCREW*.

13.43 Show how the rank of a matrix can be computed from its characteristic polynomial.

13.44 Design a version of *MatInvertCREW* for the EREW PRAM.

13.45 Design a version of *MatInvertCREW* for the three-dimensional mesh of trees.

13.46 Design an algorithm for inverting a matrix on the two-dimensional mesh. (*Hint:* Parallelize Gaussian elimination and the staggered I/O technique.)

Section 13.8 Appendix: Complex Numbers and the *n*th Roots of Unity

13.47 Prove Proposition 13.8.1.

13.48 Prove Proposition 13.8.2.

13.49 Prove Proposition 13.8.3.

13.50 Given a complex number $z = x + iy$, its *complex conjugate* $\bar{z}$ is defined by $\bar{z} = x - iy$.

a. Give a geometric interpretation of $\bar{z}$.

b. Show that the complex conjugate has the following properties:

i. $\overline{-z} = -\bar{z}$,

ii. $\overline{z_1 + z_2} = \bar{z}_1 + \bar{z}_2$,

iii. $\overline{z_1 z_2} = \bar{z}_1\,\bar{z}_2$.

c. Use part (b) to show that the complex roots of a polynomial $P(x)$ with real coefficients occur in conjugate pairs; that is, if $P(z) = 0$, then $P(\bar{z}) = 0$.

13.51 Verify that the *n*th roots of z defined by (13.8.8) are all distinct.

Section 13.9 Closing Remarks

13.52 Pan's divide-and-conquer matrix multiplication algorithm is based on a partitioning scheme that assumes n is a power of 70. The complexity $T(n)$ of Pan's divide-and-conquer algorithm satisfies the recurrence relation

$$T(n) = 143{,}640T(n/70), \quad n = 70^i, \quad i \geq 1, T(1) = 1.$$

Show that this implies that $T(n)$ is approximately $n^{2.795}$.

13.53 Up to polynomial factors in the size of the matrices and the number of processors used and up to constant factor changes in the running times of the algorithms, the amount of parallel computing time needed to perform each of the following calculations is the same.

1. Computing the nth power of a matrix
2. Inverting a lower triangular matrix
3. Inverting a general matrix
4. Computing a determinant
5. Computing the characteristic polynomial

Prove the result by establishing the following sequence of implications: In Section 13.7 it was shown that 1 and 2 implies 3 and 5. Clearly, 3 implies 2.

a. Show that 2 implies 1. (*Hint:* Invert the block matrix, where A is an $n \times n$ matrix and I is the following $n \times n$ identity matrix.)

$$\begin{bmatrix} I & 0 & 0 & 0 & 0 & \dots & 0 \\ A & I & 0 & 0 & 0 & \dots & 0 \\ 0 & A & I & 0 & 0 & \dots & 0 \\ 0 & 0 & A & I & 0 & \dots & 0 \\ \cdot & & & & & & \\ \cdot & & & & & & \\ \cdot & & & & & & \\ 0 & & & & 0 & A & I \end{bmatrix}$$

b. Show that 5 implies 4.

c. Verify the following determinant formula for the inverse of an (invertible) matrix A, and use it to show that 4 implies 3. The ijth cofactor C_{ij} of A is $(-1)^{i+j}$ times the determinant of the $(n-1) \times (n-1)$ submatrix obtained by deleting the ith row and jth column. The adjoint matrix $Adj(A)$ is the matrix whose ijth entry is C_{ji}. Then

$$A^{-1} = \frac{Adj\,(A)}{\det\,(A)}\,.$$

d. Verify that the above implications imply the equivalence of 1, 2, 3, 4, and 5.

14

Dynamic Programming

Dynamic programming is a design strategy that involves constructing a solution S to a given problem by building it up dynamically from solutions to smaller (or simpler) problems $S_1, S_2, \ldots, S_m$ of the same type. The solution to any given smaller problem S_i is itself built up from the solutions to even smaller (simpler) subproblems, and so forth. We start with the known solutions to the smallest (simplest) problem instances and build from there in a bottom-up fashion. To be able to reconstruct S from $S_1, S_2, \ldots, S_m$, some additional information is usually required. We let *Combine* denote the function that combines $S_1, S_2, \ldots, S_m$, using the additional information to obtain S, so that

$$S = Combine(S_1, S_2, \ldots, S_m).$$

Dynamic programming is similar to divide-and-conquer in the sense that it is based on a recursive division of the problem into smaller or simpler problems of the same type. However, whereas divide-and-conquer algorithms often utilize a top-down resolution method, dynamic programming algorithms invariably proceed by solving all the simplest subproblems of a problem before combining them into more complicated subproblems in a bottom-up fashion. Also, unlike many instances of divide-and-conquer, dynamic programming algorithms typically never consider a given subproblem more than once. Dynamic programming algorithms for optimization problems also can avoid generating suboptimal subproblems when the so-called *Principle of Optimality* holds, thereby leading to increased efficiency. In this chapter we consider a number of well-known problems that are amenable to the method of dynamic programming.

14.1 Optimization Problems and the Principle of Optimality

The method of dynamic programming is most effective in solving optimization problems when the Principle of Optimality holds. Consider the set of all *feasible* solutions S to the given optimization problem, that is, solutions S satisfying the constraints of the problem. An *optimal* solution S is a solution that optimizes (minimizes or maximizes) the objective function. If we wish to obtain an optimal solution S to the given problem, then we must optimize (minimize or maximize) over *all* subproblems $S_1, S_2, \ldots, S_m$ such that $S = Combine(S_1, S_2, \ldots, S_m)$. For many problems it is computationally infeasible to examine all such subproblems $S_1, S_2, \ldots, S_m$ (often there are exponentially many such subproblems). Fortunately, we can drastically reduce the number of subproblems that we need to consider if the Principle of Optimality holds.

DEFINITION 14.1.1 Given an optimization problem and an associated function *Combine*, the *Principle of Optimality* holds if the following is always true: If $S = Combine(S_1, S_2, \ldots, S_m)$ and S is an *optimal* solution to the problem, then $S_1, S_2, \ldots, S_m$, are *optimal* solutions to their associated subproblems.

We illustrate the Principle of Optimality for two familiar examples. First, we consider the problem of finding a shortest path in a graph (or digraph) G from vertex a to vertex b. Suppose we are given a vertex v of G different from a and b. If P is a path from a to b that contains v, then v determines a (natural) splitting of P into the subpath P_1 from a to v (solution S_1) and the subpath P_2 from v to b (solution S_2). Given paths P_1 and P_2, the function $Combine(P_1, P_2)$ forms the union of the two paths. If P is a shortest path from a to b, then P_1 is a shortest path from a to v and P_2 is a shortest path from v to b. Thus, the Principle of Optimality holds.

As a second example, consider the problem of finding optimal binary search trees for a set of distinct keys. Recall that a binary search tree T for keys $K_1 < \cdots < K_n$ is a binary tree on n nodes each containing a key, such that the following property is satisfied: Given any node v in the tree, each key in the left subtree rooted at v is no larger than the key in v, and each key in the right subtree rooted at v is no smaller than the key in v (see Figure 2.17). If K_i is the key in the root, then the left subtree L of the root contains $K_1, \ldots, K_{i-1}$ and the right subtree R of the root contains $K_{i+1}, \ldots, K_n$. Given a binary search tree T for keys $K_1, \ldots, K_n$, let K_i denote the key associated with the root of T, and let L and R denote the left and right subtrees (of the root) of T, respectively. Again it follows that L (solution S_1) is a binary search tree for keys $K_1, \ldots, K_{i-1}$ and R (solution S_2) is a binary search tree for keys $K_{i+1}, \ldots, K_n$. Given L and R, the function $Combine(L,R)$ merely reconstructs the tree T using K_i as the root. In the next section, we show that the Principle of Optimality holds for this problem by showing that if T is an optimal binary search tree, then so are L and R.

The solutions to many optimization problems result from making a sequence of decisions $D_1, \ldots, D_n$, where the first decision D_1 leads to a derived *single-instance* subproblem of the same type as the original problem. The solution to the derived problem is determined by the sequence of decisions $D_2, \ldots, D_n$. In the latter setting, the Principle of Optimality takes the following special form.

DEFINITION 14.1.2 The *Specialized Principle of Optimality* holds if, given any sequence of decisions $D_1, \ldots, D_n$ yielding an *optimal* solution S to the given problem, the subsequence of decisions $D_2, \ldots, D_n$ yields an *optimal* solution S' to the single-instance derived problem resulting from having made the first decision D_1.

The Specialized Principle of Optimality is the special case of the previous formulation, where $m = 1$ and $Combine(S') = S$. The additional information required to reconstruct S from S' is the information associated with decision D_1.

The problem of finding a shortest path in a graph G from vertex a to vertex b can be viewed as making a sequence of decisions $D_1, D_2, \ldots, D_p = b$ for the vertices in the path, where D_1 is the choice of a vertex v adjacent to a. The path S' determined by the remaining decisions $D_2, \ldots, D_p$ must be a shortest path, so that the Specialized Principle of Optimality also holds for the shortest-path problem.

14.2 Optimal Parenthesization for Computing a Chained Matrix Product

Our first example of dynamic programming is an algorithm for the problem of parenthesizing a chained matrix product so as to minimize the number of (scalar) multiplications performed when computing the product. If A and B are matrices of dimensions $p \times q$ and $q \times r$, then the matrix product AB involves pqr multiplications. Given a sequence (chain) of matrices $M_1, M_2, \ldots, M_n$, consider the product $M_1M_2 \ldots M_n$, where the matrix M_i has dimension $d_{i-1} \times d_i$, $i = 0, \ldots, n$, for a suitable sequence of positive integers $d_0, d_1, \ldots, d_n$. Since matrix product is an associative operation, there are a number of ways to evaluate the chained product depending on how we choose to parenthesize the expression. It turns out that the manner in which the expression is parenthesized can make a major difference in the total number of multiplications performed when computing the chained product. In this section we consider the problem of finding an *optimal parenthesization*, that is, a parenthesization that minimizes the total number of multiplications performed using ordinary matrix products.

We illustrate the problem with an example that commonly occurs in multivariate calculus. Suppose A and B are $n \times n$ matrices, X is an $n \times 1$ column vector, and we wish to evaluate ABX. The product ABX can be parenthesized in two ways, namely, $(AB)X$ and $A(BX)$, resulting in $n^3 + n^2$ and $2n^2$ multiplications, respectively. Thus, the two ways of parenthesizing make a rather dramatic difference in the number of multiplications performed (order $\Theta(n^3)$ versus order $\Theta(n^2)$).

The following is a formal (recursive) definition of a fully parenthesized chained matrix product and its associated first cut.

DEFINITION 14.2.1 Given the sequence of matrices $M_1, M_2, \ldots, M_n$, P is a *fully-parenthesized* matrix product for $M_1, M_2, \ldots, M_n$, which for convenience we simply call a *parenthesization of* $M_1M_2 \ldots M_n$, if P satisfies

$$P = M_1, \quad n = 1,$$
$$P = (P_1P_2), \quad n > 1,$$

where for some k, P_1 and P_2 are parenthesizations of the matrix products $M_1M_2 \ldots M_k$ and $M_{k+1}M_{k+2} \ldots M_n$, respectively. We call P_1 and P_2 the *left* and *right* parenthesizations of P, respectively. We call the index k the *first cut index of P*.

no. mult.	no. mult.	no. mult.	no. mult.
M_1 0	(M_1M_2) 10000	$(M_1(M_2M_3))$ 3500	$(M_1(M_2(M_3M_4)))$ 28500
		$((M_1,M_2)M_3)$ 15000	$(M_1((M_2M_3)M_4))$ 10000
			$((M_1M_2)(M_3M_4))$ 47500
			$((M_1(M_2M_3))M_4)$ 6500
			$(((M_1M_2)M_3)M_4)$ 18000

Figure 14.1 Number of multiplications performed for each full parenthesization are shown for matrices M_1, M_2, M_3, M_4 having dimensions 20×10, 10×50, 50×5, 5×30, respectively. The optimal parenthesizations are shaded.

The table in Figure 14.1 shows all the different parenthesizations of the matrices M_1, M_2, M_3, M_4 having dimensions 20×10, 10×50, 50×5, 5×30, respectively, with the optimal parenthesizations highlighted.

There is a one-to-one correspondence between parenthesizations of $M_1M_2 \dots M_n$ and 2-trees having n leaf nodes. Given a parenthesization P of $M_1M_2 \dots M_n$, if $n = 1$ its associated 2-tree $T(P)$ consists of a single node corresponding to the matrix M_1; otherwise, $T(P)$ has left subtree $T(P_1)$ and right subtree $T(P_2)$, where P_1 and P_2 are the left and right parenthesizations of P. The 2-tree $T(P)$ is the *expression tree* for P (see Figure 14.2).

Thus, the number of parenthesizations p_n equals the number t_n of 2-trees having n leaf nodes, so that by Exercise 7.39, we have

$$p_n = \frac{1}{n}\binom{2n-2}{n-1} \geq \frac{4^{n-1}}{2n^2 - n} \in \Omega\left(\frac{4n}{n^2}\right). \tag{14.2.1}$$

Figure 14.2 Associated expression 2-tree for parenthesization: $(((M_1M_2)(M_3M_4))(M_5M_6))$. The label (i,j) inside each node indicates that the matrix product associated with that node involves matrices $M_i, M_{i+1}, \dots, M_j$.

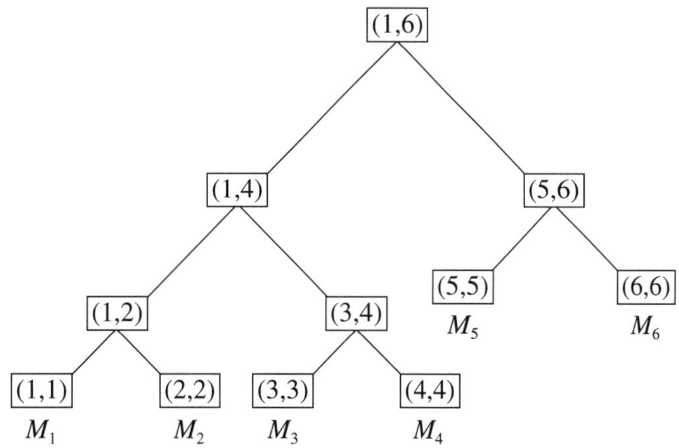

Hence, a brute-force algorithm that examines all possible parenthesizations is computationally infeasible.

We are led to consider a dynamic programming solution to our problem by noting that the Principle of Optimality holds for optimal parenthesizing. Indeed, consider any optimal parenthesization P for $M_1M_2 \ldots M_n$. Clearly, both the left and right parenthesizations P_1 and P_2 of P must be optimal for P to be optimal.

For $1 \leq i \leq j \leq n$, let m_{ij} denote the number of multiplications performed using an optimal parenthesization of $M_iM_{i+1} \ldots M_j$. By the Principle of Optimality, we have the following recurrence for the numbers m_{ij} based on making an optimal choice for the first cut index:

$$m_{ij} = \min_k \{m_{ik} + m_{k+1,j} + d_{i-1}d_kd_j : 1 \leq i \leq k \leq j \leq n\} \qquad (14.2.2)$$
$$\textbf{init. cond. } m_{ii} = 0, i = 1, \ldots, n.$$

The value m_{1n} corresponds to the minimum number of multiplications performed when computing $M_1M_2 \ldots M_n$. A divide-and-conquer algorithm *ParenthesizeRec* can be based directly on a top-down implementation of the recurrence relation (14.2.2). Unfortunately, a great many recalculations are performed by *ParenthesizeRec*, and it ends up doing $\Omega(3^n)$ multiplications to compute the minimum number m_{1n} corresponding to an optimal parenthesization.

A straightforward dynamic programming algorithm proceeds by computing the values m_{ij}, $1 \leq i \leq j \leq n$ in a bottom-up fashion using (14.2.2) (and thereby avoiding recalculations). Note that the values m_{ij}, $1 \leq i \leq j \leq n$, occupy the upper-right triangular portion of an $n \times n$ table. Our bottom-up resolution proceeds throughout the upper-right triangular portion diagonal by diagonal, starting from the bottom diagonal consisting of the elements $m_{ii} = 0$, $i = 1, \ldots, n$. The qth diagonal consists of the elements $m_{i,i+q}$, $q = 0, \ldots, n-1$. In Figure 14.3 we illustrate the computation of the m_{ij} for the example given in Figure 14.1. When computing m_{ij}, we also generate a

Figure 14.3 Table showing values m_{ij}, $1 \leq i \leq j \leq 4$, computed diagonal by diagonal from $q = 0$ to $q = 3$ using the bottom-up resolution of (14.2.2) for matrices M_1, M_2, M_3, and M_4 of dimensions 20×10, 10×50, 50×5, and 5×30, respectively. The values of c_{ij} are also shown underneath each m_{ij}, $1 \leq i \leq j \leq 4$.

i \ j	1	2	3	4	
1	0	10000 1	3500 1	6500 3	$q = 3$
2		0	2500 2	4000 3	$q = 2$
3			0	7500 3	$q = 1$
4				0	$q = 0$

table c_{ij} of indices k where the minimum in (14.2.2) occurs, that is c_{ij} is where the first cut in $M_iM_{i+1} \ldots M_j$ is made in an optimal parenthesization. The values c_{ij} can then be used to actually compute the matrix product according to the optimal parenthesization.

The following procedure *OptimalParenthesization* accepts as input the *dimension sequence* $d[0{:}n]$, where matrix M_i has dimension $d_{i-1} \times d_i$, $i = 1, \ldots, n$. Procedure *OptimalParenthesization* outputs the matrix $m[1{:}n,1{:}n]$, where $m[i,j] = m_{ij}$, $1 \le i \le j \le n$, is defined by recurrence (14.2.2). *OptimalParenthesization* also outputs the matrix $FirstCut[1{:}n,1{:}n]$, where $FirstCut[i,j] = c_{ij}$, $1 \le i \le j \le n$, which is the first-cut index in an optimal parenthesization for $M_i \ldots M_j$.

```
procedure OptimalParenthesization(d[0:n],m[1:n,1:n],FirstCut[1:n,1:n])

Input:  d[0:n] (dimension sequence for matrices M1, M2, . . . , Mn)
Output: m[1:n,1:n]   (m[i,j] = number of multiplications performed in an optimal
                     parenthesization for computing Mi . . . Mj, 1 ≤ i ≤ j ≤ n)
        FirstCut[1:n,1:n]  (index of first cut in optimal parenthesization of Mi
                           . . . Mj, 1 ≤ i ≤ j ≤ n)
   for i := 1 to n do        {initialize M[i,i] to zero}
         m[i,i] := 0
   endfor
   for diag := 1 to n − 1 do
      for i := 1 to n − diag do
         j := i + diag       {compute mij according to (14.2.2)}
         Min := m[i + 1,j] + d[i − 1]*d[i]*d[j]
         TempCut := i
         for k := i + 1 to j − 1 do
            Temp := m[i,k] + m[k + 1,j] + d[i − 1]*d[k]*d[j]
            if Temp < Min then
               Min := Temp
               TempCut := k
            endif
         endfor
         m[i,j] := Min
         FirstCut[i,j] := TempCut
      endfor
   endfor

end OptimalParenthesization
```

A simple loop counting shows that the complexity of *OptimalParenthesization* is in $\Theta(n^3)$.

It is now straightforward to write pseudocode for a recursive function *ChainMatrixProd* for computing the chained matrix product $M_1 \ldots M_n$ using an optimal parenthesization. We assume that the matrices $M_1, \ldots, M_n$ and the matrix

FirstCut[1:*n*,1:*n*] are global variables to the procedure *ChainMatrixProd*. The chained matrix product $M_1 \ldots M_n$ is computed by initially invoking the function *ChainMatrixProd* with $i = 1$ and $j = n$. *ChainMatrixProd* invokes a function *MatrixProd*, which computes the matrix product of two input matrices.

```
function ChainMatrixProd(i,j) recursive
Input:  i,j (indices delimiting matrix chain M_i, . . . , M_j)
        M_1, . . . , M_n (global matrices)
        FirstCut[1:n,1:n] (global matrix computed by OptimalParenthesization)
Output: M_1M_2 . . . M_n (matrix chain)
   if j > i then
      X := ChainMatrixProd(i,FirstCut[i,j])
      Y := ChainMatrixProd(FirstCut[i,j] + 1,j)
      return(MatrixProd(X,Y))
   else
      return(M_i)
   endif
end ChainMatrixProd
```

For the example given in Figure 14.1, invoking *ChainMatrixProd* with M_1, M_2, M_3, M_4 computes the chained matrix product $M_1M_2M_3M_4$ according to the parenthesization $((M_1(M_2M_3))M_4)$.

14.3 Optimal Binary Search Trees

We now use dynamic programming and the Principle of Optimality to generate an algorithm for the problem of finding optimal binary search trees. Given a search tree T and a search element X, the following recursive strategy finds an occurrence (if any) of a key X. First, X is compared to the key K associated with the root. If X is found there, then we are done. Otherwise, if X is less than K, then we search the left subtree, else we search the right subtree.

Consider, for example, the binary search tree given in Figure 14.4 involving the four keys 'Ann', 'Joe', 'Pat', 'Ray'. The internal nodes correspond to the successful searches X = 'Ann', X = 'Joe', X = 'Pat', X = 'Ray', and the leaf nodes correspond to the unsuccessful searches X < 'Ann', 'Ann'< X < 'Joe', 'Joe' < X < 'Pat', 'Pat'< X < 'Ray', 'Ray' < X. Suppose, for example, that X = 'Ann'. Then *BinSrchTreeSearch* makes three comparisons, first comparing X to 'Pat', then comparing X to 'Joe', and finally comparing X to 'Ann'. Now suppose that X = 'Pete'. Then *BinSrchTreeSearch* makes two comparisons, first comparing X to 'Pat' and then comparing X to 'Ray'. *BinSrchTreeSearch* (implicitly) branches to the left child of the node containing the key 'Ray', that is, to the leaf (implicit node) corresponding to the interval 'Pat' < X < 'Ray'. Let p_1,p_2,p_3,p_4 be the probability that X = 'Ann', X =

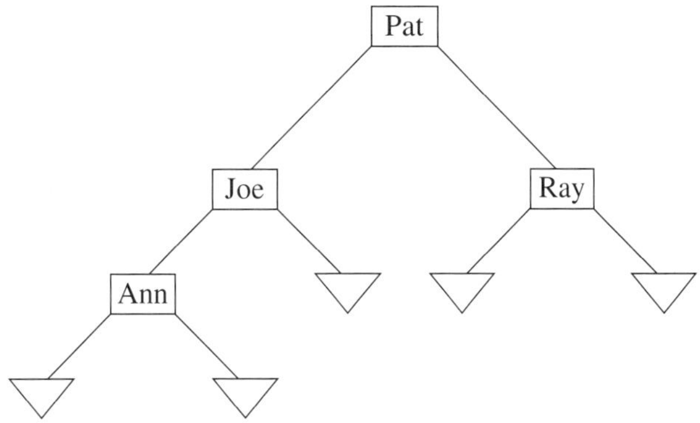

Figure 14.4 Search tree with leaf nodes drawn representing unsuccessful searches

'Joe', X = 'Pat', X = 'Ray', respectively, and let q_0,q_1,q_2,q_3,q_4, denote the probability that X < 'Ann', 'Ann'< X < 'Joe', 'Joe' < X < 'Pat', 'Pat'< X < 'Ray', 'Ray' < X, respectively. Then, the average number of comparisons made by *BinSrchTreeSearch* for the tree T of Figure 14.4 is given by

$$3p_1 + 2p_2 + p_3 + 2p_4 + 3q_0 + 3q_1 + 2q_2 + 2q_3.$$

Now consider a general binary search tree T whose internal nodes correspond to a fixed set of n keys $K_1, K_2, \ldots, K_n$ with associated probabilities $\mathbf{p} = (p_1, p_2, \ldots, p_n)$, and whose $n + 1$ leaf (external) nodes correspond to the $n + 1$ intervals I_0: $X < K_1$, I_1: $K_1 < X < K_2, \ldots, I_{n-1}$: $K_{n-1} < X < K_n$, I_n: $X > K_n$ with associated probabilities $\mathbf{q} = (q_0, q_1, \ldots, q_n)$. (When implementing T, the leaf nodes need not actually be included. However, when discussing the average behavior of *BinSrchTreeSearch*, it is useful to include them.) We now derive a formula for the average number of comparisons $A(T,n,\mathbf{p},\mathbf{q})$ made by *BinSrchTreeSearch*. Let d_i denote the depth of the internal node corresponding to K_i, $i = 1, \ldots, n$. Similarly, let e_i denote the depth of the leaf node corresponding to the interval I_i, $i = 0, 1, \ldots, n$. If $X = K_i$, then *BinSrchTreeSearch* traverses the path from the root to the internal node corresponding to K_i. Thus, it terminates after performing $d_i + 1$ comparisons. On the other hand, if X lies in I_i, then *BinSrchTreeSearch* traverses the path from the root to the leaf node corresponding to I_i and terminates after performing e_i comparisons. Thus, we have

$$A(T,n,\mathbf{p},\mathbf{q}) = \sum_{i=1}^{n} p_i(d_i + 1) + \sum_{i=0}^{n} q_i e_i. \tag{14.3.1}$$

We now consider the problem of determining an *optimal* binary search tree T, optimal in the sense that T minimizes $A(T,n,\mathbf{p},\mathbf{q})$ over all binary search trees T. This problem was solved in Chapter 9 for the case where all the p_i's are equal and all the q_i's are equal. Here we use dynamic programming to solve the problem for general probabilities p_i and q_i. In fact, we solve the slightly more general problem, where we relax the condition that $p_1, \ldots, p_n$ and $q_0, \ldots, q_n$ are probabilities (we actually need this additional generality when looking at the subproblems). That is, we solve the problem

$$\underset{T}{\text{minimize}}\ A(T,n,\mathbf{p},\mathbf{q}) \tag{14.3.2}$$

over all binary search trees T of size n, where $p_1, \ldots, p_n$ and $q_0, \ldots, q_n$ are given fixed positive real numbers. For convenience we sometimes refer to $A(T,n,\mathbf{p},\mathbf{q})$ as the *cost* of T. We define $\sigma(\mathbf{p},\mathbf{q})$ by

$$\sigma(\mathbf{p},\mathbf{q}) = \sum_{i=1}^{n} p_i + \sum_{i=0}^{n} q_i. \tag{14.3.3}$$

Note that we have removed the probability constraint that $\sigma(\mathbf{p},\mathbf{q}) = 1$.

Similar to our discussion of chained matrix products, we could obtain an optimal search tree by enumerating all binary search trees on the given identifiers and choosing the one with minimum $A(T,n,\mathbf{p},\mathbf{q})$. However, the number of different binary search trees on n identifiers is the same as the number of binary trees on n nodes, which is given by the nth Catalan number

$$b_n = \frac{1}{n+1}\binom{2n}{n} \in \Omega\left(\frac{4^n}{n^2}\right).$$

Thus, a brute force algorithm for determining an optimal binary search tree using simple enumeration is computationally infeasible. Fortunately, the Principle of Optimality holds for the optimal binary search tree problem, so we look for a solution using dynamic programming.

Let K_i denote the key associated with the root of T, and let L and R denote the left and right subtrees (of the root) of T, respectively. As we remarked earlier, L is a binary search tree for the keys $K_1, \ldots, K_{i-1}$, and R is a binary search tree for the keys $K_{i+1}, \ldots, K_n$. For convenience, let $A(T) = A(T,n,\mathbf{p},\mathbf{q})$, $A(L) = A(L,i-1,p_1, \ldots, p_{i-1},q_0, \ldots, q_{i-1})$, and $A(R) = A(R,n-i,p_{i+1}, \ldots, p_n,q_i, \ldots, q_n)$. Clearly, each node of T different from the root corresponds to exactly one node in either L or R. Further, if N is a node in T corresponding to a node N' in L, then the depth of N in T is exactly one greater then the depth of N' in L. A similar result holds if N corresponds to a node in R. Thus, it follows immediately from (14.3.1) that

$$A(T) = A(L) + A(R) + \sigma(\mathbf{p},\mathbf{q}). \tag{14.3.4}$$

We now employ (14.3.4) to show that the Principle of Optimality holds for the problem of finding an optimal search tree. Suppose that T is an optimal search tree; that is, T minimizes $A(T)$. We must show that L and R are also optimal search trees. Suppose there exists a binary search tree L' with $i-1$ nodes involving the keys $K_1, \ldots, K_{i-1}$ such that $A(L') < A(L)$. Clearly, the tree T' obtained from T by replacing L with L' is a binary search tree. Further, it follows from (14.3.4) that $A(T') < A(T)$, contradicting the assumption that T is an optimal binary search tree. Hence, L is an optimal binary search tree. By symmetry, R is also an optimal binary search tree, which establishes that the Principle of Optimality holds for the optimal binary search tree problem.

Since the Principle of Optimality holds, when constructing an optimal search tree T we need only consider binary search trees L and R, both of which are optimal. This observation, together with recurrence relation (14.3.4), is the basis of the following dynamic programming algorithm for constructing an optimal binary search tree. For $i,j \in \{1, \ldots, n\}$, we let T_{ij} denote an *optimal* search tree involving the consecutive keys $K_i, K_{i+1}, \ldots, K_j$, where T_{ij} is the null tree if $i > j$. Then, $T = T_{1n}$ is an optimal search tree involving all n keys. We define $A(T_{ij})$ by

$$\mathrm{A}(T_{ij}) = \mathrm{A}(T_{ij}, j - i + 1, p_i, p_{i+1}, \ldots, p_j, q_{i-1}, q_i, \ldots, q_j). \qquad (14.3.5)$$

Since the keys are sorted in nondecreasing order, it follows from the Principle of Optimality and (14.3.4) that

$$\mathrm{A}(T_{ij}) = \min_k \{\mathrm{A}(T_{i,k-1}) + \mathrm{A}(T_{k+1,j})\} + \sigma(p_i, \ldots, p_j, q_{i-1}, \ldots, q_j), \qquad (14.3.6)$$

where the minimum is taken over all $k \in \{i, i+1, \ldots, j\}$. Recurrence relation (14.3.6) yields an algorithm for computing an optimal search tree T. The algorithm begins by generating all single-node binary search trees, which are trivially optimal, namely, $T_{11}, T_{22}, \ldots, T_{nn}$. Using (14.3.6), the algorithm can then generate optimal search trees $T_{12}, T_{23}, \ldots, T_{n-1,n}$. In general, at the kth stage in the algorithm, the recurrence relation (14.3.6) is applied to construct the optimal search trees $T_{1k}, T_{2,k+1}, \ldots, T_{n-k+1,n}$, using the previously generated optimal search trees as building blocks. Figure 14.5 illustrates the algorithm for a sample instance involving $n = 4$ keys. Note that there are two possible choices for T_{13} in Figure 14.5, each having a minimum cost of 1.1. The tree with the smaller root key was selected.

Figure 14.5 Action of algorithm to find an optimal binary search tree using dynamic programming

i	0	1	2	3	4
p_i		.15	.1	.2	.3
q_i	.05	.05	0	.05	.1

T_{11} = [1] T_{22} = [2] T_{33} = [3] T_{44} = [4]

$\mathrm{A}(T_{11}) = .25$ $\mathrm{A}(T_{22}) = .15$ $\mathrm{A}(T_{33}) = .25$ $\mathrm{A}(T_{44}) = .45$

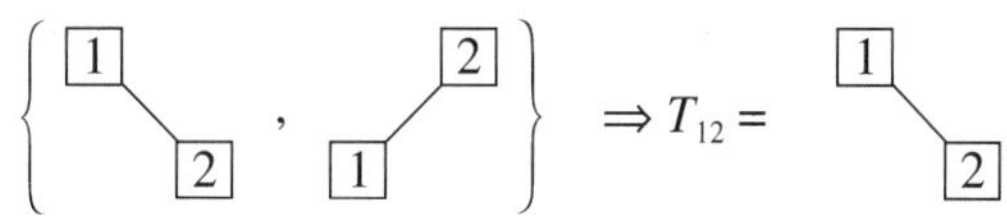

$\mathrm{A}(T_{12}) = .5$

min {0 + .15 = .15, .25 + 0 = .25} + .35 = .5

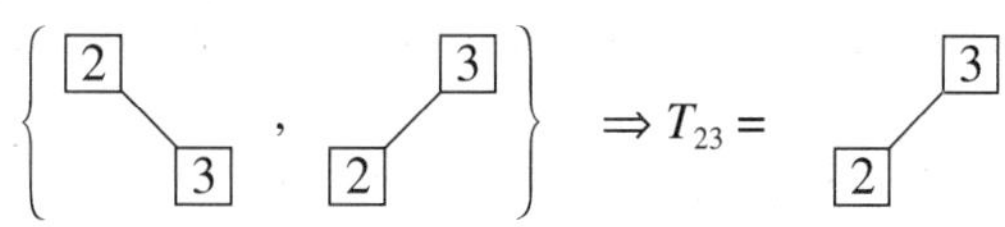

$\mathrm{A}(T_{23}) = .55$

min {0 + .25 = .25, .15 + 0 = .15} + .4 = .55

Continued

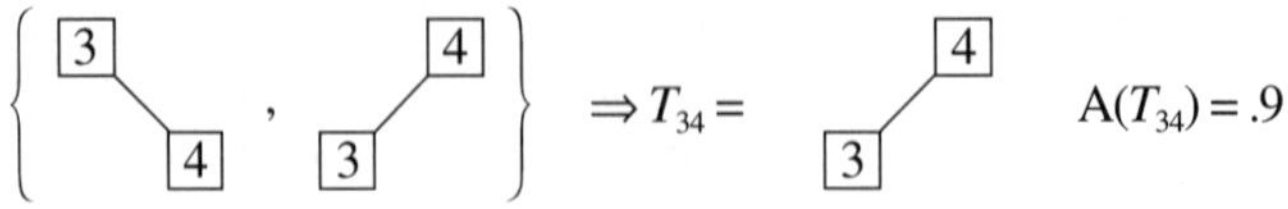

min {0 + .45 = .45, .25 + 0 = .25} + .65 = .9

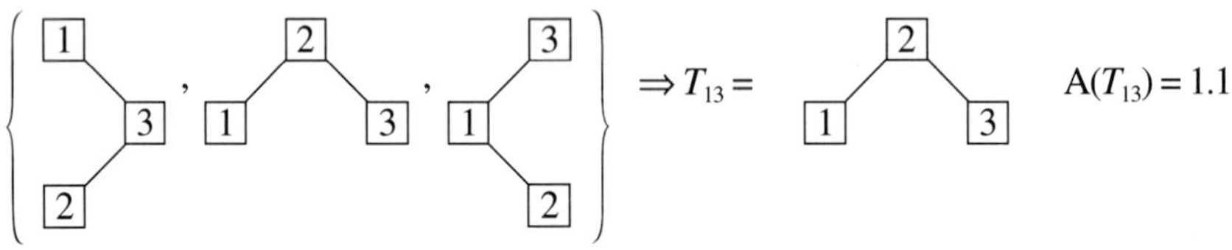

min {0 + .55 = .55, .25 + .25 = .5, .5 + 0 = .5} + .6 = 1.1

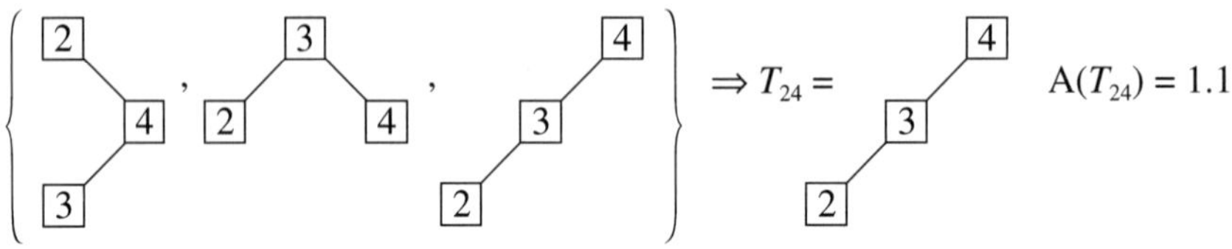

min {0 + .55 = .55, .25 + .25 = .5, .5 + 0 = .5} + .6 = 1.1

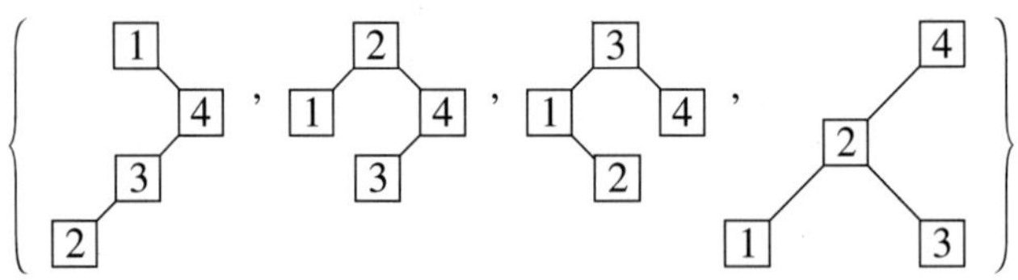

min {0 + 1.35 = 1.35, .25 + .9 = 1.15, .5 + .45 = .95, 1.1 + 0 = 1.1} + 1 = 1.95

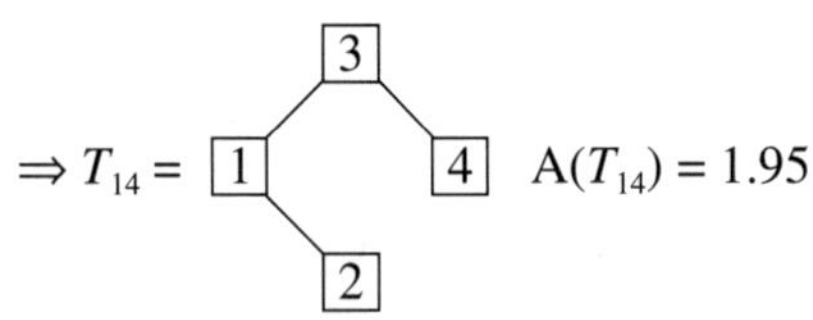

Figure 14.5 Continued

We now give pseudocode for the algorithm *OptimalSearchTree*, which implements the preceding strategy. *OptimalSearchTree* computes the root, *Root*[i,j], of each tree T_{ij}, and the cost, A[i,j], of T_{ij}. An optimal binary search tree T for all the keys, namely T_{1n}, can be easily constructed using recursion from the two-dimensional array *Root*[1:n,1:n].

```
procedure OptimalSearchTree(P[1:n],Q[0:n],Root[1:n,1:n],A[1:n,1:n])
Input:  P[1:n] (an array of probabilities associated with successful searches)
        Q[0:n] (an array of probabilities associated with unsuccessful searches)
Output: Root[1:n,1:n] (Root[i,j] is the key of the root node of T_ij)
        A[1:n,1:n] (A[i,j] is the cost A(T_ij) of T_ij)

  for i := 1 to n do
    Root[i,i] := i
    Sigma[i,i] := p[i] + q[i - 1] + q[i]
    A[i,i] := Sigma[i,i]
  endfor
  for TreeSize := 1 to n - 1 do
    for i := 1 to n - TreeSize do
      j := i + TreeSize
      Sigma[i,j] := Sigma[i,j - 1] + p[j] + q[j] {Compute σ(p_i, ... , p_j, q_{i-1}, ... , q_j)}
      Root[i,j] := i
      Min := A[i + 1,j]
      for k := i + 1 to j do
        Sum := A[i,k - 1] + A[k + 1,j]
        if Sum < Min then
          Min := Sum
          Root[i,j] := k
        endif
      endfor
      A[i,j] := Min + Sigma[i,j]
    endfor
  endfor
end OptimalSearchTree
```

Since *OptimalSearchTree* does the same amount of work for any input $P[1:n]$ and $Q[0:n]$, the best-case, worst-case, and average complexities are all equal. Clearly, the number of additions made in computing *Sum* has the same order as the total number of additions made by *OptimalSearchTree*. Therefore, we choose the addition made in computing *Sum* as the basic operation. Since *TreeSize* varies from 1 to $n-1$, i varies from 1 to $n - TreeSize$, and k varies from $i+1$ to $i + TreeSize$, it follows that the total number of additions made in computing *Sum* is given by

$$\sum_{t=1}^{n-1} \sum_{i=1}^{n-t} \sum_{k=i+1}^{i+t} 1$$

$$= \sum_{t=1}^{n-1} (n-t)\,t$$

$$= n\sum_{t=1}^{n-1} t - \sum_{t=1}^{n-1} t^2$$

$$= n\left[(n-1)\frac{n}{2}\right] - \left[(n-1)n\frac{(2n-1)}{6}\right] \in \Theta(n^3).$$

14.4 All-Pairs Shortest Paths

Both Dijkstra's and the Bellman-Ford shortest-path algorithms find paths from a single source vertex r to all other vertices of a weighted digraph $D = (V,E)$ and have complexities $\Theta(n^2)$ and $\Theta(nm)$, respectively. By repeated calls to either of these algorithms we obtain algorithms for a shortest path between every pair of vertices u and v, having complexities $\Theta(n^3)$ and $\Theta(n^2m)$, respectively. We now use dynamic programming to obtain a $\Theta(n^3)$ algorithm, *Floyd's algorithm*, for finding a shortest path between every pair of vertices of a weighted digraph D. Unlike the all-pairs shortest paths algorithm based on repeated applications of Dijkstra's algorithm, Floyd's algorithm works even if some of the edges have negative weights, provided that there is no negative cycle (sum of the weights of the edges in the cycle is negative).

Consider a digraph D with vertex set $V = \{1, \ldots, n\}$ and edge set E, whose edges have been assigned a weighting w. Let P be a path joining two distinct vertices $i, j \in V$, and suppose k is an interior vertex of P (that is, any vertex of P different from either i or j). As discussed earlier, the Principle of Optimality holds, so that the subpath P_1 from i to v is a shortest path from i to v, and the subpath P_2 from v to j is a shortest path from v to j.

For k a nonnegative integer, $k \le n$, let V_k denote the subset of vertices $\{1, \ldots, k\}$ (by convention, $V_0 = \varnothing$). Let $S_k(i,j)$ denote the length of a shortest path P from i to j, whose interior vertices (if any) all lie in V_k (if no such path P exists, then $S_k(i,j) = \infty$). Note that $S_n(i,j)$ is the length of a shortest path in G from i to j. If k is not an interior vertex of P, then P is a shortest path from i to j whose interior vertices lie in V_{k-1}, so that $S_k(i,j) = S_{k-1}(i,j)$. On the other hand, if k is an interior vertex of P, then the interior vertices (if any) of the path P_1 from i to k and the interior vertices (if any) of the path P_2 from k to j all lie in V_{k-1}. Since P_1 is a shortest path from i to k and P_2 is a shortest path from k to j, we have $S_k(i,j) = S_{k-1}(i,k) + S_{k-1}(k,j)$. Thus, either $S_k(i,j) = S_{k-1}(i,j)$ or $S_k(i,j) = S_{k-1}(i,k) + S_{k-1}(k,j)$, depending on whether or not P contains vertex k. Since P is a shortest path from i to j, it follows that the length of P is equal to the minimum of these two values, yielding the following recurrence relation for $S_k(i,j)$.

$$S_k(i,j) = \min\{S_{k-1}(i,j),\ S_{k-1}(i,k) + S_{k-1}(k,j)\} \tag{14.4.1}$$

Since, by definition, $S_0(i,j)$ is the length of a shortest path from i to j containing no interior vertices, S_0 is equal to the *weight matrix* W defined as follows: $W(i,j) = w(e)$ if D contains an edge e from vertex i to vertex j, 0 if $i = j$, and ∞ otherwise. Floyd's algorithm is based on recurrence relation (14.4.1). The matrix S_k keeps track of the length of the shortest paths and not the shortest paths themselves. To keep track of the paths, we maintain another matrix $P_k(i,j)$ defined by

$$P_k(i,j) = \begin{cases} P_{k-1}(i,j) & \text{if } S_k(i,j) = S_{k-1}(i,j) \\ k & \text{otherwise} \end{cases}, \tag{14.4.2}$$

init. cond. $P_0(i,j) = 0$ for all $i,j \in V$.

Floyd's algorithm is illustrated in Figure 14.6 (page 504) for a sample digraph D and weight function w.

A shortest path from i to j in G can then be reconstructed from P_n as follows: If $P_n(i,j) = 0$, then a shortest path from i to j is directly along the edge (i,j); otherwise, if $P_n(i,j) = k$, then k is an intermediate vertex of a shortest path from i to j and any other intermediate vertices in this shortest path can be obtained by recursively examining $P_n(i,k)$ and $P_n(k,j)$. For example, in Figure 14.6, a shortest path S from vertex 3 to vertex 5 can be reconstructed from the final matrix P_5 as follows. Since $P_5(3,5) = 2$, vertex 2 is an internal vertex of S. Now, $P_5(2,5) = 0$, so that there are no intermediate vertices of S between 2 and 5. Since $P_5(3,2) = 1$, vertex 1 is an intermediate vertex of S between 3 and 2. Now, $P_5(3,1) = 0$ and $P_5(1,2) = 0$, so that there are no intermediate vertices between 3 and 1 or between 1 and 2. Thus, a shortest path S from 3 to 5 is given by 3, 1, 2, 5.

We now give pseudocode for Floyd's algorithm.

```
procedure Floyd(W[1:n,1:n],P[1:n,1:n],S[1:n,1:n])
Input:  W[1:n,1:n]   (weight matrix for a weighted digraph D)
Output: P[1:n,1:n]   (matrix implementing shortest paths)
        S[1:n,1:n]   (distance matrix where S[u,v] is the length (cost) of
                      a shortest path from u to v in G)
  for i := 1 to n do       {initialize P and S}
    for j := 1 to n do
      P[i,j] := 0
      S[i,j] := W[i,j]
    endfor
  endfor
  for k := 1 to n do     {update S and P using (14.4.1) and (14.4.2)}
    for i := 1 to n do
      for j := 1 to n do
        if S[i,j] > S[i,k] + S[k,j] then
          P[i,j] := k
          S[i,j] := S[i,k] + S[k,j]
        endif
      endfor
    endfor
  endfor
end Floyd
```

$D =$

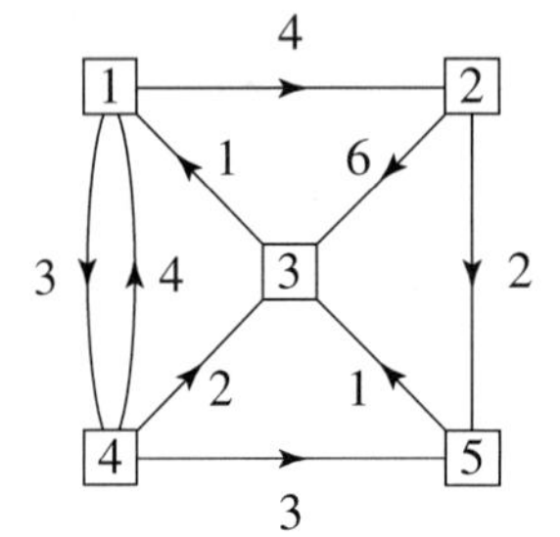

$$S_0 = W = \begin{pmatrix} 0 & 4 & \infty & 3 & \infty \\ \infty & 0 & 6 & \infty & 2 \\ 1 & \infty & 0 & \infty & \infty \\ 4 & \infty & 2 & 0 & 3 \\ \infty & \infty & 1 & \infty & 0 \end{pmatrix}$$

$$S_1 = \begin{pmatrix} 0 & 4 & \infty & 3 & \infty \\ \infty & 0 & 6 & \infty & 2 \\ 1 & 5 & 0 & 4 & \infty \\ 4 & 8 & 2 & 0 & 3 \\ \infty & \infty & 1 & \infty & 0 \end{pmatrix} \qquad P_1 = \begin{pmatrix} 0 & 0 & 0 & 0 & 0 \\ 0 & 0 & 0 & 0 & 0 \\ 0 & 1 & 0 & 1 & 0 \\ 0 & 1 & 0 & 0 & 0 \\ 0 & 0 & 0 & 0 & 0 \end{pmatrix}$$

$$S_2 = \begin{pmatrix} 0 & 4 & 10 & 3 & 6 \\ \infty & 0 & 6 & \infty & 2 \\ 1 & 5 & 0 & 4 & 7 \\ 4 & 8 & 2 & 0 & 3 \\ \infty & \infty & 1 & \infty & 0 \end{pmatrix} \qquad P_2 = \begin{pmatrix} 0 & 0 & 2 & 0 & 2 \\ 0 & 0 & 0 & 0 & 0 \\ 0 & 1 & 0 & 1 & 2 \\ 0 & 1 & 0 & 0 & 0 \\ 0 & 0 & 0 & 0 & 0 \end{pmatrix}$$

$$S_3 = \begin{pmatrix} 0 & 4 & 10 & 3 & 6 \\ 7 & 0 & 6 & 10 & 2 \\ 1 & 5 & 0 & 4 & 7 \\ 3 & 7 & 2 & 0 & 3 \\ 2 & 6 & 1 & 5 & 0 \end{pmatrix} \qquad P_3 = \begin{pmatrix} 0 & 0 & 2 & 0 & 2 \\ 3 & 0 & 0 & 3 & 0 \\ 0 & 1 & 0 & 1 & 2 \\ 3 & 3 & 0 & 0 & 0 \\ 3 & 3 & 0 & 3 & 0 \end{pmatrix}$$

$$S_4 = \begin{pmatrix} 0 & 4 & 5 & 3 & 6 \\ 7 & 0 & 6 & 10 & 2 \\ 1 & 5 & 0 & 4 & 7 \\ 3 & 7 & 2 & 0 & 3 \\ 2 & 6 & 1 & 5 & 0 \end{pmatrix} \qquad P_4 = \begin{pmatrix} 0 & 0 & 4 & 0 & 2 \\ 3 & 0 & 0 & 3 & 0 \\ 0 & 1 & 0 & 1 & 2 \\ 3 & 3 & 0 & 0 & 0 \\ 3 & 3 & 0 & 3 & 0 \end{pmatrix}$$

$$S_5 = \begin{pmatrix} 0 & 4 & 5 & 3 & 6 \\ 7 & 0 & 3 & 7 & 2 \\ 1 & 5 & 0 & 4 & 7 \\ 3 & 7 & 2 & 0 & 3 \\ 2 & 6 & 1 & 5 & 0 \end{pmatrix} \qquad P_5 = \begin{pmatrix} 0 & 0 & 4 & 0 & 2 \\ 3 & 0 & 5 & 5 & 0 \\ 0 & 1 & 0 & 1 & 2 \\ 3 & 3 & 0 & 0 & 0 \\ 3 & 3 & 0 & 3 & 0 \end{pmatrix}$$

Figure 14.6 Stages of Floyd's algorithm for a sample weighted digraph D

14.5 The Traveling Salesman Problem

Suppose that a salesperson must visit n cities and that there is a cost associated with traveling from one city to another. Starting from a home city, the salesperson wishes to visit each of the other $n - 1$ cities once and return home in such an order that the total cost incurred is a minimum among all such tours (see Figure 14.7).

The most straightforward algorithm is an *exhaustive search* that enumerates all $(n - 1)!$ tours and keeps track of the shortest tour generated. When attempting to solve the Traveling Salesman problem in practice, many salespeople resort to the solution shown in Figure 14.8 (page 506). We now discuss a *dynamic programming* algorithm for the Traveling Salesman problem having worst-case complexity in $\Theta(n2^n)$. While it is certainly an improvement over $(n - 1)!$, this exponential complexity is disappointing. However, the Traveling Salesman problem belongs to a class of problems (*NP*-hard) for which it is believed that no polynomial-time algorithm exists for any problem in the class.

The Traveling Salesman problem is equivalent to finding a minimum cost Hamiltonian cycle in a weighted digraph D. We may assume without loss of generality that D is the complete digraph $\hat{K}_n = (V, \hat{E})$, where $|V| = n$ and $\hat{E}$ consists of all pairs of distinct vertices (u,v), $u,v \in V$. Given a weighted digraph $D = (V,E)$, we merely extend the cost weighting c to $\hat{K}_n$ by setting $c(uv) = \infty$ for every (u,v) not in E. The

Figure 14.7 A minimum length tour of five cities amongst the 24 possible tours

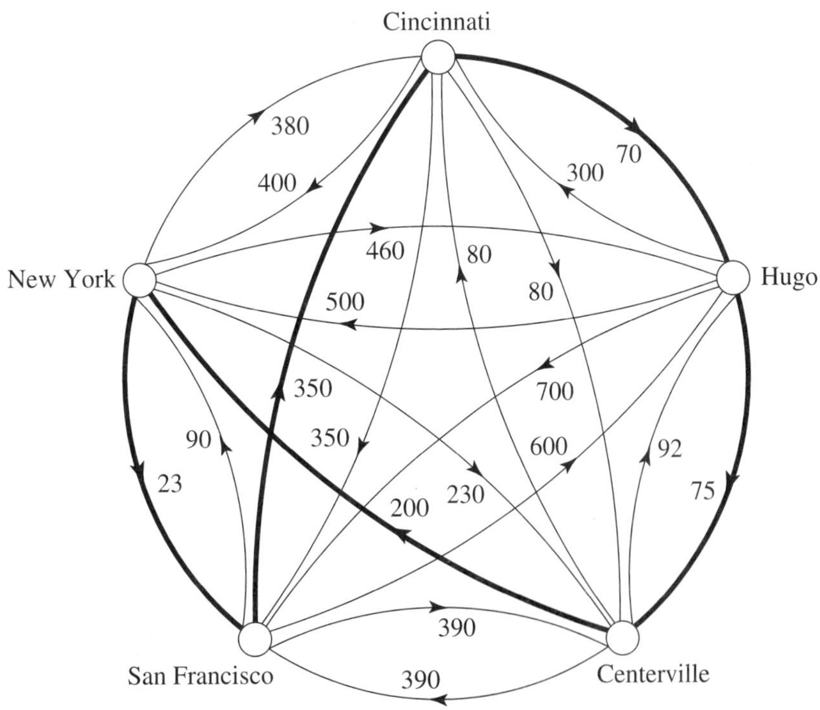

Figure 14.8 (© 1996 by Sidney Harris. Reprinted with permission. All rights reserved.)

Traveling Salesman problem for the (undirected) complete graph K_n with cost weighting c is a special case of the Traveling Salesman problem for the digraph $\hat{K}_n$: For each edge e of K_n assign both of the directed edges e^- and e^+ of $\hat{K}_n$ corresponding to e the weight $c(e)$.

Consider a Hamiltonian cycle H of $\hat{K}_n$, where without loss of generality we assume that vertex 1 is the initial and terminal vertex of H. Let i denote the first vertex visited by H after leaving vertex 1, and let H_i denote the subpath of H from vertex i to vertex 1, which passes through each vertex of $\hat{K}_n$ once (such a path is called a *Hamiltonian path*). If H is a minimum cost Hamiltonian cycle, then H_i must be a shortest path from i to 1 whose interior vertices consist of the set $V - \{1,i\}$. (A shortest path P joining two vertices i and j is a path that minimizes $Cost(P)$, where $Cost(P)$ is the sum of the costs over all the edges of P.) Now consider any subset U of V. For i a vertex not in U, let $MinCost(i,U)$ denote the cost of a shortest (minimum cost) path P from i to 1 whose interior vertices consist of the set U. Note that since P is a shortest path, P must pass through each vertex of U exactly once. Note also that $MinCost(1,V - \{1\})$ is the minimum cost of a Hamiltonian cycle, that is, the cost of an optimal salesman tour.

A solution to the problem of finding a shortest path $P = i\, v_1 \ldots v_p\, 1$, from i to 1, where $v_1, \ldots, v_p \in U$, can be expressed as a sequence of decisions, where the kth decision is to choose vertex v_k as the $(k + 1)$st vertex in the path, $k = 1, \ldots, p$. The derived problem, resulting from having made the first decision (to choose vertex v_1), is to find a shortest path from v_1 to 1, whose interior vertices consist of the set $U -$

$\{v_1\}$. It is easily verified that if P is a shortest path from vertex i to vertex 1 whose interior vertices consist of the set U, then the path $Q = v_1 \ldots v_p\ 1$ is a shortest path from vertex v_1 to vertex 1 whose interior vertices consist of the set $U - \{v_1\}$. Hence, the Specialized Principle of Optimality holds. Thus, observing that $Cost(P) = c(i,v_1) + Cost(Q)$, we obtain the following recurrence relation for $MinCost(i,U)$.

$$MinCost(i,U) = \min_{j \in U} \{c(i,j) + MinCost(j,U - \{j\})\}, \qquad (14.5.1)$$
$$\textbf{init. cond. } MinCost(i, \varnothing) = c(i,1), \quad 2 \le i \le n$$

Recurrence relation (14.5.1) yields an algorithm for solving the Traveling Salesman problem. Using (14.5.1), we compute the values of $MinCost(i,U)$ for all pairs i,U, where $|U| = 1$. We then use (14.5.1) to compute the values of $MinCost(i,U)$ for all pairs i, U, where $|U| = 2$. In general, once we have computed $MinCost(i,U)$ for all pairs i, U, where $|U| = k - 1$, we can employ (14.5.1) to compute $MinCost(i,U)$ for all pairs, where $|U| = k$, $k = 1, \ldots, n - 2$. We then use (14.5.1) to compute $MinCost(1,V - \{1\})$, thereby obtaining the minimum cost of a Hamiltonian cycle. Algorithm *TravelingSalesman* is based on this strategy and uses somewhat unrefined pseudocode. In particular, we do not make explicit how *MinCost* is maintained. (However, it is the cost of maintaining *MinCost* that contributes the major portion of the work done by the algorithm.)

```
procedure TravelingSalesman(C[1:n,1:n],Tour[1:n],Cost)
Input:  C[1:n,1:n]  (cost matrix)
Output: Tour[1:n]   (Tour[i] is the ith vertex of a minimum cost Hamiltonian
                                            cycle, i = 1, . . . , n)
        Cost (minimum cost of a Hamiltonian cycle)
  for i := 2 to n do
     MinCost[i,∅] = C[i,1]
  endfor
  for SetSize := 1 to n – 2 do
     for i := 2 to n do
        for each U ⊆ V – {i,1} such that |U| = SetSize do
           Min := ∞
           for j := 2 to n do
              if j ≠ i then
                 Sum := C[i,j] + MinCost[j,U – {j}]
                 if Sum < Min then
                    Min := Sum
                    MinVertex := j
                 endif
              endif
           endfor
           MinCost[i,U] := Min
           Path[i,U] := MinVertex
```

Continued

```
        endfor
      endfor
    endfor
    Min := C[1,2] + MinCost[2,V – {1,2}]
    for j := 3 to n do
      Sum := C[1,j] + MinCost[j,V – {1,j}]
      if Sum < Min then
        Min := Sum
        MinVertex := j
      endif
    endfor
    {MinCost[1,V – {1}] = Min}
    Path[1,V – {1}] := MinVertex
    Cost := Min
    Tour[1] := 1
    U := V – {1}
    for i := 1 to n – 1 do
      j := Path(Tour[i],U)
      Tour[i + 1] := j
      U = U – {j}
    endfor
  end TravelingSalesman
```

In Figure 14.9, we illustrate *TravelingSalesman* for a sample cost matrix $C[1{:}4,1{:}4]$.

We now analyze the worst-case complexity W(n) of *TravelingSalesman.* We choose as our basic operation an assignment to *MinCost*. For *SetSize* = k and i fixed, the number of sets $U \in V - \{1,i\}$ of size k is $\binom{n-2}{k}$. Thus, the total number of assignments to $MinCost[i,U]$ that are performed is given by

$$\sum_{k=0}^{n-2} (n-1)\binom{n-2}{k} = (n-1)2^{n-2}.$$

Figure 14.9 Action of *TravelingSalesman* for a sample digraph having four vertices

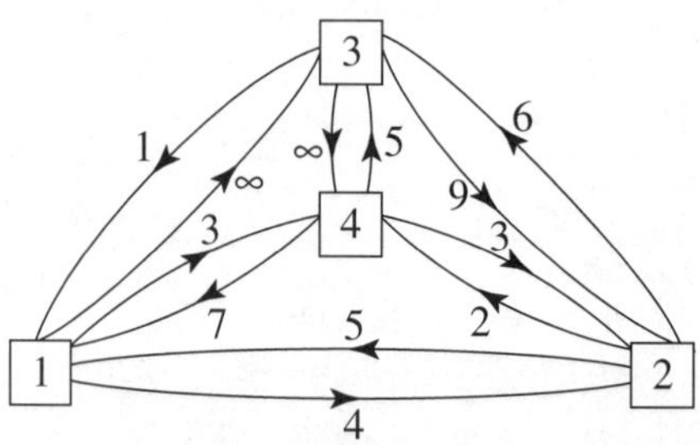

$$C = \begin{pmatrix} 0 & 4 & \infty & 3 \\ 5 & 0 & 6 & 2 \\ 1 & 9 & 0 & \infty \\ 7 & 3 & 5 & 0 \end{pmatrix}$$

$|U| = 0$
$MinCost[2,\emptyset] = C[2,1] = 5$
$MinCost[3,\emptyset] = C[3,1] = 1$
$MinCost[4,\emptyset] = C[4,1] = 7$

$|U| = 1$
$MinCost[2,\{3\}] = C[2,3] + MinCost[3,\emptyset] = 6 + 1 = 7 \qquad Path[2,\{3\}] = 3$
$MinCost[3,\{2\}] = C[3,2] + MinCost[2,\emptyset] = 9 + 5 = 14 \qquad Path[3,\{2\}] = 2$
$MinCost[4,\{2\}] = C[4,2] + MinCost[2,\emptyset] = 3 + 5 = 8 \qquad Path[4,\{2\}] = 2$
$MinCost[2,\{4\}] = C[2,4] + MinCost[4,\emptyset] = 2 + 7 = 9 \qquad Path[2,\{4\}] = 4$
$MinCost[3,\{4\}] = C[3,4] + MinCost[4,\emptyset] = \infty + 7 = \infty \qquad Path[3,\{4\}] = 4$
$MinCost[4,\{3\}] = C[4,3] + MinCost[3,\emptyset] = 5 + 1 = 6 \qquad Path[4,\{3\}] = 3$

$|U| = 2$

$$\begin{aligned}
MinCost[2,\{3,4\}] &= \min\{C[2,3] + MinCost[3,\{4\}],\ C[2,4] + MinCost[4,\{3\}]\} \\
&= \min\{6 + \infty,\ 2 + 6\} = \min\{\infty, 8\} = 8 \\
\Rightarrow Path[2,\{3,4\}] &= 4 \\
MinCost[3,\{2,4\}] &= \min\{C[3,2] + MinCost[2,\{4\}],\ C[3,4] + MinCost[4,\{2\}]\} \\
&= \min\{9 + 9,\ \infty + 8\} = \min\{18, \infty\} = 18 \\
\Rightarrow Path[3,\{2,4\}] &= 2 \\
MinCost[4,\{2,3\}] &= \min\{C[4,2] + MinCost[2,\{3\}],\ C[4,3] + MinCost[3,\{2\}]\} \\
&= \min\{3 + 7,\ 5 + 14\} = \min\{10, 19\} = 10 \\
\Rightarrow Path[4,\{2,3\}] &= 2
\end{aligned}$$

$U = V - \{1\}$

$$\begin{aligned}
MinCost[1,\{2,3,4\}] &= \min\{C[1,2] + MinCost[2,\{3,4\}], \\
&\qquad \{C[1,3] + MinCost[3,\{2,4\}] \\
&\qquad \{C[1,4] + MinCost[4,\{2,3\}]\}\} \\
&= \min\{4 + 8,\ \infty + 18,\ 3 + 10\} \\
&= \min\{12,\ \infty,\ 13\} = 12 \\
\Rightarrow Path[1,\{2,3,4\}] &= 2
\end{aligned}$$

$Cost = MinCost[1,\{2,3,4\}] = 12$

Tour[1] = 1
Tour[2] = Path[1,{2,3,4}] = 2
Tour[3] = Path[2,{3,4}] = 4
Tour[4] = Path[4,{3}] = 3

Figure 14.9 Continued

Thus,

$$W(n) = (n-1)2^{n-2} \in \Theta(n2^n).$$

Hence, *TravelingSalesman* has exponential worst-case complexity. Another serious drawback with *TravelingSalesman* is that it has space complexity belonging to $\Theta(n2^n)$.

14.6 Dynamic Programming in Parallel

Since dynamic programming is a design strategy that usually involves resolving a recurrence relation in a bottom-up level-by-level fashion, a natural parallelization results from performing the relevant operations at a given level concurrently. Similar to the corresponding parallelization of divide-and-conquer algorithms, the straightforward level-by-level parallelization on a PRAM of a sequential dynamic programming algorithm by itself does not usually yield a parallel algorithm with the best speedup for the given problem. For example, for the optimal matrix product parenthesization, optimal binary search tree and all-pairs shortest paths problems, we have a linear number of levels, so that their level-by-level parallelizations have complexities in $\Omega(n)$. The question remains as to whether these problems belong to the class NC. In other words, can they be solved on the PRAM in polylogarithmic time using a polynomial number of processors? It turns out that the optimal matrix product parenthesization, optimal binary search tree and all-pairs shortest paths problems all belong to NC. In this section, we discuss a parallel algorithm for the all-pairs shortest paths problem based on the parallel algorithm for matrix powers discussed in Chapter 11.

Consider a digraph D with set $V = \{1, \ldots, n\}$ and edge set E. Using n^2 processors on a EREW PRAM, we can compute the matrix D_k of recurrence relation (14.4.1) from the matrix D_{k-1} in constant time. The resulting parallelization of Floyd's algorithm has complexity n and cost n^3. Although we obtain a parallel algorithm that is cost optimal (when compared to Floyd's algorithm), our goal is to obtain a polylogarithmic parallel algorithm using a polynomial number of processors. We can achieve this goal by using a different recurrence relation, which is more amenable to parallelization.

Given a weight function w for the edges, let $A = (a_{ij})_{n \times n}$ be the matrix such that $a_{ij} = w(ij)$ if there is an edge from i to j, 0 if $i = j$, and ∞ otherwise. Define the matrix $A^{[q]} = (a_{ij}^{[q]})$ such that $A^{[q]}(i,j)$ is the length of a shortest path, with respect to w, from vertex i to vertex j that contains at most q edges, or ∞ if there is no such path, $i,j = 1, \ldots, n$, $q = 1, \ldots, n-1$. Note that $A^{[1]} = A$. Using the Principle of Optimality it follows that

$$a_{ij}^{[q]} = \min_{1 \le k \le n} (a_{ik}^{[q-1]} + a_{kj}^{[1]}), \quad \textbf{init. cond. } A^{[1]} = A. \tag{14.6.1}$$

Since a shortest path is necessarily a simple path and the number of edges of any simple path in D is at most $n-1$, it follows that the length d_{ij} of a shortest path in D from i to j (with respect to w) is the ijth coefficient $a_{ij}^{[n-1]}$ of the matrix $A^{[n-1]}$.

Now, consider the problem of computing the qth power $A^q = (a_{ij}^{(q)})_{n \times n}$ of a matrix A. It is immediate from the definition of matrix multiplication that

$$a_{ij}^{(q)} = \sum_{1 \le k \le n} (a_{ik}^{(q-1)} * a_{kj}^{(1)}), \quad \textbf{init. cond. } A^1 = A. \tag{14.6.2}$$

Note that recurrence relation (14.6.1) is identical to (14.6.2) except that the operation of addition is replaced with the operation min, and the operation of multiplication is replaced with the operation of addition. It follows that we can compute $A^{[q]}$ using an algorithm, such as *MatMultCREW* discussed in Chapter 11, for computing the qth power A^q of matrix A, with the operation of addition replaced with min and the operation of multiplication replaced with addition. Recall that the nth power of a matrix can be computed in time $O(\log^2 n)$ using n^3 processors on the CREW PRAM. Pseudocode for procedure *AllPairsShortestPathsCREW* based on the algorithm just described is left as an exercise.

14.7 Closing Remarks

The optimal matrix product parenthesization, optimal binary search tree, and all-pairs shortest paths problems all belong to class NC. In fact, using a technique called the *pebble game* (see references), these and other dynamic programming problems based on a certain general recurrence relation can be solved in $O((\log n)^2)$ time on an EREW PRAM with $O(n^6/\log n)$ processors.

The Traveling Salesman problem is not merely an idle curiosity (indeed, entire books have been devoted to it). The special case of finding a Hamiltonian cycle in a graph is representative of a large class of important problems (NP-complete) for which no (worst-case) polynomial algorithms are known, nor have super-polynomial lower bounds been found for any of the problems. It has been shown that if any NP-complete has a polynomial-time solution, then they all have polynomial-time solutions (see Chapter 20).

References and Suggestions for Further Reading

Bellman, R. E. *Dynamic Programming*. Princeton University Press, 1957. The first systematic study of dynamic programming.

Two other classic references on dynamic programming are:

Bellman, R. E. and S. E. Dreyfus. *Applied Dynamic Programming*. Princeton University Press, 1962.

Nemhauser, G. *Introduction to Dynamic Programming*. New York: Wiley, 1966.

Gibbons, A. and W. Rytter. *Efficient Parallel Algorithms*. Cambridge, England: Cambridge University Press, 1988. The pebble game and its application to solving a general dynamic programming recurrence can be found in this text.

Lawler, E. L., J. K. Lenstra, A. H. G. Rinnooy Kan, and D. B. Shmoys. *The Traveling Salesman Problem: A Guided Tour of Combinatorial Optimization*. New York: Wiley, 1985. A book entirely devoted to the Traveling Salesman problem.

Exercises

Section 14.1 Optimization Problems and the Principle of Optimality

14.1 Verify that the Principle of Optimality holds for shortest paths in digraphs by showing that if P is a shortest path from a to b, then for any vertex v of P, the subpath P_1 from a to v is a shortest path from a to v, and the subpath P_2 from v to b is a shortest path from v to b.

14.2 Does the Principle of Optimality hold for longest paths in digraphs? Discuss.

Section 14.2 Optimal Parenthesization for Computing a Chained Matrix Product

14.3 Given the matrix product $M_1M_2 \ldots M_n$ and a 2-tree T with n leaves, show that there is a unique parenthesization P such that $T = T(P)$.

14.4 Give pseudocode for *ParenthesizeRec* and analyze its complexity.

14.5 Show that the complexity of *OptimalParenthesization* is in $\Theta(n^3)$.

14.6 Using *OptimalParenthesization*, find an optimal parenthesization for the chained product of five matrices with dimensions 6×7, 7×8, 8×3, 3×10, 10×6.

Section 14.3 Optimal Binary Search Trees

14.7 Use dynamic programming to find an optimal binary search tree for the following probabilities, where we assume that the search key is in the search tree, that is, $q_i = 0$, $i = 0, \ldots, n$:

keys	i	1	2	3	4
probabilities	p_i	.4	.3	.2	.1

14.8 Use dynamic programming to find an optimal search tree for the following probabilities:

i	0	1	2	3	4	5
p_i		.2	.1	.2	.05	.05
q_i	.05	0	.25	0	.1	0

14.9 **a.** Design and analyze a recursive algorithm that computes an optimal binary search tree T for all the keys from the two-dimensional array *Root*[1:n,1:n] generated by *OptimalSearchTree*.

b. Show the action of your algorithm from part (a) for the instance given in Exercise 14.8.

14.10 **a.** Give a set of probabilities $p_1, \ldots, p_n$ (assume a successful search so that $q_0 = q_1 = \cdots = q_n = 0$), such that a completely right-skewed search tree T (the left child of every node is nil) is an optimal search tree with respect to these probabilities.

b. More generally, prove the following induction on n: If T is *any* given binary search tree with n nodes, then there exists a set of probabilities $p_1, \ldots, p_n$ such that T is the (unique) optimal binary search tree with respect to these probabilities.

Section 14.4 All-Pairs Shortest Paths

14.11 Show the action of Floyd's algorithm for the following graph.

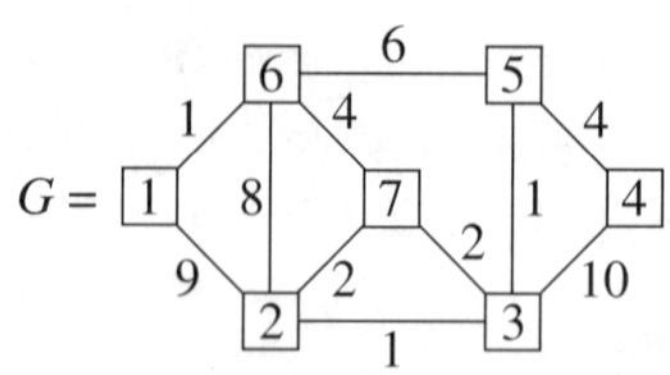

14.12 Given the following weighted digraph D, show the matrices $S_k[1{:}6,1{:}6]$, $P_k[1{:}6,1{:}6]$, $k = 1, \ldots, 6$ as computed by Floyd's algorithm. Describe how the matrix P_6 can be used, for example, to find the shortest path from vertex 2 to vertex 4.

$D =$ 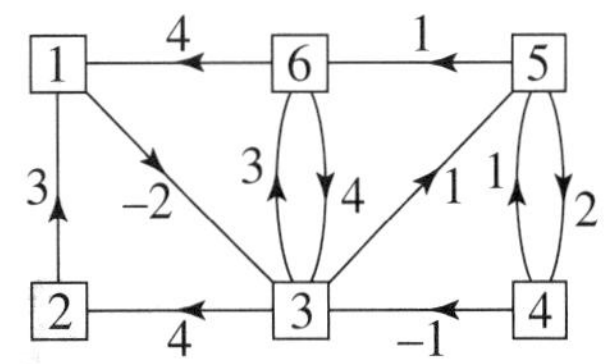

14.13 Give pseudocode for an algorithm that computes the shortest path from i to j from the matrix $P[1{:}n,1{:}n]$ generated by Floyd's algorithm.

14.14 Prove that Floyd's algorithm works even if some of the edges have negative costs, provided that there is no negative cost cycle.

14.15 Show how Floyd's algorithm can be used to obtain longest paths between every pair of vertices in a dag.

Section 14.5 The Traveling Salesman Problem

14.16 Show that if $P = i\, v_1 \ldots v_p\, 1$ is a shortest path from vertex i to vertex 1 whose interior vertices consist of the set $U = \{v_1, \ldots, v_p\}$, then the path $Q = v_1 \ldots v_p\, 1$ is a shortest path from vertex v_1 to vertex 1 whose interior vertices consist of the set $U - \{v_1\}$, so that the Specialized Principle of Optimality holds.

14.17 Show the action of procedure *TravelingSalesman* for the cost matrix

$$C = \begin{bmatrix} 0 & \infty & 9 & 3 \\ 5 & 0 & 6 & 2 \\ 9 & 6 & 0 & 7 \\ 7 & 3 & 5 & 0 \end{bmatrix}$$

Section 14.6 Dynamic Programming in Parallel

14.18 Give pseudocode for the procedure *AllPairsShortestPathsCREW.*

14.19 Adapt the procedure *AllPairsShortestPathsCREW* of Exercise 14.18 so that it is suitable for the EREW PRAM.

14.20 The *transitive closure of a digraph D* is the matrix $B = (b_{ij})$, such that $b_{ij} = 1$, if there is a directed path from vertex i to vertex j, or 0, otherwise. Let $A^{(q)} = (a_{ij}^{(q)})$ denote the matrix, such that $a_{ij}^{(q)} = 1$ if there exists a path of length at most q from i to j, or 0 otherwise.

a. Show that $B = A^{(n-1)}$.

b. Show that $A^{(q)}$ satisfies the recurrence relation:

$$a_{ij}^{(q)} = \bigwedge_{1 \le k \le n} (a_{ij}^{(q-1)} \vee a_{kj}^{(1)}),$$

where the binary operations $\vee$ and $\wedge$ are defined as follows:

$\wedge$	0	1
0	0	0
1	0	1

$\vee$	0	1
0	0	1
1	1	1

c. Using the results of parts (a) and (b), design and analyze an algorithm *TransClosureCREW* on the CREW PRAM with n^3 processors for computing the transitive closure of a digraph D.

14.21 Paths with repeated edges allowed are called *walks*. Let $A = (a_{ij})_{n \times n}$ be the adjacency matrix of a digraph D, and consider the qth-power matrix $A^q = (a_{ij}^{(q)})_{n \times n}$.

a. Prove by induction that $a_{ij}^{(q)}$ equals the number $w_{ij}^{(q)}$ of walks of length q from vertex i to vertex j in D.

b. Using the result in part (a), design and analyze an algorithm *NumWalksCREW* on the CREW PRAM with n^3 processors for computing the number $w_{ij}^{(q)}$.

***14.22** Design and analyze a polylogarithmic algorithm for solving the optimal binary search tree problem on a PRAM.

Supplementary Exercises

14.23 A *multistage digraph* is a digraph $D = (V,E)$, where $V = \{1, \ldots, n\}$ is partitioned into $k > 2$ disjoint sets $V_1, V_2, \ldots, V_k$. For each edge $e = uv$, if $u \in V_i$, then $v \in V_{k+1}$. Furthermore, the set V_1 consists of a single (source) vertex s, and V_k consists of a single (sink) vertex t. Suppose each edge has been assigned a cost. Let *Dist*(i,j) denote the length of a minimum cost (shortest) path from vertex $j \in V_i$ to t.

a. Obtain a recurrence relation for *Dist*(i,j).

b. Using the recurrence relation in part (a), design and analyze a dynamic programming algorithm for finding a minimum cost path from s to t.

14.24 Consider a sequence of n distinct integers. Design and analyze a dynamic programming algorithm to find the length of the longest increasing subsequence. For example, consider the sequence:

45 23 9 3 99 108 76 12 77 16 18 4

The longest increasing subsequence is 3 12 16 18, having length 4.

14.25 Given n integers, the partition problem is to find a bipartition of the integers into two subsets having the same sum or determine that no such bipartition exists. Design and analyze a dynamic programming algorithm for solving the partition problem.

***14.26** Design and analyze a dynamic programming solution to the coin-changing problem.

***14.27** Design a dynamic programming solution to the 0/1 Knapsack problem.

15

Backtracking and Branch-and-Bound

Suppose you find yourself entering a maze of garden hedges as shown in Figure 15.1. You want to follow some strategy that is guaranteed to lead you to the exit of the maze (or to determine that there is no exit). How can this be done? Simply walk along the trail with your left hand always touching the hedge. If you hit a dead-end, turn around and backtrack, still keeping your left hand touching the hedge. This left-hand backtracking strategy even works blindfolded! You will always find the exit, or return to the entrance if there is no exit.

Of course, by symmetry, an algorithm for generating a path through a maze can also be based on the right-hand backtracking strategy. Note that the right-hand backtracking strategy would generate a rather shorter path in Figure 15.1. In general, however, neither backtracking strategy generates the shortest path. (In fact, Figure 15.1 provides an example.) Thus, although backtracking always gets you through the maze, perhaps some good heuristics will yield a shorter path (see Figure 15.2).

Figure 15.1 Garden maze and path to exit generated by the left-hand backtracking algorithm

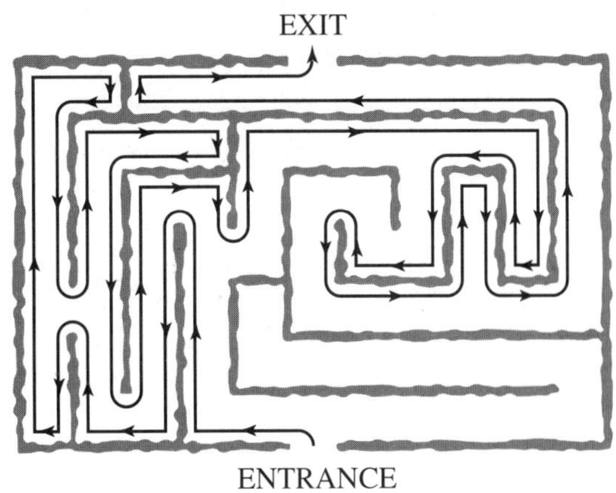

Figure 15.2 The left-hand backtracking algorithm versus heuristic search (JIM BORGMAN, © 1995 *The Cincinnati Enquirer.* Reprinted with permission. All rights reserved.)

We can associate an implicit (state space) T tree with the maze problem, where a node (state) in T corresponds to a *sequence of decisions* made that led you to a junction point in the maze. The children of a node correspond to the various branches you can take from that junction point, where we restrict the choice of branches to those not leading to a junction already visited (this avoids cycles, yielding a finite tree). The left-hand backtracking algorithm is based on performing a depth-first search of the state space tree.

15.1 State Space Trees

Backtracking and branch-and-bound design strategies are applicable to problems whose solutions can be expressed as sequences of decisions. Both backtracking and branch-and-bound are based on a search of an associated state space tree modeling all possible sequences of decisions. There may be several different ways to model decision sequences for a given problem, with each model leading to a different state space tree. We assume in our model for the decision-making process that the decision x_k at stage k must be drawn from a finite set S_k of choices. For each $k > 1$, the choices available for decision x_k may be limited by the choices that have already been made for $x_1, \ldots, x_{k-1}$. In other words, decision x_k may be restricted to a *strict* subset of S_k.

For a given problem instance, suppose n is the maximum number of decision stages that can occur. For $k \le n$, we let P_k denote the set of all possible sequences of k decisions, represented by k-tuples $(x_1, x_2, \ldots, x_k)$. Elements of P_k are called *problem states*, and problem states that correspond to solutions to the problem are called *goal* states.

Given a problem state $(x_1, \ldots, x_{k-1}) \in P_{k-1}$, we let $D_k(x_1, \ldots, x_{k-1})$ denote the *decision set* consisting of the set of all possible choices for decision x_k. More precisely, for $(x_1, \ldots, x_{k-1}) \in P_{k-1}$,

$$D_k(x_1, \ldots, x_{k-1}) = \{x_k \in S_k \mid (x_1, \ldots, x_k) \in P_k\}. \tag{15.1.1}$$

Letting $\varnothing$ denote the null tuple (), note that $D_1(\varnothing)$ is the set of choices for x_1; that is, $D_1(\varnothing) = S_1$.

The decision sets $D_k(x_1, \ldots, x_{k-1})$, $k = 1, \ldots, n$, determine a decision tree T of depth n, called the *state space tree*. The nodes of T at level k, $0 \le k \le n$, are the problem states $(x_1, \ldots, x_k) \in P_k$ (P_0 consists of the null tuple). For $1 \le k < n$, the children of $(x_1, \ldots, x_{k-1})$ are the problem states $\{(x_1, \ldots, x_k) \mid x_k \in D_k(x_1, \ldots, x_{k-1})\}$.

15.1.1 An Example

Our first illustration of state space trees is for the *Sum of Subsets* problem. An input to the Sum of Subsets problem is a multiset $A = \{a_1, \ldots, a_n\}$ of n positive integers, together with a positive integer *Sum*. A solution to the Sum of Subsets problem is a subset of elements $a_{i_1}, \ldots, a_{i_k}$ of A, $i_1 < \cdots < i_k$, such that

$$a_{i_1} + \cdots + a_{i_k} = Sum. \tag{15.1.2}$$

The Sum of Subsets problem can be interpreted as the problem of making correct change, where a_i represents the denomination of the ith coin, $i = 1, \ldots, n$, and *Sum* represents the desired change. This differs from the version of the coin-changing problem discussed in Chapter 12, since now a limited number of coins of each denomination are available. For example, consider the multiset $A = \{25,1,1,1,5,10,1,10,25\}$. The denominations are 1, 5, 10, 25, which occur with multiplicities 4, 1, 2, 2, respectively.

There are two natural ways to model a decision sequence leading to a solution to the Sum of Subsets problem. In the first model, a problem state consists of choosing k elements $a_{i_1}, \ldots, a_{i_k}$ of A, $i_1 < \cdots < i_k$ in succession, for some $k \in \{1, \ldots, n\}$. The decision sequence can be represented by the k-tuple $(x_1, \ldots, x_k) = (i_1, \ldots, i_k)$, where x_j corresponds to the decision to choose element a_{i_j} at stage j, $1 \le j \le k$. For example, consider the instance $n = 5$, and suppose that we have decided to choose the second, fourth, and fifth elements. Then, $x_1 = 2$, $x_2 = 4$, and $x_3 = 5$, so that the problem state associated with this decision sequence is the 3-tuple (2, 4, 5).

Since a higher-indexed element is chosen at each stage, it follows that $x_k \ge k$, so that $S_k = \{k, \ldots, n\}$. Further, the set P_k of all problem states of size k, $0 \le k \le n$, is given by

$$P_k = \{(x_1, \ldots, x_k) \mid \{x_1, \ldots, x_k\} \subseteq \{1, \ldots, n\} \text{ and } x_1 < x_2 < \cdots < x_k\}, \tag{15.1.3}$$

where x_i corresponds to choosing the element a_{x_i}, $i = 1, \ldots, k$.

Given that problem state $(x_1, \ldots, x_{k-1})$ has occurred (that is, the decision has been made to choose elements $a_{x_1}, \ldots, a_{x_{k-1}}$), then the available choices for decision x_k are $a_{x_{k-1}+1}, \ldots, a_n$, yielding

$$D_k(x_1, \ldots, x_{k-1}) = \{x_{k-1} + 1, x_{k-1} + 2, \ldots, n\}, \quad 1 \le k \le n. \tag{15.1.4}$$

For example, suppose that problem state (1,3) has occurred. Then the only elements available for the third decision are a_4 and a_5, so that $D_3(1,3) = \{4,5\}$. Note that $S_3 = \{3,4,5\}$, so that $D_3(1,3)$ is a strict subset of S_3. Figure 15.3 illustrates the state space tree T determined by $D_k(x_1, \ldots, x_{k-1})$ for the Sum of Subsets problem, where $n = 5$. The goal states (nodes) are not determined until a particular instance of the problem is specified. For example, for the instance $A = \{1,4,5,10,4\}$ and $Sum = 9$, (1,2,5), (2,3), and (3,5) are the goal states. On the other hand, for the same set A, if $Sum = 10$, then (1,2,3), (4) and (1,3,5) are the goal states. State space trees like this where the size of the goal states can vary for the same input size are called *variable-tuple* state space trees.

The second natural way to model the Sum of Subsets problem is an example of a *fixed-tuple* model, where goal states are always n-tuples. The decision at stage k is whether or not to choose element s_k, $1 \le k \le n$. Thus, $S_k = \{0,1\}$, where $x_k = 1$ if element a_k is chosen, and $x_k = 0$, otherwise. In the fixed-tuple model, the problem states P_k corresponding to making k decisions are k-tuples:

$$P_k = \{(x_1, \ldots, x_k) \mid x_1, \ldots, x_k \in \{0,1\}\}. \tag{15.1.5}$$

Figure 15.3 Variable-tuple state space tree T modeling the decision set D given by (15.1.4) for the Sum of Subsets problem with $n = 5$. Edges labeled with index of the chosen element. Index values of tuple shown outside some sample nodes.

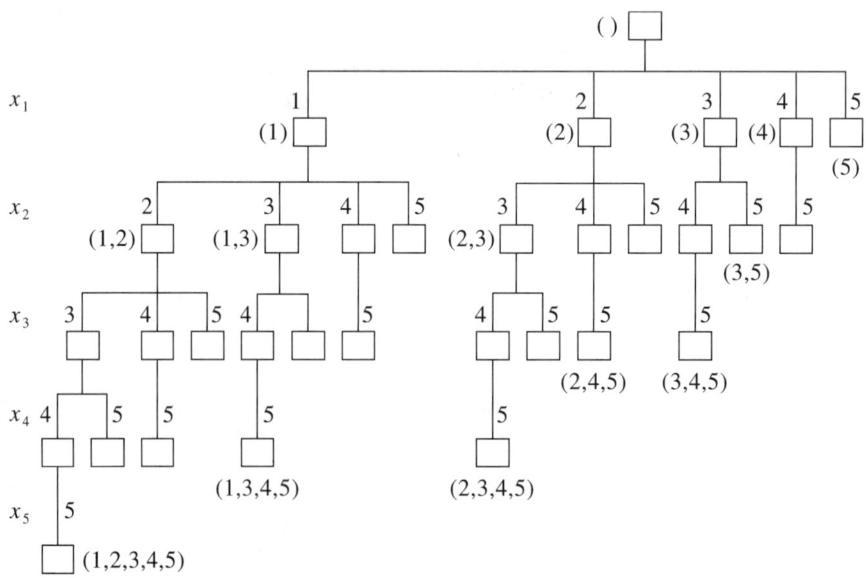

The decision at stage k is independent of the previous decisions, so that

$$D_k(x_1, \ldots, x_{k-1}) = \{0,1\}, \quad 1 \le k \le n. \tag{15.1.6}$$

Thus, the state space tree T associated with the decision sets $D_k(x_1, \ldots, x_{k-1})$ is the full binary tree on $2^{n+1} - 1$ nodes, with a left child of a node at level $k - 1$ corresponding to choosing a_k ($x_k = 1$) and a right child corresponding to omitting a_k ($x_k = 0$). Figure 15.4 shows the fixed-tuple state space tree for the same Sum of Subsets problem illustrated in Figure 15.3. For the same instance $A = \{1,4,5,10,4\}$ and $Sum = 9$ considered earlier, the goal states are now represented by the 5-tuples (1,1,0,0,1), (0,1,1,0,0) and (0,0,1,0,1).

15.1.2 Searching State Space Trees

In a general state space tree, if there are m_k choices available for x_k, then $S = S_1 \times \cdots \times S_n$ contains $m_1 m_2 \ldots m_n$ elements. Even though $D_k(x_1, \ldots, x_n)$ is often a strict subset of S_k, the state space tree for most problems is large (exponential or worse in the input size). Hence, algorithms based on a brute-force traversal of the entire state space tree (searching for goal states) are extremely inefficient. However, we can often determine that there is no goal node in the subtree rooted at a given node X in

Figure 15.4 Fixed-tuple state space tree T modeling the decision sets D_k given by (15.1.4) for the Sum of Subsets problem with $n = 5$. Edges ending at level k are labeled 0 or 1, depending on whether s_k is chosen or not. 0/1 values of tuple shown for sample leaf nodes.

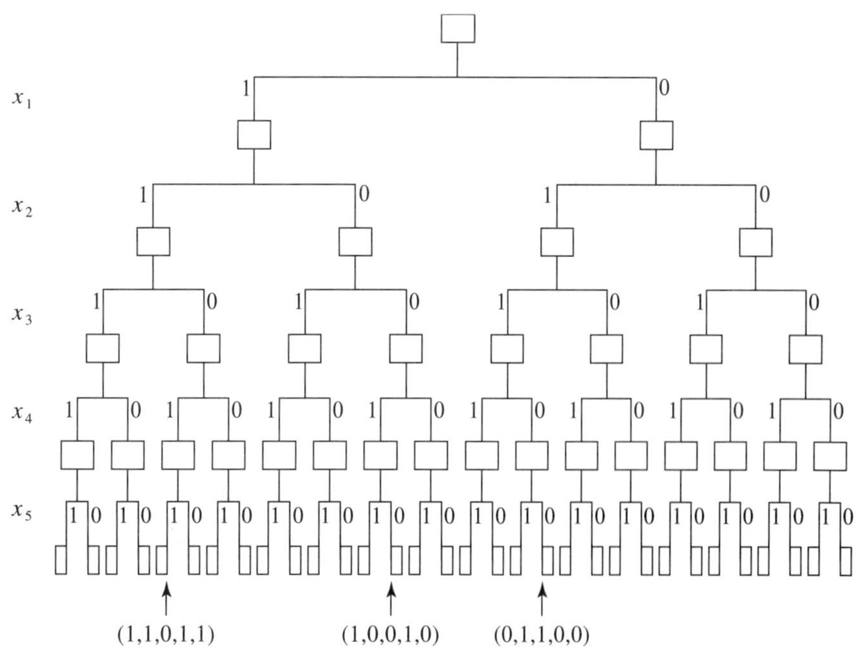

the state space tree. In this case, we say that X is *bounded*, and we can prune the state space tree by eliminating the descendants of the node X. Thus, when searching state space trees, we look for good bounding functions *Bounded*, where *Bounded* is a Boolean function such that if *Bounded*(X) is **.true.**, then there is no descendant of X that is a goal node.

In this chapter we discuss two general-purpose design strategies based on searching the state space tree associated with a given problem: *backtracking* and *branch-and-bound*. The state space tree is usually implicit to backtracking algorithms, whereas branch-and-bound algorithms usually require that the state space tree be explicitly implemented. Both backtracking and branch-and-bound utilize bounding functions to limit the number of nodes of the state space tree that need to be examined.

Backtracking is based on a depth-first search of the state space tree. When a node is accessed during a backtracking search, it becomes the current node being expanded (*E-node*), but immediately its first child not yet visited becomes the new E-node. On the other hand, branch-and-bound algorithms are based on searches of the state space tree that generate all the children of the E-node when the node is first accessed. Thus, a node can be the E-node many times during a backtracking search, but is the E-node at most once during a branch-and-bound algorithm. There are various versions of branch-and-bound, differing only in the manner in which the next E-node is chosen. For example, the children might be placed on a queue (FIFO branch-and-bound), a stack (LIFO branch-and-bound), or a priority queue (least-cost branch-and-bound).

Unless good bounding functions can be found, backtracking and branch-and-bound tend to be inefficient in the worst case. However, they can be applied in a wider variety of settings than the other major design strategies that we have discussed.

15.2 Backtracking

Before stating the general backtracking design strategy, we illustrate the method by applying it to the Sum of Subsets problem discussed in the previous section.

15.2.1 A Backtracking Algorithm for the Sum of Subsets Problem

Initially, we will not assume that the set $A = \{a_1, \ldots, a_n\}$ is ordered. (Later, we see that by sorting A in nondecreasing order, we can obtain an improved bounding function.) We use the decision sequence formulation corresponding to the variable-tuple state space tree, so that the decision sets $D_k(x_1, \ldots, x_{k-1})$ are given by (15.1.1).

To motivate the definition of a bounding function for the problem states, we consider the instance of the Sum of Subsets problem where $n = 5$, $A = \{1,4,5,10,4\}$, and $Sum = 9$. The state space tree for this instance, and the three goal states $(x_1,x_2,x_3) = (1,2,5)$, $(x_1,x_2) = (2,3)$, and $(x_1,x_2) = (3,5)$ are shown in Figure 15.5.

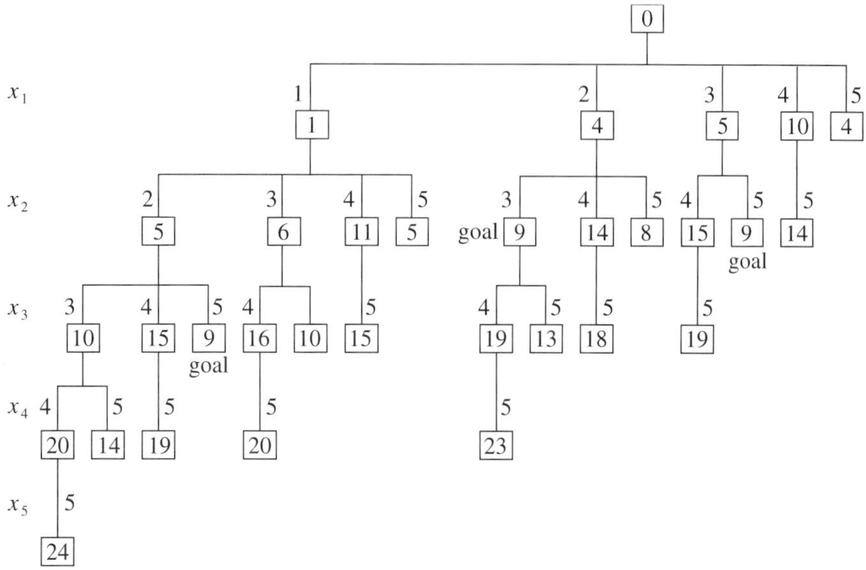

Figure 15.5 Variable-tuple state space tree for Sum of Subsets problem with $A = \{1,4,5,10,4\}$ and $Sum = 9$. The value of the sum of the elements chosen is shown inside each node. Edges are labeled with the index of the chosen element.

For example, consider the problem state (1,2,3) in the state space tree in Figure 15.5 corresponding to the choice of elements a_1,a_2,a_3. Note that $a_1 + a_2 + a_3 = 10 > Sum = 9$. Furthermore, any extension of (1,2,3) corresponds to a set of elements whose sum is even greater than 10. Thus, there is no path in the state space tree from (1,2,3) to a goal state. In other words, "*You can't get there from here!*" (See Figure 15.6, page 522.) We *bound* node (1,2,3), since there is no need to examine any of its descendants.

For the general problem state $(x_1, \ldots, x_k) \in P_k$, we define the bounding function $Bounded(x_1, \ldots, x_k)$ by

$$Bounded(x_1, \ldots, x_k) = \begin{cases} \textbf{.true.} & \text{if } s_{x_1} + \cdots + s_{x_k} \geq Sum, \\ \textbf{.false.} & \text{otherwise.} \end{cases} \tag{15.2.1}$$

Clearly, if the elements corresponding to $(x_1, \ldots, x_k)$ have a sum greater than or equal to *Sum*, then any extension of $(x_1, \ldots, x_k)$ corresponds to a set of elements whose sum is strictly greater than *Sum*. Thus, if $Bounded(x_1, \ldots, x_k) =$ **.true.**, then no descendant of $(x_1, \ldots, x_k)$ can be a goal state. For the problem instance $A = (1,4,5,10,4)$, Figure 15.7 shows the state space tree T (from Figure 15.5) after it has been pruned at all the nodes bounded by (15.2.1). The bounded nodes are labeled B, except for those bounded nodes that are also goal nodes, which are so labeled. The unlabeled leaves correspond to nodes that were leaves in the original state space T, before pruning.

Figure 15.6 A bounded node

The *backtracking* strategy performs a depth-first search of the state space tree T, utilizing an appropriate bounding function. By convention, when moving from an E-node to the next level of the state space tree, we select the left-most child not already visited. If no such child exists, or if the E-node is bounded, then we backtrack to the previous level. If only one solution to the problem is desired, then the backtracking

Figure 15.7 Pruned state space tree for Sum of Subsets $A = \{1,4,5,10,4\}$ and $Sum = 9$. Edges labeled with index of element chosen. Bounded nodes labeled by B.

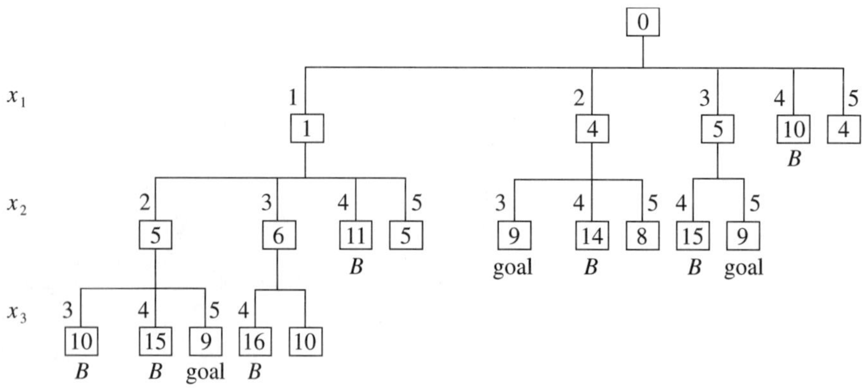

algorithm terminates once a goal state is found. Otherwise, the algorithm continues until all the nodes have been exhausted, outputting each goal state when it is reached.

We now give pseudocode for nonrecursive and recursive backtracking procedures for solving the Sum of Subsets problem. These procedures perform a depth-first search of the variable-tuple state space tree *T*, utilizing the bounding function given in (15.2.1). The procedures output all goal nodes but can be trivially modified to terminate once the first goal state is found.

```
procedure SumOfSubsets(A[1:n],Sum,X[1:n])
Input:  A[1:n] (an array of positive integers)
        Sum (a positive integer)
        X[0:n] (an array of positive integers)
Output: print all goal states; that is, print all (X[1], . . . , X[k]) such that
        A[X[1]]+ · · · +A[X[k]] = Sum

  for i := 0 to n do
     X[i] := 0
  endfor
  k := 1
  while k ≥ 1 do                          {E-node is (X[1], . . . , X[k – 1]). Initially
                                           E-node = ( ) corresponding to root}
     Searching := .true.
     while Searching do                   {searching for unbounded child of E-node}
        if X[k] = 0 then
           X[k] = X[k – 1] + 1            {visit first child of E-node}
        else
           X[k] := X[k] + 1               {visit next child of E-node}
        endif
        if X[k] > n then                  {no more children of E-node}
           Searching := .false.
        else
           if A[X[1]]+ · · · +A[X[k]] ≥ Sum then     {(X[1], . . . , X[k]) is bounded}
              if A[X[1]]+ · · · +A[X[k]] = Sum then
                 call Print(X[1], . . . , X[k])      {print goal state}
              endif
           else                                      {(X[1], . . . , X[k]) is not bounded}
              Searching := .false.
           endif
        endif
     endwhile
     if X[k] > n then
        X[k] = 0                          {backtrack to previous level; all children
        k := k – 1                         of E-node have already been visited}
```
Continued

```
      else
         k := k + 1                    {move on to next level}
      endif
   endwhile
end SumOfSubsets
```

The following recursive backtracking algorithm *SumOfSubsetsRec*(k) for the Sum of Subsets problem is called initially with $k = 0$. We assume that $A[1{:}n]$, *Sum*, and $X[1{:}n]$ are global variables. When calling *SumOfSubsetsRec* with input parameter k, it is assumed that $X[1], \ldots, X[k]$ have already been assigned values, so that $k = 0$ on the initial call.

```
procedure SumOfSubsetsRec(k) recursive
Input:  k  (a nonnegative integer, 0 on initial call)
        A[1:n] (global array of positive integers)
        Sum (global positive integer)
        X[1:n] (global array, and where X[1], . . . , X[k] have been assigned)
Output: print all descendant goal states of (X[1], . . . , X[k]); that is,
        print all (X[1], . . . , X[k],X[k+1], . . . , X[q]) such that
        A[X[1]] + ··· + A[X[k]] + A[X[k+1]] + ··· + A[X[q]] = Sum
   k := k + 1                                   {move on to next level}
   for Child := X[k – 1] + 1 to n do
      X[k] := Child
      if A[X[1]]+ ··· +A[X[k]] ≥ Sum then       {(X[1], . . . , X[k]) is bounded}
         if A[X[1]]+ ··· +A[X[k]] = Sum then
            call Print(X[1], . . . , X[k])      {print goal state}
         endif
      else                                      {(X[1], . . . , X[k]) is not bounded}
         call SumOfSubsetsRec(k)
      endif
   endfor
end SumOfSubsetsRec
```

If the elements $a_1, \ldots, a_n$ are first sorted in nondecreasing order (the reverse of the ordering used by the greedy algorithm for the Coin-Changing problem), then the following bounding function can be used for the problem states $(x_1, \ldots, x_k)$. The bounding function (15.2.2) is stronger than that given by (15.2.1).

$$Bounded(x_1, \ldots, x_k) = \begin{cases} \textbf{.true.} & \text{if } a_{x_1} + \cdots + a_{x_k} + a_{x_k+1} > Sum, \\ \textbf{.false.} & \text{otherwise.} \end{cases} \tag{15.2.2}$$

Since $a_1 \le a_2 \le \cdots \le a_n$, whenever the elements corresponding to problem state $(x_1, \ldots, x_k, x_k + 1)$ have a sum strictly greater than *Sum*, then the elements corresponding to any problem state $(x_1, \ldots, x_k, x_{k+1})$ also have a sum strictly greater than *Sum*. Thus, (15.2.2) is a valid bounding function. *SumOfSubsets* and *SumOfSubsetsRec* can be easily modified to utilize the bounding function given in 15.2.2.

15.2.2 The General Backtracking Paradigm

The following general backtracking paradigm, *Backtrack*, follows the backtracking strategy we just described for the Sum of Subsets problem. *Backtrack* finds all solutions to a given problem by searching for all goal states in a state space tree associated with the problem. *Backtrack* invokes a bounding function *Bounded* for the problem states. The definition of *Bounded* depends on the particular problem being solved. We assume that an implicit ordering exists for the elements of $D_k(x_1, \ldots, x_{k-1})$.

```
procedure Backtrack
Input:  implicit state space tree T associated with the given problem and
        decision set D_k, where D_k = ∅ for k ≥ n
        bounding function Bounded
Output: all goal states
  k := 1
  while k ≥ 1 do                {E-node is (X[1], . . . , X[k – 1]). Initially
                                  E-node = ( ) corresponding to root.}
    Searching := .true.
    while Searching do          {searching for unbounded child}
      X[k] := first of the remaining untried values from D_k(X[1], . . . , X[k – 1]),
              where this value is ∅ if all values in D_k(X[1], . . . , X[k – 1]) have
              been tried
      if X[k] := ∅ then
        Searching := .false.
      else
        if (X[1], . . . , X[k]) is a goal state then
          call Print(X[1], . . . , X[k])
        endif
        if .not. Bounded(X[1], . . . , X[k]) then
          Searching := .false.
        endif
      endif
    endwhile
    if X[k] = ∅ then
      k := k – 1                {backtrack to previous level; all children
                                  of E-node have already been visited}
```

Continued

```
      else
         k := k + 1                          {move on to next level}
      endif
   endwhile
end Backtrack
```

The procedure *BacktrackRec* is the recursive version of the procedure *Backtrack*. Since we are essentially performing a depth-first search of the state space tree T starting at the root, *BacktrackRec*(k) is initially called with $k = 0$. Note how elegantly the recursion implements the backtracking process. We assume that $D_k(X[1], \ldots, X[k])$ is empty for $k \geq n$.

```
procedure BacktrackRec(k)
Input:  k (a nonnegative integer, 0 on initial call)
        implicit state space tree T associated with the given problem and
        decision set D_k, where D_k = ∅ for k ≥ n
        explicit global array X[1:n] maintaining problem states of T, where
        X[1], . . . , X[k] are assumed already defined
        bounding function Bound
Output: all goals that are descendants of (X[1], . . . , X[k])
   k := k + 1
   for each x_k ∈ D_k(X[1], . . . , X[k − 1]) do
      X[k] := x_k
      if (X[1], . . . X[k]) is a goal state then
         call Print(X[1], . . . , X[k])
      endif
      if .not. Bounded(X[1], . . . , X[k]) then
         call BacktrackRec(k)
      endif
   endfor
end BacktrackRec
```

Remarks

1. Often, computing only one goal state is required. The procedures *Backtrack* and *BacktrackRec* can be easily modified to halt once the first goal state is reached.
2. Note that the pseudocode for procedures *SumOfSubsets* and *SumOfSubsetsRec* vary slightly from the paradigms *Backtrack* and *BacktrackRec*, in that they only check for a goal state after determining that a problem state is bounded. Putting off this check for a goal state applies to any problem where all the goal states are bounded.

15.2.3 Tic-Tac-Toe

Consider the problem of finding a tie board (result of a "cat's game") in the familiar game of tic-tac-toe. Here we are trying to determine whether tie games are possible, not to devise a strategy for playing the game. (In Section 16.4 we consider the problem of designing strategies for various perfect-information games such as tic-tac-toe.)

The game of tic-tac-toe involves two players A and B, who alternately place Xs and Os into unoccupied positions on a board such as the one shown in Figure 15.8. We assume that player A starts by placing an X in any position, and then player B places an O in any remaining position. The two players continue to alternately place Xs and Os on the board until either there are three Xs in a row (horizontally, vertically, or diagonally) so that A wins, there are three Os in a row so that B wins, or all nine positions are occupied with no three Os or three Xs in a row so that the game is tie (a cat's game).

We use backtracking to find a tie board, that is, a board corresponding to a tie game (see Figure 15.8). In fact, we solve the problem for an $n \times n$ board, where a *tie board* contains no "three in row" of either Xs or Os in any 3×3 sub-board of contiguous positions in the $n \times n$ board. Note that we have not assumed in our definition of a tie board that the number of Xs and Os differ by at most one. However, it is interesting that all tie boards in the $n \times n$ board have this property, so we don't need to check for it in our backtracking algorithm.

We refer to the cell in row i and column j of the $n \times n$ board as cell (i,j), $i,j \in \{1, \ldots, n\}$. We also refer to cell (i,j) as the cell labeled k, where $k = n(i-1) + j$, that is, k is a *row-major* labeling (see Figure 15.9). Row-major labeling of the cells is useful in defining the problem states P_k. However, when defining the bounding function, it is more convenient to use the row-column labeling. The problem of finding a board filled with Xs and Os, containing no three in a row in either Xs or Os, can be expressed as a sequence of decisions where the decision at stage k is whether to place an X or O in the cell labeled k. We set $x_k = 1$ if an X is placed in cell k in the kth stage, and $x_k = 0$, otherwise. Thus, $x_k \in S_k = \{0,1\}$, and the problem states of size k, $1 \le k \le n$, are given by

$$P_k = \{(x_1, \ldots, x_k) \mid x_1, \ldots, x_k \in \{0,1\}\}. \qquad (15.2.3)$$

Figure 15.8 A 3×3 tic-tac-toe tie board

X	X	O
O	O	X
X	X	O

Since the choice for x_k, $1 \le k \le n$, is independent of the previous $k - 1$ choices for $x_1, \ldots, x_{k-1}$, we have

$$D_k(x_1, \ldots, x_{k-1}) = \{0,1\}, \quad 1 \le k \le n. \tag{15.2.4}$$

Note that the definition of $D_k(x_1, \ldots, x_{k-1})$ given by (15.2.4) is identical to the definition for $D_k(x_1, \ldots, x_{k-1})$ given by (15.1.6) for the Sum of Subsets problem. Thus, the (fixed-tuple) state space tree T for tic-tac-toe is also identical, namely, T is the full binary tree on $2^{n+1} - 1$ nodes.

Figure 15.9*c* shows the 3×3 board configuration corresponding to the 7-tuple (1,1,0,1,0,0,1).

An obvious bounding function for tic-tac-toe is given by

$$Bounded(x_1, \ldots, x_k) = \begin{cases} \textbf{.true.} & \text{if board configuration corresponding to } (x_1,\ldots,x_k) \text{ contains 3 in a row,} \\ \textbf{.false.} & \text{otherwise.} \end{cases} \tag{15.2.5}$$

Note that *Bounded* = **.true.** for the problem state (1,1,0,1,0,0,1) in Figure 15.9. If the board configuration corresponding to the problem state $(x_1, \ldots, x_k)$ already contains three in a row (in either Xs or Os), then it obviously cannot be extended to a board configuration not containing three in a row. In particular, it cannot be extended to a goal state.

Figure 15.9 (a) Row-column labeling (i,j); (b) row-major labeling k; and (c) a 3×3 board configuration corresponding to the 7-tuple (1,1,0,1,0,0,1)

(1,1)	(1,2)	(1,3)
(2,1)	(2,2)	(2,3)
(3,1)	(3,2)	(3,3)

(a)

1	2	3
4	5	6
7	8	9

(b)

X	X	O
X	O	O
X		

(c)

Before giving pseudocode for the algorithm *TicTacToe* solving the problem of finding all $n \times n$ generalized tie board configurations, we give pseudocode for the Boolean function *BoundedBoard* based on (15.2.5). The board configuration is represented by the two-dimensional array $B[-1{:}n+2, -1{:}n+2]$, where, for convenient implementation of *BoundedBoard*, we assume the existence of "border" rows and columns indexed by $-1,0,n+1,n+2$. The cells corresponding to these rows and columns are always empty; that is, $B[i,j]$ = 'E' if either $i \in \{-1,0,n+1,n+2\}$ or $j \in \{-1,0,n+1,n+2\}$. We illustrate a bordered 4×4 board in Figure 15.10, together with an assignment of Xs and Os to the positions $k = 1, \ldots, 10$.

We assume that the two-dimensional array $B[-1{:}n+2, -1{:}n+2]$ is global to the function *BoundedBoard*. We refer to a set of three adjacent cells along a horizontal, vertical, or diagonal line as a *winning line*. Suppose the board configuration restricted to the cells labeled $1,2, \ldots, k-1$ in the row-major labeling of the $4 \times 4 \times 4$ board contains no three in a row in either Xs or Os. Then, to determine whether the board configuration restricted to the cells labeled $1,2, \ldots, k$ contains three in a row in either Xs or Os, we merely need to check all winning lines containing the cell (labeled) k. Four winning lines need to be checked: one horizontal, one vertical, and two diagonal. In Figure 15.11 (page 530) we illustrate these lines for the cell in the board of Figure 15.10 whose row-major label is $k = 11$.

The following Boolean function *BoundedBoard*(i,j) returns the value **.true.** if, and only if, the board configuration corresponding to $B[-1{:}n+2, -1{:}n+2]$ contains all Xs or Os in one of the four winning lines described above.

Figure 15.10 A bordered 4×4 tic-tac-toe board with first 10 positions filled

E	E	E	E	E	E	E	E
E	E	E	E	E	E	E	E
E	E	X	X	O	X	E	E
E	E	X	X	O	X	E	E
E	E	O	O	E	E	E	E
E	E	E	E	E	E	E	E
E	E	E	E	E	E	E	E
E	E	E	E	E	E	E	E

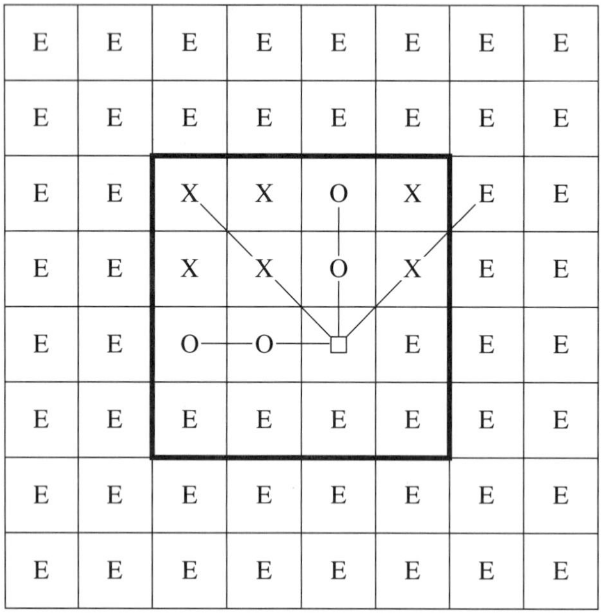

Figure 15.11 Winning lines that are checked for $k = 11$ in a $4 \times 4 \times 4$ tic-tac-toe board

```
function BoundedBoard(i,j)
Input:  B[-1:n+2, -1:n+2]  (global array corresponding to board configuration)
        i,j                (integers between 1 and n, inclusive)
Output: returns .true. if the board configuration involving the cells
        labeled 1, ..., k = n(i - 1) + j, contains three in a row in either
        Xs or Os along a line containing the cell labeled k.
   LineH := (B[i,j] = B[i,j - 1]) .and. (B[i,j] = B[i,j - 2])
   LineV := (B[i,j] = B[i - 1,j]) .and. (B[i,j] = B[i - 2,j])
   LineD1 := (B[i,j] = B[i - 1,j - 1]) .and. (B[i,j] = B[i - 2,j - 2])
   LineD2 := (B[i,j] = B[i - 1,j+1]) .and. (B[i,j] = B[i - 2,j+2])
   return(LineH .or. LineV .or. LineD1 .or. LineD2)
end BoundedBoard
```

We now give pseudocode for the algorithm *TicTacToe*. *TicTacToe* calls the procedures *Previous*(i,j) and *Next*(i,j), which accomplish the operations of backtracking to the previous cell ($k = k - 1$) and moving forward to the next cell ($k = k + 1$), respectively, in the row-major labeling. Thus, *Previous*(i,j) executes the statement

```
if j > 1 then
    j := j - 1
else
```

```
        i := i – 1
        j := n
    endif
```

and *Next(i,j)* executes the statement

```
    if j < n then
        j := j + 1
    else
        i := i + 1
        j := 1
    endif
```

```
procedure TicTacToe(n)

Input:   n (a positive integer representing size of board)
Output: all generalized tie board configurations; that is, all board
        configurations not containing three in a row in either Xs or Os

  for i := – 1 to n + 2 do
    for j := – 1 to n + 2 do
      B[i,j] := 'E';                     {initialize all positions on board to empty}
    endfor
  endfor
  i := 1                                 {k = 1}
  j := 1
  while i > 0 do
    if B[i,j] = 'O' then                 {backtrack: k = k – 1}
      B[i,j] := 'E'
      call Previous(i,j)
    else
      if B[i,j] = 'E' then
        B[i,j] := 'X'                    {visit left child of E-node}
      else
        B[i,j] := 'O'                    {visit right child of E-node}
      endif
      if .not. BoundedBoard(i,j) then
        if (i = n) .and. (j = n) then
          call PrintBoard(Board[1:n,1:n])   {print goal state}
        else
          call Next(i,j)
        endif
      endif
    endif
  endwhile

end TicTacToe
```

A recursive version, *TicTacToeRec*(i,j), of *TicTacToe* can be written as follows. *TicTacToeRec*(i,j) is initially called with $i = 0$ and $j = n$ $(k = 0)$. The augmented board *Board* is initialized to 'E'.

```
procedure TicTacToeRec(i,j) recursive
Input:  i,j                        (integers between 1 and n, inclusive, called
                                    initially with i = 0 and j = n)
        B[– 1:n + 2, – 1:n + 2]    (global array corresponding to board configuration,
                                    initialized to 'E', and B[1,1], . . . , B[i,j] filled
                                    with Xs and Os with no three in a row)
Output: all extensions of B[1,1], . . . , B[i,j] to goal states; that is, board
        configurations not containing three in a row in either Xs or Os

  call Next(i,j)   {k = k + 1}
  for Child := 1 to 2 do
     if Child = 1 then
        B[i,j] := 'X'
     else
        B[i,j] := 'O'
     endif
     if .not. BoundedBoard(i,j) then
        if (i = n) .and. (j = n) then
           call PrintBoard(Board[1:n,1:n])   {print goal state}
        else
           call TicTacToeRec(i,j)
        endif
     endif
  endfor
end TicTacToeRec
```

15.2.4 Graph Coloring

We now consider the problem of finding a coloring of the vertices of a graph G using c colors $(c \geq 2)$ in such a way that no two adjacent vertices have the same color. Such a coloring of G is called a *proper* vertex c-coloring. The problem of finding proper vertex c-colorings has a long history. In 1850 Guthrie asked whether or not the countries in any map drawn in the plane can be colored with only four colors so that any two neighboring countries (sharing a common border consisting of more than isolated points) are colored differently. The map-coloring problem can be translated into the equivalent problem of finding a vertex 4-coloring for any planar graph. (Vertices in the graph correspond to countries, and two vertices are adjacent if, and only if, their corresponding countries are neighbors.) This problem, known as the Four Color Conjecture, was one of the most famous unsolved problems in mathematics for over a century!

In 1880, Kempe published a "proof" of the Four Color Conjecture, which went unchallenged for 10 years. In 1890, Heawood discovered an error in Kempe's argument. However, Heawood was able to use Kempe's argument to show that five colors was sufficient to properly color any planar graph. The Four Color Conjecture remained unresolved until 1976, when Appel and Haken announced a proof that combined mathematical arguments with exhaustive searching using a large computer.

In addition to being of great theoretical interest, the problem of finding a proper vertex c-coloring of a graph for a given c has many practical applications. For example, consider the problem of assigning channels to television transmitting stations. If two transmitting stations are close enough together (neighbors), then they cannot be assigned the same channel without interference. Let G be the graph whose vertices correspond to the transmitting stations, where two vertices are joined with an edge of G whenever the corresponding stations are neighbors. Clearly, if there are 13 channels available, then the problem of assigning channels to the stations without interference is equivalent to the problem of finding a proper vertex 13-coloring of G.

The following greedy-like algorithm finds a proper c-coloring when $c \geq \Delta(G) + 1$, where $\Delta(G)$ is the maximum degree of the graph. We simply visit the vertices of the graph in some sequence, coloring each new vertex using a color not already used by any of the adjacent vertices previously visited. Such a color is always available since there are at most $\Delta(G) < c$ vertices adjacent to any vertex.

To describe a canonical greedy coloring, suppose for definiteness that G has vertices $v_1, v_2, \ldots, v_n$ and that a greedy coloring is generated by visiting the vertices in this sequence. Moreover, suppose we use colors that belong to $\{1,2,\ldots,c\}$. We then specify a greedy coloring as follows: When coloring vertex v_j we choose the smallest integer (color) that has not been used for any of the vertices adjacent to v_j that have already been colored, $j = 1, \ldots, n$. In Figure 15.12, we show outside each vertex the four colors generated by this greedy coloring.

There is no known simple strategy like the greedy strategy for finding a proper c-coloring of a graph G for $c \geq 3$. Of course, one (exponential complexity) algorithm for finding a proper c-coloring of a given graph G on n vertices is to examine each of the c^n possible vertex colorings of G. We now describe a backtracking algorithm for this problem, which improves on the c^n worst-case complexity of the exhaustive search algorithm, but nevertheless remains super-polynomial in the worst case. A coloring of G is obtained by first coloring vertex v_1, then coloring vertex v_2, and so

Figure 15.12 The greedy coloring of G generated by the vertex sequence v_1,v_2,v_3,v_4,v_5

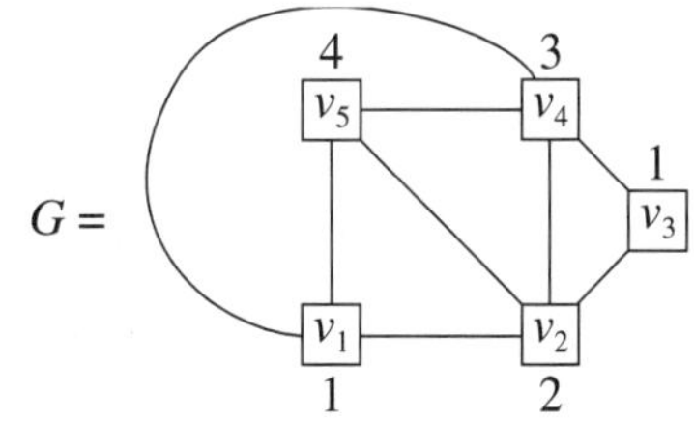

on. Thus, at stage k, we decide the color of vertex v_k from the colors $1,2, \ldots, c$. We indicate the decision to color v_k with color $i \in \{1, \ldots, c\}$ by setting $x_k = i$, so that $S_k = \{1, \ldots, c\}$. The set P_k of all problem states of size k is given by

$$P_k = \{(x_1, \ldots, x_k) \mid x_1, \ldots, x_k \in \{1, \ldots, c\}\}, \quad 0 \le k \le n. \tag{15.2.6}$$

The problem state $(x_1, \ldots, x_k)$ corresponds to the partial coloring of G where vertex v_i is colored x_i, $i = 1, \ldots, k$. Given $(x_1, \ldots, x_{k-1}) \in P_{k-1}$, the decision set $D_k(x_1, \ldots, x_{k-1})$ of possible choices for x_k is given by

$$D_k(x_1, \ldots, x_{k-1}) = \{1,2, \ldots, c\}, \quad 1 \le k \le n. \tag{15.2.7}$$

The state space tree T determined by (15.2.7) is the complete c-ary tree having depth c and c^n leaf nodes. The leaf nodes correspond to all possible vertex c-colorings of G. In particular, the goal states (if any) corresponding to proper vertex c-colorings are leaf nodes of T.

Clearly, if a problem state corresponds to a partial coloring of the vertices of G where two adjacent vertices are colored the same, then it cannot be extended to a proper vertex coloring of the entire graph G. Thus, we have the following bounding function for the problem states $(x_1, \ldots, x_k)$.

$$Bounded(x_1, \ldots, x_k) = \begin{cases} \textbf{.true.} & \text{if } i \in \{1,\ldots, k-1\} \text{ exists with } x_i = x_k \\ & \text{and } v_i v_k \in E(G) \\ \textbf{.false.} & \text{otherwise.} \end{cases} \tag{15.2.8}$$

Note that the bounding function given by (15.2.8) only checks whether $x_i = x_j$ for $j = k$. If $x_i = x_j$ for $j < k$, then problem state $(x_1, \ldots, x_j)$ is bounded by (15.2.8), so that problem state $(x_1, \ldots, x_k)$ would never have been reached by the backtracking algorithm.

Figure 15.13 illustrates the portion of the state space tree T that is generated by backtracking when finding a proper 3-coloring ($c = 3$) of a sample graph G, where the backtracking algorithm terminates once the first goal state is found.

15.2.5 Solving Optimization Problems Using Backtracking

Backtracking is frequently used to solve optimization problems, where we wish to optimize (maximize or minimize) an objective function. For example, consider the

Figure 15.13 Portion of state space tree T generated by backtracking for the problem of finding a single proper vertex 3-coloring of G

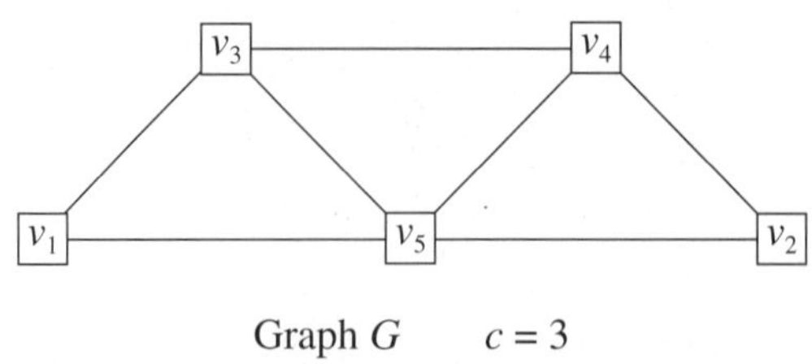

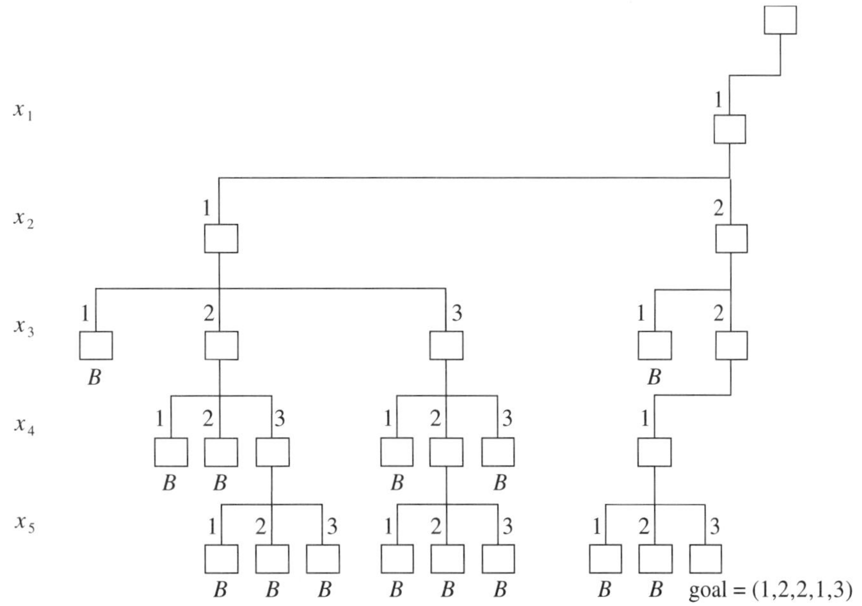

Figure 15.13 Continued

Sum of Subsets problem, interpreted as a coin-changing problem, where we now wish to make correct change using the fewest coins (the objective function is the number of coins). We will refer to this optimization problem as the *Coin-Changing* problem. The backtracking algorithm for the Sum of Subsets problem given earlier can be modified to solve the Coin-Changing problem by introducing a variable *UB* that dynamically maintains the current best *solution state*. Solution states for the Coin-Changing problem are the goal states for the Sum of Subsets problem (that is, they make correct change), and goal states for the Coin-Changing problem are solution states that use the fewest number of coins. Thus, *UB* is the fewest number of coins used by any solution state currently generated. The variable *UB* is initialized to infinity, since initially no solution state has been generated. Note that a problem state $(x_1, \ldots, x_k)$ corresponding to a choice of k coins can be bounded if k is not smaller than the current value of *UB*. This is called *dynamic bounding*, as opposed to the *static bounding* that occurs for problem states whose sum is not smaller than the correct change *Sum*.

procedure *CoinChangingRec*(*k*,*CurrentBest*) **recursive**

Input: k (a nonnegative integer, 0 on initial call)
$A[1{:}n]$ (global array of positive integers), *Sum* (global positive integer)
$X[0{:}n]$ (global array initialized to 0s)
Output: *CurrentBest* (solution state $(X[1], \ldots, X[k])$ such that k is minimum over all solution states)

Continued

```
dcl UB (global variable, initialized to ∞)
  k := k + 1                                  {move on to next level}
  for Child := X[k − 1] + 1 to n do
    X[k] := Child
    if A[X[1]]+ ··· +A[X[k]] ≥ Sum then {(X[1], . . . , X[k]) is statically bounded}
      if A[X[1]]+ ··· +A[X[k]] = Sum then      {(x[1], . . . , x[k]) is a
        if k < UB then                           solution state}
          UB := k
          CurrentBest := (X[1], . . . , X[k])
        endif
      endif
    else                         {(X[1], . . . , X[k]) is not statically bounded}
      if k < UB then             {(X[1], . . . , X[k]) is not dynamically bounded}
        call CoinChangingRec(k,CurrentBest)
      endif
    endif
  endfor
end CoinChangingRec
```

The portion of the state space tree generated by procedure *CoinChangingRec* is illustrated in Figure 15.14 for a sample set $A = \{2,3,5,2,2,6,3\}$ and $Sum = 9$.

Figure 15.14 Portion of state space tree generated by procedure *CoinChangingRec* for the set $A = \{2,3,5,2,2,6,3\}$ and $Sum = 9$. Nodes that are statically bounded ($A[X[1]] + \cdots + A[X[k]] \geq Sum$) are labeled by *SB*. Nodes that are not statically bounded but are dynamically bounded ($k \geq UB$) are labeled by *DB*. Those solution states resulting in updates to *UB* and *CurrentBest* are labeled with updated value. Upon termination, *CoinChanging* outputs the final value *CurrentBest* = (2,6).

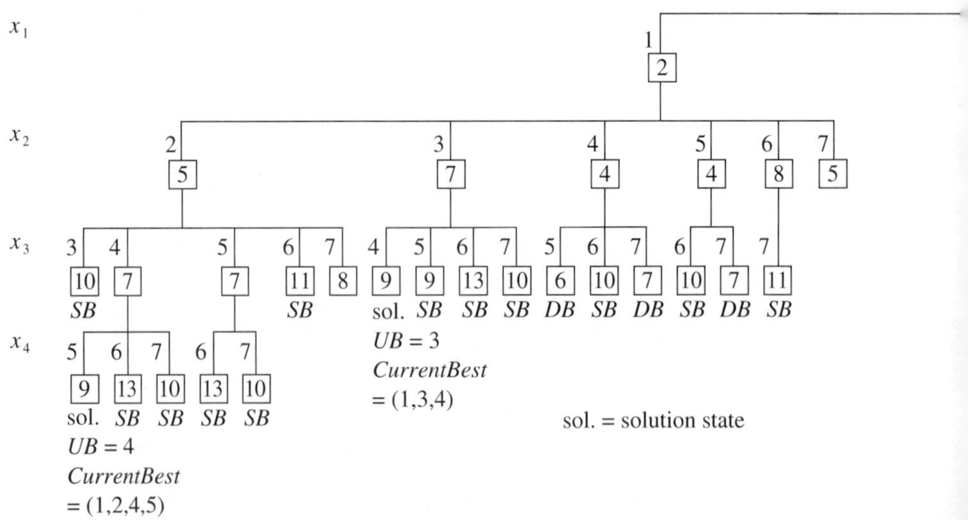

The ideas discussed in the previous examples can be applied to optimization problems in general, where we wish to optimize an objective function f over all *solution states*. A solution state is a problem state that satisfies the constraints of the particular input to the problem. It is important to differentiate between solution states and goal states. For example, in the Coin-Changing problem, only those solution states using the fewest number of coins are the goal states.

First consider the case where we wish to minimize the objective function f, and let f^* denote that minimum of f over all solution states. A solution state X such that $f(X)$ is minimum over all solution states is a goal state. Note that for *any* solution state $X = (x_1, \ldots, x_k)$, the value $f(X)$ is an upper bound for f^*. We maintain a variable *UB*, initialized to infinity. At each stage of the backtracking algorithm, a solution state *CurrentBest* is maintained such that *UB* = f(*CurrentBest*) is the minimum value of f over all solution states generated so far. For many problems, we can efficiently compute a function *LowerBound*$(x_1, \ldots, x_k)$ that is not larger than the value of f on any solution state belonging to the subtree of the state space tree rooted at $(x_1, \ldots, x_k)$. We can then dynamically bound a problem state $(x_1, \ldots, x_k)$ if *LowerBound*$(x_1, \ldots, x_k) \geq UB$.

The paradigm *BacktrackMin* for minimizing an objective function is based on the strategy just outlined. The following recursive procedure *BacktrackMin* is called initially with $k = 0$. During its resolution, *BacktrackMin* calls a function *StaticBounded*, which plays the same role as the function *Bounded* used for nonoptimization problems. For example, recall that in the Coin-Changing problem, a problem state was statically bounded if its sum was not smaller than *Sum*, whereas it was dynamically bounded if it used as many coins as a previously generated solution state. In general, a problem state is either bounded dynamically (*LowerBound*$(x_1, \ldots, x_k) \geq UB$), or bounded statically (*StaticBounded* = **.true.**).

Figure 15.14 Continued

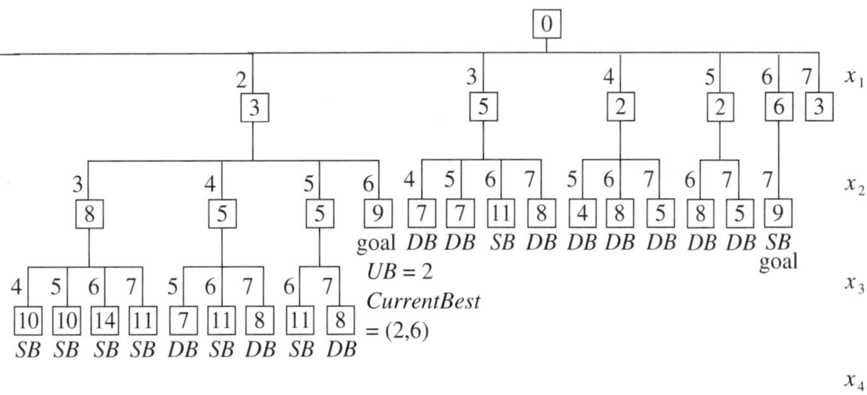

```
procedure BacktrackMin(k,CurrentBest) recursive
Input:  implicit state space tree T associated with the given problem, and
        decision function D_k, with D_k = ∅ for k ≥ n
        objective function f
        explicit global array X[1:n] maintaining problem states of T
        static bounding function StaticBounded
        a function LowerBound defined on the problem states
        k (a nonnegative integer, 0 on initial call)
        CurrentBest (solution state with the current minimum value of f)
Output: CurrentBest (upon returning to original call, CurrentBest is an
                     optimal goal state)

dcl UB (global variable, initialized to ∞)
  k := k + 1
  for each X[k] ∈ D_k(X[1], . . . , X[k − 1]) do
    if (X[1], . . . , X[k]) is a solution state then
      if f(X[1], . . . , X[k]) < UB then
        UB := f(X[1], . . . , X[k])
        CurrentBest := (X[1], . . . , X[k])
      endif
    endif
    if LowerBound(X[1], . . . , X[k]) < UB .and.
                        .not. StaticBounded(X[1], . . . , X[k]) then
      call BacktrackMin(k,CurrentBest,UB)
    endif
  endfor

end BacktrackMin
```

Remark The paradigm *BacktrackMin* ends up returning the first optimal goal state that was encountered (of course, unlike nonoptimization problems solved using backtracking, the algorithm has no way of checking that it was an optimal goal state until all goal states have been examined or eliminated). If all optimal goal states are desired, then *CurrentBest* must maintain *all* the goal states that have the current minimum value of *f*. For example, the sample set of coins shown in Figure 15.14 has two optimal goal states, represented by the tuples (2,6) and (6,7). Procedure *CoinChangingRec* only outputs the goal state (2,6).

The algorithm *BacktrackMin* is easily altered to apply to optimization problems where we wish to maximize the objection function f. For such problems, LB is the current maximum value of f, and $UpperBound(x_1, \ldots, x_k)$ is an upper bound for the maximum value of f over all solution states in the subtree of the state space tree rooted at $(x_1, \ldots, x_k)$.

Alternatively, a problem involving maximizing an objection function f can be canonically transformed into an equivalent problem of minimizing the associated objective function $g = M - f$, where M is a suitable constant. By replacing f by $M - f$ in a maximization problem, *BacktrackMin* can be applied directly to solve both minimization and maximization problems.

Our next example, the 0/1 Knapsack problem, illustrates the transformation of maximization problems into minimization problems. In Chapter 12, an efficient greedy algorithm was given for solving the Knapsack problem. The greedy method does not necessarily yield an optimal solution to the 0/1 Knapsack problem. In fact, there is no known worst-case polynomial algorithm for solving the 0/1 Knapsack problem. However, the greedy algorithm for the Knapsack problem helps us to define a useful function *LowerBound* for dynamic bounding in the transformed minimization problem.

Since the 0/1 Knapsack problem involves looking at subsets of a set of size n, we may solve the problem using backtracking by searching the same state space tree as in the Sum of Subsets problem. Consider the variable-tuple state space tree determined by (15.1.4) (see Figure 15.3). An obvious static bounding function for the problem state $(x_1, \ldots, x_k)$ is given by

$$StaticBounded(x_1, \ldots, x_k) = \begin{cases} \textbf{.true.} & \text{if } w_{x_1} + \cdots + w_{x_k} \geq C, \\ \textbf{.false.} & \text{otherwise.} \end{cases} \tag{15.2.9}$$

For $(x_1, \ldots, x_k)$ a problem state, let

$$\begin{aligned} Value(x_1, \ldots, x_k) &= \sum_{i=1}^{k} v_{x_i}. \\ Weight(x_1, \ldots, x_k) &= \sum_{i=1}^{k} w_{x_i}. \end{aligned} \tag{15.2.10}$$

For the 0/1 Knapsack problem, the solution states consist of all tuples not statically bounded, that is, all tuples $(x_1, \ldots, x_k)$ such that $Weight(x_1, \ldots, x_k) \leq C$. Goal states maximize the objective function *Value* over all solution states. The maximization problem is transformed into a minimization problem by letting

$$M = \sum_{i=1}^{n} v_i,$$

$$LeftOutVal(x_1, \ldots, x_k) = M - Value(x_1, \ldots, x_k) \tag{15.2.11}$$

In other words, $LeftOutVal(x_1, \ldots, x_k)$ is the total value of all the objects left out of the knapsack corresponding to solution state $(x_1, \ldots, x_k)$. In the transformed problem, which we now consider, the objective is to minimize the objective function *LeftOutVal*. To dynamically bound problem states, we maintain a variable *UB*, which at each stage of the backtracking algorithm keeps track of the minimum value of $LeftOutVal(x_1, \ldots, x_k)$ over all the solution states generated so far.

Consider a solution state $(x_1, \ldots, x_k)$ corresponding to the partial filling of the knapsack with the subset of objects $B_k = \{b_{x_1}, \ldots, b_{x_k}\}$. Suppose $(x_1, \ldots, x_k)$ is not bounded by (15.2.9), and let $C' = C - Weight(x_1, \ldots, x_k)$ denote the remaining capacity of the knapsack. Let $LeftOutVal^*(x_1, \ldots, x_k)$ denote the minimum value of *LeftOutVal* over all solution states in the state space subtree rooted at $(x_1, \ldots, x_k)$. In other words, $LeftOutVal^*(x_1, \ldots, x_k)$ is the smallest value of *LeftOutVal* that can be achieved by placing additional objects in the knapsack from the remaining set of objects $B' = B \setminus B_k$. Clearly, if $LeftOutVal^*(x_1, \ldots, x_k) \geq UB$, then $(x_1, \ldots, x_k)$ can be dynamically bounded. Unfortunately, there is no known efficient method to compute $LeftOutVal^*(x_1, \ldots, x_k)$. In fact, the general problem of computing $LeftOutVal^*(x_1, \ldots, x_k)$ is equivalent to the original 0/1 Knapsack problem.

Fortunately, we can efficiently compute a useful lower bound for $LeftOutVal^*(x_1, \ldots, x_k)$ by applying the greedy algorithm *Knapsack*. For *B* a given set of objects (with associated values and weights), and *C* a given capacity for the knapsack, we let *Greedy*(*C*,*B*) denote the value of the optimal placement of objects in the knapsack, where fractions of objects are permitted (that is, the value of the knapsack generated by *Knapsack*). We define $LowerBound(x_1, \ldots, x_k)$ by

$$LowerBound(x_1, \ldots, x_k) = LeftOutVal(x_1, \ldots, x_k) - Greedy(C',B'). \qquad (15.2.12)$$

Clearly, we have

$$LeftOutVal^*(x_1, \ldots, x_k) \geq LowerBound(x_1, \ldots, x_k). \qquad (15.2.13)$$

Thus, we can dynamically bound a problem state $(x_1, \ldots, x_k)$ if $LowerBound(x_1, \ldots, x_k) \geq UB$. Figure 15.15 shows the portion of the state space tree *T* generated by backtracking for a sample instance of the Knapsack problem.

15.3 Branch-and-Bound

Cowards die many times before their deaths
The valiant never taste of death but once

Julius Caesar, Act II, Scene II
W. Shakespeare

As with backtracking algorithms, branch-and-bound algorithms are based on searches of an associated state space tree for goal states. However, in a branch-and-bound algorithm, *all* the children of the *E*-node (the node currently being expanded) are generated before the next *E*-node is chosen. When the children are generated, they become *live* nodes and are stored in a suitable data structure *LiveNodes*. *LiveNodes* is typically a queue, a stack, or a priority queue. Branch-and-bound algorithms using the latter three data structures are called *FIFO* (*First-In, First-Out*) *branch-and-bound*, *LIFO* (*Last-In, First-Out*) *branch-and-bound*, and *least cost branch-and-bound*, respectively.

Immediately upon expanding the current *E*-node, this *E*-node becomes a *dead* node and a new *E*-node is selected from *LiveNodes*. Thus, branch-and-bound is quite different from backtracking, where we might backtrack to a given node many times, making it the *E*-node each time until all its children have finally been generated or the algorithm terminates. The nodes of the state space tree at any given point

i	1	2	3	4	5
v_i	16	6	5	12	4
w_i	4	2	2	6	5
v_i/w_i	4	3	2.5	2	.8

$, C = 11$
$M = 43$

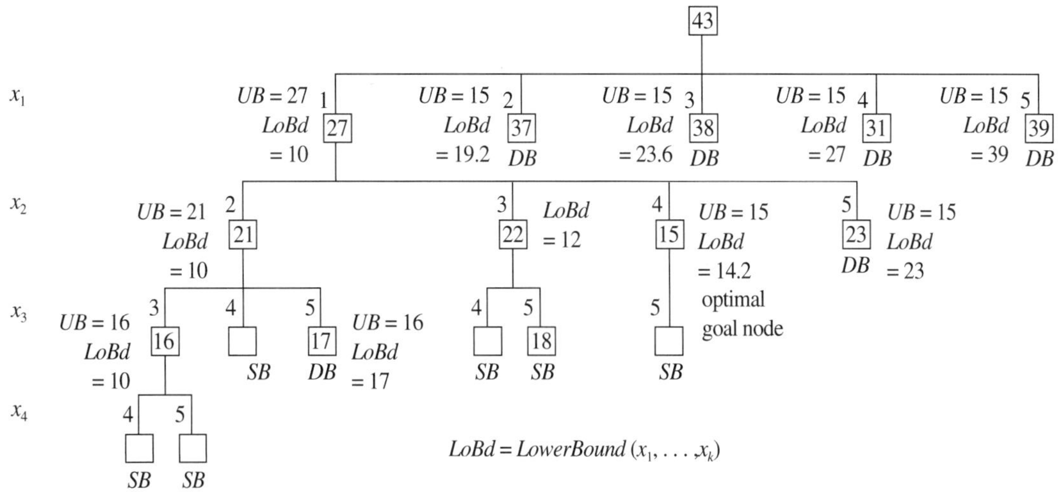

Figure 15.15 Portion of the variable-tuple state space tree T generated by the procedure *BackTrackMin* for a sample input to the 0/1 Knapsack problem. Problem states that are statically bounded ($Weight(x_1, \ldots, x_k) \geq C$) are labeled *SB*, whereas problem states that are not statically bounded but are dynamically bounded ($LowerBound(x_1, \ldots, x_k) \geq UB$) are labeled *DB*. $LeftOutValue(x_1, \ldots, x_k)$ is shown inside each solution state. $LowerBound(x_1, \ldots, x_k)$ and the current value of *UB* are shown outside each problem state where *UB* is updated, or where the problem state is dynamically bounded.

in a branch-and-bound algorithm are therefore in one of the following four states: *E-node*, *live node*, *dead node*, or *not yet generated.*

As with backtracking, the efficiency of branch-and-bound depends on the utilization of good bounding functions. Such functions are used in the attempt to determine solutions by restricting attention to small portions of the entire state space tree. When expanding a given *E*-node, a child can be bounded if it can be shown that it cannot lead to a goal node.

We illustrate branch-and-bound by revisiting the Sum of Subsets problem, where the data structure *LiveNodes* is a queue. Such a branch-and-bound, called *FIFO branch-and-bound*, involves performing a breadth-first search of the state space tree. Initially the queue of live nodes is empty. The algorithm begins by generating the root node of the state space tree and enqueueing it in the queue *LiveNodes.* At each stage of the algorithm a node is dequeued from *LiveNodes* to become the new *E*-node. All the children of the *E*-node are then generated. The children that are not bounded are enqueued (as they are generated from left to right). If only one goal state is desired, then the algorithm terminates after the first goal state is found. Otherwise, the algorithm terminates when *LiveNodes* is empty. Figure 15.16 illustrates FIFO branch-and-bound for the Sum of Subsets problem for the instance $A = (1,11,6,2,6,8,5)$ and $Sum = 10$. The action of the

Figure 15.16 Action of queue *LiveNodes* and a portion of the variable-tuple state space tree generated by FIFO branch-and-bound for the Sum of Subsets problem, $A = (1,11,6,2,6,8,5)$ and $Sum = 10$. The sum of the elements chosen is shown inside each node.

queue *LiveNodes*

generate ()	enqueue ()
dequeue	*E*-node := ()
generate (1)	enqueue (1)
generate (2)	bounded
generate (3)	enqueue (3)
generate (4)	enqueue (4)
generate (5)	enqueue (5)
generate (6)	enqueue (6)
generate (7)	enqueue (7)
dequeue	*E*-node := (1)
generate (1,2)	bounded
generate (1,3)	enqueue (1,3)
generate (1,4)	enqueue (1,4)
generate (1,5)	enqueue (1,5)
generate (1,6)	enqueue (1,6)
generate (1,7)	enqueue (1,7)
dequeue	*E*-node := (3)
generate (3,4)	enqueue (3,4)
generate (3,5)	bounded
generate (3,6)	bounded
generate (3,7)	bounded
dequeue	*E*-node := (4)
generate (4,5)	enqueue (4,5)
generate (4,6)	goal node

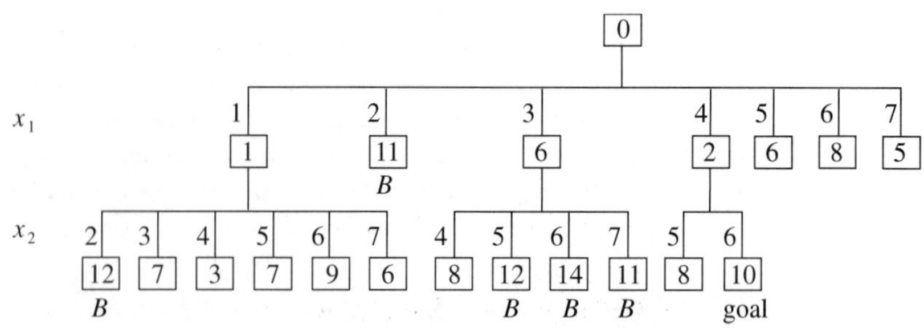

queue *LiveNodes* and the portion of the state space tree generated in reaching the first goal state are given. For this instance of the Sum of Subsets problem, FIFO branch-and-bound generates fewer nodes of the state space tree before reaching a goal state than are generated by backtracking.

Figure 15.17 illustrates *LIFO branch-and-bound*, where *LiveNodes* is a stack, for the same instance of the Sum of Subsets problem given in Figure 15.16. LIFO branch-and-bound is similar to backtracking, except that a move is made to the right-most child of a node first instead of the left-most. However, unlike backtracking, all the children of a node are generated before moving on.

LIFO branch-and-bound generates fewer nodes than FIFO branch-and-bound for the input considered in Figure 15.16, but for other inputs the opposite can be true. In general, since FIFO branch-and-bound is based on a breadth-first search of the state

Figure 15.17 Action of queue *LiveNodes* and a portion of the variable-tuple state space tree generated by LIFO branch-and-bound for the Sum of Subsets problem, $A = (1,11,6,2,6,8,5)$ and $Sum = 10$. The sum of the elements chosen is shown inside each node.

stack *LiveNodes*

generate ()	push ()
pop	*E*-node := ()
generate (1)	push (1)
generate (2)	bounded
generate (3)	push (3)
generate (4)	push (4)
generate (5)	push (5)
generate (6)	push (6)
generate (7)	push (7)
pop	*E*-node := (7)
pop	*E*-node := (6)
generate (6,7)	bounded
pop	*E*-node := 5
generate (5,6)	bounded
generate (5,7)	bounded
pop	*E*-node := (4)
generate (4,5)	push (4,5)
generate (4,6)	goal node

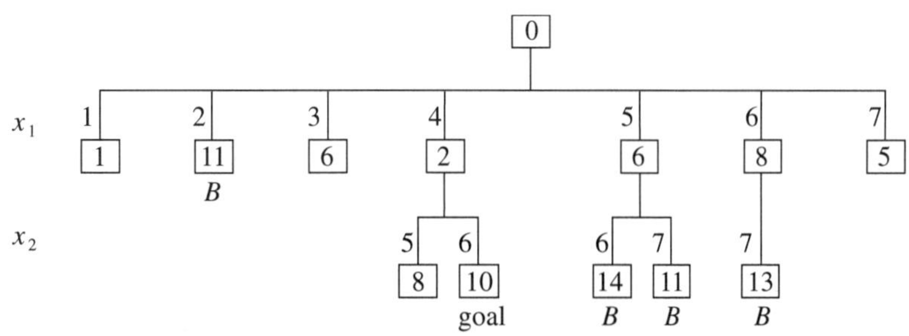

space tree, it is more efficient than LIFO branch-and-bound when goal nodes are not very deep in the state space tree.

We now consider the problem of finding a proper vertex c-coloring of a graph G. Figure 15.18a and Figure 15.18b show the portion of the state space tree T that is generated by FIFO branch-and-bound and LIFO branch-and-bound, respectively, for the same instance of the graph coloring problem as given in Figure 15.13 for backtracking. Because goal nodes are at the deepest level in the state space tree, FIFO branch-and-bound is much less efficient than either backtracking or LIFO branch-and-bound for the problem of c-coloring a graph.

15.3.1 General Branch-and-Bound Paradigm

When we use backtracking, we do not explicitly implement the state space tree. However, in branch-and-bound algorithms we must explicitly implement the state space tree and maintain the data structure storing the live nodes. In the general branch-and-bound paradigm *BranchAndBound*, the state space tree T is implemented using the parent representation.

The nodes of T that are generated by paradigm *BranchAndBound* are represented as follows.

```
TreeNode = record
    Info: InfoType
    Parent: → TreeNode
end
```

Only the value x_k need be stored in the information field *Info* of the node N corresponding to problem state $(x_1, \ldots, x_k)$. Given a pointer *PtrNode* to N, the entire tuple $(x_1, \ldots, x_k)$ is recovered by following the path in T from N to the root. For convenience, we denote $D_k(x_1, \ldots, x_{k-1})$ by $D_k(PtrNode)$, where *PtrNode* is a pointer to the problem state $(x_1, \ldots, x_{k-1})$.

The next E-node is chosen from the elements of *LiveNodes* by calling the procedure *Select(LiveNodes,E-node,k)*, where *E-node* is a pointer to the E-node and k is the size of the E-node. The definition of procedure *Select* is dependent on the type of branch-and-bound being implemented. For example, *Select* may choose the next E-node from a queue *LiveNodes* (FIFO branch-and-bound), a stack *LiveNodes* (LIFO branch-and-bound), a priority queue *Livenodes* (least cost branch-and-bound), and so forth.

Paradigm *BranchAndBound* adds a node to *LiveNodes*, by calling the procedure *Add(LiveNodes,PtrNode)*. *BranchAndBound* also invokes the Boolean functions *Answer(PtrNode)* and *Bound(PtrNode)*. *Answer(PtrNode)* assumes the value **.true.** if the node pointed to by *PtrNode* is a goal state. *Bound(PtrNode)* returns the value **.true**. if the node pointed to by *PtrNode* is bounded. Similar to backtracking, the definition of the bounding function depends on the particular problem being solved.

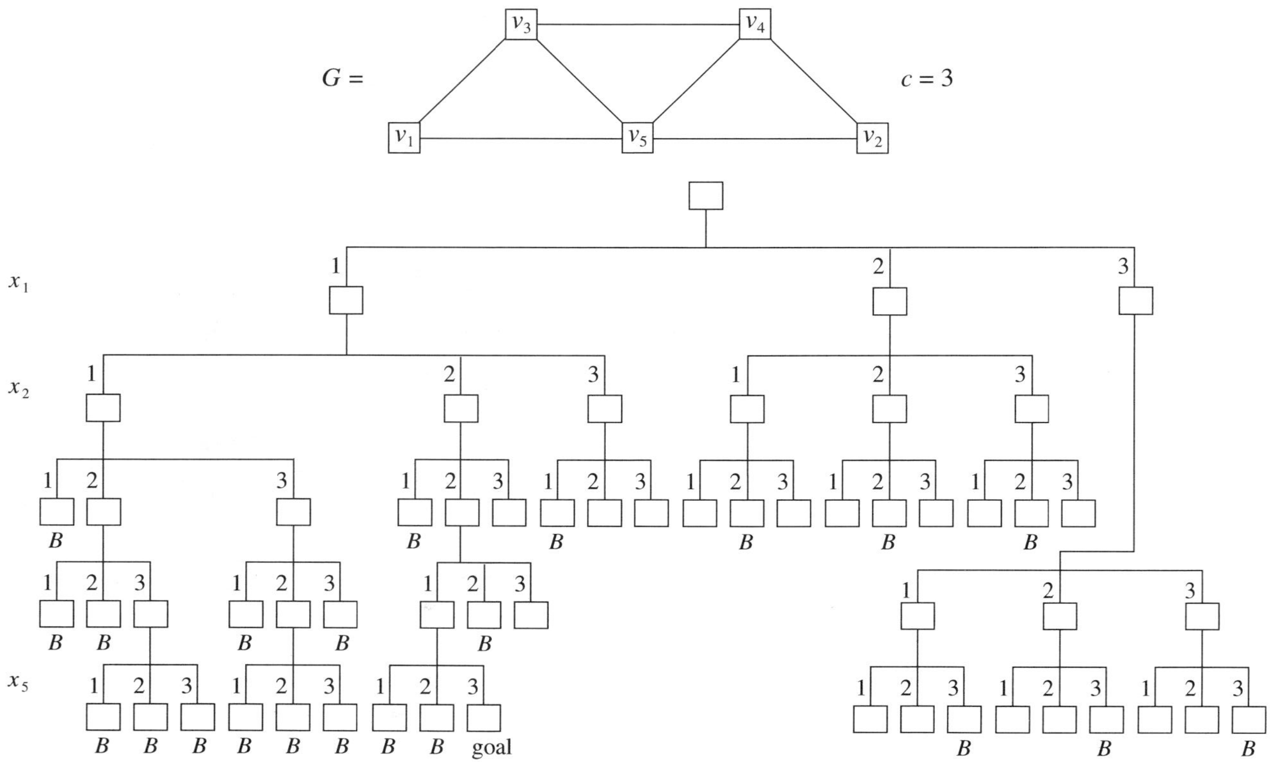

(a) Portion of state space T generated by FIFO branch-and-bound

Figure 15.18 Finding a proper vertex 3-coloring of G

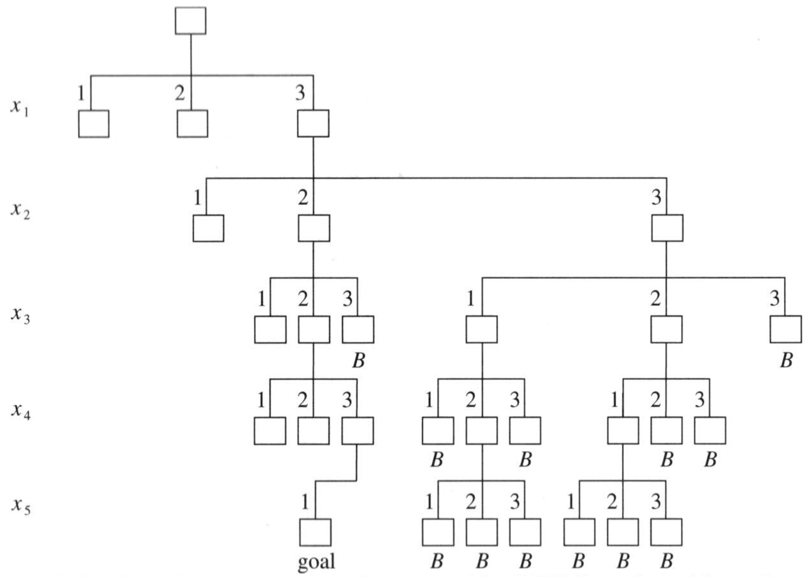

(b) Portion of state space tree T generated by LIFO branch-and-bound

Figure 15.18 Continued

```
procedure BranchAndBound
Input:   function D_k(x_1, . . . , x_{k-1}) determining state space tree T associated
                                           with the given problem)
         Bounding function Bounded
Output: All goal states to the given problem

  LiveNodes is initialized to be empty
  call AllocateTreeNode(Root)
  Root → Parent := nil
  call Add(LiveNodes,Root)                  {add root to list of live nodes}
  while LiveNodes is not empty do
     call Select(LiveNodes,E-node,k)        {select next E-node from live nodes}
     for each X[k] ∈ D_k(E-node) do         {for each child of the E-node do}

        call AllocateTreeNode(Child)
        Child → Info := X[k]
        Child → Parent := E-node
        if Answer(Child) then               {if child is a goal node then
           call Path(Child)                  output path from child to root}
        endif
        if .not. Bounded(Child) then
```

```
            call Add(LiveNodes,Child)     {add child to list of live nodes}
         endif
      endfor
   endwhile
end BranchAndBound
```

As with *Backtracking*, there is a corresponding version of the general paradigm *BranchAndBound* for problems involving minimizing or maximizing objective functions. For example, the paradigm *BranchAndBoundMin* maintains in *CurrentBest* the solution state with the current minimum value of the objective function f, and the value $f(CurrentBest)$ is used to dynamically bound nodes. Again, this dynamic bounding is done utilizing a suitable function *LowerBound*, whose value at a given node X is a lower-bound estimate on the value of the objective function at all goal states in the subtree rooted at X. A node X can be (dynamically) bounded if $f(CurrentBest) \leq LowerBound(X)$. As with the 0/1 Knapsack problem discussed earlier, the function $LowerBound(X)$ is often expressed in the form $f(X) + h(X)$, where $h(X)$ is a lower-bound estimate of the smallest incremental increase in f incurred in going from X to a descendant goal state. Sometimes $h(X)$ is a heuristic estimate that might not be provably a lower bound, but which nevertheless has been shown to work well in practice. *LiveNodes* is often maintained as a priority queue, where $LowerBound(X)$ is taken as the priority of a node X. The next *E*-node chosen by *Select* is the node in *LiveNodes* with the least value of *LowerBound*, and the strategy is called *least cost* branch-and-bound. A more detailed discussion of least cost branch-and-bound will be given in the next chapter, where it is shown to be a special case of a more general search strategy.

15.4 Closing Remarks

Both LIFO and FIFO branch-and-bound are blind searches of the state space tree T in the sense that they search the nodes of T in the same order regardless of the input to the algorithm. Thus, they tend to be inefficient for searching the large state space trees that often arise in practice. Utilizing heuristics can help narrow the scope of otherwise blind searches. The least cost branch-and-bound strategy discussed above utilizes a heuristic cost function associated with the nodes of the state space tree T, where the set of live nodes is maintained as a priority queue with respect to this cost function. In this way, the next node to become the *E*-node is the one that is the most promising to lead quickly to a goal.

Least cost branch-and-bound is closely related to the general heuristic search strategy called A*-search. A*-search can be applied to state space digraphs, rather than just state space trees. A*-search is one of the most commonly used search strategies in artificial intelligence. Both A*-search and least cost branch-and-bound are discussed in the next chapter.

References and Suggestions for Further Reading

Golumb, S. and L. Baumert. "Backtracking programming," *Journal of the ACM*, 12 (1965): 516–524. An early general description, and applications, of the backtracking method.

Walker, R. J. "An enumerative technique for a class of combinatorial problems," *Proceedings of Symposia in Applied Mathematics,* vol. X. Providence, RI: American Mathematical Society, 1960. The backtracking and branch-and-bound design strategies have been studied for a long time. (The name backtrack was coined by D. H. Lehmer in the 1950s.) Walker's article is one of the first accounts of the backtracking method.

For two early survey articles on the branch-and-bound paradigm, see:

Lawler, E. L. and D. W. Wood. "Branch-and-bound methods: a survey," *Operations Research*, 14 (1966): 699–719.

Mitten, L. "Branch-and-bound methods: general formulation and properties," *Operations Research*, 18 (1970): 24–34.

Jensen, T. R. and B. Toft. *Graph Coloring Problems.* New York: Wiley, 1995. A recent book that surveys graph coloring and related problems and includes a section on tie positions in n^m tic-tac-toe.

Exercises

Section 15.1 State Space Trees

15.1 Consider the backtracking solution to following instance of the 0/1 Knapsack problem:
Capacity of knapsack = $C = 15$

i	1	2	3	4	5	6	7
v_i	25	45	12	7	6	10	5
w_i	5	11	3	2	2	7	4

a. Give S_k, P_k, and $D_k(x_1, x_2, \ldots, x_{k-1})$.

b. Draw the *variable-tuple* state space tree (first three levels).

15.2 Repeat Exercise 15.1 for the *fixed-tuple* state space tree.

15.3 Show that the number of nodes of the state space tree T for the Sum of Subsets problem is exponential in n.

Section 15.2 Backtracking

15.4 Modify the pseudocode for the procedure *SumOfSubsets* to utilize the bounding function given in 15.2.2.

15.5 **a.** Give pseudocode for a nonrecursive backtracking procedure for solving the Sum of Subsets problem, based on the state space tree given in Figure 15.4.

b. Repeat part (a) for a recursive backtracking procedure.

15.6 **a.** Reformulate procedure *Backtrack* so that it halts once the first goal is reached.

b. Repeat part (a) for the procedure *BacktrackRec*.

15.7 Give pseudocode for versions of the procedures *Backtrack* and *BacktrackRec* that only check for a goal state after determining that a problem state is bounded. These modified algorithms only apply to problems where all the goal states are bounded.

15.8 Show that the total number of colors used in the coloring generated by the canonical greedy coloring for a given graph G can be different for different orderings of the vertices.

15.9 Give pseudocode for a nonrecursive version of the paradigm *BacktrackMin* for minimizing an objective function.

15.10 Show the portion of the state space tree that is generated during backtracking for the instance of the 0/1 Knapsack problem given in Exercise 15.1 using the bounding functions (15.2.9) and (15.2.12).

15.11 Give pseudocode for a backtracking algorithm that solves the 0/1 Knapsack problem using the bounding functions (15.2.9) and (15.2.12).

15.12 **a.** Write a computer program using backtracking that proves there are no tie boards for the $3 \times 3 \times 3$ tic-tac-toe game, even if we relax the condition that the number of Xs and the number of Os differ by one.

b. Write a computer program using backtracking that outputs all tie boards for the $3 \times 3 \times 3$ board minus the center position (when playing the game of tic-tac-toe, no X or O is placed in the center position), where we relax the condition that the number of Xs and Os are equal. Is there a tie board where the number of Xs and Os are equal?

15.13 Two queens in the ordinary chessboard are *nonattacking* if they are not in the same row, column, or diagonal. A classical problem known as the *8-queens problem* is to place eight queens on the board so that each pair of queens is nonattacking. One solution to the 8-queens problem is shown in Figure 15.19, page 550.

The *n-queens problem* is to place n queens on the $n \times n$ chessboard so that each pair of queens is nonattacking. Design a backtracking algorithm that generates all solutions to the n-queens problem.

15.14 Another classical problem associated with chess is the *Knight's Tour* problem. A knight can make (up to) eight moves as shown in Figure 15.20, page 550. Starting at an arbitrary position in the $n \times n$ board, a knight's tour is a sequence of $n^2 - 1$ moves such that every square of the board is visited once. Write a backtracking algorithm that either produces a knight's tour or determines that no such tour exists.

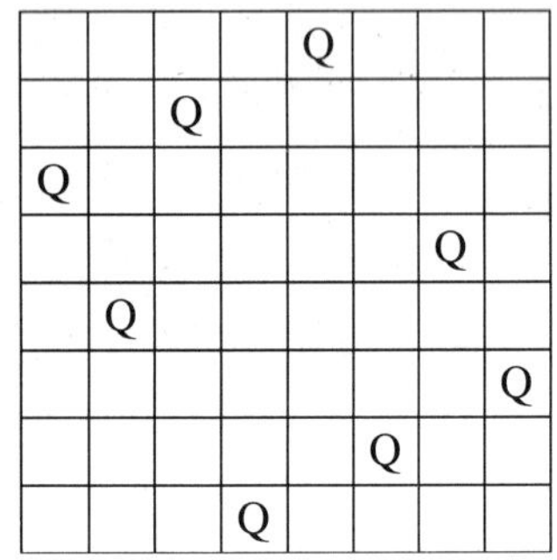

Figure 15.19 A solution to the 8-queens problem

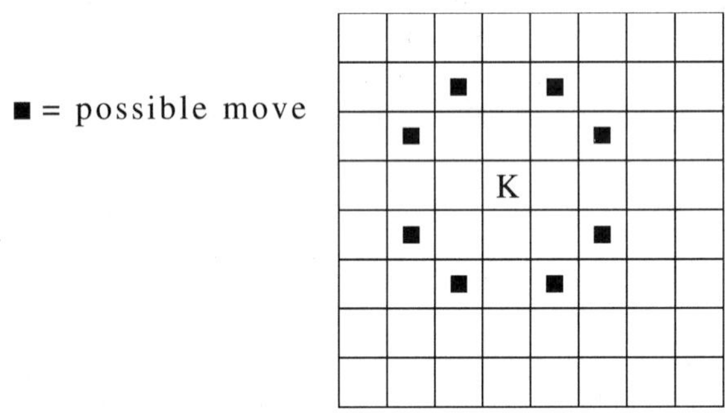

Figure 15.20 The eight possible moves for a knight in the given position

15.15 Let *Maze*[1:*n*,1:*n*] be a 0/1 two-dimensional array. Design a backtracking algorithm that either finds a path from *Maze*[1,1] to *Maze*[*n*,*n*] or determines that no such path exists. Adjacent vertices in the path correspond to adjacent cells in the matrix. You are not allowed to move to a cell that contains a 1.

15.16 Write a computer program that uses backtracking to solve the game of Hi-Q. Hi-Q is a popular game that can be found in many toy stores. Thirty-two pieces are arranged on a board as shown in Figure 15.21, with the center position left empty. The goal is to remove all the pieces but one by jumping and have the last piece end up in the middle position. A piece is allowed to jump a neighbor in either a horizontal or vertical direction (diagonal jumps are not permitted). When a piece is jumped, it is removed from the board. Output the 32 board configurations showing the solution: the initial board configuration and the board configuration after each jump is performed.

15.17 Design a backtracking algorithm that generates all Hamiltonian cycles in a digraph *D*.

15.18 Design a backtracking algorithm solving the Traveling Salesman problem.

15.19 Given a graph *G*, an *independent* set of vertices is a set of vertices no two of which are adjacent. Design a backtracking algorithm that outputs an independent set of vertices of largest size.

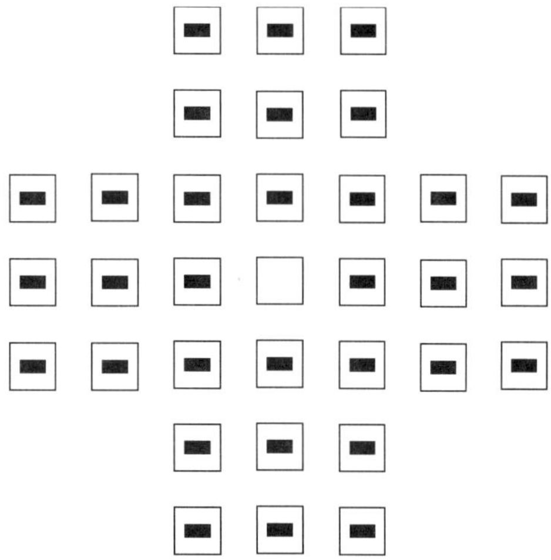

Figure 15.21 Initial board configuration for Hi-Q

15.20 Given a graph G, the *chromatic number* of G is the minimum number of colors required to properly color the vertices. Design a backtracking algorithm for computing the chromatic number of a graph.

Section 15.3 **Branch-and-Bound**

15.21 Consider the Sum of Subsets problem, for instance $(c_1, \ldots, c_7) = (1,11,6,2,6,8,5)$ and *Change* = 10. Show that for this instance of the Sum of Subsets problem, FIFO branch-and-bound generates fewer nodes of the state space tree before reaching a goal state than backtracking does.

15.22 Give pseudocode for a version of the general procedure *BranchAndBound* that terminates as soon as a goal is found.

15.23 Give pseudocode for the procedure *Path*(*PtrNode*).

15.24 Consider the following graph G.

$G =$

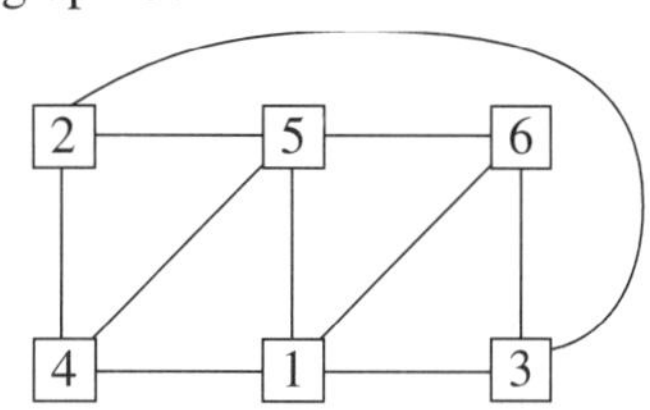

a. Determine the 3-coloring of G generated by the backtracking algorithm.

b. Determine the 3-coloring of G generated by LIFO branch-and-bound.

15.25 Draw that portion of the variable-tuple state space tree generated by FIFO branch-and-bound for the 0/1 Knapsack problem given by the following chart. Label the nodes with appropriate values of *UB*, *LoBd*, *SB*, *DB*, and indicate optimal goal node as in Figure 15.15. Trace the action of the queue *LiveNodes* as illustrated in Figure 15.16.

i	1	2	3	4	5	Capacity $C = 12$
v_i	28	15	18	5.5	1	
w_i	8	5	6	4	1	

15.26 Repeat Exercise 15.25 for least cost branch-and-bound.

15.27 Repeat Exercise 15.25 using the fixed-tuple state space tree.

15.28 Repeat Exercise 15.26 using the fixed-tuple state space tree.

15.29 Give pseudocode for a LIFO branch-and-bound algorithm that accepts a graph G on n vertices and outputs a proper vertex coloring using the minimum number of colors (thereby determining the chromatic number of G). Of course, your algorithm will certainly be horribly inefficient in the worst case. You may assume that the first vertex is colored 1. You may also assume *Push* and *Pop* routines for pushing and popping a stack. Use a suitable parent array implementation for the state space tree of problem states.

15.30 Design a LIFO branch-and-bound algorithm for the Traveling Salesman problem.

Part Four

Special Topics

16

Heuristic Search: A*-Search, Game Trees

heuristic *... providing aid or direction in the solution of a problem, but otherwise unjustified or incapable of justification ... of or relating to exploratory problem-solving techniques that utilize self-learning techniques to improve performance. ...*

—Webster's New Collegiate Dictionary

Search strategies such as backtracking, LIFO and FIFO branch-and-bound, breadth-first search, and depth-first search are blind in the sense that they do not look *ahead* beyond a local neighborhood when expanding a node. In this chapter we show how utilizing heuristics can help narrow the scope of otherwise blind searches. We introduce a type of heuristic search strategy known as A*-search, which is widely used in artificial intelligence (AI). We then discuss strategies for playing two-person games. The alpha-beta heuristic for two-person games is based on assigning a heuristic value to positions reached by looking ahead a certain fixed number of moves. Then an estimate for the best move is obtained by working your way back to the current position using the so-called *minimax* strategy.

16.1 Artificial Intelligence: Production Systems

The subject of AI is concerned with designing algorithms that allow computers to emulate human behavior (see Figure 16.1). The major areas of AI include natural language processing, automatic programming and theorem proving, robotics, machine vision and pattern recognition, intelligent data retrieval, expert systems, and game playing. Certain activities that are child's play, such as using a natural language, present theoretically and computationally difficult problems that are beyond

Figure 16.1

the reach of current technology. It is true that voice recognition computer programs are currently available that can properly interpret a limited set of spoken instructions. However, the time when we can carry out ordinary conversations with a computer, such as those between the spaceship crew and the computer *Hal* in the movie 2001—*A Space Odyssey*, appears to be somewhat farther off than the year 2001.

Many problems in AI involve production systems. An AI production system is characterized by *system states* (also called *databases*), *production rules* allowing the system to change from one state to another, and a *control system* that manages the execution of production rules allowing the system to evolve according to some desired scenario. For example, a system state might be the position of a robotic arm. A production rule allows the robotic arm to change its position. A control strategy is an algorithm that controls the movement of the arm from a given initial position to a final goal position. Here again, a two-year-old child can move his or her arms without much thinking, but designing an algorithm that allows a robot to accomplish the same thing is a complicated task.

Given any AI production system, there is a positive cost associated with the application of each production rule. A production system can be modeled as a positively weighted digraph, called a *state space digraph*, where a node in the graph is the state of the system, and a directed edge from node v to node w is assigned the cost $Cost(v,w)$ of the production rule that transforms state v into state w. Given some initial state r (root vertex in the directed graph), we are interested in whether or not we can find a (directed) path from r to a goal state. A control system for the problem is then simply a search strategy for reaching a goal state starting from r. As usual, we wish to find control systems that perform searches efficiently.

16.2 The 8-Puzzle Game

We illustrate the ideas of an AI production system by the 8-puzzle game. The 8-puzzle game is a smaller version of the 15-puzzle game invented by Sam Loyd in 1878. In the 8-puzzle game, there are eight tiles numbered 1 through 8 occupying eight of the nine cells in a 3×3 square board. The object is to move from a given initial state in the board to a goal state. The only moves (production rules) allowed are to move a tile into an adjacent empty cell. It is convenient to characterize such a move as a movement of the empty cell. Thus, there are exactly four rules for moving the empty cell: move left, move right, move up, move down. Of course, the only states allowing all four rules to be applied are when the empty cell is in the center. When the empty cell is at a corner location in the board, then only two of the four rules can be applied. The remaining locations allow the application of three of the four rules.

We will let D be the state space digraph whose vertex set consists of all possible board configurations in the 8-puzzle game, where a (directed) edge (v,w) exists if a move in the game transforms v into w. Note that if (v,w) is an edge, then (w,v) is also an edge. Thus D can be considered as a state space graph, where the two directed edges (v,w) and (w,v) are replaced with the single undirected edge $\{v,w\}$. A portion of the state space graph is shown in Figure 16.2 (page 558). Suppose we use the breadth-first search control strategy to search for a path leading from the initial state to the goal state. We assume that our control strategy always generates the children of a node in the order move left, move right, move up, move down. In Figure 16.3 we have taken an initial position that is only four moves from the goal state. However, note that 28 states (not counting the initial state) are generated by breadth-first search!

In general, the number of nodes generated by breadth-first search for the n-puzzle game is exponential in the minimum number of moves required to reach a goal. Thus, we must look for a better search strategy to solve the problem for initial states requiring many moves to reach the goal. We now describe such a strategy.

16.3 A*-Search

Given a root vertex r and a set of goal states in a state space digraph, A*-search is a strategy for finding a shortest path from r to a nearest goal. Such a path is called an *optimal path*. A*-search finds an optimal path using a generalization of Dijkstra's algorithm. In the treatment of Dijkstra's algorithm given in Chapter 12, we maintained an array $Dist[1{:}n]$. At each stage, the vertex v minimizing $Dist[v]$ over all vertices not in the tree was added to T. The values of $Dist[w]$ were then updated for all

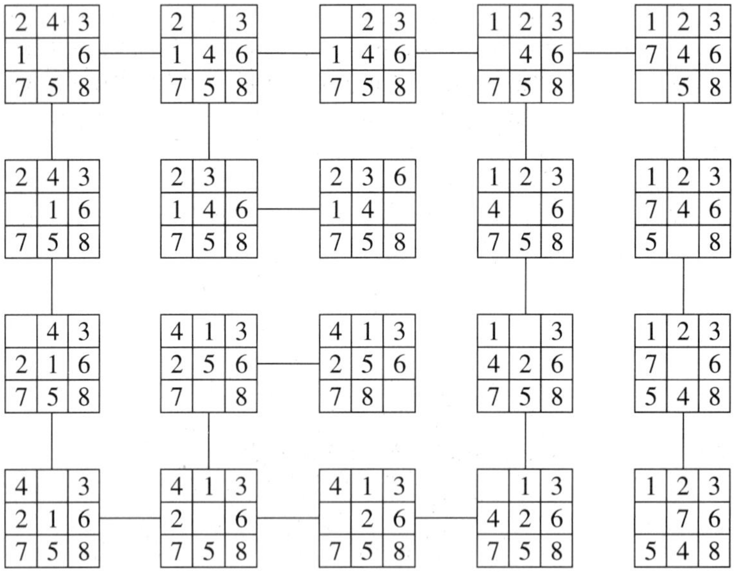

Figure 16.2 A portion of the state space graph for the 8-puzzle game

vertices w in the out-neighborhood of v. The operations performed on the array $Dist[1{:}n]$ were essentially those of a priority queue. To aid in the description of A*-search, we now give a high-level description of Dijkstra's algorithm based on maintaining a priority queue of vertices. We also modify Dijkstra's algorithm to terminate once a goal is dequeued.

We denote the priority of a vertex v in the queue by $g(v)$, where smaller values of g have higher priority. At initialization, the root vertex r is enqueued with priority $g(r) = 0$. When a vertex v is enqueued, a parent pointer from v to $Parent(v)$ is also stored. The parent pointers determine a tree. We call the subtree of this tree spanned by the vertices that have been dequeued the *dequeued tree T*. The tree T contains a shortest path in D from r to each vertex in T.

At each stage in the algorithm, a vertex v in the priority queue is dequeued, and the out-neighbors of v are examined. If an out-neighbor w of v is already in the tree T, then nothing is done to w. If the out-neighbor w has not been enqueued, then it is enqueued with priority $g(w) = g(v) + c(v,w)$, and a parent pointer from w to v is set. Finally, if the out-neighbor w is already on the queue, then the priority of vertex w is updated so that $g(w) = \min\{g(w), g(v) + c(v,w)\}$. If $g(w)$ is changed to $g(v) + c(v,w)$, then the parent pointer of w is reset to point to v. The algorithm terminates once a goal is dequeued. The path in the final tree T from r to the goal is an optimal path. The action of the algorithm is shown in Figure 16.4 (page 560) for a sample digraph.

Dijkstra's algorithm is too inefficient for most AI applications, since the shortest-path tree can grow to be quite huge, even exponentially large. The reason for its inefficiency is that only local information is assumed when looking ahead to a goal. No global information is utilized that can help the shortest-path tree send out

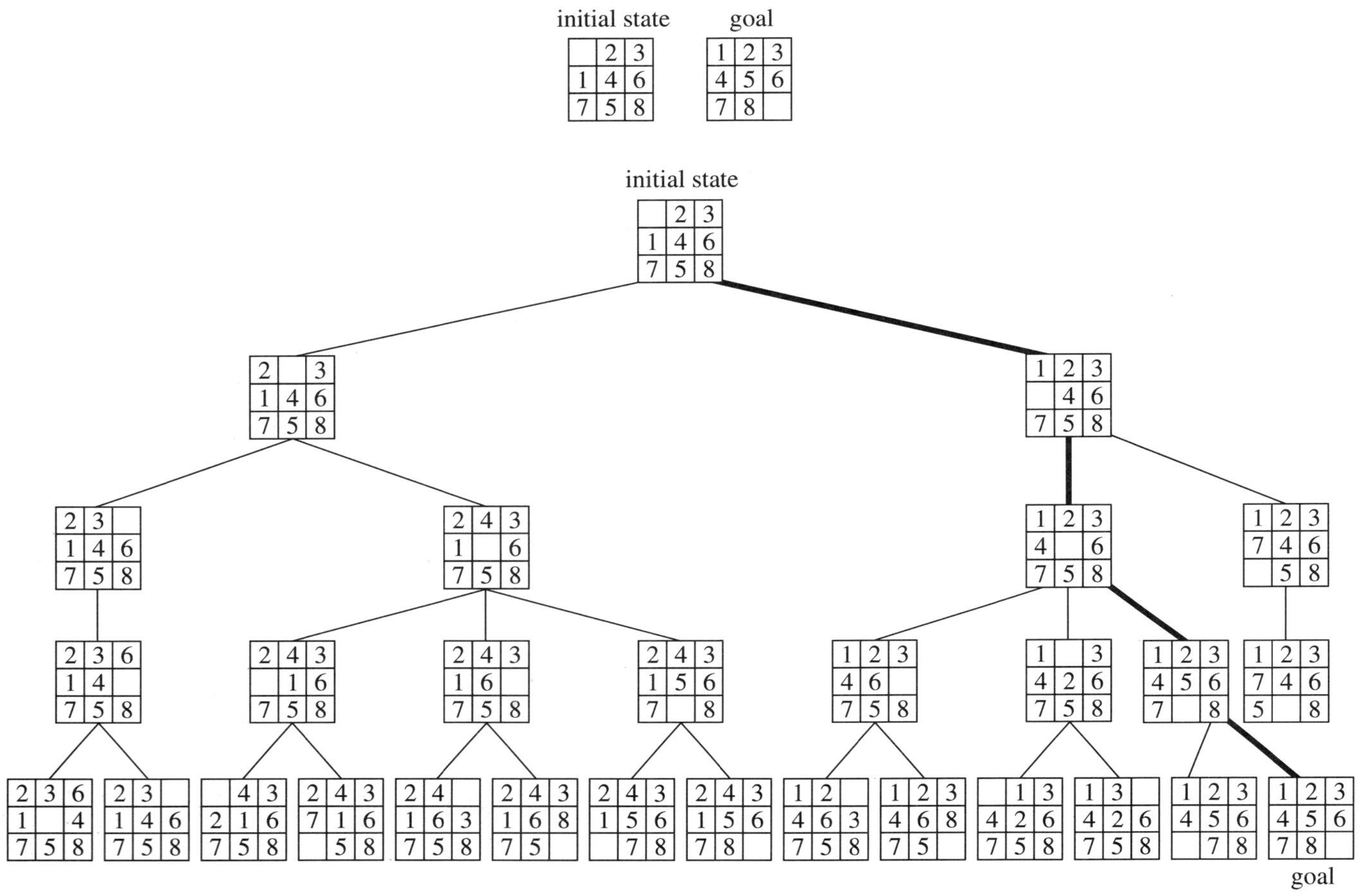

Figure 16.3 The states generated by breadth-first search in finding a path from the given initial state to the goal state in the 8-puzzle game. The path to the goal shown in bold.

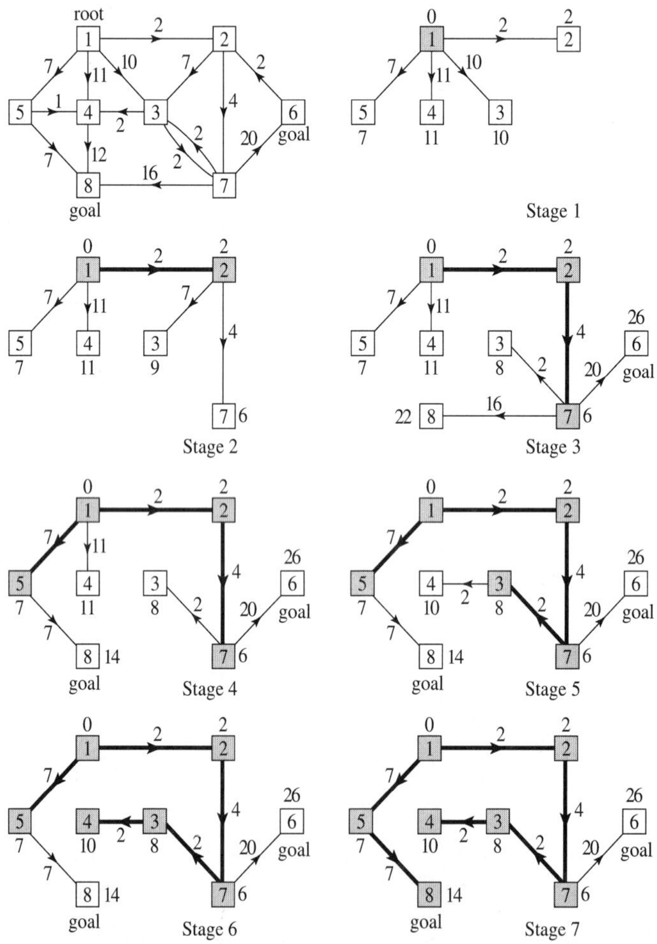

Figure 16.4 The action of Dijkstra's algorithm is shown for a sample weighted digraph *D*. The distance $g(v)$ is shown outside each node *v*. The vertices and the edges of the dequeued (shortest path) tree are shaded. The priority queue at each stage consists of vertices *w* not in the dequeued tree, where the priority of *w* is $g(w)$.

branches in a promising direction. In other words, in Dijkstra's algorithm the shortest-path tree tends to grow fat (a "shotgun approach" to a goal), rather than grow skinny (a "bee-line" approach to a goal).

16.3.1 Heuristics

When additional information other than merely the costs of the edges in the digraph is available, Dijkstra's algorithm can be improved so that the shortest path tree can be less expansive and the search can be made more efficient. The idea is that the priority value $g(v)$ of a vertex v in the queue, which is the cumulative distance (cost) from the root r via the current path determined by the parent pointers, can be replaced by an overall estimate of the cost of the shortest path from r to a goal constrained to go through v. In Dijkstra's algorithm the priority value of v is $g(v)$, but now we define the priority value of v to be the *cost function*

$$f(v) = g(v) + h(v), \qquad (16.3.1)$$

where $h(v)$ is some estimate of the cost of a shortest path from v to a goal vertex. Since the shortest path from v to a goal has not been found, the best that we can do is utilize a *heuristic* value for $h(v)$.

When no restriction is placed on the heuristic h in (16.3.1), h is merely a heuristic for a greedy algorithm. When the vertices in the dequeued tree T are re-examined and their parent pointers updated if shorter paths for them are found, the algorithm based on (16.3.1) is called *A-search*. There is no guarantee that the first path found to a goal is optimal using A-search. When A-search uses a heuristic $h(v)$ that is a lower bound of the cost of the shortest path from v to a goal, then the algorithm is called *A*-search*. We assume that A*-search terminates when it dequeues a goal, or when the queue is empty. The proof of the following theorem is left to the exercises.

THEOREM 16.3.1 Given a positively weighted digraph G (finite or infinite), if a goal is reachable from a root vertex r, then A*-search terminates by finding an optimal path from r to a goal. ❑

In A*-search, when a node v is dequeued some of its neighbors may already be in the tree T. Unlike Dijkstra's algorithm, in A*-search these neighbors must be checked to see if shorter paths now exist to them via the vertex v just dequeued. If shorter paths are found, then T must be adjusted to take them into account (see Figure 16.5).

Figure 16.5 The number shown outside each node v in T is the value $g(v)$. This example shows how the dequeued tree T can change during an A*-search. Note that the values of g at vertices a, b and c required updating.

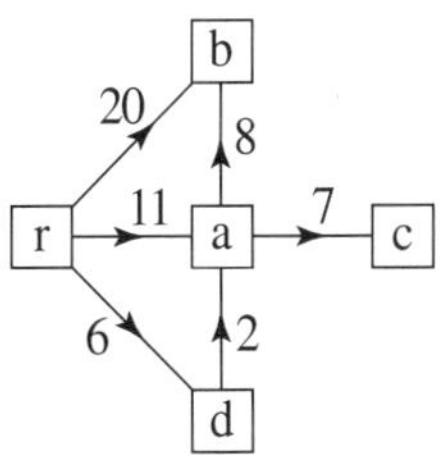

v	r	a	b	c	d
$h(v)$	–	23	20	15	29

Portion of state space induced by vertices r, a, b, c, d, and an associated function h

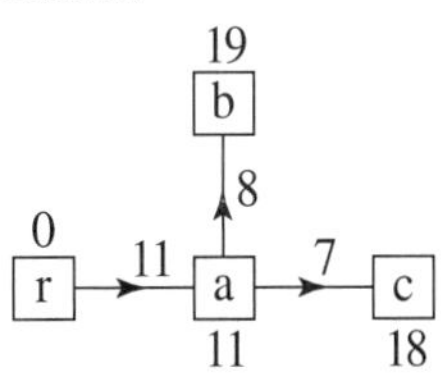

The dequeued tree T after vertices r, a, b, c have be dequeued

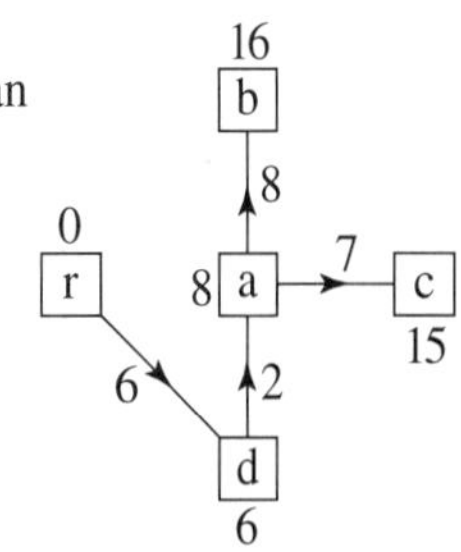

The dequeued tree T after d has be dequeued

Considerable computational cost may be incurred by A*-search when adjusting the dequeued tree T. This computational cost can be avoided by placing a rather natural and mild restriction on the heuristic h used by A*-search. The heuristic value $h(v)$ is an estimate of the cost of going from v to the nearest goal. If the edge (v,w) exists, then one estimate is $c(v,w) + h(w)$. The restriction on a heuristic, called the *monotone restriction*, says that $h(v)$ should be at least as good as this estimate.

DEFINITION 16.3.1 A heuristic $h(v)$ for A*-search for a given digraph with cost function c on the edges is said to satisfy the *monotone restriction*, if

$$\begin{aligned} h(v) &\le c(v,w) + h(w), && \text{whenever the edge } (v,w) \text{ exists,} \\ h(v) &= 0, && \text{whenever } v \text{ is a goal.} \end{aligned} \tag{16.3.2}$$

If h satisfies the monotone restriction, then we merely say that h is monotone. ||||

If h is a monotone heuristic, then $h(v)$ is a lower bound of the cost of the shortest path from v to a goal. The following proposition states that an A*-search using a monotone heuristic does not need to update parent pointers of a vertex v already in the dequeued tree T, since the path in T from r to v is already a shortest path in D from r to v. The result is consistent with the fact that the A* algorithm reduces to Dijkstra's algorithm when $h(v) \equiv 0$ (and the identically zero function trivially satisfies the monotone restriction).

PROPOSITION 16.3.2 Suppose an A*-search utilizes a monotone heuristic. Then the dequeued tree T is a shortest path tree in the state space digraph D. In particular, parent pointers for vertices in the dequeued tree T never need updating.

Proof For any vertex $v \in V$, let $g^*(v)$ denote the shortest path in D from r to v (so that $g(v) \ge g^*(v)$ at every stage in the execution of *A*-search*). We wish to show that $g(v) = g^*(v)$ at the time when v is dequeued. If $v = r$, we have $g(v) = g^*(v) = 0$, so we can suppose that $v \ne r$. Let $P = v_0, v_1, \ldots, v_j$ be a shortest path in D from $r = v_0$ to $v = v_j$. Let vertex v_k be the last vertex in P such that $v_0, v_1, \ldots, v_k$ were all in the tree T when v was dequeued (v_k exists since $r = v_0 \in T$). Then v_{k+1} was in the queue Q at the time when v was dequeued. For any pair of consecutive vertices v_i, v_{i+1} in P, using the monotone restriction, we have

$$g^*(v_i) + h(v_i) \le g^*(v_i) + h(v_{i+1}) + c(v_i,v_{i+1}). \tag{16.3.3}$$

Now v_i and v_{i+1} are in a shortest path in G, so that

$$g^*(v_{i+1}) = g^*(v_i) + c(v_i,v_{i+1}). \tag{16.3.4}$$

Substituting (16.3.4) in (16.3.3) we obtain

$$g^*(v_i) + h(v_i) \le g^*(v_{i+1}) + h(v_{i+1}). \tag{16.3.5}$$

Iterating (16.3.5) and using the transitivity of $\le$ yields

$$g^*(v_{k+1}) + h(v_{k+1}) \le g^*(v_j) + h(v_j) = g^*(v) + h(v). \tag{16.3.6}$$

Now v_{k+1} is on a shortest path P, and $v_0, v_1, \ldots, v_k$ all belong to T, so that $g(v_{k+1}) = g^*(v_{k+1})$. Hence, (16.3.6) implies

$$f(v_{k+1}) = g(v_{k+1}) + h(v_{k+1}) \le g^*(v) + h(v) \le g(v) + h(v) = f(v). \qquad (16.3.7)$$

Thus, we must have had $g(v) = g^*(v)$ when v was dequeued, otherwise $f(v_{k+1}) < f(v)$, and v would not have been dequeued in preference to v_{k+1}. ■

The following proposition helps explain the terminology *monotone restriction*. We leave the proof of Proposition 16.3.3 as an exercise.

PROPOSITION 16.3.3 The f-values of the vertices dequeued by A*-search using a monotone heuristic are nondecreasing. ❑

In problems using A*-search, the digraph and associated cost function are either implicitly or explicitly input to the algorithm. We will give examples of both scenarios. For very large digraphs, the digraph is usually implicitly defined and only that part of the digraph that is generated by the execution of A*-search is made explicit. The following is a high-level description of A*-search using a monotone heuristic. At any given point in the execution of procedure *A*-SearchMH*, T is a subtree of D rooted at the root vertex r containing a path from r to each vertex that has been dequeued by the algorithm. Assuming that a path from r to a goal exists, Theorem 16.3.1 and Proposition 16.3.2 show that when *A*-SearchMH* terminates after dequeuing a goal, the corresponding path to the goal is optimal.

```
procedure A*-SearchMH(D,c,r,GoalSet,h,T)
Input:  D = (V,E) (a digraph, either implicitly or explicitly defined)
        c (a positive cost function on E)
        r (a root vertex in D)
        h (a heuristic function satisfying monotone restriction)
        GoalSet (a set of goal vertices in D)
Output: a shortest-path out tree T rooted at r containing an optimal path
        to a goal vertex, if one exists
dcl     Q (a priority queue of vertices, with v having priority value f(v) =
           g(v) + h(v), where g(v) is cost of shortest path P(v) from r to v
           currently generated. Q also contains a parent pointer from v to
           w ∈ T where edge (w,v) belongs to P(v))
   while Q is not empty do
      dequeue vertex v in Q with minimum priority value f(v)
      add vertex v to T using parent pointer
      if v ∈ GoalSet then
         return
      endif
      for all vertices w ∉ T and adjacent to v do
```
Continued

```
            if w ∉ Q then
               enqueue w with parent v and priority value f(w) = g(w) + h(w)
               where g(w) = g(v) + c(v,w)
            else
               if f(w) > g(v) + c(v,w) + h(w) then
                  reset parent pointer of w to v and update priority value of
                  w to f(w) = g(w) + h(w), where g(w) = g(v) + c(v,w)
               endif
            endif
         endfor
      endwhile
      return "failure"
end A*-SearchMH
```

The action of procedure *A*-SearchMH* is illustrated in Figure 16.6 for a sample digraph *D*.

Figure 16.6 The action of A*-search with root vertex $r = 1$ and a monotone heuristic $h(v)$ is shown for the same weighted digraph D as in Figure 16.4.

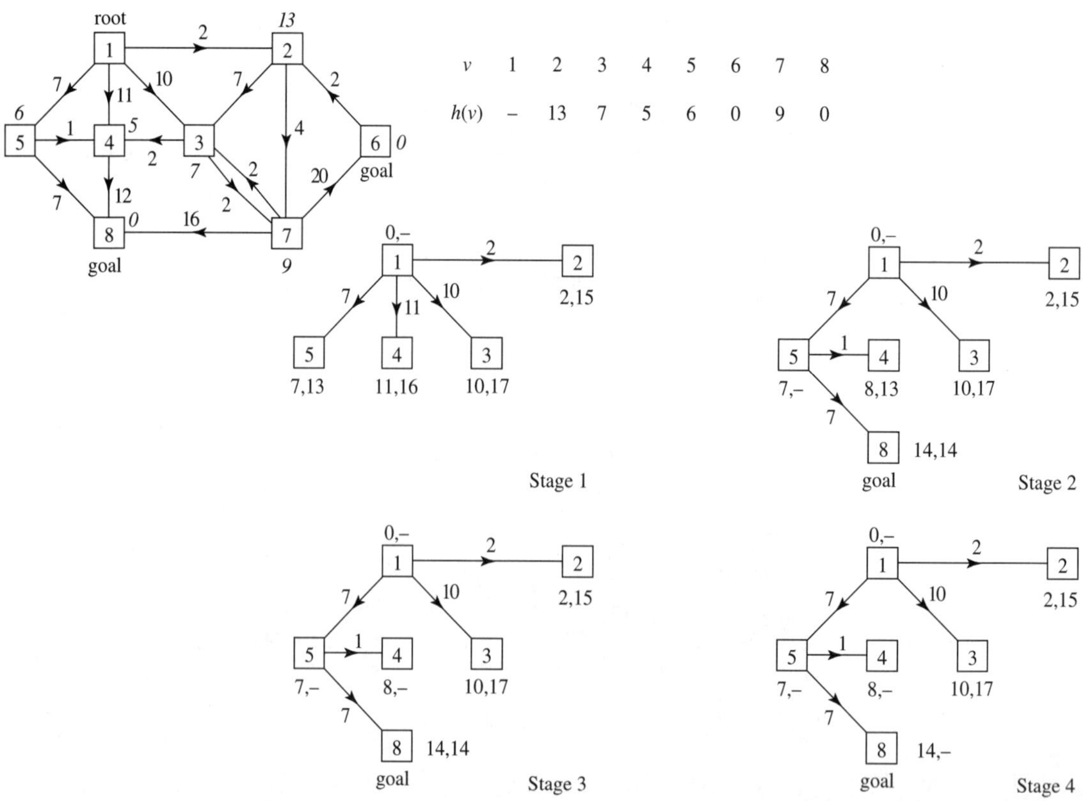

v	1	2	3	4	5	6	7	8
$h(v)$	–	13	7	5	6	0	9	0

We now illustrate procedure *A*-SearchMH* with two examples. First, we revisit the 8-puzzle problem. Then we consider the problem of finding shortest paths between cities in the United States using the freeway system. In the 8-puzzle problem, the state space graph *G* is implicitly defined. In the freeway problem, the state space graph is explicitly input to the algorithm.

16.3.2 A*-Search and the 8-Puzzle Game

Consider the heuristic

$$h(v) = \text{the number of tiles not in correct cell in the state } v.$$

It is easy to verify that $h(v)$ satisfies the monotone restriction. Using $h(v)$, Figure 16.7 shows the shortest path tree generated by A*-search for the same input as shown in Figure 16.3.

Note that A*-search only generated 9 states compared to the 28 states generated by breadth-first search for the same input. Other monotone heuristics exhibit even better behavior in general than the one used in Figure 16.7. For example, for any given state, the sum of the Manhattan distances (vertical steps plus horizontal steps) from the tiles to their proper positions in the goal state exhibits good performance.

Figure 16.7 Shortest path tree generated by A*-search for the 8-puzzle game with the same initial state and goal state as in Figure 16.3.

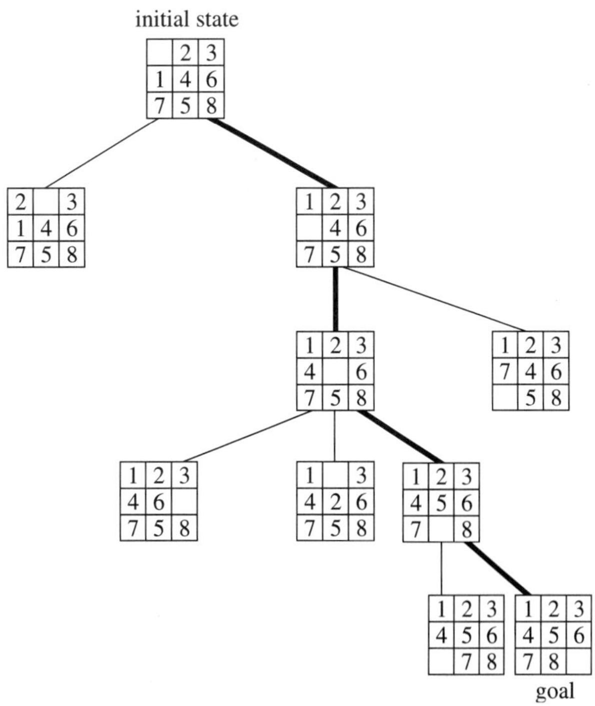

16.3.3 Shortest Paths in the Freeway System

Our second example of A*-search is for the problem of finding a shortest path on freeways between two cities in the continental United States (see Figure 16.8, page 567). The heuristic $h(v)$ we use will be a lower bound of the geographical (great-circle) distance between v and the destination (goal) city t. We assume that the distances between adjacent cities is available to the algorithm via a suitable adjacency cost matrix. The lower bound estimate is computed using the longitude and latitude of each city, which we assume are both input to the algorithm as additional information. A lower bound of 50 miles is used for the longitude distance of one degree apart. A lower bound of 70 miles is used for the latitude distance of one degree apart. The square root of the sum of the squares of the longitude distance and the latitude distance between cities v and t is used as $h(v)$. The heuristic $h(v)$ is monotone because the cost (mileage on a freeway between adjacent cities) cannot be smaller than the geographical distance between them. Indeed, if the cost between adjacent cities v and w is $c(v,w)$, then $h(v)$, $h(w)$, and $c(v,w)$ form a (almost planar) triangle, and we have $h(v) \leq h(w) + c(v,w)$, which is the monotone restriction.

Figure 16.8 shows the graph of the United States as input to a Prolog program, as well as a shortest path from Cincinnati to Houston. Figure 16.9 shows the shortest path tree generated by Dijkstra's algorithm ($h \equiv 0$), whereas Figure 16.10 shows the shortest path tree generated by A*-search. The number of vertices in the shortest path between Cincinnati and Houston is 9. The number of vertices expanded when the heuristic is used is 34, compared to 213 vertices when expanded by Dijkstra's algorithm. Thus, the portion of the graph expanded using A*-search is only 16% of that expanded using Dijkstra's algorithm for this example. The contrast between the trees grown in Figure 16.9 and Figure 16.10 shows rather nicely the difference between the "shotgun" approach to a goal made by Dijkstra's algorithm versus the "bee-line" approach made by A*-search.

The selection of a good heuristic is crucial to the success of A*-search. The closer $h(v)$ is to the actual cost of the shortest path from v to a goal, the fewer nodes will be expanded during A*-search. However, determining heuristics close to the actual cost is usually too expensive computationally, since determining close estimates is as hard as the original problem! It sometimes speeds the search to use a function for h that does not have the lower bound property, that is, using A-search instead of A*-search. For example, there are better heuristics leading to A-searches for the 8-puzzle game than the monotone heuristic $h(v)$ that was the basis for our A*-search.

16.4 Least Cost Branch-and-Bound

Least cost branch-and-bound is basically an A*-search applied to a state space tree with an additional utilization of a bounding function. We use the same notation when describing the state space tree T as used in the previous chapter. Least cost branch-and-bound applies to problems involving minimizing an objective function

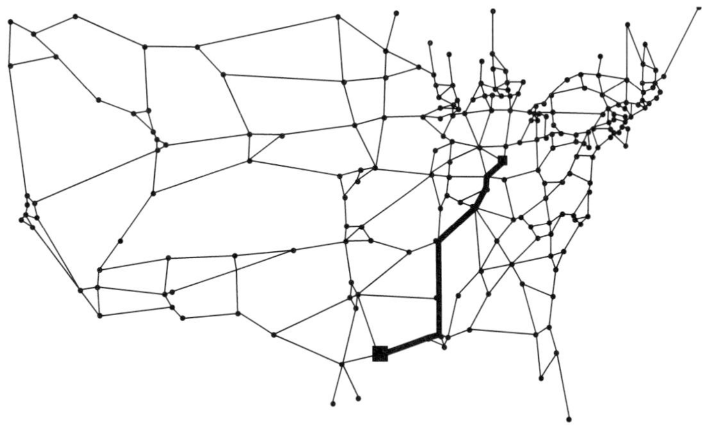

Figure 16.8 United States freeway system

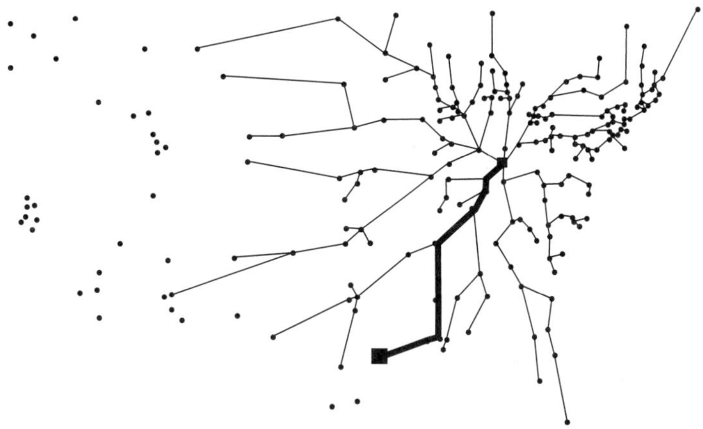

Figure 16.9 Shortest path tree generated by Dijkstra's algorithm

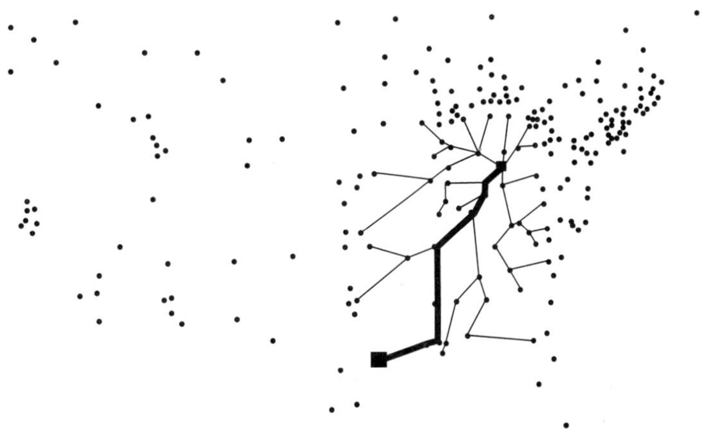

Figure 16.10 Shortest path tree generated by A*-search

φ over each solution state in the state space tree. The cost $c(v)$ of a given node $v = (x_1, \ldots, x_k)$ in the state space tree is taken to be a lower bound estimate for

$$\varphi^*(v) = \min \{\varphi(w) : w \in T_v \text{ and } w \text{ is a solution state}\}, \tag{16.4.1}$$

where T_v is the subtree of the state space tree rooted at v, so that

$$c(v) \leq \varphi^*(v). \tag{16.4.2}$$

Often, the cost function c has the same form as with A*-search,

$$c(v) = g(v) + h(v), \tag{16.4.3}$$

where $g(v)$ is the cost associated with going from the root to the node v, and $h(v)$ is a heuristic lower bound estimate of the incremental cost of going from v to a solution state v^* in T_v where φ is minimized (over T_v). Typically, $g(v)$ is $\varphi(v)$, where $\varphi(v)$ is an extension of the objective function φ to all problem states.

For example, in the Coin-Changing problem (see Chapter 12), $g(v)$ is the number of coins used in problem state v. A natural heuristic $h(v)$ is obtained in a manner similar to the greedy method. Let $r(v)$ denote the remaining change required. We use as many of the largest denomination coin as possible, then as many of the next-largest denomination coin as possible, and continue in this manner as long as we do not exceed $r(v)$. The number of coins so obtained is our lower bound estimate $h(v)$.

As a second example, consider the 0/1 Knapsack problem formulated as a minimization problem (see Chapter 15). We then have

$$g(x_1, \ldots, x_k) = \mathit{LeftOutValue}(x_1, \ldots, x_k),$$

and

$$h(x_1, \ldots, x_k) = -\,\mathit{Greedy}(C',B'),$$

where $\mathit{LeftOutValue}(x_1, \ldots, x_k)$ is the sum of the values of the objects not in the set $\{b_{x_1}, \ldots, b_{x_k}\}$, B' is the set of objects $\{b_{x_k+1}, \ldots, b_n\}$, $C' = C - (w_{x_1} + \cdots + w_{x_k})$, and $\mathit{Greedy}(C',B')$ is the value of the greedy solution to the (C',B') Knapsack problem.

In least cost branch-and-bound, the live nodes in the state space tree are maintained as a priority queue with respect to the cost function c. Unlike A*-search we maintain a global variable UB, which is the smallest value of the objective function over all solution states already generated. Then a node v can be bounded if $c(v) \geq UB$. *Moreover, assuming the lower bound cost function, if a node of least cost among the live nodes is bounded, then the algorithm can terminate having already generated an optimal solution state (goal node).*

The following paradigm for least cost branch-and-bound uses the same notation and implementation details for the state space tree as in Chapter 15.

procedure *LeastCostBranchAndBound*

Input: function $D_k(x_1, \ldots, x_{k-1})$ (determining state space tree T associated with the given problem)

```
        objective function φ defined on the solution states of T
        cost function c(v) such that:
        c(v) ≤ φ*(v) = min{φ(w) : w ∈ T_v and w is a solution state}
Output: a solution state (goal) where φ is minimized

  LiveNodes is initialized to be empty
  call AllocateTreeNode(Root)
  Root → Parent := nil
  call AddPriorityQueue(LiveNodes,Root) {add root to priority queue of live nodes}
  Goal := Root   {initialize goal to root}
  UB := ∞
  Found := .false.
  while LiveNodes is not empty .and. .not. Found do
     call Select(LiveNodes,E-node,k) {select E-node of smallest cost from
                                                  live nodes}
     if c(E-node) ≥ UB then   {Goal points to optimal solution state}
        Found := .true.
     else
        if E-node is a solution state and φ (E-Node) < UB then {update UB}
           UB := φ (E-Node)
           Goal := E-Node
        endif
        for each X[k] ∈ D_k(E-node) do   {for each child of the E-node do}
           if c(X[k]) < UB .and. .not. StaticBounded (X[1], . . . , X[k]) then
              call AllocateTreeNode(Child)
              Child → Info := X[k]
              Child → Parent := E-node
              call AddPriorityQueue(LiveNodes,Child) {add child to list of
                                                       live nodes}
           endif
        endfor
     endif
  endwhile
  call Path(Goal)   {output path from goal node to root}
end BranchAndBound
```

16.5 Game Trees

Ever since the invention of the electronic computer there has been interest in computerized strategies for playing two-player games. For example, particular interest has been focused on designing computer programs to play chess (see Figure 16.11). The first computer programs written for playing chess were not very sophisticated, partly because early computers were not powerful enough to store the vast amounts of information necessary to play a good game. Computer programs implemented on

Figure 16.11 Computerized game playing (Jim Borgman, © 1995 *The Cincinnati Enquirer.* Reprinted with special permission of King Features Syndicate. All rights reserved.)

levels for such games as backgammon, and recently computer programs for playing bridge have exhibited excellent performance.

Computerized game-playing strategies are usually based on efficient partial searching of the enormous game tree modeling all possible legal moves for a given (two-player) game. The size of the game tree of all possible moves is generally much too large to admit complete searches. Thus, computerized game-playing strategies use heuristics to estimate the value of various moves based on looking down a limited number of levels in the game tree.

Game trees can become quite large even for simple games. For example, consider the game tree associated with tic-tac-toe on the 3×3 board (see Chapter 15). Suppose we have two players, A and B, who are placing X's and O's, respectively, and player A moves first. Note that player A has 9 possible choices to place the first X. (Of course, using symmetries of the square, only three of these moves are nonequivalent. We choose to ignore this reduction for the moment.) After player A moves, then player B has 8 possible choices for placing the first O. Continuing in this fashion, we see that the game tree has an upper bound of $9! = 362{,}880$ nodes, although the actual number of nodes is smaller since the game terminates whenever three X's or three O's in a row is achieved. Figure 16.12 shows all the nodes at levels 0 and 1 of the game tree, but only that portion of the nodes at level 2 corresponding to the eight possible moves from a particular single node at level 1. In Figure 16.13 (page 571) we show all the nodes in the levels 0, 1, and 2 of the pruned game tree that results by pruning symmetric board configurations.

Now consider the game tree modeling an arbitrary game between two players A and B, who alternately make moves, with each player having complete knowledge of the moves of the other (a *perfect information* game). The root of the game tree

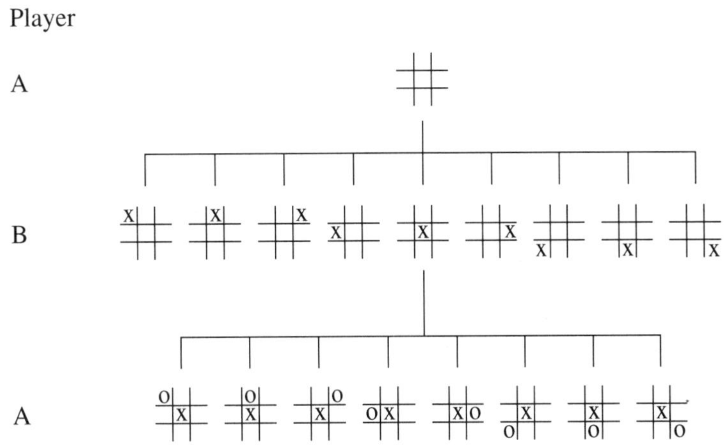

Figure 16.12 A portion of the game tree for tic-tac-toe on the 3×3 board. Children of a single node at level 1 are shown.

corresponds to the opening move in the game, which is assumed to be made by player A. Nodes at even levels in the game tree correspond to configurations where it is A's move and are called *A-nodes*. Nodes at odd levels correspond to configurations where it is B's move and are called *B-nodes*. The children of an A-node (respectively, B-node) correspond to all admissible moves available to player A (respectively, player B) from the node. Similar to the game tree for tic-tac-toe, for general games we assume a particular ordering of all admissible moves from a given node, so that our game trees are always ordered trees. Leaf nodes in the game tree are called *terminal* nodes and correspond to an ending (outcome) of the game. In general, terminal nodes correspond to a win, loss, or tie for player A, although certain games such as nim cannot end in a tie.

Figure 16.13 A portion of the game tree for tic-tac-toe on the 3×3 board pruned by symmetric board configurations. All nodes at first three levels are shown.

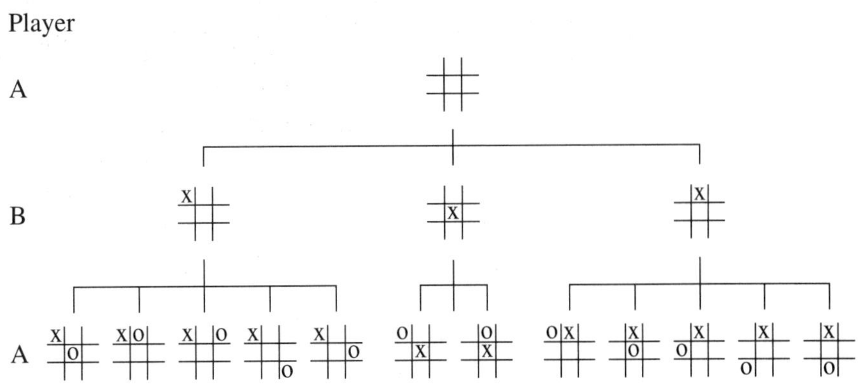

Given a game tree, the value to player A is assigned to each terminal node (outcome of the game). We also assume that the game is a *zero-sum* game, so that the value to player B of a terminal node is the negative of its value to player A. *However, when we speak of the value of a node, until further notice it is always the value to player A.* We wish to design an algorithm that determines player A's optimal first move. In other words, we wish to determine the move that player A should make so that the game will end at a terminal node of maximum value for player A, assuming that each player plays perfectly. Of course, if player B does not play perfectly, then player A can perhaps even do better.

To analyze the entire game, it is enough to design a procedure to determine the optimal opening move. Indeed, after player A makes the opening move, we would simply repeat (from player B's point of view) the optimal strategy at the game subtree rooted at the node corresponding to this move, and so forth.

If the game tree is small enough to be completely traversed in a reasonable amount of time, then there is a simple *minimax procedure* for player A to determine the optimal opening move. We simply perform a postorder traversal of the game tree, where a visit at an A-node corresponds to identifying a child of maximum value for a move from the A-node and assigning this value to the A-node. Similarly, a visit at a B-node corresponds to identifying a child of minimum value for a move from the B-node and assigning this value to the B-node. *The minimax strategy is nothing more than the definition of perfect play for the two players.* Note that postorder traversal is necessary, since the value of each child of a node must be determined before the value of the node itself can be determined. When the postorder traversal is complete, the opening move, together with the value of the game to player A, will be determined.

We illustrate the minimax procedure for the game tree corresponding to a small instance of the game of nim. In the general game of nim, there are n piles of sticks, where the ith pile contains m_i sticks, $i = 1, \ldots, n$. Each player alternately chooses a nonempty pile and removes some or all of the sticks from this pile. There is usually a restriction made on how many sticks a player is allowed to remove in a given move. The last player to remove a stick loses. For large $m_1 + m_2 + \cdots + m_n$, the game tree modeling nim would be enormous. To keep things in sight, consider the instance $n = 2$, $m_1 = 3$, $m_2 = 2$. The game tree for this instance is shown in Figure 16.14, where the numbers inside each node correspond to the number of sticks left in each pile. Thus, terminal nodes correspond to 0,0. We assign the value of +1 to a terminal A-node (A wins) and – 1 to a terminal B-node (B wins). In Figure 16.14 we have done some pruning of the complete game tree to eliminate generating symmetric children configurations of the two nodes [1,1] and [2,2]. For example, in the complete game tree, the node [2,2] generates the four nodes [0,2], [1,2], [2,0], and [2,1]. Using symmetry, we need only display the first two nodes in Figure 16.14 when drawing the (pruned) game tree.

Figure 16.15 shows the results of a postorder traversal of the game tree in Figure 16.14, where visiting a node executes the minimax procedure described previously. In Figure 16.15 we show the value (to player A) of each node outside the node. Note that the [3,2] game of nim is a win for player A. The complete traversal

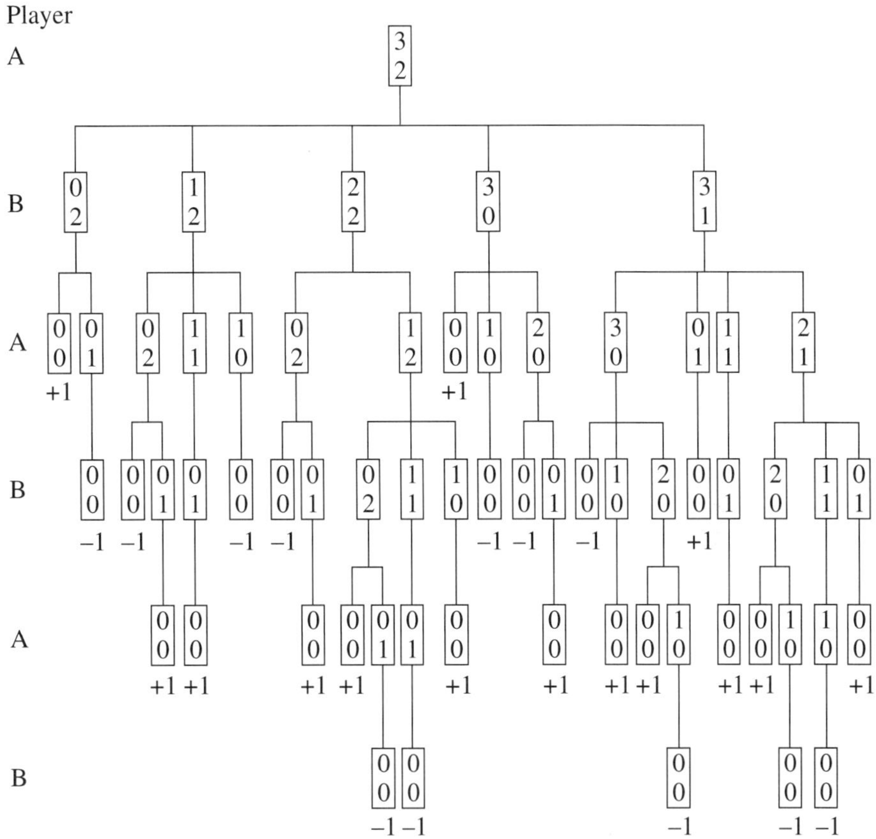

Figure 16.14 Game tree for [3,2] nim, pruned to eliminate symmetric children of [1,1] and [2,2]. Terminals node values are +1 for A wins, –1 for B wins.

of the game tree shows that both [2,2] and [3,1] are opening moves that guarantee a win for player A. If a single winning strategy is desired, then the postorder traversal could be terminated as soon as it is determined that the root node has value +1 (with the opening move [2,2]). Of course, similar termination can be done in any game where we simply have the values +1, 0, and –1 for win, tie, and loss, respectively.

Games for which a complete traversal of the game tree is feasible are usually too small to be very interesting. Even the ordinary game of 3×3 tic-tac-toe has a rather large game tree. For games like chess, the game tree is truly enormous. The game tree for chess has been estimated to contain over 10^{100} nodes! Thus, rather than attempting to traverse the entire game tree when determining an optimal move, in practice the minimax procedure is usually limited to looking ahead a fixed number r of levels in the game tree (*r-level search*). Terminal nodes encountered within r levels are assigned the value of the outcome of the game corresponding to this node.

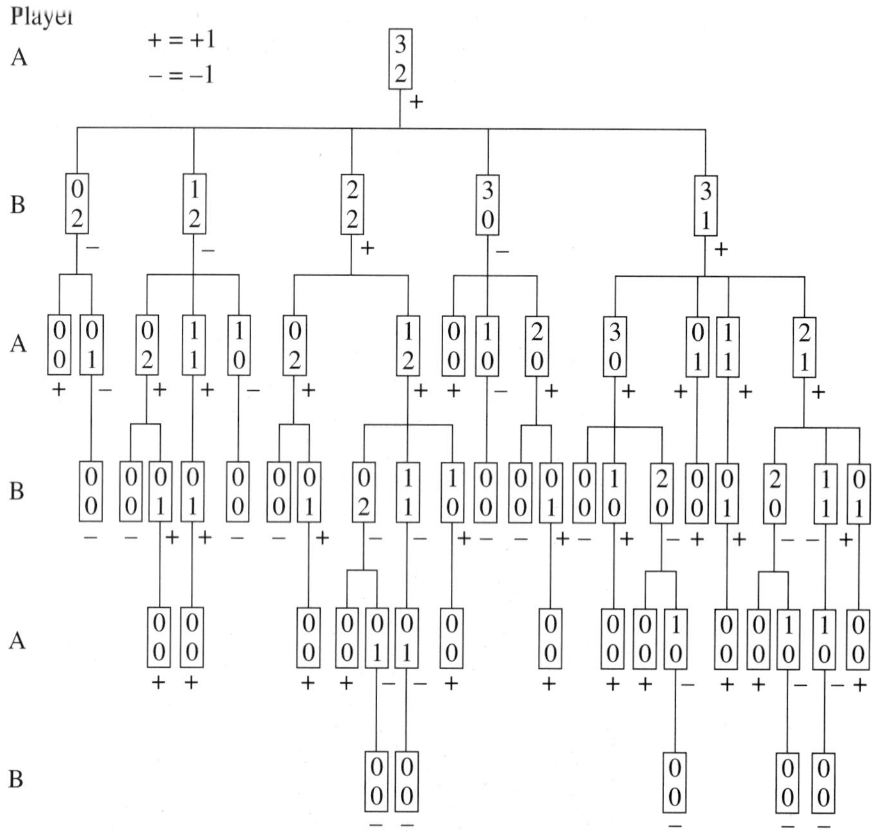

Figure 16.15 Values + = +1 and – = –1 assigned to each node by the minimax postorder traversal of the [3,2] game of nim

Nonterminal nodes at level r are assigned some estimate of the value of the node based on the best available knowledge, typically using some heuristic. Nodes that look more promising are given higher values. Of course, the better the estimate, the better the strategy generated by the r-level search.

Suppose we consider 3×3 tic-tac-toe with a two-level search. None of the nodes in the first two levels are terminal nodes, so we need to come up with some estimate of the value of each node at level 2. A natural choice would be to assign to a given node (configuration of the board) the number of winning lines completable in X's minus the number of winning lines completable by O's. For example, in Figure 16.16 we show the value of each of the nodes at level 2 in the game tree (pruned by symmetries). We also show in Figure 16.16 the result of applying the two-level search (minimax procedure) to the game tree for 3×3 tic-tac-toe, which gives the values of –1, 1, and –2 to the nodes C_1, C_2, and C_3 at level 1, respectively, and gives the root node a value of 1.

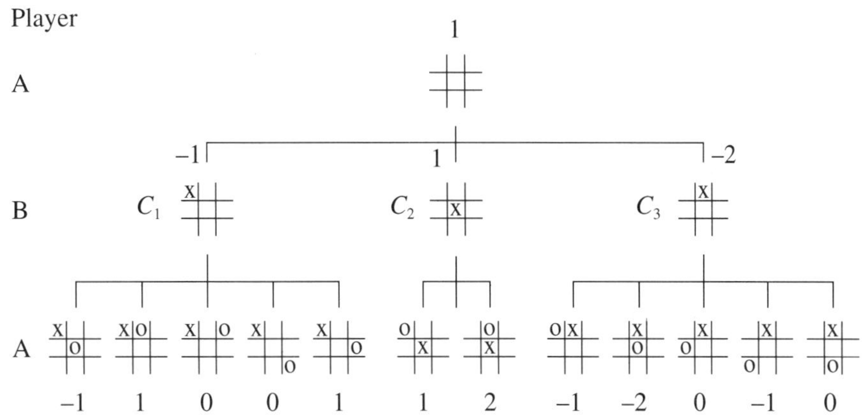

Figure 16.16 Value of each node at level 2 as computed using the number of winning lines completable by X minus number of winning lines completable by O. The values of nodes at levels 1 and 0 are computed using the minimax strategy.

We see from Figure 16.16 that player A's opening move would be to place an X in the center position. Then the minimax procedure is continued from the subtree rooted at the latter node. Unfortunately, continuing with the same two-level heuristic search method may lead to a loss for player A. (See Exercise 16.25.) The fairly obvious fix to the latter problem is to assign an appropriately large value to terminal positions when encountered in the search. We simply give terminal nodes that are wins for player A (that is, three X's in a row) any value greater than 8, which is the total number of winning lines for the 3×3 game. For example, we could assign the value 9 to terminal nodes that are wins for player A and – 9 to terminal nodes that are wins for player B. Terminal nodes corresponding to tie games (cats boards) are assigned the value 0. With the values 9, 0, and – 9 so assigned to terminal nodes, a two-level search leads to a tie game. A tie game for the 3×3 board is the best that either player can hope for when both players play perfectly.

There is a heuristic strategy called *alpha-beta pruning* that can result in a significant reduction in the amount of nodes required to visit during an n-level search and still correctly compute the value of a given node. The easiest way to explain alpha-beta pruning is by example. Consider again the two-level search made in the game tree in Figure 16.16. After returning to the root from the middle child C_2, we know that player A can make a move to a node having value 1. Then we move to the third child C_3 of the root and begin visiting the children of C_3 (grandchildren of the root). The first child of C_3 has value – 1, so we can immediately cut off our examination of the remaining children of C_3. The reason is simple: The value of C_3 is the minimum value of its children, so that the value – 1 of the first child of C_3 places an upper bound of – 1 on the value of C_3. Since a lower bound on the value of the root is already known to be 1, the value of C_3 cannot possibly affect the value of the root. The cutoff just described is illustrated in Figure 16.17. A cutoff of the search of the grandchildren of an A node (respectively, B node) is called *alpha cutoff* (respectively, *beta-cutoff*).

We formalize the notion of alpha-beta pruning as follows. A lower bound for the value of an A-node is called an *alpha value* of the A-node. Note that during an r-level search, an alpha value of a parent A-node is determined when we return to the A-node from its first child, and the alpha value can be updated, when appropriate, when we return from subsequent children. For example, after returning to the root from child C_1 in Figure 16.17, –1 was known to be an alpha value of the root. However, when returning to the root from the second child C_2, the alpha value of the root can be updated to 1.

In general, suppose during an r-level search we are examining the children of the ith child of C_i, where the parent of C_i is an A-node X. If we encounter a child of C_i (grandchild of X) whose value is not larger than an alpha value of X, then we can cut off (alpha-cutoff) our search of the remaining children of C_i, since the value of C_i cannot affect the value of X.

An entirely symmetric discussion holds for B-nodes. More precisely, an upper bound for the value of a B-node is called a *beta value* of the B-node. Given any grandparent B-node Y, if during an r-level search we encounter a child (grandchild of Y) of the ith child D_i of Y whose value is not smaller than a beta value of Y, then we can cut off (beta-cutoff) our search of the remaining children of D_i, since the value of D_i cannot affect the value of Y.

In Figure 16.18, we illustrate a sample game tree and the effect of alpha-beta pruning for a complete search (that is, a three-level search) of the tree. To illustrate the dynamic nature of alpha and beta values, in Figure 16.18a we show the indicated alpha and beta values of nodes just after returning to the node X from its second child. The value inside a given node is either the actual value of the node or an alpha or beta value as appropriate. Those nodes containing an alpha or a beta value are flagged as such. A value in a node that is shown as * means that the value is irrele-

Figure 16.17 Illustration of alpha-cutoff

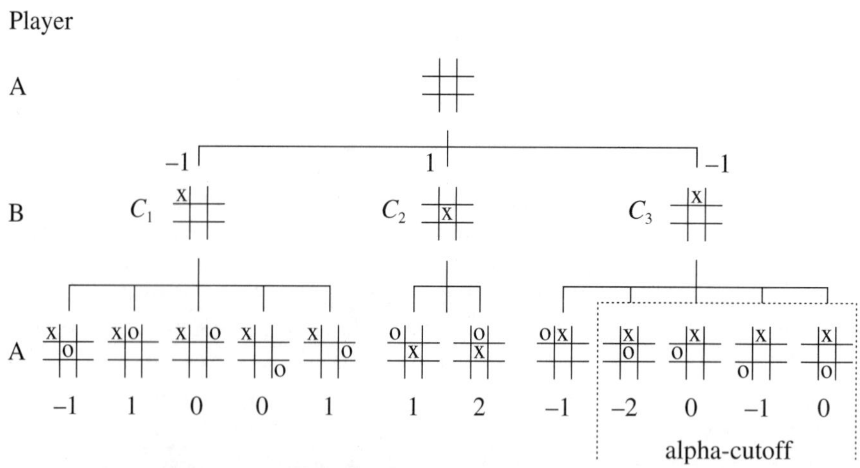

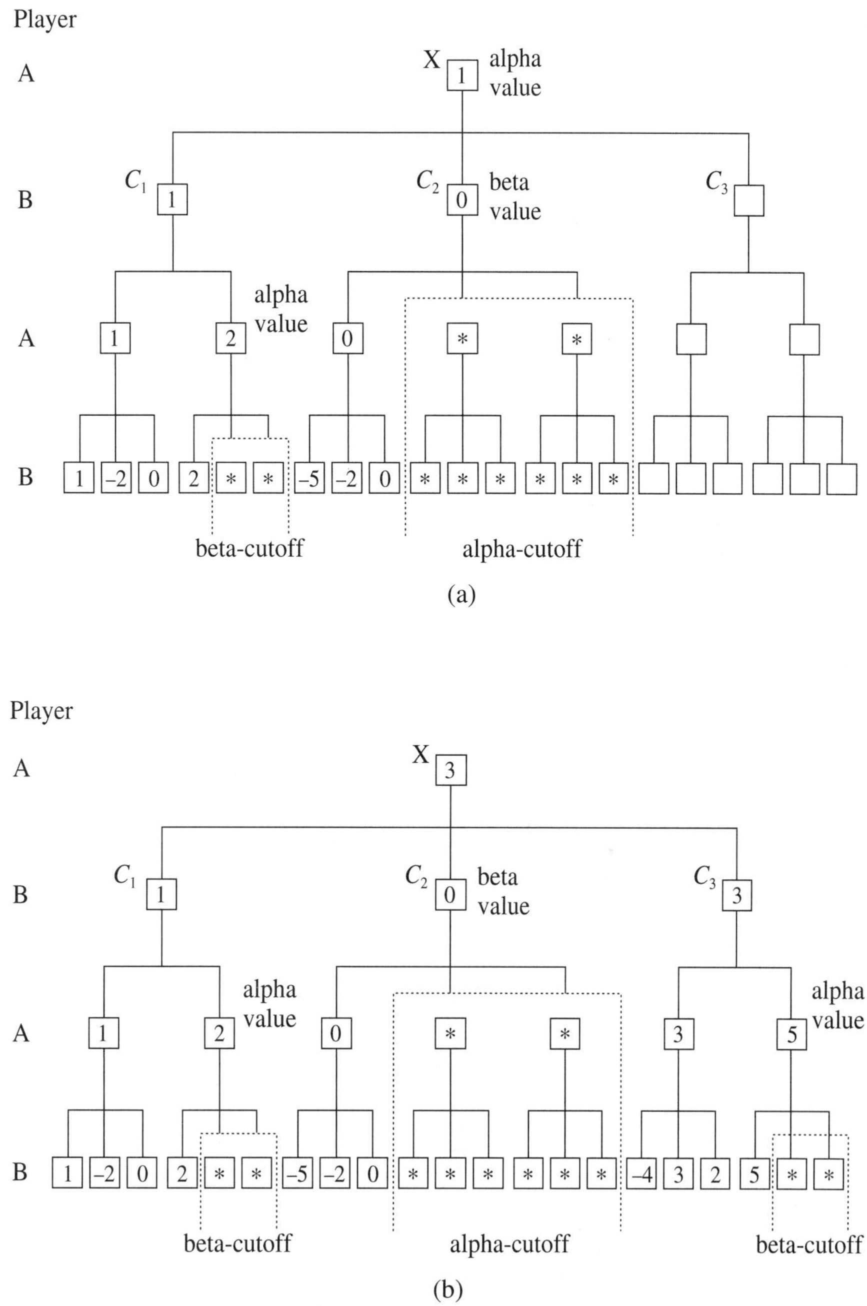

Figure 16.18 (a) The results of the search just after returning to X from its second child C_2; (b) the results of the completed search.

vant, since the node is never reached due to alpha-beta pruning. To emphasize the stage of the search in Figure 16.18a, no values are shown inside the nodes of the subtree rooted at the third child of X. In Figure 16.18b values are supplied for the nodes in the latter subtree, where we show the results of the completed search.

When writing pseudocode implementing the minimax procedure, it is convenient to consider the value of a B-node to be the value to player B, not A. In other words, we simply change sign of the values given to B-nodes in our previous discussion. This change simplifies the pseudocode by turning the minimax procedure into simply a max procedure. Note that in the max procedure, the value of either an A-node or a B-node is the maximum of the negatives of the values of their children. Thus, the identical max procedure is executed at an A-node or a B-node.

In the new scenario, our cutoff rule takes the same form whether we are examining the children of an A-node or a B-node. In either case, suppose LB is a lower bound for the value of the node v whose children are being evaluated, and suppose *ParentValue* is a lower bound for the value of the parent node. If we ever determine that $-LB \le ParentValue$, then we can cut off further examination of the children of v, since the value of v cannot affect the value of the parent of v.

The following pseudocode for the recursive function *ABNodeValue* returns the value of a node X in the game tree using a *NumLevels*-search, where the parameter *ParentValue* is a lower bound for the value of the parent of X. If X is a terminal node, or if $NumLevels = 0$, then we assume that X has been given an appropriately defined value (denoted by $Val(X)$) as described earlier. Given a node X and an integer r, the value of X would be calculated using an r-search by invoking *ABNodeValue* initially with arguments X,r,∞.

```
function ABNodeValue(X,NumLevels,ParentValue) recursive
Input:  X (a node in the game tree having children C_1,C_2, . . . , C_k),
        NumLevels (number of levels to search)
        ParentValue (lower bound on value of parent of X)
Output: returns the value of X
  if X is a terminal node .or. NumLevels = 0 then
    return(Val(X))
  else
    LB := – NodeValue(C_1,NumLevels – 1,∞) {initial lower bound for value of X}
    for i := 2 to k do
      if LB ≥ – ParentValue then   {cutoff}
        return(LB)
      else
        LB := max(LB, – NodeValue(C_i,NumLevels – 1, – LB)
      endif
    endfor
  endif
  return(LB)
end ABNodeValue
```

When measuring the efficiency of *ABNodeValue*, the quantity of interest is the number of nodes cut off from the straight minimax *r*-search of the game tree that does not use the cutoff rule. Of course, not much can be said in general unless some assumptions are made about the regularity of the game tree. Even with strong regularity conditions imposed on the game tree, the analysis is difficult. We merely state one result in this direction. Pearl has shown that for game trees in which each parent has the same number of children, and in which the terminal nodes are randomly ordered, *ABNodeValue* permits a search depth greater by a factor of 4/3 than that allowed by the straight minimax procedure in the same amount of time.

Alpha-beta pruning can be enhanced by adding an additional parameter *NodeValueLowBnd* into the algorithm *ABNodeValue*. *NodeValueLowBnd* is maintained as a lower bound on the value on the input parameter *X*. Additional pruning of the game tree results from the following observation:

> *If the value of a grandchild of X is not larger than NodeValueLowBnd, then all remaining children of this grandchild can be pruned.*

Whereas alpha-beta pruning only uses information from the parent, *NodeValueLowBnd* carries information deep into the tree, so that the resulting evaluation of the game tree is called *deep* alpha-beta pruning. We leave the design of the recursive function for deep alpha-beta pruning as an exercise.

16.6 Closing Remarks

Seeking solutions to the 8-puzzle game or finding a shortest-length trip along a freeway system are examples of what has been called *single-agent* problems. In general, A*-search is well suited to large-scale problems where the *entire solution* is sought in a reasonable length of time (and then saved for future reference) rather than for real-time problems where the first step (and each successive step) in the solution must be computed very quickly. For example, it might be acceptable for a computer to take weeks or even months to solve a highly important single-agent problem, since its solution would then be known and usable *in real time* thereafter.

Ordinary A*-search as applied to single-agent problems usually finds the entire path to a goal before even the first move from the starting position is known for sure. Hence, using A*-search for single-agent problems becomes too costly for large-scale applications where the optimal decisions along the way must be decided quickly and in advance of the final solution. For example, you might be under a short time constraint to make each move in a game such as the 8-puzzle game, rather than simply wanting to determine the *entire solution* in a larger but more reasonable length of time.

While most single-agent problems are not subject to intermediate real-time constraints, two-player games usually are. In chess, for example, there is usually a time restriction on the amount of time a player has to make the next move. Moreover, the game tree for chess is so enormous that generating complete solutions is out of the question. To make real-time decisions, the alpha-beta heuristic is based on attempting to evaluate moves in a limited *search horizon*, that is, looking ahead a fixed number of moves.

For single-agent problems, a heuristic search method called *real-time A*-search* combines the A*-search strategy with a limited look-ahead search horizon. Real-time A*-search utilizes an analog of (minimax) alpha-beta pruning called (minimin) *alpha pruning*. Alpha pruning drastically improves the efficiency of A*-search without affecting the decisions made. Real-time A*-search, like A*-search, can find the entire solution to fairly large-scale problems in a reasonable amount of time. However, real-time A*-search has the advantage that it also generates the optimal moves along the way quickly and before the entire solution is known. Refer to the references for further information on real-time A*-search.

Suppose that a perfect-information game involving alternate moves by two players A and B must end in a finite number of moves with a win for one of the two players. Then one of the two players must have a *winning strategy*, that is, a strategy that guarantees a win regardless of the moves made by the other player. The reason is simple: If neither player had a winning strategy, there would be a sequence of alternate moves made by A and B that never ends in a loss for either player. Since that sequence would not terminate, we obtain a contradiction of the finiteness assumption of the game.

A two-person *positional* game is determined by a collection of sets A_i, $i = 1, \ldots, m$. The players alternately choose an element (which they keep) from $\bigcup_{i=1}^{n} A_i$. The first player to choose *all* the elements from one of the sets wins. Tic-tac-toe is an example of a positional game, where the sets A_i are the winning lines in the board. For positional games that cannot end in a tie, such as the $3 \times 3 \times 3$ game of tic-tac-toe, the first player always has a winning strategy. Indeed, we have just seen that one of the two players must have a winning strategy, so suppose it is the second player. Then the first player makes a random opening move, and thereafter assumes the role of the second player (basically ignoring the opening move). More precisely, when the first player moves, he chooses the move dictated by the second player's winning strategy, or moves randomly if he has previously made this move. Since having made an extra move in a positional game cannot possibly hurt, the first player is thus led to a win! This contradicts the assumption that the second player has a winning strategy and shows that the first player has a winning strategy. The same argument shows that if there is a winning strategy for a positional game, then it must belong to the first player.

For positional games that cannot end in a tie, the fact that the first player has a winning strategy does not mean that there is an efficient algorithm to generate the strategy. Also, for positional games that *can* end in a tie (given, perhaps, imperfect play), there still might exist a winning strategy for the first player. For example, tie positions exist for the $4 \times 4 \times 4$ game of tic-tac-toe (winning sets being 4 in a row). However, it was conjectured for a long time that the first player has a winning strategy in $4 \times 4 \times 4$ tic-tac-toe. This conjecture was finally established by Patashnik using clever bounding arguments that allowed a pruning of the enormous game tree for $4 \times 4 \times 4$ tic-tac-toe, reducing it to a size that was amenable to computer search.

References and Suggestions for Further Reading

Kanal, L. and V. Kumar, eds., *Search in Artificial Intelligence*. New York: Springer-Verlag, 1988. A book containing numerous articles on searching in artificial intelligence including a discussion on the optimality of A*-search.

Korf, R. E. "Real-time heuristic search," *Artificial Intelligence*, 42 (1990): 189–211. A paper devoted to real-time A*-search.

Nilsson, N. J. *Principles of Artificial Intelligence*. Palo Alto, CA: Tioga Publishing, 1980. A detailed account of A*-search, which was originally developed by Hart, Nilsson, and Raphael.

Pearl, J. *Heuristics: Intelligent Search Strategies for Computer Problem Solving*. Reading, MA: Addison-Wesley, 1984. A text devoted to heuristic searching.

For two recent books on artificial intelligence that contain extensive discussions of search strategies and game playing, see:

Rich, E. and K. Knight. *Artificial Intelligence*, 2nd Ed. New York: McGraw-Hill, 1991.

Russell, S. J. and P. Norvig. *Artificial Intelligence: A Modern Approach*. Englewood Cliffs, NJ: Prentice Hall, 1995.

Patashnik, O. "Qubic: 4×4×4 tic-tac-toe," *Math. Mag.*, 53 (1980): 202–216. A nice survey discussion of *n*-dimensional tic-tac-toe, as well as the proof that the 4 × 4 × 4 is a first player win.

Two papers containing detailed analyses of alpha-beta and deep alpha-beta pruning are:

Knuth, D. "An analysis of alpha-beta cutoffs," *Artificial Intelligence*, 6 (1975): 293–316.

Baudet, G. "An analysis of the full alpha-beta pruning algorithm," *Proc. 10th Ann. ACM Symp. on Theo. of Comput.*, San Diego, CA, 1978, pp. 296–313.

Berlekamp, E. R., J. H. Conway, and R. K. Guy. *Winning Ways, for Your Mathematical Plays*, Vol. I, II. New York: Academic Press, 1982. A book covering strategies for a host of games.

Exercises

Section 16.2 The 8-Puzzle Game

16.1 Draw the first three levels of the state space tree generated by breadth-first search for the 8-puzzle game with the following initial and goal states.

goal

1	2	3
4	5	6
7	8	

initial state

4	2	7
1		6
3	5	8

16.2 The $(n^2 - 1)$-puzzle is a generalization of the 8-puzzle to the $n \times n$ board. The goal position is where the tiles are in row-major order (with the empty space in the lower right-hand corner). For $k \in \{1, \ldots, n^2\}$, let $L(k)$ denote the number of tiles t, $t < k$, such that the position of t comes after k in the row-major order in the initial arrangement (the empty space is considered as tile n^2). Show that a necessary and sufficient condition that the goal can be reached is that

$$\sum_{k=1}^{n^2} L(k) \equiv i + j \ (\mathbf{mod}\ 2), \tag{16.3.4}$$

where (i,j) is the position of the empty space in the initial arrangement.

Section 16.3 A*-Search

16.3 Show that if h is a monotone heuristic, then $h(v)$ is a lower bound of the cost of the shortest path from v to a (nearest) goal.

16.4 Prove Proposition 16.3.3.

16.5 Design an algorithm for A*-search using a heuristic that is not necessarily monotone.

16.6 For the 8-puzzle game, consider the following heuristic.

$h(v)$ = the number of tiles not in correct cell in the state v.

Show that $h(v)$ satisfies the monotone restriction.

16.7 For the 8-puzzle game, consider the following heuristic.

$h(v)$ = the sum of the Manhattan distances (vertical steps plus horizontal steps) from the tiles to their proper positions in the goal state.

Show that h satisfies the monotone restriction.

16.8 For the 8-puzzle game, let h be the monotone heuristic defined in Exercise 16.6. For the following initial and goal states, draw the states generated by making the first three moves in the game using A*-search with the priority function $f(v) = g(v) + h(v)$, ($g(v)$ is the number of moves made from the initial state to v). When enqueuing states, assume that the (possible) moves of the empty tile are ordered: move left, move right, move up, move down. Label each state v with its f-value.

goal

1	2	3
4	5	6
7	8	

initial state

4	2	7
1		6
3	5	8

16.9 Repeat Exercise 16.8 for the heuristic h defined in Exercise 16.7.

16.10 Write a program for the n-puzzle game using the Manhattan distance heuristic. Test your program for $n = 8$ and $n = 15$.

16.11 Can you find better heuristics (not necessarily lower bounds) for the n-puzzle game than the Manhattan distance heuristic? Test your heuristic empirically for $n = 8$ and $n = 15$.

16.12 Prove Theorem 16.3.1.

Section 16.4 **Least Cost Branch-and-Bound**

16.13 Show that the heuristic $h(v)$ given in Section 16.4 for the Coin-Changing problem is a lower bound for the minimum number of additional coins required to make correct change from the given problem state v.

16.14 Write a program implementing a least cost branch-and-bound solution to the Coin-Changing problem.

16.15 Design a heuristic and a least cost branch-and-bound algorithm for the variation of the Coin-Changing problem where we have a limited number of coins of each denomination. Assume the number of coins of each denomination is input along with the denominations.

16.16 Draw the portion of the variable-tuple state space tree generated by least cost branch-and-bound for the instance of the 0/1 Knapsack problem given in Figure 15.15, using the heuristic given in Section 16.4. Label each node with the value of $c(v)$ and the current value of UB.

16.17 Repeat Exercise 16.16 for the fixed-tuple state space tree.

16.18 Write a program implementing a least cost branch-and-bound solution to the 0/1 Knapsack problem.

16.19 Given the complete digraph $\hat{K}_n$ with vertices $1, 2, \ldots, n$ and a nonnegative cost matrix $C = (c_{ij})$ for its edges (we set $c_{ij} = \infty$ if $i = j$ or if the edge ij does not exist), a traveling salesman tour starting at vertex 1 corresponds to a sequence of vertices $1, i_1, i_2, \ldots, i_n, 1$, where $i_1, i_2, \ldots, i_n$ is a permutation of $2, \ldots, n$. Consider a state space tree T for the traveling salesman problem (finding a minimum cost tour) where a node at level k in T corresponds to a simple path containing $k + 1$ vertices, starting with vertex 1. Thus, T has depth n, and leaf nodes correspond to a sequence of choices $i_1, i_2, \ldots, i_n$ determining the tour $1, i_1, i_2, \ldots, i_n, 1$.

We now describe a cost function $c(v)$ for a least-cost branch-and-bound algorithm for the traveling salesman problem. The definition of $c(v)$ is based on the notion of a reduced cost matrix. A row (or column) of a (nonnegative) cost matrix is said to be *reduced* if it contains at least one zero. A (nonnegative) cost matrix is *reduced* if each row and column of the matrix is reduced (except for rows and columns whose elements are all equal to ∞). Given the cost matrix C, an associated reduced cost matrix C_r is constructed as follows. First, reduce each row by subtracting the minimum entry in the row from each element in the row. In the resulting matrix, repeat this process for each column. We define $c(r)$ to be the total amount subtracted. The following example illustrates C_r for a sample C.

$$C = \begin{pmatrix} \infty & 23 & 9 & 32 & 12 \\ 21 & \infty & 2 & 16 & 4 \\ 4 & 8 & \infty & 20 & 6 \\ 15 & 10 & 4 & \infty & 2 \\ 9 & 5 & 8 & 10 & \infty \end{pmatrix} \qquad C_r = \begin{pmatrix} \infty & 14 & 0 & 18 & 3 \\ 19 & \infty & 0 & 9 & 2 \\ 0 & 4 & \infty & 11 & 2 \\ 13 & 8 & 2 & \infty & 0 \\ 4 & 0 & 3 & 0 & \infty \end{pmatrix} \qquad c(r) = 27$$

More generally, we define (inductively on the levels of T) a reduced cost matrix for each nonleaf node by suitably reducing the cost matrix C_u associated with the parent node u of v. Suppose u corresponds to a path ending at vertex i, and v corresponds to adding the edge ij to this path. We then change all the entries in row i and column j of C_u to ∞, as well as the entry in the jth row and first column. We then perform the same subtracting operation on the resulting matrix as we did when comput ing C_r. Let s_v denote the total amount subtracted, and define $c(v) = c(u) + C_u(i,j) + s_v$.

For leaf nodes v, $c(v)$ is defined as the cost of the tour determined by v.

a. Show that $c(r)$ is a lower bound for the minimum cost of a tour.

b. More generally, show that $c(v) \leq \phi^*(v)$, where $\phi^*(v)$ = the minimum cost over all tours determined by the leaf nodes of the subtree of T rooted at v.

c. Part (b) shows that $c(v)$ is suitable for *LeastCostBranchAndBound*. Design and give pseudocode for *LeastCostBranchAndBound* implementing $c(v)$.

16.20 Draw the portion of the state space tree T generated by the least-cost branch-and-bound discussed in the previous exercise for the cost matrix illustrated in that exercise. Label each node v with its cost value $c(v)$. Also, write out the reduced matrix associated with each node generated.

16.21 Discuss other state space trees and associated cost functions $c(v)$ for the traveling salesman problem.

Section 16.5 Game Trees

16.22 Consider the two-person zero-sum game shown on page 585. The values in the leaf nodes are values to player A. Use the minimax strategy (postorder traversal) to determine the value of the game to player A. Show clearly where alpha-cutoff and beta-cutoff occur, as well as (final) actual values, alpha values, and beta values of all nodes reached in the traversal.

16.23 Rewrite the recursive function *NodeValue* as a recursive procedure that has the same input parameters *Y,NumLevels,ParentValue*, but now returns in output parameters the value V of Y and the child C_i whose value is $-V$.

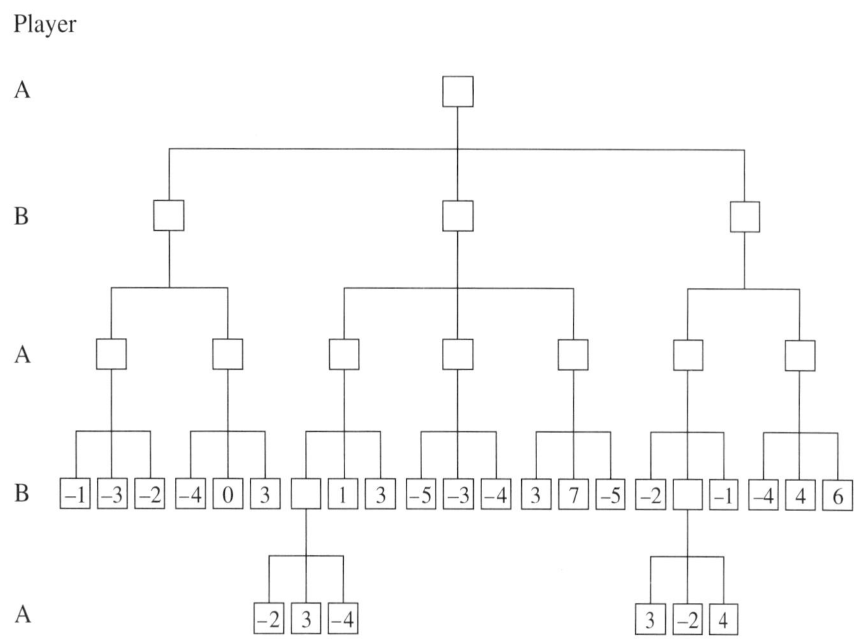

16.24 Find a sequence of admissible moves for the two-level heuristic search illustrated in Figure 16.16 that leads to a loss for player A in 3×3 tic-tac-toe.

16.25 Show that by assigning the values 9, 0, and -9 to terminal nodes that are wins, ties, or losses, respectively, for player A, the two-level search illustrated in Figure 16.16 never leads to a loss for player A.

16.26 There are no tie positions in the $3 \times 3 \times 3$ tic-tac-toe game so that the first player has a winning strategy. Find a winning strategy for the first player.

16.27 Design a recursive function *DABNodeValue*(*X*,*NumLevels*,*ParentValue*, *NodeValueLowBnd*) for deep alpha-beta pruning. The initial invocation of *DABNodeValue* should have *ParentValue* $= \infty$ and *NodeValueLowBnd* $= -\infty$.

16.28 Redo Exercise 16.28 for deep alpha-beta pruning. Indicate any pruned nodes that were not pruned by alpha-beta pruning.

17

The Dictionary Problem: Hashing and Balanced Trees

The *dictionary operations* of inserting, deleting, or searching for keys stored in some type of data structure occur in a variety of applications. Database programs must support rapid location of customer records, word processing programs must contain search and replace operations, compilers must quickly recognize keywords, and so forth. Binary search trees, multiway search trees, and hash tables are commonly used to store keys. In this chapter we discuss these ADTs and their associated algorithms for efficiently implementing dictionary operations.

17.1 The Dictionary Problem

Given a totally ordered set S (elements of S will be called *keys* or *identifiers*), the *dictionary problem* is the problem of designing an ADT to maintain a collection of items drawn from S. The classic dictionary problem restricts attention to the dictionary operations of inserting a key, deleting a key, and searching for a key, but we also might consider performing additional operations, such as finding the maximum or minimum elements, accessing the keys in sorted order, and so forth. We seek ADTs that perform dictionary operations efficiently.

One of the simplest ADTs for the dictionary problem is a sorted array. Searching can be done in logarithmic time using *BinarySearch*. However, insertion and deletion are inefficient due to the large number of array reassignments that are required on average. Another ADT for the dictionary problem is an array indexed by the keys themselves. Using this latter ADT, the dictionary operations can be performed in *constant* time! The problem, however, is that such an array requires an index range large enough to contain the entire universe from which the keys are drawn, and this universe is usually too large to make this approach feasible. A compromise is to map the keys to some reasonably large interval of integral indices $\{1, \ldots, n\}$, using a suitably defined function (called a *hash function*). Good choices of hash functions will avoid excessive "collisions" among the keys when they are mapped to the corresponding array *HashTable*[1:n]. We discuss the method of hash functions and hash tables in the next section.

Other commonly used ADTs for the dictionary problem are balanced binary and multiway search trees. In fact, we have already discussed the dictionary operations for binary and multiway search trees in Chapter 2. Search trees not only support the dictionary operations, they also have the advantage that they support accessing keys in sorted order using inorder traversals. However, they have the disadvantage that insertions and deletions might make them unbalanced, leading to linear worst-case behavior for the operations of inserting and searching for elements. By keeping the search trees balanced we can reduce the complexity of these operations to logarithmic complexity.

The ideally balanced binary tree on n nodes is the complete binary tree T_n, which has depth $\lfloor \log_2 n \rfloor$. Recall from Proposition 7.3.1 that T_n has minimum depth over all binary trees on n nodes. However, maintaining minimum depth binary trees in the presence of dictionary operations is too costly. What we seek is a compromise between a strong balance that guarantees minimum depth and an efficiently maintainable balance that is still strong enough to keep the depth logarithmic in the number of nodes.

Two well-known types of balanced binary search trees are AVL trees and red-black trees. Both AVL trees and red-black trees on n nodes have depths belonging to O(log n). Insertion and deletion of elements in AVL trees and red-black trees begin just as for ordinary binary search trees. However, after inserting or deleting an element in an AVL tree or a red-black tree, rebalancing needs to be done to maintain their respective notions of balance.

A type of balanced multiway search tree known as a B-tree is commonly used to maintain the keys associated with records in a database stored on secondary (external) memory devices such as magnetic disks. Accessing and processing information in a node of a B-tree corresponds to a disk access I/O operation. Disk accesses are much slower than accesses to main (internal) memory, so that algorithms for B-trees are designed to minimize node accesses (disk I/O operations). Since moves between adjacent levels in a B-tree require disk I/O operations, it is not unusual for B-trees to contain thousands of keys in each node to minimize the depth of the tree. Finding a key in a B-tree requires additional comparisons in each node, but comparisons are performed in internal memory and are orders of magnitude faster than disk I/O operations.

17.2 Hashing

The method called *hashing* for solving the dictionary problem utilizes a *hash function* and a *hash table* implemented using an array. Clearly, if we utilize an array for storing n keys whose index range is as large as the size of the ordered set S consisting of all possible keys, then we can reserve a unique position in the array for each key. In the so-called *direct-addressing* scheme, the ordered set S itself is used as the indexing for the array. For (a very simple) example, if S consists of integers between

1 and 99, and the actual keys in current use are 7, 23, 55, 61, and 80, then an array *DirAddr*[1:99] supporting direct-addressing would have positions (slots) 7, 23, 55, 61, and 80 occupied with key values, and the remaining 94 slots would be unoccupied. If the keys are used to index records in a database file, then it would be assumed that the element in the ith slot of *DirAddr*[1:99], $i = 1, \ldots, 99$, contains the address of the record having the key i in this file.

In a direct-addressing scheme, all the dictionary operations (inserting, deleting, and searching for a key) can be performed in constant time. Unfortunately, the direct-addressing scheme is not often feasible in practice, since the size of S is simply too large. For example, suppose we have at most ten thousand records corresponding to students at UHash College, where the key is a student's 9-digit social security number. We would need to maintain an array of size 999,999,999 to have enough positions for every possible social security number, even though at most 10,000 of these positions would be actually utilized.

The idea behind hashing is to define an appropriate mapping h from the set S of all possible keys to a much smaller set S' consisting of consecutive integers. For example, in our UHash College example, suppose h maps each social security number s onto the number determined by the last four digits of s. Then h maps the set S of all social security numbers onto the much smaller set S' of integers between 0 and 9999, inclusive. The associated hash table can then be an array H[0:9999], and a nine-digit social security number s would be stored in position $h(s)$ of the array H.

Although hashing has the advantage of saving memory, it has the drawback that multiple keys in S are mapped by h to the same integer in S'. For example, two or more students from UHash College may have social security numbers with the same last four digits. After the first student has been inserted in the hash table, the attempted insertion of a second student whose social security number has the same last four digits results in a *collision*. In the next two subsections we discuss the two standard methods used to resolve collisions, *chaining* and *open addressing*. Of course, it is best to avoid collisions as much as possible by a good choice of hash function. In our UHash College example, the choice of hash function was not very good, since it completely ignored the first five digits of the social security number. It would be better to use a hash function h that depends on all the digits of s, since there might be some bias present in the last four digits.

The word *hashing* derives from the fact that $h(s)$ is often obtained, as in our UHash College example, by chopping off or otherwise altering some aspect of the key value. Also, good hash functions should mix things up pretty well, thereby keeping collisions to a minimum.

17.2.1 Hash Functions

As just mentioned, a good hash function is one that minimizes the number of collisions. Of course, a good hash function should also be easily computable. The number of collisions is minimized if each key k from S is equally likely to be mapped by the hash function h to any given index (*slot*) in the hash table. In other words, if there are m slots $0, \ldots, m-1$ in the hash table, and we independently draw keys from S according to a given probability distribution P, then the number of collisions is minimized if

$$P\{h(k) = i\} = \frac{1}{m}, \quad i = 0, 1, \ldots, m-1. \tag{17.2.1}$$

A hash function satisfying (17.2.1) is termed a *uniform hash function*, and such a hash function together with its associated hash table is called *uniform hashing*. Unfortunately, in most instances the probability distribution P on S is not known. Even when P is known, uniform hash functions are difficult to obtain. However, heuristic techniques can be used to find hash functions that come close to satisfying the assumption of uniform hashing and therefore perform quite well in practice.

To simplify our discussion, from now on we assume that the keys are drawn from the set of integers. There is usually a natural way in which to satisfy this assumption in practice. For example, keys made up of alphabetical characters could be replaced by their ASCII (decimal) equivalents. Given integer keys, two standard hashing techniques are the division method and the multiplication method.

In the *division method*, h maps k onto k **mod** m, where m is the size of the hash table:

$$h(k) = k \textbf{ mod } m. \tag{17.2.2}$$

We must be careful in the choice of m so as not to bias the hash function. For example, if m is divisible by two, then h would have the bias that even (odd) integers would be hashed to even (odd) positions in the hash table (goodbye uniform hashing). Similar biases would exist if m has other nontrivial factors. For instance, considering again our UHash College example, note that our hash function was precisely (17.2.2) with $m = 10^4$. To avoid these types of biases, a good choice of m is a prime not too close to an exact integer power.

In the *multiplication method*, we map the keys to slots in $H[0{:}m-1]$ by the formula

$$h(k) = \lfloor m(kA - \lfloor kA \rfloor) \rfloor, \tag{17.2.3}$$

where A is a constant such that $0 < A < 1$. Note that the term $(kA - \lfloor kA \rfloor)$ in (17.2.3) is merely the fractional part of kA. While the method works for any choice of the constant A, for a particular set of keys some choices of A will tend to yield a more uniform behavior for h than others. It turns out that the choice of A given by the inverse of the golden ratio $\phi^{-1} = (\sqrt{5} - 1)/2$ works well in many situations because of the especially uniform way that the points 0, $\phi^{-1} - \lfloor \phi^{-1} \rfloor$, $2\phi^{-1} - \lfloor 2\phi^{-1} \rfloor$, $3\phi^{-1} - \lfloor 3\phi^{-1} \rfloor, \ldots, k\phi^{-1} - \lfloor k\phi^{-1} \rfloor$ divide up the unit interval $[0, 1]$ (so that $h(0)$, $h(1)$, $h(2)$, $\ldots$, $h(k)$ map to $0, 1, \ldots, m-1$ in a correspondingly uniform way).

The choice of m in (17.2.3) is guided, of course, by such things as the anticipated size of the set of keys to be stored in the hash table. Fortunately, the choice of m is not critical to the good behavior of the hashing function given by (17.2.3). Thus, m is typically taken to be a power of two to make the implementation of $h(k)$ easier on most computers.

17.2.2 Collision Resolution by Chaining

In the method of collision resolution by *chaining*, the ith entry in the hash table $H[0{:}m-1]$, $i = 0, \ldots, m-1$, does not contain a key, but instead contains a pointer to a linked list of keys all of which satisfy $h(k) = i$. We illustrate collision resolution by chaining in Figure 17.1 for our UHash College example. Recall that the keys are nine-digit social security numbers, and $h(k)$ is the integer represented by the last four digits.

When searching for a given key, we first hash the key to the slot that points to the appropriate linked list (*chain*), and then we perform a search of this chain. When inserting a key, we hash the key to the appropriate slot, and then insert the key in the corresponding linked list. Deletion of a key also amounts to simple deletion in the appropriate linked list.

Searching, insertion, and deletion each have worst-case behavior bounded by the longest chain pointed to by the hash table. Of course, in the worst case, all the keys hash to the same list, so that the dictionary operations are basically identical to that performed on an ordinary linked list (with the additional overhead of computing the hash function). Thus, the dictionary operations for hashing with chaining have linear worst-case complexity, which is certainly unattractive. However, as we will see, the average behavior of hashing with chaining is much better.

To discuss the average behavior of hashing with chaining, we need to make some assumptions and provide some notation. We focus on searching, since insertion and deletion have the same complexity as searching. Given a hash table $H[0{:}m-1]$ of m slots containing n keys, we measure the average complexity of searching a hash

Figure 17.1 A chained hash table for students' social security numbers at UHash College, where the hash function h maps a social security number onto the number determined by its last four digits

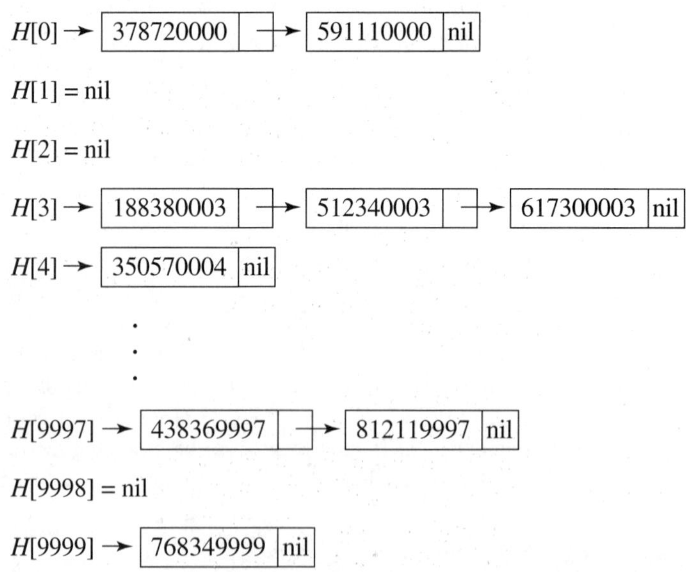

table with chaining in terms of the *load factor* $\alpha = n/m$. The ratio α can be thought of as the average number of keys in a chain. However, it is only when our hash function h is close to uniform that each chain holds approximately α keys.

We now analyze the average behavior of both successful search and unsuccessful search of a hash table with chaining under the assumption that h is a uniform hash function. Let U_n and S_n denote the average number of key comparisons done in an unsuccessful search and a successful search, respectively. Note that U_n is nothing more than the average length of a chain. Moreover, since h is a uniform hash function, the average length of a chain is given by α. Hence, under our assumptions, the average complexity of unsuccessful search is given by

$$U_n = \alpha = \frac{n}{m}. \tag{17.2.4}$$

We now consider a successful search, where we assume that any key in the hash table is equally likely to be the search element. Under our assumptions of uniform hashing, the expected number of key comparisons made when searching for a given key is one more than that made when the key was inserted into the hash table. Thus, S_n is equal to one plus the average number of key comparisons made when inserting the ith key, $i = 1, \ldots, n$. Since we have assumed uniform hashing and each key is equally likely, when the ith key was inserted, each chain had average length given by $(i - 1)/m$. Thus, we have

$$\begin{aligned} S_n = \frac{1}{n}\sum_{i=1}^{n}\left(1 + \frac{i-1}{m}\right) &= 1 + \frac{1}{nm}\sum_{i=1}^{n}(i-1) \\ &= 1 + \frac{1}{nm}\left(\frac{(n-1)\,n}{2}\right) \\ &= 1 + \frac{\alpha}{2} - \frac{1}{2m}. \end{aligned} \tag{17.2.5}$$

Note from (17.2.5) that S_n is approximately equal to $1 + \alpha/2$ for large m, which is entirely reasonable since this is the average number of comparisons performed by linear search on a list of size α.

What happens in the formulas for U_n and S_n as both m and n increase? If the number of slots m in the hash table is proportional to the number of keys n in the table, then $\alpha = n/m \in O(1)$. Then both successful search and unsuccessful search (and the other dictionary operations as well) can be done in constant time, on average. The dramatic improvement over the linear worst-case behavior is the reason that finding good hash functions is so important.

17.2.3 Collision Resolution with Open Addressing

The second major method of resolving collisions in hash tables is to use the technique of *open addressing*. When a collision is detected upon inserting a key, a search of the hash table is made to find the first available unoccupied (open) slot in the hash table. The key is then inserted at this open position. In the open addressing scheme we can never store more keys than available slots in the hash table. Thus, in contrast to the chaining scheme, the loading factor $\alpha = n/m$ for hashing with open addressing is never more than one.

There are various methods of determining the first available open position for inserting a key. Each method generates a certain *probe sequence*, which, starting from the slot given by the hash function, visits the other slots in the hash table in some determined order. We require that the probe sequence generate a permutation of the slots, to guarantee that all m slots are eventually reached by unsuccessful searches. We discuss two commonly used probing methods, *linear probing* and *double hashing*.

When inserting a key we follow the appropriate probe sequence until the first unoccupied slot is found (if any). The key is then inserted at this first available slot. Searching for a key also works by executing the appropriate probe sequence until either the key is found, an unoccupied slot is encountered, or all m slots have been examined without finding the key. Both of the latter cases correspond to unsuccessful search. The reason that encountering unoccupied slots signals unsuccessful search is that each probe sequence is completely determined by the key and the specific probing method. In particular, if an unoccupied position is ever encountered when searching for a given key, the key is not in the table since it would have been already found or originally inserted at the unoccupied position.

The previous discussion highlights a problem that arises with open addressing. Namely, deletions really confound subsequent searches. How do we tell whether an unoccupied slot was vacated by a previous deletion? Clearly, we must be able to recognize this, otherwise searches might end prematurely and erroneously signal unsuccessful search. One way around this dilemma is to keep a bit field around to signal whether or not a slot was ever occupied. While this solves the problem, it tends to generate long searches in the presence of many deletions. Thus, hashing with chaining is preferred over hashing with open addressing when there is the likelihood of a large number of deletions being performed.

Linear Probing Perhaps the simplest probing method is *linear probing*, where starting from the slot given by the hash function, we perform a linear search of the table. We assume that the indices are stored circularly (that is, modulo m), so that the probe resumes at the beginning of the table after it reaches the end. Thus, for a given k, the linear probing sequence corresponding to k is given by

$$(h(k) + i) \textbf{ mod } m, \quad i = 0, \ldots, m - 1. \tag{17.2.6}$$

In Figure 17.2 we show the contents of a hash table $H[0{:}22]$ after the insertion of the keys 3, 4, 5, 26, 6, 7, 23, 16, 39, 17, 22, and 55, where $h(k) = k \textbf{ mod } 23$.

Unfortunately, while linear probing is easy to implement, it does not come very close to uniform hashing, since long runs (called *primary clusters*) of occupied slots

Figure 17.2 Contents of $H[0{:}22]$ after inserting 3, 4, 5, 26, 6, 7, 23, 16, 39, 17, 22, and 55 in initially empty table using the hash function $h(k) = k$ **mod** 23 and open addressing with linear probing

slot	0	1	2	3	4	5	6	7	8	9	10	11	12	13	14	15	16	17	18	19	20	21	22
key	23			3	4	5	26	6	7	55							16	39	17				22

tend to build up even in relatively sparse tables. The reason for clustering is clear: A given empty slot is twice as likely to be filled by an insertion if it is preceded by a nonempty slot than if it is preceded by an empty slot. In fact, if i slots are filled preceding a given open slot, then the empty slot has probability $(i + 1)/m$ of being filled by the next insertion, whereas it has probability of $1/m$ of being filled if its preceding slot is empty. (These calculations assume, as usual, that any slot is equally likely to be the hash value of the inserted key.)

Double Hashing *Double hashing* gets its name from the fact that it utilizes two hash functions to generate its probing sequences. Given two hash functions h_1 and h_2, the double hash function generated by h_1 and h_2 has a probe sequence given by

$$(h_1(k) + ih_2(k)) \textbf{ mod } m, \quad i = 0, \ldots, m - 1, \tag{17.2.7}$$

where it is required that $\gcd(h_2(k),m) = 1$.

In particular, the double hash function has its initial value given by $h_1(k)$. The condition that $h_2(k)$ and m are relatively prime is needed to guarantee that the probe sequence generates a permutation of $0, \ldots, m - 1$ (that is, all slots are reached). To ensure relative primality, we can define $h_2(k) = 1 + (k \textbf{ mod } m')$, where m' is slightly smaller than m. In Figure 17.3 we illustrate a double hash function based on $h_1(k) = k \textbf{ mod } 23$ and $h_2(k) = 1 + (k \textbf{ mod } 21)$ for the same keys as in Figure 17.2.

Double hashing outperforms linear probing since it does not suffer from the problem of clustering. In fact, in practice double hashing usually yields a rather good approximation to uniform hashing.

Average Behavior of Hashing with Open Addressing When analyzing the average behavior of hashing with collision resolution by open addressing, a very strong form of uniform hashing is assumed. The strong form takes into account the entire probing sequence generated by the hash function, not just its initial value. A *strongly uniform hash function* h is a hash function with the property that each of the $m!$ permutations of $0, \ldots, m - 1$ are equally likely to be the probe sequence generated by h for any key. In particular, note that a strongly uniform hash function is also a uniform hash function in our previous sense, since its initial value can be any of the slots $0, \ldots, m - 1$ with equal probability.

Of course, strongly uniform hash functions are even harder to find than ordinary uniform hash functions. Double hashing is closer to being strongly uniform than linear probing. Linear probing generates a total of only m probe sequences, since a linear probe is simply a cyclic shift of the identity permutation. On the other hand,

Figure 17.3 Contents of $H[0{:}22]$ after inserting 3, 4, 5, 26, 6, 7, 23, 16, 39, 17, 22, and 55 in initially empty table using the double hash function based on the hash functions $h_1(k) = k \textbf{ mod } 23$ and $h_2(k) = 1 + (k \textbf{ mod } 21)$ and open addressing

slot	0	1	2	3	4	5	6	7	8	9	10	11	12	13	14	15	16	17	18	19	20	21	22
key	23			3	4	5	6	7		26			39		55		16	17					22

double hashing generates m^2 probe sequences, since each pair $(h_1(k),h_2(k))$ yields a distinct probe sequence.

The following theorem expresses the average behavior of unsuccessful and successful searches for hashing with collision resolution by open addressing. The proof of Theorem 17.2.1 is left to the exercises. Theorem 17.2.1 expresses the average behavior in terms of the loading factor $\alpha = n/m < 1$.

THEOREM 17.2.1 Suppose we have an open-address hash table, with a loading factor $\alpha = n/m < 1$, and a strongly uniform hashing function. Let U_n and S_n denote the expected number of probes in an unsuccessful and successful search, respectively. Then we have

$$U_n \le \frac{1}{1-\alpha}, \qquad (17.2.8)$$

and,

$$S_n \le \frac{1}{\alpha}\left(1 + \log_2\left(\frac{1}{1-\alpha}\right)\right). \qquad (17.2.9)$$

❑

Note that inserting an element has the same complexity as that for an unsuccessful search. Again, supposing that α is held roughly constant as m and n grow large, then Theorem 17.2.1 implies that both unsuccessful and successful searching can be done in constant time.

17.3 Rotations in Binary Search Trees

Rotations are the primary tool used to restore balance in balanced search trees after insertion or deletion of an element. Left and right rotations involving two nodes in a binary tree are illustrated in Figure 17.4. In Figure 17.4, α, β, and γ represent subtrees. Note that performing a left or right rotation preserves the search tree property and does not alter the inorder traversal of the binary tree, since the inorder traversal of either subtree is $\alpha, A, \beta, B, \gamma$.

In our previous discussion of algorithms for maintaining binary search trees, we did not include a parent field in each node. Algorithms for balanced search trees typically contain parent pointers to facilitate insertion and deletion of nodes. We

Figure 17.4 Left and right rotations involving nodes A and B in a binary tree T. The inorder traversal of T is preserved by either rotation.

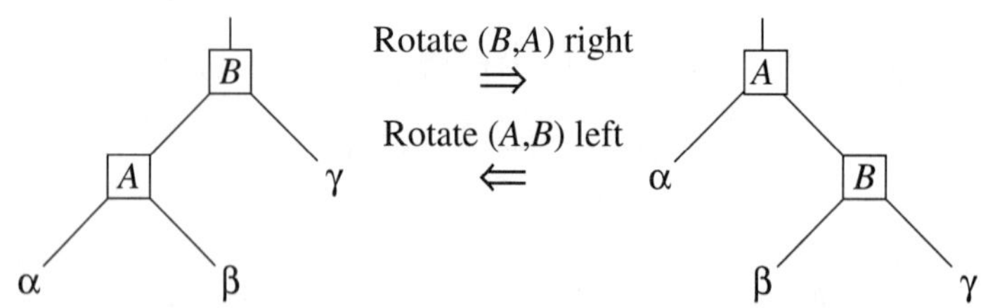

assume that the nodes in a balanced search tree have the following structure (*Info* may contain multiple fields).

```
BinSrchTreeNode = record
    Info: InfoType
    Key: KeyType
    LeftChild:   → BinSrchTreeNode
    RightChild:  → BinSrchTreeNode
    Parent:      → BinSrchTreeNode
end BinSrchTreeNode
```

The following procedure *RotateRight* performs a right rotation of the nodes pointed to by *Node*, *Node* → *LeftChild* in a binary tree pointed to by *Root*. We leave the pseudocode for the symmetric case of a left rotation as an exercise.

```
procedure RotateRight(Root,Node)
Input:  Root (→ BinaryTreeNode)   {points at root of a binary tree T}
        Node (→ BinaryTreeNode)   {points at node where rotation takes place}
Output: Root (→ BinaryTreeNode),  {points at root of the binary search tree
                                   T′ obtained by performing right rotation}
   B := Node                      {simplify notation and tie to Figure 17.4}
   A := Node→ LeftChild           { "    "    "    "    "    "    " }
             {move A up to B's position}
   if Root = B then
      Root := A
   else
      if B = (B→Parent)→LeftChild then   {B is left child of its parent}
         (B→Parent)→LeftChild := A
      else
         (B→Parent)→RightChild := A   {B is right child of its parent}
      endif
   endif
             {make A's right subtree B's left subtree}
   B→LeftChild := A→RightChild
   if A→RightChild ≠ nil then
      (A→RightChild)→Parent := B
   endif
             {make B into A's right child}
   A→RightChild := B
end RotateRight
```

In Figure 17.5, we show how two rotations are used to restore balance in an *AVL tree*, which is a binary search tree such that the depths of the left and right subtrees

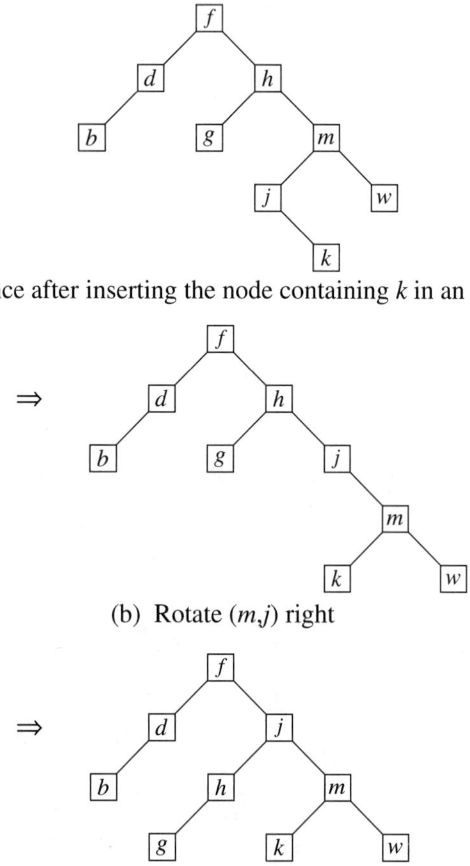

(a) Imbalance after inserting the node containing k in an AVL tree

(b) Rotate (m,j) right

(c) Rotate (h,j) left

Figure 17.5 Restoring balance after inserting k in an AVL tree

rooted at any node differ by at most one. AVL trees are one of the earliest types of balanced search trees and are named after Adel'son-Velsk'ii and Landis, who introduced them in 1962.

Algorithms implementing the dictionary operations for red-black trees (defined in the next section) are simpler than those for AVL trees, so red-black trees are more commonly implemented today. Thus, we focus our attention on red-black trees.

17.4 Red-Black Trees

About 10 years after the introduction of AVL trees, Bayer introduced what are now known as red-black trees. (Bayer gave them the name "symmetric binary B-trees.") A red-black tree is balanced in the sense that no path from the root to a leaf has length more than twice the length of any other such path. However, this balance is not part of the definition of a red-black tree, but rather is a consequence of the balance restrictions on the node coloring. The coloring balance restrictions for red-black trees are simpler to maintain than other balance restrictions such as AVL balance, with correspondingly simpler algorithms for the dictionary operations.

17.4.1 Definition and Depth of Red-Black Trees

The color balance properties of red-black trees are easier to state if we assume that augmented leaf nodes corresponding to unsuccessful searches are included as part of the tree. In particular, with augmented leaf nodes included, red-black trees are 2-trees. In what follows, for red-black trees we simply refer to the augmented leaf nodes as leaf nodes.

DEFINITION 17.4.1 A *red-black tree* is a binary search tree, augmented by (external) leaf nodes corresponding to unsuccessful search, where each node has been assigned either the color red or black subject to the following three properties.

i. Every leaf node is black.

ii. If a node is red, then both its children are black.

iii. Any two paths from a given node v down to a leaf node contain the same number of black nodes.

Remark Including augmented leaf nodes in red-black trees makes the properties (i) and (iii) in Definition 17.4.1 have a powerful balancing effect. Indeed, if augmented leaf nodes were not included, then an all black, completely skewed binary search tree would be a red-black tree.

In Figure 17.6 (page 598) we show a red-black tree on 12 internal nodes. We have indicated the color outside each node.

The following proposition states that red-black trees have small depth, which makes them good search trees.

PROPOSITION 17.4.1 Let T be a red-black tree on n internal nodes. Then

$$\text{depth}(T) \leq 2\log_2(n+1). \qquad (17.4.1) \quad \square$$

We establish Proposition 17.4.1 with the aid of Lemma 17.4.2. To state Lemma 17.4.2, for any red-black tree T, we let $b(T)$ denote the number of black nodes, excluding the root, on any path from the root of T to a leaf. Property (iii) of Definition 17.4.1 shows that $b(T)$ is well defined.

LEMMA 17.4.2 Let T be a red-black tree. Then T has at least $2^{b(T)} - 1$ internal nodes.

Proof We prove Lemma 17.4.2 using induction on the number of internal nodes n in T.

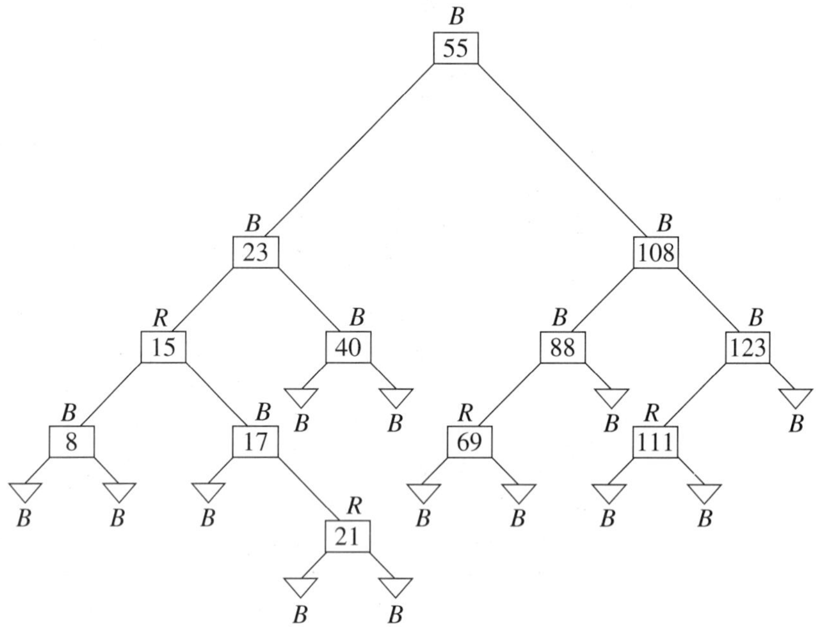

Figure 17.6 A red-black tree on 12 internal nodes

Basis step: $n = 1$. Then $b(T) = 1$. Since T has $1 = 2^1 - 1$ internal nodes, the basis step is established.

Induction step: Assume that Lemma 17.4.2 has been verified for all red-black trees having fewer than $n > 1$ internal nodes. Note that the subtree rooted at any node v of a red-black tree is also a red-black tree. Hence, the left and right subtrees L and R, respectively, of the root of T are both red-black trees having fewer than n internal nodes. Since T is a 2-tree, both L and R are nonempty. Thus, by induction hypothesis, L and R both have at least $2^{b(R)} - 1$ internal nodes. Now the values of $b(L)$ and $b(R)$ are either $b(T)$ or $b(T) - 1$, depending on whether their roots are colored red or black, respectively. It follows that T has at least $2(2^{b(T)-1} - 1) + 1 = 2^{b(T)} - 1$ nodes, which establishes the induction step and proves Lemma 17.4.2. ■

Proof of Proposition 17.4.1 Conditions (i) and (ii) of Definition 17.4.1 imply that at least half of the nodes along any path from the root to a leaf, excluding the root, must be black. In other words, $b(T) \geq \text{depth}(T)/2$. Thus, using Lemma 17.4.2, we have

$$n \geq 2^{b(T)} - 1 \geq 2^{\text{depth}(T)/2} - 1. \qquad (17.4.2)$$

Proposition 17.4.1 now follows directly from (17.4.2). ■

17.4.2 Implementation of Red-Black Trees

We assume that the nodes in a red-black tree have the following structure.

RedBlackTreeNode = **record**
Info: InfoType
Key: *KeyType*
Color: *Boolean* {0 = red, 1 = black}
LeftChild: →*RedBlackTreeNode*
RightChild: →*RedBlackTreeNode*
Parent: →*RedBlackTreeNode*
end *RedBlackTreeNode*

The augmented leaf nodes of the red-black tree are assumed to have the same structure as the nodes in the original tree, except that their information and key fields do not contain meaningful values. However, the parent fields of the augmented leaf nodes would be appropriately defined. In actual practice, allocating memory locations for each augmented leaf node is simply too wasteful of space. Instead, a single augmented node is added, called a *sentinel node*. All child pointers in the original tree that pointed to **nil** augmented leaf nodes now hold the address of the sentinel node. Again, the sentinel node is assumed to have the same structure as the other nodes in the tree. In Figure 17.7 we show the same red-black tree as in Figure 17.6, but now a single sentinel node is used for the augmented leaf nodes.

Using a sentinel node to handle **nil** child pointers simplifies boundary conditions in algorithms that restore balance by suitably moving up the tree, sometimes starting from a **nil** leaf. For convenience, in some figures, we still represent **nil** augmented nodes as separate nodes as in Figure 17.6.

Figure 17.7 Red-black tree with sentinel node

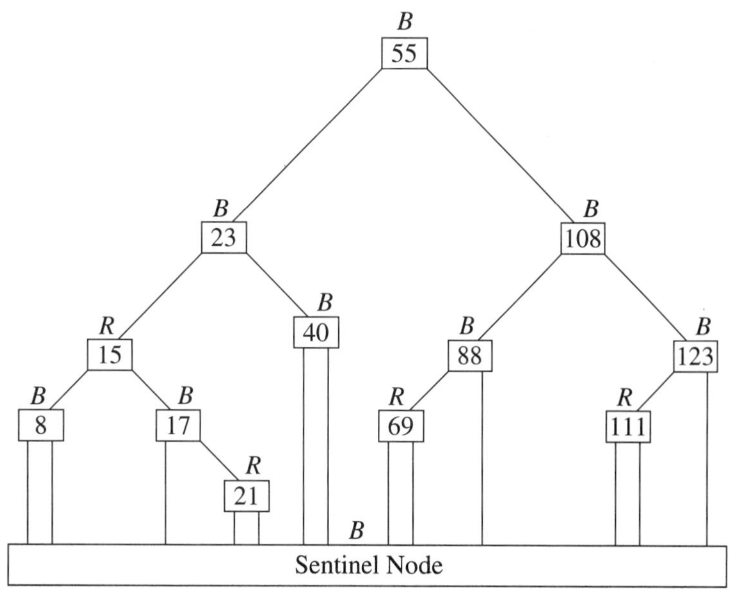

A search algorithm for a red-black tree is the same as that for an ordinary binary search tree. We now consider the operations of inserting and deleting nodes.

17.4.3 Inserting a Node into a Red-Black Tree

When inserting a node into a red-black tree, we call the procedure *BinSrchTreeInsert* (modified to maintain parent pointers) and color the newly inserted node red. If the parent of the newly inserted node is black, we are done, since the tree is already a red-black tree. Otherwise, red-black property (ii) is violated. (It is easily seen that the other properties of a red-black tree remain satisfied.) At most two rotations and some recoloring restores property (ii), without violating the other red-black tree properties.

For any node v in a binary tree T, we define the *local neighborhood* of v to consist of v together with its parent, grandparent, and uncle. Each *correct-or-move-up* step in the eventual restoration of property (ii) involves recoloring and (possibly) rotating the nodes in the local neighborhood of the current node. As implied by the name, one of two things are accomplished by a correct-or-move-up step: Either the violation of property (ii) is corrected, or the violation of property (ii) is moved up two levels in the tree. During each correct-or-move-up step, no violations of the other two red-black properties are introduced. The current node is initially set to the newly inserted element.

For each correct-or-move-up step, there are six possible cases. Figure 17.8 illustrates three of these cases, which correspond to the parent of the current node being a red left child. The other three cases are mirror images of these cases.

Case 1: The current node has a red uncle, and its parent node is a red left child.

In case 1 it doesn't matter whether the current node is a left or a right child. For definiteness, in Figure 17.8 we have shown the current node as a left child. In case 1 we simply recolor the parent, grandparent, and uncle of the current node. If the great-grandparent of the current node is black, then the simple recoloring restores a red-black tree. Otherwise, the recoloring moves the violation of property (ii) up two levels, and the current node is moved up to the grandparent. In both cases, the recoloring does not violate red-black property (iii).

Case 2: The current node is a right child, whose parent is a red left child, and whose uncle is black.

In case 2, we perform a left rotation of (*Parent*, *CurNode*) and update the current node as in Figure 17.8, thereby transforming case 2 into case 3.

Case 3: The current node is a left child, whose parent is a red left child, and whose uncle is black.

In case 3, we recolor the parent and grandparent and then perform a right rotation of (*GrandParent*, *Parent*). These actions restore a red-black tree.

We assume that each input to the algorithm *RedBlackTreeInsert* is a red-black tree where the root is black, so that our correct-or-move-up process always terminates with the restoration of a red-black tree.

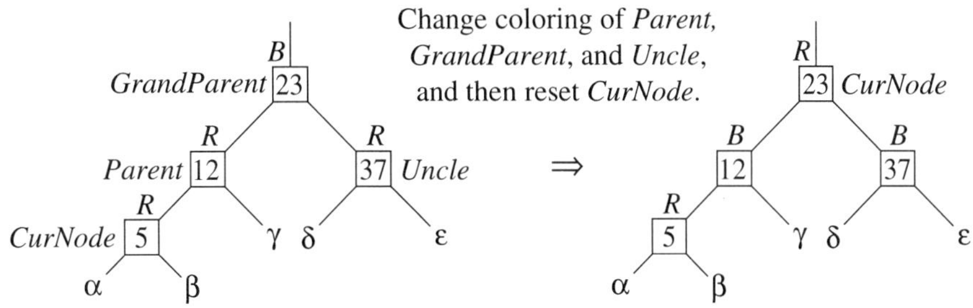

(a) Case 1: Parent of current node is a red left child. Uncle of current node is also red.

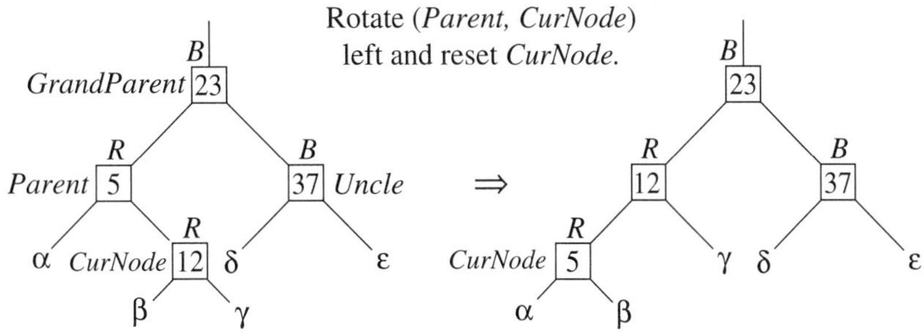

(b) Case 2: Parent of current node is red left child. Uncle of current node is black. Current node is right child.

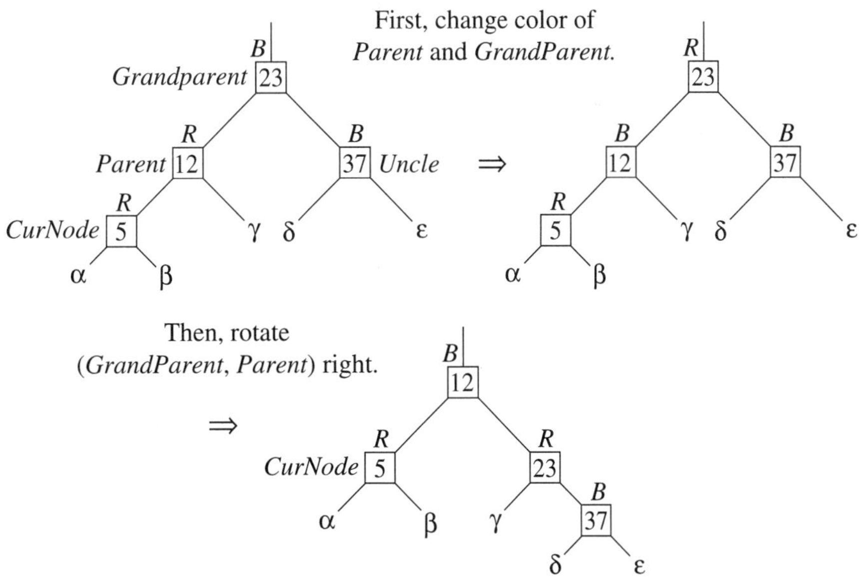

(c) Case 3: Parent of current node is red left child. Current node is left child. Uncle of current node is black.

Figure 17.8 Actions leading to restoration of a red-black tree after insertion

The pseudocode for the procedure *RedBlackTreeInsert* is a straightforward implementation of the cases illustrated in Figure 17.8. We assume that *RedBlackTreeInsert* is called with a red-black tree where the root is black. During the course of the algorithm, the root may be colored red, but the last statement before the end of *RedBlackTreeInsert* restores its color to black. For readability, we use the terms *red* and *black* to correspond to the values 0 and 1, respectively, of the Boolean field *Color*. *RedBlackTreeInsert* calls the procedure *RBCorrect(ii)OrMoveUpL* when we encounter cases 1–3, and calls the procedure *RBCorrect(ii)OrMoveUpR* for the other three symmetric cases.

```
procedure RedBlackTreeInsert(Root,X,Node)
Input:  Root (→RedBlackTreeNode) {points at black root of a red-black tree T}
        X (KeyType)                {key to be inserted in T}
Output: Root (→RedBlackTreeNode)   {root of the red-black tree T′
                                    obtained by inserting X}
        Node (→RedBlackTreeNode)   {points to node where X inserted}
   call BinSrchTreeInsert(Root,X,Node) {ordinary search tree insertion,
                          where Node now points to inserted node containing X}
   Node→Color := red
   CurNode := Node
   while CurNode ≠ Root .and. (CurNode→Parent)→Color = red do
      Parent := CurNode→Parent
      GrandParent := (CurNode→Parent)→Parent
      if GrandParent→LeftChild = Parent then
         call RBCorrect(ii)OrMoveUpL(CurNode,Parent,GrandParent)
      else
         call RBCorrect(ii)OrMoveUpR(CurNode,Parent,GrandParent)
      endif
   endwhile
   Root→Color := black
end RedBlackTreeInsert
```

We now give the pseudocode for *RBCorrect(ii)OrMoveUpL*. (The pseudocode for *RBCorrect(ii)OrMoveUpR* is obtained from *RBCorrect(ii)OrMoveUpL* by interchanging right and left.)

```
procedure RBCorrect(ii)OrMoveUpL(Root,CurNode,Parent,GrandParent)
Input:  CurNode (→RedBlackTreeNode) {points at node of a tree T
                                     whose parent violates property (ii)}
Output: CurNode (→RedBlackTreeNode) {root of the red-black tree T′ arising
                  from T by performing action according to Case 1, 2, or 3
                  and then moving up CurNode}
```

```
Uncle := GrandParent→RightChild
if Uncle→Color = red then                   {Case 1}
   Parent→Color := black
   GrandParent→Color := red
   Uncle→Color := black
   CurNode :→ GrandParent                   {move current node up two levels}
else
   if Parent→RightChild = CurNode           {Case 2}
      call RotateLeft(Root,Parent)
      CurNode := CurNode→LeftChild          {reset current node}
   endif
   Parent→Color := black                    {Case 3}
   GrandParent→Color := red
   call RotateRight(Root,GrandParent)
endif
end RBCorrect(ii)OrMoveUpL
```

We illustrate the insertion of the sequence of keys 18, 16, 55 into a red-black tree in Figure 17.9. For convenience, in Figure 17.9 we have omitted pointers to the sentinel leaf node.

Figure 17.9 Inserting the sequence of keys 18, 16, 55 into a red-black tree

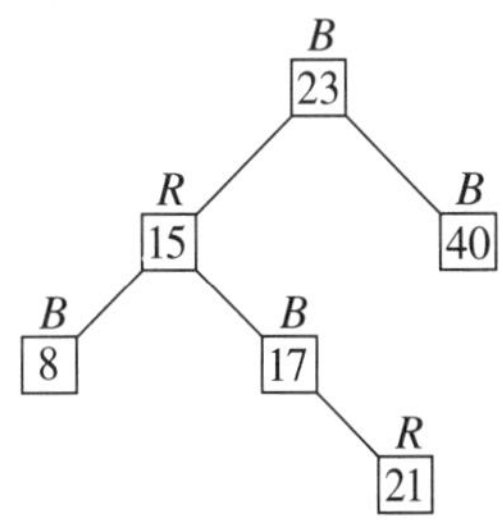

(a) Original red-black tree

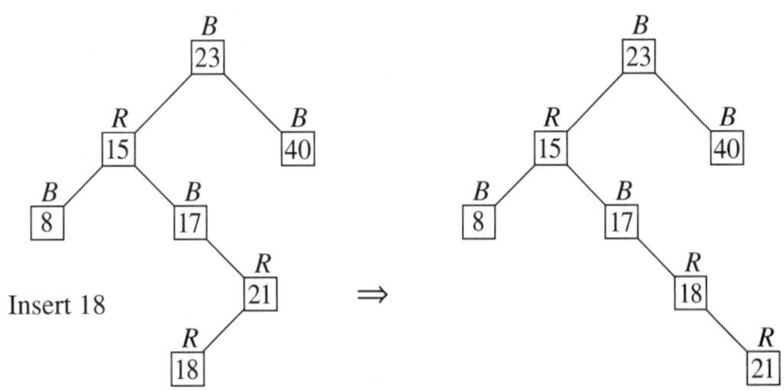

Case 2′ Case 3′ **Continued**

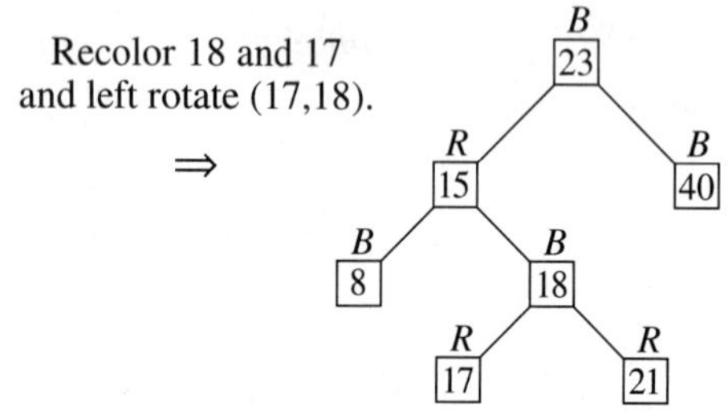

(b) Final red-black tree after inserting 18

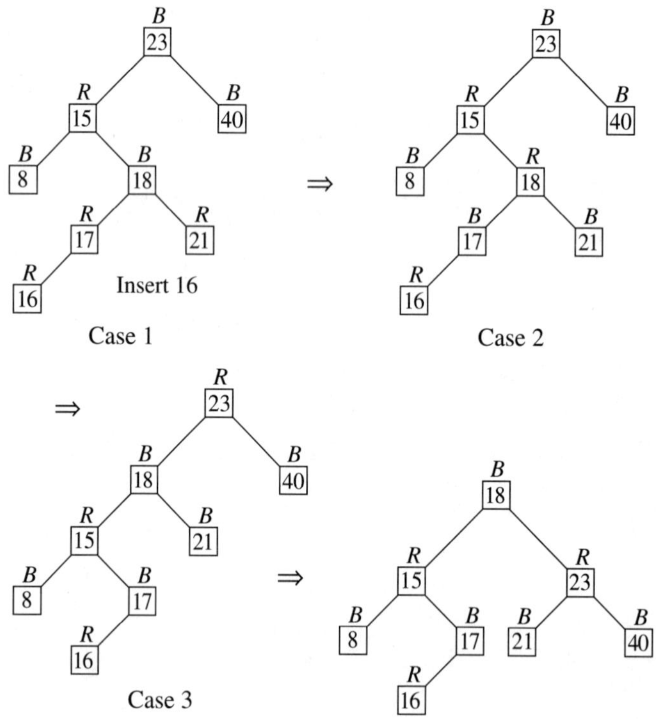

(c) Final red-black tree after inserting 16

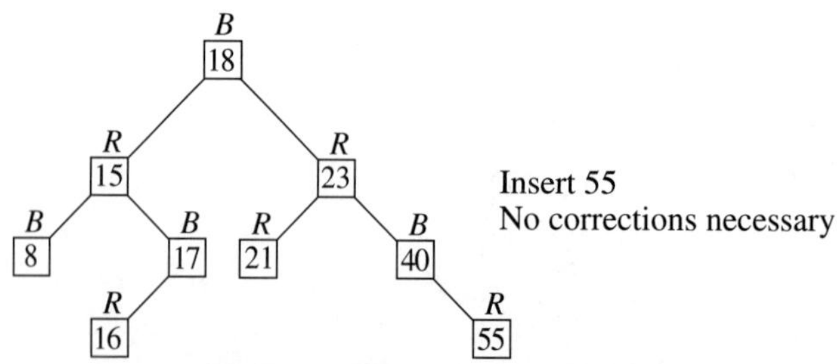

(d) Insert 55

Figure 17.9 Continued

17.4.4 Deleting a Node from a Red-Black Tree

Deleting a node v from a red-black tree is somewhat more complicated than inserting a node, but has a similar flavor. We begin by calling *BinSrchTreeDelete* (modified to maintain parent pointers) to perform an ordinary deletion from a binary search tree. Recall that when we delete an internal node v with two non-nil children (a case (3) deletion in the terminology of Section 2.5.4), we replace the values in the information and key fields of v by the values in the information and key fields, respectively, of the inorder successor w of v. Then we delete w, which has at most one non-nil child. However, it is important to note that we do *not* replace the value of the color field of v by the value of the color field of w. Thus, for a case (3) deletion, when we talk about the color of the deleted node, we mean the color of w, not v.

The deletion of a node with at most one non-nil child is accomplished by splicing it out (see Figure 2.19). Now if the deleted node was red, clearly we still have a red-black tree and nothing else need be done. If the deleted node was black, and it has a red non-nil child, then clearly we restore a red-black tree by recoloring the non-nil child black. Hence, we are left with the problem of restoring a red-black tree when the deleted node and its children are all black. Property (iii) is then violated, since any path that starts at a given node y, goes through the parent of the deleted node, and then goes on to the "spliced-in child" *CurNode* and down to a descendant leaf, will have one too few black nodes than a path from y to a leaf that doesn't go through *CurNode*. Also, any two paths from a node to a leaf that do not go through *CurNode* have the same number of black nodes. Thus, we can consider the violation of property (iii) to be "rooted" at *CurNode*, the spliced-in child.

As before, we perform recolorations and rotations to restore a red-black tree, but our local neighborhood now consists of the current node, its parent, its sibling, and the two children of the sibling. Recoloring and rotation will either correct the violation of property (iii) or move the violation up one level in the tree (at the same time not introducing any other violations of the red-black properties). There are ten cases to consider, five cases where the current node is a left child, and five cases where it is a right child. In Figure 17.10 (page 606), we illustrate the five cases handled by *RBCorrect(iii)OrMoveUpL* (current node is a left child). We use the variable names *CurNode*, *Parent*, and *Sibling* to denote the nodes in a local neighborhood of the current node.

In case 1, *Sibling* is red, whereas in the other four cases, *Sibling* is black. Case 1 transforms into case 2b, case 3, or case 4. In case 2a, the violation is moved one level up the tree and the loop is repeated (unless the current node becomes the root). In case 2b, the violation of property (iii) is moved up one level, but a violation of property (ii) is introduced there. However, the while loop in *RedBlackTreeDelete* is terminated, so that *CurNode* is colored black, which corrects the violation of properties (ii) and (iii) in one fell swoop. Case 3 is transformed into case 4, where the violation of property (iii) is corrected and the algorithm terminates. In Figure 17.10, we show sample key values in the nodes so that the nodes can be traced during the action of *RBCorrect(iii)OrMoveUpL*.

It is important to verify in all ten cases that the transformations introduce no other violations of the red-black properties (except as noted for case 2b, which is

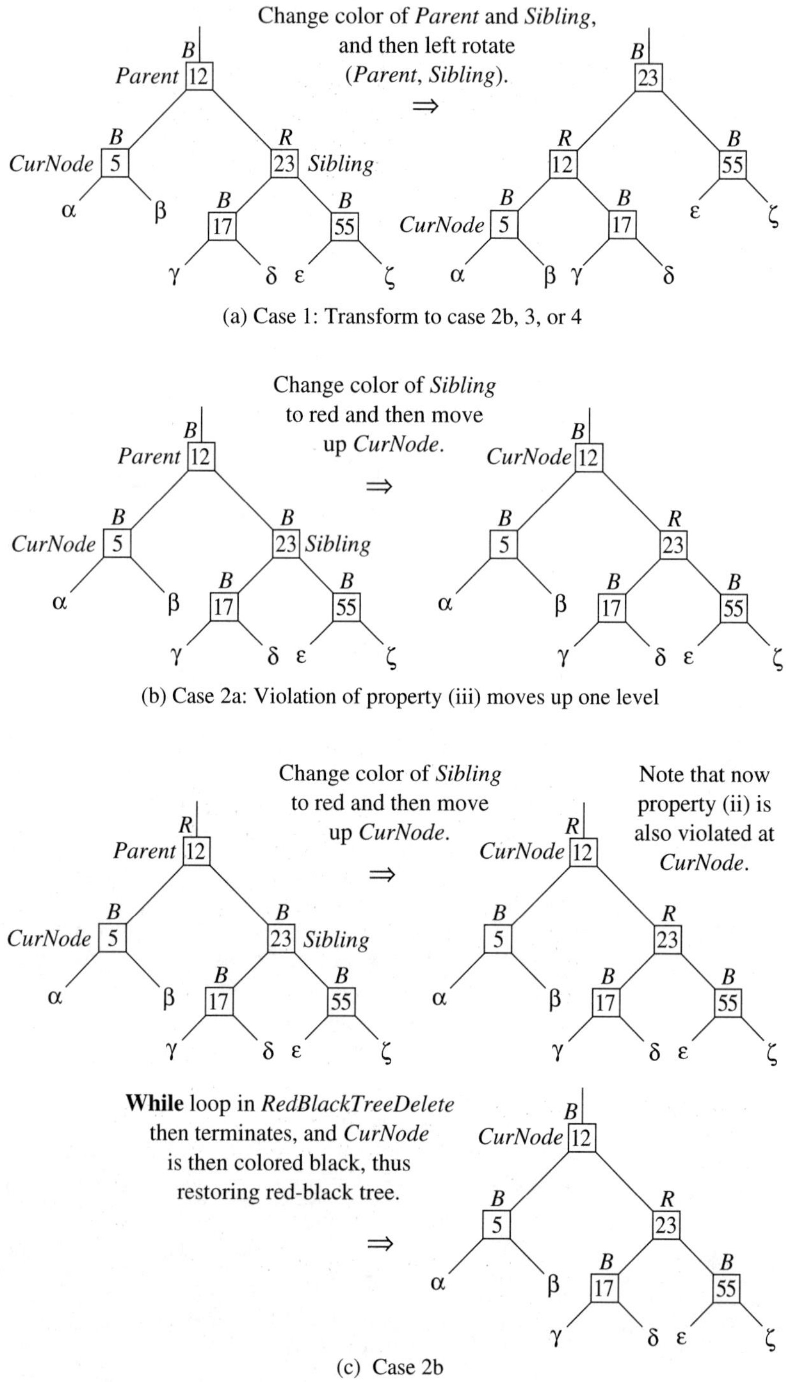

(a) Case 1: Transform to case 2b, 3, or 4

(b) Case 2a: Violation of property (iii) moves up one level

(c) Case 2b

Figure 17.10 Restoring a red-black tree after deleting a node

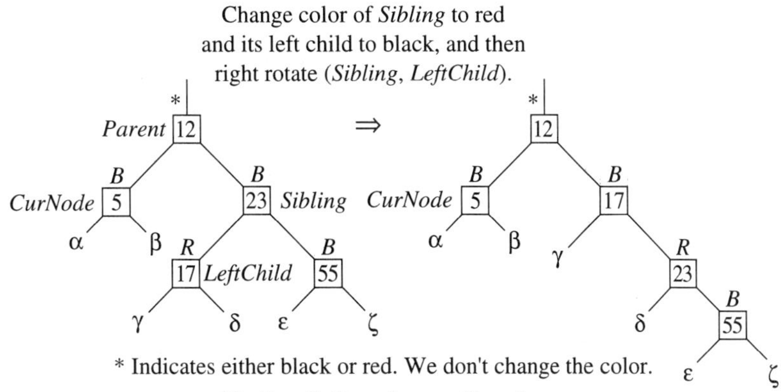

(d) Case 3: Transform to Case 4

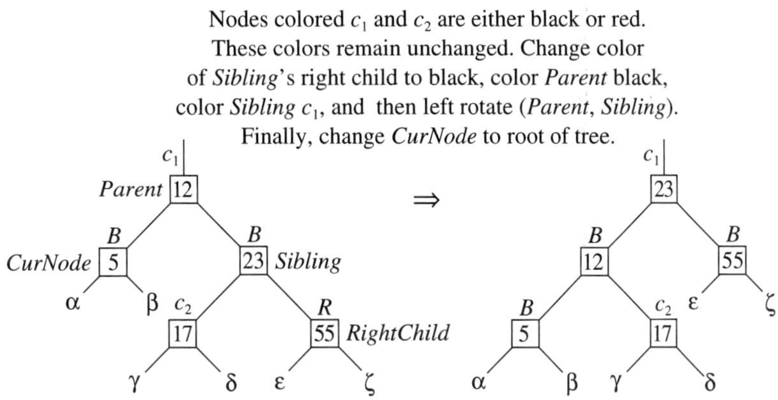

(e) Case 4

Figure 17.10 Continued

then corrected), that each case except case 2a results in restoring a red-black tree, and that case 2a moves the root of the violation of property (iii) one level up the tree. For example, in case 1, note from Figure 17.10a that any path starting a vertex v and going through either node 5, 17, or 55 on its way to a leaf node will encounter the same number of black nodes before and after the transformation. Moreover, no other paths to leaf nodes are affected by the transformation, so the violation of property (iii) remains rooted at node 5. The complete verifications of all ten cases are left as exercises.

In Figure 17.11 (page 608), we illustrate the deletion of the sequence of nodes 88, 8, 55, and 61 from a red-black tree. Again, for convenience we omit pointers to the sentinel leaf node *Sentinel*, except when the initial value of *CurNode* points at *Sentinel* as in Figures 17.11c and 17.11d.

Note that we perform at most two rotations to correct the violation of property (iii) and restore a red-black tree. When restoring the red-black properties after

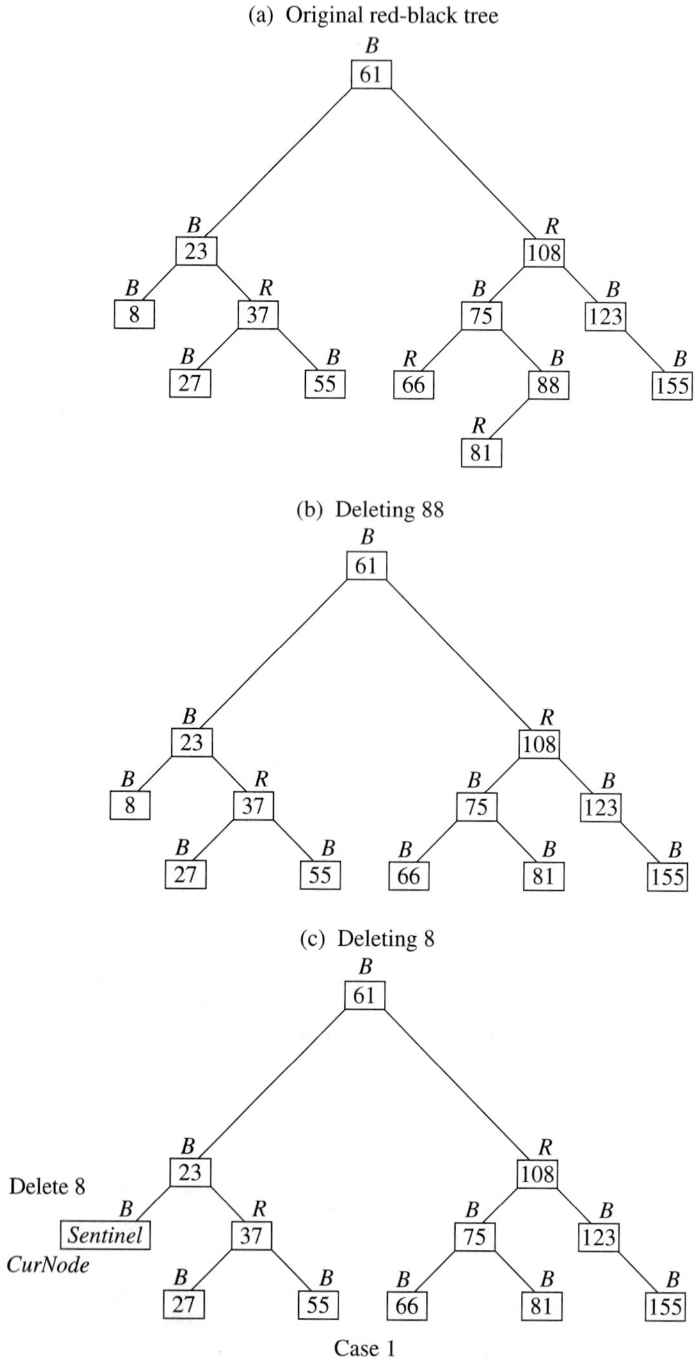

Figure 17.11 Deleting the sequence of nodes 88, 8, 55, 61 from a red-black tree

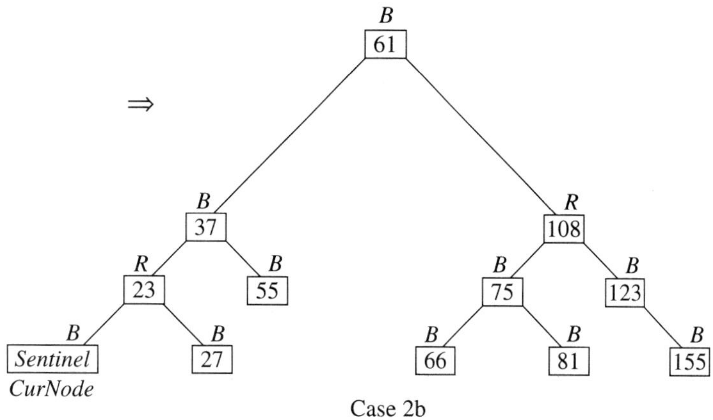

Case 2b

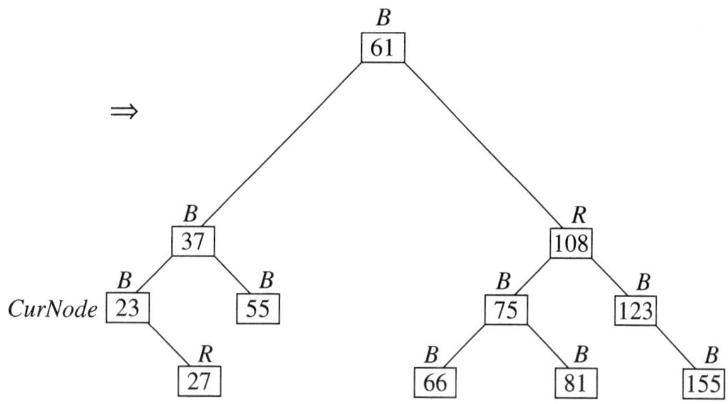

(d) Deleting 55

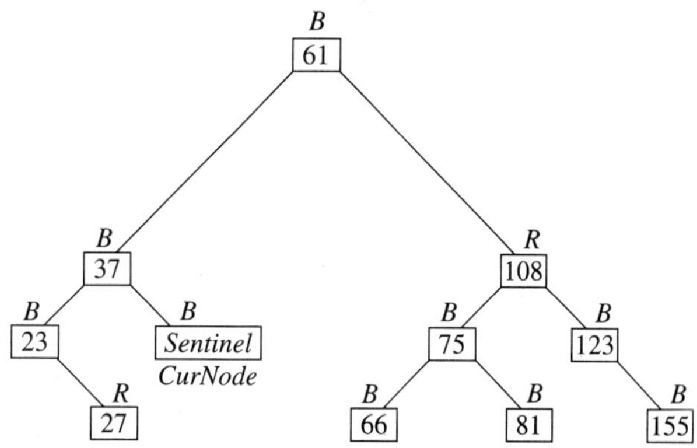

Case 3′

Figure 17.11 Continued

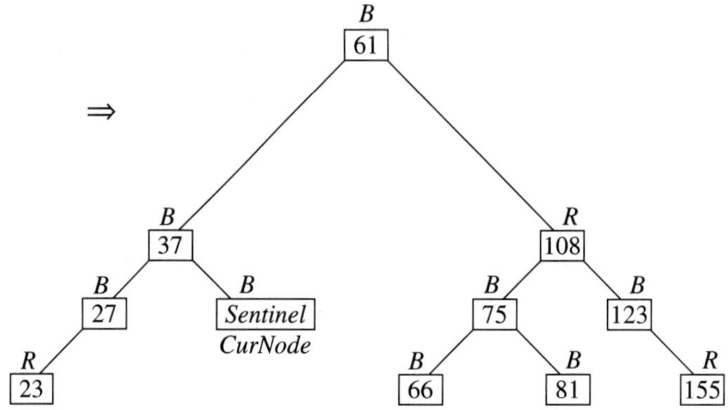

Case 4′

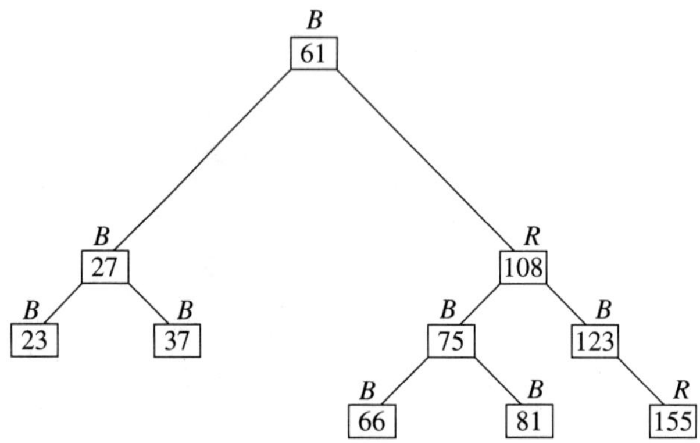

(e) Deleting 61

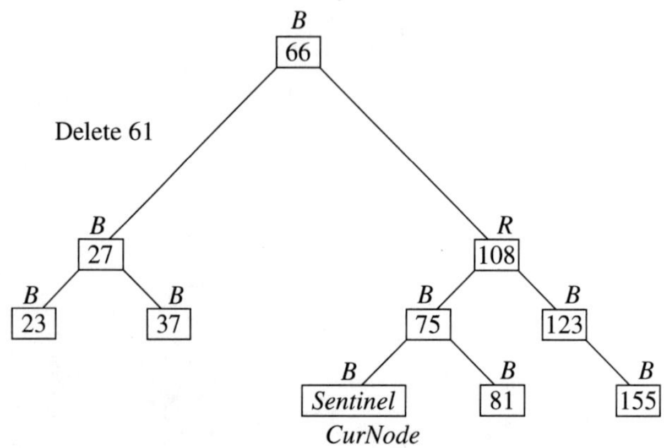

Case 2a: Replace 61 by its inorder successor 66, then delete original occurrence of 66

Figure 17.11 Continued

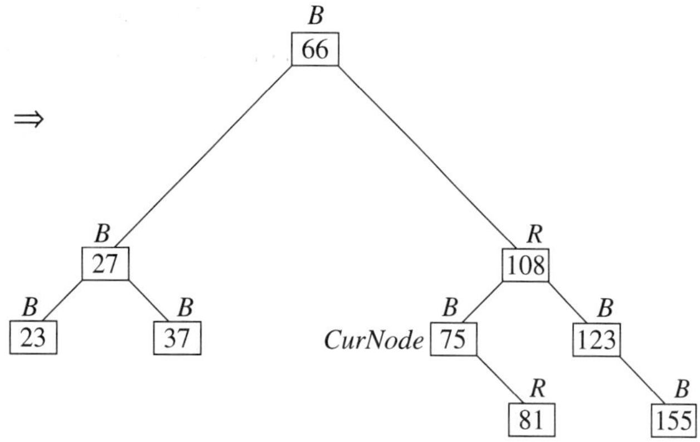

Case 2b

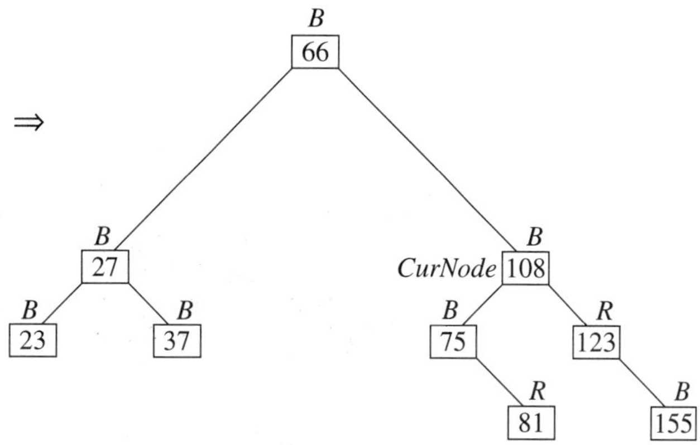

Figure 17.11 Continued

inserting or deleting node, each time we call the correct-or move-up algorithms we perform a constant number of pointer and other field reassignments. Thus, the complexity of both inserting and deleting an element in a red-black tree T having n internal nodes belongs to O(depth(T)) = O(log n). The pseudocode for deleting a node from a red-black tree is left as an exercise.

17.5 B-Trees

Introduced by Bayer and McCreight in the early 1970s, B-trees have been extensively utilized to implement the dictionary operations on database files stored on external storage devices (typically disks). Since accessing external memory through **read** and **write** operations is much more time consuming than accessing internal memory, we wish to keep external file accesses (**read**s or **write**s) to a minimum.

When using a B-tree as an index file for a database stored in external memory, we assume that each move from one level to another necessitates a **read** operation. More precisely, to access a node in a B-tree, we must **read** in all the keys in the node from external memory. We do not concern ourselves here with the memory management details associated with the paging in and paging out of various nodes.

To minimize the number of accesses to external memory performed by the dictionary operations, the depth of B-trees is kept small using various balancing techniques, as well as by storing multiple keys in each node. Storing multiple keys in each node requires extra processing in these nodes for dictionary operations, but this processing is done in fast internal memory, so that the time taken for this processing is dominated by the time taken by reading from (and writing to) an external file.

The algorithms for maintaining B-trees are straightforward, but somewhat cumbersome in their detail. We limit our discussion to defining B-trees and establishing some of their basic properties. We also illustrate and give high-level descriptions of insertion and deletion algorithms for B-trees.

17.5.1 Definition and Properties of B-Trees

The following definition imposes a very strict form of balance on B-trees. This balance, together with a restriction on the minimum number of keys contained in a node, is what makes B-trees have small depth compared to the number of keys contained in the tree (see Proposition 17.5.1).

DEFINITION 17.5.1 A *B-tree* T is a (rooted) tree satisfying the following properties:

1. T is a multiway search tree.
2. Every leaf of T is on the same level.
3. For some fixed integer $m \geq 2$, every node other than the root contains at least m keys and at most $2m$ keys. The root must contain at least one key and at most $2m$ keys.
4. An internal node containing t keys has $t + 1$ (non-nil) children.

Each node in a B-tree has at most $2m + 1$ children, so in the terminology of the previous section, B-trees are $(2m + 1)$-way search trees. For convenience, we refer to a $(2m + 1)$-way B-tree as a B-tree of *order m*. (Caution: Other names appear in the literature for B-trees of order m, such as B-trees of *minimum degree* $2m + 1$.) A node that contains the maximum number $2m$ of keys is called a *full* node. In Figure 17.12 we show a B-tree of order 2 containing 24 keys.

In practice, B-trees typically are of much higher order than shown in Figure 17.12, with orders of a thousand or more not uncommon. Since the number of nodes at level l is an exponential function of l, an enormous number of keys can be maintained by very shallow B-trees. Indeed, a B-tree of order 50 and depth 2 can store $100 + (101)(100) + (101^2)(100) = 1{,}030{,}301$ keys, which is more than enough for

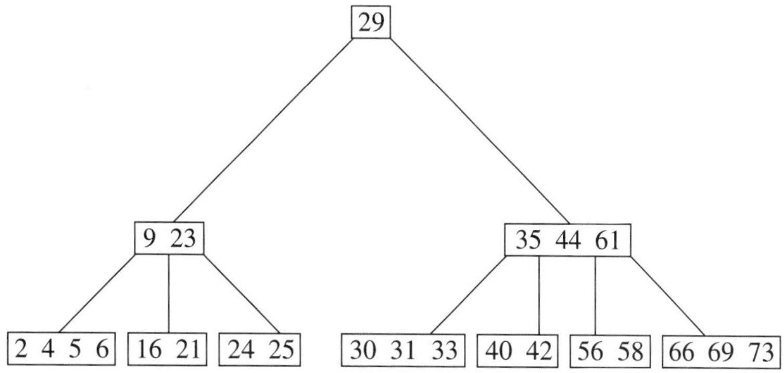

Figure 17.12 A B-tree of order 2 containing 24 keys

most databases maintained on personal computers. Since the root node can be kept in internal memory, each of the 1,030,301 keys can be found by executing at most two **read** statements.

PROPOSITION 17.5.1 Suppose T is a B-tree of order m and depth d. Then, T contains at least $2m^d - 1$ keys. Hence,

$$\text{depth}(T) < \log_m \frac{n+1}{2}. \tag{17.5.1}$$

Proof We establish the proposition by determining the minimum number $\mu(m,d)$ of keys contained in a B-tree of order m and depth d. Using conditions (3) and (4) of Definition 17.5.1, $\mu(m,d)$ is clearly obtained when the root has one key, and every other node contains m keys. We then have

$$\begin{aligned}\mu(m,d) &= 1 + m(2 + 2(m+1) + 2(m+1)^2 + \cdots + 2(m+1)^{d-1}) \\ &= 1 + m\frac{((2m+1)^d - 1)}{m} = 1 + 2((m+1)^d - 1) = 2(m+1)^d - 1. \end{aligned} \tag{17.5.2}$$

Formula (17.5.2) implies that if T is a B-tree of order m and depth d containing n keys, then

$$n \geq \mu(m,d) = 2(m+1)^d - 1. \tag{17.5.3}$$

■

Our insertion and deletion procedures for B-trees require m to be an input parameter. Thus, in our analysis of the complexity of the algorithms for performing the dictionary operations on a B-tree, we give the worst-case complexity in terms of the base-m logarithm function $\log_m n$, rather than in terms of $\Theta(\log n)$ membership. There is another reason to measure complexity in terms of m: All base-m logarithms, $m > 1$, determine the same class $\Theta(\log n)$, so that a Θ-class measure is not sufficiently precise to distinguish between the performance, say, of balanced binary trees and B-trees.

17.5.2 Searching a B-Tree

Since B-trees are multiway search trees, the search strategy described in Section 2.6 applies. Letting *B-TreeSearch* denote the resulting algorithm, clearly the worst-case complexity occurs when the search key k is either in a leaf or not in the tree at all. In both cases, we perform depth(T) **reads**. Hence, Proposition 17.5.1 implies that *B-TreeSearch* performs at most $\log_m(n + 1)/2$ **reads** and $2m \log_m(n + 1)/2$ comparisons for a B-tree of order m on n keys.

The $2m \log_m(n + 1)/2$ comparisons performed by *B-TreeSearch* in the worst case is larger than the typical upper bound of $2 \log(n + 1)$ comparisons performed by the dictionary operations on red-black trees on n keys. However, what is reduced is the depth from a worst case $2 \log_2(n + 1)$ for red-black trees to a worst case $\log_m(n + 1)/2$ for B-trees. Smaller depth trees reduce the corresponding number of time-consuming accesses to external memory.

17.5.3 Inserting a Key into a B-Tree

When inserting a key into a B-tree, we use a variation of our searching technique to descend down a path from the root of the B-tree to an appropriate leaf node for insertion. However, we might end up at a full leaf, and inserting the key there causes an overflow. There are two ways to remove the overflow at a node and thereby (eventually) restore a B-tree:

1. If the node has an immediately adjacent sibling that is not full, then send a key from the parent down to this sibling and send a key up from the node to the parent.
2. If both immediately adjacent siblings are full, then split the node into two nodes by passing the median key up to the parent.

We always try to remove overflow using method (1), since the B-tree property is then immediately restored (see Figure 17.13). Using method (2) to remove overflow, the B-tree property is immediately restored if the parent was not full, otherwise an overflow occurs when passing the median key up to the parent (see Figure 17.14). Then we must repeat the process at the parent. Continuing in this manner, we move up the tree until we eventually restore the B-tree property. In the worst case, we work all the way back to the root. If the root is full, and if method (1) is not applicable, then the root is split and a new root is created containing the median key of the old root. *Splitting the root increases the depth of the B-tree by one and is the only way that the depth of a B-tree increases.* Thus, while binary search trees grow at the bottom, B-trees always grow at the top.

We illustrate methods (1) and (2) of removing overflow in Figures 17.13 and 17.14, respectively. In both figures, the parent node is full. Hence, in Figure 17.14, the overflow gets passed up to the parent.

The example in Figure 17.14 is continued in Figure 17.15 (page 616). Figure 17.15a illustrates a case where the overflowed parent has an immediately adjacent sibling that is not full. Then method (1) is used to restore a B-tree. In Figure 17.15b, we assume both immediately adjacent siblings to the overflowed parent are full, so

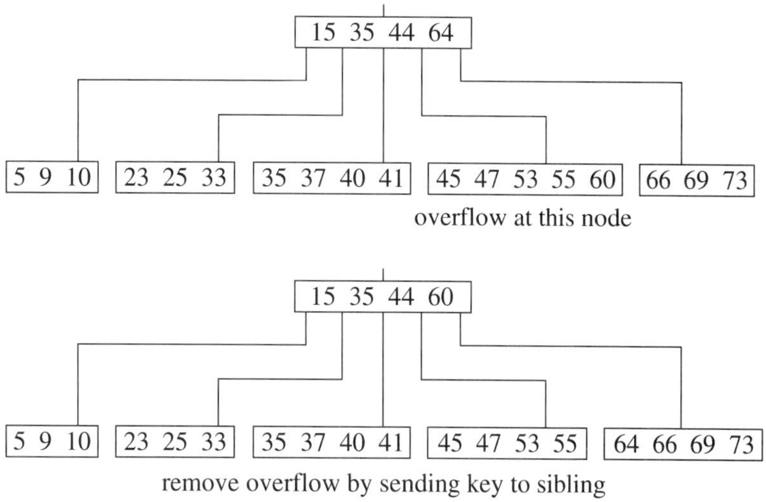

Figure 17.13 Removing overflow in a node using method (1) in a B-tree of order 2

that method (2) is used to remove overflow. We only show those nodes that are relevant to the overflow removal.

In Figure 17.16 we show the result of splitting the root node of a B-tree of order 2.

In Figure 17.17, we illustrate the process of inserting the sequence of keys into an initially empty B-tree of order 2. We only illustrate some of the intermediate steps in

Figure 17.14 Removing overflow in a node using method (2) in a B-tree of order 2. In this example, an overflow is passed up to parent.

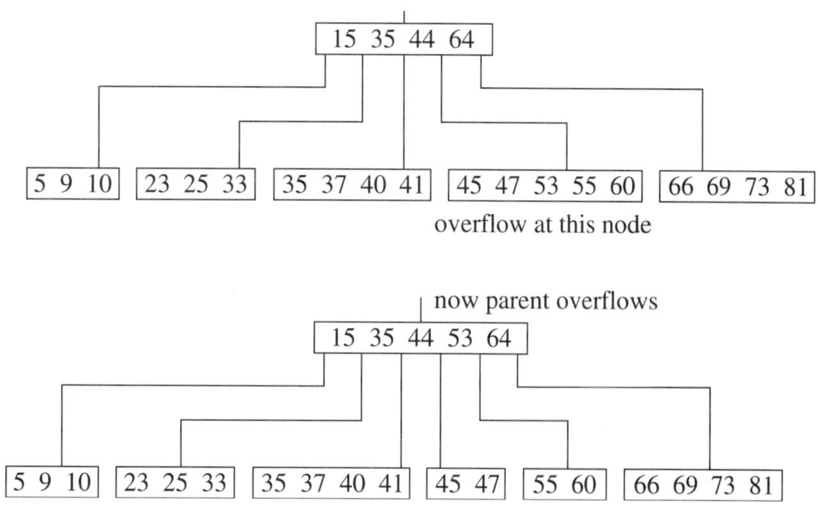

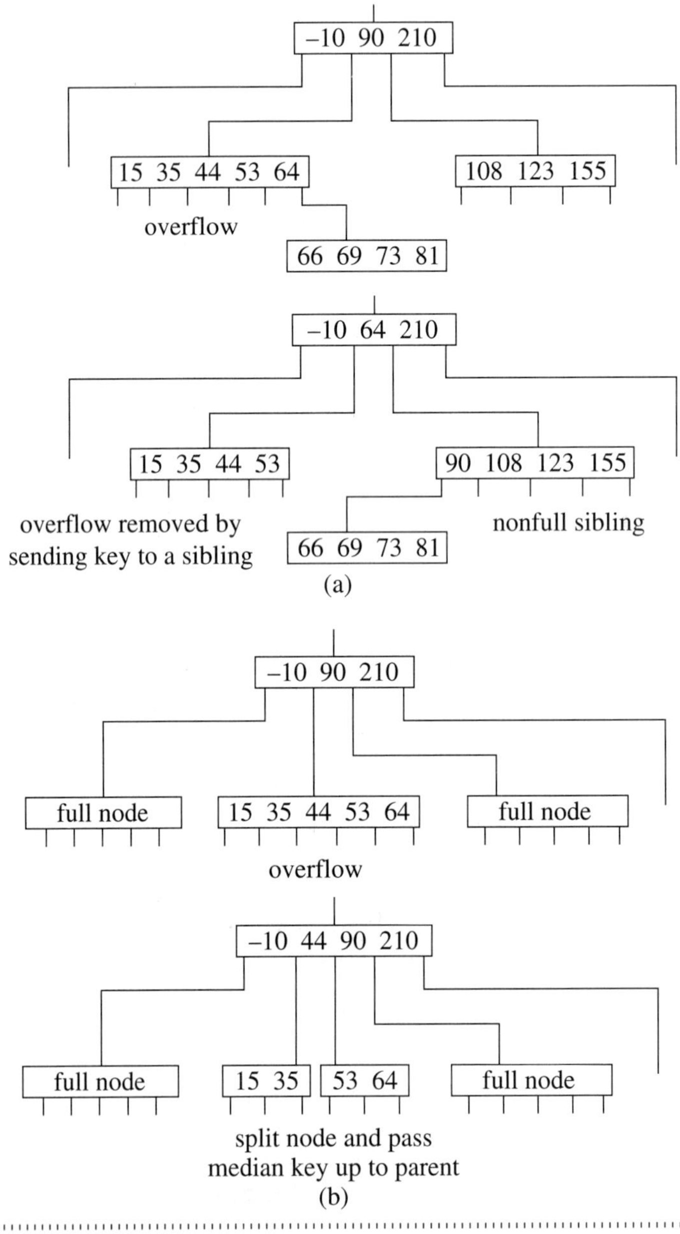

Figure 17.15 Continuation of example from Figure 17.14

the total sequence of insertions. When using method (1), if both the immediate right and left siblings are not full, we assume that we send a key to the right child.

Using the techniques illustrated in Figures 17.13–17.17, it is relatively straightforward (but cumbersome in detail) to design a recursive algorithm for

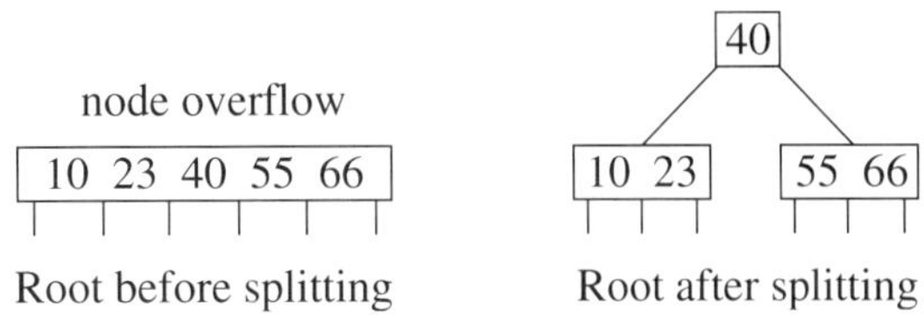

Figure 17.16 Splitting the root node of a B-tree of order 2

inserting a key in a B-tree. The recursion works as follows. If we are at a leaf node, then we search the node for a proper placement of the key. If the leaf node is not full, then we simply insert the key. If the leaf node is full, then, if possible, we use method (1) to restore the B-tree property. If we can't use method (1), then we split the node into two nodes, regarded as a left child and a right child of the median key that will be moved up to the parent node. The pointer to the original leaf node now points to the new left child leaf node, and we return the median key as well as a pointer to the newly created right child. The pointer to the right child is **nil** if no splitting occurred.

If we are not at a leaf node, then we search the node for the proper subtree in which to insert the key and recursively insert the key into this subtree. Again, if the

Figure 17.17 Inserting the sequence of keys 23, 98, 55, 37, 30, 47, 35, 15, 10, 75, 3, 53, 1, 33, 5, 28, 25, 90, 95, 56, and 8 into an initially empty B-tree of order 2

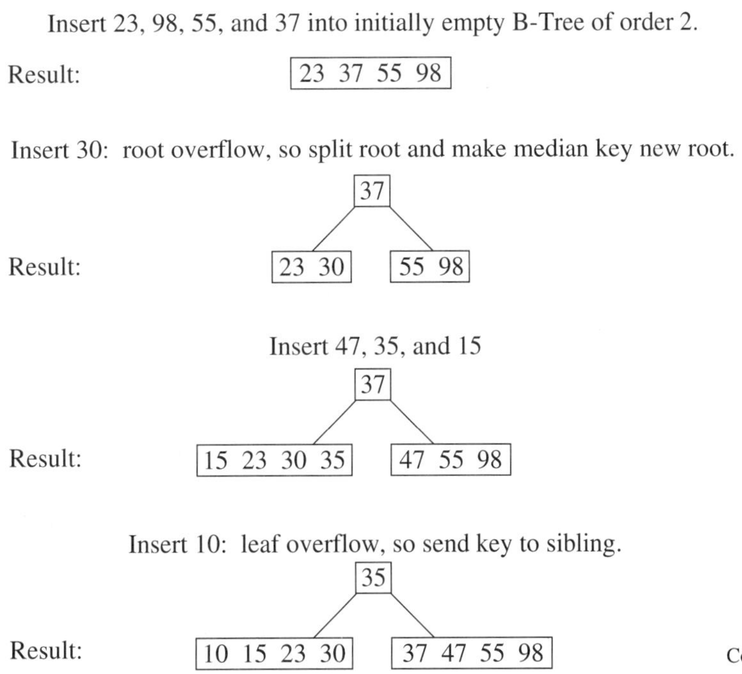

Continued

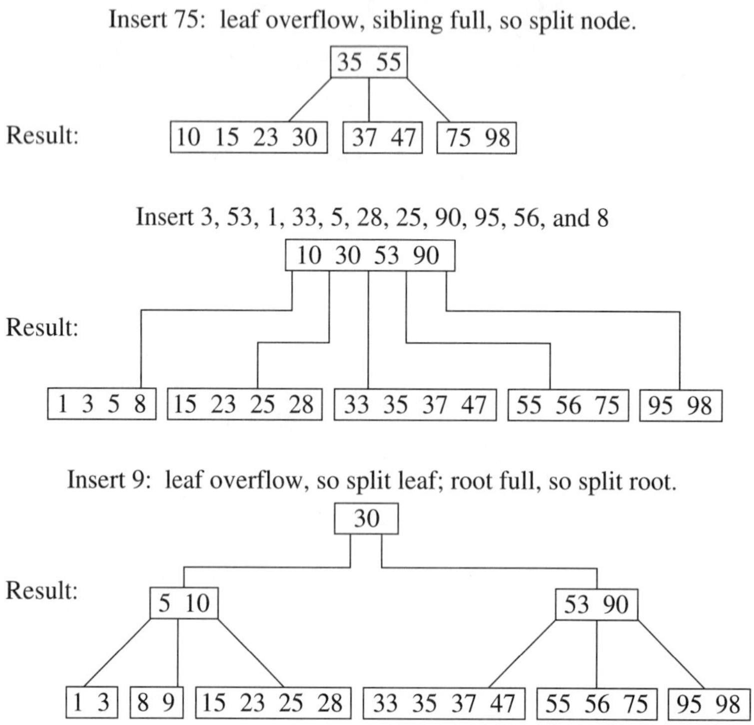

Figure 17.17 Continued

root node of the subtree has to be split, then the recursive insertion returns a median key to be inserted at the current node, together with a pointer to the node that is to be the right child of this median key.

We now give pseudocode for the procedure *B-TreeInsert*(*Root*,*k*) that inserts a key *k* into the B-tree *T* of order *m* pointed at by *Root*. We assume that the nodes in the B-tree have the following structure.

B_TreeNode = **record**
    Key[1:2*m*]: *KeyType* {an array of keys}
    NumKeys: *Integer* {keeps track of current number of keys in node, *Key*[1], . . . , *Key*[*NumKeys*]}
    Child[1:2*m*]: *Child*[*i*]→*B_TreeNode*, *i* = 1, . . . , *NumKeys* + 1 {an array of pointers to the children of the node}
    Parent: *Parent*→*B_TreeNode* {a pointer to parent node}
    Leaf: *Boolean* {a Boolean variable that is **.true**. if, and only if, the node is a leaf}
end *B_TreeNode*

Since B-trees are multiway search trees, it follows that for each internal node we have

$$k_1 \le Key[1] \le k_2 \le Key[2] \le \cdots \le k_{\text{NumKeys}} \le Key[NumKeys] \le k_{\text{NumKeys}+1},$$

where k_i is any key in the subtree pointed at by *Child*[i], $i = 1, \ldots, NumKeys + 1$.

We assume a procedure *AllocateB_TreeNode*, which when called with output parameter *Node* allocates (in internal memory) a memory location having the *B_TreeNode* record structure and whose address is placed in *Node*. In particular, all the algorithms for performing the dictionary operations on a B-tree *T* assume that *T* was initialized as an empty tree by executing the following statements.

```
call AllocateB_TreeNode(Root) {create an empty B-tree}
Root→Leaf := .true.
Root→NumKeys := 0
write(Root→B_TreeNode)
```

The nonrecursive procedure *B-TreeInsert* calls the auxiliary recursive procedure *B-TreeInsertRec*, which does most of the work.

```
procedure B-TreeInsert(Root,k,m)
Input:   Root (a pointer to root r of a B-tree T)
         k (a key to be inserted)
         m (order of T)
Output: Root (a pointer to T altered by inserting key k)
   call B-TreeInsertRec(Root,k,m,RChild,MedianKey)
   if RChild < > nil then {root r was split: create new root}
      call B-TreeNewRoot(Root,RChild,MedianKey)
   endif
end B-TreeInsert
```

```
procedure B-TreeInsertRec(Root,k,m,RChild,MedianKey) recursive
Input:   Root (a pointer to root node s of a subtree of a B-tree T)
         k (a key to be inserted)
         m (order of T)
Output: T altered by inserting key k in subtree
         RChild (a pointer to right child sibling of s if s is split, otherwise nil)
         MedianKey (median key of keys in s ∪ k if s is split, otherwise
                        undefined)
   call NodeSearch(Root,k,Root→NumKeys,Index)
   if Root→Leaf then      {insert k}
      call InsertKeyInNode(Root,k,Root→NumKeys,Index,MedianKey,RChild)
   else
      call B-TreeInsertRec(Root→Child[Index],k,m,RightChild,MedianKey)
```

```
        if RChild ≠ nil then
            call NodeSearch(Root,MedianKey,Root→NumKeys,Index)
            call InsertKeyInNode(Root,MedianKey,Root→NumKeys,Index,
                                         MedianKey,RChild)
        endif
    endif
end B-TreeInsert
```

We leave the pseudocode for procedures *B-TreeNewRoot*, *NodeSearch*, and *InsertKeyInNode* as exercises.

The number of basic operations (**read**s and **write**s) performed by *B-TreeInsert* for a B-tree T is bounded above by 4(depth(T)). Hence, by (17.4.1), *B-TreeInsert* performs at most $4(\log_m(n + 1)/2)$ basic operations for a B-tree of order m on n nodes.

17.5.4 Deleting a Key from a B-Tree

When inserting a key into a B-tree of order $2m$, we corrected node overflows by sending keys to siblings or by splitting nodes. Since we only split full nodes, the minimum number m of keys in each node was always maintained. Deleting a key from a node containing m keys leaves the node one shy of having the minimum number of keys, so a correction must be done.

As with deletion in binary search trees, deleting a key k from an internal node involves finding the inorder successor k^* of k (k^* will be in a leaf), replacing k by k^*, and then deleting k^*. The inorder successor k^* is the left-most key in the right subtree rooted at k. If the node containing k^* has more than m keys, then we merely delete k^* and are done. Otherwise, the deletion of k^* yields an underflow. One of two actions are performed to remove underflow, depending on the following two cases.

(1) If the node has an immediately adjacent sibling containing more than m keys, then send a key from the sibling up to the parent, and send a key from the parent down to the node.

(2) If both immediate siblings have m keys, then merge the node and an immediate sibling, and send down a key from the parent, placing it between the keys in the merged siblings.

Note that performing case (1) results in the restoration of a B-tree. Performing case (2) with a parent having more than m keys also results in the restoration of a B-tree. However, performing case (2) when a parent has only m nodes results in underflow at the parent. Then we must repeat case (2) at the parent. In the worst case, this underflow will propagate all the way to the root. Performing case (2) on a root with a single key destroys the root, and a new root consisting of the two merged siblings results (see Figure 17.18).

We illustrate the actions of cases (1) and (2) in Figure 17.18 for a sample B-tree of order 2.

In Figure 17.19 (page 622), we show the action of deleting the key 47 from a B-tree of order 2.

Case 1: Delete 88; send 77 from sibling to parent, and send 80 from parent to node

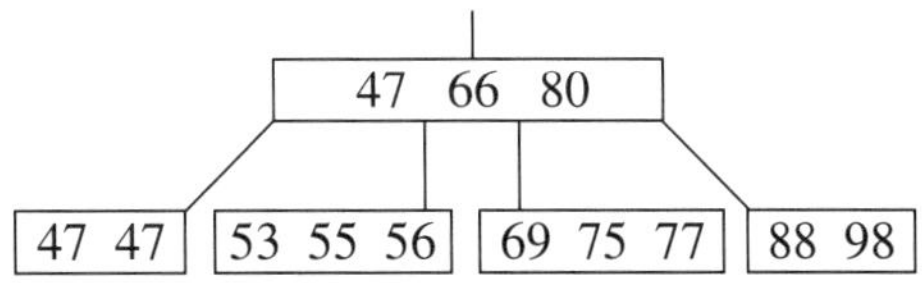

Result:

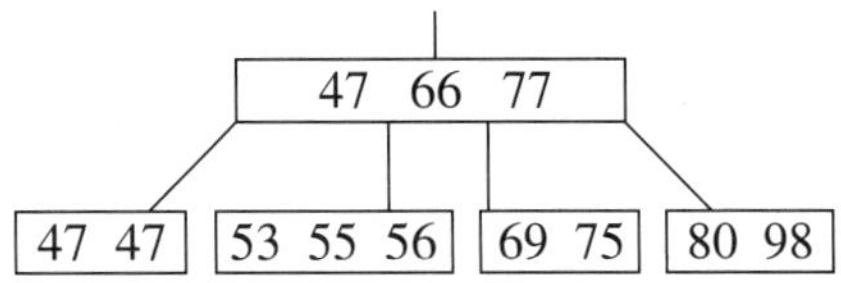

Case 2: Delete 80; merge node with sibling, and send down 77 from parent.

Result:

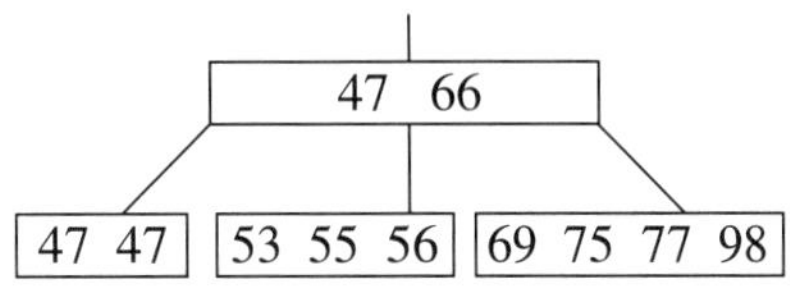

Figure 17.18 Cases (1) and (2) deletions, respectively, from a B-tree of order 2

Designing the procedure *B-TreeDelete* based on the preceding discussion is straightforward, but again somewhat cumbersome in detail. We leave the pseudocode for *B-TreeDelete* as an exercise. The number of basic operations (**read**s and **write**s) performed by *B-TreeDelete* for a B-tree T is bounded above by $4(\text{depth}(T)) < 4(\log_m(n + 1)/2)$.

17.6 Closing Remarks

For a given set of input keys S contained in some universal set U of keys, a function $h : U \Rightarrow \{0,1, \ldots, m - 1\}$ is called a *perfect hash function for S* if h is one-to-one on S. Thus, a perfect hash function avoids collisions altogether on S. If $|S| = n \leq m$, then clearly there are many perfect hash functions for S. The problem is to compute a perfect hash function efficiently for a given S, m, and U. Another requirement is that the perfect hash function should evaluate at any point in U in constant time. There has been quite a bit of research done on the problem of finding perfect hash functions. Most of the recent work in this direction involves probabilistic algorithms, which are the topic of the next chapter.

Since B-trees are so often utilized in practice to index database files, it is of interest to try to improve their performance even more. For example, two common variants of B-trees are B$^+$-trees and B*-trees (discussed in the exercises). The references contain much more detail on B-trees and their variants.

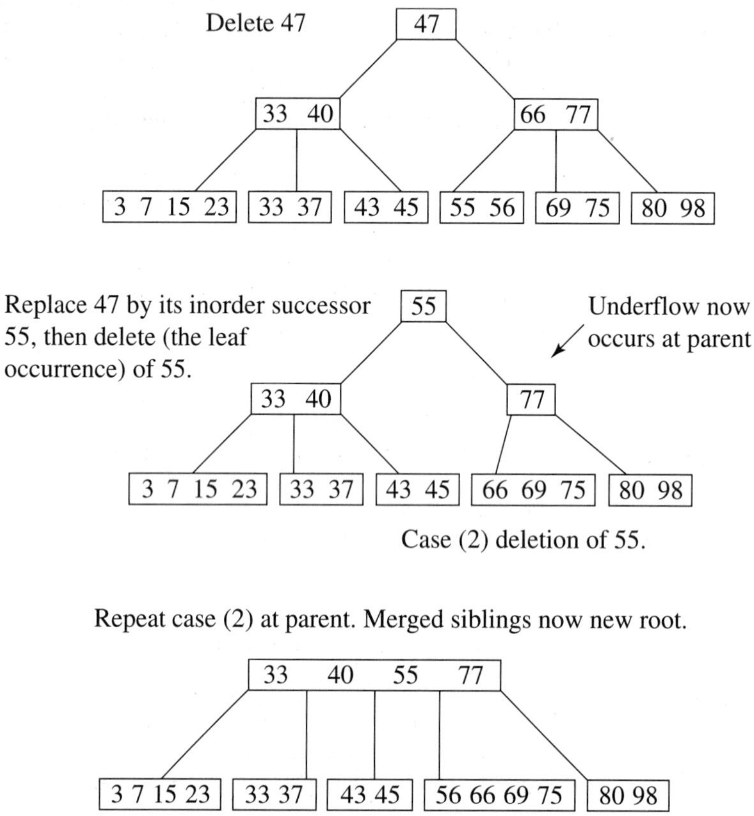

Figure 17.19 Deleting key 47 from a sample B-tree of order 2

References and Suggestions for Further Reading

Detailed discussions of hashing, as well as further information on balanced trees and B-trees, can be found in:

Aho, A. V., J. E. Hopcroft, and J. D. Ullman. *Data Structures and Algorithms*. Reading, MA: Addison-Wesley, 1983.

Gonnet, G. H. *Handbook of Algorithms and Data Structures*. Reading, MA: Addison-Wesley, 1984.

Knuth, D. E. *Sorting and Searching*, volume 3 of *The Art of Computer Programming*. Reading, MA: Addison-Wesley, 1973.

Majewski, B. S., N. C. Wormahld, G. Havas, and A. J. Czech. "Graphs, hypergraphs, and hashing," *Proceedings of the 19th International Workshop on Graph-Theoretic Concepts in Computer Science* (*WG'93*), 1993. An excellent guide to perfect hashing schemes.

Adel'son-Vel'skii, G. M. and Y. M. Landis. "An algorithm for the organization of information," *Dokl. Akad. Nauk SSR,* 146 (1962): 263–266. English translation in *Soviet Math.* (*Dokl.*), 3 (1962): 1259–1263. The notion of balancing a search tree is due to Adel'son-Vel'skii and Landis, who introduced AVL trees in this paper.

Bayer, R. "Symmetric binary B-trees: Data structure and maintenance algorithms," *Acta Informatica,* 1 (1972): 290–306. Bayer first introduced red-black trees under the name "symmetric binary B-trees."

Bayer, R. and E. M. McCreight. "Organization and maintenance of large ordered indices," *Acta Informatica,* 1 (1972): 173–189. The original reference to B-trees.

Comer, D. "The ubiquitous B-tree," *ACM Computing Surveys,* 11 (1979): 121–137. A nice survey of the topic of B-trees.

Exercises

Section 17.2 Hashing

17.1 Show the contents of an open-address hash table $H[0:22]$ after the insertion of the keys 2, 11, 55, 23, 53, 9, 61, 46, 1, 3, 25, 19, 17, 34, 10, 38, and 48 where $h(k) = k$ **mod** 23, using linear probing to resolve collisions.

17.2 Repeat Exercise 17.1, using double hashing, with hash functions $h_1(k) = k$ **mod** 23, and $h_2(k) = 1 + (k$ **mod** $21)$.

17.3 Consider the (not-so-good) hash function $h(i)$, which maps an integer i to its leading digit. Using the method of collision resolution by chaining, suppose the keys 23, 55, 123, 504, 211, 200, 88, 91, and 9 have been inserted in the hash table. Assuming each slot (leading digit) is equally likely for a given search k, determine the average behavior of successful search.

17.4 Repeat Exercise 17.3 for unsuccessful search.

17.5 Show that the probe sequence given by (17.2.7) reaches all slots in the table if, and only if, $h_2(k)$ and m are relatively prime.

17.6 Prove Theorem 17.2.1.

Section 17.3 Rotations of Binary Search Trees

17.7 Suppose we have keys $K_1 < K_2 < \cdots < K_n$ stored in a **global** array $K[1:n]$ (so that $K[i] = K_i$, $i = 1, \ldots, n$), and that T is an (abstract) binary search tree for the keys. Suppose also that the **global** array $Root[1:n,1:n]$ is such that $Root[i,j]$ is the index of the key in the root of the subtree of T containing the keys $K_i, K_{i+1}, \ldots, K_j$, $1 \le i \le j \le n$. (In particular, $Root[1,n]$ is the index of the key in the root of T.) Give pseudocode for a recursive **function** *CreateBinSrchTree*(i,j) that creates the binary search tree containing the keys $K_i, K_{i+1}, \ldots, K_j$, $1 \le i \le j \le n$, and returns a pointer to the subtree. Note that an invocation of *CreateBinSrchTree*$(1,n)$ should create the entire tree T. The nodes of the tree T are given the following structure.

```
BinSrchTreeNode = record
    Info: InfoType
    Key: KeyType
    LeftChild:    →BinSrchTreeNode
    RightChild:   →BinSrchTreeNode
    Parent:       →BinSrchTreeNode
end BinSrchTreeNode
```

CreateBinSrchTree should call the **procedure** *AllocateNode*(Q), which you can assume allocates space for the above record and returns in the output parameter Q a pointer to the allocated record.

17.8 Give pseudocode for an adaptation of *BinSrchTreeInsert* to binary search trees with parent pointers, whose nodes have the structure given in Exercise 17.7.

17.9 Give pseudocode for an adaptation of *BinSrchTreeDelete* to binary search trees with parent pointers, whose nodes have the structure given in Exercise 17.7.

17.10 Give pseudocode for a left rotation.

17.11 Recall that an AVL tree is a binary tree where the depth of the left and right subtrees of any node differ by at most one. For an AVL tree T on n nodes, show that depth(T) $\in$ O(log n).

17.12 Show that at most two rotations are needed to restore balance in an AVL tree after insertion of a node.

Section 17.4 Red-Black Trees

17.13 Prove that a red-black tree is balanced in the sense that no path from the root to a leaf has length more than twice the length of any other such path.

17.14 Describe the five cases for *RBCorrect(iii)OrMoveUpR* that are symmetric to those for *RBCorrect(iii)OrMoveUpL*

17.15 Give pseudocode for *RBCorrect(iii)OrMoveUpL* and *RBCorrect(iii)OrMoveUpR*.

17.16 Verify that the ten transformations for restoring a red-black tree introduce no other violations of the red-black properties (except as noted for case 2b, which is then corrected), that each case except case 2a results in restoring a red-black tree, and that case 2a moves the root of the violation of property (iii) one level up the tree.

17.17 Show the result of inserting the keys 9, 10, 5, 36, and 33 into the red-black tree in Figure 17.20.

17.18 Show the result of deleting the keys 27, 108, 23, 123, and 61 from the red-black tree in Figure 17.21.

Section 17.5 B-Trees

17.19 Show that most of the keys in a B-tree are in leaf nodes.

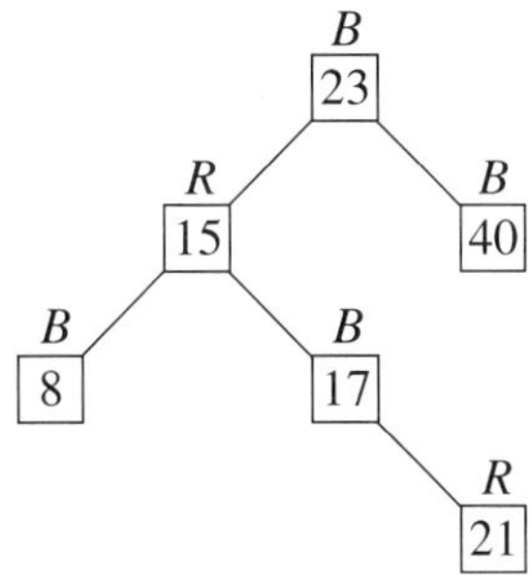

Figure 17.20

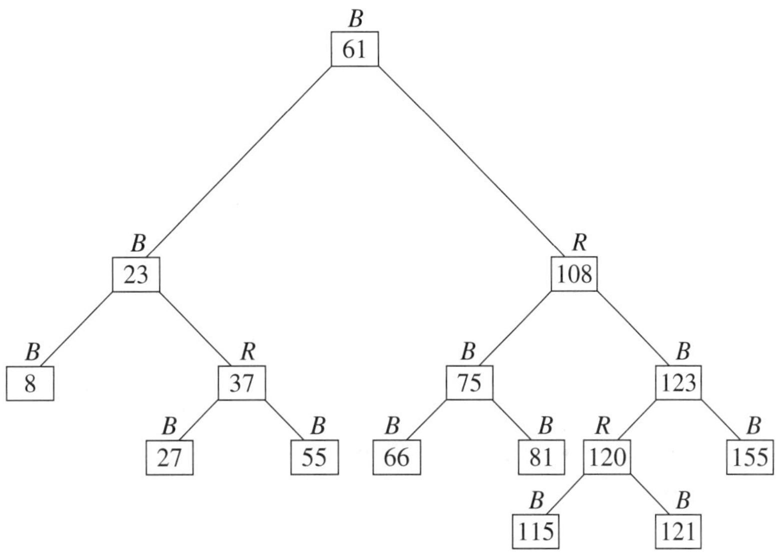

Figure 17.21

17.20 Show that for any key k in a B-tree, if k is in an internal node, then both the inorder successor and the inorder predecessor of k are in leaf nodes.

17.21 Determine the minimum depth of a B-tree of order m containing n keys, $n \geq 1$.

17.22 Show the B-tree that results from using *B-TreeInsert* to insert the elements 1, 2, and 63 into the B-tree of order 2 in Figure 17.22.

17.23 Show the B-tree that results from using *B-TreeDelete* to delete the elements 28, 29, 33, 30, 44, and 23 from the B-tree of order 2 in Figure 17.23.

For exercises 17.24–17.30: Suppose that the nodes in a B-tree T of order m have the form described in Section 17.5, and that we have procedure *AllocateB_TreeNode*,

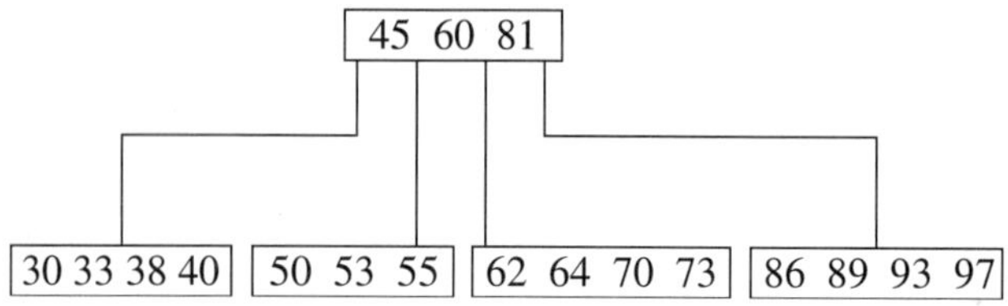

Figure 17.22

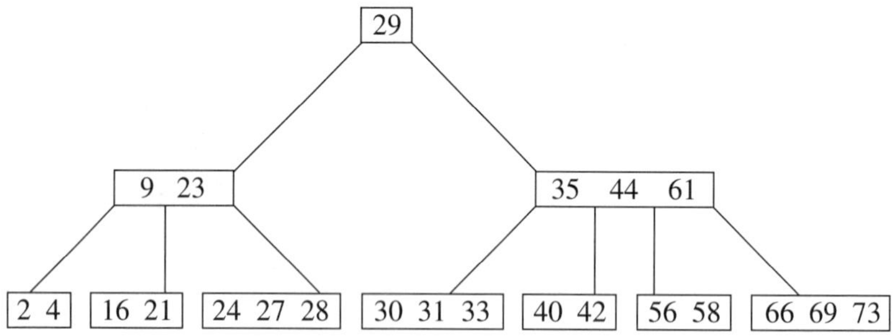

Figure 17.23

which when called with output parameter *Node* allocates (in internal memory) a memory location having the *B_TreeNode* record structure and whose address is placed in *Node*. In particular, we assume that the B-tree *T* was initialized as an empty tree by a executing the following statements.

> **call** *AllocateB_TreeNode*(*Root*) {create an empty B-tree}
> *Root* → *Leaf* **:= .true.**
> *Root* → *NumKeys* := 0
> *Parent* := **nil**
> **write**(*Root*→*B_TreeNode*)

17.24 Design and give pseudocode for a recursive algorithm *B-TreeSearch* that inputs a pointer to the root of a B-tree *T* and a search key *k* and outputs a pointer to a node *v* containing *k* (**nil** if *k* is not in *T*), as well as the index in *v*'s field *Key*[1:2*m* – 1] such that *Key*[*Index*] = *k* (0 if *k* not in *T*).

17.25 Design and give pseudocode for the procedure *B-TreeNewRoot*(*Root*,*RChild*, *MedianKey*) called by procedure *B-TreeInsert*.

17.26 Design and give pseudocode for the procedure *NodeSearch*(*Root*,*k*,*Root*→*NumKeys*,*Index*) called by *B-TreeInsert*.

17.27 Design and give pseudocode for the procedure *InsertKeyInNode*(*Root*,*k*,*Root*→*NumKeys*,*Index*,*MedianKey*,*RChild*) called by *B-TreeInsert*.

17.28 Design and give pseudocode for a nonrecursive version of *B-TreeInsert* based on its bottom-up resolution.

17.29 Design and give pseudocode for a recursive procedure *B-TreeDelete*.

17.30 Design and give pseudocode for a nonrecursive version of *B-TreeDelete* based on its bottom-up resolution.

17.31 A B*-tree is a B-tree with the additional requirement that each node (except possibly the root) is at least 2/3 full. Algorithms for maintaining B*-trees avoid splitting nodes whenever possible. For example, when inserting a key into a full node such that both immediate siblings are also full, the three full nodes are merged into two, with appropriate adjustment of the data in the parent. Design and analyze an algorithm for inserting into a B*-tree.

17.32 A B^+-tree is a B-tree where all keys in internal nodes are regarded as dummies (sign posts to guide searches), as in Figure 17.24. Keys in internal nodes are duplicated in leaves. Note that if the leaves in a B^+-tree are linked together in increasing order of the keys, then traversals are easily accomplished. Design and analyze search and insertion algorithms for B^+-trees.

Figure 17.24 A B^+-tree of order 2

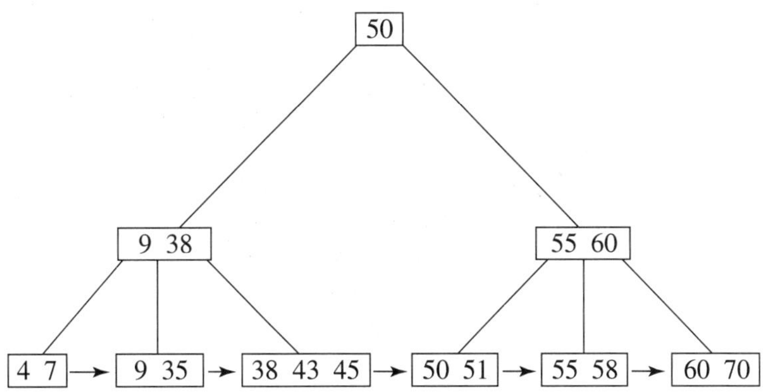

18

Probabilistic Algorithms

Probabilistic algorithms is one of the most active and rapidly growing areas of research in computer science. A probabilistic algorithm contains steps that make random choices by invoking a random (or pseudorandom) number generator. Thus, they are sometimes referred to as "coin-flipping algorithms." In particular, a probabilistic algorithm can perform differently for two runs with the same input.

Probabilistic algorithms fall into four main design categories: randomizations of deterministic algorithms, Monte Carlo algorithms, Las Vegas algorithms, and numerical probabilistic algorithms. A randomization of a deterministic algorithm results by replacing certain steps that made canonical choices by steps that make these choices in some random fashion. Randomization is done to break the connection between a particular input and worst-case behavior, and thereby homogenize the expected behavior of inputs to the algorithm.

The expected behavior of a randomized algorithm is no better than the average behavior of its associated deterministic algorithm and is usually a little worse due to such things as the overhead of calls to a random number generator. Monte Carlo algorithms often produce solutions very quickly, but only guarantee correctness with high probability. A Las Vegas algorithm never outputs an incorrect solution, but has some probability of reporting a failure to produce a solution. In practice, obtaining solutions with high probability is almost as satisfactory as the foolproof guarantee provided by a deterministic algorithm. There are many important problems, such as prime testing, for which the only known efficient algorithms for their solutions are probabilistic.

Numerical probabilistic algorithms were among the first examples of introducing randomness into the design of algorithms. A classical example is the estimation of π obtained by throwing darts at a square and recording how many darts land inside a circle inscribed in the square.

Probabilistic techniques have become very useful tool in the theory of parallel algorithms. The analog of the class NC for sequential algorithms is the class RNC for parallel algorithms. There are important examples of problems in RNC that are not known to be in NC.

18.1 Probabilistic Algorithms

Most of the probabilistic algorithms discussed in this chapter assume that we can make a random choice from a set of elements. For example, given integers m and n, $m < n$, we assume that a (*random number generator*) procedure *Random*($\{m, \ldots, n\}, j$) exists that assigns to j a random integer between m and n, with each integer being equally likely to be chosen. Moreover, successive calls to *Random* are independent of one another. Given an interval (a,b) of real numbers, we also assume a procedure *Random*($(a,b),x$) that assigns to x a randomly chosen real number between a and b. The distribution of x is uniform and independent on the interval (a,b).

In practice, random number generators are usually not available, and we must rely on *pseudorandom number generators*. A pseudorandom number generator is a deterministic procedure that from a given *seed* generates a sequence of numbers that closely resembles a sequence that might be obtained by successive calls to a genuine random number generator. However, given two identical seeds, a pseudorandom number generator generates identical sequences of numbers. Thus, most pseudorandom number generators that are implemented in programming languages utilize the system clock to generate a seed. While pseudorandom number generators closely simulate random number generators, we assume the existence of the latter in the design and analysis of our probabilistic algorithms.

The performance of a probabilistic algorithm depends not only on the input to the algorithm, but also on the results of randomized choices. Since running the algorithm twice with the same input I may result in a different number of basic operations being performed, $\tau(I)$ is no longer well defined. Instead, what is relevant is the *expected* number, $\tau_{exp}(I)$, of basic operations performed by the algorithm for input I with respect to the random choices made by the algorithm. As with any expectation, if we run the algorithm many times with *fixed input I*, then we can expect that the algorithm performs $\tau_{exp}(I)$ basic operations on average. If the algorithm performs many random choices, then even for a single run we can expect that the number of basic operations performed is very close to $\tau_{exp}(I)$. Analogous to the functions $B(n)$, $W(n)$, and $A(n)$ for the best-case, worst-case, and average complexities of a deterministic algorithm, we now have $B_{exp}(n)$, $W_{exp}(n)$, $A_{exp}(n)$ for the expected best-case, worst-case, and average complexities of a probabilistic algorithm. These are defined by

$$B_{exp}(n) = \min\{\tau_{exp}(I) \mid I \in \mathcal{I}_n\}$$

$$W_{exp}(n) = \max\{\tau_{exp}(I) \mid I \in \mathcal{I}_n\}$$

$$A_{exp}(n) = E\{\tau_{exp}\}$$

where $\mathcal{I}_n$ denotes the set of all inputs of size n to the algorithm.

In the next four sections we will design and analyze examples of each of the four main design strategies for probabilistic algorithms.

18.2 Randomizing Deterministic Algorithms

Randomized algorithms arise from deterministic algorithms by introducing randomness in some of the steps of the algorithm. Situations arise where introducing randomness into an otherwise deterministic algorithm can alleviate, in a certain sense, unacceptably bad worst-case behavior. Randomized versions of deterministic algorithms tend to cause the expected behavior of *each* input to approach the average behavior of the deterministic algorithm so that $A_{exp}(n)$ is approximately equal to $A(n)$. Thus, a randomized version of a deterministic algorithm has the most potential of being useful when the average complexity $A(n)$ of the deterministic algorithm is significantly smaller than the worst-case complexity $W(n)$. Then a suitable randomization usually results in a $W_{exp}(n)$ that is also significantly smaller than $W(n)$.

The procedure *QuickSort* is a good example of when randomization is useful. Indeed, *QuickSort* has $\Theta(n \log n)$ average complexity, but $\Theta(n^2)$ worst-case complexity. A randomization of *QuickSort* does not eliminate the possibility that an input performs $\Omega(n^2)$ basic operations, but it breaks the connection between an input and $\Omega(n^2)$ behavior. More precisely, the randomizations of *QuickSort* presented in this section will have $\tau_{exp}(I) = A(n)$ for any *any* input of size n, where $A(n)$ is the average behavior of ordinary *QuickSort*. In particular, the randomized versions of *QuickSort* satisfy the ultimate homogenization condition:

$$B_{exp}(n) = A_{exp}(n) = W_{exp}(n). \tag{18.2.1}$$

Randomized algorithms satisfying the ultimate homogenization condition (18.2.1) have been dubbed *Sherwood algorithms* in honor of Robin Hood.

18.2.1 Randomizing *LinearSearch*

LinearSearch simply scans a list $L[1{:}n]$ for the occurrence of a search element X from left to right (that is, in the order $L[1], L[2], \dots, L[n]$). Of course, it would make no difference to the efficiency of *LinearSearch* if we scan the list from right to left (that is, in the order $L[n], L[n-1], \dots, L[1]$). For successful search (X is in the list), we have $B(n) = 1$ and $W(n) = n$. If we make the additional assumption that each list element is distinct and equally likely to be X, then $A(n) = (n+1)/2$.

A simple randomization yields $B_{exp}(n) = A_{exp}(n) = W_{exp}(n) = (n+1)/2$. We merely flip a fair coin once at the beginning to decide whether we scan from left to right or from right to left. To verify that $\tau_{exp}(I) = (n+1)/2$ for any input $I \in \mathcal{I}_n$, let $L[1{:}n]$ be any input list and let k be the index in L of the occurrence of X. With probability 1/2 the scan will be from left to right, resulting in k comparisons, and with probability 1/2 the scan will be from right to left, resulting in $n - k + 1$ comparisons. Thus, the expected number of comparisons performed by the algorithm to find X is given by

$$\frac{1}{2}(k) + \frac{1}{2}(n - k + 1) = \frac{n+1}{2}.$$

Since L and X were any input I (where X is on the list), we have

$$\tau_{exp}(I) = \frac{n+1}{2}, \quad I \in \mathcal{I}_n. \tag{18.2.2}$$

LinearSearch is the simplest example we can think of that illustrates the complete homogenization effect (18.2.1) of the behavior of its randomized versions. Running the algorithm over and over again with any given input I yields an expected number of comparisons equaling $(n + 1)/2$ independently of the input. However, only a single random decision was made at the beginning, so we can't expect that the randomized version performs anything close to $\tau_{\text{exp}}(I)$ basic operations on a single run with input I. In fact, searching for the first element in the list has a 50–50 chance of making n comparisons! Thus, while the long-run behavior approaches $(n + 1)/2$, the variance of $\tau(I)$ is large for inputs I where the search element is near the beginning or the end of the list. We need to introduce more randomness into the algorithm so that the expected behavior *on a single running* is close to $(n + 1)/2$ for any input.

One way to introduce more randomness is to perform the search in two directions, using pointers *MoveRight* and *MoveLeft*, which initially point to the beginning and end of the list, respectively. Thus, initially *MoveRight* = 1 and *MoveLeft* = n. We then make a random choice (with equal probabilities) of *MoveRight* or *MoveLeft* and compare the search element X to L[*MoveRight*] or L[*MoveLeft*], depending on which one was chosen. If X is found, then we return the relevant value of *MoveRight* or *MoveLeft*. Otherwise, we increment *MoveRight* by one or decrement *MoveLeft* by one, depending on which one was chosen, and repeat the process. The process terminates when X is found (successful search) or when *MoveRight* becomes larger than *MoveLeft* (unsuccessful search). Since we have introduced randomness throughout the algorithm, we can expect a smaller variance $\tau(I)$ than our previous randomization of *LinearSearch*. We leave the coding and analysis of this probabilistic version of *LinearSearch* to the exercises.

18.2.2 Randomizing *LinkOrdLinSrch*

In Chapter 9 we designed a procedure *LinkOrdLinSrch* for searching a list $L[1{:}n]$ that was maintained in sorted order using an auxiliary array $Link[1{:}n]$. We assume that the variable *Head* contains the index of the smallest element in L. Thus, the sorted order of the list L is given by $L[Head]$, $L[Link[Head]]$, $L[Link[Link[Head]]]$, $\ldots$, $L[Link^{n-1}[Head]]$, where $Link^m$ denotes the m-fold composition of *Link*. $Link[i] = 0$ indicates the end of the list.

Given a search element X, procedure *LinkOrdLinSrch* begins by determining the best place to start a link-ordered linear search from the first $\sqrt{n}$ elements in $L[1{:}n]$. We saw that *LinkOrdLinSrch* has $\Theta(n)$ worst-case complexity, but $\Theta(\sqrt{n})$ average complexity, so it is a good candidate for randomization. Moreover, the randomization of *LinkOrdLinSrch* is straightforward. Instead of determining the best start for a link-ordered linear search from $L[1], L[2], \ldots, L[\sqrt{n}]$, the randomized version uses a random number generator to sample $\sqrt{n}$ points at random and determines the best start from these random choices. The analysis that we gave of the average behavior of *LinkOrdLinSrch* actually establishes at the same time that the randomized version has $\tau_{\text{exp}}(I) \in \Theta(\sqrt{n})$ for each $I = (L[1{:}n],X)$, $X = L[1], L[2], \ldots, L[n]$.

18.2.3 Randomizing *QuickSort*

One way to randomize *QuickSort* is to make a random choice of a pivot element for the call to *RearrangeAndPlace*. Another way to randomize *QuickSort* is to first randomly permute the elements in a given input list L and then input the permuted list L' to *QuickSort*. This latter method of randomization is an example of the general randomization technique known as *stochastic preconditioning*. Both methods of randomizing *QuickSort* accomplish the same thing; that is, the randomized versions of *QuickSort* perform the same *expected* number of comparisons for *any* input list of size n. Moreover, the expected number of comparisons performed by the randomized versions of *QuickSort* are the same as the average number of comparisons performed by the deterministic version of *QuickSort*. We now introduce some concepts and notation allowing a more precise description of the randomized versions of *QuickSort*.

QuickSort exhibits $\Theta(n^2)$ worst-case complexity, whereas its average complexity is in $\Theta(n \log n)$. Making a random choice for the pivot element in the call to *RearrangeAndPlace* leads to *RandomizedQuickSort*, which exhibits completely homogenized behavior. When making a random choice for the pivot element in *RandomizedQuickSort*, we assume that each element in the relevant interval has an equal chance of being chosen. For an interval of m elements, this is achieved by a call to *Random*($\{1, \dots, m\}, k$). Our analysis for the average complexity $A(n)$ of *QuickSort* given in Chapter 9 was based on the assumption that each element of the input list $L[1:n]$ is equally likely to be chosen as the pivot element. Thus, $\tau_{exp}(L) = A(n)$ for any input list L of size n, so that $W_{exp}(n) = A(n)$.

Note that in some theoretical sense we have not *eliminated* the worst-case $\Theta(n^2)$ behavior. Indeed, for *any* input list I of size n, there is the chance (albeit infinitesimally small for n large) that the random choice of the pivot element always belongs at an endpoint of the relevant interval, thereby forcing the algorithm to perform $\Theta(n^2)$ comparisons. However, since *RandomizedQuickSort* makes many random choices of pivot elements (one choice for each internal node in its tree of recursive calls), performing $\Theta(n^2)$ comparisons has such a small chance of occurring that for all practical purposes it can be ignored, even for relatively small lists. Thus, in practice, we can expect the number of comparisons performed in any one run of *RandomizedQuickSort* with any given input list L to be very close to $\tau_{exp}(L)$.

In Figure 18.1, we show the tree of recursive calls performed by *RandomizedQuickSort* for a sample input list of size 19.

Procedure *RandomizedQuickSort* makes $O(n)$ calls to *Random*, where we assume each call takes unit time. Thus, the overhead due to calling *Random* is dominated by the $\Omega(n \log n)$ comparisons performed by *RandomizedQuickSort*.

We now design an algorithm *RandomQuickSort2* based on stochastic preconditioning. Given an input list $I = L[1:n]$, *RandomizedQuickSort2* first calls the following procedure *Permute* with input list L, and then calls *QuickSort*.

procedure *Permute*($L[1:n]$)

Input: $L[1:n]$ (an array of list elements)
Output: $L[1:n]$ (an array of list elements randomly permuted)

```
    for i = 1 to n - 1 do
        call Random({i, . . . , n},j)
        call interchange(L[i],L[j])
    endfor
end Permute
```

Procedure *RandomQuickSort2* has the identical expected performance as *RandomizedQuickSort*.

Figure 18.1 Tree of recursive calls of *RandomizedQuickSort* for a sample run with an input list of size 19. Internal nodes are labeled with the indices of randomly chosen pivot elements in a call to *RearrangeAndPlace*. Leaf nodes are labeled with the index of single element sublists that result in immediate returns. For simplicity, we do not show leaf nodes corresponding to calls to *QuickSort* with empty lists.

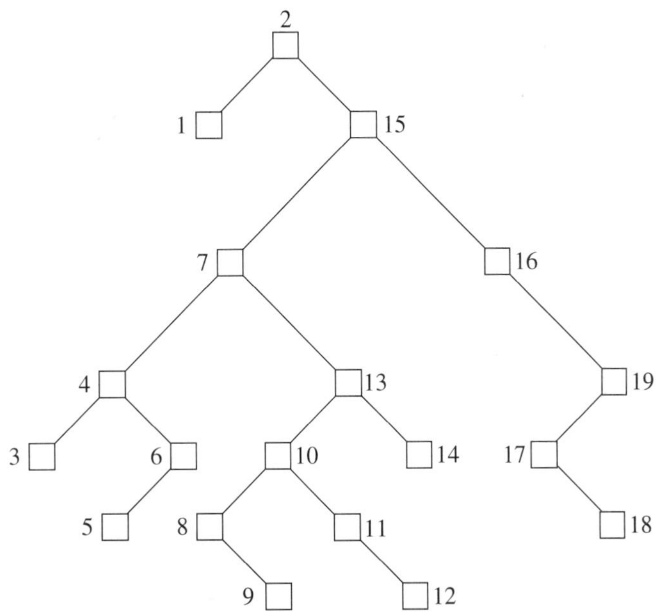

18.3 Monte Carlo Algorithms

A Monte Carlo algorithm is a probabilistic algorithm that has a certain probability of returning the correct answer whatever input is considered. The most useful class of Monte Carlo algorithms are those that have a probability of returning the correct answer greater than some fixed positive constant for any input. More precisely, for a (fixed) real number p, $0 < p < 1$, a *p-correct Monte Carlo* algorithm is a probabilistic algorithm that returns the correct answer with probability not less than *p no matter what input is considered.* Unfortunately, there is usually no efficient method available to test whether an answer returned by a Monte Carlo algorithm for a given input is correct.

18.3.1 Biased Monte Carlo Algorithms

A Monte Carlo algorithm for a decision problem is *false-biased* if it is always correct when it returns the value **.false.** and only has some (hopefully small) probability of making a mistake when returning the value **.true.**. A similar definition holds for *true-biased* Monte Carlo algorithms. For convenience, in the following discussion, we restrict our attention to Monte Carlo algorithms that are false-biased. Fundamental to the applicability of a Monte Carlo algorithm is the fact that the probability of returning the correct output increases with repeated trials. For an example, given a (false-biased) Monte Carlo algorithm *MC*, consider the following algorithm *MCRepeat*. (In a particular application, the generic name *MC* will be replaced by the specific name of the Monte Carlo algorithm.)

```
function MCRepeat(k)
Input:   k (a positive integer)
Output: .false. if MC returns .false. for any invocation, .true. otherwise
   for i := 1 to k do
      if MC returns .false. then
         return(.false.)
      endif
   endfor
   return (.true.)
end MCRepeat
```

The following proposition shows that *MCRepeat* can ramp up the correctness of *MC* to a number as close to unity as desired.

PROPOSITION 18.3.1 Suppose that we have a p-correct false-biased Monte Carlo algorithm *MC*. Then the algorithm *MCRepeat(k)* is a $(1 - (1 - p)^k)$-correct false-biased Monte Carlo algorithm.

Proof Clearly, repetition of a false-biased algorithm results in a false-biased algorithm. For a given input I, let $q_k(I)$ denote the probability that *MCRepeat(k)* outputs **.false.**. Let $p_k(I)$ denote the probability that *MCRepeat(k)* outputs the correct answer for a given input I. Then, using the formula for conditional probability given in Chapter 9, we obtain

$$\begin{aligned} p_k(I) &= q_k(I)\text{Prob}\{\textit{MCRepeat}(k) \text{ is correct } | \textit{ MCRepeat}(k) \text{ outputs } \textbf{.false.}\} + \\ &\quad (1 - q_k(I))(\text{Prob}\{\textit{MCRepeat}(k) \text{ is correct } | \textit{ MCRepeat}(k) \text{ outputs } \textbf{.true.}\} \\ &\geq q_k(I)(1) + (1 - q_k(I))(1 - (1 - p)^k) \\ &\geq q_k(I)(1 - (1 - p)^k) + (1 - q_k(I))(1 - (1 - p)^k) \\ &= (q_k(I) + (1 - q_k(I)))(1 - (1 - p)^k) = (1 - (1 - p)^k). \end{aligned}$$ ■

To illustrate the use of Proposition 18.3.1, suppose we have a Monte Carlo false-biased algorithm *MC* for primality testing that is .75-correct. If *MC* is .75-correct, then even for k as small as 8, *MCRepeat*(k) is .9999847-correct. Thus, we can be pretty confident that an integer n is prime if *MCRepeat*(8) returned **.true.** when we input n, since otherwise an event has occurred whose probability is not greater than .00000153.

Remark If we have an unbiased p-correct algorithm for a decision problem with $p \leq 1/2$, then repeating the algorithm doesn't ramp up the correctness. For unbiased p-correct Monte Carlo algorithms with $p > 1/2$, the correctness can be improved by repeating the algorithm sufficiently many times and choosing the answer occurring most often.

18.3.2 A Monte Carlo Algorithm for Testing Polynomial Equality

Given symbolic polynomials (that is, represented by their coefficient arrays) $f(x)$, $g(x)$, and $h(x)$ of degrees $2n$, n, and n, respectively, we consider the problem of testing whether $f(x)$ is identically equal to $g(x)h(x)$, written $f(x) \equiv g(x)h(x)$. (We write $f(x) \not\equiv g(x)h(x)$ to indicate the negation of $f(x) \equiv g(x)h(x)$ if there exists at least one point x such that $f(x) \neq g(x)h(x)$.) In Chapter 13 we developed a (deterministic) $O(n \log n)$ algorithm (*FFT*) for computing the symbolic product of two polynomials $g(x)$ and $h(x)$, each of degree n. Thus, an $O(n \log n)$ algorithm for testing whether or not $f(x) \equiv g(x)h(x)$ is obtained by computing $g(x)h(x)$ using *FFT* and comparing the coefficients of $g(x)h(x)$ to those of $f(x)$.

As our first illustration of a Monte Carlo algorithm, we now describe a simple false-biased (1/3)-correct Monte Carlo algorithm for testing $f(x) \equiv g(x)h(x)$ that has $O(n)$ complexity. The idea behind the algorithm is that a polynomial of degree n is completely determined by knowing its values at $n + 1$ points (see Section 7.6). Thus, two distinct polynomials of degree $2n$ could not agree at more than $2n$ integers drawn from $\{1,2, \ldots, 3n\}$. The following algorithm is therefore a false-biased (1/3)-correct Monte Carlo algorithm for testing whether or not $f(x) \equiv g(x)h(x)$.

```
function TestPolyEqual(f(x),g(x),h(x))

Input:  f(x),g(x),h(x) (symbolic polynomials of degrees 2n,n,n, respectively)
Output: .true. 100% of the time if f(x) ≡ g(x)h(x), and at most 2/3 of the
                    time if f(x) ≢ g(x)h(x)

   call Random({1, . . . , 3n},j)
   if f(j) = g(j)h(j) then
      return(.true.)
   else
      return(.false.)
   endif

end TestPolyEqual
```

The algorithm *TestPolyEqual* has $O(n)$ complexity since the evaluations of $f(j)$, $g(j)$, and $h(j)$ can be done with $2n$, n, and n multiplications, respectively, using the algorithm *HornerEval*. If $f(x) \not\equiv g(x)h(x)$, then we can have $f(j) = g(j)h(j)$ for at most $2n$ points $j \in \{1,2,\ldots,3n\}$. Hence, *TestPolyEqual* is (1/3)-correct. Thus, if we repeat *TestPolyEqual* 20 times, then Proposition 18.3.1 shows that the probability that *TestPolyEqualRepeat*(20) makes an error is less than $(2/3)^{20} < .0003$ no matter what instance is considered.

Remark By choosing a point from $\{1,2,\ldots,mn\}$, a *TestPolyEqual* would be $(2/m)$-correct. We took $m = 3$ for convenience and to keep the domain of the *Random* relatively small. Besides, 20 repetitions of *TestPolyEqual* gave us an algorithm whose correctness was very close to unity.

18.3.3 Primality Testing

You probably are familiar with the following deterministic algorithm for the primality testing problem.

```
function TestPrimality(n)
Input:  n (a positive integer)
Output: .true. if n is a prime, .false. otherwise
   for i := 2 to ⌊√n⌋ do
      if n mod i = 0 then return(.false.) endif
   endfor
   return(.true.)
end TestPrimality
```

Recall that when measuring the complexity of number-theoretic algorithms having integer input n, we take the input size to be the number of digits in n (with respect to a convenient base). The number of digits in the decimal representation of n is given by $m = \lceil \log_{10}(n+1) \rceil$. In the worst case (that is, when n is a prime), the function *TestPrimality* executes the **for** loop $\lfloor \sqrt{n} \rfloor$ times. Since $\lfloor \sqrt{n} \rfloor \approx 10^{m/2}$, we see that the worst-case complexity of *TestPrimality* is exponential in the input size m. Thus, *TestPrimality* is of no practical use for integers having, say, more than 40 digits, since it would take millions of years for today's fastest computers to execute *TestPrimality* in the worst case.

The following simple probabilistic algorithm *MCTestPrimality* for primality testing is a false-biased Monte Carlo algorithm. However, *MCTestPrimality* is still of little practical value since it is not p-correct for *any* fixed p (its correctness rapidly diminishes with the input size). Thus, the number of times that it would have to be run to achieve a given fixed confidence level would increase with the input size.

```
function MCTestPrimality(n)
Input:  n (a positive integer)
```

```
Output: .false. if a single random choice j of an integer in
          {2, . . . , ⌊√n⌋} divides n, .true. otherwise

   call Random({2, . . . , ⌊√n⌋},j)
   if n mod j = 0 then
      return(.false.)
   else
      return(.true.)
   endif
end MCTestPrimality
```

If $n = p_1p_2$ where $p_1 < p_2$ are primes, then the probability that *MCTestPrimality* returns the correct answer **.false.** is less than $1/p_1$. As mentioned earlier, this renders *MCTestPrimality* unsatisfactory in practice. What is needed is a test for primality that has a *fixed* positive probability of succeeding no matter what size the input. Such a test was discovered independently in the 1970s by Rabin and by Solovay and Strassen. We need some definitions from number theory to describe the test.

The following result, stated by Fermat in 1640, is useful for prime testing.

THEOREM 18.3.2 If n is a prime, then for all positive integers $a < n$,

$$a^{n-1} \equiv 1 \ (\text{mod } n). \tag{18.3.1}$$

□

Most composite numbers n have many integers between 2 and $n - 1$ such that $a^{n-1} \not\equiv 1$ (**mod** n), Thus, a test for primality for these composite numbers that has a high probability of being correct is to test whether $a^{n-1} \not\equiv 1$ (**mod** n) for a random choice of a. Unfortunately, there exist composite integers n, for which $a^{n-1} \equiv 1$ (**mod** n) for most $a < n$. In fact, there are composite numbers, called *Carmichael numbers*, for which (18.3.1) holds for all $a < n$. The following false-biased Monte Carlo algorithm *RSSTestPrimality* is based on a refinement of Fermat's test, and was developed independently by Rabin and Solovay and Strassen. Note that unless *RSSTestPrimality* returns an answer within the **for** loop, the final test it performs is the Fermat test (18.3.1).

```
function RSSTestPrimality(n)

Input:   n (an odd positive integer)
Output: .true. 100% of the time when n is a prime, and less than 25% of the
          time when n is composite

   call Random({2, . . . , n – 2},a)
   find positive integers k,m such that n – 1 = 2^k m, and m odd
```

Continued

```
    b := a^m mod n                      {b ≡ a^m (mod n)}
    for j := 1 to k do
       if b ≠ 1 .and. b ≠ n – 1 then    {b ≢ ± 1 (mod n)}
          return(.false.)
       else
          if b = n – 1 then             {b ≡ – 1 (mod n)}
             return(.true.)
          endif
       endif
       b := b*b mod n                   {b ≡ a^(2^j m) (mod n)}
    endfor
    if b ≠ 1 then                       {a^(n – 1) ≢ 1 (mod n)}
       return(.false.)
    else
       return(.true.)
    endif
end RSSTestPrimality
```

Rabin and Solovay and Strassen proved that *RSSTestPrimality* is false-biased and .75-correct. Note that if the binary method for computing powers is used to compute a^m (mod n), *RSSTestPrimality* is linear in the input size (number of digits of n). Thus, repeating *RSSTestPrimality* yields an efficient primality testing algorithm, whose probability of error can be made quite small even for just a few repetitions.

The *primality testing problem*, and the related *factoring problem* of finding the prime factors of a composite number, play important roles in cryptography. For example, these two problems play complementary roles in the RSA cryptosystem, which is currently one of the most widely used methods for public key cryptography and for digital signatures. The practicality of constructing RSA ciphers depends on (very high confidence) primality testing being efficient using such algorithms as *RSSTestPrimalityRepeat*, whereas the security of the cipher depends on the fact that no known (probabilistic or otherwise) algorithm exists for efficiently factoring large numbers.

18.4 Las Vegas Algorithms

Unlike a Monte Carlo algorithm, a Las Vegas algorithm never returns an incorrect answer. However, a Las Vegas algorithm is subject to two types of disasters: It may run arbitrarily long without producing an answer, or it might make a random decision from which there is no recovery (a dead end). Usually the Las Vegas algorithm signals the latter type of disaster and terminates, so that it can be restarted again from scratch. The probability that a dead end is reached on any given run is usually quite small, so that running the algorithm several times significantly increases the probability that the algorithm successfully solves the given problem. Relative to the first type of disaster, running beyond a reasonable time usually has such a small probability of occurring that it can safely be ignored in practice. Thus, in all cases Las Vegas algorithms have high probability of successfully finding the solution to a given problem.

In this section we give two examples of Las Vegas algorithms. Our first example is a simple illustration of a Las Vegas algorithm for the problem of searching a list. Our second example is a probabilistic string matching algorithm due to Karp and Rabin. The Karp-Rabin string matching algorithm illustrates how a Monte Carlo algorithm can be converted into a Las Vegas algorithm if we have a method for testing whether or not the output of the Monte Carlo algorithm is correct. Given such a test, we can repeatedly run the Monte Carlo algorithm until a correct output is generated.

18.4.1 Randomized Searching of a List with Repeated Elements

Our first example of a Las Vegas algorithm is a simple example of searching a list with repeated elements. We assume that we are searching a list $L[1{:}n]$ where the search element X occurs *at least k times* in the list. For example, suppose we are sampling a large list of students for a student having a GPA of 3.3 or higher. If we know that at least one-tenth of the students has a GPA of 3.3 or higher, then $k \geq n/10$. Note that the worst-case complexity of solving this problem using *LinearSearch* is $n - k + 1$ and is achieved for an input list where X occupies precisely the last k positions in the list L. Note also that *LinearSearch* with stochastic preconditioning would not be useful here, since it takes $n - 1$ calls to *Random* to randomly permute the list elements.

If k is sufficiently large with respect to n, say $k \geq n/d$ for some fixed constant $d > 0$, then we can *expect* to do much better than *LinearSearch* by the following simple randomized searching algorithm *RandomSearch*. Rather than making a linear scan through the list, each iteration of the **while** loop in *RandomSearch* calls a random number generator *Random* to determine which list element to compare to X. In particular, *RandomSearch* may re-examine a list element more than once. Indeed, *RandomSearch* is subject to the first type of disaster; namely, it has the potential to run arbitrarily long for a fixed input size n. However, the expected number of iterations performed by *RandomSearch* in the worst-case for an input of size n is relatively small.

```
function RandomSearch(L[1:n],X)
Input:  L[1:n] (an array of list elements), X (a search item, occurring at least
                                                     k times in L)
Output: returns index of an occurrence of X in L

   Found := .false.
   while .not. Found do
      call Random({1, . . . , n},Choice)
```

Continued

```
    if X = L[Choice] then Found := .true. endif
  endwhile
  return(Choice)
end RandomSearch
```

The expected best-case complexity is $B_{exp}(n) = 1$, since $\tau_{exp}(I) = 1$ for an input $I = (L[1{:}n],X)$ where each list element equals X. Now consider the expected worst-case complexity $W_{exp}(n)$. Clearly, $W_{exp}(n)$ is achieved by any input list in which X occurs *exactly* k times. For such an input list I_w, the search element X is found in any particular pass of the **while** loop with probability $p = k/n$. Therefore, the probability that exactly i passes are performed is given by $p_i = (1-p)^{i-1}p$, $i = 1,2, \ldots$. Thus,

$$W_{exp}(n) = \tau_{exp}(I_w) = \sum_{i=1}^{\infty} ip_i = \sum_{i=1}^{\infty} i(1-p)^{i-1}p. \tag{18.4.1}$$

Using formula (1.2.13), (18.4.1) becomes

$$W_{exp}(n) = \tau_{exp}(I_w) = p\left(\frac{1}{1-(1-p)}\right)^2 = \frac{1}{p} = \frac{n}{k}. \tag{18.4.2}$$

When k is a fixed percentage of the list elements, say $k \geq n/d$, for a fixed $d > 0$, then (18.4.2) implies that $W_{exp}(n) \leq d$, so that $W_{exp}(n) \in O(1)$.

Given any integer i, note that the algorithm *RandomSearch* performs more than i iterations with probability $(1-p)^i$, $p = k/n$. While this probability tends to zero as i tends to infinity, nevertheless *RandomSearch* has the (diminishing) potential to run arbitrarily long for a fixed input size n. However, in practice the possibility of *RandomSearch* performing unacceptably more iterations than $W_{exp}(n)$ is so remote that it can be safely ignored.

18.4.2 Probabilistic String Matching

Finding some or all occurrences of a given (pattern) string in a given (host) text is an important problem that is encountered often in computer science. For example, most word processing software packages have built-in search-and-replace functions and spell checkers, both of which depend on finding the occurrences of words in texts.

The string matching problem can be formally described as follows. An *alphabet* is a set of *characters* or *symbols* $A = \{a_1,a_2, \ldots, a_k\}$. A *string* $S = S[1{:}n]$ of length n on A is a sequence of m characters (repetitions allowed) from A. Such a string $S = s_1s_2, \ldots, s_n$ can be viewed as an array of characters $S[1{:}n]$ from A, so that the ith character s_i in the string is denoted by $S[i]$, $i = 1, \ldots, n$. More generally, we denote the substring consisting of symbols in consecutive positions i through j of S by $S[i{:}j]$, $1 \leq i \leq j \leq n$. The *null string*, denoted by ε, is the string that contains no symbols. We let A^* denote the set of all finite strings (including the null string ε) on A. The length of a string S, denoted $|S|$, is the number of characters in S. For $a \in A$, we let a^i denote the string of length i consisting of the single symbol a repeated i times.

Given a *pattern string* $P = P[1{:}m]$ of length m and *host string* $H = H[1{:}n]$ of length n, where $m \le n$, the string matching problem is to determine where P occurs in H. For convenience, we denote the substring $H[s + 1{:}s + m]$ by H_s, $s = 0, \ldots, n - m$. In our string matching algorithms, we assume that we are looking for the first occurrence (if any) of the pattern string in the host string. The algorithms can be readily modified to return all occurrences of the pattern string.

A Naive Algorithm A naive algorithm for finding the first occurrence of P in H is to position P at the start of H and simply shift the pattern P along H, one position at a time, until either a match is found or the string H is exhausted (that is, position $n - m + 1$ in H is reached without finding a match).

```
function NaiveStringMatcher(P[1:m],H[1:n])

Input:  P[1:m] (a pattern string of length m)
        H[1:n] (a host string of length n)
Output: returns the position in H of the first occurrence of P, or 0
        if P does not occur in H

  for s := 0 to n – m do {shift the pattern string P along H by s positions}
     if H_s = P then
        return(s + 1)
     endif
  endfor
  return(0)

end NaiveStringMatcher
```

When measuring the complexity of procedure *NaiveStringMatcher*, it is natural to choose comparison of symbols as our basic operation. We can test whether $H_s = P$ by using a simple linear scan. Clearly, this scan requires a single comparison in the best case ($P[1] \ne H[s + 1]$) and m comparisons in the worst case ($P[i] = H[s + i]$, $i = 1, \ldots, m - 1$). Since the **for** loop of procedure *NaiveStringMatcher* is iterated $n - m + 1$ times, *NaiveStringMatcher* never performs more than $m(n - m + 1)$ comparisons. Moreover, $m(n - m + 1)$ comparisons are performed, for example, when $P[1{:}m]$ and $H[1{:}n]$ are the strings $0^{m-1}1$ and 0^n, respectively, over the alphabet $A = \{0,1\}$. Thus, *NaiveStringMatcher* has worst-case complexity

$$W(m,n) = m(n - m + 1).$$

The action of *NaiveStringMatcher* is illustrated for a sample pattern and host string in Figure 18.2 (page 642).

Karp-Rabin String Matching Algorithm We now present a probabilistic string matching algorithm, due to Karp and Rabin. Without loss of generality we assume that A is the k-ary alphabet $\{0,1, \ldots, k - 1\}$. Each character of A can be thought of as a digit in radix-k notation, and each string $S \in A^*$ can be identified with the base k

```
P:     00100122
H:     001201001001221
        0 0 1 2 0 1 0 0 1 0 0 1 2 2 1
s = 0   0 0 1 0
s = 1     0 0
s = 2       0
s = 3         0
s = 4           0 0
s = 5             0
s = 6               0 0 1 0 0 1 2 2   a match: return 7
```

For each shift s, $P[1{:}8]$ is compared with $H = H[s + 1{:}s + 8]$ character by character from left to right until a mismatch is encountered or a match is made. The resulting false starts and match at $s = 6$ are shown.

Figure 18.2 Action of procedure *NaiveStringMatcher* for a sample pattern string $P[1{:}8]$ and a host string $H[1{:}15]$ on alphabet $A = \{0,1,2\}$

representation of an integer $\hat{S}$. For example, when $k = 10$, the string of numeric characters "6832355" can be identified with the integer 6832355. Given a pattern string $P[1{:}m]$, we can compute the corresponding integer $\hat{P}$ using m multiplications and m additions by employing Horner's Rule.

$$\hat{P} = P[m] + k(P[m-1] + k(P[m-2] + \cdots + k(P[2] + kP[1]) \ldots)) \qquad (18.4.3)$$

Similarly, we can employ Horner's Rule to compute $\hat{H}_0$. However, to compute the integer $\hat{H}_s$, $s = 0, \ldots, n - m$, we do not need to employ Horner's Rule each time. We can compute $\hat{H}_s$ from $\hat{H}_{s-1}$ using the following recurrence relation:

$$\hat{H}_s = k(\hat{H}_{s-1} - k^{m-1}H[s]) + H[s+m] \quad s = 1, \ldots, n-m. \qquad (18.4.4)$$

For example, if $k = 10$, $m = 7$, $\hat{H}_{s-1} = 7937245$, and $\hat{H}_s = 9372458$, then (18.4.4) becomes

$$\hat{H}_s = 10[7937245 - (1000000 \times 7)] + 8 = 9372458.$$

The constant $c = k^{m-1}$ in (18.4.4) can be computed in time O(log m) using the binary method for computing powers. Once c is computed it does not need to be recomputed when (18.4.4) is applied again. Thus, assuming that the arithmetic operations in (18.4.4) take constant time, each application of (18.4.4) takes constant time. Hence, the $n - m + 2$ integers $\hat{P}$ and $\hat{H}_s$, $s = 0, 1, \ldots, n - m$, can be computed in total time $O(m + n)$. Thus, a linear ($O(n + m)$) worst-case algorithm for solving the string matching problem is obtained by comparing $\hat{P}$ and $\hat{H}_s$, starting with $s = 0$ and incrementing s until $\hat{P} = \hat{H}_s$ (a match is found) or until $s = n - m + 1$ (P does not occur in H).

The problem with the preceding approach is that the integers $\hat{P}$ and $\hat{H}_s$, $s = 0, 1, \ldots, n - m$, may be too large to work with efficiently, and the assumption that (18.4.4) can be performed in constant time becomes unreasonable. To get around this difficulty, we reduce these integers modulo q for some randomly chosen integer q. To avoid multiple-precision arithmetic, q is often chosen to be a random prime number such that kq fits within one computer word.

We now let

$$\hat{P}^{(q)} = \hat{P} \textbf{ mod } q, \qquad \hat{H}_s^{(q)} = \hat{H}_s \textbf{ mod } q. \tag{18.4.5}$$

The values $\hat{H}_s^{(q)}$ and $\hat{P}^{(q)}$ can be computed in time $O(m + n)$ using exactly the same algorithm described earlier for computing $\hat{H}_s$ and $\hat{P}$, except that all arithmetic operations are performed modulo q. Clearly, if $\hat{H}_s^{(q)} \neq \hat{P}^{(q)}$ then $P \neq H_s$. However, if $\hat{H}_s^{(q)} = \hat{P}^{(q)}$, we are not guaranteed that $P = H_s$. When a shift s has the property that $\hat{P}^{(q)} = \hat{H}_s^{(q)}$, but $P \neq H_s$ we have a *spurious match*. A Monte Carlo algorithm for finding the first occurrence of P in H returns the first shift s such that $\hat{H}_s^{(q)}$ and $\hat{P}^{(q)}$ are equal or reports false to the existence of a match if no such shift exists. Clearly, the resulting algorithm is false-biased, since it always correctly reports false when no match exists. Unfortunately, unless q is sufficiently large, if a match is found there is a high probability that it is spurious. However, we can test whether or not a match is spurious by explicitly checking whether $P = H_s$, thereby turning the Monte Carlo algorithm into the following Las Vegas algorithm.

```
function KarpRabinMatcher(P[1:m],H[1:n],k,q)
Input:  P[1:m] (a pattern string of length m)
        H[1:n] (a host string of length n)
        k (A is the k-ary alphabet {1,2, . . . , k})
        q (a random prime number q such that kq fits in one computer word)
Output: returns the position in H of the first occurrence of P, or 0
        if P does not occur in H

  c := k^(m-1) mod q
  P̂^(q) := 0
  Ĥ_0^(q) := 0
  for i := 1 to m do   {apply Horner's Rule to compute P̂^(q) and Ĥ_0^(q)}
     P̂^(q) := (kP̂^(q) + P[i]) mod q
     Ĥ_0^(q) := (kĤ_0^(q) + H[i]) mod q
  endfor
  for s := 0 to n - m do
     if s > 0 then
          Ĥ_s^(q) := (k(Ĥ_(s-1)^(q) - H[s]c) + H[s + m]) mod q
     endif
     if Ĥ_s^(q) = P̂^(q) then
        if H_s = P then   {match is not spurious}
           return(s + 1)
        endif
```

```
        endif
      endfor
      return(0)
    end KarpRabinMatcher
```

In Figure 18.3 the action of function *KarpRabinMatcher* is illustrated for a sample pattern string $P[1:7] = 6832355$, host string $H[1:21] = 895732102683235544031$, and prime modulus $q = 11$. To illustrate the calculations of $\hat{H}_s^{(11)}$ in Figure 18.3, we show how *KarpRabinMatcher* computes $\hat{H}_3^{(11)}$ from $\hat{H}_2^{(11)}$. Using (18.4.4) and doing all arithmetic modulo $q = 11$, we have

$$\begin{aligned} 7321026 &\equiv 10(5732102 - 1000000 \times 5) + 6 \quad (\mathbf{mod}\ 11) \\ &\equiv 10(2 - 1 \times 5) + 6 \quad (\mathbf{mod}\ 11) \\ &\equiv 9. \quad (\mathbf{mod}\ 11) \end{aligned}$$

Since *KarpRabinMatcher* terminates after finding the first nonspurious match, the worst-case performance occurs for an input pair (P,H), where the pattern string P occurs precisely at the end ($s = n - m$) of the host string H. Since q is chosen at random, for $s \neq n - m$, we make the assumption that $\hat{H}_s^{(q)}$ takes on a particular value $i \in \{0,1, \ldots, q-1\}$ with equal probability $1/q$. Since a spurious match occurs only when $\hat{H}_s^{(q)} = \hat{P}^{(q)}$, $s = 0, \ldots, n - m - 1$, it follows that a spurious match occurs at shift s with probability $1/q$. Let r denote the expected number of spurious matches. Since a test for a spurious match involves m comparisons in the worst-case, and since $r + 1$ such tests are performed (including

Figure 18.3 Action of function *KarpRabinMatcher* illustrated for a sample pattern string $P[1:7]$, host string $H[1:21]$, and prime modulus $q = 11$

$\hat{H}[1:21]$ 895732102683235544031
$\hat{P}[1:7]$ 6832355

s	$\hat{H}_s$	$\hat{H}_s^{(11)}$
0	8 9 5 7 3 2 1	10
1	9 5 7 3 2 1 0	9
2	5 7 3 2 1 0 2	2 Spurious match
3	7 3 2 1 0 2 6	9
4	3 2 1 0 2 6 8	6
5	2 1 0 2 6 8 3	0
6	1 0 2 6 8 3 2	4
7	0 2 6 8 3 2 3	0
8	2 6 8 3 2 3 5	5
9	6 8 3 2 3 5 5	2 Match: return(10)

the test at shift $s = n - m$), the expected performance of *KarpRabinMatcher* for the input (P,H) is

$$W_{exp}(m,n) = (r + 1)m + (n - m + 1). \tag{18.4.6}$$

The value r is the expectation of the binomial distribution with $n - m$ trials (shifts), where success (a spurious match) occurs with probability $1/q$. Thus,

$$r = \frac{n-m}{q}. \tag{18.4.7}$$

Substituting (18.4.7) into (18.4.6), we obtain

$$W_{exp}(m,n) = \left(\frac{n-m}{q} + 1\right)m + (n - m + 1). \tag{18.4.8}$$

If we assume that q is bounded above by a fixed constant, then *KarpRabinMatcher* achieves a worst-case complexity in $\Theta(nm)$, which is no better than the naive algorithm. However, in practice, it is reasonable to assume that q is larger than m, in which case *KarpRabinMatcher* has complexity in $O(n)$. The Karp-Rabin string matching algorithm has the additional nice feature that it is readily adapted to the problem of finding $m \times m$ patterns in $n \times n$ hosts (see Exercise 18.8).

The string matching problem has been well studied in the literature. Two recent books entirely devoted to string matching algorithms are listed in the references. Two standard deterministic string matching algorithms are the Boyer-Moore algorithm and the Knuth-Morris-Pratt algorithm. We discuss the Knuth-Morris-Pratt algorithm in the exercises.

18.5 Probabilistic Numerical Algorithms

Historically, the first examples of probabilistic algorithms computed answers of a desired accuracy to numerical problems such as finding approximate values of definite integrals. Examples such as computing π by throwing darts discussed in the introduction to this chapter were dubbed Monte Carlo algorithms or Monte Carlo simulations. Here the term Monte Carlo algorithm is used to mean any probabilistic algorithm that approximates a numerical quantity by a random simulation process as opposed to the definition of a Monte Carlo algorithm as discussed in the previous section.

18.5.1 Numerical Integration

There are various deterministic algorithms for computing numerical approximations to the definite integral $\int_a^b f(x)dx$. For example, the interval $[a,b]$ is divided up into n subintervals of equal length $\Delta x = (b - a)/n$ using points $x_0 = a$, $x_1 = a + \Delta x$, $x_2 = a + 2\Delta x, \ldots, x_n = a + n\Delta x = b$. Then three standard approximations to $\int_a^b f(x)dx$ are given by

Left-Hand End Point Riemann Sum: $\Sigma_{i=1}^{n} f(x_{i-1})\Delta x = [\Sigma_{i=1}^{n} f(x_{i-1})](b - a)/n$,

Right-Hand End Point Riemann Sum: $\Sigma_{i=1}^{n} f(x_i)\Delta x = [\Sigma_{i=1}^{n} f(x_i)](b - a)/n$, and the average of the preceding two Riemann sums,

$$\textit{Trapezoid Rule:}\quad [f(x_0) + 2f(x_1) + 2f(x_2) + \cdots + 2f(x_{n-1}) + f(x_n)](b-a)/2n.$$

In general, for a given number of iterations, the Trapezoid Rule will achieve much better approximations to $\int_a^b f(x)dx$ than the probabilistic methods that we are about to describe, but probabilistic methods become practical for integrating functions of four or more variables.

One probabilistic method (*Monte Carlo Integration*) for finding an approximation to $\int_a^b f(x)dx$ relies on the interpretation of the integral as the (signed) area between the graph of f and the x axis. For purposes of discussion, we can assume without loss of generality that f is nonnegative on the interval $[a, b]$. We also assume that M is an upper bound for the value of $f(x)$ as x varies over $[a, b]$. Referring to Figure 18.4, we have illustrated a random collection of points chosen from the rectangle R with corners $(a,0),(a,M),(b,0),(b,M)$. We can think of the region between the graph of f and the x axis as the target in the dart board R, and the points illustrated as the result of throwing darts at R. If enough darts are thrown, we expect that the ratio of the number of darts that hit the target to the total number of darts thrown will closely approximate the ratio of the area $\int_a^b f(x)dx$ of the target to the area $M(b-a)$ of the dart board R.

The following Monte Carlo integration algorithm implements the dart throwing technique.

```
function ThrowDarts(f,a,b,M,n)
Input:  f (a function f: [a,b]→[0, + ∞))
        a,b (endpoints of an interval of real numbers)
        M (a real number such that f(x) ≤ M for all x ∈ [a,b])
        n (a positive integer equal to number of points sampled)
Output: approximation to ∫_a^b f(x)dx

  HitTarget := 0
  for Dart := 1 to n do
     call Random((a,b), x)
     call Random((0,M),y)
     if y ≤ f(x) then
        HitTarget := HitTarget + 1
     endif
  endfor
  return(M(b – a)*HitTarget/n)
end ThrowDarts
```

In practice, when f is computationally expensive to calculate, the direct test $y \le f(x)$ can sometimes be replaced by an equivalent but easier-to-compute test. For example, consider the integral $\int_0^1 \sqrt{1-x^2}dx$. Then the test $y \le \sqrt{1-x^2}$ is equivalent to the computationally easier test $x^2 + y^2 \le 1$.

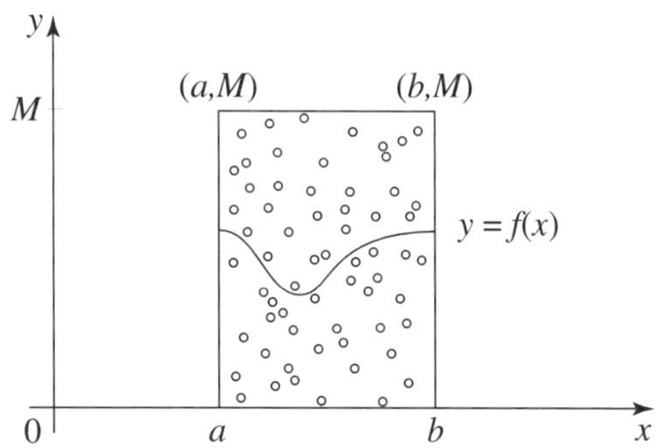

Figure 18.4 Throwing darts to approximate $\int_a^b f(x)dx$

A second method of Monte Carlo integration to approximate $\int_a^b f(x)dx$ is simply to choose n points $x_1, x_2, \ldots, x_n$ at random in the interval (a, b) and multiply the arithmetic mean $(f(x_1) + f(x_2) + \cdots + f(x_n))/n$ of f evaluated at these points by the length of the interval $b - a$. Although the second method is more efficient than *ThrowDarts*, it is still less efficient than the Trapezoid Rule. The name *ApproxMeanValueIntegral* for the following Monte Carlo integration algorithm reflects the fact that the arithmetic mean of the values of f at n randomly selected points will approximate the value $f(x^*)$ in the Mean Value Theorem for Integrals, which states that $\int_a^b f(x)dx = (b - a)f(x^*)$ for some point x^* in (a,b).

```
function ApproxMeanValueIntegral(f,a,b,n)
Input:  f (a function f: [a,b]→(−∞, +∞))
        a,b (endpoints of an interval of real numbers)
        n (a positive integer equal to number of points sampled)
Output: approximation to ∫_a^b f(x)dx

  Sum := 0
  for Count := 1 to n do
    call Random((a,b),x)
    Sum := Sum + f(x)
  endfor
  return((b − a)*Sum/n)

end ApproxMeanValueIntegral
```

For the same value of n, *ApproxMeanValueIntegral* generally gives a better approximation to $\int_a^b f(x)dx$ than does *ThrowDarts*. However, when $f(x)$ is expensive to calculate, *ThrowDarts* may be more efficient than *ApproxMeanValueIntegral*.

Monte Carlo integration starts to become computationally attractive for approximating multiple integrals in dimension greater than four. The reason is that deterministic approximations to k-fold multiple integrals are based on partitioning a rectangular parallelepiped $P = [a_1, b_1] \times [a_2, b_2] \times \cdots \times [a_k, b_k]$ by subdividing each interval $[a_i, b_i]$ into n subintervals of length $b_i - a_i$, $i = 1, \ldots, n$. For example, we then obtain an approximation to the multiple integral of $f(x_1, x_2, \ldots, x_n)$ over the original parallelepiped P by summing the products of f evaluated at a suitable point in each of the n^k rectangular subparallelepipeds times the volume of the subparallelepipeds. Since the number of points used increases exponentially with the dimension, these deterministic approximation methods rapidly become computationally infeasible. On the other hand, the precision obtained by Monte Carlo integration using n points is generally not particularly sensitive to an increase in dimension.

For approximating multiple integrals, hybrid methods have been developed to improve precision that are a mix between deterministic and probabilistic methods. Refer to the references for further discussion of Monte Carlo integrations in higher dimensions.

18.5.2 Estimating the Size of Sets

Monte Carlo integration algorithms find approximations to real numbers to within a desired accuracy. We finish our discussion of numeric probabilistic algorithms by noting that there are probabilistic algorithms that can be used to estimate the value of a (large) integer that would be otherwise difficult to compute. For example, suppose that you wish to estimate the size of a finite set X that is too large to count by a simple enumeration of its elements. Suppose also (and this is a strong assumption) that there exists a method of randomly and uniformly drawing a element from the set X. Then a probabilistic estimate of the size of X is obtained by repeatedly drawing an element from X until the first repetition occurs. Laws of probability imply that this repetition has a good chance of occurring much earlier than might be guessed intuitively.

The classical example of choosing until a repetition occurs is the so-called *birthday paradox*. The birthday paradox is not really a paradox, but rather is the somewhat surprising answer to the question: "How many people do you need to randomly sample from a given population before there is a better than a 50–50 chance that two people share the same birthday?" At first glance it might seem that there should be at least $188 = \lceil 365/2 \rceil$ randomly chosen people before they would bet even odds that two people share the same birthday. The surprise is that choosing 23 people suffices. The calculation proceeds as follows. First randomly pick a person x_1 from the population. The chance that a second randomly chosen person x_2 from the remaining population has a different birthday from the first is 364/365. Continuing in this fashion, the chance that the kth randomly chosen person x_k has a birthday different from the previously chosen people $x_1, x_2, \ldots, x_{k-1}$ is given by

$$\text{Prob}(k \text{ randomly drawn people all have different birthdays}) = \left(\frac{364}{365}\right)\left(\frac{363}{365}\right)\cdots\left(\frac{365-k+1}{365}\right). \tag{18.5.1}$$

The expression (18.5.1) begins to be less than 1/2 for $k \geq 23$.

Remark We assumed in (18.5.1) that each of the 365 days is equally likely to be a birthday for a random person. Actually, the distribution of birthdays over the entire populace is not quite uniform, and there is also the February 29th leap year anomaly. The first consideration increases the probability for common birthdays, whereas the second consideration decreases this probability. However, these small effects are probably not enough to change the threshold value of $k = 23$.

Generalizing (and slightly altering) the birthday problem, suppose that we have a set X of n objects, and we draw $k \leq n$ elements randomly from X. We assume here (as opposed to the birthday problem) that our draws are with replacement. Then the number of ways we can obtain k distinct elements is $n(n-1)\dots(n-k+1) = n!/(n-k)!$. Since there are n^k different ways to randomly draw (in order and with replacement) k elements from X, it follows that the probability that we obtain k distinct elements is given by

$$\text{Prob}(k \text{ elements drawn from } n \text{ are distinct}) = \frac{n!}{(n-k)!n^k}. \tag{18.5.2}$$

By Stirling's formula, $n! \approx \sqrt{2\pi n}\,(n/e)^n$, so that

$$\frac{n!}{(n-k)!n^k} \approx e^{-k^2/2n}. \tag{18.5.3}$$

Combining (18.5.2) and (18.5.3) yields the following theorem.

THEOREM 18.5.1 Let k be the smallest integer such that the probability is greater than 1/2 that we obtain at least one repetition in k random and uniform draws with replacement from a set of size n. Then, for large n,

$$k \approx \sqrt{2n \ln 2}\sqrt{n} \approx 1.177\sqrt{n}. \tag{18.5.4}$$ ❑

The next theorem, whose proof is left as an exercise, gives the expected value of the number of draws required before the *first* repetition occurs.

THEOREM 18.5.2 Suppose we randomly and uniformly draw elements with replacement from a set X of size n until a repetition occurs. Let ρ denote the random variable consisting of the number k of draws associated with each such event. Then, for large n, the expected value of ρ is given by

$$E(\rho) \approx \sqrt{\frac{\pi}{2}}\sqrt{n} \approx 1.253\sqrt{n}. \tag{18.5.5}$$ ❑

Theorem 18.5.2 is the basis for the following probabilistic algorithm for estimating the size n of a finite set X.

function *EstimateSize*(X)

Input: X (a finite set)
Output: an estimate of the size of X

Continued

```
  k := 1
  SetOfDraws := ∅
  call Random(X,a)
  while a ∉ SetOfDraws do
    k := k + 1
    SetOfDraws := SetOfDraws ∪ {a}
    call Random(X,a)
  endwhile
  return(2k²/π)
end EstimateSize
```

By Theorem 18.5.2, the expected number of repetitions of the **while** loop in *EstimateSize* belongs to $\Theta(\sqrt{n})$. Note that the space requirements for maintaining the set *SetOfDraws* also belongs to $\Theta(\sqrt{n})$. Thus, the expected complexity of *EstimateSize* belongs to $\Theta(\sqrt{n})$, assuming that the operations of checking set membership and adding elements to *SetOfDraws* can both be done at unit cost.

18.6 Probabilistic Parallel Algorithms

In this section we discuss the obvious parallelization that exists for many of the probabilistic algorithms discussed earlier. We also describe probabilistic parallel algorithms for solving the well-known Marriage problem and for determining a protocol for breaking communication deadlock in a completely symmetric network of processors.

18.6.1 Exploiting Obvious Parallelism

The procedure *MCRepeat*(k) for ramping up the correctness of a Monte Carlo algorithm simply repeats the algorithm k times. Since these repetitions are independent trials, there is the obvious parallelism for *MCRepeat*(k): We simply assign k processors the job of running the algorithm in parallel and then communicate the results. Hence, the k repetitions of the sequential algorithm can be reduced to a single parallel running and a parallel communication step. For example, on the EREW PRAM the communication step can be done in $O(\log k)$ time. A similar parallelization can be done for Las Vegas algorithms that are repeated to increase the certainty of correctness.

In the previous section we saw that probabilistic numerical integration involves randomly choosing n points (throwing darts) in a $(k + 1)$-dimensional parallelepiped P in Euclidean space containing the region R corresponding to the value of a definite integral of a function of k variables. The resulting algorithm then estimates the value of the integral by computing the ratio of the number of darts that hit within R to the total number of darts n thrown and then multiplying this ratio by the volume of P. The parallelization of the process for estimating a definite integral reduces the problem to adding n numbers, each of which is 0 or 1.

On the EREW PRAM having n processors P_i, $i = 1, \ldots, n$, the definite integral estimation can be done in O(log n) parallel steps (k fixed). Indeed, each of the n processors would randomly choose a point in P (each processor making k calls to a random number generator). Then processor P_i writes a 0 or 1 to the memory location *HitOrMiss*[i] in the array *HitOrMiss*[1:n], $i = 1, \ldots, n$. An invocation of the function *SumPRAM* with input *HitOrMiss*[1:n] then determines the number of points chosen within the region R, so that the definite integral can then be estimated as indicated earlier.

For interconnection network models, the processors each write the outcome of their choice of a point to a distributed variable *HitOrMiss*, and then the values in *HitOrMiss* are summed. Thus, the complexity of the parallelization reduces to the communication complexity of the network.

18.6.2 The Marriage Problem

Suppose we have a set of n boys $b_1,b_2, \ldots, b_n$ and a set of n girls $g_1,g_2, \ldots, g_n$, where each boy knows some of the girls. The so-called *Marriage problem* is to determine a necessary and sufficient condition so that each boy can marry a girl that he knows (no two boys can marry the same girl). More formally, under what conditions does there exist a permutation σ: $\{1,2, \ldots, n\} \rightarrow \{1,2, \ldots, n\}$ such that boy b_i knows girl $g_{\sigma(i)}$, $i = 1,2, \ldots, n$? If such a permutation σ exists, then the corresponding set of matches $P_\sigma = \{(b_i,g_{\sigma(i)}): i = 1,2, \ldots, n\}$ is called a *perfect matching*.

Let K_i denote the set of girls that boy b_i knows, $i = 1, \ldots, n$. For example, suppose that $n = 5$ and the sets K_i, $i = 1, \ldots, 5$, are given by

$$K_1 = \{g_1,g_2,g_4,g_5\}, K_2 = \{g_1,g_3\}, K_3 = \{g_2,g_3\}, K_4 = \{g_3,g_4,g_5\}, K_5 = \{g_3\}.$$

Then, the following set of pairs determines a perfect matching:

$$(b_1,g_4), (b_2,g_1), (b_3,g_2), (b_4,g_5), (b_5,g_3).$$

On the other hand, suppose that K_1 is replaced with the set

$$K_1' = \{g_1,g_2,g_3\}.$$

Since the four boys b_1,b_2,b_3,b_5 collectively know only the three girls g_1,g_2,g_3, a perfect matching cannot exist.

The Marriage problem is naturally modeled using a bipartite graph G, where one set X of the vertex bipartition consists of the set of n boys and the other set Y consists of the set of n girls. A vertex b_i in X is joined to a vertex g_j in Y, whenever the boy b_i knows the girl g_j (see Figure 18.5, page 652). The Marriage problem then becomes: Determine a necessary and sufficient condition for a bipartite graph to contain a *perfect matching*, that is, a set of n edges no two of which have a vertex in common.

Let E denote the edge set of the bipartite graph G. Associate the indeterminate weight x_{ij} with each edge $\{b_i,g_j\} \in E$, and let $R = (r_{ij})$ be the $n \times n$ matrix such that $r_{ij} = x_{ij}$ if $\{b_i,g_j\}$ is an edge of G, and $r_{ij} = 0$, otherwise. The matrix R for the bipartite graph G of Figure 18.5 is given in Figure 18.6. Let S_n denote the set of all permutations of $\{1, \ldots, n\}$. Given a permutation $\sigma \in S_n$, the set of

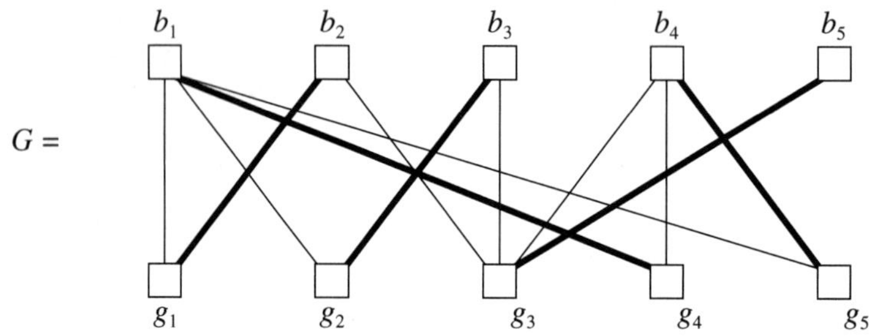

Figure 18.5 The bipartite graph G corresponding to sets $K_1 = \{g_1,g_2,g_4,g_5\}$, $K_2 = \{g_1,g_3\}$, $K_3 = \{g_2,g_3\}$, $K_4 = \{g_3,g_4,g_5\}$, $K_5 = \{g_3\}$. G contains the perfect matching (b_1,g_4), (b_2,g_1), (b_3,g_2), (b_4,g_5), (b_5,g_3).

pairs $P_\sigma = \{(b_i,g_{\sigma(i)}) \mid i = 1,2, \ldots, n\}$ is a perfect matching of G if, and only if, $r_{i\sigma(i)}$ is nonzero for each $i \in \{1, \ldots, n\}$. Define $sign(P_\sigma) = sign(\sigma)$. Let $\mathscr{P}$ denote the set of all perfect matchings of G.

The following proposition gives a solution to the Marriage problem and yields an associated parallel Monte Carlo algorithm for determining whether or not a bipartite graph has a perfect matching.

PROPOSITION 18.6.1 The bipartite graph G contains a perfect matching if, and only if, the determinant of R is not the identically zero polynomial in the indeterminates x_{ij}. ❑

Proposition 18.6.1 follows immediately from the following lemma.

LEMMA 18.6.2 $\det(R) = \sum_{P \in \mathscr{P}} sign(P) \prod_{\{b_i,g_j\} \in P} x_{ij}$.

Proof Using the definition of the determinant and the connection between permutations of $\{1, \ldots, n\}$ and perfect matchings, we obtain

$$\det(R) = \sum_{\sigma \in S_n} sign(\sigma) \prod_{i=1}^{n} r_{i\sigma(i)}$$

Figure 18.6 The matrix R for the bipartite graph G of Figure 18.5

$$\begin{array}{c} \\ b_1 \\ b_2 \\ b_3 \\ b_4 \\ b_5 \end{array} \begin{array}{c} \begin{array}{ccccc} g_1 & g_2 & g_3 & g_4 & g_5 \end{array} \\ \left[\begin{array}{ccccc} x_{11} & x_{12} & 0 & x_{14} & x_{15} \\ x_{21} & 0 & x_{23} & 0 & 0 \\ 0 & x_{32} & x_{33} & 0 & 0 \\ 0 & 0 & x_{43} & x_{44} & x_{45} \\ 0 & 0 & x_{53} & 0 & 0 \end{array}\right] \end{array}$$

$$= \sum_{\sigma \in S_n} sign(P_\sigma) \prod_{\{b_i, g_j\} \in P_\sigma} r_{ij}$$

$$= \sum_{P \in \mathscr{P}} sign(P) \prod_{\{b_i, g_j\} \in P} x_{ij}. \qquad \blacksquare$$

Determining whether the determinant of R is identically zero by computing the coefficients of R as a polynomial in the indeterminates x_{ij} is computationally infeasible. However, the determinant of R can be computed efficiently for any particular assignment of values to the indeterminates. The following proposition provides the basis for an efficient Monte Carlo algorithm for determining whether or not the determinant of R is identically zero.

PROPOSITION 18.6.3 Suppose det(R) is not the identically zero polynomial. For each edge $\{b_i,g_j\}$ of G, let R_w denote the matrix obtained from R by substituting the integer w_{ij} for x_{ij}, where the w_{ij}s are chosen uniformly and independently from the set $\{1,2, \ldots, t\}$. Then, the probability that det(R_w) is zero is at most n/t.

Proof Assume the proposition is true for matrices of size $n - 1$. Now suppose R is a matrix of size n such that det(R) is not the identically zero polynomial. Then, by Proposition 18.6.1, G contains a perfect matching P. For $\{b_i,g_j\}$ an edge of P, let R_{ij} and $(R_w)_{ij}$ denote the matrices obtained from R and R_w, respectively, by deleting the ith row and jth column. Since the bipartite graph G_{ij} obtained from G by deleting vertices b_i, g_j, and all incident edges contains the perfect matching $P\backslash\{b_i,g_j\}$, R_{ij} is nonzero. Further, it follows from the definition of a determinant that for some integer k, we have

$$\det(R_w) = w_{ij}\det((R_w)_{ij}) + k.$$

By induction hypothesis, det($(R_w)_{ij}$) is zero with probability at most $(n - 1)/t$. Let q denote the probability that both det(R_w) and det($(R_w)_{ij}$) are zero. Clearly, $q \le (n - 1)/t$. Now suppose det($(R_w)_{ij}$) is nonzero. Since w_{ij} was chosen uniformly and independently from the set $\{1,2, \ldots, t\}$, the probability that w_{ij} equals $- k/\det((R_w)_{ij})$ is at most $1/t$. Equivalently, the probability that det(R_w) is zero is at most $1/t$. Hence, the probability that det(R_w) is zero is at most

$$q + \frac{1}{t} \le \frac{n-1}{t} + \frac{1}{t} = \frac{n}{t}. \qquad \blacksquare$$

Proposition 18.6.3 yields the following Monte Carlo parallel algorithm *MCMarriage* for solving the Marriage problem. In the pseudocode for *MCMarriage*, we omit the particular parallel architecture and the number of processors, since the only portion of the code that requires this knowledge is the call to the function *Det* for computing the determinant of a matrix.

function *MCMarriage*(G,t)

Input: G (a bipartite graph with vertex set $V = \{b_1, \ldots, b_n\} \cup \{g_1, \ldots, g_n\}$ and edge set E)

t (a positive integer, $t \ge 2n$)

Continued

Output: returns **.true.** 100% of the time when G contains a perfect matching and **.false.** with probability at most n/t when G contains a perfect matching (and **.false.** 100% of the time then G does not contain a perfect matching)

```
for 1 ≤ i,j ≤ n do in parallel
    if {b_i,g_j} ∈ E then
        R[i,j] := Random({1,2, . . . , t})
    else
        R[i,j] := 0
    endif
end in parallel
if Det(R) = 0 then
    return(.false.)
else
    return(.true.)
endif
end MCMarriage
```

Since the dominant operation of *MCMarriage* is computing the determinant of an $n \times n$ matrix, *MCMarriage* can be implemented in time $O(\log^2 n)$ using $O(n^4)$ processors on the EREW PRAM or on the three-dimensional mesh-of-trees.

18.6.3 Randomized Protocols for Breaking Deterministic Deadlock

Suppose an American and a New Zealander are approaching one another on a narrow path. As they draw closer, the American moves to his right of the path to let the New Zealander pass. However, at that same instant, the New Zealander moves to her left of the path for the symmetric purpose (recall that they drive on the left in New Zealand). Then they both decide simultaneously to move to the other side of the path, and so it goes. In real life, the resulting back and forth shuffle might continue for one or two more rounds before the deadlock is broken and they end up on opposite sides of the path (or until romance breaks out).

While humans can decide on a protocol that resolves the path-passing deadlock with little more cost than a moment's embarrassment, situations arise in a communication network when processors executing completely deterministic protocols generate communication deadlocks requiring costly outside intervention by central control. The natural course of action in such cases is to introduce some type of randomness into the protocol. Indeed, randomized protocols are the basis of many communication protocols in use in the world today.

We limit our discussion of randomized communication protocols to a simple problem of choosing a distinguished (leader) processor from a set of n processors connected in a simple ring (see Figure 18.7). (The problem can be generalized to other communication networks, but the ideas are similar.) We assume that each pro-

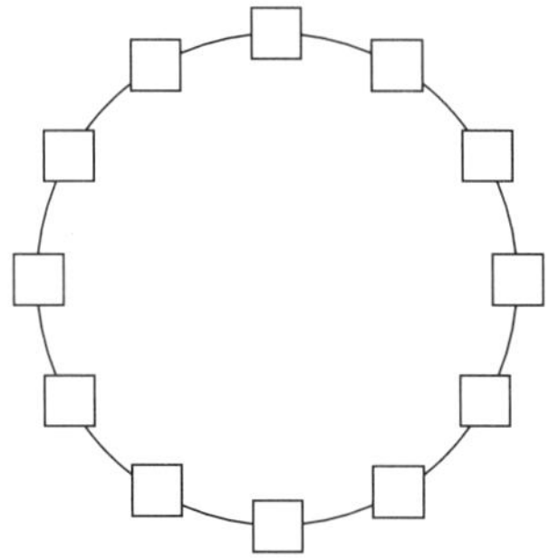

Figure 18.7 A ring of identical processors with $n = 12$

cessor knows the number n of processors in the ring. Similar to an interconnection network parallel computer, we assume that each processor has a local memory. However, unlike the situation assumed in an interconnection network parallel computer, the processors are not distinguished by an indexing scheme known to one another.

Suppose that each processor's local memory has been initialized with identical data. The processors are to decide on a single distinguished processor (choosing a leader) by performing purely local calculations and communicating the results circularly around the ring. (Since the processors are not indexed, they cannot, for example, simply choose an integer j between 1 and n and say that processor with index j is the leader.) Assuming that each processor executes the same deterministic algorithm with the same initial data, they would all communicate identical information around the ring, so that no processor could possibly be distinguished. To break the deadlock of symmetry, randomness is required.

One simple Las Vegas algorithm to choose a leader is based on the familiar "drawing straws" protocol. More precisely, the algorithm *DrawStraws*(n,m) consists of having each of the n processors draw an integer at random from $\{1, \ldots, m\}$ for some integer $m > 1$. Using $n - 1$ communication steps, the n choices are communicated to each processor in the ring and the maximum choice is determined. If the maximum was not drawn by more than one processor, then the leader is the processor who drew the maximum element. Otherwise the algorithm *DrawStraws* is repeated.

We choose bit communication as the basic operation when measuring the complexity of any algorithm for choosing a leader in our ring of processors. Thus, a single running of *DrawStraws*(n,m) has complexity $(n - 1)\log_2 m$. Hence, if $\rho(n,m)$ denotes the expected number of repetitions of *DrawStraws*(n,m) that are required for termination, we then have

$$\tau_{exp}(n,m) = \rho(n,m)(n - 1)\log_2 m. \tag{18.6.1}$$

Formula (18.6.1) shows the trade-off involved with making $\tau_{exp}(n,m)$ small: To make $\rho(n,m)$ small, we must make m (and hence $\log_2 m$) large. In particular, the expected number of basic operations performed by the Las Vegas algorithm *RepeatDrawStraws*(n,m) is not linear in n no matter what fixed value of m is used.

We now describe a Las Vegas algorithm (protocol) *RepeatDrawOne*(n) for choosing a leader whose expected behavior is linear in n. The protocol involves repeatedly performing the following operation, *DrawOne*, until a leader is determined. During the operation *DrawOne*, each processor that is still a candidate for the leader draws an integer at random from {1, . . . , *NumCand*}. The identification of those processors that have drawn the number 1 is then communicated as follows. In the first communication step, if a processor has drawn a 1, then it communicates the bit 1 to its neighbor, otherwise it communicates the bit 0. Noncandidate processors communicate bit 0 to their neighbors. Then after $n - 2$ additional communication steps all of the results are communicated to all of the processors. In particular, each processor knows how many processors drew the number 1.

If no processor draws a 1, then *DrawOne* is repeated under the same conditions (*NumCand* remains unchanged). If a single processor draws a 1, then this processor is the leader, and *RepeatDrawOne* terminates. If more than one processor drew a 1, then *NumCand* is updated to the number of such processors. Then *DrawOne* is repeated with those processors that did not draw one dropping out of the contest for leader. Thus, each processor maintains a Boolean variable *Candidate*, which is initialized to **.true.** and is set to **.false.** when it is no longer a candidate for leader. Of course, the processors that are no longer candidates for leader still participate in the communication steps.

Let $\tau_{exp}(n)$ be the expected number of parallel bit communication steps performed by *RepeatDrawOne* with input n and $\rho(n)$ be the expected number of calls to *DrawOne*. Then we have

$$\tau_{exp}(n) = (n - 1)\rho(n). \tag{18.6.2}$$

We now establish a recurrence relation for $\rho(n)$. The n choices made by the n processors during the first call to *DrawOne* can be thought of as performing n independent trials, where the probability of success (choosing a one) is $1/n$ for each trial. Thus, if $p(n,i)$ denotes the probability of i successes (exactly i processors choose one), the binomial distribution yields

$$p(n,i) = \binom{n}{i}\left(\frac{1}{n}\right)^i\left(1 - \frac{1}{n}\right)^{n-i}. \tag{18.6.3}$$

Note that the three basic (mutually disjoint) outcomes of the call to *DrawOne* described earlier correspond to $i = 0$ (repeat again from scratch), $i = 1$ (*RepeatDrawOne* terminates), and $2 \le i \le n$ (repeat again with i candidates for leader remaining). Since we call *DrawOne* at least once, we have

$$\rho(n) = 1 + p(n,0)\rho(n) + \sum_{i=2}^{n} p(n,i)\rho(i), \quad n \ge 2. \tag{18.6.4}$$

Solving for $\rho(n)$ in (18.6.4) yields the following (full history) recurrence

$$\rho(n) = \frac{1 + \sum_{i=2}^{n-1} p(n,i)\rho(i)}{1 - p(n,0) - p(n,n)}, \quad n \ge 2, \quad \textbf{init. cond. } \rho(2) = 2. \tag{18.6.5}$$

Using (18.6.5), it can be shown (see Exercise 18.19) that $\rho(n) < e \approx 2.718$, $n \ge 2$. Hence, the expected number of parallel bit communication steps performed by *RepeatDrawOne* is linear in the input size n.

18.7 Closing Remarks

Recall that a problem is in the class NC if there exists a parallel algorithm having polylogarithmic complexity for the problem on a PRAM with a polynomial number of processors. The randomized analog of this class, called RNC, is the class of algorithms for which there is a parallel Monte Carlo algorithm having polylogarithmic complexity for the problem on a PRAM with a polynomial number of processors. There are a number of important and natural problems for which NC algorithms are unknown, but for which efficient RNC algorithms are fairly readily obtained. For example, there is no known NC algorithm for determining whether or not a multivariate polynomial $P(x_1,x_2, \ldots , x_m)$ of degree n is identically zero. On the other hand, an RNC algorithm for this problem can be readily given (see Exercise 18.20). In this chapter we described an RNC parallel algorithm for determining the existence of a perfect matching in a bipartite graph. Finding an NC algorithm for this problem was open for many years and has only recently been solved.

In this chapter we described some significant applications of the use of probability in the design of sequential and parallel algorithms. An important probabilistic method for optimization problems not covered in this chapter is *simulated annealing*, which is used for finding the global minimum of a cost function that may possess several local minima. Simulated annealing emulates phenomena occurring in nature, namely, the physical process occurring when a solid is slowly cooled. Another class of probabilistic algorithms that have been widely studied are genetic algorithms. The genetic algorithm (GA) design strategy is based on *natural selection* and *survival of the fittest* laws of nature popularized by Darwin. We refer the student to the references for more on these probabilistic design strategies.

References and Suggestions for Further Reading

Alon, N., J. Spencer, and P. Erdös. *The Probabilistic Method*. New York: Wiley, 1992. Covers many of the important applications of the probabilistic method introduced by Erdös and includes a chapter devoted to the derandomization of randomized algorithms.

Brassard, G. and P. Bratley. *Algorithmics: Theory and Practice*. Englewood Cliffs, NJ: Prentice-Hall, 1988. A nice treatment of probabilistic algorithms.

Gupta, R., S. Smolka, and S. Bhaskar. "On randomization in sequential and distributed algorithms," *ACM Computing Surveys*, 26 (March 1994): 7–86. A comprehensive survey article on the use of probabilistic methods in sequential and parallel algorithms.

National Research Council. *Probability and Algorithms*. Washington, DC: National Academy Press, 1992. An outgrowth of a panel on probability and algorithms, sponsored by the National Research Council. Contains twelve independent articles that survey a number of topics on probabilistic algorithms, including simulated annealing.

Holland, J. *Adaptation in Natural and Artificial Systems*. Cambridge MA: The MIT Press, 1992. Written by the founder of the subject of genetic algorithms, this text is a revision of the 1975 version published by the University of Michigan.

A number of other books on genetic algorithms have appeared, including:

Bäck, T. *Evolutionary Algorithms in Theory and Practice*. New York: Oxford University Press, 1995.

Davis, L., ed. *Handbook of Genetic Algorithms*. New York: Van Nostrand Rheinhold, 1991.

Goldberg, D. E. *Genetic Algorithms in Search, Optimization, and Machine Learning*. Reading, MA: Addison-Wesley, 1989.

Koza, J. R. *Genetic Programming*. Cambridge, MA: The MIT Press, 1992.

Koza, J. R. *Genetic Programming* II. Cambridge, MA: The MIT Press, 1994.

Michalewicz, Z. *Genetic Algorithms + Data Structures = Evolution Programs*, 2nd Ed. Berlin, Germany: Springer-Verlag, 1994

Two recent books on string matching are:

Crochemore, M. *Text Algorithms*. New York: Oxford University Press, 1994.

Stephen, G. A. *String Searching Algorithms*. London: World Scientific Publishing, 1994.

Exercises

Section 18.2 Randomizing Deterministic Algorithms

18.1 For the randomized version of linear search where we flip a coin (once) to determine the direction of the search, compute the variance of $\tau(I)$ for $I = (L[1{:}n],X)$ where $X = L[i]$, $1 \le i \le n$.

18.2 Consider the following deterministic algorithm.

```
function TwoWayLinearSearch(L[1:n],X)

Input:  L[1:n] (an array of list elements), X (a search item)
Output: returns index of an occurrence of X in L, 0 if X not in list

   Left := 1
   Right := n
   while Left ≤ Right do
      if X = L[Left] then return(Left) endif
      if X = L[Right] then return(Right) endif
      Left := Left + 1
      Right := Right – 1
   endwhile
   return(0)

end TwoWayLinearSearch
```

a. Determine the best-case, worst-case, and average complexities of *TwoWayLinearSearch* under the assumptions that $L[1{:}n]$ consists of distinct elements and X occurs on the list, with each list element equally likely to be X.

b. Consider the randomized version of *TwoWayLinearSearch* where we flip a fair coin, and if it comes up heads we compare X to $L[Left]$ and increment $Left$ if $X \neq L[Left]$, otherwise (tails) we compare X to $L[Right]$ and increment $Right$ if $X \neq L[Right]$. The process is repeated as long as X is not found and $Left \leq Right$.

c. Find an expression for $\tau_{exp}(I)$ for $I = (L[1{:}n],X)$ where $X = L[i]$, $1 \leq i \leq n$.

d. Compute the variance $\tau(I)$ for $I = (L[1{:}n],X)$ where $X = L[i]$, $1 \leq i \leq n$.

e. Compare the variances in part (d) with those from Exercise 18.1.

18.3 Show that the variance of $\tau(L)$ for any input list L to *RandomizedQuickSort* is small.

Section 18.3 Monte Carlo Algorithms

18.4 A *majority element* x in a list $L[1{:}n]$ is an element that occurs more than $n/2$ times. Design and analyze a Monte Carlo algorithm for determining the majority element $L[1{:}n]$ in a list, if one exists.

Section 18.4 Las Vegas Algorithms

18.5 Consider the following Las Vegas algorithm for the n-Queens problem.

```
procedure QueensLasVegas(n,Column[1:n])
Input:  n (an n × n chess board)
Output: Column[1:n] (queen i is in row i and column Column[i] so that no two
                     queens attack each other)
   AvailColumns := {1, . . . , n}
   R := 1
   while AvailColumns ≠ ∅ .and. R ≤ n do
      choose a column C randomly and uniformly from AvailColumns
      Column[R] := C          {place queen R in row R and column C}
      R := R + 1
      update AvailColumns to be the set of columns available for placing the
      (R + 1)st queen in row R + 1 without creating a pair of attacking queens
   endwhile
end QueensLasVegas
```

a. Write a program implementing *QueensLasVegas*.

b. Run your program for $n = 8$. Empirically verify that the probability p for success is given by $p = 0.1293\ldots$.

18.6 Give pseudocode for a version of *QueensLasVegas* where you do not explicitly maintain the set *AvailColumns* of available columns.

18.7 Write a program similar to *QueensLasVegas* that places a number k of queens on the board in a random way and then uses backtracking to place the remaining queens (without changing the positions of the queens that were placed randomly). Test out your program for various values of k and compare its performance to *QueensLasVegas*.

18.8 Adapt the Karp-Rabin string matching algorithm to the problem of finding $m \times m$ patterns in $n \times n$ hosts.

18.9 Verify recurrence relation 18.4.4.

18.10 Given that we have a hit in the Karp-Rabin string matching algorithm, compute the conditional probability that it is spurious.

For Exercises 18.11–18.12: A string matching algorithm due to Knuth-Morris-Pratt (KMP) is based on a strategy similar to that of *NaiveStringMatcher.* Namely, the pattern string is shifted along the host string until a match occurs or until the pattern string no longer fits in the host. However, the KMP algorithm achieves $O(n + m)$ worst-case complexity by preprocessing the string P, which allows the result of a comparison of P with H_s to determine the next H_s where a match might possibly occur.

To illustrate, consider the pattern string P = 00100122 and host string H = 001201001001221. Placing P at the beginning of H, note that a mismatch occurs in position 4. The naive algorithm would then shift P by one, placing it at the second character in H, and check for a match. However, since $H[2:3] = 01 = P[2:3]$, which is not equal to $P[1:2] = 00$, a match cannot possibly occur at shift $s = 1$. A similar analysis shows that since $P[1:3] = 001 = H[1:3]$ and $P[4] = 0 \neq H[4]$, the first shift where a match could possibly occur is at $s = 3$ (we are not looking at $H[i]$, $i \geq 5$, when we consider the next possible match). What is important to notice is that we gleaned this information by comparing P against itself; this is the key to the preprocessing of P. For example, *whenever* a new scan is started at shift position s and we determine that $P[1:3] = 001 = H[s + 1:s + 3]$ whereas $P[4] \neq H[s + 4]$, then the *next* shift s' (that is, the smallest shift s' bigger than s) where a match can possibly occur is at $s' = s + 3$.

More generally, for a given $P[1:m]$ and $r \in \{0, \ldots, m - 1\}$, the preprocessing step of the KMP algorithm consists in computing the array $Next[0:m - 1]$, where $Next[r]$ is such that *whenever* a new scan is started at shift position s and we determine that $P[1:r] = H[s + 1:s + r]$ whereas $P[r + 1] \neq H[s + r + 1]$, then the *next* shift s' where a match can possibly occur is at $s' = s + Next[r]$. (Note that $Next[0] = 1$.)

In Figure 18.8, we illustrate $Next[0:7]$ and how it is used by the KMP algorithm to determine the succession of shifts in H = 001201001001221 that need to be examined when scanning for a match of P = 00100122.

18.11 Design and analyze an algorithm for computing the array $Next[0:m - 1]$ by preprocessing the pattern string $P[1:m]$.

18.12 Design and analyze the KMP algorithm for finding the first occurrence of the pattern P in the host H assuming that the arrays $H[1:n]$, $P[1:m]$, and $Next[0:m - 1]$ are input parameters.

P: 00100122		r:	0 1 2 3 4 5 6 7
		$Next\,[r]$:	1 2 1 3 5 3 2 6
	H: 001201001001221		
Sequence of possible shifts			
	001201001001221		
$s = 0$	001<u>0</u>	$r = 3$	$Next[3] = 3$
$s = 3$	0<u>0</u>	$r = 2$	$Next[2] = 1$
$s = 5$	0	$r = 0$	$Next[0] = 1$
$s = 6$	00100122	a match	

Figure 18.8 Action of the KMP algorithm for a sample pattern string $P[1{:}8]$ and a host string $H[1{:}15]$ on alphabet $\Sigma = \{0,1,2\}$

Section 18.5 Probabilistic Numerical Algorithms

18.13 When estimating the value of $\int_a^b f(x)dx$, show that we can assume without loss of generality that f is nonnegative on the interval $[a, b]$.

18.14 Show that the expression (18.5.1) begins to be less than 1/2 for $k \geq 23$.

18.15 Use Stirling's formula $n! \approx \sqrt{2\pi n}(n/e)^n$ to show that the value of (18.5.2) can be approximated for large n by

$$\frac{n!}{(n-k)\,!n^k} \approx e^{-k^2/2n} \quad . \tag{18.2.6}$$

18.16 Prove Theorem 18.5.2.

Section 18.6 Probabilistic Parallel Algorithms

18.17 Show that the expected number of basic operations performed by the Las Vegas algorithm *RepeatDrawStraws(n,m)* is not linear in n no matter what fixed value of m is used.

18.18 Show that the expected number of calls to *DrawOne* made by algorithm *RepeatDrawOne* is smaller than 2.442.

18.19 Using (18.6.5), show that $\rho(n) < e \approx 2.718$, $n \geq 2$, so that the expected number of basic operations performed by *RepeatDrawOne* is linear in the input size n.

18.20 Suppose that we are given an NC algorithm for evaluating a multivariate polynomial $P(x_1, x_2, \ldots, x_m)$ of degree n over the real numbers. For example, $P(x_1, x_2, \ldots, x_m)$ might be a determinant of matrix A whose entries involve the variables $x_1, x_2, \ldots, x_m$. Design and analyze an RNC algorithm (a parallel Monte Carlo algorithm having logarithmic complexity in n and m, which uses a polynomial number of processors in n and m) for determining whether or not P is identically zero. Use the following lemma.

LEMMA For any set of k real numbers $S = \{r_1, \ldots, r_k\}$, the number of zeros in S^m is at most $k^{m-1}n$. ❑

19

Graph Algorithms

In many applications modeled using graphs and digraphs, it is important to determine whether certain connectivity properties hold. Similarly, the problem of finding a maximum flow in a (capacitated) network has a host of applications. This chapter is devoted to a discussion of algorithms for some classical graph problems, such as finding the strongly connected components of a digraph, the biconnected components of a graph, and a maximum flow/minimum cut in a capacitated network. We also consider the problems of finding a perfect matching in a bipartite graph and a maximum (weight) perfect matching in a weighted complete bipartite graph.

19.1 Strongly Connected Components

The problem of finding the components of an undirected graph G is easily solved using depth-first traversal. The vertex set of each tree of a depth-first traversal forest is a component of G. Thus, the components of G can be computed in time $O(n + m)$ using *DFTForest* (see Chapter 8), where n and m denote the number of vertices and edges of G, respectively.

The notion of a strongly connected component in a digraph $D = (V,E)$ is a generalization of the notion of a component in a graph. Two vertices u and v in a digraph D are *strongly connected* if there is both a (directed) path from u to v and a (directed) path from v to u. Clearly, the relation *strongly connected* is an equivalence relation on the set $V = V(D)$ of vertices of D. The subdigraphs induced by these equivalence classes are called *strongly connected components* of D (see Figure 19.1).

Remark While the connected components of a graph partition the vertex set as well as the edge set, in general the strongly connected components of a digraph only partition the vertex set: An edge can be in at most one strongly connected component, but it might not be in any.

The depth-first search out-tree $T_{\text{out},v}$ of digraph D rooted at v is an out-tree that contains a path from v to every vertex u for which there exists a path in D from v to u. The depth-first search in-tree $T_{\text{in},v}$ of D rooted at v contains a path from u to v for every vertex u for which there exists a path in D from u to v. Thus, a vertex u

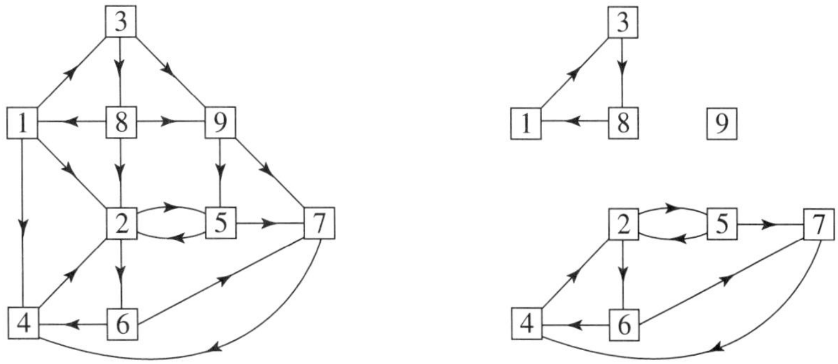

Digraph D Strongly connected components

Figure 19.1 Strongly connected components in a digraph D

belongs to the strongly connected component containing v if, and only if, u belongs to both $T_{\text{out},v}$ and $T_{\text{in},v}$. Computing $T_{\text{out},v}$ and $T_{\text{in},v}$ using *DFSOutTree* and *DFSInTree*, and then determining the vertices common to both yields an O(max $\{n,m\}$) algorithm for finding the strongly connected component containing vertex v (where m denotes the number of edges of D). Calling this algorithm for each vertex v yields an O(n max$\{n,m\}$) algorithm for obtaining all the strongly connected components of D.

We now discuss a more efficient (O(m)) algorithm for obtaining the set of strongly connected components of a digraph, which is based on the notion of the post numbering of the vertices of a digraph. For convenience, we assume that $V(D) = \{1,2, \ldots, n\}$. The post numbering of a digraph D is determined by performing an out-directed depth-first traversal of D. For definiteness, we assume that the depth-first traversal always chooses the smallest vertex whenever there is a choice. Recall that a vertex u of D becomes *explored* when the traversal accesses u having visited all vertices in the out-neighborhood of u. The *post number* of u, denoted *PostNum*(u), is the integer i where u is the ith vertex to be explored in an out-directed depth-first traversal of D (see Figure 19.2, page 664).

Remark Vertex u has post number i if u is the ith vertex to be visited in a postorder traversal of the DFT out-forest.

Clearly, *PostNum* is a permutation of $1,2, \ldots, n$. The inverse permutation of *PostNum*, which we denote by *PostNumInv*, can be com11puted in linear time. Note that *PostNumInv*(1), *PostNumInv*(2), ..., *PostNumInv*(n*59) lists the vertices in the order in which they become explored. The depth-first traversal *DFTOut* is easily modified to compute arrays *PostNum*[1:n] and *PostNumInv*[1:n], representing the permutations *PostNum* and *PostNumInv*, respectively. In Figure 19.2, we illustrate the values of the arrays *PostNum*[1:n] and *PostNumInv*[1:n] for a given digraph D, where we use vertex 3 as the root vertex.

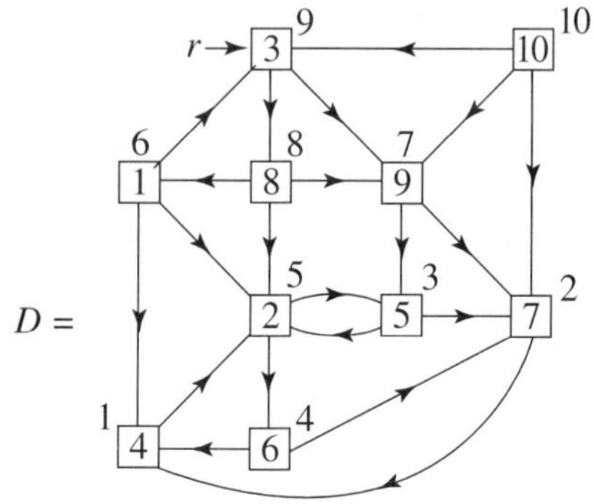

i	1	2	3	4	5	6	7	8	9	10
PostNum[i]	6	5	9	1	3	4	2	8	7	10
PostNumInv[i]	4	7	5	6	2	1	9	8	3	10

Figure 19.2 Values of the arrays *PostNum*[1:*n*] and *PostNumInv*[1:*n*] for the given digraph *D* and root vertex 3

The following proposition is the basis for an algorithm for computing the strongly connected components of a digraph.

PROPOSITION 19.1.1 Let C be the vertex set of a strongly connected component of a digraph D. Suppose u is a vertex such that $PostNum(u) < PostNum(r)$ for some vertex $r \in C$. If u is joined with an edge uv to some vertex of v of C (not necessarily equal to r), then u belongs to C.

Proof We give a proof by contraction. Assume that the hypothesis of the proposition holds, but u does not belong to C. It follows that there is no path from any vertex in C to u. Clearly, u does not become explored by the out-directed depth-first traversal until after v is accessed. Let x be the first vertex in C to be accessed by the out-directed depth-first traversal. Since C is the vertex set of a strongly connected component, it follows that every vertex of C different from x will be accessed and eventually explored before x becomes explored. Since there is no path from x to u, after having accessed x the out-directed depth-first search will not access u until after x has been explored. Thus, every vertex in C becomes explored before u does, so that every vertex in C has a smaller post number than u. But this contradicts the fact that r has a greater post number than u. ■

The following algorithm *StrongComponents* finds the strongly connected components D in two stages. In the first stage, *StrongComponents* performs an out-directed depth-first traversal to compute the array *PostNumInv*[1:*n*]. In the second stage, *StrongComponents* performs an in-directed depth-first traversal. However, rather than scanning the vertices from 1 to n when performing the depth-first searches occurring within the traversal, we scan in the order *PostNumInv*[n], *PostNumInv*[$n-1$], ..., *PostNumInv*[1]. It turns out that vertex sets of the in-trees $T_1, T_2, \ldots, T_k$ of the associated depth-first traversal in-forest are precisely the vertex sets that induce the strongly connected components of D. *StrongComponents* has the same order of complexity, $O(n + m)$, as depth-first traversal.

In the pseudocode for *StrongComponents* we use the array *InForest*[1:n] to keep track of the depth-first traversal in-forest (see Figure 19.3).

```
procedure StrongComponents(D,PostNumInv[1:n])
Input:  D (a digraph with n vertices and m arcs)
        PostNumInv[1:n] (PostNumInv[i] is the vertex u where PostNum[u] = i)
Output: InForest[1:n] (an array giving parent representation of forest of
        in-trees T_1, T_2, ... ,T_k)
dcl  Mark[1:n] {a 0 – 1 array},
     InForest[1:n]
   for v := 1 to n do
      Mark[v] := 0
      InForest[v] := 0
   endfor
   for i := n downto 1 do    {perform an in-directed depth-first
      v := PostNumInv[i]      traversal according to reverse order of post numbers}
      if Mark[v] = 0 then
         call DFSInTree(D,v,InForest)  {perform an in-directed depth-first search
                                        rooted at vertex v and store associated
      endif                             depth-first in-tree as part of InForest[1:n]}
   endfor
end StrongComponents
```

We verify the correctness of procedure *StrongComponents* by showing that the vertex sets of the output trees T_i, $i = 1, \ldots, k$, are precisely the strongly connected components of the digraph D. Given the tree T_i, let r_i denote the root vertex of T_i and let V_i denote the vertex set of T_i. We will show that the vertex set C of the strongly connected component containing r_i is precisely V_i. It is easily verified that $C \subseteq V_i$. To prove that $C \supseteq V_i$, we show inductively that the vertices of V_i having depth d in T_i belong to C, $d = 0, \ldots, depth(T_i)$. Since r_i belongs to C, this result is true for $d = 0$. Assume that all vertices of T_i having depth d belong to C and consider a vertex u of T_i having depth $d + 1$. Since we chose r_i to be the vertex not in trees $T_1, \ldots, T_{i-1}$ having the largest post number, it follows that $PostNum(u) < PostNum(r_i)$. Further, since u has depth $d + 1$ in T_i, it is joined to some vertex v of T_i at depth d. By induction hypothesis, v belongs to C, so that Proposition 19.1.1 implies that u belongs to C. Thus, by induction, all the

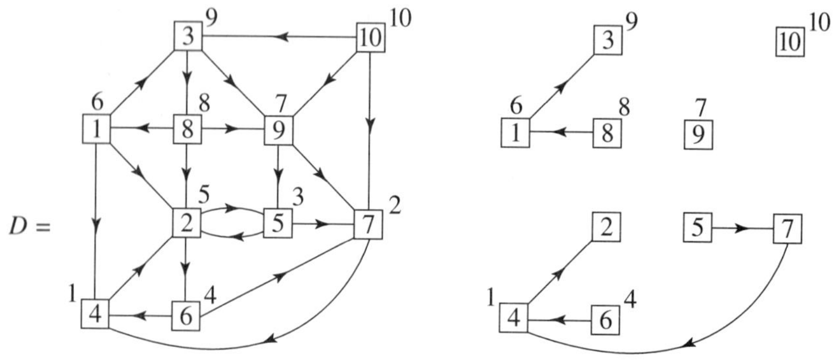

Digraph D with the post number showing outside each node

In-forest generated by *StrongComponents*

i	1	2	3	4	5	6	7	8	9	10
$InForest[i]$	3	0	0	2	7	4	4	1	0	0

Figure 19.3 Action of *StrongComponents* for the digraph D of Figure 19.1

vertices of T_i belong to C, which completes the verification of correctness of algorithm *StrongComponents*.

We illustrate in Figure 19.3 the action of the algorithm *StrongComponents* for the digraph D in Figure 19.1.

19.2 Articulation Points and Biconnected Components

In many applications of graphs, it is useful to know that stronger connectivity properties hold than merely being connected. For example, suppose G models a *communication network*, where the vertices of G might correspond to people, processors, computers, and so forth. Two vertices u and v in such a network are joined with an edge whenever direct communication is possible between them. A pair of vertices u and v that cannot communicate directly can communicate by relaying the information along a path P joining u and v. It is useful in practice that more than one path connecting vertices u and v exists, so that the communication network is more tolerant of certain failures. For example, sometimes a vertex in the network becomes "inoperable" and is no longer capable of relaying information. It then becomes of interest to find communication paths that "bypass" such a vertex.

For some vertices in a graph it might not be possible to find "bypass" paths in all cases. For example, if the deletion of a certain vertex v disconnects the graph, then vertices u,w in two different components of $G - v$ can no longer communicate with one another, since all paths connecting u and w in G must pass through v. In this sec-

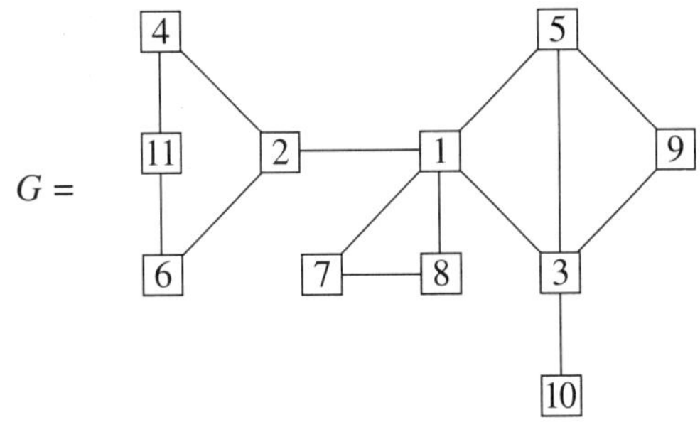

Graph G with articulation points $\{1,2,3\}$

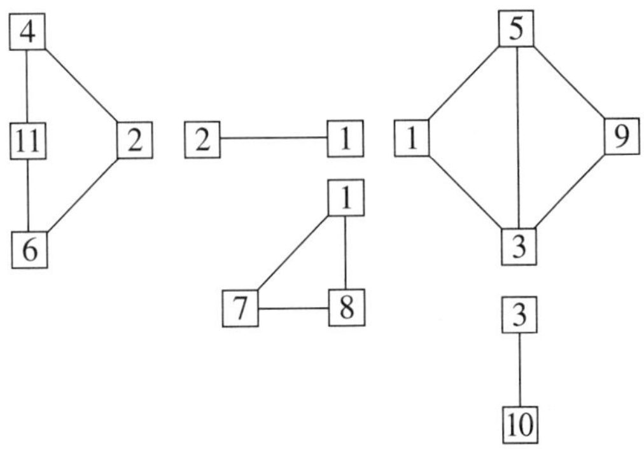

Biconnected components of G

Figure 19.4 Articulation points and biconnected components of a connected graph G

tion we consider algorithms for identifying such troublesome vertices v (articulation points).

Throughout this section we assume that the graph G is connected. A vertex v of a connected graph G is an *articulation point* (also called *cut vertex*) if the graph $G - v$ obtained from G by deleting vertex v and all incident edges is not connected. A *biconnected component* of G (also called a *block of* G) is a maximal subgraph B such that B has no articulation points (see Figure 19.4).

The algorithms we discuss for finding the articulation points of a connected graph G are based on the DFS tree T_v rooted at a vertex $v \in V(G)$. The following easily verified proposition and its corollary are useful in finding articulation points.

PROPOSITION 19.2.1 Given a vertex r of a connected graph $G = (V,E)$, consider the DFS tree T_r rooted at vertex r. Then for any edge $e \in E$, one endpoint of e is an ancestor of the other in T_r. More precisely, if $vw \in E$, and v is accessed before w in the depth-first search rooted at r, then v is an ancestor of w in T_r. ❑

COROLLARY 19.2.2 Given a vertex v of a connected graph $G = (V,E)$, consider the DFS tree T_r rooted at vertex r. Then r is an articulation point in G if, and only if, r has two or more children in T_r. ❑

Using Corollary 19.2.2, computing T_r for each vertex r in the graph G yields an $O(nm)$ algorithm for determining the set of all articulations points. We now discuss an $O(m + n)$ algorithm *ArticulationPoints* for computing the articulation points. As with the algorithm *StrongComponents*, *ArticulationPoints* is based on computing (two) numberings of the vertices based on a single depth-first search. The first numbering, *DFSNum*, is simply the order in which the vertices are (first) accessed by a depth-first search rooted at vertex r. The second numbering, *Lowest*, is computed recursively at the same time as the depth-first search tree T_r rooted at vertex r (and associated numbering *DFSNum*) is determined.

To motivate the definition of *Lowest*, we will show how the notion of bypass edges helps to determine articulation points. Given the DFS tree T_r, for a vertex $v \in V$, we let $T_{r,v}$ denote the subtree of T_r having root vertex v. For c a vertex at depth at least two in T_r, we let $P_{r,c}$ denote the path in T_r joining r to *parent*(*parent*(c)). A *bypass edge for c* is an edge $vw \in E$ such $v \in T_{r,c}$ and $w \in P_{r,c}$ (see Figure 19.5).

The following proposition and its corollary are easily verified.

PROPOSITION 19.2.3 Given a vertex r of a connected graph $G = (V,E)$, consider the DFS tree T_r rooted at vertex r. Then for any vertex c at depth at least two in T_r, there exists a path in G not containing *parent*(c) that joins c and r if, and only if, there exists a bypass edge for c. ❑

COROLLARY 19.2.4 Given a vertex r of a connected graph $G = (V,E)$, consider the DFS tree T_r rooted at vertex r. Then a vertex $v \neq r$ is an articulation point in G if, and only if, there exists a child c of v in T_r having no bypass edge for c. ❑

An algorithm for finding the articulation points in G is based on Corollary 19.2.4 and the fact that the numbering *Lowest* gives a simple test for the existence of bypass edges. For a given $v \in V(G)$, *Lowest*[v] is defined recursively as follows.

$$Lowest[v] = \min \begin{cases} \min \ \{Lowest[c] : c \text{ is a child of } v \text{ in } T_r\}, \\ \min \ \{DFSNum[w] : vw \in E, w \neq parent(v)\}, \\ DFSNum[v]. \end{cases}$$

Remark

1. We assume that min {*DFSNum*[w] : $vw \in E$, $w \neq$ *parent*(v)} = ∞ if no such w exists. Similarly, min {*Lowest*[c] : c is a child of v in T_r} = ∞ if v is a leaf node of T_r.
2. We will compute *Lowest* using a depth-first search. When computing *Lowest*[v] during such a search, if $vw \in E$, and $w \neq$ *parent*(v), then by Proposition 19.2.1 either w is an ancestor of v in T_r, so that *DFSNum*[w] is defined, or $w \in T_{r,v}$ and *DFSNum*[w] will be defined during the recursive resolution of *Lowest*[c] for an appropriate child of v. This makes min {*DFSNum*[w] : $vw \in E$, $w \neq$ *parent*(v)} well defined during our depth-first search.

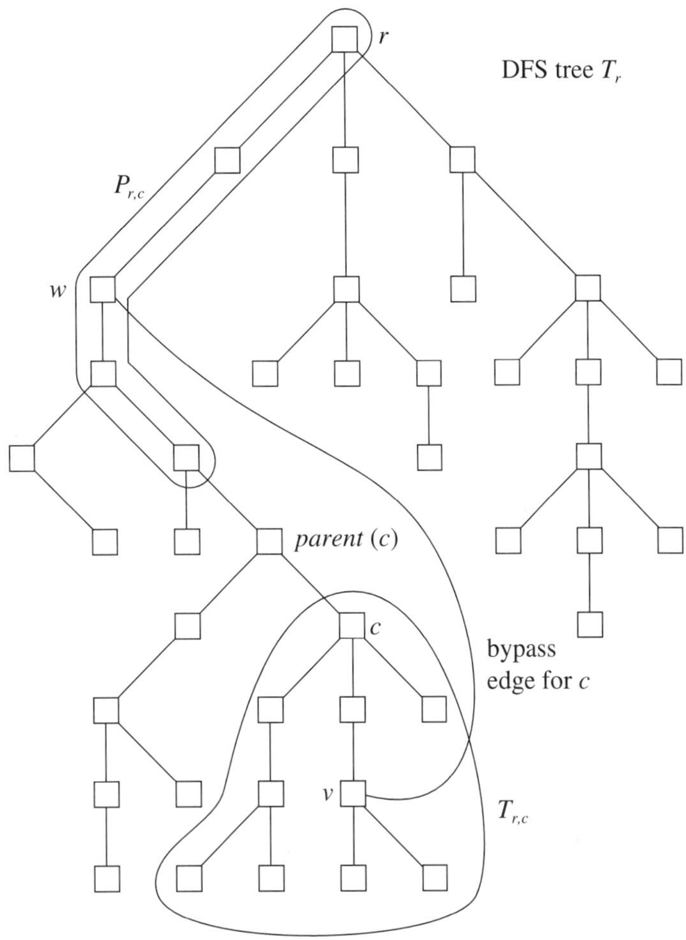

Figure 19.5 DFS-tree T_r, vertex c, subtree $T_{r,v}$, path $P_{r,v}$, and bypass edge vw for c

The following proposition is easily verified.

PROPOSITION 19.2.5 If $Lowest[v] = \min \{DFSNum[w] : vw \in E, w \neq parent(v)\}$, then there exists a bypass edge vw such that $Lowest[v] = DFSNum[w]$. ❑

In Figure 19.6 (page 670) we illustrate the values of $DFSNum[v]$ and $Lowest[v]$ for the same graph as shown in Figure 19.4, where we use root vertex $r = 1$. The pair $DFSNum[v]$, $Lowest[v]$ is shown outside each node v, and the edges of the depth-first search tree T rooted at vertex 1 are shown as dotted.

PROPOSITION 19.2.6 Given a vertex r of a connected graph $G = (V,E)$, consider the DFS-tree T_r rooted at vertex r. Then for any vertex c at depth at least two in T_r, a bypass edge for c exists if, and only if, $Lowest[c] < DFSNum[parent(c)]$.

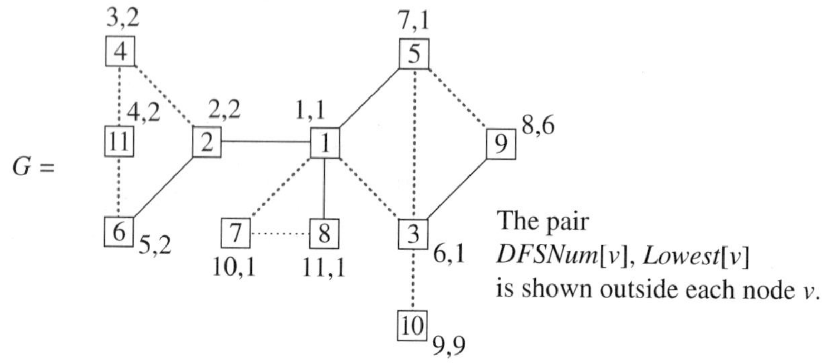

i	1	2	3	4	5	6	7	8	9	10	11
$TreeDFS[i]$	0	1	1	2	3	11	1	7	5	3	4

Figure 19.6 Values of *DFSNum* and *Lowest* for G and the array *TreeDFS*[1:10] for T_1

Proof To verify the "only if" part of the proposition, suppose vw is a bypass edge for c, so that $v \in T_{r,c}$ and $w \in P_{r,c}$. Then we have

$$
\begin{aligned}
Lowest[c] &\le Lowest[v] && \text{(by definition of } Lowest \text{ and } v \in T_{r,c}\text{)},\\
&\le DFSNum[w] && \text{(by definition of Lowest)},\\
&< DFSNum[parent(c)] && \text{(since } w \in P_{r,c}\text{)}.
\end{aligned}
$$

To verify the "if" part of the proposition, suppose no bypass edge for c exists. Consider *any* vertex $z \in T_{r,c} - c$. Note that any bypass edge for z must have $parent(c)$ as an endpoint, otherwise it would be a bypass edge for c. Thus, using Proposition 19.2.5 and the definition of *Lowest*, we have

$$\min\,\{DFSNum[w] : zw \in E,\ w \ne parent(z)\} \ge DFSNum[parent(c)]. \qquad (19.2.1)$$

Since z is a descendant of c, we have

$$DFSNum[z] \ge DFSNum[c] > DFSNum[parent(c)]. \qquad (19.2.2)$$

It follows from (19.2.1) and (19.2.2) that

$$Lowest[z] \ge DFSNum[parent(c)], \quad z \in T_{r,c}. \qquad (19.2.3)$$

By (19.2.3) and the definition of *Lowest*, $Lowest[c] \ge DFSNum[parent(c)]$. ■

The following $O(n + m)$ recursive algorithm *DFSNumberings* computes both $Lowest(u)$ and $DFSNum(u)$.

procedure *DFSNumberings*(G,v) **recursive**

Input: G (a graph with n vertices and m edges)
v (a vertex of G) {v is root vertex r of *DFS* on initial call}
Mark[1:n] (global array initialized to zeros)

```
        TreeDFS[1:n] (global array initialized to zeros)
        DFSNum[1:n] (global array)
        Lowest[1:n] (global array)
        i (global integer initialized to 0)
Output: the DFS and Lowest numberings of the vertices of G, and the parent
        implementation of the depth-first search tree TreeDFS(T_r) rooted at vertex r

  i := i + 1
  Mark[v] := 1
  DFSNum[v] := i
  for each u adjacent to v do
     if Mark[u] = 0 then
        TreeDFS[u] := v
        call DFSNumberings(G,u)
     endif
  endfor
  Min1 := min {Lowest[c] for each child c of v}
  Min2 := min {DFSNum[w] : v adjacent to w but w is not the parent of v}
  Lowest[v] := min {DFSNum[v], Min1, Min2}
end DFSNumberings
```

Remark In our description of the output of *DFSNumberings*, for simplicity we only described the output for the initial (nonrecursive) call to the procedure. The output from recursive calls also provides the proper updating of *Mark*, *DFSNum*, *TreeDFS*, and *Lowest*, thereby ensuring that the initial call returns the desired *DFS* and *Lowest* numberings of *G*.

With the aid of *DFSNumberings*, we now can present the algorithm *ArticulationPoints*, which outputs the set of articulation points of a graph *G*. For simplicity of pseudocode, we assume that the root vertex for the depth-first search is $r = 1$.

```
procedure ArticulationPoints(G,Art)
Input: G (a graph with n vertices and m edges)
Output: Art (the set of articulation points)
dcl TreeDFS[1:n], DFSNum[1:n], Lowest[1:n], Mark[1:n]
  for v := 1 to n do {initialize Mark and TreeDFS}
     Mark[v] := 0
     TreeDFS[v]:= 0
  endfor
  i := 0
  Art := ∅
  call DFSNumberings(G,1)      {compute TreeDFS[u], DFSNum[u], Lowest[u]
                                for each vertex u ∈ V}
```

Continued

```
{use Corollary 19.2.2 to test whether root r = 1 is an articulation point;
  that is, check to see if vertex 1 has more than one child in T_1}
Number := 0
c := 2
while Number < 2 .and. c ≤ n do
   if TreeDFS[c] = 1 then {c is a child of root vertex r = 1}
      Number := Number + 1
   endif
   c := c + 1
endwhile
if Number = 2 then                    {root vertex is an articulation point}
   Art := Art ∪ {1}
endif
{use Proposition 19.2.5 and Corollary 19.2.4 to determine non-root
 articulation points: whenever a child c has a parent v ≠ r such that
 Lowest[c] ≥ DFSNum[v], then v is an articulation point}
for c := 2 to n do
   v := TreeDFS[c]
   if Lowest[c] ≥ DFSNum[v] .and. v ≠ 1 then
      Art := Art ∪ {v}
   endif
endfor
end ArticulationPoints
```

ArticulationPoints outputs $Art = \{1,2,3\}$ for the input graph G of Figure 19.6. Note that these are precisely the articulation points of G. We leave it as an exercise to write an algorithm that outputs the biconnected components of a graph G given the input parameter *Art* of articulation points of G.

19.3 Perfect Matchings in Bipartite Graphs

An *independent set of edges* or *matching* in a graph G is a set of edges that are pairwise vertex disjoint. A matching that spans the vertices is called a *perfect matching*. In Chapter 18 we gave a probabilistic algorithm for determining whether or not a bipartite graph has a perfect matching. In this section we discuss a (deterministic) algorithm known as the Hungarian algorithm for finding such a matching. We also discuss the Kuhn-Munkres algorithm, which employs the Hungarian algorithm to find a maximum-weight perfect matching in an edge-weighted complete bipartite graph.

19.3.1 The Hungarian Algorithm

Recall that the Marriage problem was to determine a necessary and sufficient condition that marriages can be arranged between a set of boys $\{b_1, \ldots, b_n\}$ and a set of girls $\{g_1, \ldots, g_n\}$. It is required that each boy b_i be married only to a girl from the set K_i of girls that he knows, $i = 1, \ldots, n$. For example, suppose $n = 5$ and the sets K_i, $i = 1, \ldots, 5$ are given by

$$K_1 = \{g_1, g_2, g_3\}, K_2 = \{g_1, g_3\}, K_3 = \{g_2, g_3\}, K_4 = \{g_3, g_4, g_5\}, K_5 = \{g_3\}.$$

Since the four boys b_1, b_2, b_3, b_5 collectively know only the three girls g_1, g_2, g_3, a perfect matching does not exist. More generally, if there exists a set of k boys who collectively know strictly less than k girls, then a perfect matching does not exist. Surprisingly, the converse is also true, by a theorem of Hall:

> *A solution to the Marriage problem exists if, and only if, every set of k boys collectively knows at least k girls, $k \in \{1, 2, \ldots, n\}$.*

We can restate Hall's Theorem for bipartite graphs as follows. For S a set of vertices of G, let $\Gamma(S)$ denote the set of all vertices that are adjacent to those in S; that is,

$$\Gamma(S) = \{v \in V(G) : \text{there exists a vertex } u \text{ in } S \text{ such that } uv \in E(G)\}. \quad (19.3.1)$$

THEOREM 19.3.1 **P. Hall** A bipartite graph with vertex bipartition $V = X \cup Y$ contains a perfect matching if, and only if, for every subset S of X,

$$|S| \leq |\Gamma(S)|. \quad (19.3.2)$$

Proof First, suppose that there exists a perfect matching M in G. For $u \in V$, let $M(u)$ denote the *mate* of u in M, that is, the unique vertex v such that $uv \in M$. For $S \subseteq X$, let

$$M(S) = \{M(x) : x \in S\}. \quad (19.3.3)$$

Clearly, $M(S) \subseteq \Gamma(S)$. Hence,

$$|\Gamma(S)| \geq |M(S)| = |S|. \qquad \blacksquare$$

We complete the proof of Theorem 19.3.1 by presenting an algorithm, called *Hungarian*, which accepts an initial matching M (possibly empty) and repeatedly augments M, until either a perfect matching is generated or a set $S \subseteq X$ is found such that $|\Gamma(S)| < |S|$.

Before presenting the algorithm *Hungarian*, we need some definitions and preliminary results. We say that a vertex v of G is *M-matched* if it is incident with an edge of M. Otherwise, we say that v is *M-unmatched.* An *M-alternating path* is one where the edges alternately belong to $E(G) - M$ and M. An *M-augmenting path* is an alternating path where the initial and terminal vertices are both M-unmatched. The *symmetric difference* $M \oplus E(P)$ of M and the edges $E(P)$ of P is given by

$$M \oplus E(P) = (M \cup E(P)) - (M \cap E(P)). \quad (19.3.4)$$

The algorithm *Hungarian* searches for augmenting paths. To find an M-augmenting path in G, algorithm *Hungarian* grows a tree T, called an *M-alternating tree*, having the following properties:

1. The root r of T is an M-unmatched vertex belonging to X.

2. For each odd integer i less than the depth of T, each edge of T joining a vertex at level i to a vertex at level $i + 1$ belongs to M.
3. The leaf nodes of T all belong to X.

An M-alternating tree is illustrated in Figure 19.7. Clearly, all the paths in an M-alternating tree T are M-alternating paths. Let X_T and Y_T denote the subset of vertices of X and Y, respectively, belonging to T.

The proofs of the next three lemmas are easy and left as exercises.

LEMMA 19.3.2 If M is a matching and P is an M-augmenting path, then $M \oplus E(P)$ is a matching of size one greater than M. ❑

LEMMA 19.3.3 If T is an M-alternating tree, then

$$|X_T| = |Y_T| + 1. \tag{19.3.5}$$ ❑

LEMMA 19.3.4 Given an M-alternating tree T, if there exists a vertex $x \in X_T$ that is adjacent to an M-unmatched vertex y (not in T), then the path from the root r of T to x, together with the edge xy, determine an M-augmenting path. ❑

Figure 19.7 An M-alternating tree T rooted at vertex r

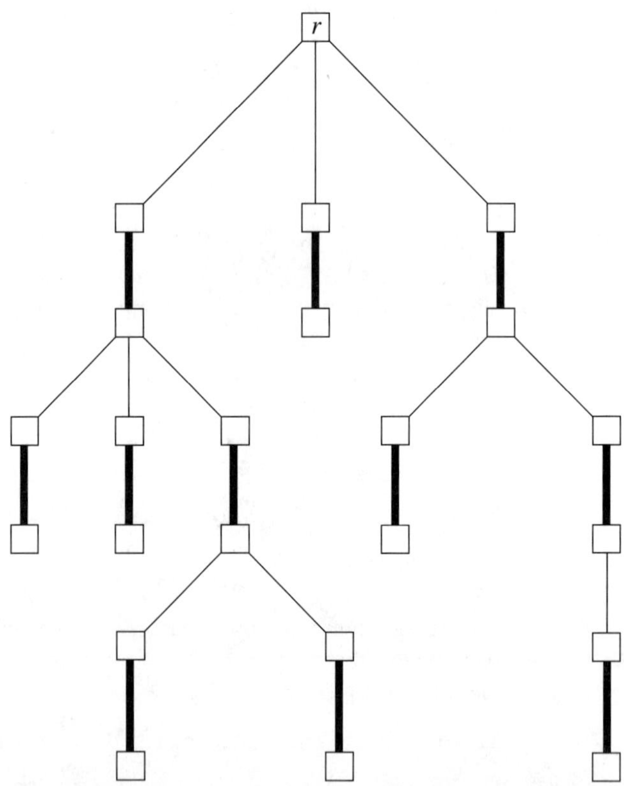

Clearly, if all the vertices in X are M-matched, then M is a perfect matching. In this case, algorithm *Hungarian* returns the perfect matching M and terminates. In the case where the vertices in X are not all matched, algorithm *Hungarian* looks for an augmenting path by growing an M-alternating tree T. The tree T may be initialized to consist of the single vertex r, where r is chosen arbitrarily from the set of unmatched vertices in X. When growing the tree T, at each stage we encounter one of the following three cases.

1. $\Gamma(X_T) - Y_T$ is empty. (Action: Algorithm terminates.)

Then by Lemma 19.3.3, $|\Gamma(X_T)| = |Y_T| = |X_T| - 1$. Algorithm *Hungarian* then returns $S = X_T$ and terminates, since by Theorem 19.3.1 no perfect matching exists.

In the next two cases, $\Gamma(X_T) - Y_T$ is nonempty, and we choose y to be any vertex in $\Gamma(X_T) - Y_T$ and x to be any vertex in X_T adjacent to y.

2. Vertex y is matched. (Action: T is augmented.)

Then we augment T to obtain a new M-alternating tree by adding the edges xy and yz, where z is the mate of y in M.

3. Vertex y is unmatched. (Action: M is augmented.)

Then an augmenting path P in T from r to x together with the edge xy has been found, so that we replace M by the augmented matching $M \oplus E(P)$ containing one more edge.

In case 3, we either have found a perfect matching, or we look for another augmenting path by growing another M-alternating tree rooted at a new unmatched vertex in X. Clearly, M can be augmented at most n times. Since at each stage the alternating tree T can be grown in time $O(n^2)$, the worst-case complexity of procedure *Hungarian* is $O(n^3)$. Pseudocode for the Hungarian algorithm follows.

```
procedure Hungarian(G,M,S)

Input:  G (a bipartite graph with vertex bipartition (X,Y))
        M (an initial matching, possibly empty)
Output: M (a perfect matching if one exists)
        S (a set of vertices with the property that |Γ(S)| < |S| if no
          perfect matching exists)

  AugmentingM := .true.
  while AugmentingM do
    if all the vertices in X are M-matched then        {M is a perfect matching}
      AugmentingM := .false.
    else {Grow M-alternating tree T}
      r := any M-unmatched vertex in X
      T := tree consisting of the single vertex r
      GrowingTree := .true.
      while GrowingTree do
```

Continued

```
            if Γ(X_T) = Y_T then
               S := X_T                                                  { |Γ(S)| < |S| }
               GrowingTree := false
               AugmentingM := false
            else
               y := any vertex in Γ(X_T) − Y_T
               x := any vertex in X_T adjacent to y
               if y is M-matched then                                    {augment T}
                  z := M(y)                                              {z is mate of y in M}
                  T := T ∪ xy ∪ yz
               else                                 {an M-augmenting path has been found}
                  P := path in T from r to x together with the edge xy
                  M = M ⊕ E(P)                                           {augment M}
                  GrowingTree := false
               endif
            endif
         endwhile
      endif
   endwhile
end Hungarian
```

In Figure 19.8 the action of procedure *Hungarian* is illustrated for a sample bipartite graph. For definiteness, in Figure 19.8 we show the perfect matching generated by always choosing vertices in order of their vertex number. More precisely, the phrase "any . . . vertex" occurring in three statements in the pseudocode for *Hungarian* is replaced by phrase "the . . . vertex of smallest index" when generating the perfect matching in Figure 19.8. We leave the intermediate steps in finding the augmenting paths as an exercise.

19.3.2 Maximum Perfect Matching in a Weighted Bipartite Graph

Suppose n workers $x_1, x_2, \ldots, x_n$ are to be assigned to n jobs $y_1, y_2, \ldots, y_n$, where each worker is qualified to perform any of the jobs. Associated with each worker/job pair (x_i, y_j) is a weight ω_{ij} measuring how effectively worker x_i can perform job y_j. The natural problem arises of finding an assignment of workers to jobs so that the total effectiveness of the workers is optimized. This problem can be modeled using a weighted complete bipartite graph $G = (V,E)$, with vertex bipartition $X = \{x_1, \ldots, x_n\}$ and $Y = \{y_1, \ldots, y_n\}$. Each edge x_iy_j of G is assigned the weight ω_{ij}, $i, j \in \{1, \ldots, n\}$. Clearly, a perfect matching in G corresponds to an assignment of each worker to a job, so that no two workers are assigned the same job. We define the *weight* of a perfect matching M, denoted by $\omega(M)$, to be the sum of the weights of its edges:

$$\omega(M) = \sum_{e \in M} \omega(e). \tag{19.3.6}$$

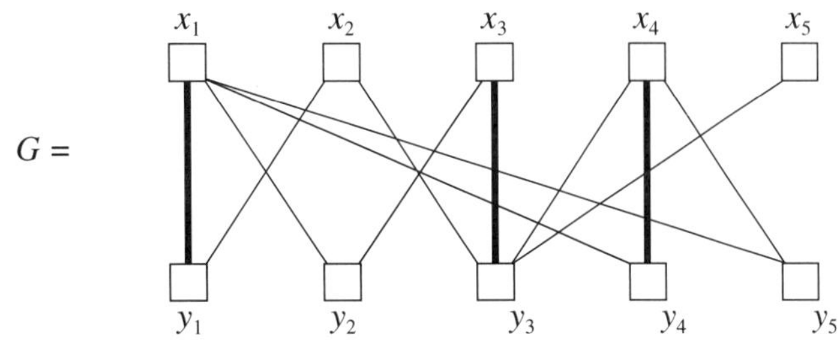

Sample bipartite graph G and initial matching (x_1,y_1), (x_3,y_3), (x_4,y_4)

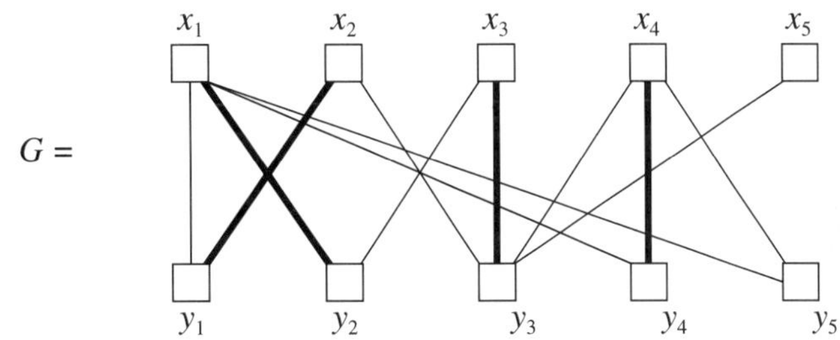

A first M-augmenting path P: $x_2y_1x_1$ found by *Hungarian* with starting M-unmatched vertex x_2 yielding augmented matching $M \oplus E(P) = \{x_1y_2, x_2y_1, x_3y_3, x_4y_4\}$

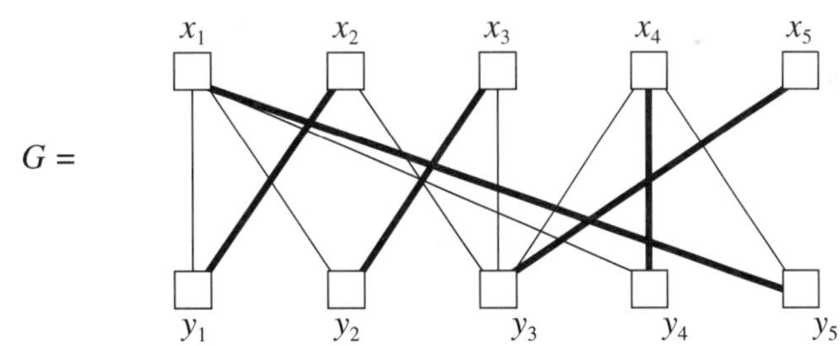

A second M-augmenting path P: $x_5y_3x_3y_2x_1y_5$ found by *Hungarian* with starting M-unmatched vertex x_5 yielding augmented perfect matching $M \oplus E(P) = \{x_1y_5, x_2y_1, x_3y_2, x_4y_4, x_5y_3\}$

Figure 19.8 Action of algorithm *Hungarian* for a sample bipartite graph with initial matching (x_1,y_1), (x_3,y_3), (x_4,y_4). Vertices are considered in the order of their indices.

A *maximum perfect matching* is a perfect matching of maximum weight over all perfect matchings of G. Since G is complete, any permutation π of $\{1,2, \ldots, n\}$ determines a perfect matching $M_\pi = \{x_iy_{\pi(i)} \mid i \in \{1, \ldots, n\}\}$, and conversely. Thus, a brute-force algorithm that enumerates all $n!$ perfect matchings and chooses one of maximum weight is hopelessly inefficient.

We now describe an $O(n^3)$ algorithm due to Kuhn and Munkres for finding a maximum perfect matching in a weighted complete bipartite graph. The Kuhn-Munkres algorithm utilizes the Hungarian algorithm, together with the notion of a *feasible vertex weighting*.

DEFINITION 19.3.1 A *feasible vertex weighting* is a mapping ϕ from V to the real numbers such that for each edge $xy \in E$ (each $x \in X$ and $y \in Y$)

$$\phi(x) + \phi(y) \geq \omega(xy). \tag{19.3.7}$$

Clearly, any sufficiently large ϕ is a feasible vertex weighting. For example, the following vertex weighting is feasible:

$$\phi(v) = \begin{cases} \max \ \{\omega(vy) : vy \in E(G)\} & \text{if } v \in X, \\ 0 & \text{otherwise} \end{cases} \tag{19.3.8}$$

The following proposition is easily verified.

PROPOSITION 19.3.5 Let ϕ be any feasible vertex weighting and M any perfect matching of G. Then,

$$\omega(M) \leq \sum_{v \in V} \phi(v). \tag{19.3.9}$$

❑

Given a vertex weighting ϕ, the *equality subgraph* $G_\phi = (V_\phi, E_\phi)$ is a subgraph of G such that $V_\phi = V(G)$ and E_ϕ consists of all the edges xy such that $\omega(xy) = \phi(x) + \phi(y)$. Note that it is possible for some of the vertices of G to be isolated vertices in G_ϕ.

PROPOSITION 19.3.6 Let ϕ be any feasible vertex weighting. If the equality subgraph G_ϕ contains a perfect matching M, then M is a maximum perfect matching in G. ❑

Observe that if M is a perfect matching in G_ϕ, then $\omega(M) = \Sigma_{v \in V}\phi(v)$. Hence, Proposition 19.3.6 follows from Proposition 19.3.5.

Starting with an initial feasible vertex weighting (such as the one given by (19.3.8)), the Kuhn-Munkres algorithm applies the Hungarian algorithm to the subgraph G_ϕ. If a perfect matching M_ϕ is found, then by Proposition 19.3.5, M_ϕ is a maximum perfect matching in G. On the other hand, suppose the Hungarian algorithm terminates by finding a matching M and an M-alternating tree T such that $\Gamma(X_T) - Y_T$ is empty. Then, we have found a set $S = X_T$ such that $|\Gamma_\phi(S)| = |S_T| + 1 < |S_T|$. Instead of terminating at this point, we replace ϕ with a new feasible weighting ϕ' and continue the Hungarian algorithm in the equality graph $G_{\phi'}$ of ϕ'. The new feasible weighting ϕ' is constructed as follows. Set

$$\varepsilon = \min\{\phi(x) + \phi(y) - \omega(xy) : x \in S, y \in Y - \Gamma_\phi(S)\}. \tag{19.3.10}$$

For each $v \in V(G)$, the weighting ϕ' is defined by

$$\phi'(v) = \begin{cases} \phi(v) - \varepsilon & v \in S, \\ \phi(v) + \varepsilon & v \in \Gamma_\phi(S), \\ \phi(v) & \text{otherwise.} \end{cases} \tag{19.3.11}$$

It is easily verified that the vertex weighting ϕ' given by (19.3.11) is a feasible vertex weighting whose equality subgraph $G_{\phi'}$ contains the M-alternating tree T. Further, $G_{\phi'}$ contains a least one edge $\{x,y\}$, where $x \in S$ and $y \in Y - \Gamma_{\phi'}(S)$. Since y is unmatched by M, the path in $G_{\phi'}$ consisting of the path in T from its root to x together with edge xy is an M-augmenting path. Thus, by continuing the Hungarian algorithm in $G_{\phi'}$, the matching M will be augmented by at least one edge. Hence, after at most $n/2$ steps in which ϕ is replaced with ϕ', we will obtain a perfect matching M and a feasible weighting ϕ^* such that M is contained in the equality subgraph G_{ϕ^*} of ϕ^*. By Proposition 19.3.6, such a perfect matching will necessarily be a maximum perfect matching G.

In the following pseudocode for the Kuhn-Munkres algorithm, we call the procedure *Hungarian*2, which is identical to *Hungarian* except that the alternating tree T is added as an input/output parameter.

```
procedure KuhnMunkres(G,ω,M)
Input:  G (a weighted complete bipartite graph with vertex bipartition
          (X,Y), |X| = |Y| = n)
        ω(a positive edge weighting of G)
Output: M (a maximum perfect matching)

  ϕ := any feasible vertex weighting {in particular, the one given by 19.3.8}
  Gϕ := equality subgraph for ϕ
  M := any matching in Gϕ  {in particular, M can be chosen to be empty}
  PerfectMatchingFound := .false.
  while .not. PerfectMatchingFound do
     call Hungarian2(Gϕ,M,S,T)
     if M is a perfect matching then
        PerfectMatchingFound := .true.
     else
        ε := min {ϕ(x) + ϕ(y) − ω(xy) : x ∈ S, y ∈ Y − Γϕ(S)}
        for all x ∈ S do
           ϕ(x) = ϕ(x) − ε
        endfor
        for all y ∈ Γϕ(S) do
           ϕ(y) = ϕ(y) + ε
        endfor
     endif

end Kuhn-Munkres
```

We leave it as an exercise to show that the worst-case complexity of procedure *KuhnMunkres* is $O(n^3)$.

19.4 Maximum Flows in Capacitated Networks

There are a myriad of practical interpretations of flows in networks, such as current flow through an electrical network, fluid flow through a network of pipelines, data flow through a computer network, traffic flow through a network of highways, and so forth. Usually, each edge in a network has a certain flow capacity (a *capacitated* network), and the problem arises of finding the maximum flow subject to the capacity constraints of the edges. One of the most celebrated theorems about capacitated networks is the Max-Flow Min-Cut theorem of Ford and Fulkerson. In this section we present this theorem and an associated algorithm for finding a maximum flow in a capacitated network.

Our study of flows in networks begins by formally defining the notion of a flow in a digraph (uncapacitated network) and establishing some elementary results about flows.

19.4.1 Flows in Digraphs

Let $D = (E,V)$ be a digraph with vertex set V and (directed) edge set E. A *real weighting* ω of the edges of D is a mapping from E to the set $\mathbb{R}$ of real numbers. We refer to $\omega(e)$ as the *ω-weight* of edge e. For $v \in V$, we let $\sigma_{in}(\omega,v)$ and $\sigma_{out}(\omega,v)$ denote the sum of the ω-weights over all the edges having head v and tail v, respectively. That is,

$$\sigma_{in}(\omega,v) = \sum_{uv \in E} \omega(uv),$$

$$\sigma_{out}(\omega,v) = \sum_{uw \in E} \omega(vw).$$

By convention, $\sigma_{in}(\omega,v) = 0$ if there are no edges having head v. Similarly, $\sigma_{out}(\omega,v) = 0$ if there are no edges having tail v.

The following proposition is easily verified.

PROPOSITION 19.4.1 Given any real weighting ω of the edges in a digraph D,

$$\sum_{v \in V} (\sigma_{in}(\omega,v)) - \sigma_{out}(\omega,v)) = 0.$$

Now suppose we are given two vertices s and t such that there are no edges having head s or tail t. A *flow* f from s to t is a weighting of the edges such that

$$\sigma_{in}(f,v) = \sigma_{out}(f,v), \quad \forall v \in V - \{s,t\}. \tag{19.4.1}$$

Given any vertex v in D, we refer to $\sigma_{in}(f,v)$ and $\sigma_{out}(f,v)$ as the *flow into* v and the *flow out of* v, respectively. Equation (19.4.1) is called a *flow conservation equation*. The *value of flow* f, denoted by $val(f)$, is defined to be the flow out of s. It follows easily from Proposition 19.4.1 that the flow out of s equals the flow into t. Hence,

$$val(f) = \sigma_{out}(f,s) = \sigma_{in}(f,t). \tag{19.4.2}$$

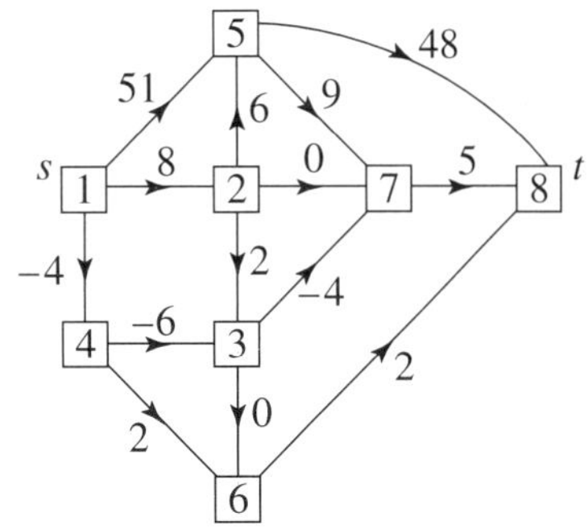

Figure 19.9 A flow of value 55

A *unit flow* is a flow f such that $val(f) = 1$. In Figure 19.9 we illustrate a flow f of value 55 on a sample digraph D. ❑

The following proposition is easily verified.

PROPOSITION 19.4.2 The set of all flows is closed under linear combinations, that is, for any flows f_1 and f_2 and real numbers λ_1 and λ_2, $\lambda_1 f_1 + \lambda_2 f_2$ is a flow. Moreover,

$$val(\lambda_1 f_1 + \lambda_2 f_2) = \lambda_1 val(f_1) + \lambda_2 val(f_2).$$

Now consider a (directed) path $P = se_1u_1e_2u_2 \ldots u_{p-1}e_pt$ from s to t (not necessarily a simple path). Associated with P is the flow χ_P from s to t given by

$$\chi_p(e) = \begin{cases} 1 & e \in E(P) \\ 0 & e \notin E(P), \quad e \in E. \end{cases}$$

We call χ_P the *characteristic flow* of P. Note that the characteristic flow is a *unit* flow.

In the next section, we compute maximum flows in a capacitated network by utilizing the notion of a semipath, which is a path where the orientation of the edges is ignored. More precisely, a *semipath S* from s to t is an alternating sequence of vertices and edges $se_1u_1e_2u_2 \ldots u_{p-1}e_pt$ such that either e_i has tail u_{i-1} and head u_i (e_i is a *forward edge of S*) or e_i has tail u_i and head u_{i-1} (e_i is *backward edge of S*), $i = 1, \ldots, p$, where $u_0 = s$ and $u_p = t$. Associated with semipath S is the flow χ_S, called the *characteristic flow* of S, given by

$$\chi_S(e) = \begin{cases} 1 & e \text{ is a forward edge of } S, \\ -1 & e \text{ is a backward edge of } S, \\ 0 & \text{otherwise}, \qquad e \in E. \end{cases} \tag{19.4.3}$$

It is easily verified that χ_S is a unit flow.

19.4.2 Flows in Capacitated Networks

A *capacitated network N* (sometimes called a *transportation* or *flow* network) consists of the 4-tuple (D,s,t,c), where $D = (V,E)$ is a digraph, s and t are two distinguished vertices of D called the *source* and *sink*, respectively, and c is a positive real weighting of the edges, called the *capacity weighting* of D. We assume that all the edges incident with s are directed out of s and all the edges incident with t are directed into t. We refer to $c(e)$ as the *capacity* of edge e.

A *flow f in a capacitated network N* is a flow in D from s to t such that for each $e \in E$, $f(e)$ is nonzero and does not exceed the capacity of e, that is

$$0 \le f(e) \le c(e), \quad \forall e \in E.$$

Figure 19.10 shows a flow of value 23 in a sample capacitated network N on 8 vertices. In Figure 19.10 and all subsequent figures, when illustrating a flow f, all edges e such that $f(e) = 0$ are omitted.

A *maximum (value) flow* is a flow f in N whose value is maximum over all flows in N. A naive attempt to finding a maximum flow might proceed as follows. First we find a path P_1 from s to t in N using an algorithm such as breadth-first search. Let λ_1 denote the minimum capacity among all the edges of P_1. Then $f = \lambda_1 \chi_{P_1}$ is a flow in N having value λ_1. Next, we adjust the capacities of the edges of N by subtracting λ_1 from each edge belonging to P_1 and deleting all the edges from N whose capacity becomes zero. We then find a path P_2 (if one exists) in the new network. We now augment the flow f by $\lambda_2 \chi_{P_2}$, where λ_2 is the minimum capacity (in the new network) among all the edges of P_2. The current flow $f = \lambda_1 \chi_{P_1} + \lambda_2 \chi_{P_2}$ has value $\lambda_1 + \lambda_2$. Again we subtract λ_2 from every edge belonging to P_2 and delete all the edges whose capacity becomes zero. Continuing in this way we find paths $P_3, \ldots, P_k$ and associated real numbers $\lambda_3, \ldots, \lambda_k$, respectively, until there are no paths left from s to t in the final network.

We illustrate the action of our naive attempt in Figure 19.11 for a sample capacitated network. Unfortunately, the final flow f attained has value 14, whereas a maximum flow shown in Figure 19.12 (page 684) has value 24. However, any flow generated by our naive attempt does have a maximality property, namely, if g is any flow different from f, then $g(e) < f(e)$ for some edge e.

Figure 19.10 A sample capacitated network N and a flow f having value $val(f) = 23$

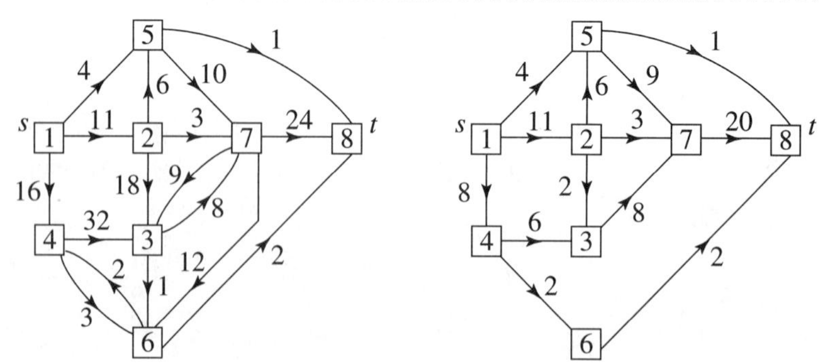

A capacitated network N A flow f in N

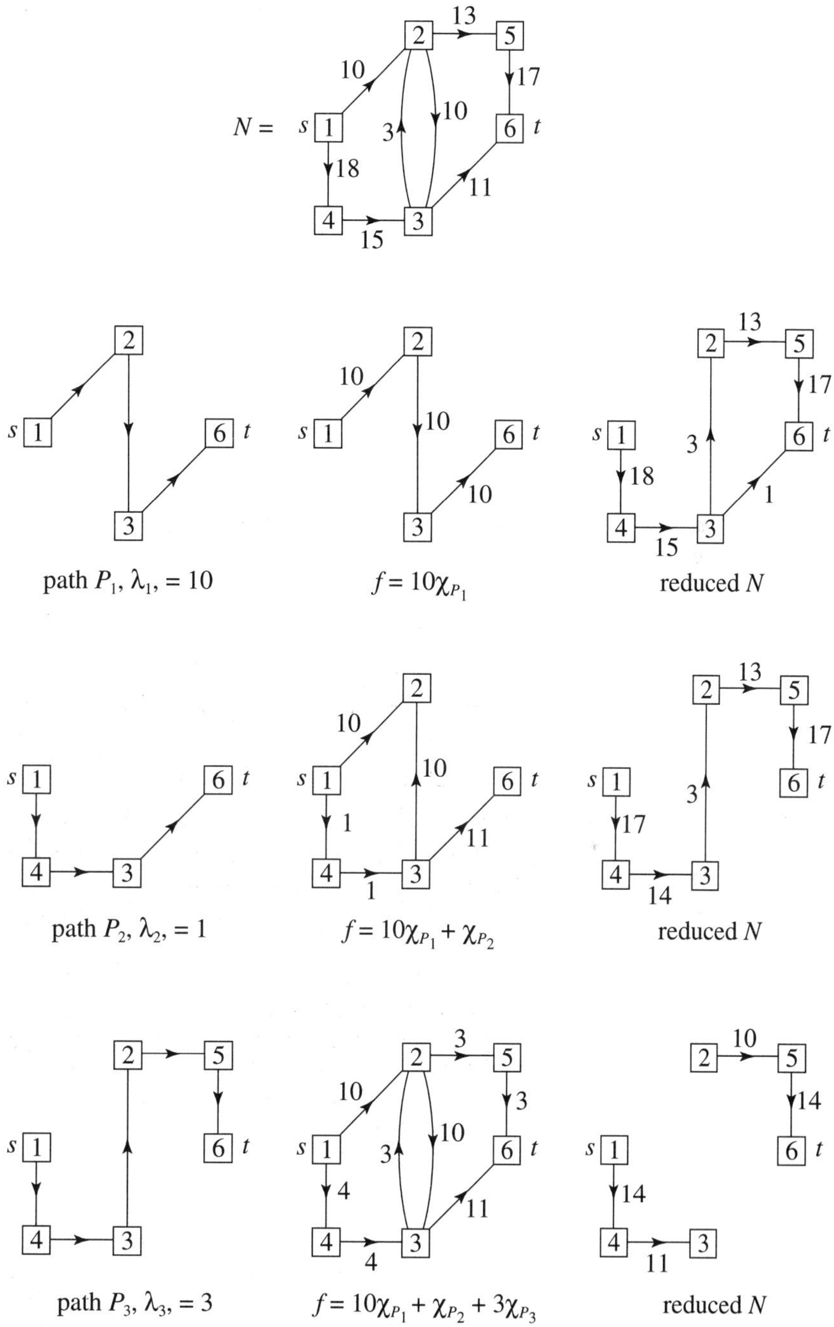

Figure 19.11 Action of naive attempt to find maximum flow for a sample capacitated network N, yielding flow $f = 10\chi_{P_1} + \chi_{P_2} + 3\chi_{P_3}$ having value 14. This flow is suboptimal, since a maximum flow f^* shown in Figure 19.12 has value 24.

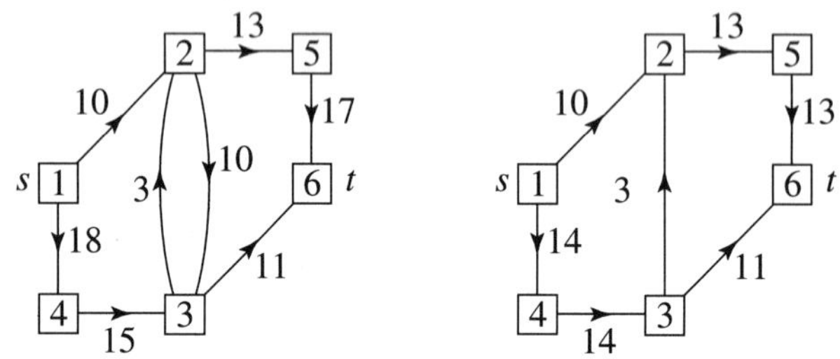

A capacitated network N A maximum flow f^* on N

Figure 19.12 A maximum flow f^* having value 24 in the capacitated network N of Figure 19.11

The flow f of Figure 19.11 generated by our naive algorithm is a dead end in our search for a maximum flow in the sense that we can no longer augment f by simply finding (directed) paths from s to t. However, Ford and Fulkerson showed that by using semipaths we can continue to augment f by adding flow to forward edges while removing flow from backward edges, until a maximum flow f is found.

19.4.3 Finding an Augmenting Semipath

The *residual capacity* with respect to a flow f is $c(e) - f(e)$. An edge e is *f-saturated* if the residual capacity is zero; otherwise, edge e is *f-unsaturated.* A semipath S is an *f-augmenting semipath* if every forward edge of S is f-unsaturated and $f(e) > 0$ for every backward edge e of S. For $e \in E(S)$, let

$$c_f(S,e) = \begin{cases} c(e) - f(e) & e \text{ is a forward edge of } S, \\ f(e) & e \text{ is a backward edge of } S. \end{cases}$$

Let $c_f(S)$ denote the minimum value of $c_f(S,e)$ among all the edges of S; that is,

$$c_f(S) = \min \{c_f(S,e) \mid e \in E(S)\}.$$

Now consider the edge weighting $\hat{f}$ given by

$$\hat{f} = f + c_f(S)\chi_S, \tag{19.4.4}$$

where χ_S is the characteristic flow of S given by (19.4.3). Thus, for each $e \in E(G)$

$$\hat{f}(e) = \begin{cases} f(e) + c_f(S) & e \text{ is a forward edge of } S, \\ f(e) - c_f(S) & e \text{ is a backward edge of } S, \\ f(e) & \text{otherwise.} \end{cases}$$

Proposition 19.4.2 implies that the edge weighting $\hat{f}$ given by (19.4.4) is a flow in the digraph D. Further, it is immediate from the definition of $c_f(S)$ that for all $e \in E$,

$$0 \le \hat{f}(e) \le c(e).$$

Hence, $\hat{f}$ is a flow in the capacitated network N. Further, since χ_S is a unit flow, it follows from Proposition 19.4.2 that

$$val(\hat{f}) = val(f) + c_f(S),$$

so that $\hat{f}$ has a strictly greater value than f.

A semipath S and the value $c_f(S)$ can be computed by finding a path from s to t in the *f-derived network* N_f constructed from the network N and the flow f. The network N_f is obtained by starting with vertex set V and adding edges as follows. For each edge uv of N that is f-unsaturated, we add an edge uv to N_f having weight $c(uv) - f(uv)$. For each edge uv of N such that $f(uv) > 0$, we add an edge vu to N_f having weight $f(uv)$. In the case when uv and vu are both edges of N, then it is possible for N_f to have two edges joining vertex u to vertex v. In our search for a maximum flow, allowing such multiedges does not present a problem. In fact, we can even eliminate the possibility of multiedges in N_f by replacing f with a new flow f obtained by subtracting min $\{f(uv), f(vu)\}$ from both $f(uv)$ and $f(vu)$ so that one of them becomes equal to zero. The latter operation does not affect the value of the flow f. Note that if $val(f) = 0$, then $N_f = N$. The f-derived network N_f for a sample network N and flow f is illustrated in Figure 19.13.

For P a path from s to t in N_f, we define the weight $\mu(P)$ to be the minimum weight over all the edges of P; that is,

$$\mu(P) = \min\{w(e) : e \in E(P)\}.$$

The following proposition follows immediately from the definitions of N_f and an f-augmenting semipath S.

PROPOSITION 19.4.3 If P is a path from s to t in N_f having weight $\mu(P)$, then the corresponding semipath S in N is an f-augmenting semipath. Further,

$$c_f(S) = \mu(P).$$

Figure 19.13 The f-derived network N_f for a sample network N and flow f

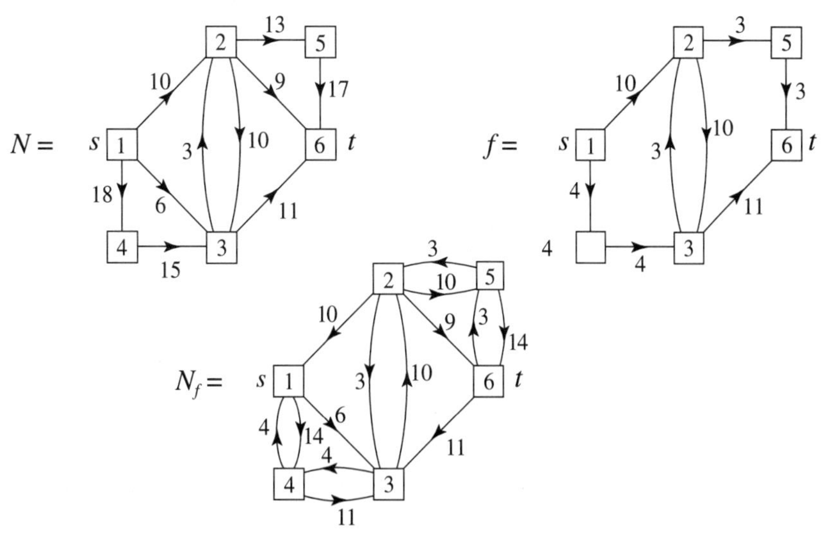

It follows from (19.4.4) that if there exists an f-augmenting semipath S, then we can find a flow whose value is strictly greater than f. The Ford-Fulkerson algorithm is based on the fact that the converse is also true; that is, *if there is no f-augmenting semipath, then f is a maximum flow.* To prove this important result, we introduce the concept of a cut.

19.4.4 Bounding Flow Values by Capacities of Cuts

Consider a bipartition of the vertex set V into two (disjoint) sets X and Y. We denote this bipartition by (X,Y). The cut associated with the bipartition (X,Y), denoted $cut(X,Y)$, is defined by

$$cut(X,Y) = \{xy \in E(D) \mid x \in X, y \in Y\}.$$

We say a set of edges Γ is a *cut*, if $\Gamma = cut(X,Y)$ for some bipartition (X,Y) of V. For $u, v \in V$, if $u \in X$ and $v \in Y$, then we say that Γ *separates u and v*. Unless otherwise stated, we will assume that Γ separates the source s from the sink t. The *capacity of* Γ, denoted by $cap(\Gamma)$, is the sum of the capacities of all the edges in Γ, that is,

$$cap(\Gamma) = \sum_{e \in \Gamma} c(e).$$

A *minimum capacity cut* (or simply *minimum cut*) is a cut Γ whose capacity is minimum over all cuts separating s and t.

Since deleting all the edges of a cut disconnects the source s from the sink t, intuitively we would expect that the value of any flow f is not greater than the capacity of any cut Γ. The following proposition affirms this intuition.

PROPOSITION 19.4.4 Let f be any flow from s to t in N and let $\Gamma = (X,Y)$ be any cut separating s and t. Then, the value of f is bounded above by the capacity of Γ, that is,

$$val(f) \le cap(\Gamma).$$

Proof Given a nonnegative weighting ω of E, we extend ω to a mapping of all $V \times V$ as follows

$$\omega(u,v) = \begin{cases} \omega(uv) & v \in E, \\ 0 & \text{otherwise.} \end{cases}$$

For $A, B \subseteq V$, let

$$\omega(A,B) = \sum_{a \in A} \sum_{b \in B} \omega(a,b). \tag{19.4.5}$$

In the notation of (19.4.5), the flow conservation equation (19.4.1) can be rewritten as follows

$$f(u,V) - f(V,u) = 0, \quad u \in V - \{s,t\}. \tag{19.4.6}$$

Equation (19.4.2) then becomes

$$f(s,V) = f(V,t) = val(f). \tag{19.4.7}$$

It follows immediately from (19.4.6) and (19.4.7) that

$$f(X,V) - f(V,X) = val(f). \tag{19.4.8}$$

Clearly, for U any subset of V, we have

$$\begin{aligned} f(U,V) &= f(U,X) + f(U,Y), \\ f(V,U) &= f(X,U) + f(Y,U). \end{aligned} \tag{19.4.9}$$

Substituting $U = X$ in both equations of (19.4.9), subtracting the second equation from the first, and employing (19.4.8), yields

$$f(X,Y) - f(Y,X) = f(X,V) - f(V,X) = val(f). \tag{19.4.10}$$

Using (19.4.10), we have

$$\begin{aligned} val(f) &= f(X,Y) - f(Y,X) \\ &\le f(X,Y) \\ &= \sum_{xy \in \Gamma} f(xy) \\ &\le \sum_{xy \in \Gamma} c(xy) \quad (\text{since } f(e) \le c(e) \text{ for all } e \in E) \\ &= cap(\Gamma). \end{aligned}$$

■

COROLLARY 19.4.5 If f is a flow from s to t and Γ is a cut separating s and t such that $val(f) = cap(\Gamma)$, then f is a maximum flow and Γ is a minimum cut.

Proof Let f' be any flow from s to t and let Γ' be any cut separating s and t. Then by Proposition 19.4.6, we have

$$\begin{aligned} val(f') &\le cap(\Gamma) = val(f), \\ cap(\Gamma') &\ge val(f) = cap(\Gamma). \end{aligned}$$

■

Corollary 19.4.5 states a condition guaranteeing that f is a maximum flow and Γ is a minimum cut, but it does not tell us whether such a flow f and cut Γ actually exist. In fact, a seminal result in flow theory is that such a flow f and cut Γ always exist.

THEOREM 19.4.6 **Max-Flow Min-Cut** Let $N = (D,c,s,t)$ be a capacitated network. Then the maximum value of a flow from s to t equals the minimum capacity of a cut separating s and t. ❑

To prove Theorem 19.4.6, by Corollary 19.4.5 it is sufficient to exhibit a flow f and cut Γ such that $val(f) = cap(\Gamma)$. In the next subsection we give an algorithm for computing such a flow f and cut Γ.

19.4.5 The Ford-Fulkerson Method

The following procedure computes a maximum flow and minimum cut by repeatedly augmenting the current flow f using an f-augmenting semipath.

```
procedure FordFulkerson(N,f,Γ)

Input:   N = (D,s,t,c) (a capacitated network)
Output:  f (maximum flow)
         Γ(minimum cut)

  f := 0
  N_f := N
  while there is a directed path from s to t in the f-derived network N_f do
     P := a path from s to t in N_f
     S := the f-augmenting semipath in N corresponding to path P in N_f
     c_f(S) := μ(P)        {μ(P) is the minimum weight over all the edges of P}
     f := f + c_f(S)χ_S    {augment f}
     update N_f
  endwhile
  X := set of vertices that are accessible from s in N_f          {s ∈ X}
  Y := set of vertices that are not accessible from s in N_f      {t ∈ Y}
  Γ := cut(X,Y)

end FordFulkerson
```

The correctness of procedure *FordFulkerson* is established with the aid of the following two lemmas, whose proofs are left as exercises.

LEMMA 19.4.7 Let f and (X,Y) be the flow and vertex bipartition, respectively, generated by procedure *FordFulkerson*. Then every edge xy of the cut $\Gamma = cut(X,Y)$ is saturated, that is,

$$f(xy) = c(xy). \qquad \square$$

LEMMA 19.4.8 Let f and (X,Y) be the flow and vertex bipartition, respectively, generated by procedure *FordFulkerson*. Then for each edge $yx \in E(D)$, where $y \in Y$ and $x \in X$, we have

$$f(yx) = 0.$$

Lemmas 19.4.7 and 19.4.8 imply that

$$f(X,Y) = cap(\Gamma),$$
$$f(Y,X) = 0. \qquad \square$$

Thus, by (19.4.10) we have

$$val(f) = f(X,Y) - f(Y,X) = cap(\Gamma).$$

By Corollary 19.4.5, f is a maximum flow and Γ is a minimum cut, which completes the correctness proof of procedure *FordFulkerson*.

The Ford-Fulkerson augmenting semipath algorithm is a general method for computing a maximum flow and minimum cut. In procedure *FordFulkerson* we did not specify how the path P is generated in the derived network N_f at each stage. In general, there may be many such paths, and the efficiency of *FordFulkerson* is dependent on which augmenting semipath S is chosen at each stage.

For a poor choice of augmenting semipaths S, procedure *FordFulkerson* may never terminate. However, in the case when the capacities on the edges are all integers, procedure *FordFulkerson* terminates after having performed at most $val(f)$ iterations of the **while** loop. Since each iteration can be performed in time $O(m)$, the worst-case complexity $W(n,m)$ of procedure *FordFulkerson* belongs to $O(m\ val(f))$. Since $val(f)$ depends on the capacities of the edges, it can be arbitrarily large. Examples are not hard to find showing that for a poor choice of augmenting semipaths S, $W(n,m)$ can also be arbitrarily large.

Edmonds and Karp showed that a good choice for the augmenting semipath S at each stage of procedure *FordFulkerson* is one which is shortest (has a minimum number of edges) over all such semipaths. At each stage, a shortest augmenting semipath S can be found by performing a breadth-first search of the f-derived network N_f to find a shortest path P from s to t. The Edmonds-Karp algorithm has worst-case complexity $W(n,m) \in O(nm^2)$. (The proof of this complexity result is beyond the scope of this book.) The Edmonds-Karp algorithm is illustrated in Figure 19.14 (page 690) for a sample flow network N having 8 vertices and 17 edges. The shortest path generated at each step is indicated with a dotted (as opposed to solid) line.

19.4.6 Maximum Flows in 0/1 Networks: Menger's Theorem

An *integer flow* in a digraph $D = (V,E)$ is a flow f, such that $f(e)$ is an integer for every edge $e \in E$.

PROPOSITION 19.4.9 Let $D = (V,E)$ be a digraph and f be any integer flow from s to t, where $s,t \in V$. Then there exists a set $\mathcal{P}$ of (simple) paths from s to t and positive integers $\{\lambda_P : P \in \mathcal{P}\}$ such that

$$val(f) = \sum_{P \in \mathcal{P}} \lambda_P . \qquad (19.4.11)$$

❑

We leave the proof of Proposition 19.4.9 as an exercise. A set of paths $\mathcal{P}$ and associated positive integers $\{\lambda_P : P \in \mathcal{P}\}$ satisfying (19.4.11) can be computed in time $O(m\ val(f))$.

A 0/1 flow is an integer flow such that, for each edge e, $f(e)$ is either 0 or 1. The following result is an immediate corollary of Proposition 19.4.9.

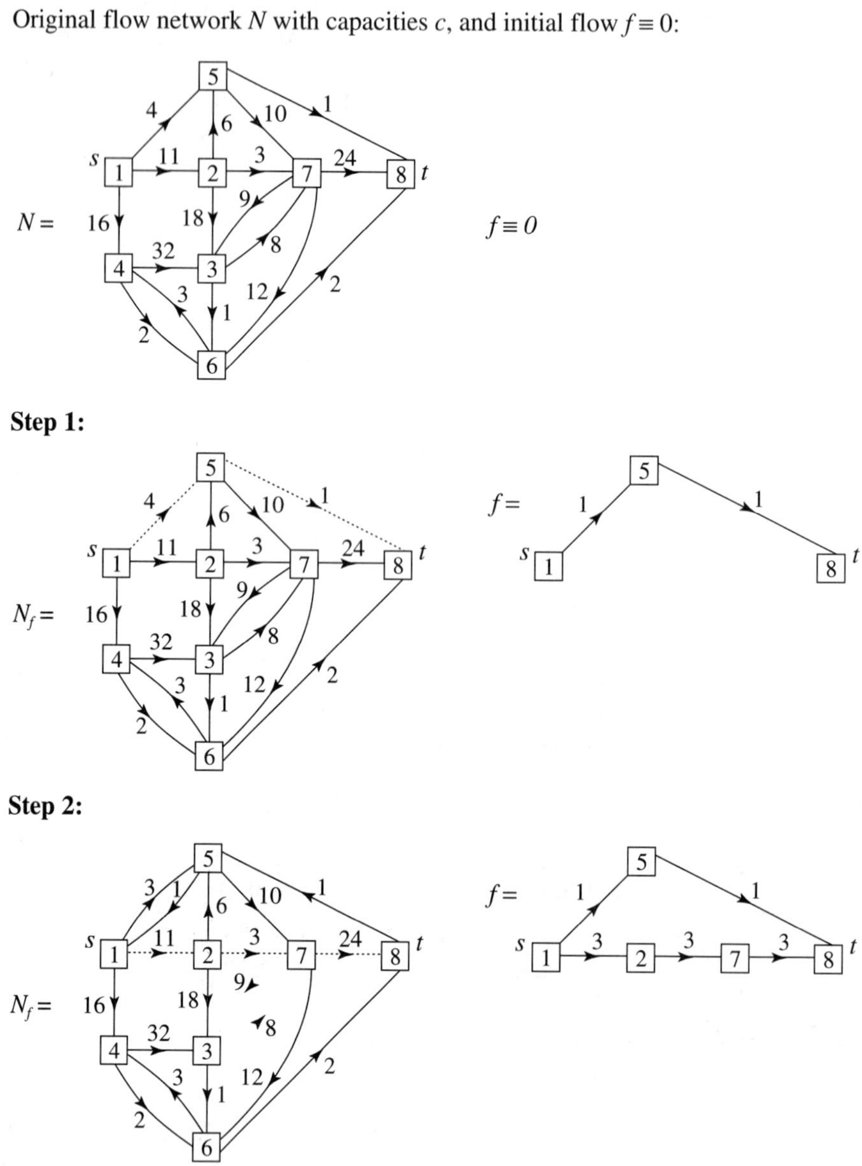

Figure 19.14 Action of the Edmonds-Karp algorithm for a sample capacitated network N

COROLLARY 19.4.10 Let f be any 0/1 flow from u to v. Then there exists a set $\mathscr{P}$ of pairwise edge-disjoint paths from u to v such that

$$val(f) = |\mathscr{P}|. \qquad \square$$

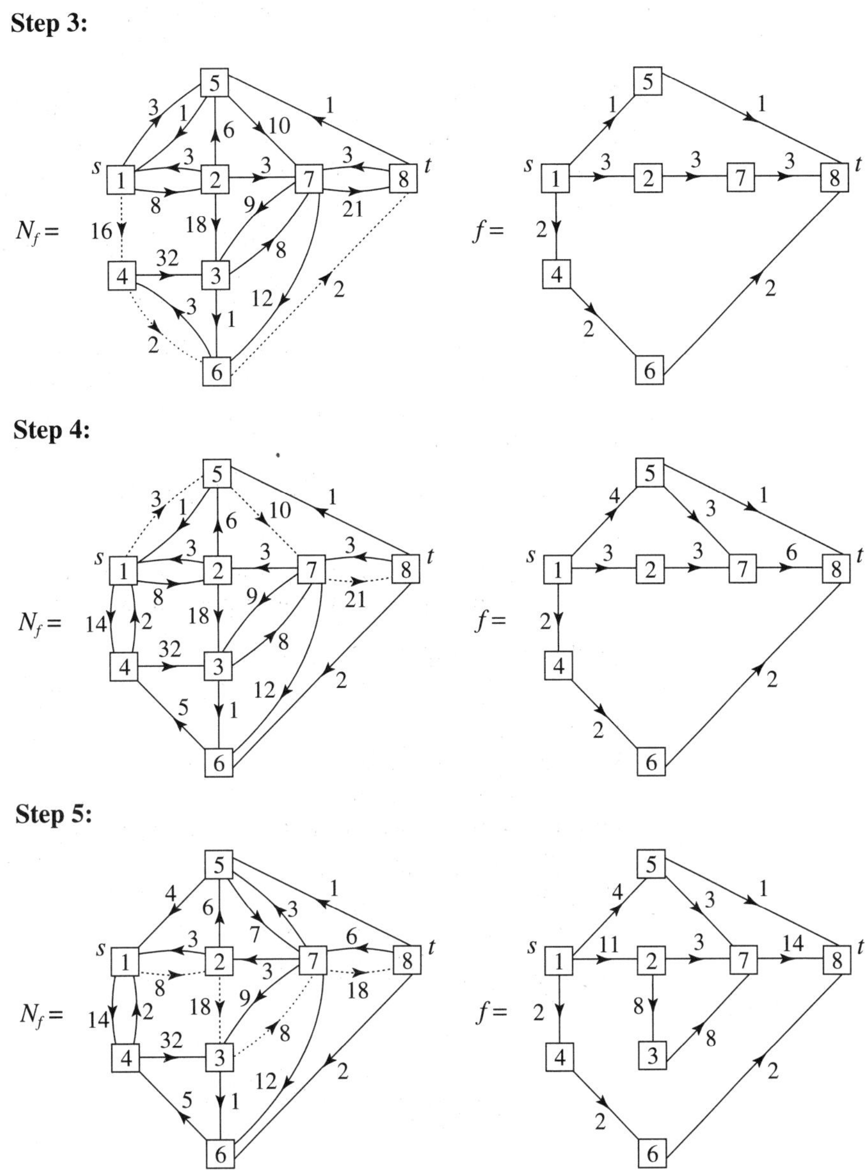

Figure 19.14 Continued

When the capacities are all integers, $c_f(S)$ is an integer at each stage of the procedure *FordFulkerson*. Thus, for integer capacities, *FordFulkerson* generates a maximum integer flow f. In particular, if each edge has unit capacity, then a maximum 0/1 flow f is generated. By Corollary 19.4.10, if f is a 0/1 flow, then there

Step 6:

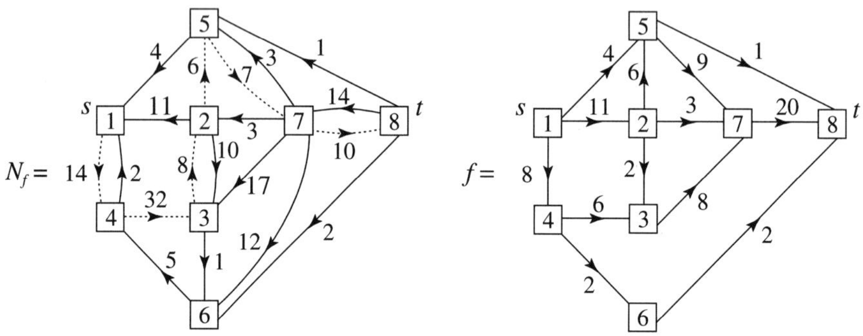

There are no more augmenting semipaths. The final flow f has value 23.

Step 7: Compute the final f-derived network N_f and minimum cut $cut(X,Y)$.

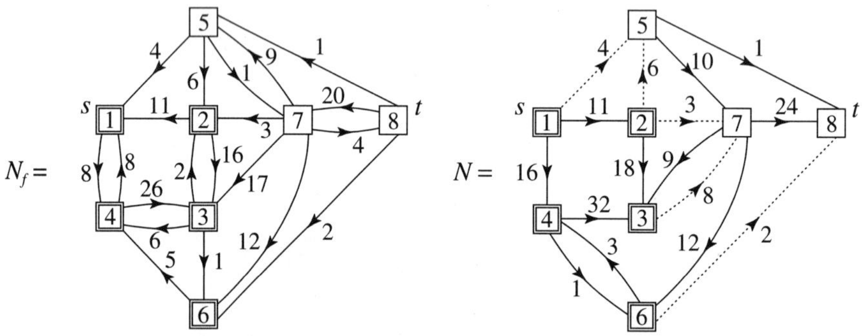

The set $X = \{1,2,3,4,6\}$ of vertices that are accessible in N_f from the source s (marked with a ☐) and the set $Y = \{5,7,8\}$ of vertices that are not accessible from s determine a cut $\Gamma = cut(X,Y)$ of capacity $c(X,Y) = 4 + 6 + 3 + 8 + 2 = 23$. Hence, we have $val(f) = 23 = cap(\Gamma)$, so that f is a maximum flow and Γ is a minimum cut.

Figure 19.14 Continued

exists a set $\mathcal{P}$ of pairwise edge-disjoint paths from u to v such that $val(f) = |\mathcal{P}|$. If $\mathcal{P}$ is a set of pairwise edge-disjoint paths from u to v then, $\Sigma_{P\in\mathcal{P}}\chi_P$ is a 0/1 flow. Thus, a set $\mathcal{P}$ of pairwise edge-disjoint paths from s to t has maximum cardinality over all such sets if, and only if, $\Sigma_{P\in\mathcal{P}}\chi_P$ is a maximum flow.

It follows that we can compute a maximum size set $\mathcal{P}$ of pairwise edge-disjoint paths from s to t in a given digraph D by first applying procedure *FordFulkerson* to the capacitated network $N = (D,c,s,t)$, where each edge e has capacity $c(e) = 1$ to obtain a maximum flow f, and then computing a set $\mathcal{P}$ of pairwise edge-disjoint paths from s to t such that $val(f) = |\mathcal{P}|$. Now consider the cut Γ generated by *Ford-Fulkerson*. Since every edge has unit capacity, the capacity of Γ equals the size of (number of edges in) Γ. Since the capacity of Γ equals the value of f, it follows that the size of Γ equals the size of $\mathcal{P}$, which yields a classical theorem of Menger.

THEOREM 19.4.11 **Menger** Let $D = (V,E)$ be a digraph. Then, for $s, t \in V$, the maximum size of a set $\mathscr{P}$ of pairwise edge-disjoint paths from s to t equals the minimum size of a cut Γ separating s and t. ❑

The following corollary is the analog of Theorem 19.4.11 for graphs.

COROLLARY 19.4.12 **Menger** Let $G = (V,E)$ be an (undirected) graph. Then, for $s, t \in V$, the maximum size of a set $\mathscr{P}$ of pairwise edge-disjoint paths from s to t equals the minimum size of a cut Γ separating s and t. ❑

19.4.7 Maximum Size Matching

Procedure *FordFulkerson* can be applied to obtain a maximum size matching in a bipartite graph G as follows. Let (X,Y) denote the associated bipartition of the vertex set V of G. Construct a digraph D as follows. The vertex set of D consists of the vertex set of G together with two new vertices s and t; that is,

$$V(D) = V(G) \cup \{s,t\}.$$

The edge set $E(D)$ of D consists of pairs xy such that $xy \in E(G)$, $x \in X$, and $y \in Y$, together with all pairs sx, $x \in X$, and all pairs yt, $y \in Y$; that is,

$$E(D) = \{xy : xy \in E(G), x \in X, y \in Y\} \cup \{sx : x \in X\} \cup \{yt : y \in Y\}.$$

Now consider the capacitated network $N = (D,c,s,t)$ where every edge e has capacity $c(e) = 1$. For f a 0/1 flow from s to t, let $\mu(f)$ denote the set of all edges xy of G such that xy is an edge of D and $f(xy) = 1$. Clearly, $\mu(f)$ is a matching M in G, and $val(f)$ is equal to the size of M. Conversely, given any matching M of G there is a flow f in N such that $M = \mu(f)$. Let f^* be the flow generated by procedure *FordFulkerson*. Since f^* is a maximum value flow in N, $\mu(f^*)$ is a maximum size matching in G. The algorithm just described for finding a maximum size matching is illustrated in Figure 19.15 (page 694).

19.5 A Monte Carlo Parallel Algorithm for Computing a Perfect Matching

In Chapter 18 we discussed a Monte Carlo parallel algorithm for determining whether or not a bipartite graph contains a perfect matching. In this section we discuss a Monte Carlo parallel algorithm due to Mulmuley, Vazirani, and Vazirani (MVV) for finding a perfect matching in a general graph. The MVV algorithm creates a candidate perfect matching M by randomly assigning chosen integer values to the edges and then taking the edges of M to be those edges satisfying a simple algebraic condition related to the determinant and the adjoint of a matrix associated with the resulting weighted graph. Since the adjoint and determinant of a matrix can be computed in time $O(\log^2 n)$ on a PRAM with $O(n^4)$ processors, the MVV algorithm shows that the problem of finding a perfect matching in general graph lies in the class RNC.

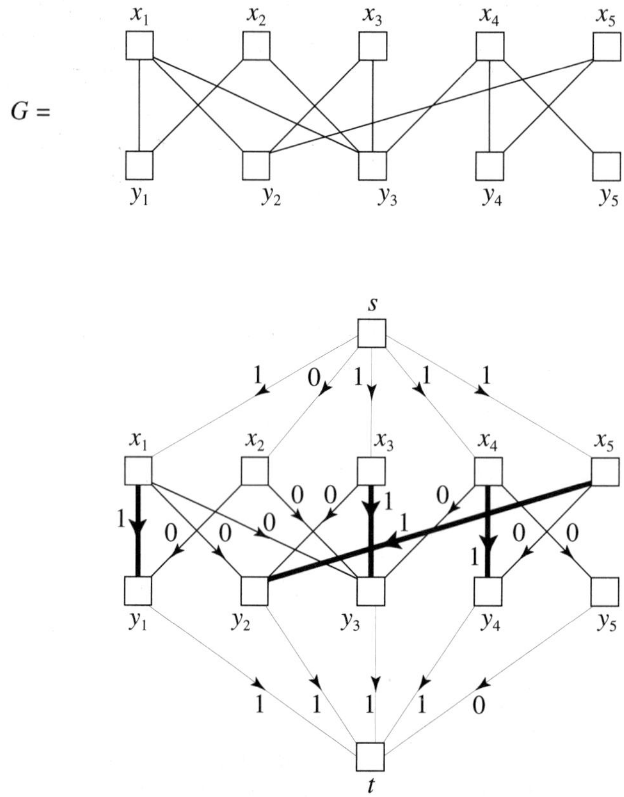

Figure 19.15 A maximum flow in an associated network where all edges have capacity 1 yielding a maximum matching in a bipartite graph G

Let $G = (V,E)$ be a graph with vertex set $V = \{1, \ldots, n\}$ and edge set $E = \{e_1, \ldots, e_m\}$. Suppose we associate with each edge e_i a weight w_i, where w_i is a positive integer. The *exponentially weighted adjacency matrix* A with respect to the weights $w_1, \ldots, w_n$ is the $n \times n$ matrix $A = (a_{ij})$ given by

$$a_{ij} = \begin{cases} 2^{w_k} & \text{if } i < j \text{ and } \{i,j\} = e_k \in E, \\ -2^{w_k} & \text{if } i > j \text{ and } \{i,j\} = e_k \in E, \\ 0 & \text{if } \{i,j\} \notin E. \end{cases} \tag{19.5.1}$$

The *adjoint* of A, denoted $adj(A)$, is the matrix whose jith entry is $(-1)^{i+j}\det(A_{ij})$, where A_{ij} is the submatrix of A obtained by stroking out row i and column j and $\det(A)$ denotes the determinant of A. A classical result in linear algebra states that

$$adj(A) = \det(A)A^{-1}.$$

Given a perfect matching M in the weighted graph G, the weight *weight*(M) of M is the sum of the weights of all the edges in M. A perfect matching M is called a *minimum-weight* matching if its weight is not larger than the weight of any perfect matching in G. The following two theorems form the basis for the MVV algorithm.

THEOREM 19.5.1 Given a graph G with vertex set $V = \{1, \ldots, n\}$ and edge set $E = \{e_1, \ldots, e_m\}$, suppose that we choose for each edge e_i a weight w_i uniformly and independently from $\{1, \ldots, 2m\}$, $i = 1, \ldots, m$. If G has a perfect matching, then with probability at least 1/2, the minimum-weight perfect matching is unique. ❑

The following theorem gives a connection between perfect matchings and the exponentially weighted adjacency matrix.

THEOREM 19.5.2 Given a graph G with vertex set $V = \{1, \ldots, n\}$ and edge set $E = \{e_1, \ldots, e_m\}$, suppose we associate with each edge e_i a positive integer weight w_i, $i = 1, \ldots, m$. Let A be the exponentially weighted adjacency matrix given by (19.5.1). Suppose G has a unique minimum-weight perfect matching M. Then,

1. M has weight L, where 2^{2L} is the highest power of 2 that divides det(A).
2. The set of edges of M are precisely those edges $e_k = \{i,j\}$ such that $(\det(A_{ij})2^{2w_k})/2^{2L}$ is an odd integer. ❑

Theorems 19.5.1 and 19.5.2 yield the following MVV algorithm for finding a perfect matching in a graph G.

```
procedure MCPerfectMatch(G,M,Matching)
Input:  G (graph with vertex set V = {1, ..., n} and edge set E = {e_1, ..., e_m})
Output: M (a candidate perfect matching)
        Matching (a Boolean variable that is .true. if M is a perfect
                  matching, otherwise .false.)
        {if a perfect matching exists, then Matching is .true. with
            probability of at least 1/2}
   for each edge e_k ∈ E choose a weight w_k uniformly and independently from
                                                                 {1, ..., 2m}
   A := exponentially weighted adjacency matrix given by (19.5.1)
   compute det(A) and adj(A) = det(A)A^(-1)
   compute the integer L such that 2^(2L) is the largest power of 2 dividing det(A)
   M := set of edges e_k = {i,j} such that (det(A_ij)2^(2w_k))/2^(2L) is an odd integer
   if M is a perfect matching then
      Matching := .true.
   else
      Matching := .false.
   endif
end MCPerfectMatch
```

By Theorems 19.5.1 and 19.5.2, *MCPerfectMatch* is a true-biased (1/2)-correct Monte Carlo algorithm. In particular, we can ramp up exponentially the probability of obtaining a correct answer by repeatedly running *MCPerfectMatch*.

We now prove Theorems 19.5.1 and 19.5.2. The proof of Theorem 19.5.1 employs the following useful lemma, known as the *Isolating Lemma*. An *m-element set system* $(S,\mathcal{F})$ consists of a finite universe set $S = \{x_1, \ldots, x_m\}$ and a family $\mathcal{F} = \{F_1, \ldots, F_k\}$ of subsets of S. Given an assignment of a weight w_i to element x_i, $i = 1, \ldots, m$, we define the weight of a subset F of S to be the sum of the weights of the elements in F; that is,

$$weight(F) = \sum_{x_i \in F} w_i.$$

LEMMA 19.5.3 **Isolating Lemma** Let $(S,\mathcal{F})$ be an m-element set system whose elements are assigned integer weights chosen uniformly and independently from $\{1, \ldots, 2m\}$. Then, the probability that there is a unique minimum-weight set in $\mathcal{F}$ is at least 1/2. ❑

The Isolating Lemma is somewhat surprising. Since there are 2^m possible subsets of S, but only $2m^2$ possible weights, there are on average $2^{m-1}/m^2$ subsets per weight. For this reason, one might expect the probability that there is a unique minimum-weight subset to be considerably smaller than 1/2.

Before proving the lemma, we first establish a preliminary claim. For $i = 1, \ldots, m$, let A_i denote the minimum weight over all subsets $F \in \mathcal{F}$ that do not contain x_i and let B_i denote the minimum weight of $F - \{x_i\}$ over all subsets $F \in \mathcal{F}$ that contain x_i. That is,

$$A_i = \min\ \{weight(F) : F \in \mathcal{F}, x_i \notin F\},$$

$$B_i = \min\ \{weight(F - \{x_i\}) : F \in \mathcal{F}, x_i \in F\}.$$

We set $A_i = 0$ $(B_i = 0)$ if x_i belongs to each set in $\mathcal{F}$ (if x_i belongs to no set in $\mathcal{F}$). The *threshold of* x_i is the random variable α_i defined by

$$\alpha_i = A_i - B_i.$$

Note that α_i is independent of w_i.

CLAIM 19.5.4 For any $i \in \{1, \ldots, m\}$, there exist two sets from $\mathcal{F}$ having minimum weight, one containing x_i and the other not containing x_i, only if $w_i = \alpha_i$.

Proof of Claim 19.5.4 First suppose $\alpha_i < w_i$, so that

$$A_i < B_i + w_i.$$

The definition of B_i then implies that the weight of every set in $\mathcal{F}$ that contains x_i is strictly greater than the minimum weight A_i of some set not containing x_i. Thus, every minimum-weight set does not contain x_i.

Now suppose $\alpha_i > w_i$, so that

$$B_i + w_i < A_i.$$

Clearly, $B_i + w_i$ is at least as great as the minimum weight of a set in $\mathscr{F}$ that contains x_i. Thus, the minimum weight A_i of a set in $\mathscr{F}$ that does not contain x_i is strictly greater than the minimum weight of a set in $\mathscr{F}$ containing x_i. It follows that all the minimum-weight sets in $\mathscr{F}$ contain x_i.

We have just shown that if $\alpha_i \neq w_i$, then either every minimum-weight set in $\mathscr{F}$ contains x_i or every minimum-weight set in $\mathscr{F}$ does not contain x_i. ■

Proof of Lemma 19.5.3 Since any two distinct minimum-weight sets must differ in at least one variable x_i, Claim 19.5.4 implies that $\mathscr{F}$ does not contain a unique minimum-weight set only if $w_i = \alpha_i$ for some $i \in \{1,2, \ldots, m\}$. Since w_i is chosen at random from $2m$ possible values and w_i is independent from α_i, the probability that $w_i = \alpha_i$ is at most $1/(2m)$. Thus, the probability that there are two minimum-weight sets in $\mathscr{F}$ is no more than

$$\begin{aligned} \text{Prob}\{\text{for some } i, \alpha_i = w_i\} &\leq \sum_{i=1}^{m} \text{Prob}\{w_i = \alpha_i\} \\ &\leq m\left(\frac{1}{2m}\right) = \frac{1}{2}. \end{aligned}$$

■

Proof of Theorem 19.5.1 Consider the set system $(S,\mathscr{F})$, where the universe S is the edge set E of G and the family $\mathscr{F}$ is the set of all perfect matchings of G. The Isolating Lemma implies that with probability of at least 1/2, G has a *unique* minimum-weight perfect matching. ■

Proof of Theorem 19.5.2 Assume that G has a unique minimum-weight perfect matching M with respect to the weights w_i, $i = 1, \ldots, m$. Consider the determinant $\det(A)$ of the exponentially weighted adjacency matrix A. By definition,

$$\det(A) = \sum_{\pi} sign(\pi) \prod_{i=1}^{n} a_{i\pi(i)},$$

where the product is taken over all permutations π: $\{1, \ldots, n\} \rightarrow \{1, \ldots, n\}$. For convenience, let

$$value(\pi) = \prod_{i=1}^{n} a_{i\pi(i)},$$

so that

$$\det(A) = \sum_{\pi} sign(\pi)\ value(\pi).$$

Associate with the (undirected) graph G the digraph $\hat{G}$ obtained from G by replacing every edge $\{v_i,v_j\}$ with the two directed edges (v_i,v_j) and (v_j,v_i). A *trail* R is a subdigraph having in-degree = out-degree = 1 at every vertex. Note that R consists of the union of disjoint directed cycles that span the vertices. There is a natural bijection ρ between trails R and permutations π with value$(\pi) \neq 0$, where $\rho(\pi)$ is the trail R consisting of the directed edges $(v_i,v_{\pi(i)})$, $i = 1, \ldots, n$.

Define sign(R) = sign($\rho^{-1}(R)$) and $value(R) = value(\rho^{-1}(R))$. Let $\gamma(R)$ denote the number of circuits of even length in R. From the definition of the sign of a permutation, we have

$$sign(R) = -1^{\gamma(R)}.$$

We tripartition the set $\mathscr{R}$ of all trails into classes $\mathscr{A}$, $\mathscr{B}$, and $\mathscr{C}$, where $\mathscr{A}$ consists of all trails such that each cycle of the trail has size 2, $\mathscr{B}$ consists of all trails such that each cycle of the trail has even length with one cycle having length at least 4, and $\mathscr{C}$ consists of the set of all trails such that at least one cycle of the trail has odd length. Let

$$\sigma(\mathscr{A}) = \sum_{A \in \mathscr{A}} sign(A)\ value(A).$$

The values $\sigma(\mathscr{B})$ and $\sigma(\mathscr{C})$ are defined similarly. Then,

$$\det(A) = \sum_{R \in \mathscr{R}} sign(R)\ value(R) = \sigma(\mathscr{A}) + \sigma(\mathscr{B}) + \sigma(\mathscr{C}).$$

Note that a *perfect matching* P corresponds to a trail A from $\mathscr{A}$. Each edge $e_k = \{i,j\}$ of P corresponds to two directed edges (i,j) and (j,i), and therefore contributes $2^{w_k}(-2^{w_k})$, so that

$$sign(A)value(A) = 2^{2w(P)},$$

where $w(P)$ denotes the weight of P. Since by assumption M is the unique minimum-weight perfect matching, it follows that, for some integer j,

$$\sigma(\mathscr{A}) = 2^{2w(M)} + 2^{2w(M)+1}j.$$

Now consider a trail B from $\mathscr{B}$. Since each cycle in B has even length, B can be partitioned into two distinct perfect matchings P_1 and P_2. Further,

$$value(B) = 2^{w(P_1) + w(P_2)}.$$

It follows that $2^{2w(M)+1}$ divides $\sigma(\mathscr{B})$.

Finally, consider a trail C from $\mathscr{C}_v$. Choose the odd cycle containing the lowest-numbered node. Reversing the direction of the odd cycle yields another trail C' such that $value(C) = -value(C')$, but $sign(C) = sign(C')$. It follows that

$$\sigma(\mathscr{C}) = 0.$$

The preceding arguments show that, for some integer k,

$$\det(A) = 2^{2w(M)} + 2^{2w(M)+1}k.$$

Hence, M has weight L, where 2^{2L} is the highest power of 2 that divides det(A), which completes the proof of part (1) of Theorem 19.5.2.

Part (2) of Theorem 19.5.2 can be proved by applying part (1) of Theorem 19.5.2 to the graph G_k obtained by deleting vertices i and j, together with all incident edges, for each edge $e_k = \{i,j\}$. ■

19.6 Closing Remarks

Papers by Tarjan and by Hopcroft and Tarjan in the early 70's popularized the use of depth-first search in graph algorithms. Although RNC algorithms have been found for computing the depth-first forest of a graph or digraph, the problem is not known to be in NC. Nevertheless, NC algorithms have been designed for the problems of finding strongly connected components in a digraph and articulation points in a graph. For example, the algorithm given in Chapter 14 for finding the transitive closure of a digraph yields an algorithm for finding the strongly connected components.

In addition to the Edmonds-Karp algorithm for finding a maximum flow, which uses the Ford-Fulkerson augmenting path method, other efficient algorithms for finding maximum flows have been designed. These algorithms include the Dinac Maximum Flow algorithm and, more recently, the Preflow Push algorithm due to Goldberg and Tarjan. The problem of finding a maximum flow from a single source s to a single sink t generalizes to the problem of finding a multicommodity flow from a set of sources to a set of sinks. The theory of multicommodity flows is an active research area, with applications to such areas as routing in communication networks and VLSI layout.

In this chapter we discussed the Hungarian algorithm for finding a perfect matching in a bipartite graph and the Kuhn-Munkres algorithm for finding a maximum-weight perfect matching in a weighted bipartite graph. We also showed how a maximum flow can be used to find a maximum size matching in a bipartite graph. Sequential algorithms for finding maximum perfect matchings in general graphs can be found in the references.

The algorithm *MCPerfectMatch* can be modified to find a maximum matching in graph G. Using a classical matching theorem of Berge and Tutte, Karloff gave a Las Vegas algorithm for finding a maximum size matching in a graph. Note that most of the steps in the algorithm *MCPerfectMatch* were deterministic. The only randomized part of the algorithm was assigning weights to ensure with high probability that the minimum-weight perfect matching is unique. An interesting open question is whether there exists a deterministic polylogarithmic algorithm for finding a perfect matching in a general graph G.

References and Suggestions for Further Reading

For good references for network optimization algorithms, including flow and matching algorithms, see:

Ahuja, R. K. *Network Flows: Theory, Algorithms, and Applications*. Englewood Cliffs, NJ: Prentice-Hall, 1993.

Even, S. *Graph Algorithms*. Rockville, MD: Computer Science Press, 1979.

Gordon, M. and M. Minoux. *Graphs and Algorithms* (trans. by S. Vajda). New York: Wiley, 1984.

Lawler, E. L. *Combinatorial Optimization: Networks and Matroids*. New York: Holt, Rinehart and Winston, 1976.

Papadimitriou, C. H. and K. Steiglitz. *Combinatorial Optimization: Algorithms and Complexity*. Englewood Cliffs, NJ: Prentice-Hall, 1982.

Tarjan, R. E. *Data Structures and Network Algorithms*. Philadelphia, PA: Society for Industrial and Applied Mathematics, 1983.

For two recent books on graph algorithms, see:

Chartrand, G. and O. R. Oellermann. *Applied and Algorithmic Graph Theory*. New York: McGraw-Hill, 1993.

Thulasiraman, K. and M. N. S. Swamy. *Graphs: Theory and Algorithms*. New York: Wiley, 1992.

Lovász, L. and M. D. Plummer. *Matching Theory*. Amsterdam: North Holland, 1986. A nice reference to results and algorithms on matchings.

Exercises

Section 19.1 Strongly Connected Components

19.1 **a.** Give pseudocode for the O(m) *naive* algorithm *StrongNaive* discussed in Section 19.4 for finding the strongly connected component containing vertex v.

b. Show how to obtain all the strongly connected components of D by repeatedly calling *StrongNaive*.

19.2 Let F be the depth-first out-forest of a digraph D consisting of trees $T_1, \ldots, T_k$, listed in the order they are generated by the out-directed depth-first traversal. Show that a vertex u has post number i if u is the ith vertex to be visited in the *postorder traversal* of F, that is, the traversal of F consisting of successive postorder traversals of T_i, $i = 1, \ldots, k$.

19.3 Give a linear time algorithm for computing the array *PostNumInv*[1:n] from the array *PostNum*[1:n].

19.4 Modify the depth-first traversal procedure *DFTOut* to compute arrays *PostNum*[1:n] and *PostNumInv*[1:n].

19.5 Show that an out-directed depth-first search starting at a vertex u of a digraph D visits all the vertices in the strongly connected component containing u.

Section 19.2 Articulation Points and Biconnected Components

19.6 Prove Proposition 19.2.1.

19.7 Prove Proposition 19.2.3.

19.8 Prove Proposition 19.2.5.

19.9 Design and analyze an algorithm that outputs the biconnected components of a graph G given the input parameter *Art* of articulation points of G.

19.10 Consider the following graph G.

$G =$ 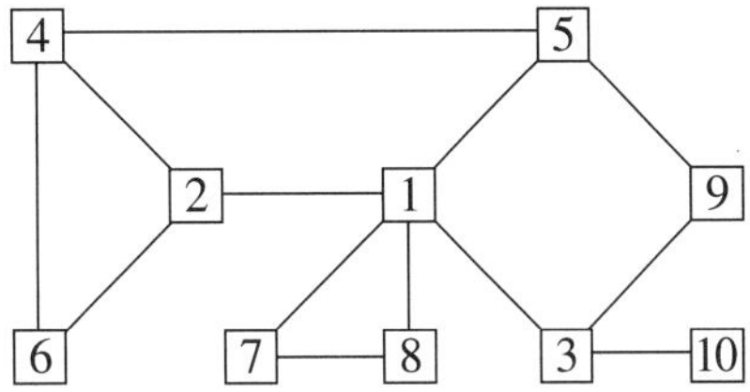

a. Using a depth-first search starting at vertex $r = 3$, compute the global arrays *TreeDFS*[1:10], *DFSNum*[1:10], and *Lowest*[1:10] output by *DFSNumberings*.

b. Demonstrate how the articulation points are obtained from the arrays computed in part (a).

19.11 Let G be graph that is not biconnected and suppose B_1 and B_2 are any two biconnected components of G. Show that B_1 and B_2 have at most one vertex in common.

Section 19.3 Perfect Matchings in Bipartite Graphs

19.12 Prove Lemma 19.3.2.

19.13 Prove Lemma 19.3.3.

19.14 Prove Lemma 19.3.4.

19.15 Show that at each stage of the procedure *Hungarian* the alternating tree T can be grown in time $O(n^2)$, so that the worst-case complexity of procedure *Hungarian* is $O(n^3)$.

19.16 Refine the pseudocode of the procedure *Hungarian* to include details for finding the augmenting paths.

19.17 Consider the following bipartite graph G.

$G =$ 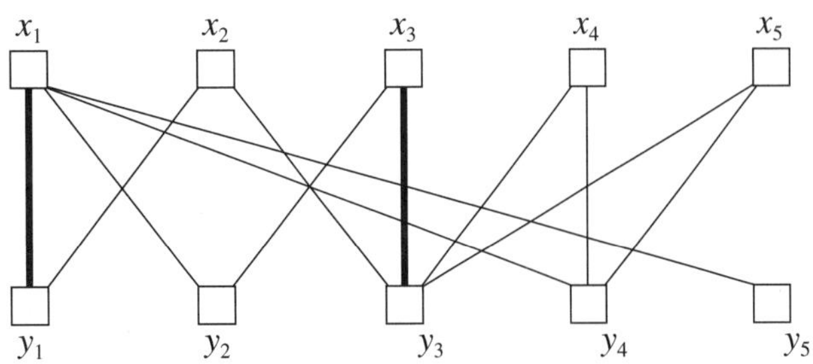

Starting with the given matching $M = \{x_1y_1, x_3y_3\}$, determine the perfect matching output by *Hungarian*. Show each augmenting path that is generated.

19.18 Prove Proposition 19.3.5.

19.19 Verify that the vertex weighting given by (19.3.8) is a feasible vertex weighting.

19.20 **a.** Verify that the vertex weighting ϕ' given by (19.3.11) is a feasible vertex weighting whose equality subgraph $G_{\phi'}$ contains the M-alternating tree T.

b. Show that $G_{\phi'}$ contains at least one edge $\{x,y\}$, where $x \in S$ and $y \in Y - \Gamma_{\phi'}(S)$.

19.21 Show that the worst-case complexity of procedure *KuhnMunkres* is $O(n^3)$.

Section 19.4 Maximum Flows in Capacitated Networks

19.22 Let f be any flow from vertex s to vertex t in a digraph. Using Proposition 19.4.1, show that the flow out of s equals the flow into t.

19.23 Prove Proposition 19.4.2.

19.24 Verify that χ_S given by (19.4.3) is a unit flow.

19.25 Verify that the final flow $f = \lambda_1\chi_{P_1} + \cdots + \lambda_k\chi_{P_k}$ generated by the naive algorithm illustrated in Figure 19.11 is maximal in the sense that if g is any other flow, then $g(e) < f(e)$ for some edge e.

19.26 Prove Lemma 19.4.7.

19.27 Prove Lemma 19.4.8.

19.28 Show that for a poor choice of augmenting semipaths S, procedure *FordFulkerson* may never terminate.

19.29 Assuming that the capacities of the network are positive integers, show that for a poor choice of augmenting paths S, the worst-case complexity $W(n,m)$ of procedure *FordFulkerson* can also be arbitrarily large.

19.30 Using the Edmonds-Karp algorithm, find a maximum flow f and a minimum cut Γ in the following capacitated network N.

$N =$ 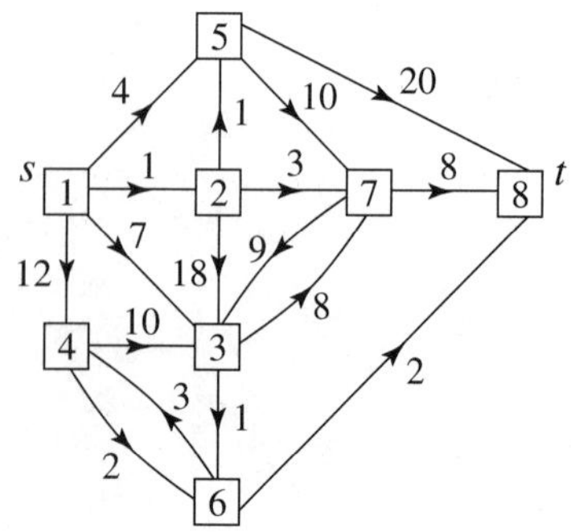

19.31 Design and analyze an algorithm for computing a set of paths $\mathscr{P}$ and associated positive integers $\{\lambda_P : P \in \mathscr{P}\}$ satisfying (19.4.11).

19.32 Derive Menger's Theorem for undirected graphs (Corollary 19.4.12) from Menger's Theorem for digraphs (Theorem 19.4.11).

19.33 Using the Edmonds-Karp algorithm, find a maximum matching in the following bipartite graph G by computing a maximum flow in the associated capacitated network.

$G =$

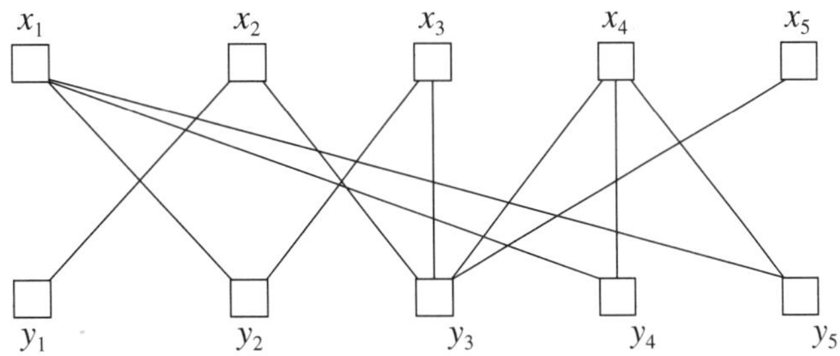

19.34 Derive Konig's Theorem, which states that the size of a maximum matching in a bipartite graph G equals the size of a minimum cover (a *cover* is a set of vertices such that every edge of G is incident with at least one vertex in the cover) from Menger's Theorem for digraphs (Theorem 19.4.11).

Section 19.5 A Monte Carlo Parallel Algorithm for Computing a Perfect Matching

19.35 Prove the second part of Theorem 19.5.2 using the first part of Theorem 19.5.2.

19.36 **a.** State and prove a generalization of the Isolating Lemma, where the integer weights are chosen uniformly and independently from $[1,t]$, where $t \geq 2m$.

b. Design and analyze a version of *MCPerfectMatch* that utilizes the generalized Isolating Lemma given in part (a).

19.37 Design an RNC algorithm similar to *MCPerfectMatch* for finding a maximum-size matching in a graph G.

Section 19.6 Closing Remarks

19.38 Design and analyze an NC parallel algorithm for finding the strongly connected components of a digraph.

20

NP-Complete Problems and the Class NC

It is a curious phenomenon that the (worst-case) complexity of sequential algorithms for most of the commonly encountered problems in computer science fall into the following two main categories:

1. bounded above by low degree polynomials,
2. bounded below by super-polynomial functions. (A function f is *super-polynomial* if $f \in \Omega(n^k)$ for *all* integers $k \geq 1$.)

Problems in graph theory illustrate this situation well. For example, we have seen $O(n^2)$ and $O(n^3)$ algorithms, respectively, for the problems of finding minimum cost spanning trees and finding the shortest paths between every pair of vertices in weighted graphs, whereas the best-known algorithms for the Hamiltonian cycle problem and the vertex-coloring problem have super-polynomial complexity. On the other hand, no one has been able to show that super-polynomial lower bounds for these problems exist (that is, lower bounds on the complexity of *any* algorithms for these problems). Problems such as the Hamiltonian cycle and graph coloring problems (or, more precisely, their decision versions as defined in the next section) belong to a large class of fundamental decision problems called *NP-complete problems* (see Section 20.3). It turns out that if any NP-complete problem is solvable by a polynomial (time) algorithm, then they all are. It has long been conjectured that no polynomial algorithm exists for any of the NP-complete problems. This conjecture is arguably the most important open question in theoretical computer science.

In this chapter we give a brief introduction to the class of NP-complete problems. Our treatment will be somewhat intuitive and informal. In particular, we will not establish the formal machinery necessary to prove the famous result of S. A. Cook that NP-complete problems exist, but refer the interested student to the references for a completely rigorous treatment.

We also expand our earlier discussion of the class NC and introduce the notion of P-completeness. P-completeness can be viewed as the analog for parallel computation of NP-completeness for sequential computation.

20.1 The Classes P and NP

We consider a problem to be *tractable* if it can be solved by a worst-case polynomial (time) algorithm. Problems having superpolynomial worst-case complexity are considered *intractable*. Throughout this chapter when we use the term *computing time* (or *complexity*) we will always mean the worst-case computing time (complexity).

Superpolynomial algorithms are computationally infeasible to implement in the worst case. Even though we have called a problem tractable if it can be solved by a polynomial algorithm, such an algorithm may still be computationally infeasible if it has complexity $\Omega(n^k)$ where k is large. For example, an algorithm having complexity n^{64} will not finish in our lifetime even for $n = 2$. Nevertheless, it is standard in the theory of algorithms to regard problems solved by polynomial algorithms as tractable, and it becomes important to identify those problems. Besides, if a polynomial algorithm has been shown to exist for a problem (even one of high degree), then there may be hope that a more practical polynomial algorithm of relatively small degree can be found for the problem.

In the theory of complexity, it is convenient and useful to restrict attention to decision problems, that is, problems whose solutions simply output yes or no (0 or 1). Let P denote the class of decision problems that are solvable by algorithms having polynomial (worst-case) complexity.

20.1.1 Input Size

When considering membership in P, we must be very careful about how input size is measured. For example, consider the problem of determining whether a given integer n is prime. A simple testing of all possible divisors between 2 and $\sqrt{n}$ yields a sublinear algorithm if n itself is taken as the input size. However, a more realistic measure of the input size is the number of digits of n with respect to a given base $b \geq 2$, in which case the input size is given by $m = \lceil 1 + \log_b n \rceil \in \Theta(\log n)$. With this measure of input size, the inefficiency of the simplistic algorithm for prime testing is revealed: It has complexity $\sqrt{n} \approx b^{m/2}$, which is exponential in the input size m. (Incidentally, it is not known whether prime testing is in P.)

More generally, when considering a decision problem having input parameters that are numeric quantities, the size of these numeric quantities is taken into account when measuring input size. We usually use the number of digits in the binary representation of the sum of the appropriate sizes of these numeric quantities. For example, consider an input $(w_1, w_2, \ldots, w_n)$, $(v_1, v_2, \ldots, v_n)$, and C to the 0/1 Knapsack problem. If we let $m = \Sigma_{i=1}^{n}(w_i + v_i)$ and d denote the number of digits of m, then the size of this input is $n + d \in \Theta(n + \log m)$.

20.1.2 Optimization Problems and Their Decision Versions

Decision problems occur naturally in many contexts. For example, an important decision problem in graph theory is to determine whether a clique of size k occurs in a given input graph. In mathematical logic, a fundamental decision problem is whether the Boolean variables in a given Boolean formula can be assigned truth values that make the entire formula true. In addition to such decision problems that are of interest in their own right, optimization problems usually give rise to associated decision problems. Moreover, restricting attention to the decision version of an optimization problem sometimes can be done without loss of generality (up to polynomial factors). In other words, the decision version of an optimization problem may have the property that a polynomial solution to the optimization problem exists if, and only if, a polynomial solution to the associated decision problem exists (the "only if" part is always true).

To illustrate, consider the optimization problem of determining the chromatic number $\chi = \chi(G)$ of a graph G on n vertices, that is, finding the minimum number of colors necessary to (properly) color the vertices of a graph G such that no two adjacent vertices get the same color. Given an integer k, the associated decision problem asks whether a proper k-coloring of G exists. Clearly, a solution to the optimization problem immediately implies a solution to the decision problem: A proper k-coloring exists if, and only if, $k \geq \chi(G)$. On the other hand, a polynomial solution to the decision problem also implies a polynomial solution to the original problem: Simply call the decision problem for $k = 1, 2, \ldots$ until a *yes* is returned (which will happen after at most n calls).

As another example, consider the 0/1 Knapsack problem with input weights $w_1, w_2, \ldots, w_n$, values $v_1, v_2, \ldots, v_n$, and capacity C. The decision version of this problem adds an integer input parameter k and asks whether a collection of the objects can be placed in the knapsack whose total value is at least k. Here again, it is obvious that the ordinary version of the 0/1 Knapsack problem yields a solution of the decision version with no extra work. However, in the case of noninteger (rational) weights and values, there is no obvious polynomial number of decision problems whose answers yield the optimal solution to the 0/1 Knapsack problem. The situation improves if we restrict attention to the case of *integer* weights and values. Then each possible solution has an integer value, with the largest possible value equal to $m = v_1 + v_2 + \cdots + v_n$. Unlike the graph-coloring problem, we cannot simply call the decision problem for $k = m, m - 1, \ldots$ until a *yes* is returned, since this will take an exponential number (with respect to input size) of calls in general. However, using a binary search strategy, by making $\log_2 m$ calls to the decision problem for suitable values of k we can determine the optimal solution in polynomial time (see Exercise 20.1).

Remark It is not known in general whether a polynomial solution to the 0/1 Knapsack decision problem implies a polynomial solution to the 0/1 Knapsack (optimization) problem.

20.1.3 The Class NP

Informally, the class NP (nondeterministic polynomial) consists of those decision problems for which *yes* instances can be solved in polynomial time by a nondeterministic algorithm. A *nondeterministic algorithm* is an algorithm that for any input instance I is allowed to make guesses to produce a structure S that will serve as a "certificate" for whether I is a *yes* instance or a *no* instance. The structure S can be thought of as providing a *candidate solution* for instance I. For example, if the decision problem is whether a graph can be properly colored using k colors for a given instance graph G, a nondeterministic algorithm would simply guess a color from $\{1,2, \ldots, k\}$ for each vertex. A nondeterministic algorithm for a decision problem contains two phases: In phase 1 nondeterministic choices (guesses) can be made, but phase 2 is deterministic.

1. For a given instance I, a certificate S is produced using nondeterminism (guessing).
2. A deterministic verification algorithm that, when given inputs I and S, outputs a "yes" or "no" to the question of whether I is a *yes* instance.

For a nondeterministic algorithm A to "solve" a given decision problem, the following two correctness conditions must be satisfied:

a. If I is a *yes* instance, then A must be able to produce a certificate S in phase 1 that causes the verification algorithm to return "yes" when given inputs I and S.

b. If I is a *no* instance, then A cannot produce a certificate S in phase 1 that causes the verification algorithm to return "yes" when given inputs I and S.

Condition (a) can be interpreted as saying that if there are solutions to the problem, then by guessing "right" during phase 1 the algorithm can produce a solution to the problem.

The class NP consists of those decision problems for which there exists a nondeterministic algorithm solving the problem where phases 1 and 2 are performed in polynomial time. Note that $P \subseteq NP$, since if A is a polynomial algorithm for a decision problem, then a nondeterministic algorithm for the problem is obtained by guessing any succinct (computable in polynomial time) certificate S for a given input I in phase 1, and then using A in phase 2 to solve the problem instance I completely ignoring S.

20.1.4 Some Sample Problems in NP

We illustrate the definition of NP with some sample problems in NP. In each of these examples, it will be obvious that if a *yes* instance is input, then a solution can be produced by the right collection of guesses in the production step of a nondeterministic algorithm. Our first example is in P, whereas the other examples are conjectured not to be in P.

List Searching Given an unordered list $L[1{:}n]$ and an element X, does X occur in the list?

The best deterministic algorithm for answering this question has complexity $\Theta(n)$. Consider the nondeterministic algorithm that guesses a position $i \in \{1,2, \ldots, n\}$ in the list and then returns the result of comparing X to the list element $L[i]$ (in time

O(1)). If X occurs in the list, then clearly the algorithm can correctly guess a position of such an occurrence, and the comparison step will result in the return of a "yes" answer. Moreover, no guess will result in the return of a "yes" answer if X is not in the list.

Clique Given the graph $G = (V,E)$ and the integer k, does there exist a clique of size k in G?

Recall that a clique V' on k vertices has the property that $uv \in E$ for all vertices u and v in V'. A nondeterministic algorithm guesses a candidate solution by choosing a set of k vertices. The verification that the set of vertices is distinct and forms a clique can be done in polynomial time.

Sum of Subsets Given the input integers $\{s_1, s_2, \ldots, s_n\}$, and the target sum C, is there are subset of the integers whose sum equals C?

The nondeterministic algorithm guesses a candidate solution by choosing a subset of the integers $\{s_{i_1}, s_{i_2}, \ldots, s_{i_m}\}$. The verification that $s_{i_1} + s_{i_2} + \cdots + s_{i_m} = C$ can be done in polynomial time.

Hamiltonian Cycle Given the graph $G = (V,E)$, does G have a Hamiltonian cycle?

A nondeterministic algorithm guesses a Hamiltonian cycle by choosing a sequence of vertices $v_1,v_2, \ldots, v_n,v_1$. Verification that the sequence of vertices is a Hamiltonian cycle can be done in polynomial time.

Showing that the problems in this section were in NP was relatively easy. However, there are problems whose membership in NP is quite difficult to establish. A notable example is prime testing, which asks whether a given integer n is prime. Note that a *yes* instance means that the answer is *no* to the question of whether n has factors, and the nonexistence of factors has no obvious certificate. However, it was shown by Pratt that a succinct certificate can be produced by a nondeterministic algorithm that can be verified in polynomial time (see Exercise 20.2).

The question of whether P = NP can be viewed as asking whether adding nondeterminism allows us to solve a class of problems in polynomial time that are not otherwise so solvable. It seems reasonable that adding the luxury of making good guesses should really help, and most computer scientists believe that P ≠ NP. Reinforcement of this belief can be found in the fact that there are hundreds of varied problems in NP that have resisted all attempts to find polynomial solutions.

20.2 Reducibility

One reason that polynomial complexity is such a convenient measure is that polynomials have nice closure properties. For example, the composition $f(x) = g(h(x))$ of two polynomials g and h is yet another polynomial; in other words, a polynomial of a polynomial is a polynomial. Polynomials have other nice closure properties: The sum (difference, product) of a polynomial and a polynomial is a polynomial. These closure properties of polynomials, together with the fact that polynomial complexity is the standard measure of efficiency in the theory of complexity, allows a nice description of when a given decision problem A is not harder to solve than another decision problem B; namely, the solution to B should yield the solution to A with at most a polynomial factor of additional work. We make this precise in the following definition.

DEFINITION 20.2.1 Given two decision problems A and B, we say that A is (polynomially) *reducible* to B, denoted $A \propto B$, if there is a mapping f from the inputs to problem A to the inputs to problem B, such that

1. f can be computed in polynomial time (that is, there is a polynomial algorithm that for input I to problem A outputs the input $f(I)$ to problem B), and
2. the answer to a given input I to problem A is *yes* if, and only if, the answer to the input $f(I)$ to problem B is *yes*.

In particular, condition (1) of Definition 20.2.1 implies that there exists a polynomial $p(n)$ such that $|f(I)| \in O(p(|I|))$, for any input I to algorithm A, where $|I|$ denotes the size of I.

The following two propositions are easy consequences of the closure properties of polynomials.

PROPOSITION 20.2.1 The relation $\propto$ is transitive; that is, if $A \propto B$ and $B \propto C$ then $A \propto C$.

PROPOSITION 20.2.2 If $A \propto B$ and B has polynomial complexity, then so does A.

We now illustrate the notion of reducibility by several examples. Further examples are given in Section 20.4.

Example 20.2.1: Hamiltonian Cycles $\propto$ Traveling Salesman Given the graph $G = (V,E)$, $|V| = n$, define the following weighting w on K_n on the complete graph on n vertices: $w(e) = 1$ if $e \in E$, otherwise $w(e) = 2$. Clearly, w can be computed in polynomial time. We now map G to the instance $f(G) = (K_n,w,k = n)$ of Traveling Salesman (so that $p(n) = n^2$). Clearly, G has a Hamiltonian cycle if, and only if, K_n with weighting w has a tour of cost no more than n. ■

Example 20.2.2: Clique $\propto$ Vertex Cover $\propto$ Independent Set $\propto$ Clique Since the relation $\propto$ is transitive, Example 20.2.2 says that the three problems are each polynomial reducible to one another. Given the graph $G = (V,E)$, recall that the complement $\overline{G} = (V, \overline{E})$ is the graph with the same vertex set V, such that $e \in \overline{E}$ if, and only if, $e \notin E$. Given G, clearly the complement $\overline{G}$ can be computed in polynomial time. The polynomial reductions are consequences of the following proposition, whose proof is left as an exercise. ■

PROPOSITION 20.2.3 Given the graph $G = (V,E)$, then the following three conditions are mutually equivalent conditions on a subset of vertices $V' \subseteq V$.

1. V' is a clique in G.
2. V' is an independent set of vertices in the complement $\bar{G} = (V, \bar{E})$.
3. The set $V - V'$ is a vertex cover of the complement $\bar{G} = (V, \bar{E})$. ❑

The reducibility illustrated in the previous examples was easy to show and not very surprising, since the problems involved were highly related. In Section 20.4 we give examples of disparate problems A and B where $A \propto B$.

The following definition makes precise the notion of two problems being polynomially equivalent to one another.

DEFINITION 20.2.2 Two decision problems A and B are *polynomially equivalent to one another* if $A \propto B$ and $B \propto A$.

Since the relation $\propto$ is clearly reflexive, it follows from Proposition 20.2.1 that the relation of polynomial equivalence is an equivalence relation on the set of decision problems.

20.3 NP-Complete Problems: Cook's Theorem

S. A. Cook raised a fundamental question: Is there a problem in NP that is as hard (up to polynomial factors) as every other problem in NP? In other words, is there a problem C in NP such that if A is *any* problem in NP, then $A \propto C$? If such a problem exists, then it is called *NP-complete*. (Note that all NP-complete problems are automatically polynomially equivalent to one another.) In one of the most celebrated results in theoretical computer science, Cook showed in 1971 that the satisfiability problem for Boolean expressions (as defined presently) is NP-complete. The following proposition follows easily from the transitivity of the relation $\propto$.

PROPOSITION 20.3.1 Suppose A is in NP, and B is NP-complete. If $B \propto A$, then A is NP-complete. ❑

In the time since Cook's result appeared, hundreds of NP-complete problems have been identified. Despite considerable effort, no polynomial algorithm has been found for any NP-complete problem. This lack of success reinforces the belief that $P \neq NP$. The following corollary of Proposition 20.3.1 shows that it is unlikely that a polynomial algorithm for any NP-complete problem will ever be found.

COROLLARY 20.3.2 If any NP-complete problem A also belongs to P, then $P = NP$. ❑

A problem A is *NP-hard* if $B \propto A$ for every problem B in NP. Equivalently, A is NP-hard if $B \propto A$ for some NP-complete problem B. For example, if the decision version of an optimization problem is NP-complete, then the optimization problem itself is NP-hard. Note that the NP-complete problems are precisely those problems in NP that are NP-hard.

To describe Cook's result, we need to define some terms from first-order logic. As with our pseudocode conventions, a *Boolean* (logical) variable is a variable that can only take on two values, true (T) or false (F). Given a Boolean variable x, we denote by $\bar{x}$ the variable has that the value T if, and only if, x has the value F. Boolean variables can be combined using logical operators *and* (denoted by $\wedge$ and called *conjunction*), *or* (denoted by $\vee$ and called *disjunction*), *not* (denoted by $\neg$ and called *negation*), and parenthesization, to form Boolean expressions (formulas). A *literal* in a Boolean formula is of the form x or $\bar{x}$, where x is a Boolean variable. Note that $\neg x$ has the same truth value as $\bar{x}$. A Boolean formula is said to be in *Conjunctive Normal Form* (CNF) if it has the form $C_1 \wedge C_2 \wedge \cdots \wedge C_m$, where each clause C_i is a disjunction of literals $x_1^{(i)} \vee x_2^{(i)} \vee \cdots \vee x_{n(i)}^{(i)}$, $i = 1, \ldots, m$.

A Boolean formula is called *satisfiable* if there is an assignment of truth values to the literals occurring in the formula that makes the formula evaluate to true. For example, the CNF formula

$$(x_2 \vee \bar{x}_3) \wedge (x_1 \vee \bar{x}_2) \wedge (x_1 \vee x_3) \wedge (\bar{x}_1 \vee x_3)$$

is satisfiable ($x_1 = x_2 = x_3 = T$), whereas the CNF formula

$$(x_2 \vee x_3) \wedge (\bar{x}_1 \vee \bar{x}_2) \wedge (x_1 \vee \bar{x}_2) \wedge (x_1 \vee \bar{x}_3) \wedge (\bar{x}_1 \vee \bar{x}_2 \vee \bar{x}_3)$$

is not satisfiable.

The *CNF-Satisfiability* problem (CNF SAT) is the problem of determining whether a given CNF formula is satisfiable. CNF SAT is a fundamental problem in mathematical logic and computer science, with numerous applications.

The input size of a CNF formula can be defined as the total number of literals (counting repetitions) occurring in the formula. Clearly, CNF SAT $\in$ NP, since a linear scan of the formula determines whether a candidate assignment to the literals occurring in the formula results in the formula evaluating to true. CNF SAT was the first example shown to be NP-complete.

THEOREM 20.3.3 Cook **CNF SAT is NP-complete.** ❑

Given any (decision) problem A in NP, there is a nondeterministic algorithm that produces a candidate solution for A that can be checked in polynomial time. Cook proved Theorem 20.3.3 by constructing a CNF formula that, roughly speaking, modeled the action of the nondeterministic algorithm. The length of the formula constructed by Cook is polynomial in the size of an instance I of A, and is satisfiable if, and only if, the problem A is true for I.

20.4 Some Sample NP-Complete Problems

As soon as it was established that CNF SAT is NP-complete, the door was opened to show that a host of other NP-complete problems exist. Proposition 20.3.1 gives a recipe for proving that a problem A is NP-complete, namely:

a. show that A is in NP, and

b. find *any* NP-complete problem B such that $B \propto A$.

Then (a) and (b) imply that A is NP-complete. Condition (a) is usually easy to show, so that (b) becomes the issue. There are cases, however, when (i) is difficult to verify. A notable example is the prime testing problem, which is in NP, but the proof is far from obvious.

Hundreds of natural and practical problems have been shown to be NP-complete. If you are working on a problem that you think might be NP-complete, you can check the vast list of NP-complete problems for one that might resemble your problem. The following subsections provide a small sample list of NP-complete problems. We have already shown that the clique problem and the sum of subsets problems are in NP. For each of the other examples, we will leave it as an exercise to show the problem is in NP. For each of the sample problems in this section, we will give a reduction that shows the problem is NP-complete.

Example 20.4.1: 3-CNF SAT An instance of the 3-CNF SAT problem is a CNF Boolean formula in which every clause contains *exactly* three literals. Since general CNF SAT is in NP, 3-CNF SAT is also in NP. Surprisingly, 3-CNF SAT is NP-complete, even though 2-CNF SAT (each clause consists of two literals) is in P (see Exercise 20.9).

We show that CNF SAT $\propto$ 3-CNF SAT by showing how any CNF formula $I = C_1 \wedge C_2 \wedge \cdots \wedge C_m$ can be mapped onto a 3-CNF formula $f(I)$ with no more $12\,|I|$ literals, such that I is satisfiable if, and only if, $f(I)$ is satisfiable. More precisely, we construct $f(I)$ by replacing each clause C_i, $i = 1, \ldots, m$, by conjunctions of three-literal clauses according to the following three cases.

1. C_i has exactly three clauses.
2. C_i has more than three clauses.
3. C_i has less than three clauses.

In case 1, we leave C_i unaltered. In case 2, suppose C_i contains $k > 3$ literals x_1, $x_2, \ldots, x_k$, so that $C_i = x_1 \vee x_2 \vee \cdots \vee x_k$. We then replace C_i by the $k - 2$ clauses

$$(x_1 \vee x_2 \vee y_1) \wedge (\bar{y}_1 \vee x_3 \vee y_2) \wedge (\bar{y}_2 \vee x_4 \vee y_3) \wedge \cdots \wedge (\bar{y}_{k-3} \vee x_{k-1} \vee x_k),$$

where $y_1, y_2, \ldots, y_{k-3}$ are new Boolean variables. We leave it as an exercise to show that C_i is satisfiable if, and only if, the conjunction of the $k - 2$ clauses replacing C_i is satisfiable.

Now consider case 3. If C_i consists of a single literal x, then we can replace x by $(x \vee y \vee z) \wedge (x \vee y \vee \bar{z}) \wedge (x \vee \bar{y} \vee z) \wedge (x \vee \bar{y} \vee \bar{z})$, where y and z are new Boolean variables. If C_i consists of the two-literal clause $x_1 \vee x_2$, then we replace C_i by $(x_1 \vee x_2 \vee y) \wedge (x_1 \vee x_2 \vee \bar{y})$, where y is a new Boolean variable. In both cases, it follows easily that C_i is satisfiable if, and only if, the conjunction of the clauses replacing it are satisfiable. ■

Example 20.4.2: Clique This example establishes a connection between Boolean formulas and graph theoretic problems, thereby extending the realm of NP-complete problems into the latter domain. We show that CNF SAT $\propto$ Clique. Consider any CNF formula $I = C_1 \wedge C_2 \wedge \cdots \wedge C_m$. Writing clause C_i as $y_1 \vee \cdots \vee y_j \vee \overline{y_{j+1}} \vee \cdots \vee \overline{y_k}$ (the j and k depend on i), where $y_1, \ldots, y_j \in \{x_1, \ldots, x_n\}$ and $\overline{y_{j+1}} \vee \cdots \vee \overline{y_k} \in \{\overline{x_1}, \ldots, \overline{x_n}\}$, let $V_i = \{(y_1,i), \ldots, (y_j,i), (\overline{y_{j+1}},i), \ldots, (\overline{y_k},i)\}$, $i = 1, \ldots, m$. We map I to the graph $G = f(I)$ as follows. The vertex set V of G is the union of the V_i's, $i = 1, \ldots, m$. We define two vertices (a,i) and (b,j) to be adjacent if, and only if, $i \neq j$ and the literal a is not the negative of the literal b. For example, if $(x_1,3)$, $(x_5,7)$, $(\overline{x_8},2)$, $(x_2,4)$, $(\overline{x_3},1)$, $(\overline{x_8},1)$, $(x_6,2)$, and $(\overline{x_6},5)$ are vertices of G, then $\{(x_1,3),(x_5,7)\}$ and $\{(\overline{x_8},2),(x_2,4)\}$ are edges of G but $\{(\overline{x_3},1),(\overline{x_8},1)\}$ and $\{(x_6,2),(\overline{x_6},5)\}$ are not. Note that G can be constructed in polynomial time.

Claim I is satisfiable if, and only if, G has a clique of size m.

Proof Suppose that G has a clique W of size m. By definition of adjacency in G, it follows that, for each clause C_i in I, there is exactly one vertex in W of the form (a,i). If a is the positive literal x_q, then assign x_q the value T. If a is the negative literal $\overline{x_q}$, then assign $\overline{x_q}$ the value T (that is, x_q is F). This assignment is well defined, since by definition of adjacency in G, none of the literals (in the second component) of a vertex is the negative of another. Moreover, this assignment makes each of the clauses C_i evaluate to T, so that I evaluates as true.

Now suppose that I is satisfiable. Let $Z = \{z_1, \ldots, z_n\}$ denote the n literals from $\{x_1, \ldots, x_n\} \cup \{\overline{x_1}, \ldots, \overline{x_n}\}$ that have the value T. Since each clause C_i must have the value T, $i = 1, \ldots, m$, C_i contains at least one literal z_i from Z. But, by the definition of G, (z_i,i) is a vertex of G. Further, the set of vertices $\{(z_i,i) : i \in \{1, \ldots, m\}\}$ forms a clique of size m. ■

Example 20.4.3: Vertex Cover and Independent Set That these problems are both NP-complete follows from Example 20.4.2 and Proposition 20.2.3. ■

Example 20.4.4: Three-Dimensional Matching Recall that the perfect matching problem in a bipartite graph interprets the classical Marriage problem, which asks whether marriages can be arranged between a set of n boys and a set of n girls so that each boy marries a girl that he knows, and no girl marries more than one boy. The three-dimensional matching problem generalizes the Marriage problem by adding n houses to the mix. Not only do we seek to arrange marriages, we also wish to match each married couple with a house from a specified subcollection of the houses that are available to the couple, and so that no two couples share the same house. Adding houses turns a problem solved by an $O(n^3)$ algorithm into an NP-complete problem.

More formally, given three sets H, B, and G (houses, boys, and girls, respectively), each having the same cardinality $|H|$, and a subset $M \subseteq H \times B \times G$, the three-dimensional matching problem asks whether there is a subset (matching) $M' \subseteq M$ of cardinality $|M'| = |H|$, such that none of the triples in M' agree in any coordinate. We refer to the triples in M' as *households*. The three-dimensional matching problem is in NP, since a nondeterministic algorithm guesses a set of $|H|$ triples, and the verification that no two triples agree in any coordinate can be done in polynomial time. We now show that the three-dimensional matching problem is NP-complete by showing that CNF SAT reduces to it.

Suppose $I = C_1 \wedge C_2 \wedge \cdots \wedge C_m$ is any instance of CNF SAT, and suppose I contains the Boolean variables $x_1, x_2, \ldots, x_n$. We will construct an instance of the three-dimensional matching problem H,B,G, $M \subseteq H \times B \times G$, where $|H| = |B| = |G| = 2mn$, such that a matching $M' \subseteq M$ exists if, and only if, I is satisfiable. H consists of m copies of the literals $\{x_1, x_2, \ldots, x_n\}$, $\{\bar{x}_1, \bar{x}_2, \ldots, \bar{x}_n\}$. We use subscripts and superscripts to denote the elements in H, where subscripts reference literals, and superscripts reference clauses in I; that is

$$H = \{x_i^{(j)}, \bar{x}_i^{(j)} : 1 \le i \le n,\ 1 \le j \le m\}.$$

We now create sets B and G of $2mn$ boys and $2mn$ girls, respectively. The sets B and G will be divided into three classes, and marriages are restricted to take place only between boys and girls in the same class. Thus, the division of the boys and girls induces a division of M into three classes of households, which will be called "truth setting," "satisfaction testing," and "remaining" households, respectively. As the names imply, truth setting households force an assignment of truth values to the Boolean variables $x_1, x_2, \ldots, x_n$, whereas satisfaction testing households determine whether this truth assignment makes the formula I true. The remaining households in M are made up of a set of additional couples created merely to occupy the remaining houses not covered by the households in the first two classes. We now describe the division of the boys and girls into three classes together with the resulting three classes of households. ■

Truth Setting Households The truth setting households will be divided into n components T_i, $i = 1, \ldots, n$, where each component holds m households. For each i, $1 \le i \le n$, we create a set of m boys and m girls

$$B_{T_i} = \{b_i^{(1)}, b_i^{(2)}, \ldots, b_i^{(m)}\},\ G_{T_i} = \{g_i^{(1)}, g_i^{(2)}, \ldots, g_i^{(m)}\},\ i = 1, \ldots, n.$$

Marriages are restricted so that a boy in the ith component B_{T_i} must marry a girl in the ith component G_{T_i}. The marriages are further restricted within each component T_i to take place between people who are adjacent when arranged in a circle, as in Figure 20.1.

In particular, there are only two marriage schemes possible in T_i, a clockwise scheme, $\{b_i^{(1)}, g_i^{(1)}\}$, $\{b_i^{(2)}, g_i^{(2)}\}$, ... , $\{b_i^{(m)}, g_i^{(m)}\}$, and a counterclockwise scheme, $\{b_i^{(1)}, g_i^{(m)}\}$, $\{b_i^{(m)}, g_i^{(m-1)}\}$, $\{b_i^{(m-1)}, g_i^{(m-2)}\}, \ldots, \{b_i^{(2)}, g_i^{(1)}\}$.

In each component T_i, only one house will be made available to a given couple. If girl $g_i^{(j)}$ marries boy $b_i^{(j)}$, then she must live in house $\bar{x}_i^{(j)}$. Otherwise (girl $g_i^{(j)}$ marries boy $b_i^{(j+1)}$, where we set $b_i^{(m+1)} = b_i^{(1)}$), she must live in house $x_i^{(j)}$. The housing restrictions are illustrated in Figure 20.2, which helps explain the "truth setting" nomenclature. We see from Figure 20.2 that any matching M' must intersect the households in the ith component T_i in precisely one of two ways: Either $M' \cap T_i$ consists of the set $T_i(t)$ of shaded households, or $M' \cap T_i$ consists of the set $T_i(f)$ of

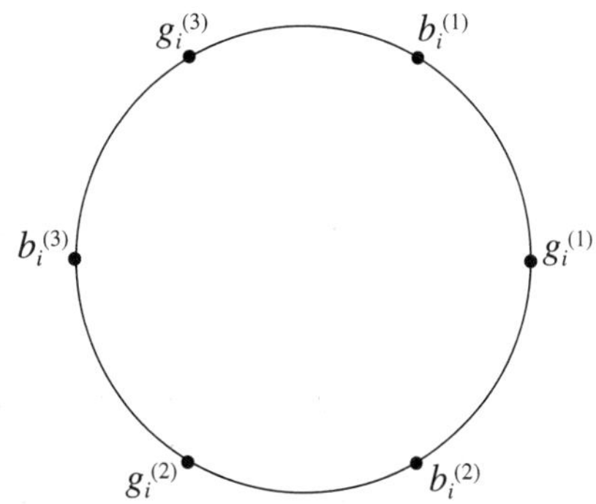

Figure 20.1 In each component, marriages must take place between adjacent people in the given circle arrangement, as illustrated for $m = 3$.

unshaded households. Thus, the component T_i forces a matching M' to set a truth value for x_i, $i = 1, \ldots, n$, where $M' \cap T_i = T_i(t)$ is considered as setting x_i to T, whereas $M' \cap T_i = T_i(f)$ is considered as setting x_i to F.

In summary, the truth setting households in M consist of n components T_i, $i = 1, \ldots, n$, where each T_i is the union of two sets of triples:

Figure 20.2 The households of T_i corresponding to $m = 3$. Any matching must intersect T_i precisely in either the shaded households $T_i(t)$ or the unshaded households $T_i(f)$.

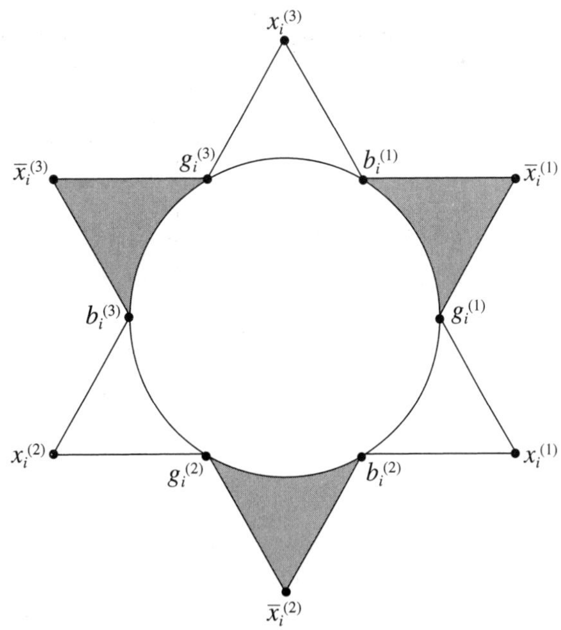

$$T_i(t) = \{(\bar{x}_i^{(j)}, b_i^{(j)}, g_i^{(j)}) : 1 \le i \le n, 1 \le j \le m\},$$
$$T_i(f) = \{(x_i^{(j)}, b_i^{(j+1)}, g_i^{(j)}) : 1 \le i \le n, 1 \le j < m\} \cup (x_i^{(j)}, b_i^{(1)}, g_i^{(m)}).$$

Moreover, no household in M other than T_i contains a boy $b_i^{(j)}$ or a girl $g_i^{(j)}$, $1 \le i \le n$, $1 \le j \le m$.

Satisfaction Testing Households We now create m boys $s_b^{(j)}$ and m girls $s_g^{(j)}$, respectively, $j = 1, \ldots, m$, and match them by prearranged marriages $\{\{s_b^{(j)}, s_g^{(j)}\} : j = 1, \ldots, m\}$. Moreover, the only houses available to couple $\{s_b^{(j)}, s_g^{(j)}\}$ are literals (x_i or $\bar{x}_i$) that appear in clause C_j. Thus, the satisfaction testing households in M consist of the triples:

$$S_j = \{(x_i^{(j)}, s_b^{(j)}, s_g^{(j)}): x_i \in C_j, i = 1, \ldots, n\} \cup \{(\bar{x}_i^{(j)}, s_b^{(j)}, s_g^{(j)}): x_i \in C_j, i = 1, \ldots, n\}.$$

No other households in M contain a boy $s_b^{(j)}$ or a girl $s_g^{(j)}$, $j = 1, \ldots, m$. Hence, any matching $M' \subseteq M$ must contain exactly one household from each S_j, $j = 1, \ldots, m$. Moreover, if the household chosen in S_j uses a house corresponding to the variable x_i, then that house must not be included in $M' \cap T_i$. In other words, we must choose a house whose truth setting for x_i as determined by M' satisfies clause C_j. *Thus, a matching $M' \subseteq M$ cannot exist unless I is satisfiable.*

Remaining Households Any matching M' uses nm houses to accommodate the truth setting households and m houses to accommodate the satisfaction testing households. Thus, there are $m(n-1)$ houses that are not covered by the truth setting or satisfaction testing households making up a *partial matching* M'. We create a set of $m(n-1)$ boys $r_b^{(j)}$ and $m(n-1)$ girls $r_g^{(j)}$, respectively, $j = 1, \ldots, m(n-1)$, so that they can occupy the remaining $m(n-1)$ houses not covered by such a partial matching M', and thereby extend M' to a matching. These boys and girls are matched by prearranged marriages consisting of couples $\{\{r_b^{(j)}, r_g^{(j)}\} : j = 1, \ldots, m(n-1)\}$. While these couples had no choice in their marriages, the set M places no housing restrictions on them. Thus, the remaining households R in M consist of the triples:

$$R = \{(x_i^{(j)}, r_b^{(k)}, r_g^{(k)}) : i = 1, \ldots, n, \quad j = 1, \ldots, m, \quad k = 1, \ldots, m(n-1)\}$$
$$\cup \{(\bar{x}_i^{(j)}, r_b^{(k)}, r_g^{(k)}) : i = 1, \ldots, n, \quad j = 1, \ldots, m, \quad k = 1, \ldots, m(n-1)\}.$$

Given the instance I of CNF SAT, the set $H \times B \times G$ and the subset of households $M \subseteq H \times B \times G$ can clearly be constructed in polynomial time. We have already noted that a matching $M' \subseteq M$ cannot exist unless I is satisfiable. We complete the proof that CNF SAT $\propto$ three-dimensional matching by showing that if I is satisfiable, then a matching $M' \subseteq M$ exists. Consider a truth assignment to the variables $x_1, x_2, \ldots, x_n$ that makes I true. For each clause C_j, we choose a literal l_j occurring in C_j having the value T. Such a choice must be possible since I is assumed true with the given truth assignment. We then set

$$M' = \Big(\bigcup_{x_i = T} T_i(t)\Big) \cup \Big(\bigcup_{\bar{x}_i = T} T_i(t)\Big) \cup \Big(\bigcup_{j=1}^{m} \{(l_j, s_b^{(j)}, s_g^{(j)})\}\Big) \cup R',$$

where R' is a suitably constructed collection of remaining households that includes all the couples $\{r_b^{(k)}, r_g^{(k)}\}$, $k = 1, \ldots, m(n-1)$, and all the remaining houses that were not covered by the truth setting or satisfaction testing households in M'. We leave it as an exercise to verify that R' can be constructed and that the resulting set M' is a matching. ■

Example 20.4.5: 3-Exact Cover Given a family $F = \{S_1, S_2, \ldots, S_n\}$ of n subsets of $S = \{x_1, x_2, \ldots, x_{3m}\}$, each of cardinality three, the *3-exact cover* problem asks if there is a subfamily $\{S_{i_1}, S_{i_2}, \ldots, S_{i_m}\}$ of m subsets of F that covers S, that is, $S = \bigcup_{j=1}^{m} S_{i_j}$. The 3-exact cover problem is in NP, since a nondeterministic algorithm guesses a subfamily, and the verification that the subfamily covers S can be done in linear time. (We can use characteristic bit vectors of length $3m$ to represent subsets of F.)

Note that three-dimensional matching $\propto$ 3-exact cover. In fact, three-dimensional matching is actually just a special case of 3-exact cover where we ignore the ordering of the subsets. More precisely, we simply associate the instance $M \subseteq H \times B \times G$ of the three-dimensional matching problem with the instance $S = H \cup B \cup G$ and $F = \{\{h,b,g\} : (h,b,g) \in M\}$ of the 3-exact cover problem.

Example 20.4.6: Sum of Subsets We now show that 3-exact cover $\propto$ sum of subsets. Given an instance $F = \{S_1, S_2, \ldots, S_n\}$, $S = \{x_1, x_2, \ldots, x_{3m}\}$ of 3-exact cover, as noted previously we can consider the sets in F as represented by their characteristic bit vectors of length $3m$. For example, if $m = 3$ and $S_1 = \{x_2, x_3, x_8\}$, then S_1 is represented by $(0,1,1,0,0,0,0,1,0)$. We map an instance of 3-exact cover to an instance of sum of subsets by interpreting each characteristic vector as an integer in the base-$(n + 1)$ system. Thus, for each $S_j \in F$, we associate with S_j the integer $s_j = \Sigma_{x_i \in S_j} (n+1)^{i-1}$. We let C be the integer corresponding to the bit vector of all ones; that is, $C = \Sigma_{i=0}^{3m-1} (n+1)^i$. We leave it as an exercise to show that F, S is a *yes* instance of 3-exact cover if, and only if, $s_1, s_2, \ldots, s_n$ and target sum C is a *yes* instance of sum of subsets. ■

Example 20.4.7: Graph Coloring We show that 3-CNF SAT $\propto$ graph coloring. Suppose $I = C_1 \wedge C_2 \wedge \cdots \wedge C_m$ is an instance of 3-CNF SAT involving the Boolean variables $\{x_1, x_2, \ldots, x_n\}$. We may assume that $n \geq 4$, since otherwise the satisfiability of I can be checked in polynomial time. We construct an instance $G = (V,E)$, $k = n + 1$, of graph coloring as follows.

$$V = \{x_1, x_2, \ldots, x_n\} \cup \{\bar{x}_1, \bar{x}_2, \ldots, \bar{x}_n\} \cup \{y_1, y_2, \ldots, y_n\} \cup \{c_1, c_2, \ldots, c_m\},$$

$$\begin{aligned} E = \{x_i\bar{x}_i &: 1 \leq i \leq n\} \cup \{y_i y_j : 1 \leq i < j \leq n\} \cup \{x_i y_j : 1 \leq i \neq j \leq n\} \\ &\cup \{\bar{x}_i y_j : 1 \leq i \neq j \leq n\} \cup \{x_i c_j : x_i \notin C_j, 1 \leq i \leq n, 1 \leq j \leq m\} \\ &\cup \{\bar{x}_i c_j : x_i \notin C_j, 1 \leq i \leq n, 1 \leq j \leq m\}. \end{aligned}$$

Clearly, G can be constructed from I in polynomial time. We leave it as an exercise to show that I is satisfiable if, and only if, the graph G is $(n + 1)$-colorable. ■

20.5 The Class co-NP

Given a decision problem Π, the complementary problem $\overline{\Pi}$ is the decision problem with the same instances, but where an instance I of the problem Π is true if, and only if, I is false for $\overline{\Pi}$. For example, suppose $\overline{\Pi}$ is the decision problem "Does a graph G contain a Hamiltonian cycle?" The complementary decision problem $\overline{\Pi}$ asks: "Is there *no* Hamiltonian cycle in G?" A certificate for a Hamiltonian cycle is easy to verify, but what would a certificate be for the complementary problem? One would be an enumeration of all possible Hamiltonian cycles, but there are far too many candidates to check in polynomial time. In fact, it is believed that the non-Hamiltonian problem does not belong to NP.

Clearly, the complement $\overline{\Pi}$ of a problem in P also belongs to P. Indeed, a polynomial algorithm for Π also serves as a polynomial algorithm for $\overline{\Pi}$ by outputting *yes* for an instance I of $\overline{\Pi}$ if, and only if, the algorithm outputs *no* for instance I of problem Π. However, the same argument does not work for problems in NP. Given a problem $\Pi \in$ NP, a *yes* instance for Π corresponds to a *no* instance for $\overline{\Pi}$, and we can't be assured of the existence of certificates that can be verified in polynomial time for *no* instances of problems in NP.

The class *co-NP* is the class of problems that are complements of problems in NP. As an example, consider again the problem of prime testing, which asks whether a given integer n is prime. Prime testing is in co-NP, since its complement belongs to NP. (A nondeterministic algorithm merely guesses a factor, and it can be checked in polynomial time whether the factor divides n evenly.)

It turns out that prime testing is in NP, so that it is in NP $\cap$ co-NP. A natural question is whether co-NP $\neq$ NP. Most computer scientists believe the answer is yes and conjecture that any NP-complete problem provides the answer. The following theorem provides the motivation for this belief.

THEOREM 20.5.1 Let Π be any NP-complete problem. If the complement $\overline{\Pi}$ belongs to NP, then NP = co-NP. ❑

We leave the proof of Theorem 20.5.1 as an exercise.

The Venn diagram in Figure 20.3 shows the conjectured relationships between P, NP, and co-NP. However, there still is the (very unlikely) possibility that P = NP, in which case the whole diagram would collapse to P.

20.6 Approximation Problems

Even though it would be ideal to find an optimal solution to a given optimization problem such as TSP, there are many situations where we would be satisfied with suitably good approximations to the optimal solution. There are various ways to measure the degree of approximation of a solution to the optimal solution. We restrict attention here to approximations that are within multiplicative factors of the optimal solution. More precisely, let $C^*(n)$ denote the value of an optimal solution

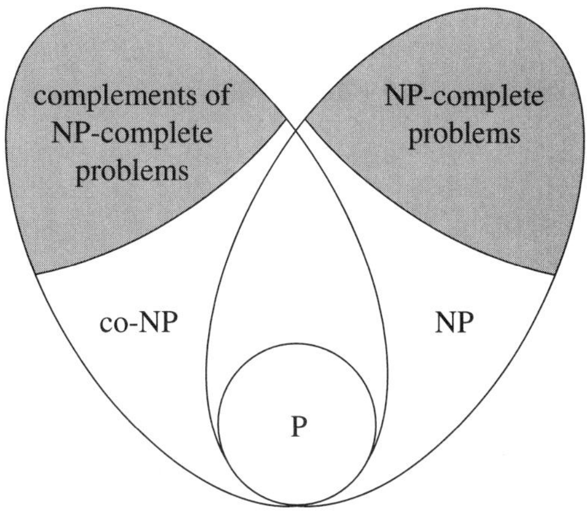

Figure 20.3 Conjectured relationships between P, NP, and co-NP

to a given optimization problem for an input of size n (we assume $C^*(n) > 0$). If the problem is a maximization problem, then for a given real-valued positive function $\rho(n)$, we say that a solution $C(n)$ is a $\rho(n)$-*approximation* to $C^*(n)$ if

$$\frac{C^*(n)}{C(n)} \leq \rho(n), \quad n = 1,2, \ldots .$$

If the problem is a minimization problem, then $C(n)$ is a $\rho(n)$-*approximation* to $C^*(n)$ if

$$\frac{C(n)}{C^*(n)} \leq \rho(n), \quad n = 1,2, \ldots .$$

When $\rho(n)$ is a constant ρ, we say that we have a ρ-*approximation* to the optimal solution $C^*(n)$. For example, if we have an algorithm whose output is a 2-*approximation* to $C^*(n)$, then the solution it produces is never more than twice as bad as the optimal solution. For many NP-hard optimization problems it is possible to find polynomial algorithms that are ρ-approximations to the optimal solutions. For other NP-hard optimization problems, it can be shown that polynomial-time ρ-approximations to the optimal solutions cannot exist unless P = NP.

As an illustration, we now design an $O(n^2)$ algorithm for a 2-approximation to the undirected version (restricted to the undirected complete graph K_n) of TSP, where the weights on the edges satisfy the triangle inequality. Given a positive weighting ω on the complete graph K_n, we say that ω satisfies the triangle inequality if for any three vertices u,v,w, we have

$$\omega(uv) \leq \omega(uw) + \omega(wv), \quad u,v,w \in V(K_n).$$

When the triangle inequality holds for the edge weighting, the following simple algorithm generates a tour whose length is no more than twice that of an optimal tour. Starting at an arbitrary initial vertex r, we generate a minimum-cost spanning tree $MC(T)$ rooted at r. $MC(T)$ can be generated by Prim's algorithms in time $O(n^2)$.

Then we perform a preorder traversal of $MC(T)$, where the vertices are visited (first accessed) in the order $v_0 = r, v_1, \ldots, v_{n-1}$. Consider the tour with edges

$$v_0v_1, v_1v_2, \ldots, v_{n-2}v_{n-1}, v_{n-1}v_n, \quad v_0 = v_n = r. \tag{20.6.1}$$

PROPOSITION 20.6.1 If the weighting ω satisfies the triangle inequality, then the cost of the tour given by (20.6.1) is no more than twice the cost of $MC(T)$.

Proof Consider any edge v_iv_{i+1}, $0 \le i \le n-1$, in the tour (20.6.1) that is *not* in $MC(T)$. Then the path in $MC(T)$ joining v_iv_{i+1} is of the form $v_iw_1, w_1w_2, \ldots, w_{j(i)}v_{i+1}$ for suitable intermediate vertices $w_1, w_2, \ldots, w_{j(i)}$. By the triangle inequality, we have

$$\omega(v_iv_{i+1}) \le \omega(v_iw_1) + \omega(w_1w_2) + \cdots + \omega(w_{j(i)}v_{i+1}). \tag{20.6.2}$$

Since each edge uv in $MC(T)$ is traversed exactly twice (in opposite directions) in a preorder traversal, each edge of $MC(T)$ occurs on the right-hand side of (20.6.2) for *exactly one* edge v_iv_{i+1} in the tour (20.6.1). Thus, the sum of the weights of those edges in the tour (20.6.1) that are *not* in $MC(T)$ is bounded above by the cost of $MC(T)$. On the other hand, clearly the sum of the weights of those edges in the tour (20.6.1) that *are* in $MC(T)$ is also bounded above by the cost of $MC(T)$. ■

COROLLARY 20.6.2 If the weighting ω satisfies the triangle inequality, then the cost of the tour given by (20.5.1) is less than twice the cost of an optimal tour.

Proof If we delete any edge from an optimal tour, then we obtain a spanning tree whose cost is less than the cost of $MC(T)$. ■

We illustrate the tour given by (20.6.2) in Figure 20.4 for a sample minimum cost spanning tree. We leave it as an exercise to find the optimal tour.

For the general TSP, it is not hard to show that the existence of a polynomial-time algorithm yielding a ρ-approximation to the optimal tour would imply that a polynomial-time algorithm would exist for the Hamiltonian cycle decision problem (see Exercise 20.20). Hence, no ρ-approximation exists for the general TSP unless P = NP.

20.7 The Classes NC and P-Complete

We considered a sequential algorithm to be good (tractable) if it has polynomial worst-case complexity. Analogously, we consider a parallel algorithm on a PRAM to be good if it uses a polynomial number of processors and has polylogarithmic worst-case complexity. The class of problems solvable by such an algorithm is called the class NC. (Since any algorithm using p processors on an EREW PRAM or CREW PRAM can be simulated on the CRCW PRAM using p processors with at most logarithmic slowdown, the definition of the class NC is independent of the model of PRAM chosen.) In this section we introduce the notion of P-completeness

$$\begin{pmatrix} 0 & 19 & 14 & 16 & 17 & 13 & 18 & 20 \\ 19 & 0 & 17 & 12 & 11 & 19 & 15 & 15 \\ 14 & 17 & 0 & 18 & 16 & 12 & 16 & 16 \\ 16 & 12 & 18 & 0 & 10 & 20 & 11 & 20 \\ 17 & 11 & 16 & 10 & 0 & 16 & 10 & 14 \\ 13 & 19 & 12 & 20 & 16 & 0 & 18 & 15 \\ 18 & 15 & 16 & 11 & 10 & 18 & 0 & 17 \\ 20 & 15 & 16 & 20 & 14 & 15 & 17 & 0 \end{pmatrix}$$

Cost matrix $C = (c_{ij})$ satisfying triangle inequality,
where c_{ij} is the cost of edge $\{i,j\}$ in K_8

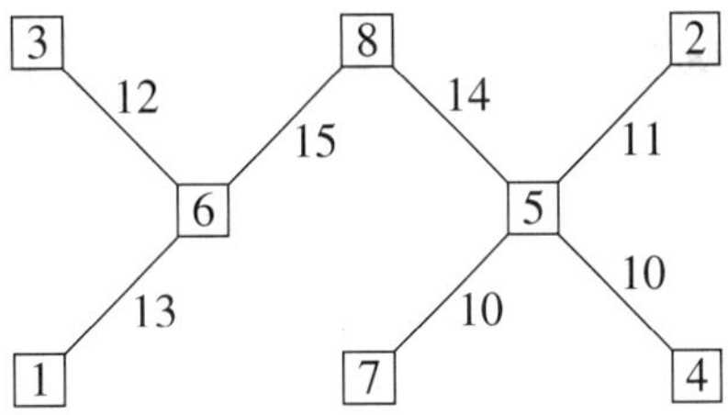

A minimum spanning tree T in weighted graph K_8 having preorder traversal

1 6 3 8 5 2 4 7

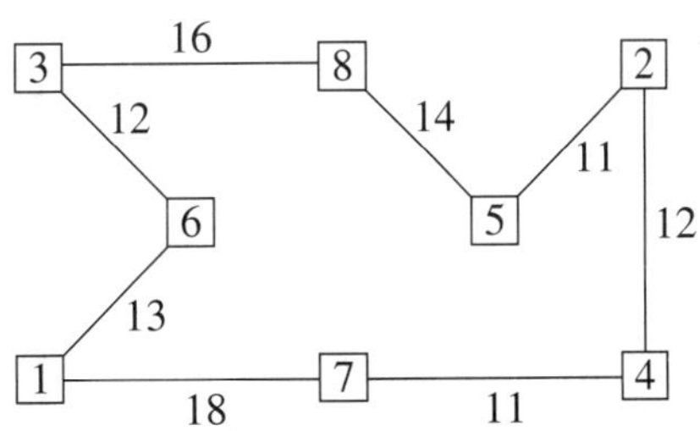

Tour generated by preorder traversal of T

Figure 20.4 Cost of the tour generated by the preorder traversal of the minimum spanning tree T is less than twice the optimal tour for TSP.

with respect to NC, which can be viewed as the analog for parallel computation of NP-completeness.

DEFINITION 20.7.1 Given two decision problems A and B, we say that A is NC *reducible* to B if there is a mapping f from instances of A to instances of B such that

1. f can be computed in polylogarithmic time using a polynomial number of processors on a PRAM,
2. the answer to a given instance I of problem A is *yes* if, and only if, the answer to instance $f(I)$ of B is *yes*.

Using the fact that the composition of two polynomials is a polynomial, it follows that if B is in NC then A is also in NC. We say that a problem B is *P-hard* if for any problem A in P, A is NC reducible to B. We say that a problem B is *P-complete* if it belongs to P and is P-hard. The P-complete problems can be viewed as the "hardest" problems in P with respect to the class NC since they have the following property: If a P-complete problem B is in NC then every problem in P is in NC. The question of whether there exists a P-complete problem that is in NC—that is, whether NC = P—is an important open problem, comparable to the problem of whether P = NP for sequential computation.

It is not obvious that there exist P-complete problems, but it turns out that many natural problems are P-complete. In this section we give several examples of such problems.

20.7.1 Straight-Line Code Problem for Boolean Expressions

A Boolean expression is an expression in the Boolean variables $x_1, \ldots, x_n$, which is formed using the three operations disjunction (or), conjunction (and), and negation, denoted by $\vee$, $\wedge$, $\neg$, respectively. In the notation used earlier, the literal $\bar{x}$ has the same truth value as $\neg x$. A constant Boolean expression is one that consists solely of the symbol T or F. An example of a Boolean expression in the variables x_1, x_2, x_3 is

$$\neg(x_1 \vee x_2) \wedge x_3 \vee \neg x_1.$$

As usual, in a Boolean expression the order of precedence is (), $\neg$, $\wedge$, $\vee$.

Now consider the following set of Boolean equations:

$$\begin{aligned} x_1 &= B_0 \\ x_2 &= B_1(x_1) \\ x_3 &= B_2(x_1, x_2) \\ &\cdot \\ &\cdot \\ &\cdot \\ x_n &= B_{n-1}(x_1, \ldots, x_{n-1}), \end{aligned}$$

where B_0 is a constant and $B_i(x_1, \ldots, x_{i-1})$ is a Boolean expression involving only the Boolean variables $x_1, \ldots, x_{i-1}$, $i = 2, \ldots, n$. Regarding = as the assignment operator, we call these equations *straight-line code*. The problem of computing the value of x_i, $i = 1, \ldots, n$, is called the *Straight-Line Code (SLC) problem*.

Clearly, the straightforward sequential algorithm for solving the SLC problem, which successively evaluates x_i, $i = 1, \ldots, n$, has linear computing time, where the input size is the total number m of symbols over all n equations (a symbol is counted k times if it occurs in k equations). Thus, the SLC problem is in P.

The SLC problem can be formulated as a decision problem, where we ask the question "Given $i \in \{1, \ldots, n\}$, does x_i have the value T?" A solution to the SLC problem can then be obtained by solving this decision problem concurrently for each x_i, $i = 1, \ldots, n$. The SLC problem is our first example of a P-complete problem. The proof that the SLC problem is P-complete is beyond the scope of this book. In the remainder of this section we illustrate NC reductions from the SLC problem to several other problems in P, thereby showing that they too are P-complete.

20.7.2 Straight-Line Code Without Disjunction

Straight-line code without disjunction (SLCWD) is straight-line code where the operation of disjunction is not used. We now describe an NC reduction of the SLC problem to the SLCWD problem, thereby showing that the SLCWD problem is P-complete. We employ the following useful identities known as DeMorgan's Laws.

PROPOSITION 20.7.1 **DeMorgan's Laws** For $x, y \in \{T, F\}$,

1. $\neg(x \vee y) = \neg x \wedge \neg y$,
2. $\neg(x \wedge y) = \neg x \vee \neg y$. ❑

Using the fact that $\neg(\neg x) = x$ and DeMorgan's Law (i) we have

$$x \vee y = \neg(\neg x \wedge \neg y).$$

We now consider the transformation f that maps an instance I of the SLC problem to the instance $f(I)$ of the SLCWD problem, where $f(I)$ is constructed from I by concurrently replacing each occurrence of $x \vee y$, where x and y are Boolean expressions, with $\neg(\neg x \wedge \neg y)$. Each such replacement involves changing $\vee$ to $\wedge$ and inserting three NOT operations and a left and right parenthesis in the appropriate place. We leave it as an exercise to design an NC algorithm for computing $f(I)$ from I. In Figure 20.5 this transformation is illustrated for a sample instance of the SLC problem. Since $f(I)$ can be computed in polylogarithmic time using a polynomial number of processors, the SLC problem is NC reducible to the SLCWD problem, thereby proving that the SLCWD problem is P-complete.

20.7.3 NOR Straight-Line Code

The operation NOR is defined by

$$x \text{ NOR } y = \neg(x_j \vee x_k)$$

NOR straight-line code is straight-line code where all assignments are of the following two forms:

$$x_1 = T$$
$$x_2 = \neg x_1$$
$$x_3 = \neg x_1 \wedge (x_1 \vee x_2)$$
$$x_4 = (x_3 \wedge \neg(x_1 \vee x_2)) \vee (x_1 \wedge \neg x_3)$$

Instance I of the general SLC problem

$$x_1 = T$$
$$x_2 = \neg x_1$$
$$x_3 = \neg x_1 \wedge (\neg(\neg x_1 \wedge \neg x_2))$$
$$x_4 = \neg(\neg(x_3 \wedge \neg(\neg(\neg x_1 \wedge \neg x_2))) \wedge \neg(x_1 \wedge \neg x_3))$$

Instance $f(I)$ of the SLCWD problem

Figure 20.5 The NC reduction transformation f from the general SLC problem to the SLCWD problem shown for a sample instance I of the SLC problem

$$\begin{aligned} x_i &= T, \\ x_i &= x_j \text{ NOR } x_k, \quad j,k < i. \end{aligned} \tag{20.7.1}$$

We now show that the problem of solving NOR straight-line code (the NOR SLC problem) is P-complete. We first describe a transformation g from an instance I of the general SLC problem to a set of equations involving only NOR operations and assignments to T, but not necessarily of the form given in (20.7.1). We construct $g(I)$ from I by concurrently performing the following replacement operations:

1. F is replaced with T NOR T,

2. $\neg x$ is replaced with x NOR x, (20.7.2)

3. $x \wedge y$ is replaced with (x NOR x) NOR (y NOR y),

where x and y are Boolean expressions. We leave it as an exercise to describe an NC transformation h that converts $g(I)$ into an instance of the NOR SLC problem by replacing each equation with a set of equations of the form given in (20.7.1). The transformation $f(I) = h(g(I))$ yields an NC reduction from the general SLC problem to the NOR SLC problem, showing that the NOR SLC problem is P-complete.

20.7.4 Lexicographically First Maximal Independent Set

Let $G = (V,E)$ be a graph with vertex set $V = \{1, \ldots, n\}$ and edge set E of size m. A *maximal independent set* is an independent set U (no two vertices of U are adjacent), such that $U \cup \{v\}$ is not an independent set for any vertex $v \in V$. The *lexicographical ordering* of 2^V (the *power set* of V, consisting of all subsets of V) is a linear ordering such that a subset $A < B$ if either $A \subset B$, or the minimum element in $A \backslash B$ is smaller than the minimum element in $B \backslash A$. For example,

$$\{2,4,5\} < \{2,4,5,9\} < \{3\} < \{5,7\}.$$

The *lexicographically first maximal independent set* is the maximal independent set U whose lexicographical order is smallest among all the maximal independent sets of G. The lexicographically first maximal independent set is illustrated for a sample graph in Figure 20.6.

In this subsection we will show that the problem of finding a lexicographically first maximal independent set (the LFMIS problem) is P-complete. (As with the SLC problem, the LFMIS problem can be formulated as a decision problem, where we ask the question: "Given a vertex $i \in V$, does i belong to the lexicographically first maximal independent set?")

First, we show that the LFMIS problem is in P, by designing a polynomial-time procedure *SequentialLFMIS* for computing the lexicographically first maximal independent set U. Procedure *SequentialLFMIS* maintains a set R of remaining vertices where R is initially the set V of all vertices. At each stage, procedure *SequentialLFMIS* (*i*) chooses the smallest vertex u from R, (*ii*) adds u to U, and (*iii*) removes vertex u and all vertices adjacent to u from R. Procedure *SequentialLFMIS* terminates when R is empty. We leave the pseudocode for procedure *SequentialLFMIS* as an exercise. It is easily verified that *SequentialLFMIS* has complexity $\Theta(m + n)$, where m denotes the number of edges of G.

We now show how to NC reduce an instance I of the NOR SLC problem to an instance $f(I)$ of the LFMIS problem. Instance $f(I)$ is a graph $G = (V,E)$ with vertex set $V = \{1,2, \ldots, n\}$, where vertex i corresponds to variable x_i of I, $i = 1,2, \ldots, n$. For each assignment $x_i = x_j$ NOR x_k of I there is an edge $\{i,j\}$ and an edge $\{i,k\}$. It is easily verified that G can be computed in polylogarithmic time on a PRAM using a polynomial number of processors. It can also be verified that a node i of G belongs to the lexicographically first maximal independent set if, and only if, the value of x_i is T in I. In Figure 20.7 (page 726) the NC transformation f is shown for a sample instance I of the NOR SLC problem.

Figure 20.6 Lexicographically first maximal independent set of G is {1,5,6,9,10}.

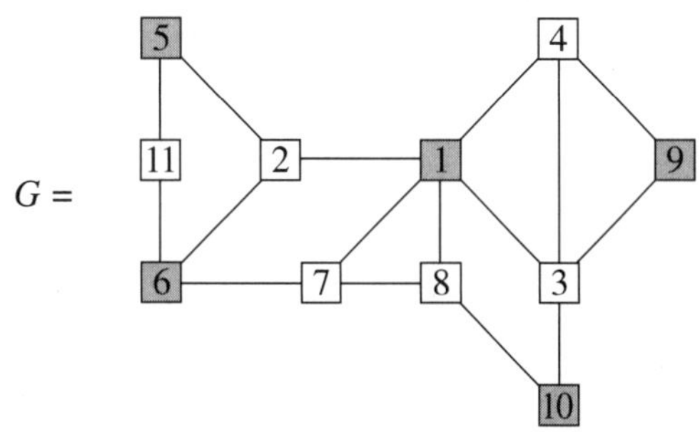

$$x_1 = T$$
$$x_2 = x_1 \text{ NOR } x_1$$
$$x_3 = x_2 \text{ NOR } x_2$$
$$x_4 = x_1 \text{ NOR } x_2$$
$$x_5 = T$$
$$x_6 = x_3 \text{ NOR } x_5$$
$$x_7 = x_1 \text{ NOR } x_5$$
$$x_8 = x_1 \text{ NOR } x_5$$
$$x_9 = x_4 \text{ NOR } x_8$$

Sample instance I of NOR SLC problem

$G =$

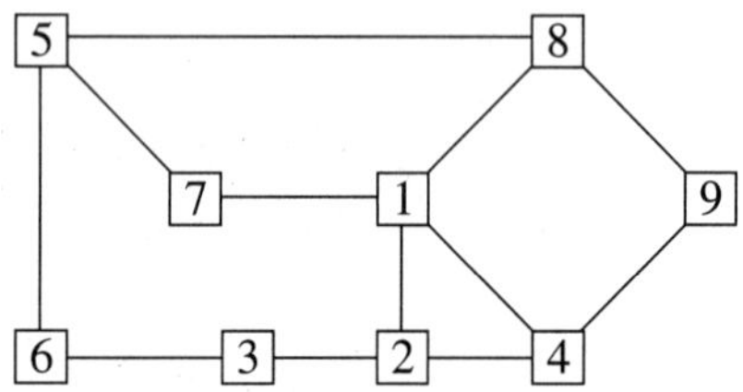

Graph $G = f(I)$ of the LFMIS problem

Figure 20.7 The transformation from a sample instance I of the NOR SLC problem to the instance $G = f(I)$ of the LFMIS problem

20.8 Closing Remarks

Cobham introduced the class P in 1964. Edmonds, who independently proposed the class P in 1965, also introduced the class NP and conjectured that P $\neq$ NP. Cook, who introduced the notion of NP-completeness in 1971, showed that CNF SAT and 3-CNF SAT are NP-complete. Shortly thereafter, Levin independently discovered the notion and also proved the existence of NP-complete problems. Karp popularized the notion of NP-completeness by establishing that a wide variety of important and practical problems are NP-complete.

Our discussions of the notions of NP, NP-completeness, and P-completeness were somewhat informal in the sense that we did not establish a formal model of computation such as a Turing Machine. The interested student can find formal treatments in the references.

References and Suggestions for Further Reading

Garey, M. and D. Johnson. *Computers and Intractability: A Guide to the Theory of NP-Completeness*. New York: W. H. Freeman & Co., 1979. A classic reference to the theory of NP-completeness, which contains a list of hundreds of important NP-complete problems.

For books containing extensive discussions of NP-completeness, see:

Horowitz, E. and S. Sahni. *Fundamentals of Computer Algorithms*. Computer Science Press, 1978.

Papadimitriou, C. and K. Steiglitz. *Combinatorial Optimization: Algorithms and Complexity*. Englewood Cliffs, NJ: Prentice-Hall, 1982.

Greenlaw, R., H. J. Hoover, and W. L. Ruzzo. *Limits to Parallel Computation: P-Completeness Theory*. New York: Oxford University Press, 1995. A recent text devoted to P-completeness.

Exercises

Sections 20.1 and 20.2 The Classes P and NP and Reducibility

20.1 Suppose that the weights and values of the objects in the 0/1 Knapsack problem are all integers. Show that 0/1 Knapsack problem and its associated decision version are polynomially equivalent. (*Hint*: Use a binary search technique.)

20.2 Given an integer n, it was shown by Pratt that n is prime if, and only if, there exists an integer a such that

a. $a^{n-1} \equiv 1 \pmod{n}$, and

b. $a^{(n-1)/p} \not\equiv 1 \pmod{n}$ for all prime divisors p of $n - 1$.

Based on this result, show that the prime testing problem is in NP.

20.3 Prove a result for TSP analogous to that of Exercise 20.1.

20.4 Prove Proposition 20.2.1.

20.5 Prove Proposition 20.2.2.

20.6 Prove Proposition 20.2.3.

Sections 20.3 and 20.4 NP-Complete Problems

20.7 Verify that each of the NP-complete problems in Section 20.4 belong to NP.

20.8 Prove Proposition 20.3.1

20.9 Show that 2-CNF SAT is in P.

20.10 This exercise refers to the mapping f from CNF SAT to 3-CNF SAT described in Example 20.4.1.

a. Show that $|f(I)| \leq 12|I|$ for any instance I of CNF SAT.

b. Show in cases 2 and 3 that each clause C_i of I is satisfiable if, and only if, the conjunction of the clauses replacing C_i is satisfiable.

20.11 In Example 20.4.4, verify that the set R' of remaining households can be constructed and that the resulting set M' is a matching.

20.12 Show that the instance F,S of the 3-exact cover problem is a *yes* instance if, and only if, the instance $s_1,s_2, \ldots, s_n$ and C constructed in Example 20.4.6 is a *yes* instance of the sum of subsets problem.

20.13 Show that the problem of determining whether a graph G is bipartite (has a proper 2-coloring) is in P.

20.14 In Example 20.4.7, show that the graph G constructed from the instance I of 3-CNF SAT is $(n + 1)$-colorable if, and only if, I is satisfiable.

20.15 Given a multiset $S = \{s_1,s_2, \ldots, s_n\}$ of positive integers, the *partition problem* asks whether S can be partitioned into two subsets having the same sum. Show that the partition problem is NP-complete. (*Hint*: Show that sum of subsets $\propto$ partition.)

20.16 Show that the 0/1 Knapsack problem is NP-complete. (*Hint*: Show that partition $\propto$ 0/1 Knapsack [partition is defined in previous exercise].)

20.17 Suppose we have a set of n objects $\{x_1,x_2, \ldots, x_n\}$ of sizes $\{s_1,s_2, \ldots, s_n\}$, where $1 < s_i < 1$, $i = 1, \ldots, n$. The *bin packing* optimization problem is to place the objects into bins having unit size using the minimum number of bins. Each bin can contain any subset of objects whose total size does not exceed one. The decision version asks for a given integer k whether we can pack the objects using no more than k bins. Show that the bin problem is NP-complete. (*Hint*: Show that sum of subsets $\propto$ bin packing.)

Section 20.5 The Class co-NP

20.18 Prove Theorem 20.5.1.

Section 20.6 Approximation Problems

20.19 Show that the 2-approximation algorithm provided by the minimum spanning tree argument does not yield a ρ-approximation for any $\rho < 2$.

20.20 Show that if a polynomial algorithm A exists yielding a ρ-approximation for the general TSP problem, then a polynomial algorithm exists for the Hamiltonian cycle decision problem. (*Hints*: Let $G = (V,E)$ be any instance of the Hamiltonian cycle decision problem. Consider the instance of TSP on the complete graph $G' = (V,E')$ with the following weighting: $\omega(u,v) = 1$ if $uv \in E$, otherwise $\omega(u,v) = \rho|V| + 1$. Show that G' has a tour of cost $\leq \rho|V|$ if, and only if, G has a Hamiltonian cycle.)

20.21 Consider the following algorithm for the vertex cover problem.

procedure *VertexCoverApprox*(*G*,*Cover*)

Input: G (a graph with vertex set V and edge set E)
Output: *Cover* (a vertex cover containing no more than twice the number of vertices in a minimum vertex cover)

```
    Cover := ∅
    RemainEdges := E
    while RemainEdges ≠ ∅ do
       choose any edge uv ∈ RemainEdges
       Cover := Cover ∪ {uv}
       remove from RemainEdges all edges incident with either u or v
    endwhile
end VertexCoverApprox
```

Show that *VertexCoverApprox* is a 2-approximation algorithm.

20.22 Consider the bin packing optimization problem defined in Exercise 20.17. Assuming an ordering of the bins, an algorithm using the *first-fit* heuristic successively places the objects into the first bin that can accommodate the object. Show that the first-fit algorithm is a 2-approximation.

Section 20.7 The Classes NC and P-Complete

20.23 Prove Proposition 20.7.1 (DeMorgan's Laws).

20.24 Use induction to prove the following generalization of DeMorgan's Laws. For $x_1, x_2, \ldots, x_k \in \{T, F\}$,

a. $\neg(x_1 \wedge x_2 \wedge \cdots \wedge x_k) = \neg x_1 \vee \neg x_2 \vee \cdots \vee \neg x_k$,

b. $\neg(x_1 \vee x_2 \vee \cdots \vee x_k) = \neg x_1 \wedge \neg x_2 \wedge \cdots \wedge \neg x_k$.

20.25 Let f be the transformation that maps an instance I of the SLC problem to the instance $f(I)$ of the SLCWD problem, where $f(I)$ is constructed from I by concurrently replacing each occurrence of $x \vee y$, where x and y are Boolean expressions, with $\neg(\neg x \wedge \neg y)$. Design an NC algorithm for computing $f(I)$ from I.

20.26 **a.** Design an NC algorithm for performing the replacement operations given in (20.7.2) to obtain $g(I)$ from an instance I of the SLCWD problem.

b. Design an NC algorithm for converting $g(I)$ to an instance I of the NOR SLC problem.

20.27 **a.** Give pseudocode for the (sequential) procedure *SequentialLFMIS* for computing the lexicographically first maximal independent set of a graph G.

b. Analyze the complexity of procedure *SequentialLFMIS*.

20.28 **a.** Design an NC algorithm for performing the reduction from an instance I of the NOR SLC problem to the instance $G = f(I)$ of the LFMIS problem shown in Figure 20.7.

b. Show that x_i has the value T in instance I of the NOR SLC problem if, and only if, vertex i is in the lexicographically maximal independent set in $G = f(I)$.

Index